Field Crop Diseases
Third Edition

Field Crop Diseases

Third Edition

Robert F. Nyvall

Iowa State University Press / Ames

Robert F. Nyvall, Professor of Plant Pathology, University of Minnesota, specializes in diseases of cultivated wild rice and in the development of mycoherbicides to control weeds. After receiving his Ph.D. in plant pathology from the University of Minnesota, he did research on vegetable diseases at Washington State University. While an Extension Plant Pathologist for 15 years at Iowa State University, he specialized in diseases of maize and soybean. Dr. Nyvall also served as Superintendent of the North Central Experiment Station of the University of Minnesota.

© 1999 Iowa State University Press
All rights reserved

Iowa State University Press
2121 South State Avenue, Ames, Iowa 50014

Orders: 1-800-862-6657
Office: 1-515-292-0140
Fax: 1-515-292-3348
Web site: www.isupress.edu

♾ Printed on acid-free paper in the United States of America

First edition © 1979 AVI. *Field Crop Diseases Handbook*

Second edition © 1989 Van Nostrand Reinhold, New York. *Field Crop Diseases Handbook*

Third edition, 1999. *Field Crop Diseases*

Library of Congress Cataloging-in-Publication Data

Nyvall, Robert F.
    Field crop diseases / Robert F. Nyvall.—3rd ed.
      p. cm.
    Rev. ed. of: Field crop diseases handbook. 2nd ed. c1989.
    Includes bibliographical references (p.   ) and index.
    ISBN 0-8138-2079-0
    1. Field crops—Diseases and pests—Handbooks, manuals, etc. 2. Plant diseases—Handbooks, manuals, etc. 3. Phytopathogenic microorganisms—Control—Handbooks, manuals, etc. I. Nyvall, Robert F. Field crop diseases handbook. II. Title.
SB731.N94   1999
632—dc21   99-18995

The last digit is the print number: 9 8 7 6 5 4 3 2 1

To

my grandchildren, Emily and Grant

my children, Martha and Chris, Tracey and Nathan

my wife, Sandra

my mother, Gertrude

and especially to the memory of my father, Robert

Contents

Contents

Contents

Contents

6. Diseases of Cotton
(Gossypium barbadense and
G. hirsutum)   183

Contents   xiii

Contents   xxi

18. Diseases of Soybean
(Glycine max)   621

Contents

20. Diseases of Sugarcane

Contents   xxvii

Contents

Preface

The purpose of *Field Crop Diseases, Third Edition,* remains the same as it has been in previous editions: to provide a basic knowledge of the diseases of many of the world's important field crops. Each disease description is meant to stand alone, or to be a self-contained story, and to give a good working knowledge of a disease. However, each description is not intended to be comprehensive. If further information is desired, in-depth knowledge may be found in sources such as research journals or in compendia that are published by the American Phytopathological Society (APS).

Some things about the book have changed:

The title is now *Field Crop Diseases* instead of *Field Crop Diseases Handbook* because the word "handbook" lends a different connotation to the contents than is intended.

The common names of host plants, such as rapeseed and corn, have changed. Since the term "rape" has awkward connotations today, "canola" is now commonly used to identify this crop. "Corn" has been changed to "maize," which most of the world uses, even though my American mentality often found it awkward to use "maize" throughout the book.

The authority behind each Latin binomial of the causal microorganism has been dropped to simplify reading by nonplant pathologists.

More than 400 diseases and 1900 references have been added. This edition discusses 1614 plant diseases of 24 field crops.

There are frustrations associated with writing such a book. One is the variety and overlap of common names of diseases. The APS has attempted to alleviate this problem by publishing a standard list of common disease names. Where possible, I have used the names on this list. However, the names of diseases commonly used in the literature do not always coincide with the APS list, thereby contributing to the general confusion. I have attempted to list other names of a disease in the description, but these lists may not always be comprehensive.

Field Crop Diseases
Third Edition

1. Diseases of Alfalfa *(Medicago stiva)*

Diseases Caused by Bacteria

Alfalfa Sprout Rot

Cause. *Erwinia chrysanthemi* survives for up to 2 weeks on inoculated dried seeds. Air, the water supply, or greenhouse workers may introduce bacteria into a sprouting house. Spread within a tray is by seed-to-seed contact. Rot is most severe at high moisture and temperatures of 28°C and above. Little disease occurs below 21°C.

Distribution. The United States (California).

Symptoms. Radicles are a translucent yellow as they emerge from the seed. In 24 to 48 hours, seeds stop growing and turn into a yellowish, odiferous mass that contains numerous bacteria. Disease initially occurs in a few trays, but it may spread throughout all trays in a few days.

Management
1. Control temperature in the sprouting house.
2. Practice sanitation. Bacteria likely survive in water remaining in tanks used for soaking seeds.
3. Soak seeds for 2 hours in 0.5% sodium hypochlorite or calcium hypochlorite. However, hypochlorite is not currently registered for alfalfa seed treatment in the United States.

Bacterial Leaf Spot

Cause. *Xanthomonas campestris* pv. *alfalfae* (syn. *X. alfalfae* and *Phytomonas alfalfae*) survives in infected residue that has been incorporated into soil or is lying on the soil surface, in hay, and in debris associated with seed. During warm, wet weather, bacteria are splashed or blown onto leaves and enter them through small wounds that have been made by any means. During dry weather, bacteria may enter leaves through wounds made by

3

windblown soil particles. The bacteria multiply inside the leaf and frequently ooze to the leaf surface, where they may be splashed by rain or
rubbed by leaf-to-leaf contact onto adjacent healthy leaves.

Distribution. Wherever alfalfa is grown under warm, wet conditions.

Symptoms. Diseased seedlings are often killed or stunted at high temperatures. Initially, small, water-soaked spots occur in chlorotic areas on
leaflets. The spots enlarge to irregular-shaped lesions (2–3 mm in diameter)
with chlorotic margins that are pronounced on the underside of leaflets.
Eventually, the lesions become light yellow to tan, often with a lighter center, and have a translucent, papery texture. Lesions often glisten because of
the dried bacterial exudate on their surface. Diseased leaves defoliate prematurely. Stem lesions are water-soaked initially, then turn brown or black.

Management
1. Grow resistant cultivars. Plants within a cultivar differ in susceptibility
2. Sow in the spring.

Bacterial Stem Blight

Cause. *Pseudomonas syringae* pv. *syringae* (syn. *P. medicaginis* and *Phytomonas
medicaginis*) survives in infested residue in soil. During wet weather, bacteria enter a host plant through stem wounds previously caused by frost.

Distribution. Western North America.

Symptoms. Bacterial stem blight occurs from early spring until the first harvest. Normally, only the lower five internodes are diseased.
    Initially, lesions are water-soaked and chlorotic at the point of leaf at-
tachment to the stem, leaflet midribs, and petioles. Lesions become linear,
are discolored light to dark brown or black, and extend down one side of
the stem for one to three internodes and into the crown and roots of older
plants. Bacterial exudate may be present on leaf surfaces. Leaves attached
to diseased areas on the stem become chlorotic and die. Diseased stems are
thin, brittle, and shorter than healthy stems.

Management
1. Hardy cultivars are less prone to frost injury; therefore, they will be less
   likely to become infected.
2. Harvest after all danger of frost is past.

Bacterial Wilt

Cause. *Clavibacter michiganense* subsp. *insidiosus* (syn. *Corynebacterium michiganense* subsp. *insidiosum, C. insidiosum,* and *Aplanobacter insidiosum).* The
Clavibacter subspecies *insidiosus* sometimes is spelled "insidiosum" in the
literature and considered a pathovar.

Clavibacter michiganense subsp. *insidiosus* survives in plant tissue and soil for several years and is seedborne. Bacteria are spread from plant to plant by mowing machinery, irrigation water, or surface rainwater. Long-distance spread is by hay and seed.

Plants are normally infected the second or third year of growth. Bacteria enter roots through wounds made by a variety of causes, including winter injury and insects. Bacteria then proceed into the xylem tissue of stems, crowns, and roots, where they rapidly increase in number and prevent water from moving up the plant. Eventually, diseased tissues disintegrate and release bacteria into the surrounding soil, where, over time, they are distributed to adjacent healthy host plants.

The greatest incidence of bacterial wilt occurs in low, poorly drained areas of fields and over larger geographic areas during periods of continuous wet weather. Presence of the root knot nematode, *Meloidogyne hapla*, increases the severity of bacterial wilt.

Distribution. Wherever alfalfa is grown, but alfalfa wilt is considered to be more important in the United States than in other countries.

Symptoms. Bacterial wilt is first noticed as scattered dead plants in a field. Initially, diseased plants are stunted; stems and leaves remain small after cutting and give plants a dwarfed, bunchy appearance. Leaves are somewhat thickened, rounded at the tip, and tend to cup upward. Plants initially wilt during moisture stress caused by daytime heat but recover their turgidity at night. Later, permanent wilt symptoms develop as leaflet margins become yellowish or bleached, followed by entire leaflets becoming chlorotic, then necrotic. Eventually, the entire plant dies.

The outer vascular tissue of crowns and taproots is yellow to tan and appears as a discolored ring. The entire root then becomes rotted. Diseased plants are more prone to winterkill than healthy plants.

Management. Grow resistant cultivars.

Crown Gall

Cause. *Agrobacterium tumefaciens*.

Distribution. It is likely that crown gall is widespread, but it is not a serious disease of alfalfa.

Symptoms. Galls present on crowns are irregular-shaped and light green when actively growing but are discolored light to dark tan when not growing.

Management. Not necessary.

Dwarf

Cause. *Xylella fastidiosa* is a nutritionally fastidious, gram-negative, xylem-in-habiting bacterium. Bacterial cells are single, straight rods (1.0–4.0 × 0.25–0.5 μm).

Several species of leafhoppers carry the bacterium to susceptible hosts and inject it into the plants during feeding. In California, *Draeculacephala minerva* and *Carneocephala fulgida* are important vectors. Other potential vectors include xylem-feeding insects with piercing, sucking mouthparts. Leafhoppers of the subfamily Cicadellinae and spittlebugs (family Cercopidae) are known vectors in which bacteria multiply in their foreguts, but the bacteria are not retained after molting and are not transovarial. Grassy weeds apparently provide a favorable habitat for development of vector populations.

Bacterial cells are limited to the xylem tissue of host plants. Once bacteria are established in the xylem tissue, they can systemically colonize both roots and shoots. Disease expression is caused by vascular occlusions in the xylem tissue, which cause water stress and, eventually, plant wilting.

Xylella fastidiosa also causes Pierce's disease of grape and almond leaf scorch.

Distribution. The United States (California, Georgia, Mississippi, and Rhode Island). Dwarf occurs sporadically and is considered to be an uncommon disease of alfalfa.

Symptoms. Diseased plants are normally stunted and have smaller-than-normal, dark bluish-green leaves. Blooming may be retarded or inhibited and fewer buds develop after each harvest, which results in a diminished number of stems. Infrequently, stem tips may wilt.

Diseased roots initially have a slight yellow vascular discoloration beneath the bark of the taproot. Later, the entire stele may be diseased with a yellowish-brown discoloration extending into the surrounding tissue. Taproot xylem tissue becomes brown when exposed to air.

Management
1. Grow resistant varieties. 'California Common 49' is resistant to dwarf but is susceptible to the spotted alfalfa aphid.
2. Control grassy weeds.

Root and Crown Rot Complex

This disease is sometimes called *Pseudomonas viridiflava* root and crown rot or *P. viridiflava* root and crown rot complex.

Cause. *Pseudomonas viridiflava.*

Distribution. Not known.

Symptoms. The crown and upper taproot have a light brown "dry rot." Light brown streaks extend beyond the rot into the vascular system for about one-third the length of the diseased root.

Management. Not reported.

Diseases Caused by Fungi

Acrocalymma Root and Crown Rot

Cause. *Acrocalymma medicaginis,* teleomorph *Massarina walkeri,* presumably overseasons as pycnidia on residue.

Distribution. Australia (southeastern Queensland), where the disease is common in plants more than 1 year old.

Symptoms. Red-flecked cortical and wood tissues occur at the extremity of a wedge-shaped lesion that commences in the crown branches and extends into the taproot. Bark of diseased areas is often fissured. Later a "dry rot" develops and older, diseased tissue darkens.

Management. Not known.

Anthracnose

Anthracnose is also called southern anthracnose.

Cause. *Colletotrichum trifolii. Colletotrichum destructivum, C. dematium* f. *truncata,* and *C. gloeosporioides* have also been reported as causal agents of anthracnose.

 Colletotrichum trifolii survives as acervuli and mycelia in residue associated with fields, feedlots, storage areas, and machinery, and on live stems and crowns. During warm, moist or humid weather, spores are produced in acervuli and disseminated by wind and splashing water to stems and crowns. Secondary inoculum is provided by spores borne in acervuli that are produced on new lesions.

 Two physiologic races, labeled as race 1 and race 2, of *C. trifolii* occur. In general, disease caused by both races is more severe as temperatures increase. Light quality and duration have no effect on disease incidence or severity.

Distribution. Canada and the United States. Race 1 is present wherever alfalfa is grown. Race 2 has been reported from Maryland, North Carolina, and Virginia. *Colletotrichum destructivum* is present primarily in Ontario.

Symptoms. Anthracnose is first detected in summer or autumn by the presence of wilted and dead, yellow or light brown shoots that are scattered throughout a field. The disease is considered to be a major cause of mid- and late-summer yield reductions and plant mortality as a result of the general reduction of shoot and crown bud numbers and of overall crown size and health.

Large, diamond-shaped lesions with bleached or white centers form on the lower portions of diseased stems. Under magnification, several small, black acervuli, appearing as upright "hairy" structures due to the presence of numerous dark setae, can be seen in the center of most lesions. Young, dead shoots may droop to form "shepherd's crooks," which is a useful diagnostic symptom.

Colletotrichum trifolii grows from diseased stems into crowns, causing tissue to become bluish black or black. Diseased crowns are weakened, resulting in plants that are predisposed to winter injury or are killed prematurely. In South Africa, *C. trifolii* has been reported to cause a root rot primarily in areas under irrigation. Both *C. destructivum* and *C. dematium* f. *truncata* are mildly pathogenic on petioles and leaves but are primarily secondary invaders in stem lesions caused by *C. trifolii*.

Management. Anthracnose cannot be managed in established stands.
1. Grow resistant cultivars.
2. Cut alfalfa before losses become too severe.
3. Under experimental conditions, the fungicides benomyl, copper hydroxide, and mancozeb reduced disease severity.

Aphanomyces Root Rot

This disease is sometimes called Aphanomyces seedling blight and root disease.

Cause. *Aphanomyces euteiches* and *A. cochlioides*. Infection of plants is caused by zoospores that are motile in wet soils over a wide temperature range (3° to 39°C). However, lower temperatures are most favorable to infection. Zoospores have been reported to be more motile for up to 24 hours in water at temperatures of 3° to 5°C.

Distribution. Canada (Ontario and Quebec) and the United States.

Symptoms. Typical symptoms include postemergence death of seedlings, root and hypocotyl browning, and chlorosis of foliage. Sometimes, seedlings are only stunted. Later, a root rot of mature plants may occur. Dry weight of live diseased plants that survive the first growing season is reduced.

Management. Grow the least susceptible cultivars. Resistance has been identified in several alfalfa cultivars.

Black Patch

Cause. *Rhizoctonia leguminicola* is seedborne and survives as mycelia in infested residue.

Distribution. The United States (West Virginia).

Symptoms. Lesions on foliage vary in color from brown to gray-black and may have concentric rings. Seedlings are often blighted and overgrown with coarse, black mycelia. Similar-appearing aerial mycelial growth later occurs on stems and petioles and frequently girdles the stems. Dark lesions and similar dark mycelia may occur on flowers and seeds.

Management. Not reported.

Black Root Rot

Cause. *Thielaviopsis basicola,* synamorph *Chalara elegans,* survives as chlamydospores in residue and soil. When soil conditions are moist during the growing season, chlamydospores germinate to form mycelia that penetrate roots. Later in the season, chlamydospores form within the diseased tissues and either remain in infested residue that has fallen to the soil surface at harvest time or return to the soil upon the disintegration of residue.

Distribution. The United States (Maine, Maryland, New Jersey, and Texas).

Symptoms. The cortex of the hypocotyl below the soil line and taproots and fibrous roots become discolored dark brown and necrotic. A severe root rot may develop, resulting in wilting of the foliage and eventual plant death.

Management. Not reported.

Brown Root Rot

Brown root rot is also called Plenodomus root rot.

Cause. *Phoma sclerotioides* (syn. *Plenodomus meliloti*) overwinters as pycnidia in soil residue. Roots are infected when soils either thaw out in the spring or before they freeze in the winter. Later in the growing season, plants become resistant to infection.

Distribution. Canada, Finland, and the United States (Alaska).

Symptoms. Slightly sunken, brown lesions occur on both lateral roots and the taproot. Numerous dark pycnidia develop on or just below the lesion surface. Plants die when rot proceeds to the crown. However, plants may recover if rot stops and enough taproot remains to produce new branch roots below the crown.

Management. Rotate alfalfa with resistant crops, such as small grains.

Common Leaf Spot

Common leaf spot is also called Pseudopeziza leaf spot.

Cause. *Pseudopeziza medicaginis. Pseudopeziza trifolii* f. sp. *medicaginis-sativae* is sometimes considered a synonym of *P. medicaginis* but is not reported in Farr et al. (1989).

Pseudopeziza medicaginis survives either as mycelia or apothecia in leaf residue on the soil surface. During abundant moisture and temperatures of 15° to 24°C, ascospores are produced on apothecia and "ejected" into the air, where they are carried by wind currents to healthy leaflets. Seedling stands grown under a thick cover crop that retains moisture, such as oat, can be severely diseased.

Distribution. Wherever alfalfa is grown.

Symptoms. Plants are normally not killed but defoliation reduces hay quality and yield. Circular, dark brown spots (2 mm in diameter) develop on leaflets. There is a sharp line delineating spots and healthy-appearing tissue. When spots are fully developed, the centers become thickened and tiny, light brown, cup-shaped apothecia form on the upper leaflet surface.

Management
1. Grow resistant cultivars.
2. Cut diseased stands in the prebloom or bud stage before diseased leaves defoliate. This maintains hay quality and removes diseased leaves that will be a later source of infection.

Corky Root Rot

Corky root rot is called corchosis.

Cause. *Xylaria* sp. Little is known of the life cycle of the fungus.

Distribution. Argentina.

Symptoms. Lateral roots are destroyed, then a dry, brown, sunken canker develops on the taproot and increases in size until the root is encircled. Fungus growth into roots can also occur from diseased crowns. Eventually, the entire root is invaded and the plant dies. The root retains its original shape but is light in weight and corky, with areas of white mycelia growing on the darkened areas of the root. Finally, stands are thinned out and fail to recover after harvest.

Management. Not reported.

Crown Wart

Cause. *Physoderma alfalfae* (syn. *Urophlyctis alfalfae* and *Cladochytrium alfalfae*) survives as resting spores in soil. Zoospores are liberated in warm, wet soil and "swim" to the developing buds of crown branches, where they germinate to form infection hyphae that infect buds.

Distribution. Australia, Ecuador, Europe, New Zealand, and the United States. Crown wart is common in the western United States but is less common in the eastern and northeastern United States.

Symptoms. Crown wart is most common in the spring when soils are excessively moist. Irregular-shaped, white galls are formed on the crown near the soil surface. Galls rarely develop on leaves. Upon maturity, galls become gray to brown and have mottled brown interiors (350 mm in diameter) that contain masses of resting spores. Galls decompose later in the growing season and release spores into the soil.

Management. Not practiced since crown wart is not considered to be a serious disease.

Cylindrocarpon Root Rot

Cause. *Cylindrocarpon magnusianum* (syn. *C. ehrenbergii*), teleomorph *Nectria ramulariae*. *Cylindrocarpon magnusianum* survives as sclerotia-like stromata in infested soil residue. Infection occurs through wounds at the base of branch roots during the end of winter dormancy. Plants become resistant later in the growing season.

Distribution. Canada and the United States (Minnesota).

Symptoms. A water-soaked area that eventually becomes discolored dark brown may occur only on a portion of a root or may involve the entire root. Stromata develop in cracks in the bark of diseased roots and give them a dark and rough appearance.

Management. Rotate alfalfa with resistant crops, such as small grains.

Cylindrocladium Root and Crown Rot

Cause. *Cylindrocladium crotalariae*. *Cylindrocladium clavatum* and *C. scoparium* are also reported to cause disease.

Distribution. The United States (Hawaii).

Symptoms. Black and sunken lesions occur on diseased roots and crowns. A brown root rot is sometimes present.

Management. Not reported.

Damping Off

Damping off is sometimes called Pythium seed and seedling blight.

Cause. *Fusarium acuminatum*, *Rhizoctonia solani,* and *Pythium* spp. The primary *Pythium* species reported to be involved in damping off are *P. debaryanum, P. irregulare, P. splendens,* and *P. ultimum.* Other *Pythium* species reported to be pathogenic to alfalfa include *P. aphanidermatum, P. myriotylum, P. pulchrum,* and *P. rostratum.*

Oospores and sporangia survive in infested residue and germinate either directly, by forming a germ tube, or indirectly, by producing zoospores that "swim" through soil water to roots and seeds. Disease caused by most *Pythium* species occurs in cold, wet soils; however, higher temperatures favor disease caused by *P. aphanidermatum* and *P. myriotylum.*

Distribution. Wherever alfalfa is grown.

Symptoms. Preemergence damping off frequently occurs but is rarely observed because rapid seed and seedling decay below the soil surface "obscures" symptoms. If seeds germinate, the seedling radicle and cotyledons turn brown and soft.

Postemergence seedling blight is characterized by hypocotyls and roots becoming light brown and water-soaked. Eventually, plants collapse and dry up. Surviving seedlings are stunted. In South Africa, *P. paroecandrum* does not cause severe damping off but infects rootlets, thereby reducing shoot growth and development.

Management
1. Apply a fungicide seed treatment.
2. Seedlings become immune after a few days; plants older than 2 weeks usually do not become infected.

Downy Mildew

Cause. *Peronospora trifoliorum. Peronospora aestivalis* has been reported to be a synonym. *Peronospora trifoliorum* overwinters as perennial mycelia in crown buds and shoots and as oospores in infested residue. In the spring, conidia are formed in darkness during high relative humidity and are disseminated by wind or splashed onto leaves. Optimum germination occurs in free water at a temperature of 18°C.

Distribution. Temperate areas of the world.

Symptoms. A rapidly growing plant has few symptoms. Downy mildew is most damaging the first year of plant growth during cool, wet springs. Symptoms disappear during warm, dry weather but may return during cool weather.

Young leaflets, especially at the tips of stems, are dwarfed and twisted or rolled downward and contain light green to yellow blotches. A grayish, cottony growth consisting of mycelia, conidiophores, and conidia is often visible on the underside of diseased leaflets during moist, cool weather or during periods of high humidity.

When the entire stem is diseased, all leaves and stem tissue are chlorotic. Diseased stems are larger in diameter and much shorter than healthy stems.

Management
1. Grow cultivars that have a high percentage of resistant plants.
2. Cut alfalfa in prebloom stage to save foliage.
3. Sow in spring to reduce seedling infection.

Forked Foot Disease

Cause. *Pythium ultimum* is frequently associated with this condition.

Distribution. Not known for certain but likely wherever alfalfa is grown.

Symptoms. Six or more adventitious roots develop above lesions on seedling radicles and give roots a "forked" appearance. Seedlings are dwarfed or stunted.

Management
1. Apply a fungicide seed treatment.
2. Seedlings become immune after a few days; plants older than 2 weeks do not become infected.

Fusarium Crown and Root Rot

Cause. *Fusarium acuminatum, F. avenaceum, F. oxysporum, F. sambucinum,* and *F. solani.* Other *Fusarium* species are presumably involved.

Fungi survive as chlamydospores in infested residue and soil and as saprophytic mycelia in infested residue. *Fusarium* species enter alfalfa roots by direct penetration or through wounds caused by feeding of nematodes, insects, or other agents. Feeding injury by the insect *Sitona hispidula* has been reported to result in root infection and severe root rot caused primarily by *F. oxysporum.* The primary role of wounding is not to breach the root surface but to alter the host–pathogen interaction to favor fungal development in the root. Some *Fusarium* species, such as *F. oxysporum* f. sp. *medicaginis, F. oxysporum* f. sp. *lycopersici, F. sambucinum,* and *F. solani,* produce substances that are toxic to living plant tissues.

Distribution. Wherever alfalfa is grown.

Symptoms. Seedlings may damp off either before or after emergence, especially during warm, wet weather. Diseased plants are usually stunted. Leaves are chlorotic, have a bleached appearance, curl at the edges, and proceed to wilt.

The disease is characterized by a discolored light brown to black, wedged-shaped necrotic area that spreads from the diseased crown downward into the taproot. Necrotic areas often occur in the cortex of the branch roots and taproot, often in association with wounds.

Management. No practical means of management is practiced; however, frequent harvesting of plants late in the growing season increases the possibility of root rot.

Fusarium Wilt

Cause. *Fusarium oxysporum* f. sp. *medicaginis* is considered to be the main pathogen. *Fusarium oxysporum* f. sp. *vasinfectum,* races 1 and 2, and *F. oxysporum* f. sp. *cassia* are also reported to cause similar symptoms.

    Fusarium oxysporum f. sp. *medicaginis* survives for several years as chlamydospores in soil and residue and as mycelia in infested residue and diseased live plants. Infection occurs by fungi growing into wounds on small roots and taproots and progressing up the xylem tissue. The nematode *Meloidogyne hapla* is reported to increase the virulence of *F. oxysporum* f. sp. *medicaginis.* The clover root curculio, *Sitona hispidulus,* increased the severity of Fusarium wilt in greenhouse experiments.

    Disease is favored by high soil temperatures. Variable soil moisture is not considered an important factor in disease development.

Distribution. Wherever alfalfa is grown but Fusarium wilt is more severe in tropical and semitropical alfalfa-growing areas of the world.

Symptoms. Fusarium wilt occurs in irregular-sized and -shaped areas within a field; however, only a small percentage of plants in the affected area display symptoms at one time. The greatest loss occurs approximately 2 years after stand establishment. Scattered plants continue to wilt and die throughout a field the remaining years of the stand. Plants that have been affected by wilt, but have not died, will be dwarfed or stunted.

    The first symptom is a rapid wilting of stems on one side of a diseased plant. Later, stems and leaves appear chlorotic or bleached. Leaves at the bottom of the stem often are a light pink. Tips of stems wilt during hot days or periods of soil moisture deficiency but recover during cool nights or when soil moisture is replenished. Plants die slowly, often over a period of several months. Frequently leaves of diseased plants become dry and brittle but retain their green color and somewhat resemble the leaves of freshly cut hay.

    The interior of a diseased taproot is a cinnamon brown to red color that is often apparent only as streaks in the woody area. As the disease progresses, the entire outer portion of the woody cylinder becomes a light brown to red-brown discolored rot that extends from the crown into the taproot.

Management. There is no satisfactory management. Individual plants have been reported to have some resistance. Seedlings inoculated with vesicular-arbuscular mycorrhizal fungi had a lower incidence of wilt than nonmycorrhizal plants.

Lepto Leaf Spot

Lepto leaf spot is also called Leptosphaerulina leaf spot, halo spot, pepper spot, brown leaf spot, and Pseudoplea leaf spot.

Cause. *Leptosphaerulina trifolii* (syn. *L. briosiana, Pleosphaerulina briosiana, Pseudoplea briosiana,* and *Pseudoplea medicaginis*) overwinters as mycelia and pseudothecia in infested leaves on the soil surface. During cool, moist weather, spores are discharged from pseudothecia and disseminated by wind to young leaves. Optimum germination occurs at temperatures of 22° to 25°C in the laboratory. Following moist weather, leaf spot is most severe on young leaves that have grown back after the first cutting.

Distribution. Wherever alfalfa is grown.

Symptoms. At first, numerous small, reddish-brown flecks appear on both leaf surfaces and the petioles. The flecks enlarge to spots that have a tan center surrounded by an irregular brown border and halo. Optimum lesion development is reported to occur under high light conditions. Diseased leaves eventually die and become necrotic but continue to cling to the stem. On older growth, the upper leaves become infected and have typical symptoms but seldom die.

Management. Effective management measures are not known; however, the following may reduce disease incidence and severity:
1. Grow cultivars that are reported to have some disease resistance.
2. Sow certified disease-free seed.
3. Cut plants in the prebloom plant stage.
4. Rotate alfalfa with a resistant host, such as soybean, maize or a small grain, for at least 2 years.
5. The fungicides benomyl, copper hydroxide, and mancozeb were effective in reducing disease severity under experimental conditions.

Marasmius Root Rot

Cause. *Marasmius* sp.

Distribution. Egypt.

Symptoms. In greenhouse pathogenicity tests, roots are brown and water-soaked. *Marasmius* sp. grows in rope-like strands over the entire root system. Diseased plants are killed in the field.

Management. Not reported.

Mycoleptodiscus Crown and Root Rot

Cause. *Mycoleptodiscus terrestris* (syn. *Leptodiscus terrestris*) overwinters as sclerotia in infested residue and soil. Conidia form in summer and are disseminated by wind to plant hosts when they are forcibly ejected from the fruiting structure, called an acervulus, by setae that unfold as mucilage dries out. Disease is favored by warm, humid weather.

Distribution. The central and eastern United States.

Symptoms. Preemergence and postemergence damping off occurs. A brown to black rot occurs in lateral roots, taproots, and crowns of older plants. The margin of decayed tissue is black while tissue behind the margin is lighter in color. Numerous sclerotia are formed in decayed tissue. Small spots develop on leaves; reddish-brown lesions that occur on stems develop into crown rot.

Management. Not reported.

Myrothecium Root Rot

Cause. *Myrothecium roridum* and *M. verrucaria*. Mycelia penetrate intact roots but will enter through wounds in some situations.

Distribution. The United States (Pennsylvania and Wisconsin); however, Myrothecium root rot is probably more widely distributed.

Symptoms. In laboratory tests, brown, water-soaked rots with poorly delineated margins were initially evident on roots. Root growth, in general, was inhibited. In other greenhouse tests, entire diseased roots had a dark brown rot with occasional streaks that extended upward in the vascular system. However, rotted roots remained firm and were not water-soaked. Foliar symptoms were chlorosis; purpling of leaflet margins; leaf curling, mottling, and stunting; and death of diseased leaves and petioles.

Management. Not reported.

Oidiopsis Powdery Mildew

Cause. *Oidiopsis* sp.

Distribution. Asia, Brazil, and Egypt.

Symptoms. Initially, white, powdery patches occur on the underside of leaflets. Eventually, the whole leaf is involved and becomes necrotic when disease becomes severe.

Management. Not reported.

Oidium Powdery Mildew

Cause. *Oidium* sp. is the conidial stage of many species in the Erysiphaceae.

Distribution. The United States (Hawaii).

Symptoms. Not well known, but a symptom is generally presumed to be the typical white to gray "powderish" appearance on upper leaflet surfaces.

Management. Not reported.

Phoma Root Rot

Phoma root rot is the root rot phase of spring black stem.

Cause. *Phoma medicaginis* (syn. *Ascochyta imperfecta, P. herbarum* f. *medicaginum,* and *P. medicaginis* var. *medicaginis*) overwinters as pycnidia in infested residue and is seedborne as mycelia on the seed coat. Pycnidia are produced on lesions of diseased tissue in late summer and autumn but rarely during the growing season. Spores, which are produced in pycnidia in the spring during wet weather and temperatures of 18° to 24°C, are disseminated by wind.

Some researchers have reported that wounds are not required for fungal entry into the roots but more extensive rot occurred when roots were wounded than not wounded. However, others report wounding is necessary for infection. Stem stubble provides a suitable infection court for the pathogen to invade crown tissues. Inoculations caused larger lesions on lateral roots than on diseased main roots.

Distribution. Not reported, but the disease is presumably widespread.

Symptoms. Main and lateral roots have a black, dry necrosis and eventually disintegrate. Dry weight of foliage and roots is reduced.

Management. Not reported.

Phymatotrichum Root Rot

This disease is also called cotton root rot, Ozonium root rot, and Texas root rot.

Cause. *Phymatotrichopsis omnivora* (syn. *Phymatotrichum omnivora, P. omnivorum,* and *Ozonium omnivorum*) survives, primarily in alkaline soils, as sclerotia and brown sclerotial strands to depths of 2 m in the soil. Sclerotia germinate in early summer when soil temperatures are high and infect roots. Low soil temperatures will kill the fungus. Conidia are produced on spore mats formed on the soil surface, particularly after a summer rain or during humid weather; however, they are not thought to be important either in infection or dissemination.

Distribution. Mexico and the southwestern United States.

Symptoms. Infested areas within a field may appear as circles of dead plants, but alfalfa in the center of the circle may not be diseased. If alfalfa is diseased, dead plants are replaced by grasses. Lesions on the taproots are yellow to brown, sunken, and clearly defined, although irregular-shaped. Leaves of diseased plants become chlorotic and bronze. Eventually, diseased plants wilt and die if lesions encircle the root or crown. Irregular-shaped, white to tan spore mats sometimes appear on the soil surface.

Management
1. Plow under green manure crops or animal manure.
2. Grow resistant crops, such as a grass, for at least 3 years. However, if sclerotia persist for a long time, this management measure is relatively ineffective.
3. Add sodium salts to soil. This helps to reduce disease severity.
4. Grow cultivars whose plants vigorously produce new roots. These cultivars have some tolerance to the disease.

Phytophthora Root Rot

Phytophthora root rot is also called wet foot disease.

Cause. *Phytophthora megasperma* f. sp. *medicaginis.* In some literature, *P. medicaginis* and *P. megasperma* have been referred to synonymously with *P. megasperma* f. sp. *medicaginis,* but Farr et al. (1989) do not consider them to be synonyms. *Phytophthora drechsleri* and *P. cryptogea* have also been reported as causal fungi. Additionally, *P. sojae* f. sp. *medicaginis* has been ascribed to be a causal fungus, but a report based on mitochondrial DNA relatedness indicates that the alfalfa Phytophthora is not related to *P. sojae,* which is specific to soybean.

Survival is primarily by oospores in infested residue and soil. Mycelia in residue and chlamydospores have also been reported to be survival mechanisms; however, only oospores are capable of long-term survival. Hyphae from diseased alfalfa plants grow through soil to colonize roots of other plants including black medic, birdsfoot trefoil, and maize. These plant colonizations increase survival potential and serve as sources of inoculum to infect alfalfa.

Eventually oospores, chlamydospores, and sporogenous hyphae are produced in diseased tissue. In the presence of moisture, sporogenous hyphae bear sporangia that germinate directly, forming either mycelia or sporangia, or indirectly, producing zoospores, at temperatures of 24° to 27°C. Infection occurs at the tips of small roots and the bases of fine lateral roots.

Water-saturated soil predisposes roots to infection by increasing root damage. More damage increases the exudation of nutrients, which, subsequently, increases the chemolactic attraction of zoospores to the roots. Root rot is more severe when susceptible plants are also infected with the nematode *Meloidogyne hapla* at 28°C (the optimum temperature).

The optimum growth of a high-temperature cultivar of *P. medicaginis* from the desert areas of the southwestern United States (designated as cultivar HTI and reported as *P. megasperma)* occurs at temperatures of 27° to 33°C. In South Africa, *P. drechsleri* has been isolated from alfalfa roots with symptoms similar to those caused by *P. medicaginis.*

Distribution. Canada and the north central and eastern United States. *Phytophthora drechsleri* was reported from South Africa.

Symptoms. Symptoms occur in wet soils. Seeds and seedlings damp off. Leaves become chlorotic or reddish and defoliate from diseased plants. Eventually plants become stunted, wilt, and die.

Rootlets and taproots are rotted. There is a sharp line that delineates diseased from healthy tissue on roots. Such diseased tissue can occur on the taproot at any depth below the soil line. The taproot itself may be rotted off at any depth below the crown and has a yellowish-brown discoloration that eventually becomes black.

Rot stops as soils dry and if enough taproot remains alive, side roots are produced and the plant remains alive. Such plants have a shallow root system and produce abundant forage when surface moisture is plentiful, but little or no forage when surface moisture is depleted.

Management
1. Grow resistant cultivars.
2. Improve water drainage in soils that tend to be poorly drained.
3. Coating seed with *Bacillus cereus* significantly increased emergence.

Powdery Mildew

Cause. *Erysiphe polygoni.*

Distribution. Italy and the United States (Massachusetts and Wyoming).

Symptoms. Typical white to gray, powder-like growths that consist of conidia, conidiophores, and mycelia occur primarily on the upper surfaces of diseased leaflets.

Management. Not reported.

Rhizoctonia Foliage Blight

Rhizoctonia foliage blight is also called web blight.

Cause. *Rhizoctonia solani* AG-1 IB survives by forming sclerotia or by growing saprophytically on plant residue. During high temperatures and abundant moisture—conditions that are frequently found in thick stands of alfalfa—mycelia grow from a precolonized substrate to infect crowns or stems just below the soil line. Sclerotia, infested residue, and soil that contains propagules of *R. solani* may be splashed onto the plant foliage. During flooding or irrigation, host plants are infected above the soil line.

Distribution. Wherever alfalfa is grown in hot, humid weather; however, the disease is seldom severe.

Symptoms. Rhizoctonia foliage blight is a progression of Rhizoctonia stem blight. Lower leaves and stems are initially affected but the disease may eventually progress halfway up the plant. Diseased leaves appear watery and "bluish" in color. Eventually, leaves wilt, become brown, and shrivel. A web-like growth of fungus mycelia may be seen growing over shriveled leaves and stems.

Management. Increase air movement by proper grazing or by cutting to allow stems to dry out.

Rhizoctonia Root Rot

Rhizoctonia rot is also called Rhizoctonia root rot canker and black root canker.

Cause. *Rhizoctonia solani* AG-4 survives by forming sclerotia or by growing saprophytically in plant residue. During periods of high temperatures and moisture, *R. solani* infects the areas where the lateral roots emerge from the taproot, and the taproot itself.

Distribution. Australia, Iran, and the irrigated areas of the United States (Arizona and California).

Symptoms. Affected areas in a field are circular to irregular-shaped and have healthy plants next to diseased plants that are wilting. Taproots are covered by a large number of cankers that are oval to round (6–12 mm in diameter) and vary in color from yellow to tan, often with a slightly darker border. Taproots may be girdled, but plants are damaged most by cankers formed on the crowns. When soil temperatures decrease during the winter, diseased areas heal and lesions or cankers turn dark brown to black.

Management. Growers in irrigated areas of the southwestern United States have reestablished stands by resowing infested fields in October or November.

Rhizoctonia Stem Blight

Rhizoctonia stem blight is also called Rhizoctonia blight and stem canker.

Cause. *Rhizoctonia solani* AG-4 survives by forming sclerotia or by growing saprophytically in plant residue. During high temperatures and moisture, mycelia that originate from a precolonized substrate will infect crowns or stems just below the soil line. Sclerotia, infested residue, and soil containing propagules of *R. solani* may be splashed onto the plant foliage and stems. During flooding or irrigation, plants are infected above the soil line.

Distribution. Wherever alfalfa is grown.

Symptoms. Dead stems are scattered throughout a stand. Sunken cankers found at the base of diseased stems vary in color from tan to reddish brown to dark brown. Stems are girdled, causing leaves and tips of stems to yellow and wilt.

Management
1. Sow certified, high-quality seed in a well-prepared seedbed with good drainage.
2. Proper mowing and grazing practices are effective means of partially managing the disease since the drying effects of direct sunlight and good air circulation retard disease development.

Rhizopus Sprout Rot

Cause. *Rhizopus stolonifer.* This is the common ubiquitous bread mold fungus.

Distribution. Potentially any place where alfalfa is grown for sprouts. It is more common on trays or screens than in vats.

Symptoms. Diseased tissue is soft and translucent. Coarse gray to brown, "cottony" mycelia with large, black sporangia develop in rotted areas.

Management
1. Practice good sanitation.
2. Avoid injuring plants.
3. Change to a vat system.

Root and Crown Rot Complex

This disease complex is sometimes called crown rot.

Cause. Several bacteria and fungi, including *Cylindrocladium crotalariae, Erwinia amylovora* var. *alfalfae, Flexibacter* sp., *Fusarium acuminatum, F. avenaceum, F. solani, F. tricinctum, F. oxysporum, Myrothecium* spp., *Phoma medicaginis, Pseudomonas marginalis* var. *alfalfae, Rhizoctonia solani, Serratia marcescens,* and *Thielaviopsis basicola.*

These fungi and bacteria normally persist either as saprophytes or as resting structures in plant residue and soil. Cutting alfalfa for hay or an injury by machinery, frost, insects, and animals causes a small portion of stem to be killed, thus allowing bacteria and propagules of saprophytic fungi to begin growing on the dead tissue. Infection spreads down the dead cut stem into the crown.

Distribution. Wherever alfalfa is grown.

Symptoms. The crowns of most alfalfa plants 1 year old or older may be rotted and have a brown to black discoloration interspersed with live tissue. Tissue that produces crown buds is often killed, resulting in a plant with few live stems. Because crown rot progresses slowly for a number of years after the disease is initiated, there is a gradual thinning of the stand from year to year.

Management

1. The last cutting of alfalfa should be timed to allow the plant to build up a large enough supply of carbohydrates to overwinter successfully.
2. Grow cultivars adapted to an area.
3. Use good management, fertilize properly, and adjust soil pH between 6.2 and 7.0.
4. Avoid placing livestock in an alfalfa field after the last cutting because wounded crowns provide an entrance for pathogens to grow into the crown.

Rust

Cause. *Uromyces striatus* (syn. *U. medicaginis, U. oblongus,* and *U. striatus* var. *medicaginis*). In warmer alfalfa-growing areas of North America, *U. striatus* survives as urediniospores and perennial, systemic mycelia in infected plants. *Uromyces striatus* is capable of surviving and reproducing on a large number of plant species in the tribes Cicereae, Galegeae, Genisteae, Hedysareae, Trifolieae, and Vicieae.

In the summer, urediniospores are disseminated by wind to northern alfalfa-growing areas where healthy plants are infected during periods of high humidity. Reported optimum temperatures for disease development vary. Some researchers reported that infection occurred at 21° to 30°C; however, in Iowa, infection efficiency was greater at 17.5°C than at 28.0°C. In the midwestern United States, rust may build up on plant regrowth in the autumn and on unharvested alfalfa in fencerows.

Uromyces striatus is a heteroecious rust with the presence of the aecial and pycnial stages reported on *Euphorbia* species in Europe and Canada. Only the uredinial and telial stages have been reported on alfalfa in the United States.

Distribution. Wherever alfalfa is grown under warm or temperate conditions; however, rust is most prevalent later in the summer and autumn.

Symptoms. Rust may cause severe defoliation late in the growing season if harvest is delayed or stands are held for seed production. Autumn-sown stands may be weakened, thinned or lost due to winter injury.

Uredinial pustules are reddish brown and sometimes are surrounded by a yellow halo. Pustules are most abundant on the undersides of leaflets but also occur on petioles and stems. Rusted leaves yellow, shrivel, and defoliate prematurely or are more likely to be injured by an early frost than are healthy leaves.

Aecia develop on leaves and stems of *Euphorbia* species and may kill stem apices. Witches' brooms and hypertrophy may occur on stems.

Management
1. Grow the most resistant cultivars.
2. Harvest on time to reduce leaf loss and remove inoculum.

Sclerotinia Crown and Stem Rot

Cause. *Sclerotinia sclerotiorum* and *S. trifoliorum*. *Sclerotinia trifoliorum* is more widespread and is the better known of the two pathogens on alfalfa.

Sclerotinia sclerotiorum survives as sclerotia in infested residue and on the soil surface. Sclerotia germinate during cool, wet weather in the spring and early summer to produce either apothecia or mycelia that infect crowns, leaves, and stems. Pollen also becomes infected by ascospores, which suggests that pollen may play a significant role in the epidemiology of blossom blight in alfalfa, especially in seed-growing areas. Sclerotia are produced on the host surfaces, in the pith, and under decaying plant parts on the soil surface. Disease is most severe in fields where infested residue remains on the soil surface from the last harvest.

Sclerotinia trifoliorum survives during the winter and high summer temperatures as sclerotia on the soil surface, imbedded in diseased stems and crowns, and adhering to the plant surface. During cool, wet weather in the growing season, sclerotia germinate to form mycelial strands or apothecia in which asci and ascospores are produced. However, most primary infection of alfalfa has been reported to occur in the late autumn and early winter when oversummered sclerotia in soil germinate to form apothecia. Ascospores are then disseminated by wind to host plants, where they infect crowns, leaves, and stems. Pollen grains can also become infected and cause blossom blight, similar to *S. sclerotiorum*.

Subsequent disease development then occurs from winter through early spring in most locations. From midwinter through early spring, growth of mycelia commences within and between host plants; however, mycelia grow only a short distance in soil and normally do not cause infection.

Disease is favored by frequent rainfall, prolonged high humidity, and mild temperatures during winter months. Disease is especially severe when alfalfa is sown into untilled sod in which numerous alternative hosts for *S. trifoliorum* are growing.

Distribution. Cool, humid areas of Europe and North America. *Sclerotinia trifoliorum* is more widespread on alfalfa than *S. sclerotiorum* and is found primarily in the south central and southeastern United States. *Sclerotinia sclerotiorum* has been reported mainly as a pathogen during the summer months in Canada, the Pacific Northwest, and the state of Georgia in the United States.

Symptoms. Symptoms caused by the two fungi are similar and appear at the same time in early spring. Sclerotinia crown and stem rot is most destructive on new seedlings and can be recognized by small circular patches of dead or dying plants within a field. The first symptoms occur in the autumn as small brown spots on diseased leaves and stems. Eventually, spots expand and diseased leaves and stems wilt and die. The fungus then spreads to the crown, where disease is most destructive the following spring when the crown becomes soft and has a gray-green discoloration. A white, "fluffy" mass of mycelia, in which small, black sclerotia are formed, grow over dead plant parts.

Sclerotinia trifoliorum may cause extensive or complete loss of stands particularly on first-year, autumn-sown crops.

Management
1. Rotate alfalfa with a resistant host. Do not grow alfalfa for 3 years.
2. Plow under residue to bury sclerotia.
3. Sow seed that is free of sclerotia.
4. Autumn or winter burning of stubble reduces numbers of sclerotia of *S. sclerotiorum,* thereby reducing incidence and severity of disease. Autumn burning has been reported to be more effective in southeastern Washington in reducing the numbers of sclerotia in the dense layer of surface plant residue and in the surface soil at depths of 0 to 2 cm.
5. Sow in spring because mature plants are relatively resistant.
6. Grow the least susceptible cultivar. No cultivar is completely resistant or tolerant to disease.

Southern Blight

Southern blight is also called Sclerotium blight.

Cause. *Sclerotium rolfsii* survives for several years in the soil or in plant residue as small, brown sclerotia that are disseminated by water and wind. Sclerotia germinate under hot, humid conditions to form mycelial strands that infect plants. Southern blight is not an important disease of alfalfa except in some silty or sandy soils.

Distribution. Southern Europe and the southern alfalfa-growing areas of the United States.

Symptoms. From a distance, the symptoms of southern blight within a field appear as scattered patches of dead plants. Plants are bleached to a light tan color and have a white, cotton-like mycelial growth on stems near the soil surface. Numerous small, tan to brown sclerotia that resemble "seeds" are formed in mycelia growing on diseased stems, crowns, and infested residue on the soil surface.

Management
1. Plow under infested residue.
2. Practice proper mowing and grazing. The causal fungus does not grow well when exposed to direct sunlight and air movement.

Spring Black Stem

Spring black stem is also called Ascochyta leaf spot and spring leaf spot.

Cause. *Phoma medicaginis* (syn. *P. herbarum* f. *medicaginum* and *Ascochyta imperfecta*) overwinters as pycnidia in infested tissue and as mycelia on the seed coat. Spores are produced in the spring during wet weather and temperatures of 18° to 24°C and are disseminated by wind to host plants. Pycnidia are produced on lesions of diseased tissue in late summer and autumn but rarely during the growing season.

Distribution. Europe, North America, and South America. Spring black stem is normally not present on first-year alfalfa unless the fungus was seedborne or spores were disseminated from nearby fields by wind.

Symptoms. Yield losses caused by spring black stem are due primarily to defoliation of the first cutting of hay. Disease may also be extensive during the autumn, but symptoms are less pronounced.

The most severe disease symptoms occur on lower leaves, but upper leaves may also be diseased. Diseased leaves have dark brown or black spots with irregular borders that enlarge and merge to cover most of the leaf area. Leaves eventually become chlorotic and defoliate.

Stem and petiole lesions initially are dark green and watery. Lesions then become dark brown to nearly black and have a "watery-appearing" margin. Stem lesions enlarge and merge until most of the lower stem is blackened. Young shoots are often girdled and killed, and the fungus may grow into the crown and upper root. Seedpods may discolor and shrivel during wet, humid conditions.

Management
1. Harvest hay before defoliation is severe. This also allows plants to dry out, making fungal sporulation and infection more difficult.
2. Plow under plants in cases of severe disease.
3. Treat seed with a fungicide to prevent seedborne introduction of the disease.
4. Sow seed produced in an arid area.
5. Grow moderately resistant cultivars. Highly resistant cultivars are not available.

Stagonospora Leaf Spot

Cause. *Stagonospora meliloti* is one of three phases of the same fungus on alfalfa and other clovers. *Stagonospora meliloti* is associated mainly with the leaf spot phase during the summer. *Phoma meliloti* has been reported as one of the causal agents of the root rot phase and is found as pycnidia produced on stems during low autumn temperatures. Farr et al. (1989) consider *P. meliloti* to be a synamorph of *S. meliloti*. The teleomorph, *Leptosphaeria pratensis,* is found as perithecia on stems in the spring and occasionally in the late autumn.

The fungi overwinter only in residue as mycelia and pycnidia of *S. meliloti* and perithecia of *L. pratensis*. During wet spring weather, conidia are extruded from pycnidia and disseminated by rain and irrigation water and infrequently by wind to serve as primary inoculum. Secondary spread is also by conidia during wet summer conditions. Pycnidia of *S. meliloti* are produced on residue later in the growing season. Little is known about the role of the perithecial stage in the epidemiology of *Stagonospora* leaf spot.

Distribution. Widespread, but *Stagonospora* leaf spot is most common in warm, humid areas or where alfalfa is grown under irrigation.

Symptoms. Spots on leaves and stems are 3 to 6 mm in diameter and circular to irregular-shaped. The spot is pale buff to almost white in the center and has a light to dark brown diffuse margin. In some instances, spots have faint concentric zones the same color as the margin. Diseased leaves defoliate soon after spots form. Pycnidia, which appear as small, dark specks, are scattered throughout the older spots.

Management
1. Rotate with a resistant host for 2 to 3 years.
2. Grow the most resistant cultivars.

Stagonospora Root Rot

This disease is sometimes called Stagonospora crown and root rot.

Cause. *Stagonospora meliloti,* teleomorph *Leptosphaeria pratensis. Phoma meliloti* is also reported as a cause and Farr et al. (1989) consider *P. meliloti* to be a synamorph of *S. meliloti. Stagonospora meliloti* is associated mainly with the leaf spot phase during the summer.

The fungi overwinter only in residue as mycelia and pycnidia of *S. meliloti* and possibly *P. meliloti* and as perithecia of *L. pratensis,* although little is known about the role of the perithecial stage in the epidemiology of Stagonospora root rot. During wet spring weather, conidia are extruded from pycnidia and disseminated to host plants by rain or irrigation water and infrequently by wind. Secondary spread is also by conidia during wet conditions. The root rot phase develops slowly, over 2 to 3 years at 15° to 25°C, from stem and crown infections. Pycnidia of *P. meliloti* are produced on stems during low autumn temperatures; pycnidia of *S. meliloti* are produced on residue later in the growing season.

Distribution. Wherever alfalfa is grown.

Symptoms. A brown to black rot of the taproot and crown are common symptoms. Lesions on the taproot are reported to occur 7.25–12.50 cm below the crown. Lesions on roots and crowns are cracked and irregular in shape. Lengthwise sections through rotted areas show affected bark and xylem tissue with bright, reddish-brown flecks.

Management
1. Rotate with a resistant plant host for 2 to 3 years.
2. Grow the most resistant cultivars.

Stemphylium Leaf Spot

Stemphylium leaf spot is also called target spot.

Cause. *Stemphylium alfalfae,* teleomorph *Pleospora alfalfae; S. botryosum,* teleomorph *P. tarda; S. globuliferum,* teleomorph *Pleospora* sp.; *S. herbarum,* teleomorph *P. herbarum;* and *S. vesicarium,* teleomorph *Pleospora* spp.

Overwintering is by mycelia in infested residue and on seed, and as perithecia in residue. Spores are produced during moist weather at any time of the growing season and are disseminated by wind to leaves of host plants. *Stemphylium botryosum* is divided into two biotypes: A warm-temperature (W-T) biotype and a cool-temperature (C-T) biotype. Temperature requirements for disease development depend upon the fungal biotype. Disease caused by the W-T biotype occurs at temperatures of 23° to 27°C, while disease caused by the C-T biotype is most severe at temperatures of 8° to 16°C. Under greenhouse conditions, disease also is severe when plants are exposed to light before and after inoculation, followed by alternating dark and light periods and extended free moisture for several days. Disease is most severe on dense foliage during late summer and autumn but can occur throughout the growing season.

A *Stemphylium*-incited disease in Australia and South Africa has been attributed to a fungus that more closely resembles *S. vesicarium* than *S. botryosum. Stemphylium vesicarium* is seedborne; seed becomes infected in either wet or dry climates. Disease caused by this *Stemphylium* is more common during cooler weather.

Distribution. Africa, Australia, Europe, New Zealand, and the United States, particularly the southeastern United States. The warm-temperature (W-T) biotype occurs in central and eastern North America during warm, wet periods; the cool-temperature (C-T) biotype occurs in California during cool, wet springs.

Symptoms. The W-T biotype produces oval or circular, slightly sunken, light brown leaf spots with dark brown borders that are surrounded by a pale yellow halo. Spots enlarge and form concentric light and dark brown zones to somewhat resemble a target. During wet weather, older lesions appear sooty from the abundant production of large spores within the lesion. Severe defoliation may occur. A single large lesion may cause a leaf to become chlorotic and defoliate prematurely.

Black areas may appear on peduncles, petioles, and stems. Girdling of stems and petioles in wet weather causes foliage to wilt and die.

Symptoms of the C-T biotype are common in the interior valleys of California but occur throughout the year along the coast. The C-T biotype produces elongate lesions, of various sizes but seldom exceeding 3 to 4 mm, that cease to expand once the border is formed. Lesions have irregular outlines and light tan, almost white, centers with dark brown borders. Premature defoliation is unusual.

Stemphylium vesicarium is reported to cause symptoms similar to the C-T biotype during cooler weather in Queensland, Australia. Only leaflets on actively growing shoots are affected. The Australian *S. vesicarium* causes circular to irregular-shaped leaf lesions with white to cream centers surrounded by a sharply defined dark brown margin. On susceptible cultivars, spots coalesce to cause a leaf blight.

Symptoms caused by *S. vesicarium* reported from South Africa also differ with temperature. In warm areas under overhead irrigation, lesions are irregular or circular to oval-shaped (0.1–3.0 mm in diameter); are light to dark brown and often are lighter in the center, which sometimes results in a concentrically zonate or target effect; and have diffuse margins that are frequently surrounded by lighter-colored halos. In cooler areas, lesions are circular to oval (0.1–2.0 mm in diameter) but uniformly white or tan with a distinct black margin. Elongated black lesions up to 2 mm long occur on petioles and stems. Under both temperature regimes, lesions are usually distinct but may later coalesce. Defoliation may occur.

Symptoms on annual Medicago vary from black specks to spots (1.0–2.0 mm in diameter) with tan centers and irregular brown margins.

Management

1. Grow the most resistant cultivar. No cultivar is highly resistant. In Australia the nondormant cultivars are most susceptible.
2. Harvest stands early to save foliage and allow plants to dry out.

Summer Black Stem

Summer black stem is also called Cercospora leaf spot and Cercospora black stem.

Cause. *Cercospora medicaginis* overwinters as mycelia in infested residue and is seedborne in warm and humid alfalfa-growing areas but not under other conditions. During warm, moist weather later in the growing season, spores are produced that are disseminated by wind and rain to host plants. Secondary spread of conidia occurs as plants grow taller and form a natural humidity chamber that prevents conidia from drying out.

Distribution. Africa, Asia, Europe, South America, and the central and eastern United States.

Symptoms. Summer black stem symptoms appear during hot, moist periods late in the growing season. Initially, small, brown spots that occur on both leaf surfaces enlarge to circular lesions (3–6 mm in diameter) and turn a reddish to smoky brown color. Frequently, only one or two lesions will develop on a leaflet. During moist weather, lesions appear ashy gray due to sporulation on the lesion surface. During severe disease, entire leaflets die and defoliation occurs.

The leaf spot phase is followed by the appearance of reddish to chocolate brown, elliptical or linear lesions on diseased petioles and stems. These lesions eventually enlarge and coalesce. Under moist conditions, the entire stem may be discolored. Smaller peduncles, petioles, and stems may die, resulting in further defoliation and loss of seed.

Summer black stem is most damaging when harvesting is delayed.

Management
1. Grow the most resistant cultivar.
2. Harvest hay when plants are in early bloom stage to reduce defoliation and allow foliage to dry out.
3. Do not allow cattle to graze on fields late in season since disease is more severe in second year stands that do not complete growth before dormancy.

Verticillium Wilt

Cause. *Verticillium albo-atrum* (dark mycelial species) overwinters as thick-walled mycelia in soil, live alfalfa plants, infested residue, and parasitized weeds, and is internally seedborne. *Verticillium albo-atrum* has a limited saprophytic ability and does not survive well in dead plant tissue that becomes colonized by more aggressive saprophytes. Under humid conditions, the fungus will colonize the pod and seed coat, but it does not survive long on the outside of seed. Pollen can be infected in vitro.

During cool, moist weather, conidia are initially produced on diseased leaflets, petioles, stipules, and stubble remaining from last seasons harvest. Eventually, conidial sporulation also occurs on diseased stems. Stubble is an effective source of inoculum only before it decomposes.

Dissemination is by any means that moves infested plant residue. A unique means of dissemination between fields utilizes the sclerotia of *Sclerotinia sclerotiorum,* which are colonized by *V. albo-atrum*. Because sclerotia are approximately the same size as alfalfa seed, they are present in cleaned and sized alfalfa seed lots and disseminate *V. albo-atrum* from seed fields to other fields. Secondary spread during the growing season is by harvest machinery, windborne conidia, and plant debris. A combination of wet conditions and ideal air currents results in significant dispersal of conidia for a short time after harvest and before healing of the cut stem occurs. Some evidence suggests, however, that airborne conidia do not cause systemic infections through penetration of uninjured leaf and stem tissues. Further evidence suggests that inoculum carried on the cutter bar, rather than airborne conidia, is the most important means of dispersal within and between fields. Long-range dissemination of *V. albo-atrum* may be by internal seedborne inoculum.

Under laboratory conditions, several insects have been implicated in dissemination. Grasshoppers, *Melanophus sanquinipes* and *M. bivittatus;* alfalfa weevils, *Hypera postica;* and woolly bears, *Apantesis blakei,* were fed diseased alfalfa leaves. The fungus was able to safely pass through their digestive systems and was detected in feces about 1 day after feeding. Fungus gnats, *Bradysia impatiens,* disseminate *V. albo-atrum* in the greenhouse, and the leafcutter bee, *Megachile rotundata,* disseminates the fungus in the field. The pea aphid, *Acyrthosiphon pisum,* is a vector by carrying spores as a surface contaminant on its legs and antennae. *Verticillium albo-atrum* has been reported to pass through the digestive tract of sheep, which affords another means of spread.

Plants are infected through the roots. Mycelia grow into the xylem tissues, where conidia are produced that spread upward within the xylem tissue of the plant. The fungus is present only in xylem elements of diseased stems. Fungus and vascular occlusions eventually plug the xylem tissue, preventing movement of water up the plant and causing wilt symptoms. Optimum growth of *V. albo-atrum* occurs at temperatures of 19° to 25°C.

The alfalfa strain *V. albo-atrum* is known to only infect alfalfa. *Verticillium dahliae* (microsclerotial species) occasionally causes similar symptoms on alfalfa in England, but it is less virulent.

Distribution. Canada, Europe, and the United States, primarily the Pacific Northwest.

Symptoms. Fewer symptoms are present during drought stress than during non-drought-stressed conditions. The first symptom is a "flagging" or wilting of the upper leaves on warm days. Initially, V-shaped, pinkish-orange to brown areas occur on leaflets. Leaflets on severely diseased shoots become necrotic and twisted, forming spirals. Lower leaves eventually wilt, become yellow to white, and die.

New shoots develop from the crown, but they also become diseased and die. Entire plants become stunted and have dead, yellow stems. Frequently, stems remain green but attached leaves are bleached and dead. Conidiophores cover the bases of infected stems and give them a grayish appearance. When the stem dies, the infected area becomes black.

In advanced stages of the disease, plants are stunted and have yellow and desiccated shoots and leaves. Xylem tissue is brown and frequently the discoloration can be traced from the taproot up into the stems. This symptom can be confused with both bacterial and Fusarium wilts. Flowering may be suppressed.

A high percentage of plants in some resistant cultivars are symptomless carriers of the pathogen, but plant height, dry weight, and flowering are affected. Stem dry weight is less affected by *V. albo-atrum*. Up to 50% reduction in yield can occur in susceptible plants by the end of the second year.

Verticillium albo-atrum is often confined only to the midrib and lateral veins of infected leaflets. This pattern prevails even after browning and death of an entire leaflet because the fungus is confined to xylem tissue. There is often a discontinuous pattern in the midrib of symptomed and symptomless leaflets due to the movement of spores within the xylem and their eventual germination within the xylem tissue.

Management
1. Grow resistant cultivars.
2. Sow seed that is free of plant debris and has been treated with a fungicide seed protectant.
3. Harvest the youngest stands first.
4. Clean equipment before leaving an infested field.
5. Seedlings inoculated with vesicular-arbuscular mycorrhizal fungi are reported to have a lower incidence of wilt than nonmycorrhizal seedlings.

Violet Root Rot

Cause. *Rhizoctonia crocorum* (syn. *R. violaceae*), teleomorph *Helicobasidium brebissonii,* survives in soil as sclerotia or as a saprophyte in infested residue. Infection is most common when moisture is present in low organic matter soils, but disease may also occur in high organic soils.

Distribution. Europe and infrequently in the United States.

Symptoms. Symptoms occur in midsummer as circular patterns associated with low, flooded areas within a field and as individual plants in older stands that have other root injuries. Diseased plants in an area are brown, in contrast to surrounding green plants. Diseased roots are brown on the inside and covered with thick mats of violet to cinnamon mycelia that extend 20 or more cm below the soil surface. Later, roots are rotted and shredded and have a brown to dark violet discoloration. Barely visible, tiny, black sclerotia are seen on diseased roots.

Management
1. Grow alfalfa in well-drained soil.
2. Harvest hay in prebloom stage.
3. Rotate alfalfa with resistant crops, such as small grains and corn.

Winter Crown Rot

Winter crown rot is also called Coprinus snow mold and snow mold.

Cause. *Coprinus psychromorbidus.* However, in some reports *C. urticicola* has been tentatively identified as the causal organism. If two organisms are involved, this may account for the differences in temperature for optimum growth in cultures: 12°C according to some reports, and 13° to 15°C according to others.

In the autumn, during conditions favorable for disease development, the fungus grows in close association with alfalfa crowns and produces hydrogen cyanide (HCN), which is absorbed by crown tissue at temperatures near 0°C. Mycelia invade crown buds in March and destroy the diseased tissue. When temperatures rise above freezing, alfalfa is no longer susceptible.

Distribution. Canada (Alberta, British Columbia, Manitoba, and Saskatchewan) and the United States (Alaska).

Symptoms. Different-sized areas of dead plants occur in a field. Dark brown, rotted areas occur on crowns and infrequently on roots. Often only the crown bud and underlying tissue is diseased. The taproot may be unaffected even though the crown is dead, but it will eventually be rotted by secondary organisms. Mycelia may grow on the soil surface in early spring.

Management
1. Grow cultivars that have *Medicago falcata* parentage.
2. Grow small grains instead of alfalfa in infested soil for a minimum of 3 years.
3. Application of borax to plants in the fall prevents HCN production. However, this treatment is sometimes phytotoxic.

Yellow Leaf Blotch

Cause. *Leptotrochila medicaginis* (syn. *Pseudopeziza jonesii* and *Pyrenopeziza medicaginis*) overwinters as stromata in residue on the soil surface. During cool, wet weather in late spring, stromata give rise to apothecia in which ascospores are produced. Ascospores are disseminated by wind to host plants.

Distribution. Wherever alfalfa is grown in temperate climates.

Symptoms. Yellow leaf blotch is most severe on the lower diseased leaves in stands of rank, tall plants. Young lesions appear as yellow stripes and blotches that are elongated parallel to the leaflet veins. As lesions enlarge, they become "fan-shaped" or circular and their color changes from yellow to orange-yellow or brown.

Small, dark pycnidia develop mostly on the upper surface of the blotches. Eventually, pseudostromata, and later stromata, form in the center of the blotch, giving it a dark brown to black color. Stromata are uncommon on the under surface of the leaf until after leaf defoliation. Dead leaves frequently remain attached to the stem for an extended time, curling downward as they dry. Similar yellow blotches, which later become dark brown, may occur on the stems.

Management
1. Cut plants before leaf defoliation becomes severe.
2. Do not leave a high stubble or allow weeds to become a problem.
3. Rotate alfalfa with a resistant host, such as soybean, maize, or a small grain, for at least 2 years.
4. Grow resistant cultivars.

Diseases Caused by Nematodes

Alfalfa Stem

Alfalfa stem nematode is sometimes called the bulb and stem nematode.

Cause. *Ditylenchus dipsaci* survives for years as larvae in dry hay stems, alfalfa crowns, soil, and other plant hosts. Larvae, which feed and reproduce in shoots near the crown, move over the plant surface in a film of water and infect plants through stomates. A female nematode lays 75 to 100 eggs; optimum reproduction and infection are in heavy, wet soils at soil temperatures of 15° to 21°C. Dissemination is by machinery, rain, and irrigation water.

Distribution. Worldwide.

Symptoms. Diseased plants are stunted and have a bushy appearance because swollen nodes and shortened internodes result in abnormal proliferation and swollen, short stems. White stems are scattered throughout a field.

Alfalfa plants growing in weedy, low-lying areas of the field have increased severity of fungal and bacterial rots. Shoots from diseased buds are severely dwarfed and have buds that are swollen, spongy, and easily detached. A ruined stand and considerable yield loss may occur with 2 to 3 years of disease.

Management
1. Grow resistant cultivars.
2. Plow under infested stands.
3. Use nematicides where feasible.
4. Rotate alfalfa with other crops for 2 to 3 years.

Cyst

Cause. *Heterodera trifolii.*

Distribution. Not known.

Symptoms. Light-colored immature females may be visible on roots. Roots tend to be smaller and have reduced nodulation.

Management. Not known.

Dagger

Cause. *Xiphinema americanum* is an ectoparasite.

Distribution. Widely distributed.

Symptoms. Root growth is reduced. Necrotic areas occur where feeding has occurred. Cells next to necrotic tissue may enlarge and cause a gall-like growth.

Management. Not known.

Pin

Cause. *Paratylenchus hamatus.*

Distribution. Thought to be widespread.

Symptoms. Necrotic areas occur on roots.

Management. Not known.

Reniform

Cause. *Rotylenchulus* spp.

Distribution. Tropical and semitropical areas.

Symptoms. Enlarged cells occur at the site of nematode feeding. Feeding sites provide entrance for secondary microorganisms, specifically fungi, that may cause necrotic areas.

Management
1. Practice clean tillage.
2. Nematicides are effective where it is practical to use them.

Root Knot

Cause. Root knot nematodes *Meloidogyne arenaria, M. chitwood,* (primarily race 2), *M. hapla, M. incognita,* and *M. javanica* infect alfalfa that grows in sandy soils. The second stage, or newly hatched larvae, infect the root tip. *Meloidogyne arenaria, M. incognita,* and *M. javanica* do not survive if the temperature averages below 3°C during the coldest month. *Meloidogyne hapla* is limited by temperatures greater than 27°C.

Distribution. Widely distributed.

Symptoms. The top growth of diseased plants is chlorotic and stunted. Roots branch excessively and have galls or knots that vary in size. Bacterial wilt severity may increase.

Management
1. Use nematicides in areas of high nematode infestation.
2. Grow resistant cultivars.

Root Lesion

Cause. *Pratylenchus penetrans* is the most important root lesion nematode on alfalfa. Other root lesion nematodes associated with alfalfa are *P. crenatus*, *P. neglectus*, *P. coffeae*, *P. pratensis*, and *P. vulnus*. All life stages except the first-stage larva invade roots. The preferred area of infection is root hairs on feeder roots, where nematodes force their way through or between epidermal and cortical cells. Maximum invasion by males and larvae occurs at temperatures of 10° to 30°C, while females invade roots at temperatures of 5° to 35°C.

Gravid females deposit eggs in infected root tissue or soil. The second stage larvae emerge. Generation time is 4 to 8 weeks depending upon species.

Distribution. Wherever alfalfa is grown in the tropical and temperate areas.

Symptoms. Injury is most severe in sandy and sandy loam soils. Initially, a lesion appears on the root as a water-soaked area that becomes yellow and elliptical. Dark brown cells later appear in the center of the lesion. Yields and cold tolerance of diseased plants are decreased and infection by *Fusarium* species is increased.

Management. Nematicides are effective but impractical.

Spiral

Cause. *Helicotylenchus* spp. are ectoparasites.

Distribution. Widely distributed.

Symptoms. Necrotic spots occur at feeding sites. Roots may be severely damaged when high nematode populations are present.

Management. Nematicides control nematodes on some crops.

Stubby Root

Cause. *Paratrichodorus* spp.

Distribution. Not known for certain, but it is presumed *Paratrichodorus* species are widely spread on alfalfa.

Symptoms. Symptoms on alfalfa are not well known, but a primary symptom is reduced root growth.

Management. Not known.

Stunt

Cause. *Tylenchorhynchus* spp.

Distribution. Widely distributed.

Symptoms. The growth of foliage and roots is reduced. Diseased plants may be chlorotic.

Management. Nematicides have been effective on some crops.

Diseases Caused by Phytoplasmas

Aster Yellows

Cause. A phytoplasma transmitted by leafhoppers.

Distribution. Wherever alfalfa is grown, but aster yellows is not a serious disease of alfalfa.

Symptoms. Diseased plants are stunted and excessively branched. Flowers are chlorotic and sterile, develop leaf-like structures, and remain on the plant rather than being shed at normal senescence.

Management. Not reported.

Witches' Broom

Cause. A phytoplasma-like organism that overwinters in several perennial plants, such as *Astragalus, Hedysarum, Lathyrus, Lotus, Medicago, Melilotus,* and *Trifolium.* It is transmitted by grafting, dodder, and the leafhoppers *Orosius argentatus, Scaphytopius actus,* and *S. dubius.*

Distribution. Australia, North America, Russia, Saudi Arabia, and possibly other semiarid areas where alfalfa is grown.

Symptoms. An abnormal number of crown buds develop that form thin, short, pale green stems, which give a "broom-like" appearance to a diseased plant. Leaflets are small and have a yellowing and crinkling around their edges that give plants a yellowish cast. Flower buds develop slowly and may be green on some plants.

    Diseased plants are short-lived and may appear to recover during cool weather in the winter and spring, but recurrent symptoms develop during warm weather and moisture stress in summer.

Management. Not reported.

Unknown Phytoplasma Causing a Witches' Broom

This may possibly be the same organism described in the previous Witches' Broom description.

Cause. A phytoplasma-like organism. Pleomorphic bodies, spherical to fila-
mentous in shape and 60 to 650 nm in diameter, were observed in phloem
tissue and roots. The causal organism is transmitted by leafhoppers and
sometimes by dodder but is not transmitted mechanically.

Distribution. South Africa.

Symptoms. Diseased plants are stunted and have an abnormal proliferation
of stems. Leaves are abnormally small and chlorotic. Phyllody and negative geotropism occur.

Management. Not reported.

Diseases Caused by Viruses

Alfalfa Enation

Cause. Alfalfa enation virus (AEV). AEV is a rhabdovirus that is transmitted in
the field by the cowpea aphid, *Aphis craccivora,* and in the laboratory by
grafts.

Distribution. Eastern and southern Europe and Morocco.

Symptoms. Diseased leaflets are "crinkled." Enations several millimeters long
occur on the underside of the midvein of the crinkled leaflets. Diseased
plants may be of normal size but become "bushy."

Management. Not reported.

Alfalfa Latent

Alfalfa latent virus is also called pea streak virus.

Cause. Alfalfa latent virus (ALV) is in the carlavirus group. ALV is transmitted
in a nonpersistent manner by the pea aphid, *Acyrthosiphon pisum*. ALV is
also seed transmitted at a low rate and sap transmitted by machinery during harvesting.

Distribution. Hungary and the United States (Arizona, Nebraska, and the
northwestern United States); however, ALV is presumed to be widespread
in most alfalfa-growing areas.

Symptoms. Diseased plants are apparently symptomless.

Management. Grow alfalfa cultivars resistant to the pea aphid to prevent
spread of ALV to food legume crops.

Alfalfa Mosaic

Cause. Alfalfa mosaic virus, (AMV), overwinters in alfalfa and other perennial host plants. AMV is disseminated from diseased to healthy plants pri-

marily by the pea aphid, *Acyrthosiphon pisum,* although other aphids may also be involved. AMV is also seedborne and is transmitted through pollen and occasionally the ovules. Infected seed is the most likely source of inoculum in new alfalfa-growing areas.

The AMV complex is composed of many strains that differ in infectivity and other characteristics. It is similar to ilarviruses and is comprised of four elongate forms (18 × 31–59 nm). Three forms of the virus are bacilliform; the fourth is spheroidal.

Distribution. Wherever alfalfa is grown.

Symptoms. Older stands have the highest number of diseased plants. Symptoms are most obvious during cool weather in the spring and autumn. The most common symptoms are yellow streaks parallel to the leaf veins and yellow or light green mottling that is often accompanied by a distortion of the leaves. Stunting often occurs, and, infrequently, plants die. Infected plants may also be symptomless.

Management
1. Sow virus-free seed.
2. Control aphids where possible.
3. Grow the most resistant cultivar.

Bean Leaf Roll

Bean leaf roll was named legume yellows virus and pea leaf roll virus.

Cause. Bean leaf roll virus (BLRV) is in the luteovirus group and consists of isometric particles 28 nm in diameter. BLRV is confined to phloem and phloem parenchyma cells. BLRV is persistently transmitted by the pea aphid, *Acyrthosiphon pisum,* from mid-June to mid-July, but it is not mechanically transmitted.

Distribution. North America.

Symptoms. Diseased plants have a mild, transient yellowing of older leaves.

Management. Management measures are not necessary since BLRV has not been demonstrated to cause a yield loss in alfalfa.

Lucerne Transient Streak

Cause. Lucerne transient streak virus (LTSV) is in the sobemovirus group. The virus consists of isometric particles 30 nm in diameter. The method of transmission in the field is not known, but in the laboratory LTSV is readily transmitted by sap.

Distribution. Australia, Canada, and New Zealand.

Symptoms. Symptoms are most obvious on newly expanded leaflets and fade as leaflets age. No symptoms occur during high temperatures in the summer. Chlorotic streaks that vary in size from small spots to streaks 1 to 2 mm wide occur on the main lateral veins of leaflets. Leaflets are often distorted around the streaks. Yield losses of up to 18% have occurred in the field.

Management. Not reported.

Red Clover Vein Mosaic

Cause. Red clover vein mosaic virus (RCVMV) is in the carlavirus group and has straight to slightly flexuous, rod-shaped particles (13 × 650 nm). RCVMV is sap transmitted in an inefficient, nonpersistent manner by several aphid species, including the pea aphid, *Acyrthosiphon pisum,* and the green peach aphid, *Myzus persicae.*

Distribution. Canada (Alberta) in breeding lines maintained in the greenhouse and field and in *Trifolium* species in Europe, North America, and South America.

Symptoms. There is a general reduction in the winter hardiness of diseased plants. Initially, a yellow vein mosaic and chlorosis occur at the leaflet edges. Necrosis that occurs in interveinal areas results in premature plant death.

Management. Not known.

Selected References

Alcorn, J. L., and Irwin, J. A. G. 1987. *Acrocalymma medicaginis* gen. et sp. nov. causing root and crown rot of *Medicago sativa* in Australia. Trans. Br. Mycol. Soc. 88:163-167.

Allen, S. J., Barnes, G. L., and Caddel, J. L. 1982. A new race of *Colletotrichum trifolii* on alfalfa in Oklahoma. Plant Dis. 66:922-924.

Basu, P. K. 1981. Existence of chlamydospores as soil survival and primary infective propagules. (Abstr.) Phytopathology 71:202.

Blackstock, J. McK. 1978. Lucerne transient streak and lucerne latent: Two new viruses of Lucerne. Aust. J. Agric. Res. 29:291-304.

Carroll, R. B., Jones, E. R., and Swain, R. H. 1977. Winter survival of *Colletotrichum trifolii* in Delaware. Plant Dis. Rptr. 61:12-15.

Christen, A. A. 1982. Demonstrations of *Verticillium albo-atrum* within alfalfa seed. Phytopathology 72:412-414.

Christen, A. A. 1983. Incidence of external seedborne *Verticillium albo-atrum* in commercial seed lots of alfalfa. Plant Dis. 67:17-18.

Claflin, L. E., and Stuteville, D. L. 1973. Survival of *Xanthomonas alfalfae* in alfalfa debris and soil. Plant Dis. Rptr. 57:52-53.

Claflin, L. E., Stuteville, D. L., and Armbrust, D. V. 1973. Wind-blown soil in the epidemiology of bacterial leaf spot of alfalfa and common blight of bean. Phytopathology 63:1417-1419.

Cowling, W. A., and Gilchrist, D. G. 1981. Distinction between the "Californian" and

"Eastern" forms of Stemphylium leaf spot of alfalfa in North America. (Abstr.) Phytopathology 71:211.

Cowling, W. A., and Gilchrist, D. G. 1982. Effect of light and moisture on severity of Stemphylium leaf spot of alfalfa. Plant Dis. 66:291-294.

Cowling, W. A., Gilchrist, D. G., and Graham, J. H. 1981. Biotypes of *Stemphylium botryosum* on alfalfa in North America. Phytopathology 71:679-684.

Emberger, G., and Welty, R. E. 1981. Relationship of Fusarium wilt resistance and soil moisture to yield and persistence of alfalfa. (Abstr.) Phytopathology 71:766.

Farr, D. F., Bills, G. R., Chamuris, G. P., and Rossman, A. Y. 1989. Fungi on Plants and Plant Products in the United States. American Phytopathological Society, St. Paul, MN. 1252 pp.

Gilbert, R. G. 1985. *Sclerotinia sclerotiorum* causing crown and stem rot of alfalfa. (Abstr.) Phytopathology 75:1333.

Gilbert, R. G. 1988. Verticillium wilt of alfalfa: Dissemination via sclerotia of *Sclerotinia*. (Abstr.) Phytopathology 78:1514.

Gilbert, R. G. 1991. Burning to reduce sclerotia of *Sclerotinia sclerotiorum* in alfalfa seed fields of southeastern Washington. Plant Dis. 75:141-142.

Gilbert, R. G., and Peaden, R. N. 1988. Dissemination of *Verticillium albo-atrum* in alfalfa by internal seed inoculum. Can. J. Plant Pathol. 10:73-77.

Gossen, B. D. 1987. Initiation of crown bud rot of alfalfa. (Abstr.) Can. J. Plant Pathol. 9:277.

Gotlieb, A. R., Pellett, N. E., and Parker, B. 1987. *Sitona/Fusarium*/cold hardiness interaction in alfalfa. (Abstr.) Phytopathology 77:1615.

Graham, J. H. (Coordinator). 1979. A Compendium of Alfalfa Diseases. American Phytopathological Society, St. Paul, MN. 65 pp.

Graham J. H., Kreitlow K. W., and Falkner, L. R. 1972. Diseases, pp. 497-526. *In* Alfalfa Science and Technology. C. H. Hanson (Ed.). American Society of Agronomy, Madison, WI.

Graham, J. H., Devine, T. E., and Hanson, C. H. 1976. Occurrence and interaction of three species of *Colletotrichum* on alfalfa in the mid-Atlantic United States. Phytopathology 66:538-541.

Griffin, G. D. 1968. The pathogenicity of *Ditylenchus dipsaci* to alfalfa and the relationship of temperature to plant infection and susceptibility. Phytopathology 58:929-932.

Griffin, G. D., and Thyr, B. D. 1988. Interaction of *Meloidogyne hapla* and *Fusarium oxysporum* f. sp. *medicaginis* on alfalfa. Phytopathology 78:421-425.

Hancock, J. G. 1983. Seedling diseases of alfalfa in California. Plant Dis. 67:1203-1208.

Hancock, J. G. 1984. Prevalence and pathogenicity of *Pythium paroecandrum* on alfalfa in California. (Abstr.) Phytopathology 74:855.

Hemmati, K., and McClean, D. L. 1977. Gamete-seed transmission of alfalfa mosaic virus and its effect on seed germination and yield in alfalfa plants. Phytopathology 67:576-579.

Hiruki, C., and Miczynski, K. 1987. Severe isolate of alfalfa mosaic virus and its impact on alfalfa cultivars grown in Alberta. Plant Dis. 71:1014-1018.

Holub, E. B., and Grau, C. R. 1987. Root disease of alfalfa caused by *Aphanomyces euteiches* in Wisconsin. (Abstr.) Can. J. Plant Pathol. 9:278.

Huang, H. C. 1989. Distribution of *Verticillium albo-atrum* in symptomed and symptomless leaflets of alfalfa. Can. J. Plant Pathol. 11:235-241.

Huang, H. C., and Hyvr, A. M. 1984. Transmission of *Verticillium albo-atrum* to alfalfa via feces of leaf-chewing insects. (Abstr.) Phytopathology 74:797.

Huang, H. C., Hironaka, R., and Howard, R. J. 1986. Survival of *Verticillium albo-atrum* in alfalfa tissue buried in manure or fed to sheep. Plant Dis. 70:218-221.

Huang, H. C., Kokko, E. G., and Erickson, R. S. 1997. Infection of alfalfa pollen by *Sclerotinia sclerotiorum*. (Abstr.). Can. J. Plant Pathol. 19:111.

Irwin, J. A. G. 1984. Etiology of a new *Stemphylium*-incited leaf disease of alfalfa in Australia. Plant Dis. 68:531-532.

Jimenez Diaz, R. M., and Millar, R. L. 1988. Sporulation on infected tissues and presence of airborne *Verticillium albo-atrum* in alfalfa fields in New York. Plant Pathol. 37:64-70.

Koch, S. H., Knox-Davies, P. S., and Lamprecht, S. C. 1988. *Colletotrichum trifolii* causing root rot of lucerne in South Africa. Phytophylactica 20d:305-309.

Kuan, T. L., and Erwin, D. C. 1980. Predisposition effect of water saturation of soil on Phytophthora root rot of alfalfa. Phytopathology 70:981-986.

Laemmlen, F. F., Cooksey, D. A., and Erwin, D. C. 1990. Naturally occurring crown gall of alfalfa. (Abstr.) Phytopathology 80:890

Leath, K. T., and Hill, R. R., Jr. 1974. *Leptosphaerulina briosiana* on alfalfa: Relation of lesion size to leaf age and light intensity. Phytopathology 64:243-245.

Leath, K. T., and Hower, A. A. 1993. Interaction of *Fusarium oxysporum* f. sp. *medicaginis* with feeding activity of clover root curculio larvae in alfalfa. Plant Dis. 77:799-802.

Leath, K.T., and Kendall, W. A. 1983. *Myrothecium roridum* and *M. verrucaria* pathogenic to roots of red clover and alfalfa. Plant Dis. 67:1154-1155.

Lukezic, F. L. 1974. Dissemination and survival of *Colletotrichum trifolii* under field conditions. Phytopathology 64:57-59.

Lukezic, F. L., and Leath, K. T. 1983. *Pseudomonas viridiflava* associated with root and crown rot of alfalfa and wilt of birdsfoot trefoil. Plant Dis. 67:808-811.

Marks, G. C., and Mitchell, J. E. 1970. Penetration and infection of alfalfa roots by *Phytophthora megasperma* and the pathological anatomy of infected roots. Can. J. Bot. 49:63-67.

McVey, D. V., and Gerdemann, J. W. 1960. Host-parasite relations of *Leptodiscus terrestris* on alfalfa, red clover and birdsfoot trefoil. Phytopathology 50:416-421.

O'Neill, N. R. 1996. Pathogenic variability and host resistance in the *Colletotrichum trifolii*/*Medicago sativa* pathosystem. Plant Dis. 80:450-457.

Ooka, J. J., and Vchida, J. Y. 1982. Cylindrocladium root and crown rot of alfalfa in Hawaii. (Abstr.) Phytopathology 72:955.

Pennypacker, B. W. 1983. Dispersal of *Verticillium albo-atrum* in the xylem of alfalfa. Plant Dis. 67:1226-1228..

Pennypacker, B. W., Leath, K. T., and Hill, R. R., Jr. 1985. Resistant alfalfa plants as symptomless carriers of *Verticillium albo-atrum*. Plant Dis. 69:510-511.

Pennypacker, B. W., Leath, K. T., and Hill, R. R., Jr. 1988. Growth and flowering of resistant alfalfa infected by *Verticillium albo-atrum*. Plant Dis. 72:397-400.

Pennypacker, B. W., Leath, K. T., and Hill, R. R., Jr. 1991. Impact of drought stress on the expression of resistance to *Verticillium albo-atrum* in alfalfa. Phytopathology 81:1014-1024.

Pesic, Z., and Hiruki, C. 1987. Occurrence of red clover vein mosaic virus (RCVMV) in alfalfa breeding lines. (Abstr.) Phytopathology 77:1732.

Pfender, W. F., Hine, R. B., and Stanghellini, M. E. 1977. Production of sporangia and release of zoospores by *Phytophthora megasperma* in soil. Phytopathology 67: 657-663.

Pierce, L., and McCain, A. H. 1987. Alfalfa sprout rot caused by *Erwinia chrysanthemi*. Plant Dis. 71:786-788.

Pratt, R. G., and Rowe, D. E. 1991. Differential responses of alfalfa genotypes to stem inoculations with *Sclerotinia sclerotiorum* and *S. trifoliorum*. Plant Dis. 75:188-191.

Pratt, R. G., and Rowe, D. E. 1995. Comparative pathogenicity of isolates of *Sclerotinia trifoliorum* and *S. sclerotiorum* on alfalfa cultivars. Plant Dis. 79:474-477.

Ribeiro, O. K., Erwin, D. C., and Khan, R. A. 1978. A new high-temperature *Phytophthora* pathogenic to roots of alfalfa. Phytopathology 68:155-161.

Rodriguez, R., and Leath, K. T. 1987. *Phoma medicaginis* var. *medicaginis*: A primary root pathogen of alfalfa. (Abstr.) Phytopathology 77:1618.

Rodriguez, R., and Leath, K. T. 1992. Pathogenicity of *Phoma medicaginis* var. *medicaginis* to crowns of alfalfa. Plant Dis. 76:1237-1240.

Rodriguez, R., Leath, K. T., and Hill, R. R., Jr. 1990. Pathogenicity of *Phoma medicaginis* var. *medicaginis* to roots of alfalfa. Plant Dis. 74:680-683.

Seif El-Nasr, H. I., Abdel-Azim, O. F., and Leath, K. T. 1983. Crown and root fungal disease of alfalfa in Egypt. Plant Dis. 67:509-511.

Seif El-Nasr, H. I., Abdel-Azim, O. F., and Leath, K. T. 1984. Marasmius root rot of alfalfa and khella in Egypt. Plant Dis. 68:906-907.

Shyh-Jane Lu, N., Barnes, D. K., and Frosheiser, F. I. 1973. Inheritance of Phytophthora root rot resistance in alfalfa. Crop Sci. 13:714-717.

Simmons, E. G. 1985. Perfect states of *Stemphylium*. II. Sydowia 38:284-293.

Skinner, D. Z., and Stuteville, D. L. 1995. Host range expansion of the alfalfa rust pathogen. Plant Dis. 79:456-460.

Smith, O. F. 1940. Stemphylium leaf spot of red clover and alfalfa. J. Agric. Res. 61:831-846.

Stack, J. P., and Millar, R. L. 1985. Competitive colonization of organic matter in soil by *Phytophthora megasperma* f. sp. *medicaginis*. Phytopathology 75:1020-1025.

Stack, J. P., and Millar, R. L. 1985. Relative survival potential of propagules of *Phytophthora megasperma* f. sp. *medicaginis*. Phytopathology 75:1025-1031.

Streets, R. B., and Bloss, H. E. 1973. Phymatotrichum root rot. American Phytopathology Society, Monograph No. 8. 38 pp.

Stuteville, D. L., and Erwin, D. C. (Ed.). 1990. A Compendium of Alfalfa Diseases, Second Edition. American Phytopathological Society, St. Paul, MN. 84 pp.

Stutz, J. C., Leath, K. T., and Kendall, W. A. 1985. Wound-related modifications of penetration, development, and root rot by *Fusarium roseum* in forage legumes. Phytopathology 75:920-924.

Sundheim, L., and Wilcoxson, R. D. 1965. *Leptosphaerulina briosiana* on alfalfa: Infection and disease development, host-parasite relationships, ascospore germination and dissemination. Phytopathology 55:546-553.

Thompson, A. H. 1987. Phytophthora root rot of lucerne in Transvaal, South Africa. Phytophylactica 19:319-322.

Thompson, A. H., and Pietersen, G. 1988. Witches' broom of lucerne associated with a mycoplasma-like organism in South Africa. Phytophylactica 20:297-303.

Turner, V., and Van Allen, N. K. 1983. Crown rot of alfalfa in Utah. Phytopathology 73:1333-1337

Veerisetty, V., and Brakke, M. K. 1977. Alfalfa latent virus: A naturally occurring carlavirus in alfalfa. Phytopathology 67:1202-1206.

Vincelli, P. C. 1992. Two diseases of alfalfa caused by *Rhizoctonia solani* AG-1 and AG-4. Plant Dis. 76:1283.

Vincelli, P. C., Nesmith, W. C., and Doney, J. 1991. Seedling disease of alfalfa in Kentucky caused by *Aphanomyces euteiches*. (Abstr.) Phytopathology 81:1136.

Webb, D. H., and Nutter, F. W., Jr. 1997. Effects of leaf wetness duration and temperature on infection efficiency, latent period, and rate of pustule appearance of rust in alfalfa. Phytopathology 87:946-950.

Welty, R. E., and Rawlings, J. O. 1980. Effects of temperature and light on development of anthracnose on alfalfa. Plant Dis. 64:476-478.

Wick, R. L., and Jeschke, N. 1987. Deterioration of alfalfa sprouts by *Pseudomonas* species during production. (Abstr.) Phytopathology 77:1620.

2. Diseases of Barley *(Hordeum vulgare)*

Diseases Caused by Bacteria

Bacterial Blight

Bacterial blight is also called bacterial leaf blight, bacterial leaf streak, bacterial streak, black chaff, black chaff and bacterial streak, and Xanthomonas streak.

Cause. *Xanthomonas translucens* pv. *translucens* (syn. *X. campestris* pv. *translucens*) is seedborne and overwinters in infested residue. Some researchers have reported *X. campestris* pv. *undulosa* as a causal bacterium.

The taxonomy of the causal organism(s) continues to be a subject in flux. Bacterial blight, which is called bacterial streak, bacterial leaf streak or black chaff on other small grains, is considered to be caused by five of the former *X. campestris* pathovars: pv. *cerealis*, pv. *hordei*, pv. *secalis*, pv. *translucens*, and pv. *undulosa*. The pathovars are often grouped together under the name "translucens" or are considered a single species or a pathovar named *X. translucens* and *X. campestris* pv. *translucens*, respectively. Bragard et al. (1997) agree that *X. translucens* pv. *translucens* and *X. translucens* pv. *hordei* are true synonyms. They further report that using restriction fragment-length polymorphism and fatty acid methyl esters analysis, the pathovars *X. translucens* pv. *cerealis*, *X. translucens* pv. *translucens*, and *X. translucens* pv. *undulosa* cluster in different groups and correspond to true biological entities.

Inoculum is spread to new locations primarily by infested seed. However, bacteria are also disseminated by the physical contact of a diseased leaf rubbing against a healthy one, splashing rain, and insects. Bacteria enter the plant through natural openings such as stomata and wounds.

Distribution. Wherever barley is grown.

Symptoms. Black chaff is seldom serious, and symptoms are confined mainly to leaves. Small, water-soaked spots occur on tender green leaves and sheaths of older plants, and sometimes on seedlings. Occasionally, a spot may become large and blotchy, causing the leaf to die, shrivel, and turn light brown in color. The spots enlarge and coalesce to become glossy, yellow or brown, translucent, narrow stripes that are limited by leaf veins but frequently extend the length of a diseased leaf blade. When a newly developed stripe is cut laterally, a milky "ooze" consisting of bacteria will exude from the cut edges. Severely diseased leaves die back from the tips. If a flag leaf is infected, the head may not emerge from the boot but may break through the side of the sheath and be distorted and blighted. Under humid, early morning conditions, droplets of milky bacterial exudate may be seen on the surface of spots and stripes. Bacterial exudate eventually dries into hard, yellowish flakes that can be easily removed from the leaf surface.

Chaff may also be affected after several days of damp or rainy weather. Water-soaked areas may occur on glumes but bacterial exudate will not be evident. Seed is not destroyed but may become brown, shrunken, and carry bacteria that, potentially, can infect next year's crop. Diseased plants grow slowly and are stunted, but disease symptoms are not noticed until plants are about two-thirds grown.

Management
1. Sow disease-free seed.
2. Treat seed with a seed-protectant fungicide.
3. Rotate barley with other crops. *Xanthomonas translucens* pv. *translucens* also infects wheat.
4. After dry heat treatments of heavily infested seed at 71°, 75° or 84°C for 11 days, the reduction in germination was negligible and *X. translucens* pv. *translucens* was not detected.

Bacterial Leaf Blight

Cause. *Pseudomonas syringae* pv. *syringae*. Bacteria infect leaves through stomates or wounds during cold, wet, and windy weather.

Distribution. North America.

Symptoms. Normally, only upper leaves, particularly the flag leaves, are diseased. Initially, small, water-soaked, gray-green lesions form, often where the leaf bends. Later, these coalesce to form larger tan or white lesions that cover the entire leaf. Eventually, the whole leaf may die, become necrotic, and curl inward from the edges.

Other symptoms caused by *P. syringae* in south central Colorado have been described as typical halo blight.

Management. Not necessary.

Bacterial Stripe

Bacterial stripe is also called bacterial stripe blight.

Cause. *Pseudomonas syringae* pv. *striafaciens* (syn. *P. striafaciens*) is seedborne and survives in infested crop refuse for at least 2 years. Bacteria are splashed or blown onto leaves during cool, wet weather. Warm, dry weather stops the spread of the bacteria.

Distribution. Australia, Europe, North America, and South America. Bacterial stripe occurs infrequently.

Symptoms. Initially, sunken, water-soaked dots appear on diseased leaves and, if dots are numerous, the leaf may die. The dots enlarge into water-soaked stripes or blotches with narrow yellowish margins that extend the length of the leaf blade. Stripes become a translucent rusty brown as they age. Droplets of bacteria exude from stripes in moist weather and later dry up and form white scales on the stripe surfaces.

Management
1. Plow under infected residue.
2. Treat seed with a seed-protectant fungicide.
3. Grow resistant cultivars.

Barley Basal Kernel Blight

Barley basal kernel blight is also called bacterial kernel blight.

Cause. *Pseudomonas syringae* pv. *syringae* survives on infested residue and leaf surfaces of nonhost plants and is seedborne. Kernel infection occurs before the milky dough stage. The infection process starts with an epiphytic inoculum buildup on the kernel surface. Bacteria enter kernels primarily through stomates located on the inner epidermis of the barley lemma before the lemma becomes attached to the caryopsis. At soft dough stage, *P. syringae* might enter kernels through pits or wounds in the external epidermis.

Distribution. North America.

Symptoms. Infected kernels have tan to dark brown spots (approximately 2 mm in diameter) with distinct margins that develop at the embryo end on the posterior surface of the kernel. A lemma spot symptom caused by *P. syringae* pv. *syringae,* but possibly also by other *P. syringae* pathovars, is a well-defined tan to dark brown discoloration of the kernel lemma. Bacteria are found in intercellular spaces of the aleurone layer, within amyloplast cells, and in vascular bundles.

Management. Not known.

Basal Glume Rot

Basal glume rot may be the same disease as basal bacteriosis.

Cause. *Pseudomonas syringae* pv. *atrofaciens* (syn. *P. atrofaciens* and *Phy-*

tomonas atrofaciens) survives epiphytically on seed and in infested soil residue. Bacterial spores on dust particles are disseminated by wind and become entrapped in the water in grooves and small spaces of the spikelets. Bacteria are also disseminated by insects and splashing water. Bacteria multiply near glume joints in the presence of moisture but remain dormant when conditions are dry.

Distribution. Wherever barley is grown.

Symptoms. Symptoms occur during wet weather, particularly at heading time. The main symptom is a brown, discolored area at the base of the glumes that cover a kernel. This discoloration is darker on the inside than the outside of the glume. Usually only one-third of the glume is discolored but sometimes the entire glume may be affected. Infrequently, the only sign of disease is a dark line at the attachment of the glume to the spike.

Severely diseased spikelets are slightly dwarfed and lighter in color than healthy ones. The base of a diseased kernel has a light brown to black discoloration.

Infected leaves have small, dark, water-soaked spots. Spots eventually elongate and become yellow, then necrotic as the tissue dies.

Management
1. Seed should be thoroughly cleaned and treated with a seed-protectant fungicide.
2. Rotate barley with resistant crops, such as legumes.
3. Plow under residue.

Diseases Caused by Fungi

Anthracnose

Cause. *Colletotrichum graminicola* is seedborne and overwinters as mycelia and conidia in infested residue, cereals, and wild grasses. Optimum conidia production is during wet weather at a temperature of 25°C; conidia are disseminated primarily by wind to barley and other host plants.

Anthracnose is most severe when barley is grown on sandy soils that are low in fertility.

Distribution. Wherever barley is grown.

Symptoms. Seedling infection may occur under severe disease conditions. Symptoms generally become apparent toward plant maturity as premature ripening or whitening of diseased plants, a general reduction in plant vigor, and plant death. Elliptical-shaped, water-soaked lesions (1–2 cm long) that later become bleached and necrotic occur on the lower part of the diseased plant. The entire crown and bases of stems become bleached and then turn brown. Later, acervuli, which appear as small, black, raised spots to the unaided eye and as clumps of dark spines under magnification, develop on the lesion surfaces of diseased lower leaf sheaths and culms. Acervuli may also develop on moist leaf blades of dead plants. Kernels may be shriveled and light in weight if plants are infected early.

Management
1. Rotate barley with a noncereal crop.
2. Improve soil fertility.
3. Treat seed with a fungicide.

Arthrinium arundinis Kernel Blight

Cause. *Arthrinium arundinis* is a saprophyte of decaying grasses, but barley grown in highly humid environments may be susceptible.

Distribution. The United States (Montana).

Symptom. A brown discoloration occurs on the basal portion of the diseased kernel.

Management. Not reported.

Ascochyta Leaf Spot

Cause. *Ascochyta hordei* var. *americana* is generally accepted as the cause of Ascochyta leaf spot; however, *A. graminea, A. sorghi,* and *A. tritici* have also been reported to cause leaf spot diseases of barley.

    Ascochyta hordei var. *americana* survives as mycelia and pycnidia in infested residue. Pycnidiospores are liberated during wet weather and are disseminated by splashing rain and wind. Disease is associated with dense foliage, leaves in contact with soil, and high humidity.

Distribution. Japan and the United States, primarily the northwestern states and New York.

Symptoms. Spots, which are round (4 mm in diameter), oval (2–4 × 8–12 mm), or irregular in shape and sometimes have several rings, occur primarily on approximately one-third of the lower leaf blades nearest the tip end. Initially, spots are brown but fade to pale yellow, ashy gray, or tan necrotic centers that are surrounded by a narrow brown band.

    Lesions at leaf tips often coalesce to cause dieback but without the brown margin. The latter symptom is often masked by frost injury. Pycnidia, which appear as tiny black specks, are present in diseased tissue but are absent from frost-damaged tissue.

Management. Disease management is not necessary since Ascochyta leaf spot is generally considered to be of minor importance.

Barley Stripe

Barley stripe is also called barley leaf stripe, barley stripe disease, Helminthosporium stripe, leaf stripe, Pyrenophora leaf stripe, and stripe disease.

Cause. *Pyrenophora graminea,* anamorph *Drechslera graminea* (syn. *Helminthosporium gramineum*) is seedborne and survives up to 5 years as mycelia in hulls, pericarps, and seed coats. At seed germination, mycelia, primarily from the pericarp but also from the seed coat over the pericarp, grow into the sheath that surrounds the leaves of the seedling. The critical stage for infection of the germinating embryo is when the coleoptile reaches the apex of the seed until the seedling emerges from the soil. From the seedling leaves, mycelia grow into the first leaf and subsequently into succeeding leaves until eventually all leaves are infected. Infection is optimum at soil temperatures of 12°C or less and is reduced or prevented at soil temperatures above 15°C.

Secondary infection is caused by conidia that are produced in stripes of infected barley leaves during high moisture conditions. The conidia mature in about 16 hours at 12°C and are disseminated by wind to the heads of healthy plants. Conidia lodge near the tips of glumes and germinate, with the resulting mycelia growing either between hulls and kernels or into seed coats. Infection occurs at temperatures of 10° to 33°C. Seeds are most susceptible to infection in the early stages of plant maturation; infection levels decrease from the boot stage to milk stage of plant growth. Free water, although beneficial, is not required because high relative humidity provides adequate moisture for seed infection.

Disease spread is directly related to moisture. Compared with plants grown in a dry environment, irrigation or high relative humidity at heading will increase disease incidence up to threefold. Barley stripe is more important on winter barley than on spring barley.

Distribution. Wherever winter barley is grown.

Symptoms. The first symptom occurs at the late tillering stage when small, yellow spots appear on seedling leaves. These minuscule spots could be easily overlooked by a casual observer. Weeks before heading, one to several narrow, yellow to light tan, parallel stripes extend the length of leaf blades and sheaths. The alternate yellow and green striping of diseased leaves contrasts with the uniform green of healthy ones. The margins of stripes turn reddish or dark brown but the stripe centers remain tan or light brown. Diseased tissue dies, then splits within the stripes, causing the entire leaf to become shredded. Stripes eventually become gray to olive gray due to sporulation by *D. graminea.* The culm dies if the growing point is infected.

Diseased plants are stunted. Heads either do not emerge or emerge twisted, compressed, brown, and standing erect, in contrast to mature healthy heads that bend over slightly. Grain in diseased heads is undeveloped or shriveled and brown.

Management
1. Treat seed with a seed-protectant fungicide.
2. Grow resistant cultivars.

Blast

Cause. *Pyricularia oryzae.* However, some isolates obtained from two-rowed barley are not pathogenic to rice and have been tentatively identified as *P. grisea.* Rice likely provides an inoculum source for *P. oryzae,* and ryegrass may provide a source for *P. grisea.* See blast in the chapter on rice diseases.

Distribution. Reported in two-rowed barley, *Hordeum distichum,* from Japan; however, blast likely occurs where barley and rice or other hosts are grown in the same vicinity.

Symptoms. Spots on leaves begin as small, water-soaked, white, gray, or blue dots that quickly enlarge under moist conditions (0.3–0.5 × 10.0–15.0 mm). Mature leaf spots are elliptical and pointed at both ends and have gray to white centers that are surrounded by brown to red-brown margins.

Spots may vary in shape and color depending on cultivar, environment, and age of the spot. Resistant cultivars have only brown specks; moderately resistant cultivars have small, round lesions with necrotic centers and brown margins.

Later in the growing season, some cultivars become infected at the point where the flag leaf attaches to the sheath. The lesion continues to expand by enlarging downward on the sheath and upward on the leaf and becomes gray at the point of attachment. The flag leaf may subsequently break off and become detached from the plant.

Nodes may be diseased. The affected node itself eventually breaks apart and remains connected together only by a few vascular strands. The base of the leaf sheath, which is connected to the node, turns black due to the sporulation of conidia. The portion of the plant above the diseased node dies.

Any portion of the panicle, panicle branches, and glumes may become diseased and have brown lesions on the diseased plant part. If the base of the panicle is diseased, rotten neck or neck rot symptoms cause the stem below the panicle to snap.

Management. Grow resistant cultivars. In two-rowed barley, resistance has been found to be conditioned by a single dominant gene.

Cephalosporium Stripe

Cause. *Hymenula cerealis* (syn. *Cephalosporium gramineum*) survives as conidia, and as mycelia in residue in the top 8 cm of soil for at least 5 years. Soilborne conidia serve as primary inoculum and infect roots during winter
and early spring through mechanical injuries caused by soil heaving and
insects. Infection is most severe in wet, acid soils (pH 5.0) because of a combination of increased fungal sporulation and root growth. The subcrown
internode can also be infected just as the seed is germinating.

After infection, conidia enter xylem vessels and are carried upward in
the plant's xylem tissue, where they lodge and multiply at nodes and
leaves, subsequently preventing water from moving up the plant. *Hymenula cerealis* also produces metabolites that are harmful to the plant. At
harvest time, *H. cerealis* is returned to the soil in infested residue, where it
is a successful saprophytic competitor against other soilborne microorganisms.

Distribution. Great Britain, Japan, and in most winter barley–growing areas
of North America.

Symptoms. Stunted plants are scattered throughout a field but are more numerous in the lower and wetter areas of the field. During jointing and
heading, one or two, and rarely up to four, distinct yellow stripes, in which
infected veins appear as thin brown lines, develop the length of the plant
on leaf blades, sheaths, and stems. Stripes eventually become brown and
are highly visible in contrast to the rest of the green leaf. After harvest,
stripes remain visible on the yellow straw. At plant maturity, culms at or
below nodes become darkened because of sporulation by *H. cerealis*. Heads
are white and do not contain seed, or if seed is present, it is usually shriveled.

Management
1. Rotate barley for at least 2 years with a noncereal crop.
2. Sow winter barley later in the autumn or when the soil temperature at
   the 10-cm level in the soil is below 13°C. The plants will have limited
   root growth, which reduces the number of potential infection sites.
3. Plow residue deeper than 8 cm.

Ceratobasidium gramineum Snow Mold

Cause. *Ceratobasidium gramineum* (syn. *Corticium gramineum*). Snow mold occurs in well-drained, upland paddy fields.

Distribution. Japan. *Ceratobasidium gramineum* snow mold is most common
on barley but also occurs on wheat.

Symptoms. As the snow melts, leaves on diseased plants are rotted and dark
green. After leaves dry, they turn pale gray to pale yellow. Lesions on sur-

viving leaves are ellipsoidal and have pale gray mycelial mats in their centers and narrow brown margins. Leaf sheaths frequently are brown.

Symptom development ceases after snowmelt and the plants are exposed to light and drying wind. However, later in the growing season, ellipsoidal lesions are found on leaf sheaths near the soil level.

Management. Not reported.

Common Root Rot

Cause. Several fungi, including *Bipolaris sorokiniana,* teleomorph *Cochliobolus sativus, Fusarium culmorum,* and *F. graminearum.* Conidia of *B. sorokiniana* and chlamydospores of *Fusarium* spp. survive for several years in soil. These fungi also are excellent saprophytes and colonize plant material added to the soil. *Microdochium bolleyi* also is a secondary invader of roots and is sometimes isolated together with *B. sorokiniana* and *Fusarium* spp. In Finland, *M. bolleyi* is reported to commonly be isolated together with *B. sorokiniana* and *Fusarium* spp.

Plants under stress from drought, warm temperatures, lack of nutrition, and Hessian fly injury are most subject to infection. Moisture is required for infection, but once disease is initiated, further development requires warm temperatures and moisture stress. Disease initiation by *B. sorokiniana* begins 25 to 30 days following sowing and reaches maximum levels at about 60 days after sowing.

Sites for secondary sporulation of fungi vary, presumably because of the different sporulation requirements for different fungi. The most abundant sporulation for some fungi, especially *B. sorokiniana,* is on crowns, with the remainder occurring on crown roots and on subcrown internodes, seed pieces, and seminal roots. Other sporulation by *Fusarium* spp. occurs on mature plants; the highest numbers of conidia are present on necrotic or senescent tissues, such as coleoptiles, and the lower outside leaf sheaths.

Disease incidence increases under continuous cultivation. Isolates of *C. sativus* from fields in continuous barley are more pathogenic to barley than isolates from fields in first-year barley. Isolates of *C. sativus* from diseased wheat and barley are highly virulent to their original host species but are weakly virulent to alternate host species.

An increase in root rot levels causes an increase in net blotch severity on winter barley.

Distribution. Generally distributed wherever barley is grown, but common root rot is more frequent where plants are under moisture stress during the growing season.

Symptoms. Common root rot is usually noticed after plant heading, when it resembles drought damage. Lesions develop on the subcrown internode in 7 to 9 weeks after emergence and cause the lower leaves to die. Brown lesions of variable size may develop on the leaf base nearest the soil surface and extend partially or completely around the stem. Later in the growing season, the lower parts of leaves, tiller buds, crown roots, and internodes below and above crowns have brown lesions.

Rotted crown and crown roots can be seen when dead leaves and tillers are removed from the crown. Diseased plants are shorter, and survival of winter barley is reduced, particularly when infected by *B. sorokiniana.*

Management. The following aid in managing disease:
1. Rotate barley with noncereal crops.
2. Plow under infested residue where feasible.
3. Experimental seed treatment with the fungicide imazalil lowered disease severity in the subcrown internode, increased grain yield by 6%, and increased test weight compared to the control.

Covered Smut

Cause. *Ustilago hordei.* Smutted heads emerge at the same time as heads of healthy plants. When the semipersistent membrane covering the teliospores is ruptured by wind, rain, and harvest machinery, the teliospores are liberated and become windborne for a considerable distance. They come to rest on or under the hulls of healthy kernels and on the soil surface. Spores on the kernel surface remain dormant until the kernel is sown the following spring or autumn. Spores deposited under hulls may germinate and infect just the outer layers of the seed, where the resultant mycelia remain dormant until sowing time the following spring or autumn. Spores also overwinter in soil.

Seedlings become infected between germination and emergence by seedborne spores and mycelia or by soilborne spores. Mycelia then grow internally just behind the growing point of the infected plant. At the boot stage, mycelia grow into the flower ovary, eventually converting it into masses of spores.

The amount of infection depends on the percentage of infested kernels that are sown. Acid or neutral soils, soil temperatures of 14° to 25°C, and high soil moisture are favorable conditions for disease development.

Physiologic races of *U. hordei* are present.

Distribution. Wherever barley is grown.

Symptoms. A somewhat persistent grayish-white membrane that encloses millions of black teliospores emerges from the boot instead of a normal green head. From a distance, smutted heads will appear grayish and are easily noticed throughout a field. Each smutted seed is now a dark, powdery mass of teliospores that crushes easily and colors fingers a dark brown or black. After the membrane ruptures and teliospores are released, the remaining naked rachis contrasts with the surrounding healthy heads.

Management.
1. Treat seed with a systemic seed-protectant fungicide.
2. Grow resistant cultivars.

Crown Rust

Cause. *Puccinia coronata.* The rust is highly virulent on winter barley and is pathogenic to spring barley; foxtail barley, *Hordeum jubatum*; quackgrass, *Elytrigia repens*; common buckthorn, *Rhamnus carthartica*; *Secale* spp.; and many gramineous species. This form of *P. coronata* differs from *P. coronata* var. *avenae* by having low pathogenicity on *Avena* spp., darker urediniospores, and longer telial appendages.

Distribution. The United States (Minnesota, Nebraska, North Dakota, and South Dakota).

Symptoms. Infection occurs on leaves, leaf sheaths, awns, and peduncles. Uredinia are elongate and light orange. When viewed under magnification, the teliospore apical cell has four to six appendages that are up to 46 µm long.

Management. Not reported.

Downy Mildew

Cause. *Sclerophthora rayssiae* is an obligate parasite that survives for several years as oospores that are in residue or are released into soil when residue decays. Oospores are seedborne and are also disseminated in water and windborne residue and soil.

Leaves of newly emerged seedlings in contact with water 24 hours or longer are most susceptible to infection. Oospores germinate in water-saturated soil to produce sporangia. Zoospores are then produced within sporangia and, upon their release from the sporangium, "swim" through water to the leaves of developing seedlings, where they germinate to produce a germ tube. Oospores may also survive for months in dry soil and germinate to form a single germ tube that directly penetrates a host. Infection occurs over a wide temperature range (7° to 31°C). *Sclerophthora rayssiae* then develops systemically in the plant, particularly in meristematic tissue. Eventually, during the growing season, oospores are produced within diseased tissue to complete the life cycle of the fungus.

Distribution. Wherever winter barley is grown.

Symptoms. Diseased plants are scattered in areas where excess water was present in low areas of a field, usually after excessive rainfall. Diseased plants are dwarfed and deformed, tiller excessively, and may be "flagged" with prominent yellowed leaves. Leaves are leathery, stiff, "warty" and thickened.

Affected heads are twisted in various ways and distorted. Seed is normally not formed in severely diseased plants. Severely diseased plants may die before jointing.

In less severely diseased plants, dwarfing may be slight, with one or more of the upper leaves stiff, upright, or curled, and twisted. Heads and stems may not be deformed.

In diseased tissue, numerous round, yellow to brown oospores may be observed under magnification.

Management
1. Provide proper soil drainage.
2. Control grassy weeds that may serve as hosts.
3. Sow cleaned seed from disease-free plants.

Dwarf Bunt

Cause. *Tilletia controversa* survives as teliospores in soil or on seed. When seed is sown, teliospores on or close to the seed germinate in the presence of moisture to produce a promycelium on which 8 to 16 basidiospores are formed. A basidiospore then fuses in the middle with a compatible basidiospore to form an H-shaped structure that germinates to form yet other structures called secondary sporidia. The secondary sporidia germinate to produce mycelia that infect seedlings.

The infection process takes 35 to 105 days, depending on temperature. The optimum temperature for spore germination is 1° to 5°C, a condition that occurs under heavy snow cover over unfrozen ground. The mycelium grows internally in meristematic tissue of the plant, invades the developing head, and replaces seed with teliospores. At harvest, diseased seed is broken and teliospores are windborne and contaminate healthy kernels or soil. Isolates of *T. controversa* from barley can infect wheat and vice versa.

Distribution. Winter barley in the United States (Utah). Dwarf bunt has never been a serious disease problem on barley and is rarely or infrequently observed.

Symptoms. Dwarf bunt is restricted to areas where winter barley grows under snow cover for a long time. Infected tillers are stunted and grow to about one-fourth to one-half the height of a normal plant. Diseased heads are more compact than healthy plant heads, and sori fill the seed coat to form bunt "balls" that are rounder and shorter than normal barley seed. A foul odor, similar to rotten fish, occurs when bunt balls are broken and teliospores are moistened.

Management. Not necessary because of the low incidence of disease.

Ergot

Cause. *Claviceps purpurea,* anamorph *Sphacelia segetum,* survives for approximately 1 year as sclerotia in soil. Prior to plant blossoming, sclerotia germinate to produce stromatic heads at the end of stipes in which perithecia are formed. Ascospores form in perithecia and are primarily windborne but occasionally are splashed by rain to florets, where they germinate. The resulting mycelium grows into the young ovary and eventually produces conidia along with a sweet, sticky liquid called honeydew. Insects are attracted to honeydew and, in the process of feeding on the sweet liquid, inadvertently become contaminated with conidia. Insects disseminate the conidia to healthy flowers, where the conidia germinate and cause a secondary infection.

Eventually, the diseased ovaries enlarge and are converted into sclerotia, which are composed of hardened mycelium, from the kernel base to the kernel tip. Since infection of flowers does not occur after pollination, infection is favored by prolonged flowering, which tends to occur during cool, wet weather.

Plants infected with the barley stripe mosaic virus are more susceptible to infection by *C. purpurea* than healthy plants. Several grasses and other cereals are susceptible to infection by *C. purpurea.*

Distribution. Wherever barley is grown. However, ergot is normally not a severe disease of barley.

Symptoms. The only plant part affected is the head. Initially, infected flowers ooze droplets of a yellowish, sticky liquid called honeydew that attracts insects. Insects may be observed clustered about a diseased flower. Also, dust adheres to the sticky liquid to give it a "dirty" or dusty appearance. At maturity, diseased kernels are replaced by larger, hard, black to bluish-black sclerotia that protrude from the flume.

Management
1. Plow residue deep into the soil so sclerotia will be placed in the soil profile, where they cannot release ascospores into the air. Sclerotia are normally decomposed by soil microorganisms in about 1 year.
2. Control grassy weeds around fields that may be collateral hosts for *C. purpurea* and increase inoculum that will infect cereal crops, including barley.
3. Rotate barley with resistant hosts.
4. Separate sclerotia from seed. Sclerotia may be harmful when fed to livestock because of the potential for production of mycotoxins.

Eyespot

Eyespot is also called culm rot, foot rot, straw breaker, and straw breaker foot rot.

Cause. *Pseudocercosporella herpotrichoides* (syn. *Cercosporella herpotrichoides*), teleomorph *Tapesia yallundae*. Two varieties of the fungus are recognized. *Pseudocercosporella herpotrichoides* var. *herpotrichoides* produces four-septate conidia measuring 35.0–80.0 × 1.5–2.5 μm; *P. herpotrichoides* var. *acuformis* produces four- to six-septate conidia measuring 43.0–120.0 × 1.2–2.3 μm. Races have not been reported, but pathogenic specialization to host species does occur. There are two main pathogenic types: W-type (wheat-type) isolates are most virulent on wheat and less virulent on barley and rye. R-type (rye-type) isolates are equally virulent on wheat, barley, and rye.

    Pseudocercosporella herpotrichoides survives as mycelium in infested crop residue that was previously colonized during the parasitic stage of the fungus. Conidia are produced during cool, damp weather in the autumn or spring and subsequently disseminated, primarily by rain, to crown and basal culm tissue of nearby plants. Roots are normally not infected.

    Infection is favored by high soil moisture, dense crop canopy, and high humidity near the soil surface. Winter barley is more likely to be infected than spring barley because of predisposition by spring frosts and excessive nitrogen fertilization. *Pseudocercosporella herpotrichoides* has a wide host range and infects other cereals.

Distribution. It is generally assumed that eyespot is widespread wherever barley is grown.

Symptoms. Eye-shaped or ovate lesions with white to tan centers and brown margins develop first on the basal leaf sheath and darken with age. Similar spots form on the stem directly beneath those on the sheath and cause lodging. Under moist conditions, lesions enlarge and a black stroma-like mycelium develops over the crown surface and base of the culms, giving them a charred appearance. The stems then shrivel and collapse, causing diseased plants to fall in all directions at maturity. Roots are not affected but a necrosis occurs around roots in the upper crown nodes. Chlorotic plants with white heads or heads reduced in size and number are also present.

Management
1. Rotate barley with legumes.
2. Sow spring barley or delay sowing winter barley.

False Loose Smut

False loose smut is also called nigra loose smut, black smut, black loose smut, intermediate loose smut, and semi-loose smut.

Cause. *Ustilago nigra* (syn. *U. avenae*) is seedborne, surviving as teliospores (sometimes called chlamydospores) on the seed surface, under the hull, and in soil. Smutted heads emerge at approximately the same time as heads of healthy plants. A mass of teliospores that has replaced the kernels is covered by a semipersistent membrane which may remain intact until harvest time unless it is ruptured by wind, rain, or other means. Teliospores are disseminated by wind and are deposited on or under hulls of healthy kernels. Spores on the seed hull surfaces remain dormant until kernels are sown the following spring or autumn. Spores deposited under seed hulls germinate immediately and the resultant mycelia grow into the kernel layers that are just under the hull. There, the mycelium remains dormant until sowing time the following spring or autumn.

Seedlings become infected between germination and emergence by seedborne spores and mycelia or by soilborne spores. Soil temperatures of 15° to 21°C and dry moisture conditions are most favorable for infection. Following infection, the mycelium grows just behind the growing point of the seedling. As diseased plants enter the boot stage, the mycelium grows into the young flowers and converts them to masses of teliospores. The smutted head then emerges from the boot at approximately the same time as a healthy head.

Physiologic races of *U. nigra* exist.

Distribution. Africa, Asia, Australia, Europe, Central America, North America, and South America.

Symptoms. Symptoms vary between those of loose smut and covered smut, but more closely resemble those of loose smut. False loose smut can be distinguished from true smut with certainty only by observing smut spores and their germination under a microscope. The most reliable distinguishing feature is the sporidial germination type of *U. nigra* compared to the mycelial germination type of *U. nuda.*

Ustilago nigra teliospores are spherical to subspherical, 5–9 μm in diameter, and dark brown to black with echinulations varying from slight to pronounced. Chlamydospores germinate to form a three- or four-celled basidium bearing sporidia from each cell.

Ustilago nuda teliospores are globose to subglobose or elongate, 5–7 μm in diameter, olivaceous brown, lighter on one side, and minutely echinulate. Spores germinate to form a slender germ tube that branches and rebranches to form mycelium.

Smutted heads appear at approximately the same time as healthy heads. The dark brown to black spore mass and variation in looseness of the mass are gross characteristics. The membrane around the spore mass varies from easily broken to stronger persistent. When membranes break, spores are quickly washed or blown away, leaving a bare rachis. Persistent membranes retain their spores until harvest.

Management. Treat seed with a systemic seed-protectant fungicide.

Fusarium Head Blight

Fusarium head blight is also called Fusarium blight, head blight, and scab.

Cause. *Fusarium graminearum,* teleomorph *Gibberella zeae,* is generally considered to be the principal pathogen. *Fusarium acuminatum, F. avenaceum, F. culmorum, F. equiseti,* and *F. nivale* are also reported to be pathogenic to barley.

Fusarium graminearum overwinters as mycelia and spores on seed, mycelia in infested residue, and perithecia on maize, wheat, and other crop residue infected the previous season. Primary inoculum is a combination of ascospores that are produced in perithecia and conidia produced on mycelium during warm, moist weather and disseminated by wind to flowers and developing seed. Secondary inoculum is primarily conidia produced on diseased heads. Disease severity is increased by wet, warm weather during anthesis and ripening of kernels, and when grain is lodged or swathed.

Scabby grain is toxic, particularly to swine. Several toxins are produced: zearalenone (F-2 toxin) is involved in causing hyperestrogenism, and vomitoxin is involved in the refusal factor or vomiting syndrome. Feed containing 3% or more scabby kernels may be poisonous to dogs, humans, and swine. However, cattle and sheep are normally not affected by ingesting scabby grain. The highest numbers of Fusarium propagules and mycotoxin concentrations occur in hulls.

Distribution. Wherever barley is grown during warm, moist summer weather.

Symptoms. The number of diseased kernels on a head may vary. Either only one or more spikelets or the entire head may be affected. Infection begins in flowers and spreads to other parts of the head, causing it to be dwarfed and giving kernels the appearance of premature ripening. Initially, diseased spikelets are water-soaked, starting at the base, then die and become light brown. Eventually, hulls change from a light to a dark brown color. When infection occurs late in kernel development, hulls may be brown only at their bases. In severe cases, entire kernels become shrunken, brown, and lightweight and have a "floury" and discolored interior. During wet weather, a pink mold consisting of mycelia and conidia grows on diseased spikelets. Later, small, black perithecia grow in the same area.

Sowing infected seed may result in seed and seedling damping off.

Management
1. Do not grow barley following barley, maize, or wheat.
2. Plow under infected residue where feasible.
3. Do not spread manure containing infested straw or maize stalks on soil in which barley is growing.
4. Sow early to allow grain to escape much of the warm, moist weather of summer.

5. Treat seed with a fungicide seed protectant to reduce damping off.
6. Separate scabby kernels from healthy kernels with cleaning equipment.
7. Grow hull-less cultivars as a means of managing mycotoxins.

Gibberella Seedling Blight

Cause. *Gibberella zeae,* anamorph *Fusarium graminearum,* overwinters as mycelia and spores on seed, and as mycelia and perithecia on barley, maize, rye, and wheat residue. Seedling infections are primarily from seed-borne inoculum. Disease is favored by temperatures of 15°C and above and dry soil.

Distribution. Wherever barley is grown.

Symptoms. Initially, seedlings are stunted, become chlorotic, and die. The root system has a reddish-brown rot and may be covered with white to pink mycelium.

Management
1. Treat seed with a fungicide seed protectant.
2. Sow certified seed.
3. Sow when the soil temperature is 15°C or below.

Gray Snow Mold

Typhula incarnata is considered one of the causal agents for speckled snow mold of wheat.

Cause. *Typhula incarnata* (syn. *T. itoana* and *T. graminum*). *Typhula incarnata* survives as sclerotia under deep snow cover. Sclerotia germinate during wet weather in the autumn to produce either basidiocarps on which basidiospores are borne or hyphae. Basidiospores are released in the Pacific Northwest of the United States in mid to late October and are airborne to seedlings. Basidiospores may remain viable for up to 60 days. The optimum temperature range for disease development is 1° to 5°C.

Distribution. Canada (Quebec), central Europe, Japan, and the mountain valleys of the western United States. Gray snow mold is associated with snow cover.

Symptoms. Winter barley plants are often infected without causing a reduction in survival or yield. However, if crowns become infected, plants normally die. Chlorotic plants are observed to be scattered throughout a field after the snow melts in the spring. A few or many necrotic leaves are covered with a thick, gray-white mycelium. Diseased leaves dry up and disintegrate easily when the snow cover disappears and temperatures rise. Sclerotia, which appear as hard, black structures of varying sizes, form on lower leaf sheaths.

Management. Do not grow barley in areas where prolonged snow cover may occur.

Halo Spot

Cause. *Selenophoma donacis* (syn. *Phyllosticta stomaticola* and *Septoria donacis*) and *Selenophoma donacis* var. *stomaticola*. *Selenophoma donacis* var. *stomaticola* reportedly can be differentiated from *S. donacis* primarily by spore size. *Selenophoma donacis* var. *stomaticola* spores are falcate, aseptate, and variable in size (1–3 × 10–20 µm but occasionally up to 25 µm long). *Selenophoma donacis* spores are stoutly falcate to boomerang-shaped and somewhat larger (2.0–4.5 × 18–35 µm).

Ever since *Selenophoma donacis* overwinters as pycnidia, pycnidiospores, and mycelia in residue, seed, and cereals. During cool, moist weather, pycnidiospores are exuded from pycnidia and disseminated by wind and splashing rain to host plants.

Distribution. In cool, moist climates of Australia, Canada, Great Britain, northern Europe, New Zealand, South Africa, and the United States. Although halo spot rarely causes severe damage, epiphytotics have been reported in southwest England and Norway.

Symptoms. In the spring, spots appear on leaves and sometimes on culms of winter barley. Spots are less than 4 mm long, are elliptical, rectangular, or square, and have purple-brown margins that fade as the margins age. Centers of spots become gray and speckled with small, black pycnidia. Spots may become so numerous that much of the leaf surface is destroyed.

Management. Since halo spot does little damage to barley, disease management is not necessary.

Kernel Blight

Kernel blight is sometimes called black point and seed discoloration.

Cause. Some reported causes of seed discoloration are *Alternaria alternata*, *Arthrinium arundinis*, *Bipolaris sorokiniana*, *Cladosporium herbarum*, *Fusarium culmorum*, and *F. graminearum*. Discoloration is associated with excess rainfall between anthesis and harvest and is enhanced by late harvest.

Distribution. Generally, wherever barley is grown.

Symptoms. The embryo is dark brown to black. Germination is decreased if the embryo is invaded.

Management. Treat seed with a seed-treatment fungicide.

Leaf Rust

Leaf rust is also called dwarf leaf rust and brown leaf rust.

Cause. *Puccinia hordei* (syn. *P. anomala*). The wheat leaf rust fungus, *P. recondita* f. sp. *tritici,* infects barley in some geographical areas but is generally considered a weak pathogen to barley.

Puccinia hordei is a heteroecious rust fungus that overwinters as urediniospores on volunteer and winter barley in the southern United States. In the spring, urediniospores are windborne to the northern United States and initiate infection just before heading. Leaf rust is most severe under warm, humid conditions where plants mature later in the growing season.

Star-of-Bethlehem, *Ornithogalum* spp., is the alternate host for *P. hordei* in Europe, Israel, and parts of Asia. In Israel, *O. brachystachys* and *O. trichophyllum,* together with *Dipcadi erythraeum* and *Leopoldia eburnea,* have been reported to be alternate hosts of *P. hordei.* Basidiospores do not infect *O. umbellatum* and *O. nutans,* and no pycnia or aecia have been found on these two *Ornithogalum* species in the United States except under experimental conditions.

Distribution. Wherever barley is grown.

Symptoms. Symptoms are inconspicuous until uredinial development is abundant. Lower leaves are infected first, with disease subsequently progressing upward on plants during warm, humid weather.

Uredinia are small, oval, light yellow-brown pustules scattered irregularly on both sides of diseased leaves. Severely diseased leaves eventually become chlorotic, making it difficult to observe the tiny, yellow-brown uredinia.

Telia, which form as plants near maturity, are brown, oblong to round, and covered by plant epidermis. Telia tend to coalesce to form gray patches on leaves, but they are less abundant than uredinia.

Seeds and their accompanying structures are normally not infected.

Management
1. Grow resistant cultivars. Most of the commercial barley cultivars grown in the United States are susceptible to *P. hordei* with the exception of those bred in Virginia.
2. Apply foliar fungicides.

Leaf Speckle

Leaf speckle is also called Septoria blotch, Septoria leaf blotch, Septoria leaf spot, Septoria speckled leaf blotch, speckled blotch, and speckled leaf blotch. Speckled blotch is sometimes considered a separate disease caused by *Septoria passerinii.* However, it is included in this discussion because of the similarity in the life cycles of the two fungi and the disease symptoms they produce.

Cause. *Stagonospora avenae* f. sp. *triticea* (syn. *Septoria avenae* f. sp. *triticea*), teleomorph *Phaeosphaeria avenaria* f. sp. *triticea* (syn. *Leptosphaeria avenaria* f. sp. *triticea*), and *Septoria passerinii.*

Stagonospora avenae f. sp. *triticea* and *S. passerinii* survive as mycelia for 2 to 3 years in live plants and as pycnidia on residue that is buried in soil or lying on the soil surface. Pycnidiospores produced on the previous year's residue are the most important source of inoculum during the growing season. During temperatures of 15° to 25°C and 6 or more hours of moisture in the autumn or spring, pycnidiospores are exuded from pycnidia in a cirrhi and disseminated by splashing and blowing rain to the lower leaves of poorly growing plants. Pycnidiospores then germinate immediately in the presence of moisture and infect leaves. The resulting mycelia in the leaves of winter barley will become inactive during the winter but will resume growth during warm weather.

Ascospores produced in perithecia as plants mature may infrequently serve as primary inoculum when they are disseminated by wind in the spring or autumn. Pycnidia are produced late in the growing season.

Distribution. Wherever barley is produced, but disease is most severe in cool, wet climates.

Symptoms. Yellow to light brown lesions of various sizes occur primarily on leaves and leaf sheaths. *Septoria passerinii* produces long linear lesions with definite margins parallel to the leaf veins.

Symptoms produced by *S. avenae* f. sp. *triticea* have indefinite margins with the yellowish-brown area blending into the green of the leaf blade and sheath.

Eventually, lesions containing numerous small, black pycnidia merge and cover large areas of the diseased leaf. Leaf margins often pinch or roll and become dry.

Management
1. Grow resistant cultivars.
2. Apply foliar fungicides to susceptible cultivars when weather favors disease development.
3. Plow under or burn infested residue, where feasible.
4. Rotate barley with resistant crops, such as legumes.

Leaf Spot

Both fungi were isolated from leaf spots.

Cause. *Monographella nivalis* (syn. *M. nivalis* var. *mayor*) and *Fusarium culmorum.* Both fungi are seedborne and are isolated from different plant parts during plant development. Therefore, endophytic growth as a latent phase of the fungi is hypothesized.

Distribution. Germany.

Symptoms. Spots first appear on lower leaves early in the growing season. Later, spots do not occur on intermediate leaves but will reappear on upper leaves of plants in their late tillering stages. Spots are 3 to 5 mm long and grayish green but later become black on the upper leaves.

Management. Not reported.

Leptosphaeria Leaf Spot

Cause. *Leptosphaeria herpotrichoides* (syn. *Phaeosphaeria herpotrichoides*) overwinters as mycelia and as ascospores in pseudothecia on infested residue. Free water must be present on leaves for 48 hours or more for infection to occur.

Distribution. Canada, Europe, and the United States in areas that have long periods of wet weather. Leptosphaeria leaf spot is considered a minor disease.

Symptoms. Irregular, diffuse, yellow to tan spots occur on diseased leaves. Under extended wet conditions, up to 50% of the leaf may be affected.

Management. Grow the most resistant cultivars.

Loose Smut

Loose smut is also called true loose smut, nuda loose smut, and brown loose smut.

Cause. *Ustilago nuda.* Farr et al. (1989) consider *U. nuda* to be a synonym of *U. tritici. Ustilago nuda* differs from other smut fungi on barley in that it infects flowers.

Ustilago nuda is seedborne, surviving as dormant mycelia in embryos. When diseased seed germinates, mycelia grow systemically within plants toward the shoot apex and inhabit seed primordia. Eventually, smutted heads consisting of masses of teliospores are produced in place of healthy kernels.

Smutted heads emerge from boots a day or two earlier than the heads of healthy plants. Each smutted head is enclosed in a fragile membrane that soon disintegrates, allowing the teliospores to be released. Teliospores are disseminated by wind to the flowers of healthy barley plants and germinate at temperatures of 16° to 22°C in moisture provided by dews or light showers. The germ tube penetrates the flower ovary and possibly the stigma with subsequent mycelial growth occurring in the germ or embryo of the developing seed. The mycelium, which is both intercellular and intracellular, develops within gall tissues derived mostly from the rachilla cortex and becomes dormant in the embryo until the following season.

Several physiologic races of *U. nuda* exist.

Distribution. Wherever barley is grown.

Symptoms. Infected seed does not have obvious external symptoms and germination is ordinarily not affected. Before heading, diseased plants have dark green, erect leaves with chlorotic or yellowish streaks. Smutted heads consist of a brown to dark brown mass of teliospores enclosed by a fragile, gray membrane consisting of plant epidermis. The membrane soon ruptures, leaving an erect, naked rachis that protrudes slightly above the reclining heads of healthy plants.

Sori have been reported in flag leaves of greenhouse-grown 'Larker' barley, where they appeared prior to or occasionally at the same time as the emergence of smutted or healthy heads from leaf sheaths. Frequently, heads did not appear nor were any present in the leaf sheath. Infrequently, auricles, ligules, and sheaths were also involved.

Losses are generally directly related to the percentage of infected heads: 10% infected heads normally yields a 10% yield loss.

Management
1. Treat seed with a systemic fungicide seed protectant.
2. Grow resistant cultivars.

Net Blotch

Cause. *Pyrenophora teres,* anamorph *Drechslera teres. Pyrenophora teres* has been divided into two forms based on symptomology. *Pyrenophora teres* f. *teres* is the name for the net-causing isolates, and *P. teres* f. *maculata* for the spot-causing isolates. Similarly, the anamorphs are identified as *D. teres* f. *teres* and *D. teres* f. *maculata.* However, some workers do not utilize the taxonomic forms but refer to the organism either as *D. teres* or *P. teres.* Barley is the only known host.

The source of primary inoculum may vary with geographical area and form of the causal fungus. Some researchers have reported the causal fungus overwinters either as seedborne mycelia or as pseudothecia in infested residue. Wild grass species may also be a potential source of primary inoculum. Cross-inoculation experiments in Israel have shown that isolates from *Hordeum murinum, H. murinum* subsp. *leporinum, H. marinum,* and *H. vulgare* subsp. *spontaneum* are capable of infecting cultivated barley. In western Australia, *P. teres* f. *teres* has been isolated from *Hordeum* species and *Bromus diandrus.* In California, *P. teres* f. *teres* was found to have a wide host range: susceptible species were found in 15 of 16 host genera that were artificially inoculated.

During temperatures of 10° to 15°C and 100% relative humidity, spores are produced on residue and disseminated by wind to healthy plants. It is not known for certain if primary inoculum consists of conidia or ascospores. Later in the growing season, conidia produced in pycnidia on primary lesions cause secondary infections during cool and moist weather conditions. Production of secondary conidia occurs only on senescent or dead host tissue and requires leaf surface wetness, high relative humidity, and appropriate temperature regimes. The duration of leaf surface wetness required for secondary infection ranges from 3.0 to 8.5 hours at temperatures of 8° to 20°C. Moisture on leaf surfaces in the form of dew is more conducive to infection than splashing rain.

As plants mature, *P. teres* will grow into sheath and culm tissue, where pseudothecia are eventually produced later in the growing season. Diseased winter barley is more prone to freezing injury of crowns.

van den Berg and Rossnagel (1991) reported that spot blotch caused by *P. teres* f. *maculata* in Saskatchewan closely followed plant development and that each leaf was infected shortly after its emergence. Only conidia produced on infested stubble were reported to be responsible for the infection of the lower leaves. Conidia produced on both stubble and the lower leaves caused most of the infections on the upper leaves. They concluded that infested crop debris acts as an inoculum reservoir for the lower leaves and that infested crop debris and the lower leaves act as an inoculum reservoir for the upper leaves.

In Pennsylvania, winter barley has the greatest disease severity in the earliest plantings; however, if the weather becomes warm and humid, previous differences in disease severity attributable to dates of planting are eliminated. Net blotch severity of winter barley also increases as root rot levels increase.

Distribution. Wherever barley is grown.

Symptoms. Leaves, stems, and seeds are affected. Disease severity is greatest on the oldest leaves of winter barley. The net-like symptoms caused by *P. teres* f. *teres* initially are brown spots or blotches near the tips of seedling leaves. Dark brown, narrow lines that run longitudinally and transversely in the lighter brown area give spots a netted or crosshatched appearance. This symptom is best seen when leaves are held up to a light source. Spots increase in length to form short, narrow streaks or they coalesce to form long, brown stripes with irregular margins. On older spots, netting or crosshatching is visible only at the margins.

Small, brown streaks that develop on glumes and indistinct brown lesions that develop on kernel bases cause reduced yields and shriveled seed. Diseased stems are light brown and lack strength.

Pyrenophora teres f. *maculata* causes a "spot form" of net blotch that may be confused with symptoms caused by *Cochliobolus sativus* on leaves and leaf sheaths. Symptoms, which may involve the entire leaf, consist of dark brown, elliptical or fusiform lesions (3 × 6 mm) that may be surrounded by a chlorotic zone of varying width, depending upon the cultivar.

Management
1. Treat seed with a fungicide seed protectant.
2. Rotate barley with resistant crops.
3. Grow resistant cultivars. Resistance is simply inherited and conditioned by one or two genes. It has been reported that mixing barley cultivars in a field reduces severity of the net blotch/scald complex; however, disease is reduced because of moderately resistant genotypes in the mixture and no yield benefits are apparent.
4. Apply foliar fungicides to high-yield-potential malt-quality or seed-production fields when weather conditions favor disease development. A biological control consisting of a bacterium applied to leaves has shown promise in experiments.
5. Increasing in-row seed spacing from 2 to 4 cm between plants and row spacing from 12 to 28 cm has been reported to decrease disease severity.
6. Application of nitrogen as ammonium nitrate at Zadok's growth stage 30 has been reported to decrease disease incidence.

Pink Snow Mold

Pink snow mold is also called Fusarium patch.

Cause. *Microdochium nivale* (syn. *Fusarium nivale, Microdochium nivalis,* and *Gerlachia nivalis*). The teleomorph is *Monographella nivalis* (syn. *Calonectria nivalis, C. graminicola, Griphosphaeria nivalis, Micronectriella nivalis,* and *Monographella nivalis* var. *mayor*). Overseasoning of the fungi is primarily by mycelia within diseased live plants that are covered by snow.

Leaf sheaths and blades are infected near the soil line during cool, wet weather in the autumn by mycelia growing from residue and by ascospores and conidia produced on infested residue and disseminated by wind to barley seedlings. Mycelia will then grow from these primary infections when plants are covered by snow.

Secondary infection is caused primarily by conidia produced in the spring and infrequently by ascospores. Perithecia develop in lower leaf sheaths in late spring and early summer during cool, wet weather.

Distribution. Canada (Quebec), central Europe, Japan, and the mountain valleys of the western United States.

Symptoms. The most obvious symptom is observed after the snow melts, when infested residue, leaf sheaths, and leaf blades have a pinkish color caused by the fungal mycelia and sporodochia. Leaves are chlorotic and necrotic but remain dry and intact when the snow cover disappears and temperatures rise. In contrast, diseased leaves that crumble after snowmelt are a characteristic symptom of both gray snow mold and speckled snow mold.

Management. Avoid growing barley where there is a persistent snow cover.

Powdery Mildew

Cause. *Erysiphe graminis* f. sp. *hordei* (syn. *Blumeria graminis* f. sp. *hordei*) overwinters as cleistothecia on residue and as mycelia and conidia in live leaves. Primary inoculum is ascospores that form in cleistothecia in the spring and are disseminated by wind to plants. Secondary inoculum is conidia that are produced when mycelia are established on leaf surfaces during temperatures of 15° to 22°C and 100% relative humidity but without free water. Cleistothecia are formed on leaf surfaces as plants approach maturity.

Barley is very susceptible during periods of rapid growth. Such rapid growth is frequently aided by heavy nitrogen fertilization that results in a dense stand of tender, rank plants.

Distribution. Wherever barley is grown.

Symptoms. Superficial mycelia, conidia, and conidiophores appear as a light gray powder or white spots on the upper surfaces of leaf blades, leaf sheaths, and floral bracts. Most fungal growth occurs on upper leaf surfaces and, infrequently, on lower surfaces. The powdery-appearing leaf areas enlarge, eventually causing leaves to become chlorotic, then necrotic. In some cultivars, leaf tissue adjacent to mycelia becomes brown and necrotic. Numerous small, round, dark cleistothecia develop on diseased areas and can be readily observed with a hand lens. Powdery mildew ceases to be a problem when the weather becomes dry and warm later in the growing season.

Management
1. Grow resistant cultivars.
2. Apply a foliar fungicide if economically feasible.

Pythium Root Rot

Pythium root rot is also called browning root rot.

Cause. *Pythium arrhenomanes* (syn. *P. arrhenomanes* var. *canadense*), *P. graminicola*, *P. heterothallicum*, *P. irregulare*, *P. tardicrescens*, *P. torulosum*, and *P. ultimum* have been reported to be pathogenic to barley. *Pythium irregulare* was reported to be the most pathogenic *Pythium* species to barley at temperatures of 10° to 25°C.

Most *Pythium* species survive as oospores in residue and soil for 5 or more years. Infection occurs in 3 to 5 hours, usually in the autumn or spring, when oospores germinate to form sporangia in which zoospores are produced.

Zoospores are released into moist soil and swim a short distance to root tips. Optimum infection is in soils with a high water potential of −1 bar or less and low temperatures. At warmer soil temperatures, oospores may germinate directly to form only one germ tube that penetrates roots. Oospores are produced in diseased tissue and are released into soil when residue decomposes, or oospores may survive in tissue when it remains intact.

Distribution. Wherever barley is grown. Disease is most common under wet soil conditions.

Symptoms. Pythium root rot occurs throughout the season. Early infections may kill seed and seedlings before or after they have emerged, resulting in uneven stands. Later infections reduce plant vigor, tillering, stem elongation, and head size. Heads of diseased plants also are unable to emerge from the boot.

Crown or basal stem tissues are ordinarily not infected. Lower leaves may become chlorotic, then necrotic, and dry up due to poor root growth caused by the fungi pruning root tips. Rootlets may be soft and a light to dark yellow or brown color that occurs mainly in the stele. Brown lesions may be present on larger roots. If soils become dry, new root growth may occur.

Management
1. Treat seed with a seed-protectant fungicide.
2. Drain areas in fields that tend to remain wet after the rest of the field has dried up.
3. Apply the recommended amounts of phosphorous fertilizer in a band or broadcast at planting time.

Rhizoctonia Root Rot

Rhizoctonia root rot is also called bare patch, barley stunt disorder, purple patch, Rhizoctonia patch, and stunting disease.

Cause. *Rhizoctonia solani,* teleomorph *Thanatephorus cucumeris. Rhizoctonia solani* isolates causing root rot may belong in AG-3, while those causing sharp eyespot belong primarily in AG-4 and a few in AG-1. In the United Kingdom, *R. solani* isolates were found to belong in AG-8 instead of AG-3. *Rhizoctonia solani* survives as sclerotia in soil and as mycelia in residues of a large number of host plants.

In Scotland, disease is restricted to light, sandy soils within 40 km of Elgin, Scotland, and occurs on barley following heavily fertilized grass that has been grazed by cattle. In Australia, the disease is found on calcareous mallee soils in regions around Salmon Gums. In the United States, disease occurs where barley was either direct drilled into stubble, sown with a minimum of preparatory tillage, or sown the same day soil was tilled. The herbicide chlorsulfuron has been reported to increase disease.

Distribution. Australia, Canada, England, South Africa, and the United States (Oregon and Washington).

Symptoms. At 12°C, seedling emergence and growth are reduced. However, at 18°C and 26°C, only seedling growth is reduced after emergence, but not the emergence itself of germinated seedlings. Symptoms start 6 weeks after sowing as circular patches with distinct borders that vary from 20 to 300 cm in diameter and persist in the same location for up to 4 years. Plants inside the circular patch are stunted and chlorotic, but those outside the patch are a normal green. Plant shoots tend to be spindly, stunted, and abnormally erect. Leaf development is retarded, and after leaves form, they senesce prematurely and become chlorotic, starting at the leaf tip. Basal leaf sheaths near the soil surface are often discolored brown.

Lesions on diseased primary and secondary roots and subcrown internodes are brown and often girdle the root. Cortices collapse and roots have pinched-off pointed tips or are frequently rotted back to brown stumps. Surviving roots often produce large numbers of lateral roots, which results in a heavily branched root system with a large amount of cortical tissues and root tips that have sloughed off.

A number of plants die prematurely. Those that survive are poorly developed and yield little grain.

Management. The disease apparently does not occur in the northwestern United States if some time has elapsed between tillage and sowing. The disease is more common in barley grown in minimum-tillage systems. Compared with no-till or minimum-tillage systems, cultivation prior to sowing reduces disease incidence in Australia. Under zero-tillage conditions, nitrogen in the form of ammonium sulfate, sodium nitrate, or urea reduces disease.

Scald

Scald is also called barley scald, leaf scald, and Rhynchosporium scald.

Cause. *Rhynchosporium secalis* (syn. *Marsonia secalis*) overwinters primarily as stroma either in lesions on winter barley that was infected in the autumn or in residue on the soil surface. Primary inoculum is conidia that are produced in the spring on stroma during cool temperatures (10°–18°C optimal) and humid weather. Conidia are disseminated primarily by splashing water and, for a short distance, by wind. Moisture in the form of rain is more conducive to infection than dew. Secondary inoculum consists of conidia produced in both new and old lesions during cool and humid weather.

Rhynchosporium secalis can be seedborne as mycelia in the pericarp and hull. Greenhouse tests demonstrated the transmission of the fungus from the seed to the seedling: at a soil temperature of 16°C, hyphae invaded the coleoptile as it emerged from the embryo. However, the importance of seedborne inoculum in the field is not well known.

Scald is more severe under reduced tillage conditions and irrigation. Different races have been identified.

Distribution. Wherever barley is grown.

Symptoms. Lesions occur on coleoptiles, leaves, leaf sheaths, glumes, floral bracts, and awns. Young lesions occur as oval to irregular-shaped blotches that are a dark to pale gray or bluish-gray color. Later, lesions become water-soaked and, as tissues die, change to brown, then to light tan bordered by a brown margin that is sometimes surrounded by a chlorotic area. Often, lesions have a zonate appearance. Inconspicuous brown spots may occur on the tips of glumes.

An unusual symptom developed on Ethiopian barley cultivars inoculated with an *R. secalis* isolate from Montana. Leaf blades were free of symptoms but brown discoloration and subsequent wilting of leaf sheaths occurred.

Management
1. Rotate barley with other crops. Barley is the only known host to *R. secalis.*
2. Plow under infected crop residues.
3. Grow resistant cultivars. It has been reported that mixing barley cultivars in a field reduces severity of the net blotch/scald complex. Disease is reduced because of moderately resistant genotypes in the mixture, however, and no yield benefits are apparent.

Sclerotinia Snow Mold

Cause. *Sclerotinia borealis* (syn. *Myriosclerotinia borealis*) overseasons as sclerotia in residue. In the autumn, sclerotia germinate to produce apothecia on which ascospores are formed. Ascospores are disseminated by wind to healthy plants. *Sclerotinia borealis* has been identified on cocksfoot, *Dactylis glomerata.*

Distribution. Canada (Quebec), central Europe, Japan, and the mountain valleys of the western United States.

Symptoms. Symptoms are observed in isolated patches throughout a field after the snow melts. Dead leaves on plants are gray and appear "thready" or decomposed with only the vascular tissue remaining. Numerous black sclerotia (2–4 mm long) are produced in diseased tissue during late spring and summer.

Management. Do not grow barley in areas of prolonged snow cover.

Septoria Speckled Leaf Blotch

Cause. *Stagonospora avenae* f. sp. *triticea* (syn. *Septoria avenae* f. sp. *triticea*), teleomorph *Phaeosphaeria avenaria* f. sp. *triticea* (syn. *Leptosphaeria avenaria* f. sp. *triticea*). *Septoria passerinii* is also reported to be a causal fungus. Survival, for at least 2 years, is by pycnidia and pseudothecia in infested residue. Primary inoculum is either pycnidiospores or ascospores that are disseminated by splashing water to host plants. Pycnidiospores from over-wintering pycnidia also are suspected to provide continuous inoculum throughout the growing season.

Extended periods of moisture at 97% relative humidity and above, to-gether with day temperatures of 16° to 23°C and night temperatures of 10° to 20°C, are optimal for pycnidiospore germination. Optimum tempera-tures for infection are 16° to 18°C during the day and 11° to 14°C at night. Pycnidia form on diseased tissue later in the growing season.

Distribution. Worldwide.

Symptoms. Grayish-green lesions of variable sizes and shapes occur on leaves and leaf sheaths. Later in the growing season, the lesions turn light tan. *Stagonospora avenae* f. sp. *triticea* causes lesions with indefinite margins, whereas *S. passerinii* causes linear lesions with definite margins that are parallel to the leaf veins.

Young lesions are generally lens-shaped but later may expand, merge, and cover much of the leaf. Leaf margins roll and become dry. Pycnidia be-gin to appear later in the growing season as small, black dots within le-sions. Later, large blotches occur as leaves and leaf sheaths senesce, and nu-merous pycnidia are produced.

Management
1. Plow under residue.
2. Rotate barley with nonhosts.

Sharp Eyespot

Sharp eyespot is also called Rhizoctonia root rot in some literature.

Cause. *Rhizoctonia cerealis.* However, *R. solani* is still reported in some litera-ture as a causal fungus although *R. cerealis* and *R. solani* were separated into two species in 1977. Where *R. solani* is attributed as a cause, isolates are re-ported to belong in the AG-4 group and a few in AG-1.

Rhizoctonia cerealis survives as sclerotia in soil or as mycelia in residues of a large number of plant hosts. Sclerotia germinate or mycelia grow from residue to infect roots and culms, particularly in cool soils with 20% or less moisture. Barley may be infected any time during the growing season, but damage is most severe when seedlings are infected. Barley is not as suscep-tible as oat, rye, and wheat.

Distribution. Wherever barley is grown.

Symptoms. Serious losses rarely occur. Diamond-shaped lesions that resem-ble those of eyespot, but are more superficial, occur on lower leaf sheaths. Lesions are light tan with dark brown margins and frequently have dark mycelium on them. Sclerotia may develop in lesions and between culms and leaf sheaths.

Seedlings may be killed, but when roots are infected, new ones may be produced. Plants are stiff and grayish; they are delayed in maturity, lodge, and produce white heads.

Management. No management is effective. Vigorous plants growing in well-fertilized soil are not as likely to become as severely diseased as plants growing in nutrient-deficient soil.

Southern Blight

Southern blight is also called foot rot and white rot.

Cause. *Sclerotium rolfsii,* teleomorph *Athelia rolfsii,* presumably overseasons as sclerotia and mycelial strands that also function as the main infective propagules. Dissemination is by any means that moves infested soil. Disease is favored by high temperatures of around 30°C and periods of high relative humidity and moisture.

Distribution. Brazil, India, Puerto Rico, and the United States (California). Southern blight is a rare disease on barley.

Symptoms. Groups of plants wilt and appear chlorotic or "blighted." Small, irregular-shaped or elongated, necrotic spots develop on the base of the stem near the soil surface. Spots eventually develop into lesions that further enlarge, coalesce, become sunken, and ultimately girdle the plant. Subsequently, stem tissues become soft and necrotic for approximately 2.50 to 3.75 cm above and below the soil line. The entire plant becomes chlorotic and, eventually, white or "bleached" in appearance. Diseased plants are stunted, and tillering and root length are reduced. Eventually, sclerotia develop on necrotic tissues near the soil line.

Similar symptoms have been reported from Brazil. White mycelium and sclerotia occur on necrotic areas of roots, crowns, and lower portions of stems. Plants become white in appearance and die.

Management
1. Treat seed with a seed-treatment fungicide.
2. Do not grow barley in fields in which susceptible crops have grown.
3. Plow under infested residue.
4. Maintain proper plant population to prevent spread of the pathogen from plant to plant.

Speckled Snow Mold

Cause. *Typhula ishikariensis* (syn. *T. ishikariensis* var. *idahoensis, T. ishikariensis* var. *ishikariensis,* and *T. ishikariensis* var. *canadensis*). *Typhula ishikariensis* survives as sclerotia under a thick snow cover. Sclerotia germinate during wet weather in the autumn to produce either basidiocarps on which basidiospores are borne, or hyphae. Basidiospores are released in early to mid-November in the northwestern United States and are disseminated by wind to seedlings. The optimum temperature for infection is 1°C. *Typhula ishikariensis* has been identified on cocksfoot, *Dactylis glomerata*.

Distribution. Canada (Quebec), central Europe, Japan, and the mountain valleys of the western United States. Speckled snow mold is associated with snow cover.

Symptoms. A few or many necrotic leaves, which are covered with a thick, gray-white mycelium, dry up and disintegrate when the snow cover has melted and temperatures rise. The numerous dark sclerotia (0.2–2.0 mm in diameter) produced under leaf sheaths, within plant tissue, or scattered throughout the mycelium give a "speckled" appearance to diseased plants. Plants die when crowns become diseased.

Management. Do not grow barley in areas of prolonged snow cover.

Spot Blotch and Associated Seedling and Crown Rots

This disease syndrome is also called black point, head blight, Helminthosporium blight, Helminthosporium head blight, and kernel blight. Spot blotch and associated seedling and crown rots also may be considered part of the common root rot syndrome.

Cause. *Bipolaris sorokiniana* (syn. *Helminthosporium sativum* and *H. sorokinianum*), teleomorph *Cochliobolus sativus*. *Bipolaris sorokiniana* is seedborne and overwinters as mycelia and conidia on residue and seedling leaves of winter barley, and as conidia in soil. Conidia are produced in spring on either infested residue or wild grasses and are disseminated by wind to seedlings. Seedborne inoculum results in seedling blight and in root and crown rot in dry, warm soils.

Leaf spots will begin to develop at temperatures lower than 20°C, but optimal symptoms, including a pronounced chlorosis, develop at temperatures of 20°C and above during wet conditions or high relative humidity. High light intensities favor leaf spot development, but incubation of diseased plants under temporary low light conditions results in chlorosis and accelerated senescence of leaves.

Based on their virulence to barley cultivars, three pathotypes of *B. sorokiniana* (labeled 0, 1, and 2) currently have been identified.

Distribution. Wherever barley is grown in warm, humid conditions.

Symptoms. Seedlings may be killed after they have emerged or, less frequently, before emergence from the soil. Leaves of diseased seedlings are dark green. At their bases near the soil line, leaf sheaths have dark brown to black lesions that eventually extend into the leaf blades. Crowns and roots are dark brown and rotted, and seedlings are dwarfed, chlorotic, and excessively tillered. Later, head emergence and kernel filling of the diseased plant may be affected, depending upon how severe the disease was when the plant was a seedling.

Round to oblong (1–5 × 2–25 mm), dark brown spots with definite margins appear on lower leaves following warm, moist weather favorable for disease development. The spots may eventually coalesce to form lesions that cover large areas of the leaf. Older lesions are olive-colored due to sporulation of *B. sorokiniana.* Severely diseased leaves will completely dry up.

Lesions on floral bracts and kernels vary from small black spots to an overall dark brown discoloration of the kernel end. Browning or blackening of lemma and palea is due to discoloration of the pericarp layer and underlying embryo region. Early head blight causes sterility or killing of individual kernels soon after pollination. Infected kernels may result in seedling blight if the kernels are sown, and they are normally discounted as unacceptable for malting purposes.

Management
1. Grow resistant cultivars.
2. Treat seed with a seed-protectant fungicide.
3. Apply a foliar fungicide if weather conditions favor the spread of the disease.
4. Rotate with resistant crops, such as legumes.

Stagonospora Blotch

Stagonospora blotch is also called glume blotch, and leaf and glume blotch.

Cause. *Stagonospora nodorum* (syn. *Septoria nodorum*), teleomorph *Phaeosphaeria nodorum* (syn. *Leptosphaeria nodorum*). *Stagonospora nodorum* survives for up to 3 years as mycelium in live plants and seed, and for an indefinite time as pycnidia on residue. During temperatures of 20° to 27°C and wet weather in the autumn or spring, pycnidiospores are exuded from pycnidia in a cirrhi and disseminated by splashing and blowing rain to the lower leaves of plants. Spore germination and infection are optimum at temperatures of 15° to 25°C and 6 or more hours of wetness between flowering and grain harvest.

Secondary pycnidia and pycnidiospores are produced 10 to 20 days after infection. In Denmark, ascospores, which reportedly are produced in pseudothecia as barley matures in late summer or early autumn, are disseminated by wind to other plants. However, ascospores are not known to be produced in North America. *Stagonospora nodorum* infects both wheat and barley.

Distribution. Wherever barley is grown.

Symptoms. Glumes, culms, leaf sheaths, and leaves are infected; however, little damage occurs until plants approach maturity. The first symptoms appear in the spring as brown or gray spots on the leaves and sheaths or at the junction of the leaf and sheath. Later, oval, red-brown spots (1–2 cm long) that develop on leaf veins have chlorotic margins that may extend to the leaf tip. Irregular-shaped lesions that develop at the junction of the leaves and sheaths affect the entire leaf. Diseased leaves senesce earlier than healthy leaves.

As lesions on leaves begin to age, portions of their necrotic centers fade to light brown or gray but longitudinal brown-pigmented streaks remain in the lesion. Eventually, dark pycnidia become readily visible as black specks throughout the lesions. Diseased stem nodes, which turn brown and shrivel, are speckled with dark pycnidia, as are leaf lesions. Stem infections cause straw to lodge and bend over above the nodes.

Two to 3 weeks after the head emerges, small gray to brown spots can be observed on the top third of glumes. Eventually, these spots enlarge and become a darker chocolate brown. Later, the centers of spots become gray with numerous pycnidia that resemble tiny black specks similar to those in leaf and stem lesions. Seed may become diseased with both the attached lemma and palea, and the caryopsis becoming discolored. Pycnidia also develop on diseased seed surfaces. These symptoms may be confused with those of black chaff or basal glume rot. However, Stagonospora blotch spots do not form streaks and are not as sharply defined or as dark brown as those of black chaff. Stagonospora blotch also does not have the water-soaked appearance of basal glume rot.

A *Septoria* sp. that is different pathologically from *S. nodorum* has been reported to cause oval leaf spots with a buff center that contains numerous pycnidia. This *Septoria* sp. was also isolated from glumes.

Management
1. Sow certified, disease-free seed that has been cleaned and treated with a seed-protectant fungicide.
2. Grow the most tolerant barley cultivar. Several winter barley cultivars have been reported to have some resistance.
3. Plow under or burn infested residue where feasible.
4. Apply a foliar fungicide if conditions warrant it.
5. Rotate barley every 3 or 4 years with a nonsusceptible crop.
6. Sow winter barley after Hessian fly–free date.

Stem Rust

Cause. *Puccinia graminis* f. sp. *tritici* and *P. graminis* f. sp. *secalis* are heteroecious rust fungi that have the barberries *Berberis vulgaris, B. canadensis, B. fendleri* and some species of *Mahonia* as the alternate host. Japanese barberry, *B. thunbergii,* is thought to be immune.

Puccinia graminis overwinters as teliospores on barley and wheat residue and as urediniospores on winter wheat and barley in the southern Great

Plains. Severity of stem rust on barley in the northern Great Plains depends on the amount of inoculum, in the form of urediniospores, that overwinters in the southern United States and northern Mexico and is disseminated north by wind.

The full life cycle is as follows. The teliospore, or overwintering stage, is produced on barley as the plants mature. Teliospores germinate in spring to produce a hyaline basidium on which four hyaline basidiospores, or sporidia, develop on sterigmata. Basidiospores are then disseminated by wind to barberry plants, where infections result in pycnia. Pycnia exude slender pycniospores and receptive hyphae in sticky drops of liquid that are attractive to insects. Pycniospores are transferred, primarily by insects, from one pycnia to another, where pycniospores germinate and exchange genetic material with receptive hyphae. The result is the formation of aecia, normally on the bottom of the leaf. Aeciospores are then produced in aecia formed on the underside of barberry leaves and disseminated by wind to barley. The result of the aeciospore infection is the formation of uredinia. Urediniospores are produced in uredinia to serve as the secondary inoculum during the growing season. Urediniospores are disseminated by wind to barley plants to cause secondary infections throughout the growing season until plants reach maturity. The optimum temperature for disease development is 20°C.

Several races are known to exist. A new race called QCC has developed that is capable of attacking barleys that possess the T gene source of resistance to stem rust. At present, all barley varieties and some winter wheat varieties are susceptible, but durum wheat and hard red spring wheat are resistant to *P. graminis* f. sp. *tritici* and *P. graminis* f. sp. *secalis*.

Distribution. Wherever barley is grown.

Symptoms. Stem rust occurs on stems, leaf sheaths, leaf blades, glumes, and beards. Uredinia are characteristic brick-red pustules with ragged edges that were created as the urediniospores broke through the leaf epidermis. Dark brown to black teliospores form in old uredinial pustules. Basidia and basidiospores are colorless and microscopic and cannot be readily observed without the aid of magnification. Pycnia are small orange spots on upper barberry leaves with a drop of liquid in the center of the spot. Aecia and aeciospores are orange "bell-like" structures on the underside of barberry leaves. During severe disease development, kernels are shriveled and stems become brown, dry, and brittle and often break over.

Management
1. Grow resistant cultivars.
2. Eliminate the common barberry bushes from the vicinity of any barley fields.
3. Apply fungicides in emergency situations.

Stripe Rust

Stripe rust is also called barley stripe rust, barley yellow rust, glume rust, and yellow rust.

Cause. *Puccinia striiformis* f. sp. *hordei* (syn. *P. glumarum*) is a rust fungus whose alternate host is not known. The fungus oversummers as urediniospores that are disseminated by wind to other plants, such as rye, wheat, barley, and grasses. Optimum disease development is at temperatures of 10° to 15°C with periodic rain or dew. Little or no infection occurs during warm weather in the summer. Teliospores are formed but are not known to function as overwintering spores.

Distribution. In North America, stripe rust occurs at the higher elevations and cooler climates along the Pacific Coast and intermountain areas from Canada to Mexico. It also occurs under the same environmental circumstances in South America and in the mountainous areas of central and western Europe, South Asia, and Africa.

Symptoms. Stripe rust symptoms occur on leaves and heads during cool weather in the spring before the symptoms of other rusts are apparent. Yellow uredinia appear on barley foliage formed in the autumn and on new foliage produced early in the spring. Eventually, uredinia coalesce to form long stripes between veins of leaves and sheaths. Small, linear lesions also occur on floral bracts. Telia develop as narrow linear dark brown pustules covered by the epidermis.

Management
1. Grow resistant cultivars.
2. Apply foliar fungicides.

Take-All

Take-all is also called dead head and white head.

Cause. *Gaeumannomyces graminis* var. *tritici* survives as mycelia in live overwintering plants, mycelia and perithecia in infested residue for about 10 months, mycelia in soil outside of residue for about 5 months, and, parasitically, on the roots of a number of different grasses. Maize roots can also become infected and serve as a source of inoculum to infect subsequent barley crops.

Dissemination of the fungus is by any means that moves residue. Plant-to-plant spread occurs by hyphae growing through the soil from root to root or directly by root to root contact. Ascospores are produced in perithecia during wet weather but apparently are not disseminated either by splashing water or wind a great distance. Therefore, ascospores are not considered important in either the dissemination of the disease or the infection of susceptible plants.

Seedlings become infected in the autumn or spring when roots grow in the vicinity of residue precolonized by *G. graminis* var. *tritici*. Plants infected early in the growing season become the most severely diseased. Disease is favored by high pH soils that are deficient in nitrogen and phosphorous and that remain cool and wet (–1.2 to –1.5 bars) for prolonged periods of time.

Take-all is more severe on wheat than barley; however, barley is usually more severely affected than rye.

Distribution. Wherever barley is grown.

Symptoms. Take-all symptoms become most obvious at heading as patches or circular areas of plants in a field are uneven in height and appear to be in different stages of maturity. The first symptom on a diseased plant is light brown to dark brown necrotic lesions on roots. At the jointing stage, most roots are brown to black, sparse, dead, and brittle. Plants eventually die, or if they are still alive, they are stunted, leaves are chlorotic, and the whole plant is easily broken free of crowns when pulled from the soil.

A dark discoloration, which is a good diagnostic characteristic, occurs on the stem above the soil line, together with a mat of dark brown mycelium that grows under the lower sheath between the stem and the inner leaf sheaths. Plants have few tillers and heads are bleached (white heads), sterile, and ripen prematurely.

Management
1. Rotate barley with a resistant crop.
2. Sow winter barley as late as feasible.

Verticillium Disease

Cause. *Verticillium dahliae.* Oat and wheat are also susceptible but do not develop severe symptoms.

Distribution. United States (Idaho).

Symptoms. Leaves, usually along the leaf margin, have longitudinal chlorotic stripes in which the vascular bundles have become discolored brown. Chlorotic leaves are often flaccid and eventually become necrotic. Symptoms are similar to those associated with Cephalosporium stripe.

Management. Resistance is reported to exist among present cultivars.

Wirrega Blotch

Cause. *Drechslera wirreganensis.* Six other grasses are hosts. Disease is favored by reduced tillage practices.

Distribution. Australia, in areas with rainfall exceeding 350 mm.

Symptoms. Two symptoms occur on leaves: a blotch and a spot. Blotches are large, medium brown, oval lesions, sometimes with chlorotic margins. Depending on the cultivar, blotches may remain small or extend along the full length of the leaf to give the appearance of a toxin spreading from the lesion. At the center of a blotch, tissue becomes necrotic and dry until it eventually tears apart.

Spots are small and dark brown with light-colored centers, usually without a chlorotic margin. There is often a hole at the center of a spot.

Management
1. Some barley cultivars are resistant.
2. Change tillage practices from reduced tillage to conventional tillage.

Yellow Leaf Spot

Yellow leaf spot is also called blight, leaf spot, and tan spot.

Cause. *Pyrenophora tritici-repentis* (syn. *P. trichostoma*), anamorph *Drechslera tritici-repentis* (syn. *Helminthosporium tritici-repentis*).

Pyrenophora tritici-repentis overwinters as pseudothecia on infested residue. During frequent rains and cool, cloudy, humid weather early in the growing season, primary inoculum is ascospores that are released and disseminated by wind to plants. Later in the growing season, under similar conditions, secondary inoculum is provided by conidia produced in older lesions and disseminated by wind to plants. In the autumn, pseudothecia are produced on diseased culms and leaf sheaths.

Distribution. Worldwide; however, barley is rarely affected.

Symptoms. Initially, tan flecks appear on both sides of lower leaves and progress upward to upper leaves during the growing season. Flecks eventually become tan, diamond-shaped lesions up to 12 mm long, with a yellow border and a dark brown spot in the center that is caused by sporulation of *P. tritici-repentis*. Lesions eventually coalesce to cause large areas of leaf to die from the tip inward. Pseudothecia will eventually develop on residue and appear as small, dark, raised bumps.

Management
1. Apply a foliar fungicide before the disease becomes severe and if weather conditions favor spread of the disease.
2. Plow under infested residue.
3. Grow resistant barley cultivars.

Diseases Caused by Nematodes

Cereal Cyst

The cereal cyst nematode is sometimes called the oat cyst nematode.

Cause. *Heterodera avenae, H. hordecalis,* and *H. latipons* survive as eggs and as larvae within cysts in soil for several years. Dissemination is by any means that moves soil. The optimum emergence of larvae from eggs from midwinter until spring occurs at a temperature of 10°C and moist soil; larvae then migrate to nearby plant roots. Optimum infection occurs behind the growing point at 20°C. As nematodes mature, they swell and erupt through root tissues. Males return to a vermiform shape, but females become lemon-shaped as egg production begins. Within 2 months of infection, the female dies and the body hardens into a cyst that protects eggs and larvae from growing season to growing season. One generation is completed each growing season.

Distribution. Australia, Canada (Ontario), Middle East, northern Europe, India, Japan, New Zealand, and the United States (Idaho, Oregon, Washington).

Symptoms. The cereal cyst nematode is the most important nematode pathogen of barley; yield reductions up to 22% have been reported. Symptoms are most pronounced on plants growing in light, sandy soils. A severe infestation causes patches of weeds to overgrow stunted barley plants whose leaves have a chlorotic discoloration similar to a nitrogen or phosphorous deficiency. Head emergence is delayed and the number of spikelets is reduced.

Roots are stunted and knotted and have a pronounced proliferation. Immature, white cysts that are approximately the size of pinheads or smaller start to appear on roots at heading.

Management
1. Grow resistant cultivars.
2. Rotate barley with a nongrass crop.

Root Gall

The disease caused by the root gall nematode is known as "krok" in Scandinavia.

Cause. *Subanguina radicicola* (syn. *Ditylenchus radicicola* and *Anguillulina radicicola*) survives by continuous habitation in host roots. Larvae penetrate roots and develop in cortical tissue, forming a gall in 2 weeks. Mature females begin egg production within galls. Eventually, galls disintegrate and release into the soil larvae, which cause secondary infections. A generation of nematodes is completed in about 60 days.

Distribution. Canada, Europe, Russia, and the United States.

Symptoms. Top growth of seedlings is reduced, and the tips of outer leaves are chlorotic. Roots are branched, twisted, and bent at the sites of the numerous inconspicuous galls that vary in diameter from 0.5 to 6.0 mm. The centers of the larger galls have a cavity that is usually filled with nematode larvae.

Management. Rotate barley with noncereal crops.

Root Knot

The root knot nematode is also called the cereal root knot nematode.

Cause. *Meloidogyne naasi* is an endoparasitic nematode that overwinters as eggs in soil. In the spring, larvae hatch from eggs and infect roots. By the middle of summer, females inside root knot tissue release their eggs into the soil.

Distribution. Northern Europe, Iran, the United States, and Yugoslavia.

Symptoms. Stunted, chlorotic plants occur in patches that range in size from a few square meters to large areas within a field. Diseased plants have short heads with no spikelets or are completely lacking a seed head. Swellings or thickenings comprised of swollen cortical cells and bodies of nematodes containing egg masses can be found on roots, especially near the root tips in the spring and summer. When root knots are cut open, egg masses within the knot will turn dark when exposed to the air.

Management
1. Sow barley in autumn.
2. Rotate barley with root crops such as sugar beet.

Root Lesion

Cause. *Pratylenchus crenatus, P. fallax, P. neglectus,* and *P. thornei* overwinter as eggs, larvae, or adults in host tissue and soil. At temperatures of 13°C and above, both larvae and adults penetrate roots, where they move through cortical cells. The females deposit eggs as they migrate through tissue. Older roots are eventually abandoned by nematodes and new roots sought as sites for further penetration and feeding.

Distribution. Wherever barley is grown. *Pratylenchus thornei* is found in the dry regions of Australia and Israel; *P. crenatus* and *P. thornei* are common in the United States; and *P. crenatus, P. fallax,* and *P. neglectus* are common in lighter soils in Europe.

Symptoms. Plants will be stunted, chlorotic, and under moisture stress. *Rhizoctonia solani* often infects through nematode wounds and eventually rots roots and crowns. New roots may also become dark and stunted, with resultant loss in yield.

Management
1. Sow in autumn.
2. Use soil nematicides where high costs warrant them.

Stunt

Cause. *Merlinius brevidens, Tylenchorhynchus dubious,* and *T. maximus* are ectoparasites that feed on external cells and are favored by wet soils.

Distribution. Peru and the United States.

Symptoms. Damage has been reported only on winter barley. Plants in areas throughout a field of approximately 1 m in diameter have short, stubby roots, are stunted and chlorotic, and have fewer tillers and kernels than healthy plants.

Management. Not reported.

Disease Caused by Phytoplasmas

Aster Yellows

Cause. The aster yellows phytoplasma (AY-PLO) survives in several plants and leafhoppers. AY-PLO is transmitted primarily by the aster leafhopper, *Macrosteles fascifrons,* as it feeds on diseased plants. Transmission of AY-PLO is less common by *Endria inimica* and *M. laevis.*

Two strains of AY-PLO exist: the eastern strain (NAY-PLO) and the western strain (CAY-PLO). Disease development occurs at temperatures of 25° to 32°C, with the most severe symptoms occurring at the higher temperatures.

Distribution. Generally throughout eastern Europe, Japan, and North America. Aster yellows is rarely severe on barley.

Symptoms. Seedlings will either die 2 to 3 weeks after infection or, if they survive, plants will be stunted, leaves will be entirely chlorotic or have yellow blotches, and heads will be sterile with distorted awns.

Infection of older plants causes leaves to become stiff and yellow, red, or purple inward from the leaf tip or margin. Root systems may not be well developed.

Management. Not reported.

Diseases Caused by Viruses

African Cereal Streak

Cause. African cereal streak virus (ACSV) is limited to the phloem. ACSV is transmitted only by the delphacid leafhopper, *Toya catilina*. The natural virus reservoir is presumably native grasses. Disease development occurs at temperatures of 20°C and above.

Distribution. East Africa.

Symptoms. Initially, faint broken chlorotic streaks begin near the leaf base and extend upward. Eventually, definite alternate yellow and green streaks develop along the entire leaf blade, which eventually becomes completely chlorotic. Leaves formed after infection develop a "shoestring" habit and die.

Young plants become chlorotic, severely stunted, and die. Older plants become soft, flaccid, and velvety to the touch and have yellow, distorted heads that yield little seed. A phloem necrosis develops.

Management. Not reported.

American Wheat Striate Mosaic

Cause. American wheat striate mosaic virus (AWSMV) is in the rhabdovirus group. AWSMV is transmitted by the leafhoppers *Endria inimica* and *Elymana virenscens*.

Distribution. Canada and the central United States.

Symptoms. Plants are stunted and have slight to moderate leaf striations that consist of yellow to white parallel streaks.

Management. Not reported.

Australian Wheat Striate

Cause. Australian wheat striate virus (AWSV) is also known as chloris striate mosaic virus (CSMV). AWSV is transmitted by the leafhoppers *Nesoclutha obscura* and *N. pallida* but is not sap-transmissible.

Distribution. Australia. The disease is of little importance in barley.

Symptoms. Plants are dwarfed and have fine, broken yellow to gray streaks or stripes on leaves.

Management. No management measures are necessary.

Barley Mosaic

Cause. Barley mosaic virus (BMV) is seedborne and is transmitted both by the corn leaf aphid, *Rhopalosiphum maidis,* and mechanically. Oat and wheat are also hosts.

Distribution. India.

Symptoms. Diseased plants are generally stunted. Initially, leaves are chlorotic but later develop mosaic symptoms. Diseased seeds, which are small and shriveled, have poor germination.

Management. Grow resistant cultivars.

Barley Stripe Mosaic

Barley stripe mosaic is also known as barley false stripe, false stripe, lantern head, and stripe mosaic.

Cause. Barley stripe mosaic virus (BSMV) can remain viable in seed for up to 8 years and overwinters in wild oat, *Avena fatua.* When infected seed germinates, the highest numbers of resulting seedlings become diseased at temperatures of 20° to 24°C. BSMV is spread when injured leaves rub against each other and by infected pollen, but it is not known to be transmitted by insects or other means. BSMV is spread further in barley sown in the spring than in the autumn.

Distribution. Southern Asia, Australia, Europe, Japan, Mexico, western North America, and Russia.

Symptoms. Symptoms vary with the mode of infection and temperature. Leaf symptoms that result from seedborne infection are a mottling or spots that are either narrow or wide, numerous or few, and continuous or broken. The color of the mottling or spots may be light green, tan, yellowish, or bleached white, which contrasts to the rest of the green leaf. Infrequently, entire diseased leaves may be nearly white.

Virulent strains of BSMV cause brown stripes, which are continuous or broken and have irregular margins, often form a V- or chevron-shape on leaves. Sometimes plants are severely stunted, and florets, which are sterile, may have no heads or poorly developed heads and kernels. Protein synthesis is affected.

Management
1. Grow resistant cultivars.
2. Sow disease-free seed.
3. Rotate barley with nongrass crops.

Barley Yellow Dwarf

Barley yellow dwarf is also called yellow dwarf.

Cause. Several luteoviruses are grouped under the name barley yellow dwarf virus (BYDV). The viruses share the characteristic of infecting gramineous plants, but differ in various other properties, such as virulence, host range, serological behavior, and transmissibility by aphid vectors.

BYDV survives or is reservoired in perennial grasses and grains sown in the autumn. Perennial grasses are a large reservoir of BYDV but may not serve as the most important source of inoculum. In Canada, inoculum carried by aphids from elsewhere was considered to be the main source of infection rather than grasses, winter wheat, or maize. In the United States and possibly within North America, most overwintering is reported to occur in warmer barley-growing areas. In areas where oversummering is a factor, maize is a reservoir for both aphids and some isolates of BYDV.

BYDV is transmitted by several species of aphids, including the corn leaf aphid, *Rhopalosiphum maidis*; oat bird-cherry aphid, *R. padi*; rice root aphid, *R. rufiabdominelis*; Russian wheat aphid, *Diuraphis noxia*; early instars of greenbug, *Schizaphis graminum*; and the English grain aphid, *Sitobion avenae.*

Based on transmission by different insects, five types of BYDV are recognized. MAV is transmitted specifically by *S. avenae*; RMV by *R. maidis*; RPV by *R. padi*; SGV by *S. graminum*; and PAV nonspecifically by *R. padi* and *S. avenae* and infrequently by *D. noxia*. However, some BYDV isolates obtained from cereal plants in Victoria, Australia were serologically similar to MAV but distinct in being readily transmissible by the aphid *R. padi*. Some RMV isolates from Montana were transmitted efficiently by both *R. maidis* and *S. graminum,* and two RMV isolates were occasionally transmitted by *R. padi*. After virus acquisition, an aphid may transmit BYDV for the rest of its life.

Subsequent investigations of serological relationships showed a parallel separation into serotypes. Vector specificities, serotyping, and other characteristics have been used to group MAV, PAV, and SGV serotypes into BYDV subgroup 1 and RPV and RMV serotypes into BYDV subgroup 2.

BYDV is not transmitted through eggs, newborn aphids, seed, or soil, or by mechanical means. Disease development occurs during temperatures of 10° to 18°C and moist conditions that favor grass and cereal growth, and aphid reproduction and migration. Storm fronts aid flights of aphids, which may cover hundreds of kilometers from southern areas, where most overwintering is reported to occur, to Canada and the Upper Midwest in the United States.

Distribution. Africa, Asia, Australia, Europe, New Zealand, North America, and South America.

Symptoms. Symptom expression is greatest at low temperatures of around 16°C. Plant stands appear uneven due to stunting of plants that were infected early in the growing season. This latter symptom is confused with nutrient deficiency and other unfavorable growing conditions.

Initially, faint yellowish-green blotches occur near the leaf tip, the leaf margin, or on the leaf blade itself, a characteristic that helps to distinguish yellow dwarf from other maladies. Discolored areas enlarge and coalesce at the leaf base, but tissue next to the midrib remains green longer than the rest of the leaf. Eventually, diseased plants are dwarfed and leaf color becomes a bright yellow or occasionally red or purple. The bright yellow discoloration occurs on older leaves but not the youngest leaves except in late infections when only flag leaves show symptoms.

In the laboratory, root dry weight, length, and numbers were reduced the most when plants were infected early. Root development stops when shoots display symptoms and is not correlated to the degree of BYDV resistance.

Infection by BYDV increases freezing injury to crowns of winter barley. There is also a reduction in malt quality and yield.

Management
1. Grow resistant or tolerant cultivars.
2. Avoid sowing in early autumn or late spring.

Barley Yellow Mosaic

Cause. Barley yellow mosaic virus (BYMV) survives up to 5 years in air-dried soil. The soil fungus *Polymyxa graminis* is thought to be the vector. Dissemination is by any means that will move soil, particularly machinery, wind, and water. Two types of BYMV have been reported from Germany. One type is disseminated both by soil and mechanically, while the other type is disseminated only by soil.

Distribution. Belgium, England, France, Germany, and Japan.

Symptoms. Symptoms are most prominent in spring and disappear as temperatures rise above 18°C. Plants in areas throughout a field are stunted and yellow. Initially, diseased leaves have yellow spots and short streaks, then become totally yellow, beginning at the leaf tip. Necrotic spots sometimes appear on older leaves, which may die prematurely.

Management
1. Grow resistant cultivars.
2. Sow barley in spring.

Barley Yellow Streak Mosaic

Cause. Barley yellow streak mosaic virus (BaYSMV) is a novel virus transmitted by the brown wheat mite, *Petrobia latens,* a non-web-spinning spider mite that is generally believed to reproduce parthenogenetically. The mite deposits eggs on debris in surface soil layers near its host plant. BaYSMV particles are extremely large for a plant virus (64 by 1000 nm on average) and appear to be enveloped and contain ssRNA. BaYSMV has been transmitted mechanically to *Nicotiana benthamiana* and *Chenopodium quinoa*.

Distribution. Canada (Alberta) and the United States (Idaho and north central Montana).

Symptoms. Diseased leaf symptoms consist of yellow-green mosaic, streaking, color banding, and severe necrosis.

Management. Resistant genotypes have been identified.

Barley Yellow Striate Mosaic

Cause. Barley yellow striate mosaic virus (BYSMV) is transmitted by planthoppers. Wheat is also a host of BYSMV.

Distribution. France, Italy, and Morocco.

Symptoms. Diseased leaves are stunted and chlorotic and have a mosaic pattern.

Management. Not reported.

Barley Yellow Stripe

Cause. The causal agent has not been identified, but it is thought to be a virus or virus-like entity. It is transmitted by the leafhopper *Euscelis plebejus* but not by sap.

Distribution. Turkey.

Symptoms. Symptoms commonly occur along field borders and other grassy areas infested by *E. plebejus*. Fine continuous stripes that occur on leaves sometimes are followed by leaf yellowing and death.

Management. Not reported.

Brome Mosaic

Cause. Brome mosaic virus (BMV) is in the bromovirus group and is transmitted primarily by sap. The Russian aphid, *Diuraphis noxia,* has been reported as a vector. The nematodes *Xiphinema coxi* and *X. paraelongatum* are inefficient vectors, and the cereal leaf beetle has transmitted BMV under controlled conditions. Transmission by nymphs and adults of red spider mites, *Tetranychus cinnabarinus,* has also been reported.

Distribution. North central Europe, Russia, South Africa, the United States, and Yugoslavia.

Symptoms. Diseased plants are dwarfed and have shriveled heads. Leaf symptoms are most obvious on young plants and fade as plants mature, often disappearing by heading. Initially, light yellow or white spots and streaks occur that spread rapidly, giving leaves a bright yellow mosaic pattern.

Management. Not considered necessary.

Cereal Chlorotic Mottle

Cause. Cereal chlorotic mottle virus (CeCMV) is in the rhabdovirus group and is transmitted by cicadellids.

Distribution. Northern Africa and Australia.

Symptom. Severe necrotic and chlorotic streaks occur on leaves.

Management. Not reported.

Cereal Tillering

Cause. Cereal tillering disease virus (CTDV) is circulative in the planthoppers *Laodelphax striatellus* and *Dicranotropis hamata* but is not passed through eggs.

Distribution. Finland, Italy, and Sweden.

Symptoms. Plants are dwarfed and have excessive tillering. Leaves are dark green and sometimes malformed with serrated margins. Grain yields are reduced.

Management. Not reported.

Eastern Wheat Striate

Cause. Eastern wheat striate virus (EWSV) is transmitted by the leafhopper *Cicadulina mbila* but not by seed, soil, aphids, or mechanically. EMSV overseasons in the perennial Narenga grass.

Distribution. India.

Symptoms. Plants infected while very young have the most severe symptoms and usually die. Diseased leaves and leaf sheaths develop thin, chlorotic stripes that become necrotic as diseased plants mature. Heads are partially filled with shriveled, poor-quality grain. Plants are stunted.

Management. Not reported.

Enanismo

Enanismo is also called cereal dwarf.

Cause. The causal agent has not been described, but one or more viruses and a toxin produced by the leafhopper *Cicadulina pastusae* may be responsible. Leafhopper adults and nymphs transmit the causal agent in a circulative fashion; females are more efficient vectors than males.

Distribution. Colombia and Ecuador.

Symptoms. Seedlings are killed or stunted. Later infections cause less stunting, but yellow leaf blotches with dark green stripes appear. Gall-like enations develop on new leaves 1 to 3 weeks after insects have fed on older leaves. Plants infected just before heading develop distorted heads with poorly filled kernels, which reduces yields in localized areas.

Management. Sow later in the spring after vector activity declines.

Flame Chlorosis

Cause. Double-stranded RNAs ranging in size from 900 to 2800 base pairs are present in vesicles and are likely the disease agent or its replicative intermediate, rather than virions. The causal agent is considered to be soil-transmitted and has been transmitted by sowing seed or planting seedlings in soil where diseased plants had grown. There may be an association with infection by *Pythium* spp.

The grassy weeds green foxtail, *Setaria viridis,* and barnyard grass, *Echinochloa crus-galli,* are also hosts.

Distribution. Canada (Manitoba), in spring-sown barley.

Symptoms. A striking flame-like pattern of leaf chlorosis and severe stunting occurs in spring barley. Chloroplasts and mitochondria of affected cells are hypertrophied and contain an extensive proliferation of fibril-containing vesicles that form within the organellar envelope. Affected plants continue to produce leaves with symptoms after transfer to sterile potting medium.

Management. Not reported.

High Plains Virus

Cause. Since High Plains virus (HPV) is transmitted by the wheat leaf curl mite, *Aceria tosichella,* which also transmits wheat streak mosaic virus (WSMV), mixed infections by these two viruses can occur. HPV is associated with a 32-kDa protein that resembles tenuiviruses. However, it is also suggested by some researchers that HPV has some resemblance to tospoviruses.

HPV is also transmitted to maize and wheat.

Distribution. The United States (Colorado, Idaho, Kansas, Nebraska, Texas, and Utah).

Symptoms. The first symptom is small, chlorotic spots on leaves. These spots rapidly expand into a mosaic pattern and then into a general yellowing of the plant. Severely diseased plants eventually become necrotic. Field infections very often are mixed infections of WSMV and HPV.

Management. Not reported.

Hordeum Mosaic

Cause. Hordeum mosaic virus.

Distribution. Canada (Alberta).

Symptoms. Symptoms develop at temperatures of 10° to 30°C. Streaks develop near the base of young leaves and a diffuse chlorotic mottle develops on all leaves.

Management. Not reported.

Maize Yellow Stripe

Cause. Maize yellow stripe virus (MYSV) is associated with tenuivirus-like filaments. MYSV is transmitted in a persistent manner by both nymphs and adults of the leafhopper *Cicadulina chinai.* Acquisition and inoculation threshold times are each 30 minutes, with a latent period of 4.5 to 8.0 days, depending on temperature. Acquisition and inoculation occur at a minimum temperature of 14°C and a maximum temperature of 25°C. The maximum retention period is 27 days.

Wheat, sorghum, and graminaceous weeds are winter hosts. Different strains of MYSV exist.

Distribution. Egypt.

Symptoms. Symptoms on barley are not well known. Three symptom types exist on maize: a fine stripe, coarse stripe, and chlorotic stunt. Each type may represent different MYSV strains. Experimentally, fine stripes appear on the first leaves, followed by coarse stripes on younger leaves; however, some leaves have both symptom types.

Management. Not reported.

Northern Cereal Mosaic

Cause. Northern cereal mosaic virus (NCMV) is in the rhabdovirus group and is transmitted by the planthopper *Laodelphax striatellus.* Other insects that have been implicated as vectors for NCMV are *Unkanodes sapporonus* and species of *Delphacodes.*

Distribution. Japan, Korea, and Siberia.

Symptoms. Initially, yellowish-white or whitish-green specks occur on young and newly emerged leaves and leaf sheaths. Later, leaf blades are short and narrow with yellowish-white stripes parallel to veins. Plants are stunted, have increased tillering, and have a few unfilled panicles or no panicles at all.

Management. Not reported.

Oat Blue Dwarf

Oat blue dwarf is possibly the same disease as moderate barley dwarf.

Cause. The oat blue dwarf virus (OBDV) is in the marafivirus group and is transmitted primarily by the adult aster leafhopper, *Macrosteles fascifrons,* with occasional transmission by immature leafhoppers. OBDV cannot be transmitted by seed, mechanically, or soil.

Distribution. Czechoslovakia, Finland, North America, and Sweden.

Symptoms. Symptom development is greater at temperatures of 16° to 18°C than at higher temperatures. Plants are blue-green and stunted, tiller excessively, and either do not produce heads or have sterile spikes. Leaves are short and stiff and have enations along veins of the leaves and leaf sheaths.

Management. Not reported.

Oat Pseudorosette

Cause. Oat pseudorosette virus (OPV) is transmitted by the leafhopper *Laodelphax striatellus.*

Distribution. Russia.

Symptoms. Plants are stunted and tiller excessively.

Management. Grow the least susceptible cultivars.

Oat Sterile Dwarf

Oat sterile dwarf is called oat base tillering disease in Finland and oat dwarf tillering disease in Sweden.

Cause. Oat sterile dwarf virus (OSDV) is in the fijivirus group and is transmitted by delphacid planthoppers, especially *Javesella pellucida* and *Dicranotropis hamata.* OSDV is transmitted after 1 month incubation in vectors, but the virus is rarely passed through eggs.

Distribution. England and eastern and northern Europe.

Symptoms. Plants are slightly stunted and have few juvenile tillers. Leaves are darker green than normal.

Management. Control grassy weeds and oat cover crops.

Rice Black-Streaked Dwarf

Cause. Rice black-streaked dwarf virus (RBSDV) is in the fijivirus group and is transmitted by the planthoppers *Laodelphax striatellus, Unkanodes sapporonus,* and *U. albifascia* but not by seed or sap. RBSDV overwinters in winter wheat and is retained from season to season in all stages of planthoppers. Nymphs are more efficient vectors than adults, but the virus is not transmitted through eggs. Plants acquire RBSDV after 30 minutes feeding by planthoppers in which the virus has incubated 7 to 35 days.

Distribution. China, Japan, Korea, and Russia.

Symptoms. Diseased plants are severely stunted and have twisted leaves. Waxy swellings are present on veins, the under surfaces of leaves, and culms.

Management
1. Grow resistant rice cultivars to lessen the chance of virus transmission to barley.
2. Do not grow barley close to rice.

Rice Stripe

Cause. Rice stripe virus is in the tenuivirus group and is transmitted by the planthopper *Laodelphax striatellus.*

Distribution. Japan, Korea, and Taiwan.

Symptoms. Plants are dwarfed and have yellow-green to yellow-white stripes that are parallel to midribs of leaves.

Management. It is not necessary to manage rice stripe in barley since the extensive use of resistant rice cultivars has reduced the vector population.

Russian Winter Wheat Mosaic

Cause. Winter wheat mosaic virus is disseminated by the leafhoppers *Psammotettix striatus* and *Macrosteles laevis.*

Distribution. Eastern Europe and Russia.

Symptoms. Seedlings are stunted and have excessive tillering and necrosis. Surviving seedlings are rosettes without leaf mosaic symptoms. Older plants are slightly stunted, and leaves have a typical mosaic and streak-mosaic pattern with chlorotic dashes and streaks parallel to veins.

Management. Not reported.

Soilborne Wheat Mosaic

Soilborne wheat mosaic is also called wheat soilborne mosaic.

Cause. Soilborne wheat mosaic virus (SBWMV) is in the furovirus group and survives in the soilborne fungus *Polymyxa graminis*. SBWMV is spread by any method that disseminates soil containing *P. graminis*. Motile zoospores infect root hairs and epidermal cells in the autumn and infrequently in the spring during wet soil conditions and soil temperatures of 10° to 20°C. Two to 4 weeks after infection, *P. graminis* replaces plant cell contents with plasmodial bodies that either segment into additional zoospores or develop into thick-walled resting spores.

Disease is most severe on barley sown in the autumn in low, wet areas. Soils may remain infected for years. There are different strains of SBWMV.

Distribution. Argentina, Brazil, Canada, Egypt, Italy, Japan, and the eastern and central United States. Soilborne wheat mosaic is not considered an important disease of barley.

Symptoms. Symptoms, which are most prominent in spring on the youngest or lowest leaves, range from light green to yellow leaf mosaics or mottling that later develops into parallel streaks. Plants may be slightly or severely stunted and rosetted by some strains of SBWMV. Warm weather prevents disease development, thus confining symptoms to lower leaves.

Management
1. Rotate barley with noncereal crops.
2. Sow barley late in the autumn.
3. Grow tolerant cultivars.

Wheat Dwarf

Cause. The causal agent is suspected to be a virus that is transmitted by several leafhoppers, primarily *Psammotettix alienus* and *Macrosteles laevis*, but not by sap, seed, or soil.

Distribution. Bulgaria, Czechoslovakia, Russia, and Sweden.

Symptoms. Plants infected as seedlings are severely stunted and do not form heads, but plants infected later are less stunted. Leaves have small, light green to yellow-brown spots and blotches that may coalesce to cause yellowing and necrosis.

Management. Not reported.

Wheat Streak

Wheat streak is also called wheat streak mosaic.

Cause. Wheat streak mosaic virus (WSMV) is in the potyvirus group and survives in infected plants. No active virus has been detected in dead plants or in seed. WSMV is transmitted primarily by windborne wheat curl mites, *Aceria tulipae,* for up to 2.4 km but also is transmitted mechanically.

Mites overwinter on live plants. Only young mites acquire WSMV after feeding 15 or more minutes on diseased plants. They carry the virus internally for several weeks, but neither the mite nor the virus can survive longer than 1 or 2 days in the absence of a living plant.

As winter barley plants mature during late summer or early autumn, WSMV is disseminated by mites to volunteer grasses and cereals and eventually to early-sown barley. However, wheat streak mosaic is a problem on barley only when mite populations build to extremely high levels in nearby wheat fields. WSMV has a wide host range, but some grasses are hosts for the mites but not for the virus, and vice versa, and some are susceptible or resistant to both.

Distribution. Eastern Europe, western and central North America, and Russia.

Symptoms. The most severe symptoms occur in plants sown early in the autumn but do not appear until the following spring. Initially, light green to light yellow blotches, dashes, or streaks occur parallel to leaf veins and turn brown as plants mature. Feeding mites often cause leaf edges to curl tightly in toward the upper midvein. Plants become stunted, display a general yellow mottling, and develop large numbers of tillers that vary in height. Plants may die before maturity or have sterile or partially sterile heads with shriveled kernels. At harvest, stunted plants, some with sterile heads, remain that are the same height or shorter than stubble.

Synergistic effects are suspected between WSMV and other viruses, such as barley yellow dwarf virus, making field identification difficult.

Management
1. Destroy volunteer cereals and grasses in adjoining fields 2 weeks before sowing and 3 to 4 weeks before sowing in the same field.
2. Sow after Hessian fly–free date.

Wheat Yellow Leaf

Cause. Wheat yellow leaf virus is in the closterovirus group and is transmitted by the corn leaf aphid, *Rhopalosiphum maidis*.

Distribution. Japan.

Symptoms. Leaves become yellow and blight. Plants die or ripen prematurely.

Management. Not reported.

Selected References

Anikster, Y. 1981. Alternate hosts of *Puccinia hordei*. Phytopathology 72:733-735.

Bockelman, H. E., Sharp, E. L., and Bjarko, M. E. 1983. Isolates of *Pyrenophora teres* from Montana and the Mediterranean region that produce spot-type lesions on barley. Plant Dis. 67:696-697.

Boosalis, M. G. 1952. The epidemiology of *Xanthomonas translucens* (J.J. & R) Dowson on cereals and grasses. Phytopathology 42:387-395.

Bragard, C., Singer, E., Alizadeh, A., Vauterin, L., Maraite, H., and Swings, J. 1997. *Xanthomonas translucens* from small grains: Diversity and phytopathological relevance. Phytopathology 87:1111-1117.

Braun, A., Brumfield, S., Hamacher, J., and Sands, D. C. 1997. Ultrastructural localization of *Pseudomonas syringae* in barley kernel tissue. (Abstr.) Phytopatholgy 87:S11.

Brown, J. K., and Wyatt, S. D. 1981. Corn as an oversummering host of barley yellow dwarf virus and aphid vector in eastern Washington. (Abstr.) Phytopathology 71:863.

Brown, M. P., Steffenson, B. J., and Webster, R. K. 1993. Host range of *Pyrenophora teres* f. *teres* isolates from California. Plant Dis. 77:942-947.

Brumfield, S. K. Z., Carroll, T. W., and Gray, S. M. 1992. Biological and serological characterization of three Montana RMV-like isolates of barley yellow dwarf virus. Plant Dis. 76:33-39.

Burton, R. J., Coley-Smith, J. R., and Wareing, P. W. 1988. *Rhizoctonia oryzae* and *R. solani* associated with barley stunt disease in the United Kingdom. Trans. Br. Mycol. Soc. 91:409-417.

Chico, A. W. 1983. Reciprocal contact transmission of barley stripe mosaic virus between wild oats and barley. Plant Dis. 67:207-208.

Conner, R. L., and Atkinson, T. G. 1989. Influence of continuous cropping on severity of common root rot in wheat and barley. Can. J. Plant Pathol. 11:127-132.

Cook, R., and Williams, T. D. 1972. Pathotypes of *Heterodera avenae*. Ann. Appl. Biol. 71:267-271.

Crosier, W. F., Nash, G. T., and Crosier, D. C. 1970. *Ustilago nuda* on leaves of 'Larker' barley. Plant Dis. Rptr. 54:927-929.

Cunfer, B. M. 1981. *Septoria* sp. on barley and *Hordeum pusillum*. (Abstr.) Phytopathology 71:869.

Damsteegt, V. D., Gildow, F. E., Hewings, A. D., and Carroll, T. W. 1992. A clone of the Russian wheat aphid (*Diuraphis noxia*) as a vector of the barley yellow dwarf, barley stripe mosaic, and brome mosaic viruses. Plant Dis. 76:1155-1160.

Darlington, L. C., Carroll, T. W., and Mathre, D. E. 1976. Enhanced susceptibility of barley to ergot as a result of barley stripe mosaic virus infection. Plant Dis. Rptr. 60:584-587.

Davies, K. A., and Fisher, J. M. 1976. Factors influencing the number of larvae of *Heterodera avenae* invading barley seedlings in vitro. Nematologica 22:153-162.

Dehne, H. W., and Oerke, E. C. 1985. Investigations on the occurrence of *Cochliobolus sativus* on barley and wheat. 1. Influence of pathogen, host plant and environment on infection and damage. Zeitschrift fur Pflanzenkrankheiten und Pflanzenschutz 92:270-280.

Delserone, L. M., and Cole, H., Jr. 1987. Effects of planting date on development of net blotch epidemics in winter barley in Pennsylvania. Plant Dis. 71:438-441.

Delserone, L. M., Frank, J. A., and Cole, H., Jr. 1983. Net blotch epidemics on winter barley in the fall as influenced by planting date. (Abstr.) Phytopathology 73:965.

Delserone, L. M., Cole, H., Jr., and Frank, J. A. 1987. The effects of infections by *Pyrenophora teres* and barley yellow dwarf virus on the freezing hardiness of winter barley. Phytopathology 77:1435-1437.

Dewey, W. G., and Hoffmann, J. A. 1975. Susceptibility of barley to *Tilletia controversa*. Phytopathology 65:654-657.

Dubin, H. J., and Stubbs, R. W. 1986. Epidemic spread of barley stripe rust in South America. Plant Dis. 70:141-144.

Duczek, L. J. 1990. Sporulation of *Cochliobolus sativus* on crowns and underground parts of spring cereals in relation to weather and host species, cultivar, and phenology. Can. J. Plant Pathol. 12:273-278.

Esnard, J., and Hepperly, P. R. 1995. First report of southern blight of common barley in Puerto Rico. Eur. J. Plant Pathol. 101:497-501.

Fargette, D., Lister, R. M., and Hood, E. L. 1982. Grasses as a reservoir of barley yellow dwarf virus in Indiana. Plant Dis. 66:1041-1045.

Farr, D. F., Bills, G. R., Chamuris, G. P., and Rossman, A. Y. 1989. Fungi on Plants and Plant Products in the United States. American Phytopathological Society, St. Paul, MN. 1252 pp.

Frank, J. A., and Cole, H., Jr. 1987. The influence of common root rot on net blotch of winter barley. Phytopathology 77:1454-1457.

Gill, C. C., and Westdal, P. H. 1966. Effect of temperature on symptom expression of barley infected with aster yellows or barley yellow dwarf viruses. Phytopathology 56:369-370.

Haber, S., and Chong, J. 1993. Flame chlorosis induces vesiculations in chloroplasts and mitochondria: What does it mean? (Abstr.) Can J. Plant Pathol. 15:57.

Haber, S., and Hardener, D. E. 1992. Green foxtail (*Setaria viridis*) and barnyard grass (*Echinochloa crus-galli*): New hosts of the virus-like agent causing flame chlorosis in cereals. Can. J. Plant Pathol. 14:278-280.

Haber, S., and Kim, W. 1989. Flame chlorosis: A new virus-like disease of barley in Manitoba. (Abstr.) Can. J. Plant Pathol. 11:189.

Hannukkala, A., and Koponen, H. 1988. *Microdochium bolleyi*, a common inhabitant of barley and wheat roots in Finland. Karstenia 27:31-36.

Harder, D. E., and Bakker, W. 1973. African cereal streak, a new disease of cereals in East Africa. Phytopathology 63:1407-1411.

Hosford, R. M., Jr. 1978. Effects of wetting period on resistance to leaf spotting of wheat, barley, and rye by *Leptosphaeria herpotrichoides*. Phytopathology 68:591-594.

Huftalen, C. S., and Bergstrom, G. C. 1986. First report of Ascochyta leaf spot caused by *Ascochyta hordei* var. *americana* on barley in New York. Plant Dis. 70:1074.

Hyung, L. S., and Skikata, E. 1977. Occurrence of northern cereal mosaic virus in Korea. Kor. J. Plant Prot. 16:87-92.

Ingram, D. M., and Cook, R. J. 1987. Influence of temperature and plant residues on pathogenicity of *Pythium* spp. on wheat, barley, peas and lentils. (Abstr.) Phytopathology 77:1239.

Inouye, T. 1976. Wheat yellow leaf virus. Descriptions of Plant Viruses, No. 157. Commonw. Mycol. Inst., Assoc. Appl. Biol., Kew, Surrey, England.

Jenkins, J. E. E. 1974. New or uncommon plant diseases and pests: *Botrytis* diseases in barley. Plant Pathol. 23:83-84.

Jin, Y., and Steffenson, B. J. 1992. *Puccinia coronata* on barley. Plant Dis. 76:1283.

Johnson, D. A., Tew, T. L., and Banttari, E. E. 1977. Factors affecting symptoms in barley infected with the oat blue dwarf virus. Plant Dis. Rptr. 61:280-283.

Kainz, M., and Hendrix, J. W. 1981. Response of cereal roots to barley yellow dwarf virus infection in a mist culture system. (Abstr.) Phytopathology 71:229.

Kelemu, S., and Sharp, E. L. 1986. Unique symptoms induced in Ethiopian barley cultivars by *Rhynchosporium secalis*. Plant Dis. 70:800.

Lister, R. M., and Sward, R. J. 1988. Anomalies in serological and vector relationships of MAV-like isolates of barley yellow dwarf virus from Australia and the U.S.A. Phytopathology 78:766-770.

Lockhart, B. E. L. 1986. Occurrence of cereal chlorotic mottle virus in northern Africa. Plant Dis. 70:912-915.

Luzzardi, G. C., Luz, W. C., and Pierobom, C. R. 1983. Podridao branca dos cereais causada por *Sclerotium rolfsii* no Brazil. Fitpathologia Brasileira 8:371-375 (in Portugese).

MacNish, G. C. 1985. Methods of reducing Rhizoctonia patch of cereals in Western Australia. Plant Pathol. 34:175-181.

Martin, R. A. 1985. Influence of seeding rate and nitrogen top dressing on net blotch development in barley. (Abstr.) Can. J. Plant Pathol. 7:446.

Martinez-Cano, C., Grey, W. E., and Sands, D. C. 1992. First report of *Arthrinium arundinis* causing kernel blight on barley. Plant Dis. 76:1077.

Martinez-Espinoza, A. D., Bjarko, M. E., and Riesselman, J. H. 1990. Studies on *Rhynchosporium secalis* as a seedborne pathogen. (Abstr.). Phytopathology 80:1006.

Mathre, D. E. (Ed.). 1982. Compendium of Barley Diseases, First edition. The American Phytopathological Society, St. Paul, MN. 78 pp.

Mathre, D. E. 1986. Occurrence of *Verticillium dahliae* on barley. Plant Dis. 70:981.

Mathre, D. E. 1989. Pathogenicity of an isolate of *Verticillium dahliae* from barley. Plant Dis. 73:164-167.

Mathre, D. E. (Ed.). 1997. Compendium of Barley Diseases, Second Edition. The American Phytopathological Society, St. Paul, MN. 90 pp.

Metz, S. A., and Scharen, A. L. 1979, Potential for the development of *Pyrenophora graminea* on barley in a semiarid environment. Plant Dis. Rptr. 63:671-675.

Mihuta-Grimm, L., and Forster, R. L. 1989. Scab of wheat and barley in southern Idaho and evaluation of seed treatments for eradication of *Fusarium* spp. Plant Dis. 73:769-771.

Murray, D. I. L., and Nicholson, T. H. 1979. Barley stunt disorder in Scotland. Plant Pathol. 28:200-201.

Nagaich, B. B., and Sinha, R. C. 1974. Eastern wheat striate: A new viral disease. Plant Dis. Rptr. 58:968-970.

Nutter, R. W., Jr., Pederson, V. D., and Timian, R. G. 1984. Relationship between seed infection by barley stripe mosaic virus and yield loss. Phytopathology 74:363-366.

O'Brien, P. C., and Fisher, J. M. 1978. Factors influencing the number of larvae of *Heterodera avenae* within susceptible wheat and barley seedlings. Nematologica 24:295-304.

Palival, Y. C. 1982. Role of perennial grasses, winter wheat and aphid vectors in the disease cycle and epidemiology of barley yellow dwarf virus. Can. J. Plant Pathol. 4:367-374.

Peters, R. A., Timian, R. G., and Wesenberg, D. 1983. A bacterial kernel spot of barley caused by *Pseudomonas syringae* pv. *syringae*. Plant Dis. 67:435-438.

Piening, L. J., and Orr, D. 1988. Effects of crop rotation in common root rot of barley. Can. J. Plant Pathol. 10:61-65.

Raemaekers, R. H., and Tinline, R. D. 1981. Epidemic of diseases caused by *Cochliobolus sativus* on rainfed wheat in Zambia. Can. J. Plant Pathol. 3:211-214.

Richardson, M. J., and Noble, M. 1970. *Septoria* species on cereals: A note to aid their identification. Plant Pathol. 19:159-163.

Roane, C. W., Roane, M. K., and Starling, T. M. 1974. *Ascochyta* species on barley and wheat in Virginia. Plant Dis. Rptr. 58:455-456.

Robertson, N. L., and Carroll, T. W. 1991. Mechanical transmission, partial purification, and preliminary chemical analysis of barley yellow streak mosaic virus. Plant Dis. 75:839-843.

Rovira, A. D., and McDonald, H. J. 1986. Effects of the herbicide chlorsulfuron on Rhizoctonia bare patch and take-all of barley and wheat. Plant Dis. 70:879-882.

Sampson, G., and Cloug, K. S. 1979. A Selenophoma leaf spot on cereals in the Maritimes. Can. Plant Dis. Surv. 59:3.

Seifers, D. L., Harvey, T. L., Martin, T. J., and Jensen, S. G. 1997. Identification of the wheat curl mite as the vector of the High Plains virus of corn and wheat. Plant Dis. 81:1161-1166.

Shane, W. W., Baumer, J. S., and Teng, P. S. 1987. Crop losses caused by Xanthomonas streak on spring wheat and barley. Plant Dis. 71:927-930

Sinha, R. C., and Benki, R. M. 1972. American wheat striate mosaic virus. CMI/A.A.B. Descriptions of Plant Viruses. Set 6., No. 99.

Skaf, J. S., Brumfield, S. K., and Carroll, T. W. 1992. Barley yellow streak mosaic virus infection of barley in Idaho. Plant Dis. 76:861.

Slack, S. A., Shepherd, R. J., and Hall, D. H. 1975. Spread of seedborne barley stripe mosaic virus and effects of the virus on barley in California. Phytopathology 65:1218-1223.

Steffenson, B. J., Grasmick, D. L., and Webster, R. K. 1988. Comparative epidemiology of net blotch and leaf scald of barley. (Abstr.) Phytopathology 78:1538.

Takamatsu, S. 1989. A new snow mold of wheat and barley caused by foot rot fungus, *Ceratobasidium gramineum*. Ann. Phytopathol. Soc. Japan 55:233-237.

Tekauz, A., and Chico, A. W. 1980. Leaf stripe of barley caused by *Pyrenophora graminea*: Occurrence in Canada and comparison with barley stripe mosaic. Can. J. Plant Pathol. 2:152-158.

Teviotdale, B. L., and Hall, D. H. 1976. Factors affecting inoculum development and seed transmission of *Helminthosporium gramineum*. Phytopathology 66:295-301.

Thomas, P. L. 1981. Distinguishing between the loose smuts of barley. Plant Dis. 65:834.

van den Berg, C. G. J., and Rossnagel, B. G. 1991. Epidemiology of spot-type net blotch on spring barley in Saskatchewan. Phytopathology 81:1446-1452.

Velasco, V. R., Ishimaru, C. A., and Brown, W. M., Jr. 1991. Halo blight of spring wheat caused by *Pseudomonas syringae*. (Abstr.) Phytopathology 81:1159.

Veljavec-Gratian, M., and Steffenson, B. J. 1997. Pathotypes of *Cochliobolus sativus* on barley in North Dakota. Plant Dis. 81:1275-1278.

Walker, J. 1975. Take-all disease of Graminae: A review of recent work. Rev. of Plant Pathol. 54:113-144.

Weller, D. M., Cook, R. J., MacNish, G., Bassett, E. N., Powelson, R. L., and Peterson, R. R. 1986. Rhizoctonia root rot of small grains favored by reduced tillage in the Pacific Northwest. Plant Dis. 70:70-73.

Westphal, A., Tidemann, A. V., and Fehrmann, H. 1995. Occurrence of *Monographella nivalis* var. *mayor* and *Fusarium culmorum* on barley in Germany in 1992. (Abstr.). Phytopathology 85:1169.

Yaegashi, H. 1988. Inheritance of blast resistance in two-rowed barley. Plant Dis. 72:608-610.

Zamora, M. R., Burnett, P. A., Vivar, H. E., Rodriguez, R., and Navarro, M. 1988. Barley mosaic virus in Mexico. Plant Dis. 72:546.

3. Diseases of Buckwheat *(Fagopyrum esculentum)*

Diseases Caused by Fungi

Ascochyta Leaf Blight

Cause. *Ascochyta fagopyri* (syn. *A. bresadolae*) overwinters as pycnidia in infested residue. Disease is favored by wet weather.

Distribution. Widespread.

Symptoms. Round to irregular-shaped leaf spots have light tan to dark brown centers and darker margins. Spots may coalesce to form brownish-purple blotches. Pycnidia, resembling black specks, may be scattered throughout the spots.

Management. Not reported.

Blight

Cause. *Phytophthora parasitica* (syn. *P. nicotianae* var. *parasitica*).

Distribution. Russia.

Symptoms. Cotyledons are infected first. Brown spots spread in concentric circles on diseased stems. Disease results either in reduction of plant growth or plant death.

Management. Not reported.

Botrytis Rot

Cause. *Botrytis cinerea* is externally seedborne and overwinters in infested residue and soil. Disease is favored by warm, wet weather.

Distribution. Russia.

Symptoms. "Powdery" spots that are covered with a gray mold or a black film form on diseased leaves and stems. Spots on diseased lower stems cause them to break off.

Management. Apply a fungicide seed treatment.

Chlorotic Leaf Spot and Stipple Spot

Cause. Possibly *Bipolaris sorokiniana* and *Alternaria alternata.* Although *A. alternata* was isolated from lesions, it was not pathogenic alone when inoculated to buckwheat.

Distribution. Canada (Manitoba).

Symptoms. Chlorotic, circular lesions (12–26 mm in diameter) are randomly scattered on leaves on the upper half of diseased plants. Lesions are of three types: spreading and uniformly chlorotic; spreading with concentric, chlorotic bands that alternate with normal dark green tissue; and small, restricted, tan lesions with sharply defined borders. This latter symptom is referred to as stipple spot. Necrosis occurs in the older lesions and in the older chlorotic rings.

Management. Not reported.

Collar Rot

Cause. *Sclerotinia* sp. is presumed to overwinter as sclerotia.

Distribution. Not known for certain but presumed to be widespread.

Symptoms. The first symptom is the presence of prematurely ripened plants that are easily pulled from the soil and are either scattered or grouped among healthy-appearing green plants. Pale gray stem lesions develop at the soil line. Initially, lesions are water-soaked and eventually expand to girdle stems and kill plants. Stems are bleached, have a chalky-white appearance, and tend to shred longitudinally. Sclerotia appear as hard, black, grain-sized bodies inside stems near the bases of killed plants.

Management. Not reported.

Downy Mildew

Cause. *Peronospora ducometi* (syn. *P. fagopyri*) is seedborne and causes systemic infection. Seed infested with oospores gives rise either to symptomless or stunted plants, depending on which of two types of hyphae occurs in diseased plants. Type A hyphae are found in stems, leaves, flowers, and seeds but not in roots; are narrow (8.6 μm in diameter) and branched; and produce haustoria. Sexual and asexual reproductive structures originate from type A hyphae. Type B hyphae are wide (27 μm in diameter) and un-branched and do not form haustoria. Type B hyphae often grow along the outside of xylem vessels but are rarely found inside xylem tissue.

Distribution. Canada, Europe, and Japan.

Symptoms. Type A hyphae are associated with stunting, leaf mottling, and rugosity. Type B hyphae are common in petioles and flowers of symptom-less plants but occasionally are found in stunted plants. Large, circular, chlorotic lesions appear first on the leaves below the top of the diseased plant and eventually cover most or all of the leaf. Lesions sometimes have concentric chlorotic bands that alternate with normal, dark green tissue. Some badly diseased leaves have a mosaic-like appearance. Systemic symptoms occur on the tops of plants as shortened internodes, epinasty of leaves, and a reduced number of seeds. Sparse conidia and conidiophores occur on the under surfaces of diseased leaves. Conidia that are "clumped" together are purplish and can be seen without magnification.

Symptoms of seedborne infection include stunted seedlings and seedlings with a small stem diameter. Leaves of diseased seedlings are rugose and mottled.

Management. Not reported.

Fusarium Wilt

Cause. *Fusarium* sp. is externally seedborne and soilborne.

Distribution. Russia.

Symptoms. Wilting symptoms occur first at the top of diseased plants, then proceed to involve the entire plant. Lower stems become discolored, necrotic, and eventually covered with "pinkish" conidia, conidiophores, and mycelia. Small-sized seed may be formed. Roots are necrotic and easily pulled from soil.

Management. Not known.

Leaf Spot

Cause. *Phyllosticta polyconorum.*

Distribution. Not known for certain but presumed to occur under warmer buckwheat-growing conditions.

Symptoms. Brownish, irregular-shaped leaf spots sometimes have a dark brown to purplish margin. The center of the spot frequently falls out, giving the diseased leaf a ragged appearance.

Management. Not reported.

Powdery Mildew

Cause. *Erysiphe commonis* f. sp. *fagopyri* is thought to overwinter as cleistothecia.

Distribution. Not know for certain but thought to be widespread.

Symptoms. The upper surfaces of diseased leaves have a "typical" white, powdery appearance due to mycelia, conidiophores, and conidia. Eventually, cleistothecia form and appear as dark specks within the mycelium.

Management. Not reported.

Pythium palingenes Disease

Cause. *Pythium palingenes.* Disease is favored by midday air temperatures exceeding 30°C and water-saturated soils.

Distribution. The United States (Missouri).

Symptoms. Diseased plants wilt during the warmest portion of the day. Sunken, tan lesions occur on stems at the soil line. Mats of white mycelia grow on the surface of the taproot.

Management. Not reported.

Ring Spot

Cause. *Cercospora fagopyri.* Disease is favored by extended periods of high humidity and warm temperatures.

Distribution. Ring spot is presumed to occur in the warmer buckwheat-growing areas.

Symptoms. Reddish, circular to irregular-shaped lesions occur on diseased leaves. Lesions vary in size from a pinpoint to large necrotic areas formed by the growth and coalescing of lesions. Circular lesions may have a dark purplish margin.

Management. Not reported.

Diseases Caused by Viruses

Broad Bean Wilt

Cause. Broad bean wilt virus (BBWV) serotype 1 is in the fabavirus group. BBWV is transmitted mechanically and by aphids.

Distribution. Italy.

Symptoms. In a mixed infection with turnip mosaic potyvirus, diseased plants displayed a yellow or chlorotic mottle on the leaves.

Management. Not reported.

Turnip Mosaic

Cause. Turnip mosaic virus (TuMV) is in the potyvirus group. TuMV is likely transmitted by aphids.

Distribution. Italy.

Symptoms. In a mixed infection with broad bean wilt fabavirus, infected plants displayed a yellow or chlorotic mottle on the leaves.

Management. Not reported.

Selected References

Bellardi, M. G., and Rubies-Autonell, C. 1997. First report of broad bean wilt fabavirus on *Polygonum fagopyrum*. Plant Dis. 81:959.

Conners, I. L. 1967. An Annotated Index of Plant Diseases in Canada. Canada Dept. Agric. Res. Branch. Pub. No. 1251.

Farr, D. F., Bills, G. R., Chamuris, G. P., and Rossman, A. Y. 1989. Fungi on Plants and Plant Products in the United States. American Phytopathological Society, St. Paul, MN. 1252 pp.

Mihail, J. D. 1993. Diseases of alternative crops in Missouri. Can. J. Plant Pathol. 15:119-122.

Savitskiy, K. A. 1970. Getchika (Buckwheat). Moscow: 'Kolos'. 312 pp.

Zimmer, R. C. 1974. Chlorotic leaf spot and stipple spot, newly described diseases of buckwheat in Manitoba. Canada Plant Disease Survey 54:55-56.

Zimmer, R. C. 1978. Downy mildew, a new disease of buckwheat (*Fagopyrum esculentum*) in Manitoba, Canada. Plant Dis. Rptr. 62:471-473.

Zimmer, R. C. 1984. A possible new downy mildew syndrome on buckwheat seedlings. Canada Plant Disease Survey 64:7-9

Zimmer, R. C., McKeen, W. E., and Campbell, C. G. 1990. Development of *Peronospora ducometi* in buckwheat. Can. J. Plant Pathol. 12:247-254.

4. Diseases of Canola (Rapeseed) and Mustard *(Brassica campestris, B. juncea, B. napus, B. nigra,* and *B. rapa)*

Diseases Caused by Bacteria

Bacterial Black Rot

Bacterial black rot is also called black rot.

Cause. *Xanthomonas campestris* pv. *campestris* is seedborne and overwinters in soil residue. Bacteria are disseminated by splashing water and enter plants through stomates. Little is known of the life cycle of *X. campestris* as it relates to disease development on canola.

Distribution. Canada; however, bacterial black rot is not a common disease.

Symptoms. Diseased leaves become chlorotic and have dark veins in the chlorotic areas.

Management. Bacterial black rot is not important enough to warrant management.

Bacterial Pod Rot

Cause. *Pseudomonas syringae* pv. *maculicola.*

Distribution. The United States.

Symptoms. Diseased pods become water-soaked and necrotic and eventually dry up.

Management. Not reported.

Bacterial Soft Rot

Bacterial soft rot is also called bacterial stalk rot.

Cause. *Erwinia carotovora* and *Pseudomonas marginalis* pv. *marginalis.* Disease is most severe on vigorously growing plants that have become more succulent due to excessive nitrogen fertilization and on plants growing in poorly drained soil.

Distribution. India.

Symptoms. Water-soaked lesions that are frequently accompanied by a white "frothing" occur at the collar region of diseased plants. Tender branches are also affected as the lesions advance upward on the plant. Leaves are water-stressed and eventually wither. Diseased stems and branches, particularly in the pith tissues, become soft and pulpy and produce dirty, white ooze with a foul odor. The diseased collar region becomes sunken and turns buff white to pale brown. Within a short time, severely diseased plants topple down at the basal region.

Management. Not reported.

Bacterial Wilt and Rot

Cause. *Erwinia carotovora* subsp. *atroseptica (syn. E. atroseptica).*

Distribution. Mexico.

Symptoms. Rot starts where larvae of the insect *Hylemyia* sp. damage stems at the soil level or where branching occurs. At or before flowering time, the centers of stems may be partially or completely rotted, causing diseased plants to wilt. Wilting is associated with a mottled leaf appearance.

Management. Not reported.

Crown Gall

Cause. *Agrobacterium tumefaciens.* Bacteria enter the root system and stem near the soil line through wounds. Once inside the infected plant, the bacterium invades intercellularly and stimulates the surrounding cells to divide.

Distribution. Widely distributed, but the disease rarely occurs on canola and mustard.

Symptoms. Initially, small swellings occur on diseased stems or roots, particularly near the soil line. These swellings later turn into galls that become dark brown to black. Galls have no definite size and shape. Diseased plants, in general, are stunted and have small, chlorotic leaves.

Management. No practical disease management strategy is available. Most canola and mustard cultivars appear to have a high degree of resistance or tolerance. However, disease may be more severe if canola and mustard follow sunflower, another host of *A. tumefaciens,* in a rotation.

Unnamed Seedling Root Rot

Cause. A fluorescent pseudomonad designated *Pseudomonas* rp2. This is an opportunistic pathogen that infects immature roots. Surface micro-colonies penetrate tissue of immature host roots by intrusion along epidermal cell anticlinal walls.

Distribution. Canada (Alberta).

Symptoms. Diseased roots are inhibited in growth and development. Brown, water-soaked lesions also occur on roots.

Management. Not reported.

Xanthomonas Leaf Spot

Cause. *Xanthomonas campestris* pv. *armoraciae.*

Distribution. India, Turkey, and the United States.

Symptoms. Light brown spots occur on diseased leaves and veins.

Management. Not reported.

Diseases Caused by Fungi

Alternaria Black Spot

Alternaria black spot is also called gray leaf spot, pod drop, and dark leaf spot.

Cause. *Alternaria brassicae, A. brassicicola,* and *A. raphani* survive as mycelia in infested residue, and are seedborne as conidia and mycelia on the outside of seed coats. Primary inoculum is conidia produced on residue and disseminated by wind to host plants. Secondary inoculum is conidia that are produced on leaf spots. Black spot is favored by humid, warm conditions that occur at midseason.

Distribution. Canada, Europe, Mexico, and the United States.

Symptoms. Infected seed may cause preemergence or postemergence damping off. Hypocotyls and cotyledons of surviving seedlings may be infected by secondary conidia produced on damped-off seedlings. Initially, brown to black dots occur on diseased leaves, stems, and pods.

Under humid conditions, dots enlarge on diseased leaves to form gray, circular spots containing concentric rings with purple or black borders. However, under dry conditions, spots remain small and black or become gray with black borders. Tissue surrounding spots becomes chlorotic. Severe disease may result in defoliation.

On stems and pods, dots enlarge into lesions that are either completely black or black with gray centers. Pods with diseased pedicels fail to develop and drop from plants. Severely spotted pods dry, shrink, and may split open prematurely to drop shrunken, diseased seed on the soil surface.

Specific symptoms caused by *A. brassicae* on Japanese mustard, *Brassica campestris* subsp. *nipposinica*, are small (2–4 mm in diameter), circular to oblong, brown leaf spots that contain concentric rings. Elongated brown spots develop on petioles. Symptoms on tah tsai, *B. campestris* subsp. *narinosa*, are circular (3–6 mm in diameter), tan to brown leaf spots that contain concentric rings and are surrounded by yellow borders. These leaf spots are slightly larger than on Japanese mustard. Elongated brown spots develop on petioles.

Management
1. Sow clean, disease-free seed.
2. Rotate canola and mustard with noncrucifers for a minimum of 3 years.
3. Plow under infested residue.
4. Control volunteer mustard, canola, and cruciferous weeds.

Alternaria Leaf Spot

Cause. *Alternaria alternata.*

Distribution. Canada on Polish-type canola.

Symptoms. The symptoms described here were on artificially inoculated plants. Grayish to brownish necrotic spots develop on diseased leaves. Spots eventually enlarge, resulting in drying and dropping out of affected areas. Flowers are frequently blighted.

Management: Not reported.

Anthracnose

Cause. *Colletotrichum gloeosporioides,* teleomorph *Glomerella cingulata,* and *C. higginsianum.*

Distribution. India, Nigeria, and Uganda.

Symptoms. Oval, elliptical, water-soaked lesions appear on stems or leaf axis. The spots do not encircle the stem completely, but longitudinal streaks of dead tissue are formed. The cortical tissue of diseased stem and branches cracks open and exposes inner tissue.

Management. Not reported.

Ascochyta Leaf Spot

Cause. *Ascochyta* spp. The pathogens survive in infested residue and are probably seedborne. Dissemination is by splashing rain and movement of infected seed. The fungus may also be windborne.

Distribution. Wherever canola is grown.

Symptoms. Diseased seed is discolored.

Management. Not reported.

Blackleg

Blackleg is also called root and collar rot and stem canker.

Cause. *Leptosphaeria maculans,* anamorph *Phoma lingam.* Formerly, four different strains, or pathogenicity groups, of *L. maculans* were recognized that varied considerably in virulence and symptoms to a set of *B. napus* differentials as follows: A common strain, PG-1, is weakly virulent, produces ascospores late in the growing season, and infects plants as they approach maturity, thereby causing little injury and few symptoms. The second strain, PG-2, is highly virulent and sporulates early in the spring. A third strain, PG-3, has been found on stinkweed, *Thlaspi arvense,* in Canada but does not seem to be important in causing blackleg on canola or mustard. Another strain from Australia and Europe, labeled PG-4, also differs in pathogenicity.

Based on pathogenicity and a number of correlated cultural, chemical, and genetic phenotypes, *P. lingam* currently is classified broadly into aggressive (virulent) and nonagressive (avirulent) strains. Increasing evidence suggests the two strains represent different species. Variation in pathogenicity exists within the aggressive strain: isolates have been classified into three groups, PG-2, PG-3, and PG-4, based on differential pathogenicity on *Brassica napus* cultivars. Pathogenicity of *L. maculans* to canola has been reported to be related to the production of the phytotoxin sirodesmin PL.

Leptosphaeria maculans and *P. lingam* survive up to 2 years in residue as pseudothecia and pycnidia, respectively. *Phoma lingam* is also seedborne as mycelia in seed, which constitutes an important means of introducing the fungus into new areas.

During wet weather or high relative humidity and cool temperatures of 8° to 15°C in the spring or autumn, ascospores are produced in pseudothecia after a few hours and disseminated by wind for several miles. Pycnidiospores are produced in pycnidia and are disseminated by splashing rain to host plants. Pycnidiospores are also disseminated by adhering to seed after harvest and by being disseminated by wind to neighboring fields during harvesting.

Pycnidia are produced on diseased plants throughout the growing season, while perithecia develop only as diseased plants mature. Stems and crowns from the current season's spring canola crop may serve as a source of ascospores that are able to infect the winter canola crop in the autumn.

Canola is most susceptible to infection during the first few weeks of growth, while the plant is in the rosette stage. While young plant tissue is susceptible to infection, all plant parts, including leaves, stems, and roots, can be infected. Infection of petioles and stems usually occurs through an injury. Subsequent disease development is favored more by a higher temperature of 18° than a lower temperature of 12°C.

Distribution. Australia, Canada, Europe, and the United States (Kentucky and North Dakota).

Symptoms. In the spring, inconspicuous pallid areas form on diseased leaves. Eventually, the areas become irregular-shaped, gray lesions dotted with pycnidia that appear as small, black specks. Occasionally, lesions or cankers are associated with axils of leaves.

The stem canker phase can originate from injuries and leaf lesions by systemic growth of *L. maculans* through the laminae and petioles into the main stem. The first noticeable symptom may be patches of lodged plants caused by diseased stems breaking at the soil line. Stems have poorly defined white to ash-gray, brown, or black lesions, often with purplish borders, that commonly develop near the soil line. Long, black transverse streaks occur in the cortex. Pycnidia are numerous and appear as tiny, black specks on the lesions. During wet weather, conidia are extruded in a pinkish exudate from pycnidia. Severely diseased stems eventually become girdled by large cankers that cause plants to ripen prematurely and produce light, shriveled seed, and that break over the entire plant.

Lesions near the base of stems extend downward into the root system, where black cankers are formed. Diseased roots subsequently disintegrate, further weakening plants and frequently causing them to break off just below the soil line.

Management
1. Rotate canola and mustard with resistant crops, such as cereals, every 3 years.
2. Do not grow canola and mustard near fields in which the disease occurred the previous season.
3. Plow under infested residue because *L. maculans* is relatively short-lived and does not survive longer than 2 years in the soil.
4. New canola and mustard varieties are tolerant. Analyses of F1 and F2 progeny suggest resistance is determined by two dominant, independently segregating genes. In Canada, cultivars with partial resistance have been released.

5. Fungicides have been reported to be effective in managing disease when they were applied on a 7-week-old crop.
6. Treat seed with a fungicide seed treatment.
7. Sow seed certified to be free of *L. maculans.*

Black Mold Rot

Cause. *Rhizopus stolonifer.*

Distribution. Canada.

Symptoms. Seed rot and preemergence seedling blight occur. Postemergence seedling blight is scattered throughout a field and losses are usually negligible.

A hard, brown to black lesion (1–2 cm long) may be seen at the base of the diseased stem. Such diseased stems are sometimes girdled near the soil line. The taproot may be discolored.

Management. Treat seed with a seed-treatment fungicide.

Black Root

Cause. *Aphanomyces raphani* presumably survives as oospores in infested residue or soil. Zoosporangia, in which zoospores are produced, are likely formed from oospores. Oospores are eventually produced in diseased roots. Disease is most severe under wet soil conditions.

Distribution. Not known.

Symptoms. Seedlings are stunted, and roots and hypocotyls are blackened. *Aphanomyces raphani* moves up the hypocotyl into the cotyledons, causing chlorosis and blackening of the diseased tissues. Under magnification, oospores are evident in diseased root cortex.

Management. Not reported.

Brown Girdling Root Rot

Brown girdling root rot is likely similar to foot rot because of the mutual pathogens involved. However, brown girdling root rot is treated as a separate disease from foot rot because of its separate treatment in the literature and apparent dissimilarity of symptoms, which suggests the possibility of different pathovars. Brown girdling root rot is also called Rhizoctonia root rot. The postemergence seedling blight phase is called wirestem.

Cause. The most frequently isolated fungi are *Fusarium roseum, Pythium ultimum,* and *Rhizoctonia solani*; however, *R. solani* is considered the primary cause of the root rot syndrome. The following anastomosis groups of *R. solani* are known to be involved: AG-2, AG-2-1, and AG-4.

Isolates of AG-2 and AG-2-1 are more virulent than AG-4 to plants sown early in the growing season and to plants of all ages during periods of high soil moisture. Early sowing, in general, results in significantly greater seedling infection by *F. roseum, P. ultimum,* and *R. solani,* regardless of what anastomosis groups may be involved.

Distribution. Canada (Alberta and British Columbia).

Symptoms. Preemergence and postemergence damping off may occur. Later, distinct brown to red-brown lesions occur on diseased taproots. Several lesions may coalesce and girdle taproots or a single large lesion may girdle a taproot.

Management
1. Rotate canola and mustard with cereals; however, it is not certain if cereals have the same disease that occurs on canola and mustard.
2. Sow cleaned, healthy seed early in the growing season.
3. *Rhizoctonia solani* is most destructive in a loose, well-worked soil; therefore, the seedbed should have firm soil.
4. Control cruciferous weeds and volunteer plants in and adjacent to fields.

Cercospora Leaf Spot

Cause. *Cercospora brassicicola.* Disease is favored by extended periods of high humidity and temperatures of 28° to 30°C.

Distribution. Sri Lanka.

Symptoms. Red-purple, angular to irregular-shaped lesions occur on diseased leaves. Lesions vary in size and originate from a pinpoint that enlarges and coalesces to form large, necrotic areas. Veinal necrosis may also occur. Severe disease may cause rapid chlorosis and necrosis, resulting in defoliation that begins with the upper diseased leaves and progresses down the plant.

Red-purple, slightly sunken lesions may occur on petioles and stems.

Management. Not reported.

Charcoal Rot

Cause. *Macrophomina phaseolina.* Optimum infection and resulting disease symptoms occur during high temperatures (32° to 36°C).

Distribution. The United States (Kentucky and Indiana).

Symptoms. Diseased plants prematurely senesce or die. Microsclerotia are present in taproots as multiple specks that somewhat resemble charcoal. Diseased tissue is shredded.

Management. Not reported.

Clubroot

Cause. *Plasmodiophora brassicae* overwinters as resting spores, called hypnospores, that form within an infected cell. Hypnospores are liberated into soil when roots disintegrate after the growing season. The following spring, when the soil warms up and moisture is present, a hypnospore germinates to form a single biflagellate zoospore that "swims" to a root hair. The zoospore loses its flagella, becomes amoeboid, and enters the epidermal cell, presumably with the aid of enzymes. Inside the plant, the fungus produces an amoeboid thallus that soon gives rise to zoosporangia from which zoospores are liberated and move out of the diseased root. Cells of infected roots contain naked fungal plasmodium that is able to move from cell to cell. Eventually, naked protoplasm breaks into multiple cells around which walls form to become hypnospores.

   Clubroot may occur at any soil temperature between 9° and 30°C but is most severe at the cooler temperatures. Hypnospores do not germinate when the soil pH is higher than 7.2. In Canada, disease incidence and severity is greater in regions with severe winters.

   Many physiological races of *P. brassicae* exist.

Distribution. In cool, canola-growing areas of the world.

Symptoms. Diseased plants are stunted and chlorotic. The roots contain the typical galls symptomatic of clubroot. Galls are often spindle-shaped and may involve the entire diameter of the diseased root. Multiple branch roots frequently grow from galls to give them a hairy appearance.

Management. Resistance is common in canola cultivars, especially in Sweden.

Damping Off

Damping off is also called seedling blight.

Cause. Several fungi, including *Pythium* spp., *Phytophthora* spp., *Fusarium* spp., and *Rhizoctonia solani*. Some *Pythium* species involved in the preemergence phase of seedling blight are *P. paroecandrum* and *P. sylvaticum*.

   Disease severity and incidence are greater in heavy, wet soils that are common in the low-lying areas of a field, and during cool, wet weather. The causal fungi are relatively weak pathogens that infect young, succulent tissue.

Distribution. Wherever canola is grown.

Symptoms. Seeds fail to germinate or seedlings are killed either preemergence or postemergence. Postemergence damping off is characterized by constriction and tapering of the hypocotyl and brown lesions on the diseased hypocotyl and cotyledons. Seedlings rot at the soil line, causing them to fall over. Seedlings may emerge and not grow larger then the one- to four-leaf stage for a period of time, then they either die or resume growth.

Management
1. Sow seed when the soil temperature has warmed up.
2. Seed should be sown at a depth of 1 to 25 mm in a firm seedbed.
3. Treat seed with a seed-protectant fungicide.

Downy Mildew

When downy mildew occurs together with white rust caused by *Albugo candida,* it is called staghead.

Cause. *Peronospora parasitica* overwinters primarily as systemic mycelia in perennial or overwintering plants and as oospores in soil and infested residue. Conidia (sporangia) are produced on diseased leaves of systemically infected plants and disseminated by wind to other hosts.

Zoospores are not commonly formed; therefore, infection occurs by conidia germinating directly to form either germ tubes that penetrate hosts or a conidiophore on which a large conidium is formed. Secondary inoculum is provided by conidia produced on leaves. Different races of *P. parasitica* exist.

When downy mildew occurs together with white rust to cause staghead, the effect is synergistic since disease development caused by both fungi together is greater than either fungus alone. Cool, moist weather favors staghead development.

Distribution. Wherever canola is grown; however, downy mildew is usually more severe in northern canola-growing areas.

Symptoms. Spots appear on diseased leaves, stems, and seedpods as small purple, irregular-shaped areas. Leaf spots enlarge to form yellow areas in which a white, downy growth occurs on diseased upper leaf surfaces, while conidia and conidiophores form on lower leaf surfaces. The downy growth also occurs on blisters and stagheads.

Management
1. Grow resistant cultivars, such as Argentine-type cultivars. Resistant cultivars of turnip canola are available. The oilseed canola cultivar 'Cresor' is resistant to 14 isolates of *P. parasitica.*
2. Control volunteer turnip canola and wild mustard.
3. Sow only cleaned seed.
4. Rotate canola with resistant crops.

Foot Rot

Cause. *Rhizoctonia solani* AG-4 and *Fusarium roseum. Rhizoctonia solani* is considered the primary pathogen. *Rhizoctonia solani* survives as sclerotia and mycelia in infested residue. *Fusarium roseum* survives as chlamydospores in soil, mycelia in residue, and perithecia on infested residue. Both fungi have a wide host range.

Other fungi associated with this disease complex are *Alternaria alternata, F. solani, F. tricinctum, Leptosphaeria maculans, Pythium debaryanum,* and *P. polymastum.* The *Fusarium* spp. are sometimes seedborne.

Distribution. Wherever canola and mustard are grown; however, disease incidence is usually sporadic.

Symptoms. Prematurely ripened plants are found singly or in patches throughout a field during the summer. Hard, clearly defined, brown lesions with rough surfaces, sometimes with a black border, are present on diseased stem bases. During wet conditions, salmon-pink spore masses may be present on lesions. A white, mycelial growth may be present on discolored root tissue. Severe disease will girdle stems and kill plants.

Management
1. Rotate canola and mustard with cereals; however, it is not certain if cereals have the same disease that occurs on canola and mustard.
2. Sow cleaned, healthy seed early in the growing season.
3. *Rhizoctonia solani* is most destructive in a loose, well-worked soil; therefore, the seedbed should have firm soil.
4. Control cruciferous weeds and volunteer plants in and adjacent to fields.

Fusarium Seed Infection

Cause. *Fusarium moniliforme, F. oxysporum,* and *F. semitectum.*

Distribution. Not known for certain, but Fusarium seed infection is presumably widespread.

Symptoms. Qualitative and quantitative changes occur in the amino acid and sugar content of diseased seeds, and protein content declines.

Management. Not reported.

Fusarium Wilt

Fusarium wilt is also called yellows.

Cause. *Fusarium oxysporum* f. sp. *conglutinans* survives as chlamydospores in residue and soil. Different races are reported to exist.

Distribution. Widespread.

Symptoms. Symptom expression varies with plant age. Foliar symptoms progress from the base of the diseased plant upward. Leaves droop, display vein-clearing and chlorosis, followed by wilting and drying, a symptom that often precedes death of the entire plant.

Plants infected during preflowering and early flowering become defoliated. Stems of such plants develop external longitudinal ridges and furrows, a symptom that generally is not observed when older plants are infected. Plants infected early in their growth are often stunted, but plants infected later generally do not have this symptom. Frequently only one side of the plant displays symptoms. Xylem or vascular tissue is discolored brown, a symptom caused by the plugging of the tissue by a dark gummy substance.

Management
1. Do not grow canola and mustard following cabbage or other *Brassicae* spp.
2. Canola and mustard may not be susceptible to all races of Fusarium wilt.

Gray Mold

Cause. *Botrytis cinerea.* Disease is favored by rainy weather.

Distribution. Germany, Sweden, Great Britain, and the former USSR.

Symptoms. Mycelium that varies from a "dirty white" or gray to black in color occurs on diseased flowers and pods.

Management. Not reported.

Leaf Spot

Cause. *Myrothecium roridum* and *Phyllosticta brassicae.*

Distribution. India.

Symptoms. Brownish spots occur on diseased leaves. Sometimes the center of the spot falls out and gives the leaf a ragged appearance.

Management. Not reported.

Light Leaf Spot

Light leaf spot is also called leaf scorch.

Cause. *Pyrenopeziza brassicae* (syn. *Cylindrosporium concentricum*). *Pyrenopeziza brassicae* is seedborne and presumably overwinters in canola stubble and on wild crucifers and brassicas. Primary inoculum is conidia that are splashed from the overwintering sites onto nearby host plants. Disease is favored by excess moisture and temperatures of 5° to 15°C.

Distribution. France, Germany, Great Britain, and Sweden.

Symptoms. Rarely, severe infection may kill seedlings. Blanched, "papery" lesions first appear on older diseased leaves, which eventually wither. During the stem extension growth stage of the plant, infection may spread to the upper leaves and bracts to cause severe leaf scorch, gross plant distortion, and stunting. Stunted, diseased winter canola plants that survive through the winter regenerate new shoots in the spring. Siliquae of diseased plants may mature prematurely and split.

Management
1. Cultivars vary in tolerance or resistance. The more resistant cultivars are high erucic acid types.
2. Foliar fungicides have been effective in field trials.

Pod Drop

Cause. *Alternaria alternata* and *Cladosporium* sp. Dead or dying petals are infected and provide a food base for fungi to grow into the pedicels.

Distribution. Canada.

Symptoms. Pedicels are blackened and weakened at, or below, their point of attachment to pods, which sometimes causes pods to drop. Pedicel tips then display a typical black discoloration. Pods on diseased pedicels do not fill normally.

Management. Summer turnip canola is more resistant than Argentine types.

Powdery Mildew

Cause. *Erysiphe polygoni* and *E. cruciferarum* overwinter as cleistothecia or mycelia in volunteer host plants. Disease is favored by relatively dry weather conditions and moderate temperatures of 7.1° to 25.0°C.

Distribution. Worldwide, but powdery mildew is usually not considered serious on canola and mustard.

Symptoms. Dirty white, circular, floury patches occur on both sides of lower diseased leaves. These powdery mildew patches increase in size and coalesce to cover the entire stem and leaves. Severely diseased plants grow poorly and produce less siliquae. Green siliquae also show white patches in the initial stage of infection. Later, such siliquae become covered with a white mass of mycelium and conidia. Severely diseased siliquae remain small in size and produce seeds that are also small and shriveled. Such siliquae produce a few seeds at the base that have twisted sterile tips. Later in the season, cleistothecia may be formed on both sides of diseased leaves, stems, and siliquae as scattered small, black bodies visible to the unaided eye.

Management
1. Limited resistance has been identified in some canola and mustard cultivars.
2. Foliar fungicides have been effective in field trials.

Rhizoctonia Crown Rot

Rhizoctonia crown rot may be a similar disease to foot rot and brown girdling root rot. However, symptoms are sufficiently different to currently treat them as separate diseases.

> **Cause.** *Rhizoctonia solani* AG-2-1 and AG-4. Crown rot is most prevalent in wet or poorly drained areas of fields, but the disease is also reported to be severe on early-sown canola growing on sandy soils in Indiana. Early-sown plants, which make extensive growth in the autumn, are more severely diseased than smaller plants that were sown later in the growing season. By early winter, crown tissues near the soil surface are infected through direct penetration of mycelia and through leaf scars, but root tissues below the crown generally are not infected until midwinter and spring.

> **Distribution.** The United States (Indiana). However, crown rot is probably distributed wherever canola is grown.

> **Symptoms.** Severe crown rot occurs with varying degrees of cortical necrosis. Root tissues below the crown are not diseased. The bases of diseased leaves frequently have dark brown to black, necrotic lesions that result in leaf death.

> **Management.** Sowing date and cultivar selection may reduce severity of crown rot.

Ring Spot

This disease is also called black blight, black stem ring spot, and white leaf spot and gray stem.

> **Cause.** *Mycosphaerella brassicicola* overwinters as thick-walled mycelia and perithecia in infested residue. The fungus is not seedborne to any extent. In the spring, conidia, produced from mycelia that overwintered in residue, become disseminated by wind to host plants. Most secondary inoculum and spread are by conidia produced in spots during cool, wet environmental conditions.
>
> Perithecia form on diseased plant tissue later in the growing season during periods of low night temperatures and heavy dews. Ascospores are thought to be a means of spread of *M. brassicicola* in Europe and the United States but not in Canada.

> **Distribution.** Canada, Europe, and the United States.

> **Symptoms.** Circular (1–2 cm in diameter), whitish leaf spots are produced on diseased leaves. Later in the season, large, elongate, purple to gray lesions with lighter centers that are speckled with numerous perithecia occur on diseased stems and pods. Lesions may coalesce and completely blacken large portions of affected plants.

Management. Although ring spot is not considered important because it develops too late in the life of the plant, the following are aids to management:

1. Rotate canola and mustard with noncruciferous crops.
2. Control weeds.
3. Sow healthy, cleaned seed.

Sclerotinia Stem Rot

Cause. *Sclerotinia sclerotiorum* overwinters as sclerotia either in the soil or as a contaminant of seed lots. Infrequently, seed itself may be infected. Mycelia do not survive well enough in infested stems to be significant sources of pathogenic inoculum.

Sclerotia have a dormancy mechanism that requires certain physical factors to occur before germination. Sclerotia from Saskatchewan required 2 to 6 weeks incubation at 10°C, followed by 4 to 6 weeks at 20°C in diffuse light, for initiation of stipes and cap expansion. Sclerotia have a higher germination at high soil moisture than at low soil moisture before plant flowering. However, after flowering, germination is higher at lower soil moisture because of the large number of viable sclerotia.

After physical factors have been satisfied, sclerotia germinate in one of two ways. First, in the spring, sclerotia germinate to produce mycelia that grow a short distance across moist soil and infect plants at the soil line. Carpogenic germination, reported to be the most common method of sclerotia germination, occurs later in the growing season, at flowering time, when sclerotia germinate to produce one to several cup-shaped apothecia. Ascospores are borne in asci on the apothecium surface and are disseminated by wind, pollen, and insects to host plants, where infection occurs in leaf axils and, more commonly, in the dead or dying petals of blossoms that have fallen onto leaves and stems. Although most ascospores are dispersed within 5 m of their source, some may be transferred up to 200 m. Sclerotia are eventually formed from mycelium in diseased stems.

In a slight derivation from the previous method of infection, ascospores first infest canola blossoms; infection is initiated when infested petals dehisce, fall, and adhere to the stem. Ascospores germinate, penetrate the stem, and form lesions in which sclerotia develop. Sclerotia fall out of these lesions into the soil and, after physiological conditioning, including overwintering, germinate to form apothecia.

Sclerotinia stem rot is reported to be most severe under warm, wet weather conditions. Two weeks of wet weather before and during flowering favors disease development. Lodged mustard and canola provide conditions conducive to disease. In North Dakota, Sclerotinia stem rot is reported to be more prevalent during rainy seasons when sclerotia-infested field soils are wet for approximately 10 to 14 days at the beginning of blossoming and when blossoms remain wet for several days.

Distribution. Wherever canola and mustard are grown.

Symptoms. The first symptom is the presence of prematurely ripened plants that are scattered or are grouped among green, healthy-appearing plants in a field. These diseased plants are easily pulled from the soil.

The pale gray lesions with faint concentric markings that are on stems develop from the soil line or from the axils of branches or leaves. At first, lesions are water-soaked, then eventually expand to girdle stems, which kills the diseased plants. Such stems are bleached, have a chalky white appearance, and tend to shred longitudinally. Sclerotia appear as hard, black, grain-sized bodies inside stems near the bases of dead plants. Under moist conditions, sclerotia occur in all parts of stems and pods.

Management
1. Rotate canola and mustard with cereals and grasses every 4 years.
2. Sow only cleaned seed with all sclerotia removed.
3. Plow under infected residue.
4. Control volunteer canola, mustard, and susceptible weeds.
5. Apply foliar fungicides.

Smut, Root Gall

Cause. *Urocystis brassicae.* Galls from roots contain spore balls of the fungus. When galls decay, the spores that are liberated into the soil constitute the primary source of inoculum. Disease does not spread but remains in the same general area within a field.

Distribution. India.

Symptoms. Gall-like bodies that occur on roots may be as large as 2.5 to 3.5 cm in diameter. Galls may be observed in all parts of the root system. Initially, galls are small and whitish but become grayish black as they mature. Diseased plants remain stunted and branch freely. Diseased mustard plants produce flowers earlier than healthy plants but have poor seed production.

Management. Not reported.

Southern Blight

Cause. *Sclerotium rolfsii.*

Distribution. Malaya.

Symptoms. A general leaf blight occurs.

Management. Not reported.

Storage Molds

Cause. *Penicillium* spp., primarily *P. verrucosum* var. *cyclopium*; and *Aspergillus* spp., primarily *A. amstelodami, A. repens,* and *A. sejunctus.* Other reported fungi are *Alternaria* spp., *Aspergillus candidus, A. fumigatus, A. versicolor, A. wentii, Cephalosporium acremonium* (syn. *Acremonium strictum*), *Cladosporium cladosporioides,* and *Wallemia sebi.*

Verticillium Wilt

Cause. *Verticillium albo-atrum* and *V. dahliae.*

Distribution. Sweden.

Symptoms. A general wilting of diseased plants occurs.

Management. Not reported.

White Blight

Cause. *Rhizoctonia solani.* Disease is favored by high temperatures and excess moisture.

Distribution. India.

Symptoms. A general blighting of diseased leaves occurs. Blighted leaves are covered with the white to grayish mycelium of *R. solani.*

Management. Not reported.

White Leaf Spot

Cause. *Pseudocercosporella capsellae* (syn. *Cercosporella brassicae*).

Distribution. The United States (California). However, white leaf spot is likely to be wherever canola and mustard are grown.

Symptoms. Circular, light tan to off-white spots measuring 2 to 8 mm in diameter occur on diseased leaves. White downy or "fuzzy" sporulation that consists of conidia, conidiophores, and mycelium occurs in the leaf spots.

Management. Not reported.

White Rust

White rust is also called staghead, either as a synonym only for white rust or to refer to the white rust–downy mildew complex. White rust is usually associated with downy mildew caused by *Peronospora parasitica.*

Cause. *Albugo candida* (syn. *A. cruciferarum*) overwinters as oospores in residue and soil and as a contaminate of seed lots. In the spring, oospores germinate to produce sessile vesicles or short exit tubes that terminate in a vesicle. Zoospores are released from vesicles and "swim" to host tissue, where they encyst and germinate to form infective mycelium that penetrates plant tissue. The resulting mycelium growing inside the infected plant then becomes systemic.

Secondary infection is by sporangia produced in leaf blisters and disseminated by wind to other hosts. Sporangia germinate by producing zoospores that first encyst, then germinate to form infective mycelium that penetrates the host and causes localized infections. Oospores eventually develop within diseased swollen or hypertrophied tissue.

White rust is most severe during wet weather conditions. Day and night temperatures of 22°C and 17°C, respectively, are more favorable to fungal growth than 15°C and 10°C, respectively.

Different races exist. Collections from radish, *Raphanus sativus;* brown and oriental mustard, *Brassica juncea;* shepherd's purse, *Capsella bursa-pastoris;* and turnip rape, *B. campestris* subsp. *eu-campestris* represent different races. Race 7 infects *B. campestris* but not *B. napus.* Differences in susceptibility exist between *Brassica napus* and *B. rapa.*

Distribution. Wherever canola and mustard are grown.

Symptoms. The most obvious symptom is a swelling and deformation of terminal parts of flower stalks which results in the "spiny" staghead symptom that accounts for most of the yield losses. Initially, stagheads are green but later become brown and hard as they dry up.

Raised green or brown blisters occur on the undersides of diseased leaves, stems, and pods. Initially, blisters are smooth but later rupture to release white to cream-colored masses of sporangia. Similar blisters may appear on surfaces of green stagheads. Portions of individual flowers may also become distorted with sterile inflorescences.

Management
1. Grow resistant Argentine-type cultivars. Resistant cultivars of turnip canola may be available.
2. Control volunteer turnip canola and wild mustard.
3. Sow only cleaned seed.
4. Rotate canola with resistant crops.
5. Under experimental conditions, metalaxyl applied as a foliar, seed, and soil application reduced disease incidence.

Winter Decline Syndrome

Cause. Physical damage on roots, crowns, or stems caused by sublethal low temperatures occurs on plants that have bolted prematurely, are not adequately cold-hardened, or have undergone plant heaving and/or depletion of soil oxygen from water-logged soils. The physical damage provides ports of entry for invasion by a variety of secondary organisms. *Clavibacter* sp.,

Fusarium sp., *Rhizoctonia* sp., and *Xanthomonas* sp. have been inconsistently isolated from affected tissue.

Distribution. The southeastern and midsouthern states of the United States.

Symptoms. Taproots and crowns of diseased plants deteriorate and often become hollow. Diseased plants may die before or during bolting, may bolt normally but later lodge and fall down because of weakened crown tissue, or remain upright but die prematurely and reduce the seed yield. Stands may be reduced by over 90%.

Management. Not reported.

Wirestem

Wirestem is a form of Rhizoctonia root rot.

Cause. *Rhizoctonia solani.*

Distribution. Widespread.

Symptoms. A narrow, rotted root comprised primarily of the stele begins below apparently healthy tissue. Diseased tissue is a light brown lesion on the taproot and the bases of larger lateral roots. Eventually, the taproot may be girdled. Eventually, the diseased portion of the root somewhat resembles a "wire" that hangs below the healthy-appearing root tissue. Root regeneration may occur above the rotted area.

Management. Not reported.

Yeast Infection of Seed

Cause. *Nematospora sinecauda* is transmitted by the false chinch bug, *Nysius niger.* Flixweed, *Descurainia sophia,* is the overwintering host.

Distribution. Canada (Prairie Provinces) and the United States (Montana and North Dakota).

Symptoms. An undesirably high plate count of microorganisms occurs in condiment flours ground from diseased seed.

Management. Not reported.

Disease Caused by Nematodes

Root Knot

Cause. *Meloidogyne arenaria* and *M. incognita. Meloidogyne arenaria* and *M. incognita* do not survive if the temperature averages below 3°C during the coldest month.

Distribution. The United States (Georgia).

Symptoms. Typical root galls form on diseased roots.

Management. Sources of resistance are available in advanced germplasms.

Diseases Caused by Phytoplasmas

Aster Yellows

Cause. Aster yellows phytoplasma survives in several dicotyledonous plants and in leafhoppers. The phytoplasma is disseminated from collateral hosts to canola and mustard primarily by the aster leafhopper, *Macrosteles fascifrons,* and less commonly by the leafhoppers *Endria inimica* and *M. laevis.*

Distribution. Wherever canola and mustard are grown. Aster yellows is not an important disease of canola or mustard.

Symptoms. Diseased plants are infected systemically but symptoms usually occur only on inflorescences. Flowers and pods become distorted and sterile and frequently form small, blue-green, hollow, bladder-like structures instead of normal seedpods.

Management. Not reported.

Green Petal Disease

Cause. A phytoplasma that is transmitted by the insect *Euscelis plebejus* and by grafting.

Distribution. Germany and Hungary.

Symptoms. Internodes are shortened and diseased plants are stunted. Virescence, flower proliferation and deformation, and vein clearing of the leaves are prominent symptoms.

Management. Not reported.

Diseases Caused by Viruses

Cauliflower Mosaic

Cause. Cauliflower mosaic virus is transmitted by the aphids *Myzus persicae* and *Brevi coryne.*

Distribution. Bulgaria, New Zealand, and the United States.

Symptoms. Diseased leaves display a general mosaic.

Management. Not reported.

Crinkle

Cause. Turnip crinkle virus (TCV) is a single-stranded RNA virus in the carmovirus group. TCV supports the replication of at least three satellite RNAs. One satellite RNA (sat-RNA C) intensifies symptoms on a turnip cultivar, but it does not have any intensifying effect on canola.

Distribution. Not known.

Symptoms. Diseased plants are stunted. Leaves are mildly to severely crinkled and have vein clearing.

Management. Not reported.

Mosaic

Cause. Turnip mosaic virus (TuMV) is in the potyvirus group. TuMV is transmitted mechanically and, frequently, by the green peach aphid, *Myzus persicae*. TuMV can also be transmitted by several other species of aphids.

Distribution. Widespread.

Symptoms. A general mosaic pattern occurs on diseased leaves.

Management. Not reported.

Phyllody

Cause. Sesamum phyllody virus (SPV). SPV is transmitted by a jassid insect, *Orosius albicinctus*.

Distribution. India.

Symptoms. Diseased floral parts are transformed into leafy structures. The corolla becomes green and sepaloid, and stamens are green and become indehiscent. The gynoecium is borne on a distinct gynophore and produces no ovules in the ovary. There also are some leafy structures attached to the false septum. The affected parts of the raceme do not form siliquae. Some diseased plants may only have the terminal portions of branches affected; others may have entire branches that display symptoms.

Management. Not reported.

Rai Mosaic

Cause. Rai mosaic virus (RMV). RMV is sap-transmissible and is transmitted in a nonpersistent manner by the aphids *Myzus persicae* and *Lipaphis erysimi*.

Distribution. India.

Symptoms. Characteristic mosaic symptoms occur on diseased leaves. Diseased plants are stunted and have deformed lamina.

Management. Not reported.

Yellows

Cause. Beet western yellows virus (BWYV) and broccoli necrotic yellows virus. BWYV is transmitted in a nonpersistent manner by the aphids *Myzus persicae* and *Macrosiphum euphorbiae* and possibly by other aphid vectors.

Distribution. The United Kingdom.

Symptoms. Interveinal chlorosis occurs on lower and intermediate diseased leaves. Leaves are frequently purple and plants ripen prematurely.

Management. Not reported.

Diseases Caused by Unknown Factors

Brown Girdling Root Rot

Cause. Unknown.

Distribution. Western Canada.

Symptoms. Symptoms are confined only to diseased roots. Light brown lesions occur on taproots and the bases of larger lateral roots 5 cm or more below the soil surface. Dark brown, sunken lesions with vertical streaks may expand, coalesce, and girdle taproots. Lesions may expand up to the soil line but rarely beyond. If girdling occurs more than 5 cm below the soil line, some root regeneration may occur if there is sufficient moisture.

Management. Argentine cultivars have some field tolerance.

Pollen Necrosis

Cause. Pollen shed from the plant. There are several possibilities as to what causes the symptom: (1) A chemical constituent of the pollen grain is perceived as an alien agent by the leaf cells, which respond in a hypersensitive manner, or enzymes within the pollen grains may be responsible for the necrotic reaction. (2) The pollen grains may promote some microbe(s) that are responsible for the browning reaction. (3) The necrotic areas may serve as points of entry for certain pathogens.

Distribution. Canada, Germany, and India.

Symptoms. Minute, necrotic spots develop under deposits of pollen grains. Eventually, a zone of chlorosis develops around the lesion.

Management. Not known.

Unknown

Cause. Unknown, but probably of genetic or physiologic origin.

Distribution. India.

Symptoms. Diseased plants are stunted and have small, puckered, leather-like, twisted leaves that are darker green than healthy leaves. Flowers are poorly developed and few seeds are produced. Longitudinal fissures that ooze a water-soluble yellow gum are present on aerial plant parts.

Management. Not reported.

Selected References

Anonymous. 1967. Diseases of field crops in the Prairie Provinces. Canada Dept. Agric. Res. Branch, Ottawa, Pub. 1008.

Assabqui, R., and Hall, R. 1990. In vitro production of sirodesmin PL correlated with pathogenicity of *Leptosphaeria maculans* to rapeseed. (Abstr.). Phytopathology 80:117.

Baird, R. E., Hershman, E. E., and Christmas, E. P. 1994. Occurrence of *Macrophomina phaseolina* on canola in Indiana and Kentucky. Plant Dis. 78:316.

Berkenkamp, B., and Vaartnou, H. 1972. Fungi associated with rape root rot in Alberta. Can. J. Plant Sci. 52:973-976.

Burgess, L., and McKenzie, D. L. 1991. Role of the insect *Nysius niger* and flixweed, *Descurainia sophia,* in infection of Saskatchewan mustard crops with a yeast, *Nematospora sinecauda.* Can. Plant Dis. Surv. 71:37-41.

Calman, A. I., and Tewari, J. P. 1987. Involvement of *Pythium* spp. in the preemergence phase of canola seedling blight. (Abstr.) Can. J. Plant Pathol. 9:274.

Calman, A. I., Tewari, J. P., and Mugala, M. 1986. *Fusarium avenaceum* as one of the causal agents of seedling blight of canola in Alberta. Phytopathology 70:694.

Campbell, J. N., Cass, D. D., and Peteya, D. J. 1987. Colonization and penetration of intact canola seedling roots by an opportunistic fluorescent *Pseudomonas* sp. and the response of host tissue. Phytopathology 77:1166-1173.

Davidson, J. G. N. 1976. Plant Diseases in the Peace River Region in 1976. Agric. Can. Res. Sta., Beaverlodge, Alberta. Mimeo.

Davidson, J. G. N. (Undated). Disease Control in Rapeseed. Agric. Can. Res. Sta., Beaverlodge, Alberta. Mimeo.

Dueck, J. 1981. Dormancy in sclerotia of *Sclerotinia sclerotiorum.* (Abstr.) Phytopathology 71:214.

Fucikovsky, L. 1979. Bacterial disease of rape and carrot in Mexico. (Abstr.) Phytopathology 69:915.

Gladders, P., and Musa, T. M. 1980. Observations on the epidemiology of *Leptosphaeria maculans* stem canker in winter oilseed rape. Plant Pathol. 29:28-37.

Gugel, R. K., and Petri, G. A. 1992. History, occurrence, impact, and control of blackleg of rapeseed. Can. J. Plant Pathol. 14:36-45.

Gugel, R. K., Yitbarek, S. M., Verma, P. R., Morrall, R. A. A., and Sadasivaiah, R. S. 1987. Etiology of the Rhizoctonia root rot complex of canola in the Peace River region of Alberta. Can. J. Plant Pathol. 9:119-128.

Hammond, K. E., Lewis, B. G., and Musa, T. M. 1985. A systemic pathway in the infection of oilseed rape plants by *Leptosphaeria maculans*. Plant Pathol. 34:557-565.

Henry, A. W. 1974. Bacterial pod spot of rape in Alberta. Can. Plant Dis. Surv. 54:91-94.

Hill, C. B., Phillips, D. V., and Hershman, D. E. 1992. Canola winter decline syndrome. Plant Dis. 76:861.

Huber, D. M., Herr, L. J., Christmas, E. P., and Roseman, T. S. 1991. Crown rot, a serious disease of canola in the Midwest. (Abstr.) Phytopathology 81:1186.

Huber, D. M., Christmas, E. P., Herr, L. J., McCay-Buis, T. S., and Baird, R. 1992. Rhizoctonia crown rot of canola in Indiana. Plant Dis. 76:1251-1253.

Humpherson-Jones, F. M. 1985. The incidence of *Alternaria* spp. and *Leptosphaeria maculans* in commercial brassica seed in the United Kingdom. Plant Pathol. 34:385-390.

Hwang, S. F., Swanson, T. A., and Evans, I. R. 1986. Characterization of *Rhizoctonia solani* isolates from canola in west central Alberta. Plant Dis. 70:681-683.

Kohli, Y., Morrall, R. A. A., Anderson, J. B., and Kohn, L. M. 1992. Local and trans-Canadian clonal distribution of *Sclerotinia sclerotiorum* on canola. Phytopathology 82:875-880.

Koike, S. T. 1996. Japanese mustard, tah tsai, and red mustard as hosts of *Alternaria brassicae* in California. Plant Dis. 80:822.

Koike, S. T. 1996. Red mustard, tah tsai, and Japanese mustard as hosts of *Pseudocercosporella capsellae* in California. Plant Dis. 80:960.

Kolte, S. J. 1985. Diseases of Annual Edible Oilseed Crops. Volume II: Rapeseed-Mustard and Sesame Diseases. CRC Press Inc., Boca Raton, FL. 135 pp.

Kruger, W., and Wittern, I. 1985. Epidemiological investigations in the root and collar rot of soil seed rape caused by *Phoma lingam*. Phytopathol. Zeitschr. 113:125-140.

Lamey, H. A. 1995. Survey of blackleg and Sclerotinia stem rot of canola in North Dakota in 1991 and 1993. Plant Dis. 79:322-324.

Lamey, H. A., and Hershman, D. E. 1993. Blackleg of canola (*Brassica napus*) caused by *Leptosphaeria maculans* in North Dakota. Plant Dis. 77:1263.

Lefol, C., and Morrall, R. A. A. 1997. Dispersal of ascospores of *Sclerotinia sclerotiorum* in canopies of flowering canola. (Abstr.). Can. J. Plant Pathol. 19:113.

Li, X. H., and Simon, A. E. 1990. Symptom intensification on cruciferous hosts by the virulent satellite RNA of turnip crinkle virus. Phytopathology 80:238-242.

Liu, Q., and Rimmer, S. R. 1990. Effect of host genotype, inoculum concentration, and incubation temperature on white rust development in oilseed rape. Can. J. Plant Pathol. 12:389-392.

Lucas, J. A., Crute, I. R., Sherriff, C., and Gordon, P. L. 1988. The identification of a gene for race-specific resistance to *Peronospora parasitica* (downy mildew) in *Brassica napus* var. *oleifer* (oilseed rape). Plant Pathol. 37:538-545.

Mathur, R. S., and Singh, B. R. 1975. A virus-like disorder of yellow mustard in India. Plant Dis. Rptr. 59:174-175.

McGee, D. C. 1977. Blackleg (*Leptosphaeria maculans* (Desm.) Ces et de Not.) of rapeseed in Victoria: Sources of infection and relationships between inoculum, environmental factors and disease severity. Aust. J. Agric. Res. 28:53-62.

McGee, D. C., and Petrie, G. A. 1978. Variability of *Leptosphaeria maculans* in relation to blackleg of oilseed rape. Phytopathology 68:625-630.

McGee, D. C., and Petrie, G. A. 1979. Season patterns of ascospore discharge by *Lep-*

tosphaeria maculans in relation to blackleg of oilseed rape. Phytopathology 69:586-589.

Mengistu, A., Rimmer, S. R., Koch, E., and Williams, P. H. 1991. Pathogenicity grouping of isolates of *Leptosphaeria maculans* on *Brassica napus* cultivars and their disease reaction profiles on rapid-cycling brassicas. Plant Dis. 75:1279-1282.

Mills, J. T., and Sinha, R. N. 1980. Safe storage periods for farm-stored rapeseed based on mycological and biochemical assessment. Phytopathology 70:541-547.

Mithen, R. F., and Lewis, B. G. 1988. Resistance to *Leptosphaeria maculans* in hybrids of *Brassica oleracea* and *Brassica insularis*. J. Phytopathol. 123:253-258.

Petrie, G. A. 1974. Diseases in Rapeseed, Canada's "Cinderella" Crop, Third Edition. R. K. Downey, A. J. Klassen, and J. McAnsh (eds.). Pub. No. 33. Rapeseed Association, Canada.

Petrie, G. A. 1975. Diseases of Rapeseed and Mustard. *In* Oilseed and Pulse Crops in Western Canada, J. T. Harapiak (ed.). Modern Press, Saskatoon, Saskatchewan.

Petrie, G. A. 1988. Races of *Albugo candida* (white rust and staghead) on cultivated Cruciferae in Saskatchewan. Can. J. Plant Pathol. 10:142-150.

Petrie, G. A., and Dueck, J. 1977. Diseases of Rape and Mustard. *In* Insect Pests and Diseases of Rape and Mustard, L. Burgess, J. Dueck, G. A. Petri, and L. G. Putnam (eds.). Pub. No. 48. Rapeseed Association, Canada.

Ponce, F., and Mendoza, C. 1983. Fungal diseases and pests of the rapeseed *Brassica napus* L. and *B. campestris* L. in the high valleys of Mexico. (Abstr.) Phytopathology 73:124.

Rao, D. V., Hiruki, C., and Chen, M. H. 1977. A mosaic disease of rape in Alberta caused by turnip mosaic virus. Plant Dis. Rptr. 61:1074-1076.

Rempel, C. B., and Hall, R. 1993. Dynamics of production of ascospores of *Leptosphaeria maculans* in autumn on stubble of the current year's crop of spring rapeseed. Can. J. Plant Pathol. 15:182-184.

Sippell, D. W., Davidson, J. G. N., and Sadasivaiah, R. S. 1985. Rhizoctonia root rot of rapeseed in the Peace River region of Alberta. Can. J. Plant Pathol. 7:184-186.

Stelfox, D., Williams, J. R., Soehngen, V., and Topping, R. C. 1978. Transport of *Sclerotinia sclerotiorum* ascospores by rapeseed pollen in Alberta. Plant Dis. Rptr. 62:576-579.

Stone, J. R., Verma, P. R., Dueck, J., and Spurr, D. T. 1987. Control of *Albugo candida* race 7 in *Brassica campestris* cv. *Torch* by foliar, seed, and soil applications of metalaxyl. Can. J. Plant Pathol. 9:137-145.

Teo, B. K., Yitbarek, S. M., Verma, P. R., and Morrall, R. A. A. 1988. Influence of soil moisture, seeding date, and *Rhizoctonia solani* isolates (AG 2-1 and AG-4) on disease incidence and yield in canola. Can. J. Plant Pathol. 10:151-158.

Teo, B. K., Morrall, R. A. A., and Verma, P. R. 1989. Influence of soil moisture, seeding date, and canola cultivars ('Tobin' and 'Westar') on the germination and rotting of sclerotia of *Sclerotinia sclerotiorum*. Can. J. Plant Pathol. 11:393-399.

Tewari, J. P., Tewari, I., and Paul, V. H. 1994. Necrosis of *Brassica* leaves under deposits of pollen grains. (Abstr.) Can. J. Plant Pathol. 16:77.

Thomas, P. E., Hang, A. N., Reed, G. C., and Reisenauer, G. 1993. Potential role of winter rapeseed culture on the epidemiology of potato leaf roll disease. Plant Dis. 77:420-423.

Vaartnou, H., and Tewari, I. 1972. *Alternaria alternata:* Parasitic on rape in Alberta. Plant Dis. Rptr. 56:676-677.

Vanterpool, T. C. 1974. *Pythium polymastum:* Pathogenic on oilseed rape and other crucifers. Can. J. Bot. 52:1205-1208.

Vigier, B., Chiang, M. S., and Hume, D. J. 1989. Source of resistance to clubroot (*Plasmodi-ophora brassicae* Wor.) in triazine-resistant spring canola (rapeseed). Can. Plant Dis. Surv. 69:113-115.

Walker, J. T., Phillips, D. V., Denmon, L., and Melin, J. 1997. Reaction of canola cultivars to root-knot nematodes *Meloidogyne arenaria* and *Meloidogyne incognita*. (Abstr.). Phytopathology 87:S115.

Williams, J. R., and Stelfox, D. 1979. Dispersal of ascospores of *Sclerotinia sclerotiorum* in re-lation to Sclerotinia stem rot of rapeseed. Plant Dis. Rptr. 63:395-399.

Williams, P. H. 1992. Biology of *Leptosphaeria maculans*. Can. J. Plant Pathol. 14:30-35.

Yitbarek, S. M., Verma, P. R., Gugel, R. K., and Morrall, R. A. A. 1988. Effect of soil temper-ature and inoculum density on preemergence damping-off of canola caused by *Rhi-zoctonia solani*. Can. J. Plant Pathol. 10:93-98.

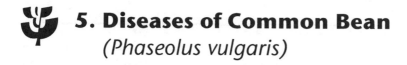

5. Diseases of Common Bean
(Phaseolus vulgaris)

Diseases Caused by Bacteria

Bacterial Brown Spot

Bacterial brown spot is also called brown spot.

Cause. *Pseudomonas syringae* pv. *syringae* is seedborne and overwinters in infested residue. Overwintering also occurs on hairy vetch, *Vicia villosa*; common vetch, *V. sativa*; kudzu, *Pueraria thunbergiana*; and probably other weeds. In New York, bacteria overwintered on weeds primarily in areas where a severe incidence of brown spot occurred the previous growing season. Otherwise the bacteria were not commonly isolated from weeds and residue was considered the primary source of inoculum.

Epiphytic phytopathogenic bacteria provide inoculum for disease, with the probability of disease increasing as epiphyte populations increase. In Wisconsin, an epiphytic population of approximately 10,000 bacteria per gram of snap bean leaf tissue was the minimum infection threshold required to produce brown spot symptoms. Higher populations of bacteria occur on leaves of susceptible cultivars than on resistant cultivars.

Disease severity increases when leaves are moist for extended periods of time.

Distribution. Wherever common bean is grown.

Symptoms. Dark brown necrotic spots of various sizes occur on diseased leaves but without the initial water-soaking symptoms common to other foliar diseases caused by bacteria. The marginal chlorosis or halos surrounding lesions also are absent. Pod symptoms are brown spots and, frequently, a pod twisting at the point of infection.

Management
1. Sow healthy seed treated with a seed-protectant fungicide.
2. Rotate common bean with resistant crops.

133

3. Do not grow common bean adjacent to lima bean *(Phaseolus limensis)* fields since lima bean is very susceptible to *P. syringae.* Inoculum may be easily disseminated from one field to the next.
4. Resistant germplasm in common bean has been identified.

Bacterial Wilt

Cause. *Curtobacterium flaccumfaciens* pv. *flaccumfaciens* (syn. *Corynebacterium flaccumfaciens)* is seedborne. Some workers report the bacterium does not overwinter in infested residue or soil but others state that it does, although poorly. Bacteria are disseminated by driving rain and hail, then enter plants through wounds caused by mechanical injuries and storms.

Distribution. Wherever common bean is grown.

Symptoms. A few water-soaked lesions occur on leaves; however, these do not tend to be as numerous as lesions of bacterial blight or halo blight. Diseased plants eventually wilt. Initially, leaves wilt when temperatures are high but become turgid again when temperatures are reduced. As the disease progresses, leaves eventually turn brown and defoliate from the plant. Sutures of pods are discolored and pods become flaccid. Depending on the mode of infection, diseased seed will have two general symptoms: When seed is invaded internally, white seed may appear yellowish due to bacteria that are visible through the seed coat. If seed is infected externally, bacterial crust forms on the outside of the seed. Plants grown from diseased seed are frequently killed while still small.

Management. Sow disease-free seed.

Bean Wildfire

Bean wildfire is also called wildfire.

Cause. *Pseudomonas syringae* pv. *tabaci* (syn. *P. tabaci*). This is a different strain from the *P. tabaci* that causes wildfire of tobacco.

Distribution. Brazil.

Symptoms. Symptoms of wildfire occur only on leaves. Lesions are water-soaked initially, then become light to dark brown and surrounded by a pronounced chlorotic halo. These symptoms resemble those of halo blight but pods and seeds do not become diseased. Chlorotic halos of bean wildfire are produced at relatively "high" temperatures, but halo blight bacteria induce the chlorotic effect only at "cooler" temperatures.

Management. Not reported.

Common Blight

Cause. *Xanthomonas campestris* pv. *phaseoli (syn. X. phaseoli* and *X. phaseoli* var. *fuscans)* is seedborne in both resistant and susceptible genotypes and overwinters in infested residue. Survival of bacteria occurs in conservation tillage systems but not in conventional tillage systems where residue is incorporated into the soil. The lack of survival is presumably due to antagonism to *X. campestris* pv. *phaseoli* by soil microorganisms. Under tropical conditions, bacteria survive on the soil surface for 5 months but cannot be detected in buried infested residue after 30 days. Generally, infested bean residue is considered to be an important source of inoculum for *X. campestris* pv. *phaseoli*. However, in Michigan, pathogenic bacteria were not isolated from residue for 10 years and did not constitute a source of primary inoculum. Survival also occurs on kudzu leaves, *Pueraria thunbergiana,* and, to a lesser extent, on infested stems and pods of common bean in the southern United States.

Internally infected seed is the main source of primary inoculum, but externally infected seed is also a source. Bacteria have been reported to survive for up to 6 years on seed. Infected seed gives rise to diseased plants and eventually bacteria are blown or splashed onto leaves, where they enter either through natural openings or wounds caused by windblown soil or other means. Bacteria exude from the initial infections and are disseminated to other host plants by rain, wind, plant-to-plant contact, and machinery. Disease development is favored by warm, humid conditions.

Distribution. Wherever common bean is grown under warm, humid conditions.

Symptoms. Seedlings that germinate but do not emerge through the soil have cotyledons with reddish lesions. Diseased plants that do emerge are stunted and the undersides of leaves have small, water-soaked spots with dried bacterial ooze in the center. The spots or lesions turn brown and coalesce to form large dead areas, causing leaves to defoliate and plants to die.

Small, water-soaked spots also occur on seedling stems. The spots eventually enlarge and become brown to reddish in color. On older plants, spots expand lengthwise along the stem. Often a lesion may encircle the stem at a lower node when pods are half mature and cause the plant to fall over.

Water-soaked spots with concentric reddish-brown or brick red rings appear on diseased pods. These spots eventually become dry and sunken and have a crust of dried, yellow bacterial ooze. Bacteria produced on pods may infect the seed through the pod hinges. Diseased seed has a discolored hilum and discolored tissue surrounding the hilum, which give the seed a "varnished" appearance. Diseased seed may sometimes appear healthy and symptomless.

Management
1. Sow certified seed.
2. Treat seed with a seed-protectant bactericide.
3. Do not cultivate when beans are wet because bacteria are spread by machinery.
4. Rotate common beans with cereals or other resistant crops every 3 to 4 years.
5. Apply copper fungicides in extraneous situations; however, fungicides are of limited value.
6. Plow under infested bean residue.
7. Disinfect seed equipment and storage facilities.
8. Grow the most tolerant cultivars. However, under tropical conditions, tolerant cultivars may become completely susceptible.

Erwinia nulandii Leaf Spot

Cause. *Erwinia nulandii.*

Distribution. The United States (Nebraska).

Symptoms. Initially, diseased leaves have small, yellow spots that enlarge and coalesce to form irregular-shaped, necrotic lesions. Eventually, diseased leaf tissue collapses. White bean seed has a pink discoloration.

Management. Not reported.

Fuscous Blight

Cause. *Xanthomonas campestris* pv. *phaseoli* (syn. *X. phaseoli* var. *fuscans)*. According to Fahy and Persley (1983), the causal bacterium for common blight and fuscous blight are considered the same. However, because other researchers consider the causal organisms to be different, common blight and fuscous blight are discussed here as two separate diseases.

Xanthomonas campestris pv. *phaseoli* survives up to 6 years on seed and overwinters in infested residue. *Xanthomonas campestris* pv. *phaseoli* also grows epiphytically on the leaves of nonhost plants, which may be a significant factor in the epidemiology of fuscous blight. Disease development is favored by warm, wet weather.

Distribution. Wherever common bean is grown under warm, wet conditions.

Symptoms. Symptoms of fuscous blight are identical to common blight. Positive identification of the disease is possible only when the causal organism is identified.

Seedlings that do not emerge through the soil have cotyledons with reddish lesions. Plants that do emerge are stunted and the undersides of leaves have small, water-soaked spots with dried bacterial ooze in the center. The spots or lesions turn brown and coalesce to form large dead areas, which cause diseased leaves to defoliate and plants to die. Small, water-soaked spots that occur on seedling stems enlarge and become brown to reddish in color.

On older plants, spots enlarge lengthwise along the stem. Often a lesion may encircle the stem at a lower node when pods are half mature and cause the plant to fall over.

Water-soaked spots with concentric reddish-brown or brick red rings appear on pods. Spots eventually become dry and sunken and have a crust of dried yellow bacterial ooze. Bacteria produced on pods may infect seed through pod hinges. Diseased seed has a discolored hilum and discolored tissue surrounding the hilum, which give the seed a "varnished" appearance. Diseased seed may sometimes appear healthy and symptomless.

Management
1. Sow certified seed.
2. Treat seed with a seed-protectant bactericide.
3. Do not cultivate when beans are wet because bacteria are spread by machinery.
4. Rotate common beans with cereals or other resistant crops every 3 to 4 years.
5. Apply copper fungicides in extraneous circumstances; however, fungicides are of limited value.
6. Plow under infested bean residue.
7. Disinfect seed equipment and storage facilities.
8. Grow the most tolerant cultivars. However, under tropical conditions, tolerant cultivars may become completely susceptible.

Halo Blight

Halo blight is also called bean halo blight.

Cause. *Pseudomonas syringae* pv. *phaseolicola* (syn. *P. phaseolicola*) is primarily seedborne and overwinters in infested residue under dry conditions. Only race 1 survives in the weed *Neonotonia wightii;* however, inoculation with all three races causes disease symptoms in this weed. Race 2 strains are reported to survive in dead standing plants in northern Tanzania and in the residue of the susceptible bean cultivar 'Canadian Wonder' that is buried 2 to 5 cm deep in the soil. Survival in standing plants may bridge the gap between two bean-growing seasons in warmer parts of the world. Survival of both races 1 and 2 is generally longer in stems than in leaves. Race 3 is presumed to survive in residue also.

Race 2 strains obtained from infested debris in northern Tanzania were reported to be consistently more aggressive than those from growing bean plants. It is suggested that virulence of race 2 strains may increase during survival in bean residue under dry environmental conditions due either to a mutation or to a selection of a subpopulation of bacteria whose increased virulence is associated with survival in residue.

Diseased seed gives rise to diseased cotyledons. Bacteria exude from these initial infections and are disseminated to other host plants by rain, wind, contact between plants, and machinery. Residue from highly susceptible cultivars constitutes a source of primary inoculum under dry conditions. Bacteria are splashed or blown from residue to leaves where they enter through natural openings or wounds caused by windblown soil or other means. Epiphytic phytopathogenic bacteria also provide inoculum for disease; susceptible cultivars are more likely to serve as inoculum sources than are resistant cultivars.

Optimum disease development is at 21°C with wet weather. Some isolates reported to be strains of race 2 are known to have optimum growth at 34°C. In northern Tanzania, bacterial populations are higher and disease severity is greater when beans are intercropped with maize than when bean is grown as a monoculture.

At least three races are known. Presumably other races also exist.

Distribution. Where common bean is grown under cool, humid conditions.

Symptoms. Plants growing from internally infected seed are stunted and gradually die. Seedlings may not emerge through the soil, and reddish lesions are normally present on cotyledons.

Small, water-soaked spots occur on the undersides of diseased leaves and turn reddish brown; the surrounding tissue forms a large yellow-green halo. Spots rapidly enlarge and coalesce to form large, brown, dead areas on leaves that eventually cause defoliation. Dried bacterial ooze that is normally a light cream color is present in the center of each spot. The upper trifoliate leaves on plants that are infected later in the growing season are chlorotic.

Reddish spots that become dry and sunken with concentric rings develop on pods. The pod hinge may become discolored also and may have bacterial exudate on it. Seed may become infected through the hinge and become discolored; however, seed from symptomless pods also may be infected and display symptoms.

Young stems have reddish lesions that extend lengthwise. Infections at nodes may girdle the stems when pods are half-mature, causing stems to break at the diseased node.

Race 3 from Africa causes a hypersensitive reaction in the cultivar 'Tendergreen,' which is susceptible to races 1 and 2. A race 2 strain isolated from residue in northern Tanzania induced systemic chlorosis and stunting on the resistant bean cultivar 'Edmund.'

Management
1. Apply copper fungicides to foliage. Copper fungicides are more helpful in managing halo blight than common blight.
2. Sow certified seed.

3. Do not cultivate soil when beans are wet because machinery may disseminate bacteria from diseased to healthy plants.
4. Rotate common bean with cereals or other resistant crops every 3 to 4 years.
5. Plow under infested bean residue.
6. Disinfect seed equipment and storage facilities.
7. Grow the least susceptible cultivars.

Pseudomonas blatchfordae Leaf Spot

Cause. *Pseudomonas blatchfordae* overwinters in infested bean straw on the soil surface. The oldest leaves are most susceptible to infection and subsequent disease development and symptom expression. Symptom development is most rapid at 24°C.

Distribution. The United States (Nebraska).

Symptoms. Initially, small yellow spots that are accompanied by water-soaking appear on the oldest leaves. Spots increase in size and become surrounded by a narrow chlorotic halo. The middle areas of the spot turn orange to brown and give the spot a "sunscalded" appearance. Eventually, the diseased area becomes necrotic and causes defoliation.

Management. Not reported.

Diseases Caused by Fungi

Alternaria Leaf Spot

Alternaria leaf spot and black pod disease are both caused by *Alternaria alternata* but are treated as two separate diseases on the basis of symptoms.

Cause. *Alternaria alternata* (syn. *A. tenuis), A. brassicae* f. *phaseoli,* and *A. brassicicola.* Disease development is most severe at high relative humidity and temperatures of 16° to 24°C. Plants increase in susceptibility as they grow older and when they are grown in nitrogen- and potassium-deficient soils.

Distribution. The United States.

Symptoms. Small, brown, irregular-shaped lesions occur on diseased leaves. Lesions expand, become gray-brown to dark brown and oval-shaped, and have concentric zones. The older portions of lesions sometimes fall out, giving a "shot-hole" appearance to diseased leaves. Lesions coalesce to form large necrotic areas and defoliation may occur.

Management
1. Ensure adequate soil fertility.
2. Apply foliar fungicides.

Angular Black Spot

Cause. *Erratomyces patelli. Protomycopsis patelii* is reported to cause disease in India, and the agent of a similar disease is known as *Entyloma vignae* in North America. However, Piepenbring and Bauer (1997) state *E. vignae* is a synonym of *P. patelii.* The fungus is related to smut fungi of the genus *Tilletia* because it produces relatively large, opaque teliospores that have a partition in their wall and that germinate with holobasidia carrying needle-shaped basidiospores. In contrast to species of *Tilletia* and related genera, the teliospores are scattered in intercellular spaces in the mesophyll without rupturing it and develop mostly intercalaryly. A new genus, *Erratomyces,* is proposed.

Spore germination occurs on water agar at 30°C or more.

Distribution. India, Central and South America.

Symptoms. Numerous polyangular, grayish, dark leaf spots (2–3 mm in diameter) occur predominantly on the lower leaves. Spots coalesce to form lesions that are 6 mm or larger in size.

Management. Not reported.

Angular Leaf Spot

Cause. *Phaeoisariopsis griseola* (syn. *Isariopsis griseola). Phaeoisariopsis griseola* is divided into two major groups. Group 1 isolates are generally recovered from Andean gene pool materials, whereas group 2 isolates are recovered from Mesoamerican materials. Group 1 isolates are more pathogenic on Andean beans, whereas group 2 isolates are more pathogenic on Mesoamerican beans. Random Amplified Polymorphic DNA and pathogenicity data suggest groups 1 and 2 may have originated in the Andes and Mesoamerica, respectively, and that coevolution of the *P. griseola* fungus and its common bean host has resulted in increased levels of disease in this host-pathogen interaction.

Phaeoisariopsis griseola is seedborne but survives primarily in infested residue on the soil surface and in infested plant residue that remains upright after harvest. Conidia are produced during wet weather and are disseminated by wind and rain to host plants. Secondary infection is by conidia produced in lesions on the undersides of diseased leaves. These conidia are also wind or rain disseminated to other host plants. The optimum temperature for infection is 24°C. Disease severity is higher in a monocultural system.

Distribution. Africa, Central America, South America, and the eastern and southern United States. Angular leaf spot is of minor importance in the United States; however, sporadic outbreaks have occurred and severe disease has been reported on dark red kidney bean in Michigan. Angular leaf spot is a major disease of dry edible bean in tropical and subtropical regions.

Symptoms. Small, brownish spots first occur on diseased leaves and may eventually become numerous enough to cause defoliation. During moist weather, *P. griseola* sporulates abundantly within a spot and gives it a grayish color.

Similar spots on pods are usually small but may enlarge and coalesce to involve a large area of the pod. There is no correlation between disease severity on pods and percentage of seeds infected with the pathogen. Seed infection normally occurs at the hilum; therefore, seeds become infected when located directly under lesions present at the pod suture.

Management
1. Rotate common bean with resistant crops.
2. Sow disease-free seed.
3. Apply foliar fungicide.
4. Grow the most resistant cultivar. Resistant germplasm has been identified.

Anthracnose

Cause. *Glomerella lindemuthiana,* anamorph *Colletotrichum lindemuthianum. Glomerella cingulata,* anamorph *C. gloeosporioides,* is also reported to be a causal fungus.

Several races of *C. lindemuthianum* occur. Previously, alpha, beta, delta, epsilon, gamma, and lambda races have been identified in Canada, while only alpha, beta, and gamma races have been reported in the United States. A larger number of physiologic races have been reported in Africa, Europe, and Latin America.

Currently, Balardin et al. (1997) have characterized 41 races based on virulence to 12 differential cultivars of *Phaseolus vulgaris.* The 41 races are categorized into two groups: those found over a wide geographic area and those restricted to a single country.

Glomerella spp. are seedborne and overwinter as perithecia in dry infested residue. *Colletotrichum lindemuthianum* also survives as mycelium under the seed coat for at least 2 years if seed is dry and stored under cool conditions at an optimal temperature of 4°C. *Colletotrichum lindemuthianum* will also survive for at least 5 years in dry residue. An alternating wet-dry cycle is detrimental to fungus survival and *C. lindemuthianum* will lose viability in several days.

In New York, *C. lindemuthianum* survived 4 months in infested bean residue placed 0, 10, and 20 cm deep in the soil, but the fungus was no longer detectable 22 months after placement at any soil depth. Bean residue infested with *C. lindemuthianum* that remains on the soil surface for only one winter may serve as an inoculum reservoir the following growing season. Disease has been reported to be most severe when tillage practices leave infested residue on the soil surface.

Infected seed gives rise to diseased seedlings that have lesions on the cotyledons. Conidia are then produced in these lesions and disseminated by wind and rain. Long-distance spread of 3 to 5 m during a rainstorm occurs when splashing raindrops are blown by gusting winds. Spores are covered by a sticky substance that enables them to adhere to whatever they touch. Therefore, conidia are disseminated by insects, machinery, and several other means.

Disease is most severe during wet weather at temperatures of 14° to 18°C. Disease development is restricted at temperatures above 25°C.

Distribution. Canada (Ontario); tropical and subtropical areas of Latin America, central and eastern Africa; and eastern, midwestern, and southern states in the United States.

Symptoms. Symptoms may occur on any part of a susceptible plant. Diseased seed has dark, sunken lesions of various sizes that frequently extend through the seed coat into the cotyledons. Initial symptoms may appear on cotyledonary leaves as small, dark brown to black lesions. Mature lesions have pinkish spore masses in their centers.

Numerous small rust-colored specks appear on hypocotyls. The specks gradually enlarge longitudinally and form sunken lesions or eyespots that cause the hypocotyls to rot off.

On older stems, oval- to eye-shaped lesions become sunken, brown cankers with purplish to brick red borders that are 5 to 7 mm long. These sunken cankers extend up and down diseased stems. Cankers may eventually weaken stems enough to cause them to break over. Pods above symptomatic stems dry prematurely and shrink, causing seed numbers and size to be reduced. Numerous pinpoint black dots, which are acervuli, develop in lesions on diseased stems and pods. Later in the growing season, perithecia develop on diseased stems.

Lesions appear first on surfaces of diseased lower leaves and leaf veins as small, angular, brick red to purple spots that eventually become dark brown to black. Later, lesions also appear on veinlets of diseased upper leaves. Brownish, angular, dead spots, which also may appear on upper leaf surfaces, give the diseased leaves a ragged appearance.

Lesions on pods start as small, elongated reddish-brown spots that become somewhat circular (6 mm in diameter) and sunken with a rusty to brown border. Spots may become numerous, and during moist weather, they become pinkish in color due to conidial sporulation.

Management
1. Grow a resistant cultivar.
2. Grow common bean in the same field every third or fourth year.
3. Sow disease-free seed grown in areas where anthracnose normally does not occur and that have a seed inspection program.

4. Apply foliar fungicides before anthracnose becomes severe.
5. Do not enter fields when plants are wet.
6. Apply a fungicide seed treatment to seed.
7. Quarantine pedigree inspection.
8. Incorporate residue in the autumn to hasten decomposition.

Aphanomyces Root and Hypocotyl Rot

Cause. *Aphanomyces euteiches* f. sp. *pisi* infects both peas and beans. *Aphanomyces euteiches* f. sp. *phaseoli* infects alfalfa and beans.

Fungi survive in soil for several years as oospores, which presumably are the primary source of inoculum. Sporangia and zoospores are produced in roots and are thought to be the source of secondary inoculum, although this has not been observed.

Aphanomyces euteiches f. sp. *phaseoli* has only been reported in sandy soils; however, it is not known if it is limited to this environment. Disease is most severe at high soil moisture levels and temperatures of 20° to 28°C.

Distribution. The United States (Minnesota, New York, Wisconsin, and possibly Michigan).

Symptoms. *Aphanomyces euteiches* f. sp. *phaseoli* causes more injury on bean than *A. euteiches* f. sp. *pisi* does. Diseased plants are stunted, and roots and lower stems are rotted. Necrosis of hypocotyls ranges from discolored streaks to complete destruction of the diseased hypocotyls. Infrequently, plant death occurs.

Management. Grow the most tolerant cultivars. Resistant germplasm has been identified.

Ascochyta Blight

Ascochyta blight is also called Ascochyta leaf spot and blight, and speckle disease, although speckle disease is reported as a separate disease caused by *Stagonosporopsis hortensis* (syn. *Ascochyta boltshauseri*). Ascochyta blight is basically the same disease as black node disease.

Cause. *Phoma exigua* var. *exigua* (syn. *Ascochyta phaseolorum*). Survival is presumably as pycnidia in infested residue on the soil surface and as mycelia in seed. The fungus is favored by a relative humidity greater than 80% and temperatures of 15° to 20°C.

Distribution. Mid- to high-altitude regions of eastern and central Africa, Central and Andean America, and western Europe.

Symptoms. All aerial plant parts may be diseased. Leaves have brown to black lesions (10–30 mm in diameter) that normally develop concentric zones and may contain numerous small, black pycnidia.

Concentric dark gray to black lesions may appear on pods. Diseased seeds become brown to black. The fungus may spread systemically throughout the plant, and defoliation and pod drop may occur. Seed yield losses are more severe following early infection and premature defoliation. Losses of about 40% have occurred under moderate disease pressure in an experimental trial in Colombia.

Management

1. Intermediate resistance has been identified in common bean germplasm. However, some researchers have reported that resistance is not available in common bean but does occur in scarlet runner bean.
2. Apply foliar fungicides.
3. Rotate common bean with resistant crops.
4. Treat seed with a seed-treatment fungicide.

Ashy Stem Blight

Ashy stem blight is also called charcoal rot.

Cause. *Macrophomina phaseolina* (syn. *Botryodiplodia phaseoli, M. phaseoli, Rhizoctonia bataticola,* and *Sclerotium bataticola*). *Macrophomina phaseolina* is seedborne. Survival is primarily as microsclerotia in soil for 3 to 36 months, but *M. phaseolina* may also survive as pycnidia in residue.

Primary inoculum is presumably from direct germination of microsclerotia, seedborne mycelia, or conidia from pycnidia. Dissemination is by any means that moves infested soil or seed. Sclerotia of *M. phaseolina* are reported to colonize beans under relatively dry conditions.

Airborne conidia, released from pycnidia, serve as one source of secondary inoculum for initiating infections on foliage of older plants. Another source of secondary inoculum is the expansion of stem infections below the soil level. These infections occur later in the growing season when diseased plants start to senesce. Infection and subsequent disease are favored by high humidity, temperatures of 27°C or higher for prolonged periods of time, and "normal" soil moisture conditions. The microsclerotia are formed in host tissues and released into the soil as the diseased tissues decay.

Macrophomina phaseolina can grow and produce large quantities of microsclerotia under low water potentials. Radial growth on potato dextrose agar (PDA) was maximum at osmotic water potential values of –1220 to –1880 J/kg, but growth was reduced at lower values. Production of sclerotia on potato dextrose broth was not affected by osmotic water potentials of –670 to –3920 J/kg but was completely inhibited from –8270 to –12,020 J/kg. Germination of sclerotia after 2 days on PDA is not significantly affected from –320 to –4760 J/kg, but germination is drastically reduced with further reductions in osmotic water potentials.

Distribution. The Caribbean, Central and South America, and the eastern, southern, and western United States.

Symptoms. Initially, dark, sunken, irregular-shaped lesions occur on seedling stems at the soil line and eventually kill the seedling. Sunken, reddish-brown lesions occur on stems at the soil level or lower. Eventually, lesions enlarge and girdle the stem, extending down into the roots and up into the branches. Lesions become gray in the center and numerous microsclerotia and pycnidia (about the size of a pinhead) appear as small, black objects that contrast with the gray centers of lesions. Young diseased plants usually die before seed is produced.

Infection of older plants produces superficial lesions that are usually not sunken. Frequently, disease is more pronounced on one side of a plant, which causes the primary leaf on that side to droop and die. Any remaining leaves turn yellow. Severely diseased plants are wilted, chlorotic, and defoliated, and mature early or die.

Management
1. Sow only healthy seed treated with a seed-protectant fungicide.
2. Rotate common bean with a resistant crop every 2 to 3 years.
3. Grow resistant cultivars. Choose drought-tolerant cultivars. The mechanism for drought tolerance confers resistance to *M. phaseolina* under field conditions.

Aspergillus nidulans Disease

Cause. *Aspergillus nidulans* (syn. *A. nidulellus*).

Distribution. Unknown.

Symptoms. Inoculation of excised tissue results in limited water-soaking and host-cell death. Sporulation occurs within several days.

Management. Not reported.

Bipolaris sorokiniana Leaf and Pod Lesions

Cause. *Bipolaris sorokiniana*.

Distribution. Canada (New Brunswick and Nova Scotia) and the United States (Wisconsin), on snap bean.

Symptoms. Small (1 mm in diameter), brown to black, round to irregular-shaped lesions occur on diseased leaves and pods. Pod lesions pose a problem for processing beans because blanching of fruit does not mask lesions.

Management. Not reported.

Black Node Disease

Black node disease is basically the same disease as Ascochyta blight.

Cause. *Phoma exigua* var. *diversispora* (syn. *P. diversispora*). Survival of *P. exigua* var. *diversispora* is presumably as pycnidia in infested residue on the soil surface and as mycelia in seed. Disease is favored by a relative humidity greater than 80% and temperatures of 15°to 20°C.

Distribution. Eastern Africa and western Europe.

Symptoms. Concentric dark gray to black lesions may appear on diseased branches, stems, and nodes, and cause stem girdling and plant death.

Management
1. Intermediate resistance has been identified in common bean germplasm. However, some researchers have reported that resistance is not available in common bean but does occur in scarlet runner bean.
2. Apply foliar fungicides.
3. Rotate common bean with nonhost crops.
4. Treat seed with a seed-treatment fungicide.

Black Pod Disease

Black pod disease is also called Alternaria pod flecking, and pod and seed coat discoloration of white bean. Alternaria leaf spot and black pod disease are both caused by the same fungus but are treated as two separate diseases on the basis of symptoms.

Cause. *Alternaria alternata* (syn. *A. tenuis*) overwinters in infested residue and seed and on leaf lesions of numerous weed species. *Alternaria alternata* is a weak parasite that colonizes stomata cavities of growing plants. In the autumn, when plant senescence begins, *A. alternata* begins to grow rapidly. Disease is most severe during cool, wet conditions that may occur in a closed foliage canopy. Autumn rains that delay harvest accentuate disease severity.

Distribution. Canada (Ontario) and the United States (New York).

Symptoms. Initially, small, irregular-shaped, water-soaked flecks appear on pods. Later, flecks coalesce to produce long streaks or irregular, large, reddish to dark brown or black discolored lesions. Lesions are either sunken or raised for just a few cells deep into the pod but have a corky layer below. Conidia are often produced in lesions. Small, irregular, brownish spots are also produced on stem and leaf tissues.

Symptoms of pod and seed coat discoloration of white bean varies. Pod discoloration varies from dark gray flecks to stipples to dark gray patches that later coalesce. Seeds, in discolored pods, have various degrees of discoloration that persists through processing.

Management
1. Grow the most resistant cultivar.
2. Apply foliar fungicides.

Black Root Rot

Cause. *Thielaviopsis basicola* survives as chlamydospores in infested residue and soil. Primary inoculum is chlamydospores that germinate to produce mycelia that penetrate intact bean tissues, wounds, or lesions caused by other pathogens. Chlamydospores are produced throughout diseased tissue. Under wet conditions, hyphae grow on the plant surface and produce endoconidia and chlamydospores. Sporulation of *T. basicola* is most prolific at temperatures of 25° to 28°C, but disease is most severe at temperatures of 15° to 20°C. In New York, however, the disease is most severe during periods of high soil moisture and high temperatures.

Distribution. Wherever common bean is grown.

Symptoms. Elongated, narrow lesions occur on diseased hypocotyl and root tissues. Initially, the lesion color is reddish purple but becomes dark gray to black. Eventually, the lesions coalesce and form large, dark areas on roots and stems. Lesions either are superficial and cause little injury or develop into deep lesions that cause plant stunting, foliage wilting, and eventual defoliation and subsequent plant death.

Management
1. Resistance has been identified in some breeding lines.
2. Grow rye in alternate years. Populations of *T. basicola* and disease incidence decrease in soil sown to rye compared to fallow soil. Disease-suppressing effects occur during growth and subsequent decomposition and are associated with microbial antagonism and antibiotic production with rye as the substrate.

Cercospora Leaf Blotch

Cause. *Cercospora canescens* and *Pseudocercospora cruenta* (syn. *Cercospora cruenta*), teleomorph *Mycosphaerella cruenta*. *Cercospora caracallae* and *C. phaseoli* are also reported to be causal fungi. The fungi are seedborne and likely overseason in infested residue.

Distribution. Latin America and the southern United States.

Symptoms. Vigorously growing leaves infrequently display symptoms. Mature leaves have brown or reddish-brown lesions that vary in shape from circular to angular and in size from 2 to 10 mm in diameter. Eventually, lesions may have a gray center, caused by fungal sporulation, with a reddish border that is the remainder of the original lesion. Lesions may coalesce and dry up, and portions of them may fall out and give the diseased leaf a ragged appearance. Severely diseased leaves become chlorotic.
Branches, stems, and pods may have lesions.

Management
1. Resistant cultivars have been identified.
2. Apply copper fungicides early.
3. Rotate common bean with nonhost crops.

Chaetoseptoria Leaf Spot

Cause. *Chaetoseptoria wellmanii* may be seedborne. Disease occurs in areas of moderately cool temperatures and moist environments.

Distribution. Central America and South America.

Symptoms. Circular to irregular-shaped lesions (up to 10 mm in diameter) that consist of light tan centers surrounded by a reddish-brown border occur on diseased primary leaves soon after plant emergence. Small, gray pycnidia, which appear as "specks," form in lesions. Defoliation and yield reduction occur.

Management
1. Rotate common bean with other crops.
2. Sow disease-free seed.
3. Apply foliar fungicide.

Floury Leaf Spot

Cause. *Mycovellosiella phaseoli* (syn. *Ramularia phaseoli)* presumably survives in infested residue. Disease occurs in areas of moderate temperatures and moisture.

Distribution. Africa, Asia, Europe, and Central and South America. In Colombia, floury leaf spot is more common at elevations of 1500 to 2000 m.

Symptoms. Initially, white or "floury," angular lesions occur on older leaves and progress upward on the diseased plant to the newer leaves. The floury appearance is due to fungal growth of mycelium and spores. These lesions may coalesce and appear irregular-shaped. The upper leaf surface may have a light green discoloration but no floury appearance of mycelium and spores. Defoliation may eventually occur.

Management
1. Rotate common bean with other crops.
2. Apply foliar fungicides.

Fusarium Root Rot

Fusarium root rot is also called dry root rot.

Cause. *Fusarium solani* f. sp. *phaseoli* survives saprophytically as mycelium in infested residue and as chlamydospores in soil. Conidia, and some mycelial cells growing on residue, are eventually converted to chlamydospores. Chlamydospores are stimulated to germinate by root excretions of susceptible plant hosts. The chlamydospores germinate to form germ tubes that enter roots directly or through wounds and grow in the outer layers of tissue, especially the cortex. *Fusarium solani* f. sp. *phaseoli* is disseminated in dust and plant debris mixed with seed, in soil on machinery, and by any means that transports soil.

Beans sown into cold soils or subjected to low temperatures after emergence tend to have severe root rot, while beans sown into warm soils with adequate moisture tend to escape serious root rot. Root growth is restricted in cold soil, and herbicide and decomposing organic matter may also be toxicogenic. Preconditioning stresses, such as soil compaction by tractor wheels, subsurface pans, and intermittent drought, especially in coarse-textured soils of low water-holding capacity, restrict root growth and precondition roots to infection and disease development.

Root rot and postemergence damping off is greater in soils that have been subsoiled or disked rather than plowed. In 1 of 3 years, postemergence damping off is greater when nitrogen is broadcast prior to sowing than when nitrogen is applied through overhead irrigation.

High disease incidence and severity are associated with high populations of *F. solani* f. sp. *phaseoli* in soil. Infection may be great at lower inoculum levels when the lesion nematode, *Pratylenchus penetrans,* is present; however, there is no effect on final disease severity.

Distribution. Wherever common bean is grown.

Symptoms. Initially, slightly red to brown areas or streaks occur on diseased taproots and hypocotyls, often around the points of secondary root emergence. This reddish discoloration increases in intensity and may eventually cover the taproot and hypocotyl. Later in the season, the red color is replaced by a brown discoloration and roots become hollow and dry. The pith area in the taproot is often a bright red. Lateral roots are often destroyed and plants develop a secondary root system near the soil line. Most diseased roots do not extend beyond the plowed layer of soil and penetrate no deeper into the soil than 7.6 to 15.2 cm.

Aboveground symptoms do not appear until roots are seriously decayed. Symptoms are similar to those caused by low soil-moisture conditions or lack of nitrogen. Young plants will be stunted and have chlorotic leaves. Older plants will have chlorotic leaves and some defoliation. Yield is reduced by a reduction of seed weight; however, the number of pods per plant is normally not reduced. Root rot has been reported to be more severe on bean grown in soil infested with both *F. solani* f. sp. *phaseoli* and *Pythium ultimum.*

Management
1. Grow the most resistant cultivars.
2. Rotate common bean with resistant crop plants every 3 to 4 years.

3. Sow seed that is free of any debris.
4. Use practices that condition soil for good moisture distribution and root penetration.
5. Sow seed rates that are high enough to ensure a good stand.
6. Do not feed infested bean straw to animals because the fungus may be spread in the manure.
7. Plow under common bean refuse.
8. Do not cut off newly formed roots during cultivation.
9. Practice subsoiling or other practices that increase root volume and depth.
10. Sow seeds into warm soil.
11. Irrigate when necessary.
12. Test soil to determine fungus population and identify fields at high risk from Fusarium root rot.

Fusarium Yellows

Fusarium yellows is also called Fusarium wilt.

Cause. *Fusarium oxysporum* f. sp. *phaseoli* survives as mycelia in infested residue and as chlamydospores in soil. Spores and chlamydospores may also be externally seedborne. Infection of bean plants is through root and hypocotyl wounds. Optimum disease development is at 20°C.

Wilt severity in common bean increases proportionally to inoculum concentration in the soil, especially chlamydospores. Disease has been reported severe where plants were stressed by high temperatures, poor water drainage, and soil compaction. Fusarium yellows affects most commercial pinto, great-northern, and navy bean cultivars of common beans.

At least two physiologic races exist. One race is in Europe and North America, and the second race is in Brazil.

Distribution. Wherever common bean is grown.

Symptoms. Initially, chlorosis and senescence occur on the lower leaves and progress to the upper leaves of the diseased plant. Leaves become increasingly chlorotic and, in some cases, eventually defoliate. A reddish-purple to reddish-orange discoloration of the vascular system of roots, stem, petioles, and peduncles occurs. Early infection results in plant stunting.

Management
1. Grow tolerant or resistant cultivars.
2. Rotate common bean with resistant crops.
3. Sow disease-free seed.
4. Apply a fungicide seed treatment.

Gray Leaf Spot

Gray leaf spot is also called gray spot and mancha gris (gray blotch).

Cause. *Cercospora vanderysti* and *C. castellanii*. Disease development is favored by cool, wet weather at elevations greater than 1500 m.

Distribution. Colombia.

Symptoms. Initially, numerous slightly chlorotic, angular spots (2–5 mm in diameter) occur on diseased upper leaf surfaces. Spots may be so numerous that diseased leaves have a mosaic-like appearance. Spots eventually coalesce and cause leaves to defoliate. Eventually, a grayish-white, powdery growth consisting of conidia and conidiophores occurs in these spots. The most distinguishing symptom of mancha gris occurs in spots on the surfaces of lower leaves, where a gray mat or cushion-like fungal growth consisting of a dense growth of flexuous conidiophores and their conidia completely covers the spots.

Management
1. Rotate common bean with other crops.
2. Apply foliar fungicides.
3. Some breeding lines show resistance.

Gray Mold

Cause. *Botrytis cinerea,* teleomorph *Botryotinia fuckeliana,* overwinters as stroma and sclerotia on infested residue and other hosts, including weeds. Survival for extended periods may also occur as mycelia and conidia. Conidia are produced early in the spring and throughout the summer when wet conditions occur and are disseminated by wind and splashing rain to plant hosts. Also, ascospores of *B. fuckeliana* are produced on apothecia that develop from stroma or sclerotia and are disseminated by wind.

Disease development is most severe under cool, humid conditions. Common bean is infected at different stages of growth but senescing cotyledons are colonized first. Young stem and leaf tissues can become diseased before plant bloom and serve as inoculum sources within a field. Infection frequently occurs where old blossoms have fallen onto the plant or have been retained at the tips of pods. Diseased stems may continuously produce inoculum into the bloom period.

Distribution. Wherever common bean is grown.

Symptoms. All aboveground plant parts may show symptoms, but pods are most commonly diseased. Economic loss is due primarily to pod rot, which reduces quality and increases processing costs. Initially, water-soaked lesions occur and, under humid conditions, a white mold develops, followed by a gray fungal growth consisting of conidia and conidiophores. Tip wilt and decay of shoots may occur.

Management
1. Sow certified disease-free seed.
2. Rotate common bean with resistant crops.
3. Apply foliar fungicides if weather conditions favor disease development.
4. Do not sow common bean too thickly because a thick canopy retards air circulation and favors disease development.

Phyllosticta Leaf Spot

Cause. *Phyllosticta phaseolina* survives as pycnidia in infested residue. Disease is favored by high humidity and rainfall at moderate temperatures.

Distribution. Europe, Japan, Latin America, South Africa, and the United States.

Symptoms. Mature leaves display small, angular, water-soaked spots that may coalesce and enlarge to 7 to 10 mm in diameter. Spots have a light-colored, necrotic center surrounded by a reddish-brown margin. Centers of older lesions may fall out and give diseased leaves a tattered appearance. Small, black specks, which are the pycnidia, develop throughout the lesion. Petioles and stems may also become infected and flower buds may become necrotic. Small lesions (1 mm in diameter) with dark centers and reddish margins infrequently develop on pods.

Management
1. Apply foliar fungicides.
2. Rotate common bean with a resistant crop.

Phymatotrichum Root Rot

Phymatotrichum root rot is also called cotton root rot.

Cause. *Phymatotrichopsis omnivora* (syn. *Phymatotrichum omnivorum)* survives as sclerotia in soil or infested residue. Fungal strands grow from sclerotia until they contact a host root. The fungal strand then encompasses the root and grows upward to the soil surface, where it grows around the upper root system and eventually penetrates the plant.

　　Disease is favored by alkaline, calcareous soils and moist, warm soils at temperatures of 15° to 35°C.

Distribution. Northern Mexico and the southwestern United States.

Symptoms. An initial slight yellowing or bronzing of leaves is followed by a sudden wilting as plants begin to flower. Cortical tissue sloughs easily and is covered by a visible network of hyphal strands. A few days after wilting, plants die in circular patterns corresponding to the radial growth of the fungus originating from dying plants. Following rains, a cottony growth or spore mat of the fungus may occur on the soil surface around the stems of dead plants.

Management
1. Rotate common bean with a resistant crop for at least 4 years.
2. Control susceptible weeds.
3. Deep chisel or deep plow green manure crops, manure, or composts.

Pink Pod Rot

Cause. *Trichothecium roseum* is a common soil fungus that is saprophytic on residue or is weakly parasitic to common bean. In Canada, *T. roseum* is frequently isolated from seeds of soybean, pea, faba bean, kidney bean, and scarlet runner bean.

Distribution. The United States (Red River Valley of North Dakota). *Trichothecium roseum* has been occasionally isolated from common bean roots showing root rot symptoms in Canada (southwestern Ontario).

Symptoms. A white, powdery mold that later turns pink occurs on senescent and mature pods. Mold occurs infrequently on stems, petioles, and dead leaves. Seeds from diseased pods are shriveled and yellow or pinkish yellow.

At first, inoculated pods have water-soaked areas that develop into a lesion whose center becomes chocolate brown. Later, a white, powdery mold develops in the lesion, then spreads outward and becomes pink, similar to field symptoms. Seeds from infected pods do not germinate.

Management. Not reported.

Pod Rot, Seed Rot, and Root Rot

Cause. *Fusarium pallidoroseum* (syn. *F. semitectum*) causes disease under humid conditions.

Distribution. Brazil.

Symptoms. Initially, pinpoint-sized, water-soaked lesions that occur on diseased pods become necrotic and rusty brown and vary in shape from circular to elongated, depending on the diseased cultivar. Under humid conditions, pods become soft and covered with fungal mycelia. Under less humid conditions, pods become brittle and dry.

White-seeded snap bean seed becomes reddish brown, shriveled, and covered with mycelia. Light brown seed has blood red, circular lesions. Rotted roots are rusty brown and "pulpy" and have a pruned appearance.

Management. Not reported.

Powdery Mildew

Cause. *Erysiphe polygoni* overwinters as cleistothecia on residue. Ascospores are formed in cleistothecia and disseminated by wind to plant hosts. Conidia form on superficial mycelia and function as secondary inoculum. Cleistothecia form later in the growing season on older infection sites or on mature diseased leaves. Mycelia on infested residue may also function as a means of overwintering, especially in geographical areas with mild winters.

Distribution. Worldwide.

Symptoms. The most damage occurs to plants that mature late in the autumn in the southern United States. Typically, a white, powdery material consisting of conidia and conidiophores is produced on all aboveground parts of plants. Diseased leaves become chlorotic and may defoliate. Pods and stems become purple. Pods are malformed, small, and poorly filled, and fall from diseased plants before seed matures.

Management
1. Grow the most disease resistant cultivar adapted to an area.
2. Apply a foliar fungicide before disease becomes severe.

Pythium Root and Hypocotyl Rot

Cause. *Pythium aphanidermatum, P. aristosporum, P. catenulatum, P. dissotocum, P. irregulare, P. myriotylum, P. splendens,* and *P. ultimum* have all been reported to be pathogenic to beans. These fungi are most pathogenic in soils with high moisture. The optimum conditions for disease are a soil temperature of 15°C and a soil water potential of 0 to −1bar.

Distribution. Not reported, but likely widespread.

Symptoms. Damping off may occur. Symptoms, which occur on hypocotyls 16 to 22 days after sowing, initially appear as elongated, tan to brown, soft, water-soaked areas that extend 3 to 5 cm above the soil line and one-quarter to three-quarters the distance around the hypocotyl circumference. These symptoms are also present below the soil line. Although frequently roots are white and healthy, rot may begin at the tips of taproots and proceed upward on hypocotyls. Within 3 weeks, water-soaked areas become dry and tan to reddish brown and have a slightly sunken surface. Later, after about 4 weeks, most of the belowground hypocotyl and fibrous roots die and appear papery and reddish-brown. In advanced cases, roots may be a hollow tube. Root pruning that results in shortened roots may also occur. Adventitious roots may be produced above or at the soil line.

    Plants wilt and die, or if adventitious roots are produced, they may be stunted or chlorotic and produce smaller and fewer pods, although the seed weight is not reduced. Secondary infection by *Fusarium* spp. may occur.

Management
1. Grow the most resistant cultivars.
2. Apply systemic fungicides to seed.
3. Plowing to a depth of 20 to 25 cm increased plant stand, vine weight, and yield in research plots. Root rot and postemergence damping off is greater in soils that have been subsoiled or disked rather than plowed.
4. Disease is suppressed by certain herbicides.
5. Throw soil up on the bases of plants during cultivation to promote the growth of adventitious roots.
6. Practice proper fertilizer application. In 1 of 3 years, postemergence damping off is more severe when nitrogen is broadcast before sowing than when nitrogen is applied through overhead irrigation.

Pythium Wilt

Cause. *Pythium aphanidermatum* (syn. *P. butleri)* survives as oospores in soil and infested residue.

Distribution. Wherever common bean is grown.

Symptoms. A water-soaked lesion occurs on the diseased stem near the soil line and progresses upward into the lower branches but rarely extends below the soil line. The outer layer of stem and branch tissue becomes soft and watery and readily separates from fibrous tissue. Leaves wilt and plants quickly die. Death of plants is usually so rapid that wilted leaves are green, then quickly turn brown.

Management. Not necessary since Pythium wilt is infrequently a serious disease.

Rhizoctonia Root Rot and Damping Off

Cause. *Rhizoctonia solani.* Several anastomosis groups occur, depending upon location. AG-4 is the primary causal fungus isolated from dry bean hypocotyls and roots from Zaire, and AG-2-2 is the primary causal fungus from Ohio. AG-1 has also been reported to be pathogenic. AG-1 and AG-2-2 are also pathogenic to common bean and to sugar beet. Foliar isolates were AG-1 IB from Zaire.

Rhizoctonia solani survives as sclerotia in soil and residue and is a vigorous saprophyte in infested residue. Roots or hypocotyls are infected by mycelium growing from a precolonized substrate or by sclerotia germinating and forming infective hyphae. Infection occurs early in the growing season at a maximum temperature of 25° and at low soil moisture. However, there are reports that *R. solani* is most active in warm, moist soil.

Distribution. Wherever common bean is grown.

Symptoms. Preemergence damping off may occur. Seedlings are infected during emergence and become twisted and stunted. Reddish, sunken cankers occur on diseased roots and main stems. If a canker girdles a stem, the seedling will die. Lesions enlarge and the taproot and the stem below ground may be destroyed. Lesions then appear more brown than red. Diseased plants are stimulated to produce numerous adventitious roots near the soil surface. Older diseased plants are stunted and chlorotic.

Management

1. Rotate common bean every 3 to 4 years but avoid rotations with sugar beet.
2. Ridge soil around the base of plants and irrigate soon thereafter to promote adventitious roots.
3. Avoid mechanical injury to seedlings.
4. Plow infested soil to a depth of 20 to 25 cm. This increases plant stand, vine weight, and yield. Root rot, hypocotyl rot, and postemergence damping off are greater in soils that have been subsoiled or disked rather than plowed because plowing reduces the population of *R. solani*.
5. Apply nitrogen fertilizer properly. In 1 of 3 years, postemergence damping off is greater when nitrogen is broadcast before planting than when nitrogen is applied through overhead irrigation.
6. Grow the most resistant bean. Black-seeded cultivars are more resistant than white-seeded cultivars because seed coats of black seeds adhere tightly to cotyledons and protect the germinating seed from infection. Black seed coats also contain phenolic compounds that inhibit growth of *R. solani*. Seed coats of white seeds crack readily before emergence. The resistance of 3-week-old red kidney bean plants to *R. solani* is associated with the inability of the fungus to form infection cushions and penetrate the hypocotyl.

Rust

Cause. *Uromyces appendiculatus* (syn. *U. phaseoli*) is a macrocyclic autoecious rust. In the spring, teliospores germinate to produce basidiospores that are disseminated by wind to bean leaves. Following infection by basidiospores, pycnia are formed on the upper surfaces of diseased leaves and aecia are formed primarily on the under leaf surfaces. Aeciospores form in aecia and infect bean leaves. Uredinia are formed in the aeciospore infections. Urediniospores are formed in uredinia and are disseminated by wind to bean leaves, thereby providing secondary inoculum and functioning as a repeating cycle. The two-spotted spider mite, *Tetranychus urticae*, is attracted to uredinia on older diseased leaves. Populations of mites will be twofold to sixfold greater on diseased leaves than on healthy ones. Mites vector the echinulate urediniospores to rust-free plants. Teliospores are eventually produced in the old uredinia as the diseased plant dies or matures.

Disease development is optimum at temperatures of 17.5° to 22.5°C with 6 to 8 hours of leaf wetness. Several physiological races of *U. appendiculatus* exist.

Distribution. Wherever common bean is grown.

Symptoms. Usually symptoms become most conspicuous later in the growing season, but epiphytotics may occur earlier. Pycnia initially appear only on leaf upper surfaces as chlorotic flecks that later become a slightly raised white bump. Aecia develop first on leaf lower surfaces as white rings or clusters of pustules; later, aecia may occasionally develop on leaf upper surfaces. Uredinia begin as small white spots that enlarge and rupture in a few days to expose orange-yellow urediniospores.

Infected leaves become chlorotic, die, turn brown, and defoliate. Pods and stems may also become diseased. As tissue dies, black telia replace urediniospores in old pustules.

Management
1. Destroy volunteer beans.
2. Rotate common bean with other crops every third year.
3. Do not grow common bean adjacent to previously infected fields.
4. Plow under common bean residue in the autumn.
5. Inspect fields during growing season to detect early symptoms of rust. Apply foliar fungicides if rust develops rapidly from early bloom to 4 weeks before harvest.
6. Grow resistant cultivars adapted to an area. Several pinto cultivars are resistant. Most black turtle and kidney beans are resistant to current rust races. Most navy beans are moderately resistant. Adult resistance in some Jamaican varieties is correlated with mean hair density or leaf pubescence on both leaf surfaces.

Smut

Smut is also called leaf blister smut and Entyloma leaf spot. Although there are disparate reports of smut and Entyloma leaf spot in the literature, the symptom descriptions appear to be similar enough to warrant them as the same disease.

Cause. *Entyloma petuniae* survives in residue.

Distribution. Dominican Republic, Guatemala, and Honduras.

Symptoms. The first or second trifoliate leaves are normally infected first. Initially, spots are water-soaked but later are reported to become a grayish-brown to black discoloration on the leaf upper surfaces and grayish-blue on the leaf lower surfaces. What are described in the literature as grayish-black blisters occur on leaf upper surfaces and contain subepidermal masses of black chlamydospores. Mature necrotic spots are dark brown, which gives leaves a burned appearance. Lesions are somewhat round or oval and sometimes are limited by leaf veins or veinlets but tend to coalesce.

Management
1. Rotate common bean with other crops.
2. Destroy plant residue.
3. Apply foliar fungicides.

Southern Blight

Cause. *Sclerotium rolfsii.*

Distribution. Generally distributed in warm common bean–growing areas.

Symptoms. Brown, water-soaked lesions occur on hypocotyls just beneath the soil line. Disease proceeds into the taproot and eventually destroys the cortex. Lower leaves become chlorotic, followed by plant wilting and death. White mycelium develops around hypocotyls and on the surrounding soil. Spherical, white sclerotia (1–2 mm in diameter) develop in the mycelium and eventually turn brown.

Management
1. Rotate common bean with resistant crops.
2. Grow tolerant or resistant cultivars.

Speckle Disease

Speckle disease is also called Ascochyta leaf spot. However, Ascochyta leaf spot is reported as a separate disease caused by *Phoma exigua* var. *exigua* (syn. *Ascochyta phaseolorum*).

Cause. *Stagonosporopsis hortensis* (syn. *Ascochyta boltshauseri*). The causal fungi are seedborne. Disease is favored by cool temperatures and abundant moisture at elevations above 1500 m.

Distribution. Higher elevations in Central and South America.

Symptoms. Dark gray to black zonate lesions that contain small, black pycnidia occur on leaves. Defoliation occurs in severe cases. Lesions also occur on peduncles, petioles, pods, and stems, where girdling can cause plant death.

Management
1. Rotate common bean with resistant crops.
2. Grow resistant cultivars.
3. Sow disease-free seed.
4. Apply foliar fungicides.

Stem Rot

Cause. An unknown Basidiomycete that resembles *Athelia* spp.

Distribution. The United States (Florida).

Symptoms. Stem lesions are dry, firm, and tan, and do not extend above the soil surface. Severity of stem rot will vary. Frequently, only small lesions or a single lesion will develop on underground stems. However, dry rot may be present on the entire underground portions of stems of plants that wilted and quickly died or still remained alive. Occasionally some plants will have only root rot symptoms but no stem lesions. Mycelial strands form on roots and underground stems of most diseased plants.

Management. Not reported.

Web Blight

Cause. *Rhizoctonia solani,* teleomorph *Thanatephorus cucumeris.* Most isolates causing web blight belong to the intraspecific groups AG-1-IB, including microsclerotial and macrosclerotial types, and AG-2-2. Isolates that cause damping off or root rot are mainly groups AG-3, AG-4, and AG-5.

Fungus survival is by sclerotia in soil and infested residue and by being seedborne. Disease is spread by sclerotia disseminated by wind, rain, machinery, and other means; airborne basidiospores; mycelial bridges between plants; and infested soil debris.

Sclerotia and infested debris splashed onto plant parts by rain serves as a primary inoculum source. Sclerotia germinate under favorable conditions to form mycelia that infect host plants. Trifoliate leaves are infected by splashed inoculum but are infected more frequently by advancing hyphae from infested tissues.

Distribution. Costa Rica and the southern United States; however, web blight is likely more widespread.

Symptoms. Small, round, water-soaked spots are produced on diseased leaves. Spots on leaves are lighter in color than healthy tissue and appear scalded. Spots become tan and are surrounded by a dark border.

Spots on young pods are light tan and irregular-shaped, but spots on mature pods are dark brown and sunken and resemble those of anthracnose. Spots may coalesce to cover the entire pod. Seeds are blemished and discolored. Seedborne fungi are pathogenic to bean seedlings and reduce seedling emergence and establishment.

During moist, warm weather in the final stages of disease, mycelial growth, resembling spider webs, grows over all plant parts, binding leaves, petioles, flowers, and pods together. Numerous small, brown sclerotia are imbedded or scattered throughout the mycelium.

Management
   1. Rotate common bean with resistant plants.
   2. Diseased plants should be destroyed soon after harvest.
   3. Apply a foliar fungicide when disease occurs.
   4. Mulch soil. This helps to prevent dissemination by splashing.
   5. Grow the most tolerant cultivar adapted to an area.

White Leaf Spot

White leaf spot is called mancha bianca in Spanish.

Cause. *Pseudocercosporella albida* (syn. *Cercospora albida)* presumably survives in infested residue. Disease is most severe under cool and wet conditions.

Distribution. Central America and South America.

Symptoms. Initially, whitish, angular spots that are restricted by veins occur first on the undersides of diseased leaves, then on the upper sides of leaves. Spots enlarge, coalesce, and cause defoliation in severe cases. All leaves of susceptible cultivars become diseased, but only the lower leaves of resistant cultivars become diseased.

Management. Grow resistant common bean cultivars.

White Mold

White mold is also called Sclerotinia white mold.

Cause. *Sclerotinia sclerotiorum* (syn. *Whetzelinia sclerotiorum)* survives 5 or more years as sclerotia in soil and up to 3 years as dormant mycelia in testae and cotyledons of seeds. Survival of *S. sclerotiorum* in seed results in failure of most seed to germinate but provides a means of dissemination. Survival of mycelia in infested stems is too low to be a significant source of pathogenic inoculum.

Sclerotia germinate either carpogenically or myceliogenically. Sclerotia germinate carpogenically to produce one or many apothecia, but will not do so until they have been "preconditioned." This preconditioning, or physiological maturation, occurs during the winter or noncrop season; however, freezing is not necessary.

In semiarid areas, carpogenic germination is initiated after the plant canopy covers the soil surface. Apothecia are produced from sclerotia in areas that stay moist for relatively long periods of time, such as the sides of irrigation furrows or next to stems shaded by the plant canopy. Optimum apothecia production occurs at a soil matric potential of –0.25 bars and soil temperatures of 15° to 18°C for 10 to 14 days. Ascospores then are produced in a layer of asci on the surface of apothecia and are disseminated by wind to old blossoms (which commonly serve as an energy source to support infection), stems, or leaves. Insects and splashing water also disseminate *S. sclerotiorum*. Conformation of senescent blossoms may influence ascospore germination through availability of free moisture with nutrients leached from blossoms and held within blossom convolutions. The most common means of disease spread after initial infection is by contact of healthy plant parts with diseased plants.

During moist, humid weather, sclerotia also germinate myceliogenically to produce mycelia that first colonize dead plant tissue, then, using dead tissue as a substrate, invade living plant tissue. However, this is not considered an effective mode of infection.

The disease can occur early in the growing season but becomes most severe later in the season, with epiphytotics occurring only after flowering during wet, humid conditions at an optimum temperature of 25°C. Disease severity is greatest under a dense plant canopy and is reduced early in the season if no plant canopy has developed.

Severe epiphytotics of white bean have occurred in Ontario when the following conditions exist: Apothecia are present, the plant surface is wet longer than 39 hours, the air temperature is 15° to 25°C, and there is a closed canopy containing petals.

Injury from Sclerotinia wilt increases with increasing irrigation in Fusarium-free soils but is negligible in Fusarium-infested soils. *Sclerotinia sclerotiorum* is a pathogen of a large number of plant genera.

Distribution. Wherever common bean is grown.

Symptoms. Symptoms appear on stems, leaves, and pods. Soft, watery, irregular-shaped areas first occur on stems just above the soil line, then on leaflet axils, leaves, and pods. These areas rapidly enlarge and, in a short time, dense, white, cottony masses of mycelia containing small, brown droplets cover the infected spots. Small, hard, black sclerotial bodies appear in the mycelium. Main stems and branches are often encircled by the infection, wilt, and die. The affected plant parts dry out and become bleached and "punky." The white masses of mycelia turn light gray, or brown, and dry out, leaving the conspicuous sclerotia. Infected bean seed is orange-colored and has a chalky texture. Sclerotia are often formed in place of seed.

Management
1. Grow resistant cultivars with open canopy structures.
2. Treat seed with a fungicide seed treatment.
3. Apply foliar fungicide at early bloom stage of growth.
4. Clean out harvesting equipment between fields.
5. Rotate common bean with small grains.

6. Space rows 76 cm or wider to encourage air movement and dry out tissues.
7. Biological control with application of *Coniothyrium minitans* to soil in the spring reduced apothecial production from sclerotia buried in the soil and increased parasitism on sclerotia produced on disease plants.

Diseases Caused by Nematodes

Awl

Cause. *Dolichodorus heterocephalus* favors wet soils.

Distribution. Widespread.

Symptoms. Pit-like lesions occur on the sides of seedling hypocotyls. Diseased roots are stubby and discolored, often with some root destruction.

Management. Not known.

Bulb and Stem

Cause. *Ditylenchus* spp. are endoparasitic nematodes capable of surviving extended periods of time in dry tissue or soil. *Ditylenchus* spp. are reported to be seedborne on other crops.

Distribution. Widespread.

Symptoms. Diseased plants are stunted and stems may be swollen and twisted.

Management. Not known.

Cyst

Cause. *Heterodera glycines.*

Distribution. Widespread.

Symptoms. Beans normally do not display symptoms, even at high nematode populations. Yields may be reduced.

Management. Not reported.

Dagger

Cause. *Xiphinema americanum* is an external parasite.

Distribution. Widespread.

Symptoms. Diseased roots have curly tips with gall-like growths on the tips.

Management. Not reported.

Lance

Cause. *Hoplolaimus* spp. are ectoparasites that feed a distance back from the root tips.

Distribution. Widespread.

Symptoms. Diseased plants are stunted and chlorotic. Necrotic areas occur on roots.

Management. Nematicides are effective on other crops.

Lesion

Cause. *Pratylenchus penetrans* and *P. scribneri.*

Distribution. In warm common bean–growing areas.

Symptoms. Diseased roots are greatly reduced and discolored.

Management
1. Rotate common bean with other crops.
2. Apply nematicides.

Nematode Angular Leaf Spot

Cause. *Aphelenchoides ritzemabosi* survives for up to 27 months in air-dried bean leaf tissue stored at room temperature. In greenhouse inoculations, symptoms developed in about 11 days at 22°C and became more pronounced in 14 to 20 days. Alfalfa is also affected and may provide a mechanism through which the nematode is able to persist.

Distribution. The United States (Wyoming).

Symptoms. Numerous dark, angular lesions occur on leaves of the pinto bean cultivar 'Othello.' An occasional superficial necrosis occurs on the upper surface of petioles.

Management. Do not rotate alfalfa with common beans.

Pin

Cause. *Paratylenchus* spp. are reported to feed either on the outside or the inside of roots.

Distribution. Widespread. Pin nematodes tend to be common in warmer climates.

Symptoms. Plants are not seriously injured. Diseased plants may be stunted and chlorotic. Necrotic spots may occur on roots.

Management. Not reported on bean, but nematicides have been effective on other crops.

Reniform

Cause. *Rotylenchulus reniformis* favors sandy soil. Only females are pathogenic.

Distribution. The United States (Florida). However, *R. reniformis* is widespread on other crops.

Symptoms. Yields are reduced.

Management. Nematicides have been effective in increasing yields.

Ring

Cause. *Criconemoides ovantus* is an ectoparasite.

Distribution. Not known.

Symptoms. Lesions occur on diseased roots.

Management. Nematicides have been effective on other crops.

Root Knot

Cause. *Meloidogyne arenaria, M. hapla, M. incognita,* and *M. javanica.* Infested soil contains eggs that hatch to produce larvae that move through soil to penetrate root tips. Larvae migrate through roots to the vascular tissues, become sedentary, and commence feeding. Excretions of larvae stimulate cell proliferation, resulting in development of syncythia, from which larvae derive food. Females lay eggs that are pushed out of roots into soil.

Distribution. The southern United States and California, primarily on light, sandy soils.

Symptoms. Fleshy, irregular-shaped galls that are larger than nodules and more irregular in shape are produced on roots. Plants are stunted, chlorotic, and unthrifty. Root growth is also suppressed; however, it has been reported that decreases in root and top growth are a temporary effect.

Nematode infestation has been reported to stimulate nodulation; however, nodules remain small and are not efficient in fixing nitrogen. Nitrogenase activity increases but leghaemoglobin decreases. These effects are reported to be temporary, and a few days difference in infestation alters the severity of the effect.

Management
1. Do not grow common bean in infested soil for at least 3 years.
2. Apply a nematicide to soil if economic conditions warrant it.

Spiral

Cause. *Helicotylenchus dihysteria* is an ectoparasite.

Distribution. Widespread, but reported to be more numerous under tropical and subtropical conditions.

Symptoms. Root lesions are present but little decay occurs. Diseased plants are stunted.

Management. Not known.

Sting

Cause. *Belonolaimus longicaudatus.* Disease is favored by sandy soils.

Distribution. Most prevalent in warm, sandy soils in the tropics and semi-tropics.

Symptoms. Yields are reduced from reduced foliage growth. Diseased roots have necrotic lesions.

Management
1. Apply nematicides.
2. Grow a summer cover crop of hairy indigo, *Indigofera hirsuta.*

Stubby Root

Cause. *Paratrichodorus christiei* is an ectoparasite.

Distribution. Not known for certain.

Symptoms. Reduced root growth results in short and thickened roots. Diseased plants are stunted.

Management. Nematicides have been effective on other crops.

Stunt

Cause. *Tylenchorhynchus* spp., *Merlinius* spp., and *Quinisulcius acutus.* Nematodes may be either internal or external parasites.

Distribution. Most prevalent in warm soils. Likely widespread in warmer climates.

Symptoms. Roots are shriveled and sparsely developed and do not elongate. Top growth is reduced.

Management. Nematicides have been effective on other crops.

Diseases Caused by Phytoplasmas

Machismo

Cause. A phytoplasma-like organism that is disseminated by the leafhopper *Scaphytopices fulginosus*. The causal organism is also graft-transmitted but is not mechanically or seed-transmitted. The average period of incubation until symptom expression is 37 days.

Distribution. Colombia.

Symptoms. Symptoms occur at flowering and pod formation. Diseased plants produce wrinkled, distorted pods with no seed. Plants may produce normal-appearing pods, but seeds germinate in the pod. In severe cases, buds proliferate, producing a witches' broom. Plants infected later in the season generally produce normal yields.

Management. Not reported.

Witches' Broom

Cause. A phytoplasma-like organism transmitted by the leafhopper *Orosius orientalis*.

Distribution. Japan.

Symptoms. Symptoms occur 1 month after leafhopper feeding. Plants are chlorotic and leaflet size is reduced. There is a shoot proliferation and phyllody of floral organs.

Management. Not reported.

Diseases Caused by Viruses

Bean Calico Mosaic

Cause. Bean calico mosaic virus (BCMoV) is in the geminivirus group. BCMoV is transmitted by the whitefly, *Bemisia tabaci,* and by mechanical means under greenhouse conditions.

Distribution. Mexico (Sonora).

Symptoms. Diseased leaves attain a bright yellow and green mosaic that resembles a calico pattern. Plants are stunted.

Management. Not reported.

Bean Common Mosaic

Cause. Bean common mosaic virus (BCMV) is in the potyvirus group. Several strains of BCMV are known to exist, and each strain may cause different symptoms. Two serotypes are known and have been suggested as two distinct potyviruses. Serotype A induces a necrosis, and serotype B induces mosaic symptoms. It has further been suggested that isolates of BCMV comprise two distinct potyviruses. On the basis of peptide profiles, McKern et al. (1992) propose that strains of one of the potyviruses be named "bean necrosis mosaic virus" since these strains induce temperature-insensitive necrosis in bean cultivars that carry the dominant I gene. The strains of the other potyvirus would retain the name BCMV. Peptide profile data also suggest that isolates of adzuki bean mosaic virus, blackeye cowpea mosaic virus, dendrobium mosaic virus, peanut stripe virus, and three potyvirus isolates from soybean are all strains of the second potyvirus, BCMV.

Some isolates of BCMV are seedborne, which constitutes the major means of overwintering. BCMV is transmitted mechanically and in a nonpersistent manner by several species of insects, including 11 known species of aphids. BCMV also may be carried in pollen.

Distribution. Wherever common bean is grown.

Symptoms. Symptoms vary between plants and are influenced by cultivar, temperature, light, soil fertility, virus strains, and moisture conditions. General symptoms include dwarfing, excessive branching, leaf cupping, and mottling.

The mosaic pattern may appear as dark bands along major veins with clearing of leaf areas along margins and between leaf veins. Also, the mosaic pattern may be dark green areas varying in size from small granular flecks to large blister-like areas in chlorotic leaves. The darker areas are raised and give leaves a warty, puckered shape. Primary leaves of plants grown from infected seed rarely have a mosaic pattern. Diseased plants growing under low temperatures and low light intensity may not have mosaic symptoms. Severe symptoms (local necrotic lesions on leaves) occur under high light intensity and at temperatures around 26.5°C. Leaves with only granular mottling have little distortion and look normal. Leaves with less pronounced symptoms may be elongated, narrow, and cupped downward.

At high temperatures, a symptom occurs called black root, which is a lethal systemic vascular necrosis resulting from hypersensitivity of some dominant I-gene-bearing resistant bean cultivars to some BCMV serotype A strains. Plants develop a systemic necrosis that causes the vascular system to be necrotic. Initially, young leaflets wilt slightly, then turn brown to black and wilt, followed by the whole plant wilting and dying. The maturity of less severely diseased plants is delayed. Plants grown from infected seed or infected early in plant growth may be spindly, have little growth, and rarely form pods.

Management
1. Sow only certified seed.
2. Grow resistant cultivars.
3. Rogue out infected plants if a small area is involved.

Bean Curly Dwarf

Cause. Bean curly dwarf mosaic virus is in the comovirus group.

Distribution. Central America, North America, and South America.

Symptoms. Diseased leaves have chlorotic local lesions. Ring spots infrequently occur. Depending upon the virus/host combination, necrosis occurs on leaves and stems.

Management. Not reported.

Bean Dwarf Mosaic

Bean dwarf mosaic was formerly called bean chlorotic mottle and infectious chlorosis of Malvaceae.

Cause. Bean dwarf mosaic virus (BDMV) belongs to the geminivirus group and is serologically related to bean golden mosaic, tomato golden mosaic, mung bean yellow mosaic, and African cassava mosaic viruses. BCMV is reservoired in several tropical weeds that serve as sources of inoculum. For example, two species of *Sida* are known to be reservoirs of BDMV. BDMV is transmitted mechanically and by the whitefly, *Bemisia tabaci.*

Distribution. Central America and South America.

Symptoms. Plants infected at the seedling stage are stunted and have malformed or curled and deformed leaves that have a mosaic pattern and chlorotic or yellow patches. Later, diseased plants may display only the variegated patches, which are irregularly distributed in the foliage. Diseased plants generally abort their flowers or produce severely distorted pods. Witches' broom symptoms have often been described as part of the disease syndrome, but these malformations may be the result of mixed virus infections.

Management. Grow resistant or tolerant varieties.

Bean Golden Mosaic

Bean golden mosaic is also called bean golden yellow mosaic.

Cause. Bean golden mosaic virus (BGMV) is in the geminivirus group. BGMV is transmitted mechanically and by the whitefly, *Bemisia tabaci,* but it is not seedborne. BGMV may be reservoired in numerous weeds.

Distribution. Tropical countries in the Western Hemisphere and the United States (Florida). Bean golden mosaic is considered the most devastating disease of beans in Latin America, particularly in Brazil, Mexico, Dominican Republic, and Guatemala.

Symptoms. Diseased plants have an intense yellow appearance. Leaves generally curl downward, and newly emerged leaves have a bright yellow general mosaic pattern but older leaves have a less distinct mosaic pattern. Plants of some cultivars are stunted and have severely malformed pods, which, together with a high incidence of flower abortion, cause yield losses of 100%.

Management
1. Grow tolerant cultivars. No cultivar is immune to BGMV. Some black-seeded cultivars are not as severely affected under moderate BGMV disease pressure.
2. Control insects through application of insecticides.

Bean Mild Mosaic

Cause. Bean mild mosaic virus.

Distribution. Central America and South America.

Symptoms. A mild mosaic occurs on diseased bean leaves.

Management. Not reported.

Bean Pod Mottle

Cause. Bean pod mottle virus is in the comovirus group.

Distribution. The United States.

Symptoms. Diseased leaves have chlorotic local lesions, and ring spots infrequently occur. Symptoms in very susceptible cultivars ranged from mild to severe mosaic and foliar malformation. Depending upon the virus/host combination, necrosis occurs on leaves and stems.

Management. Not reported.

Bean Red Node

Bean red node is also called red node.

Cause. Tobacco streak virus (TSV) is in the ilarvirus group. Two pathotypes exist. Pathotypes I and II belong to distinct serotypes and cause different symptoms. TSV is seed-transmitted and mechanically transmitted. TSV pathotype I is efficiently seed-transmitted and is also apparently transmitted by thrips *(Frankliniella* sp.). Other insects also may be implicated in transmission.

Seed transmission of TSV may depend on early movement into and replication in pollen-associated tissues.

Distribution. The United States (Florida and Washington) and possibly other bean-growing areas.

Symptoms. Pathotype I isolates induce reddening of nodes at the stem, necrosis of primary and trifoliate leaf veins, and sunken reddish lesions on bean pods. Pods are frequently shriveled and deformed. Pathotype II isolates cause a green to yellow mosaic.

Management. Avoid planting near adzuki bean, alfalfa, chickpea, fenugreek, and white sweet clover.

Bean Rugose Mosaic

Cause. Bean rugose mosaic virus (BRMV) is in the comovirus group. BRMV is transmitted mechanically and by the beetles *Cerotoma* spp. and *Diabrotica* spp.

Distribution. Central America and South America.

Symptoms. Leaf symptoms include a light and dark green mosaic pattern that is often accompanied by severe leaf blistering, curling, and malformation that gives the leaf a thickened or leathery appearance. Early infection results in severe stunting. Diseased pods may have a mosaic pattern and be malformed.

Management
1. Grow a resistant cultivar.
2. Sow virus-free seed.

Bean Southern Mosaic

Cause. Bean southern mosaic virus (BSMV) is in the sobemovirus group. BSMV is transmitted by beetles.

Distribution. Warm bean-growing areas.

Symptoms. Diseased foliage has a green mosaic and is rugose.

Management. Resistance has been identified.

Bean Yellow Mosaic

Cause. Bean yellow mosaic virus (BYMV) is in the potyvirus group. BYMV overwinters in perennial plants, such as wild sweet clover; it is transmitted mechanically and by several species of aphids, but the virus is not seed-borne. Temperatures of 28°C and higher in the growth chamber inhibit development of tip necrosis. Several strains of BYMV exist.

Distribution. Wherever common bean is grown.

Symptoms. Symptoms of bean yellow mosaic are more prevalent along borders of fields. Initially, leaflets droop at the point of attachment to the stem, followed by development of angular, pale spots (1.5–3.0 mm in diameter). In some bean cultivars, the pale, angular spots become more pronounced and are bright yellow. Spots may coalesce, which results in general chlorosis. Leaves become thick, cupped, and brittle, which causes them to be malformed and distorted or to have a thickened granular appearance. Some virus strains cause a purpling of the bases of lower leaves. Plants may be dwarfed and bunchy and have delayed maturity and reduced yield. Plants may die.

The necrotic strain of BYMV causes tip necrosis and leaf abscission.

Management
1. Grow resistant cultivars that carry the "I" gene.
2. Eliminate wild sweet clover from fence rows and ditch banks.

Bean Yellow Stipple

Cause. Bean yellow stipple virus (BYSV) is in the bromovirus group and is considered a strain of cowpea chlorotic mottle virus. BYSV is transmitted by the chrysomelid beetles *Cerotoma ruficornis* and *Diabrotica balteata*. BYSV is transferred after an acquisition period of less than 24 hours and is retained for 3 to 6 days by *C. ruficornis* and 1 to 3 days by *D. balteata*.

Distribution. Costa Rica, Cuba, and the United States (Arkansas and Illinois).

Symptoms. Inoculated plants initially display a light yellow mottle that gradually develops into clearly distinct small spots which are unevenly distributed in the trifoliate leaves. Spots may coalesce and form large, irregular-shaped, bright yellow patches. The number and intensity of spots decrease in newer leaves as plants mature. A slight stunting or vein necrosis occurs in some cultivars.

Management. Not necessary since bean yellow stipple is of no economic importance.

Clover Yellow Vein

Cause. Clover yellow vein virus (CYVV) overseasons in perennial clover hosts. CYVV is transmitted mechanically and in a nonpersistent manner by several aphids.

Distribution. Europe and North America.

Symptoms. Diseased foliage has a pronounced yellow mosaic and malformation, and plants have a pronounced stunting. Some cultivars develop apical necrosis, premature defoliation, and wilting, and plants eventually die. Pods are severely distorted and mottled. Frequently, root rot caused by soil-borne pathogens develops.

Management. Grow resistant cultivars.

Cowpea Mild Mottle

Cause. Cowpea mild mottle virus is in the carlavirus group.

Distribution. Tanzania.

Symptoms. Diseased leaves are mottled with chlorotic spots.

Management. Not reported.

Cowpea Severe Mosaic

Cause. Cowpea severe mosaic virus is in the comovirus group.

Distribution. Brazil.

Symptoms. Leaves of diseased plants have vein clearing, mottling, and chlorotic spots. Pods are mottled and distorted.

Management. Not reported.

Cucumber Mosaic

Cause. Cucumber mosaic virus (CMV) is in the cucumovirus group. CMV is seedborne.

Distribution. Europe and the United States.

Symptoms. Leaves have a mosaic appearance.

Management. Not reported.

Curly Dwarf Mosaic Disease

Cause. Bean curly dwarf mosaic virus (BCDMV) is transmitted mechanically, and by the Mexican bean beetle and the spotted cucumber beetle. Fifteen other species of legumes are also susceptible to BCDMV.

Distribution. El Salvador.

Symptoms. Diseased plants are dwarfed. Leaves have a mosaic pattern, are rugose, and curl downward. Inoculated plants have symptoms that range from mild mosaic to lethal top necrosis.

Management. Not reported.

Curly Top

Cause. The curly top virus overwinters in a large number of perennial plants and is transmitted by the leafhopper *Circulifer tenellus*.

Distribution. The western United States.

Symptoms. Plants usually die of curly top if infected in the crookneck stage. If a plant has developed only primary leaves at the time of infection, the growing point will die without further production of trifoliate leaves, followed by yellowing and drying of the whole plant.

Plants infected in early blossom stage drop blooms, become chlorotic and die. If a plant is infected in pod set, new pods will not be set but lightweight and poorly developed seed will mature.

When older plants become infected, the first symptom is a downward cupping of the youngest trifoliate leaf. The growing point then dies or subsequent internodes are shortened, causing plants to be stunted and bushy. Trifoliate leaves formed before infection become chlorotic, thickened, and cup downward. The whole plant becomes brittle, and leaves and the growing point are easily broken off.

Management. Grow resistant cultivars.

Dwarfing Disease

Cause. Soybean dwarf virus (SDV) is transmitted by the aphids *Aulacorthum solani* and *Acyrthosiphon pisum*. Subterranean clover red leaf virus is considered a strain of SDV.

Distribution. Australia (Tasmania), Japan, and New Zealand.

Symptoms. Symptoms are similar to those on soybean. Plants are severely stunted. Leaves are rugose, curl down, and may have interveinal yellowing.

Management. Not reported.

Peanut Mottle

Cause. Peanut mottle virus (PMV) is in the potyvirus group. PMV is transmitted mechanically and in a nonpersistent manner by several species of aphids. Infected peanut seed provides primary inoculum.

Distribution. Australia and the United States (New York and Texas).

Symptoms. Three types of symptoms occur: one in susceptible genotypes and the other two in resistant genotypes. Susceptible genotypes display necrotic lesions and vein necrosis of inoculated unifoliate leaves. Uninoculated trifoliate leaves in the field develop a mosaic, followed by chlorosis and necrosis of veins, petioles, and stems. Plants frequently die.

Resistant genotypes react with necrotic lesions and vein necrosis or with chlorotic lesions that eventually coalesce. In both reactions, the virus remains localized in inoculated leaves.

Few or no seeds are produced on diseased plants.

Management
1. Grow resistant cultivars.
2. Separate bean fields from peanut fields.

Peanut Stunt

Cause. Peanut stunt virus is in the cucumovirus group. For further information, *see* Chapter 12, "Diseases of Peanut."

Distribution. The United States (North Carolina, Tennessee, and Washington).

Symptoms. Diseased plants are stunted, and there is necrosis of growing tips or branches in some cultivars. Leaves are malformed, mottled, and crinkled. Pods are few, small, and malformed and have a few small seeds.

Management. Not reported.

Southern Bean Mosaic

Cause. Southern bean mosaic virus (SBMV), which is in the sobemovirus group, consists of different strains. SBMV is seedborne; the incidence of infection is higher in seeds with cracked coats than in seeds with intact coats. SBMV is also transmitted mechanically, by growing plants in virus-infested soil, and by several beetles. Accumulation of the cowpea strain of southern bean mosaic virus in bean is facilitated by coinfection with sunn-hemp mosaic, a tobamovirus.

Distribution. South America.

Symptoms. Either circular (1–3 mm in diameter), brownish-red local lesions or systemic mottling and vein banding symptoms occur on diseased leaves. Leaves also may be blistered and malformed or crumpled.

Pods have dark green, water-soaked, irregular-shaped blotches. The number and weight of seed may be reduced; however, the number of pods may be greater in diseased than in healthy plants.

Management
1. Sow virus-free seed.
2. Grow resistant cultivars.

Soybean Yellow Shoot

Cause. Soybean yellow shoot virus (VABS) is in the potyvirus group. VABS is mechanically transmitted.

Distribution. Brazil.

Symptoms. Symptoms on inoculated plants vary from interveinal mosaic, severe mosaic, and systemic or vein necrosis to top necrosis, followed by plant death.

Management. Not reported.

Stipple Streak

Cause. Tobacco necrosis virus (TNV) is transmitted in a nonpersistent manner by attachment to the surface of zoospores of the fungus *Olpidium brassicae*. *Olpidium brassicae* is disseminated primarily by splashing rain and irrigation water. Disease is most common on wet soils, particularly soils high in organic matter or peat soils. Dissemination of TNV is by plant-to-plant contact during windy weather and by man. The virus has several strains and may have a satellite virus.

Distribution. Australia, Europe, New Zealand, and the United States (New York).

Symptoms. Symptoms are most severe during warm, cloudy weather but may not occur on leaves developed during hot, sunny weather. Scattered reddish, dark brown, or black spots and streaks or bands occur on stems, petioles, and veins. Variously sized and shaped reticulate necroses of veins may develop on leaves in isolated patches. Such leaves are malformed and often wither and drop. However, tip leaves may remain normal.

Pods of plants may be symptomless or display rusty brown, irregular-shaped, sometimes concentric necrotic lesions. If plants are infected when young, they may become distorted and shrivel. Necrotic lesions may develop on pods after harvest.

Management. Not reported.

Selected References

Abawi, G. S., Potlach, F. J., and Molin, W. T. 1975. Infection of bean by ascospores of *Whetzelinia sclerotiorum*. Phytopathology 65:673-678.

Abawi, G. S., Crosier, D. C., and Cobb, A. C. 1977. Pod flecking of snap beans caused by *Alternaria alternata*. Plant Dis. Rptr. 61:901-905.

Acosta, L. O., Alegria, A., Lastra, R., and Morales, F. 1991. Partial characterization of a Colombian isolate of bean mild mosaic virus from *Phaseolus vulgaris*. (Abstr.). Phytopathology 81:690.

Adegbola, M. O. K., and Hagedorn, D. J. 1969. Symptomatology and epidemiology of Pythium bean blight. Phytopathology 59:1113-1118.

Anderson, F. N., Steadman, J. R., Coyne, D. P., and Schwartz, H. F. 1974. Tolerance to white mold in *Phaseolus vulgaris* dry edible bean types. Plant Dis. Rptr. 58:782-784.

Balardin, R. S., Jarosz, A. M., and Kelly, J. D. 1997. Virulence and molecular diversity of

Colletotrichum lindemuthianum from South, Central, and North America. Phytopathology 87:1184-1191.

Batra, L. R., and Stavely, J. R. 1994. Attraction of two-spotted spider mite to bean rust uredinia. Plant Dis. 78: 282-284.

Bernier, C. 1975. Diseases of pulse crops and their control. In Oilseed and Pulse Crops in Western Canada, J. T. Harapick (ed.). Modern Press, Saskatoon, Sask.

Blad, B. L., Steadman, J. R., and Weiss, A. 1978. Canopy structure and irrigation influence white mold disease and microclimate of dry edible beans. Phytopathology 68:1431-1437.

Boland, G. J., and Hall, R. 1987. Epidemiology of white mold of white bean in Ontario. Can. J. Plant Pathol. 9:218-224.

Brown, J. K., Jimenez-Garcia, E., and Nelson, M. R. 1988. Bean calico mosaic virus, a newly described geminivirus of bean. (Abstr.) Phytopathology 78:1579.

Burke, D. W., and Miller, D. E. 1983. Control of Fusarium root rot with resistant beans and cultural management. Plant Dis. 67:1312-1317.

Burke, D. W., Miller, D. E., Holmes, L. D., and Barker, A. W. 1972. Counteracting bean root rot by loosening the soil. Phytopathology 62:306-309.

Caesar, A. J., and Pearson, R. C. 1983. Environmental factors affecting survival of ascospores of *Sclerotinia sclerotiorum.* Phytopathology 73:1024-1030.

Cafati, C. R., and Saettler, A. W. 1980. Transmission of *Xanthomonas phaseoli* in seed of resistant and susceptible *Phaseolus* genotypes. Phytopathology 70:638-640.

Campbell, C. L., and Steadman, J. R. 1985. The relationship of *Sclerotinia sclerotiorum* ascospore germination on bean blossoms to disease development in a bean canopy microclimate. (Abstr.) Phytopathology 75:1369.

Claflin, L. E., Stuteville, D. L., and Armbrust, D. V. 1973. Windblown soil in the epidemiology of bacterial leaf spot of alfalfa and common blight of bean. Phytopathology 63:1417-1419.

Cook, G. F., Steadman, J. R., and Boosalis, M. G. 1975. Survival of *Whetzelinia sclerotiorum* and initial infection of dry edible beans in western Nebraska. Phytopathology 65:250-255.

Correa, F. J., and Saettler, A. W. 1987. Angular leaf spot of red kidney beans in Michigan. Plant Dis. 71:915-918.

Coyne, D. P., and Schuster, M. L. 1970. 'Jules,' a great northern dry bean variety tolerant to common blight bacterium *(Xanthomonas phaseoli).* Plant Dis. Rptr. 54:557-559.

Cupertino, F. P., Costa, C. L., Lin, M. T., and Kitajimi, E. W. 1982. Infeccas natural do feyoeiro *(Phaseolus vulgaris* L.) Pelo virus do mosaico severs do feijao macassar. Fitopatologia Brasileria 7:275-283 (in Portugese).

Daub, M. E., and Hagedorn, D. J. 1981. Epiphytic populations of *Pseudomonas syringae* on susceptible and resistant bean lines. Phytopathology 71:547-550.

Dean, R. A., and Timberlake, W. E. 1977. *Aspergillus nidulans:* A potential plant pathogen? (Abstr.) Phytopathology 77:1692.

Dhingra, O. D. 1978. Internally seedborne *Fusarium semitectum* and *Phomopsis* sp. affecting dry and snap bean seed quality. Plant Dis. Rptr. 62:509-512.

Dhingra, O. D., and Kushalappa, A. C. 1980. No correlation between angular leaf spot intensity and seed infection in beans by *Isariopsis griseola.* Fitopaathologia Brasileria 5:149-152.

Dickson, M. H., and Abawi, G. S. 1974. Resistance to *Pythium ultimum* in white-seeded beans *(Phaseolus vulgaris).* Plant Dis. Rptr. 58:774-776.

Dillard, H. R., and Cobb, A. C. 1993. Survival of *Colletotrichum lindemuthianum* in bean debris in New York State. Plant Dis. 77:1233-1238.

Echandi, E., and Hebert, T. T. 1970. An epiphytotic of stunt in beans incited by the peanut stunt virus in North Carolina. Plant Dis. Rptr. 54:183-184.

Echandi, E., and Hebert, T. T. 1971. Stunt of beans incited by peanut stunt virus. Phytopathology 61:328-330.

Engelkes, C. A., and Windels, C. E. 1996. Susceptibility of sugar beet and beans to *Rhizoctonia solani* AG-2-2 IIIB and AG-2-2 IV. Plant Dis. 80:1413-1417.

Fahy, P. C., and Persley, G. J. 1983. Plant Bacterial Diseases: A Diagnostic Guide. Academic Press, Inc. New York, NY.

Farr, D. F., Bills, G. R., Chamuris, G. P., and Rossman, A. Y. 1989. Fungi on Plants and Plant Products in the United States. American Phytopathological Society, St. Paul, MN. 1252 pp.

Franc, G. D., and Beaupre, C. M.-S. 1993. A new disease of pinto bean caused by *Aphelenchoides ritzemabosi* in Wyoming. Plant Dis. 77:1168.

Franc, G. D., Beaupre, C. M.-S, Gray, F. A., and Hall, R. D. 1996. Nematode angular leaf spot of dry bean in Wyoming. Plant Dis. 80:476-477.

Fuentes, A. L., and Hamilton, R. I. 1991. Sunn-hemp mosaic virus facilitates cell-to-cell spread of southern bean mosaic virus in a nonpermissive host. Phytopathology 81:1302-1305.

Galindo, J. J., Abawi, G. S., and Thurston, H. D. 1982. Variability among isolates of *Rhizoctonia solani* associated with snap bean hypocotyls and soils in New York. Plant Dis. 66:390-394.

Galindo, J. J., Abawi, G. S., Thurston, H. D., and Galvez, G. E. 1982. Source of inoculum and development of web-blight of beans in Costa Rica. (Abstr.) Phytopathology 72:170.

Galvez, G. E., Mora, B., and Alfaro, R. 1984. Integrated control of web blight of beans (*Phaseolus vulgaris*). (Abstr.) Phytopathology 74:1015.

Gilbertson, R. L., Carlson, E., Rand, R. E., Hagedorn, D. J., and Maxwell, D. P. 1986. Survival of *Xanthomonas campestris* pv. *phaseoli* in bean stubble and debris with three tillage systems. (Abstr.) Phytopathology 76:1078.

Gilbertson, R. L., Rand, R. E., and Hagedorn, D. J. 1990. Survival of *Xanthomonas campestris* pv. *phaseoli* and pectoytic strains of *X. campestris* in bean debris. Plant Dis. 74:322-327.

Godoy-Lutz, G., Arias, J., Steadman, J. R., and Eskridge, K. M. 1996. Role of natural seed infection by the web blight pathogen in common bean seed damage, seedling emergence, and early disease development. Plant Dis. 80:887-890.

Goode, M. J., Hagedorn, D. J., and Cross, J. E. 1985. Occurrence of snap bean bacterial blight pathogens on wild legumes. (Abstr.) Phytopathology 75:1287.

Granada, G. A. 1979. Machismo, a new disease of beans in Colombia. (Abstr.) Phytopathology 69:1029.

Groth, J. V., and Mogen, B. D. 1978. Completing the life cycle of *Uromyces phaseoli* var. *typica* on bean plants. Phytopathology 68:1674-1677.

Hagedorn, D. J., and Binning, L. K. 1982. Herbicide suppression of bean root and hypocotyl rot in Wisconsin. Plant Dis. 66:1187-1188.

Hagedorn, D. J., and Rand, R. E. 1979. Development of processing type beans (*Phaseolus vulgaris*) resistant to bacterial brown spot (*Pseudomonas syringae*). (Abstr.) Phytopathology 69:1030.

Hagedorn, D. J., and Rand, R. E. 1986. Development and release of WIS (MDR) 147 bean breeding line. (Abstr.) Phytopathology 76:1067.

Hagedorn, D. J., Rand, R. E., and Saad, S. M. 1972. *Phaseolus vulgaris* reaction to *Pseudomonas syringae*. Plant Dis. Rptr. 56:325-328.

Hall, R. 1991. Compendium of Bean Diseases. American Phytopathological Society, St. Paul, MN. 73 pp.

Hampton, R. O., Silbernagel, M. J., and Burke, D. W. 1983. Bean common mosaic virus strains associated with bean mosaic epidemics in the northwestern United States. Plant Dis. 67:658-661.

Hanson, P. M., Pastor-Corrales, M. A., and Kornegay, J. L. 1993. Heritability and sources of Ascochyta blight resistance in common bean. Plant Dis. 77:711-714.

Heimann, M. G., Stevenson, W. R., and Rand, R. E. 1989. *Bipolaris sorokiniana* found causing lesions on snap bean in Wisconsin. Plant Dis. 73:701.

Hoch, H. C., Hagedorn, D. J., Pinnow, D. L., and Mitchell, J. E. 1975. Role of *Pythium* spp. as incitants of bean root and hypocotyl rot in Wisconsin. Plant Dis. Rptr. 59:443-447.

Howard, C. M., Conway, K. E., and Albregts, E. E. 1977. A stem rot of bean seedlings caused by a sterile fungus in Florida. Phytopathology 67:430-433.

Hunter, J. E., Dickson, M. H., Boettger, M. A., and Cigna, J. A. 1982. Evaluation of plant introductions of *Phaseolus* spp. for resistance to white mold. Plant Dis. 66:320-322.

Hutton, D. G., Wilkinson, R. E., and Mai, W. F. 1973. Effect of two plant parasitic nematodes on Fusarium dry root rot of beans. Phytopathology 63:749-751.

Johnson, K. B., and Powelson, M. L. 1983. Influence of prebloom disease establishment by *Botrytis cinerea* and environmental and host factors on gray mold pod rot of snap bean. Plant Dis. 67:1198-1202.

Kaiser, W. J., Wyatt, S. D., and Klein, R. E. 1991. Epidemiology and seed transmission of two tobacco streak virus pathotypes associated with seed increases of legume germplasm in eastern Washington. Plant Dis. 75:258-264.

Katherman, M. J., Wilkinson, R. E., and Beer, S. V. 1980. The potential of four dry bean cultivars to serve as sources of *Pseudomonas phaseolicola* inoculum. Plant Dis. 64:72-74.

Kelly, J. D., Afanador, L., and Cameron, L. S. 1994. New races of *Colletotrichum lindemuthianum* in Michigan and implications in dry bean resistance breeding. Plant Dis. 78:892-894.

Kobriger, K., and Hagedorn, D. J. 1984. Additional *Pythium* species associated with bean rot complex in Wisconsin's Central Sands. Plant Dis. 68:595-596.

Lanter, J. M., Pastor Corrales, M. A., and Hancock, J. G. 1977. Progress of angular leaf spot of bean grown in monocultures and in bean-maize intercrops. (Abstr.) Phytopathology 77:1699.

Legard, D. E., and Schwartz, H. F. 1987. Sources and management of *Pseudomonas syringae* pv. *phaseolicola* and *Pseudomonas syringae* pv. *syringae* epiphytes on dry beans in Colorado. Phytopathology 77:1503-1509.

Lewis, J. A., Lumsden, R. D., Papavizas, G. C., and Kantzes, J. G. 1983. Integrated control of snap bean disease caused by *Pythium* spp. and *Rhizoctonia solani*. Plant Dis. 67:1241-1244.

Mabagala, R. B., and Saettler, A. W. 1992. *Pseudomonas syringae* pv. *phaseolicola* populations and halo blight severity in beans grown alone or intercropped with maize in northern Tanzania. Plant Dis. 76:687-692.

Mabagala, R. B., and Saettler, A. W. 1992. Races and survival of *Pseudomonas syringae* pv. *phaseolicola* in northern Tanzania. Plant Dis. 76:678-682.

Mabagala, R. B., and Saettler, A. W. 1992. The role of weeds in survival of *Pseudomonas syringae* pv. *phaseolicola* in northern Tanzania. Plant Dis. 76:683-687.

McFadden, W. R., Hall, R., and Phillips, L. G. 1989. Relation of initial inoculum density to severity of Fusarium root rot of white bean in commercial fields. Can. J. Plant Pathol. 11:122-126.

McKern, N. M., Mink, G. I., Barnett, O. W., Mishra, A., Whittaker, L. A., Silbernagel, M. J., Ward, C. W., and Shukla, D. D. 1992. Isolates of bean common mosaic virus comprising two distinct potyviruses. Phytopathology 82:923-929.

McLaren, D. L., Huang, H. C., and Rimmer, S. R. 1996. Control of apothecial production of *Sclerotinia sclerotiorum* by *Coniothyrium minitans* and *Talaromyces flavus*. Plant Dis. 80:1373-1378.

Meiners, J. P., Waterworth, H. E., Lawson, R. H., and Smith, F. F. 1977. Curly dwarf mosaic disease of beans from El Salvador. Phytopathology 67:163-168.

Miller, D. E., and Burke, D. W. 1986. Reduction of Fusarium root rot and Sclerotinia wilt in beans with irrigation, tillage and bean genotype. Plant Dis. 70:163-166.

Morales, F. J., and Castano, M. 1985. Effect of a Colombian isolate of bean southern mosaic virus on selected yield components of *Phaseolus vulgaris*. Plant Dis. 69:803-804.

Morales, F. J., and Niessen, A. I. 1988. Comparative responses of selected *Phaseolus vulgaris* germplasm inoculated artificially and naturally with bean golden mosaic virus. Plant Dis. 72:1020-1023.

Morales, F. J., and Niessen, A. I. 1988. Isolation and partial characterization of bean dwarf mosaic virus. (Abstr.) Phytopathology 78: 858.

Morales, F., Niessen, A., Ramirez, B., and Castano, M. 1990. Isolation and partial characterization of a geminivirus causing bean dwarf mosaic. Phytopathology 80:96-101.

Muckel, R. D., and Steadman, J. R. 1981. Dissemination and survival of *Sclerotinia sclerotiorum* in bean fields in western Nebraska. (Abstr.) Phytopathology 71:244.

Natti, J. J. 1971. Epidemiology and control of bean white mold. Phytopathology 61:669-674.

Ntahimpera, N., Dillard, H. R., Cobb, A. C., and Seem, R. C. 1997. Influence of tillage practices on anthracnose development and distribution in dry bean fields. Plant Dis. 81:71-76.

Olaya, G., and Abawi, G. S. 1996. Effect of water potential on mycelial growth and on production and germination of sclerotia of *Macrophomina phaseolina*. Plant Dis. 80:1347-1350.

Pastor-Corrales, M. A., and Abawi, G. S. 1988. Reactions of selected bean accessions to infection by *Macrophomina phaseolina*. Plant Dis. 72:39-41.

Patel, P. N., et. al. 1964. Bacterial brown spot of bean in central Wisconsin. Plant Dis. Rptr. 48:335-337.

Perry, V. G. 1953. The awl nematode, *Dolichodorus heterocephalus*, a devastating plant parasite. Proc. Helminth. Soc. Wash. 20:21-27.

Pfender, W. F., and Hagedorn, D. J. 1981. Aphanomyces root and stem rot of snap beans. (Abstr.) Phytopathology 71:250.

Pfender, W. F., and Hagedorn, D. J. 1982. *Aphanomyces euteiches* f. sp. *phaseoli*, a causal agent of bean root and hypocotyl rot. Phytopathology 72:306-310.

Pieczarka, D. J., and Abawi, G. S. 1978. Effect of interaction between *Fusarium, Pythium,* and *Rhizoctonia* on severity of bean root rot. Phytopathology 68:403-408.

Pieczarka, D. J., and Abawi, G. S. 1978. Influence of soil water potential and temperature on severity of Pythium root rot of snap beans. Phytopathology 68:766-772.

Piepenbring, M., and Bauer, R. 1997. *Erratomyces,* a new genus of Tilletiales with species on Leguminosae. Mycologia 89:924-936.

Polach, F. J., and Abawi, G. S. 1975. The occurrence and biology of *Botryotinia fuckeliana* on beans in New York. Phytopathology 65:657-660.

Prasad, K., and Weigle, J. L. 1976. Association of seed coat factors with resistance to *Rhizoctonia solani* in *Phaseolus vulgaris.* Phytopathology 66:342-345.

Raid, R. N., McDaniel, L. L., Nuessly, G. S., and Scully, B. T. 1995. First report of bean red node in the eastern United States. Plant Dis. 79:539.

Reddy, M. S., and Patrick, Z. A. 1989. Effect of host, nonhost and fallow soil on populations of *Thielaviopsis basicola* and severity of black root rot. Can. J. Plant Pathol. 11:68-74.

Reeleder, R. D., and Hagedorn, D. J. 1981. Inheritance of resistance to *Pythium myriotylum* hypocotyl rot in *Phaseolus vulgaris* L. Plant Dis. 65:427-429.

Ribeiro, R. De L. D., Hagedorn, D. J., Durbin, R. D., and Vchytil, T. F. 1979. Characterization of the bacterium inciting bean wildfire in Brazil. Phytopathology 69:208-212.

Saad, S., and Hagedorn, D. J. 1969. Symptomatology and epidemiology of Alternaria leaf spot of bean *Phaseolus vulgaris.* Phytopathology 59:1530-1533.

Saettler, A. W., and Correa, F. J. 1987. Transmission of *Phaeoisariopsis griseola* in bean seed. (Abstr.) Phytopathology 77:1707.

Saettler, A. W., and Stadt, S. J. 1981. Internal seed infection by *Pseudomonas phaseolicola* in susceptible and tolerant bean cultivars. (Abstr.) Phytopathology 71:252.

Saettler, A. W., Cafati, C. R., and Weller, D. M. 1986. Nonoverwintering of Xanthomonas bean blight bacteria in Michigan. Plant Dis. 70:285-287.

Salgado, M. O., and Schwartz, H. F. 1993. Physiological specialization and effects of inoculum concentration of *Fusarium oxysporum* f. sp. *phaseoli* on common beans. Plant Dis. 77:492-496.

Schieber, E., and Zentmeyer, G. A. 1971. A new bean disease in the Caribbean area. Plant Dis. Rptr. 55:207-208.

Schuster, M. L., Blatchford, G. J., and Schuster, A. M. 1980. A new bacterium, *Pseudomonas blatchfordae* nov. sp., pathogenic for bean. Fitopathologia Brasileria 5:283-297.

Schuster, M. L., Schuster, A. M., and Nuland, D. S. 1981. A new bacterium pathogenic for beans (*Phaseolus vulgaris* L.). Fitopatologia Brasileria 6:345-358.

Schwartz, H. F., and Steadman, J. R. 1978. Factors affecting sclerotium populations of, and apothecium production by, *Sclerotinia sclerotiorum.* Phytopathology 68:383-388.

Schwartz, H. F., Galvez, G. E., Schoonhoven, A. Van, Howeler, R. H., Graham, P. II., and Flor, C. 1978. Field Problems of Beans in Latin America. Centro Internacional de Agricultura Tropical, Cali, Colombia. 136 pp.

Schwartz, H. F., Katherman, M. J., and Thung, M. D. T. 1981. Yield response and resistance of dry beans to powdery mildew in Colombia. Plant Dis. 65:737-738.

Shaik, M. 1985. Race-nonspecific resistance in bean cultivars to races of *Uromyces appendiculatus* var. *appendiculatus* and its correlation with leaf epidermal characteristic. Phytopathology 75:478-481.

Shukla, D. D., McKern, N. M., Ward, C. W., and Ford, R. E. 1991. Molecular parameters suggest that cowpea aphid-borne mosaic virus is a distinct potyvirus and that bean common mosaic virus consists of at least three distinct potyviruses. (Abstr.) Phytopathology 81:1166.

Silbernagel, M. J., and Mills, L. J. 1990. Genetic and cultural control of Fusarium root rot in bush snap beans. Plant Dis. 74:61-66.

Sippell, E. W., and Hall, R. 1981. Effects of *Fusarium* and *Pythium* on yield components of white bean. (Abstr.) Phytopathology 71:564.

Skiles, R. L., and Cardona-Alvarez, C. 1959. Mancha gris, a new leaf disease of bean in Colombia. Phytopathology 49:133-135.

Steadman, J. R. 1983. White mold, a serious yield-limiting disease of bean. Plant Dis. 67:346-350.

Steadman, J. R., Kerr, E. D., and Mumm, R. F. 1975. Root rot of bean in Nebraska: Primary pathogen and yield loss appraisal. Plant Dis. Rptr. 59:305-308.

Stockwell, V., and Hanchey, P. 1984. The role of the cuticle in resistance of beans to *Rhizoctonia solani*. Phytopathology 74:1640-1642.

Sumner, D. R., Smittle, D. A., Threadgill, E. D., Johnson, A. W., and Chalfant, R. B. 1986. Interactions of tillage and soil fertility with root diseases in snap bean and lima bean in irrigated multiple-cropping systems. Plant Dis. 70:730-735.

Teakle, D. S., and Moris, T. J. 1981. Transmission of southern bean mosaic virus from soil to bean seeds. Plant Dis. 65:599-600.

Tu, J. C. 1981. Anthracnose (*Colletotrichum lindemuthianum*) on white bean (*Phaseolus vulgaris* L.) in southern Ontario: Spread of the disease from an infection focus. Plant Dis. 65:477-480.

Tu, J. C. 1981. Etiology of pod and seed coat discoloration of white beans. (Abstr.) Phytopathology 71:909.

Tu, J. C. 1983. Epidemiology of anthracnose caused by *Colletotrichum lindemuthianum* on white bean (*Phaseolus vulgaris*) in southern Ontario: Survival of the pathogen. Plant Dis. 67:402-404.

Tu, J. C. 1984. Biology of *Alternaria alternata*, the causal fungus of black pod disease of white beans in southwestern Ontario. (Abstr.) Phytopathology 74:820.

Tu, J. C. 1985. Pink pod rot of bean caused by *Trichothecium roseum*. Can. J. Plant Pathol. 7:55-57.

Tu, J. C. 1988. Control of bean anthracnose caused by the delta and lambda races of *Colletotrichum lindemuthianum* in Canada. Plant Dis. 72:5-8.

Tu, J. C. 1988. The role of white mold-infected white bean (*Sclerotinia sclerotiorum* (Lib.) de Bary). J. Phytopathology 121:40-50.

Tu, J. C. 1989. The role of temperature and inoculum concentration in the development of tip necrosis and seedling death of beans infected with bean yellow mosaic virus. Plant Dis. 73:405-407.

Tu, J. C. 1990. First report of anthracnose caused by *Glomerella cingulata* on white beans in Ontario, Canada. Plant Dis. 74:394.

Tu, J. C., Sheppard, J. W., and Laidlaw, D. M. 1984. Occurrence and characterization of the epsilon race of bean anthracnose in Ontario. Plant Dis. 68:69-70.

Van Bruggen, A. H. C., Arneson, P. A., and Whalen, C. H. 1983. Emergence of dry beans as affected by *Rhizoctonia solani*, soil moisture and temperature. (Abstr.) Phytopathology 73:1347.

Van Bruggen, A. H. C., Whalen, C. H., and Arneson, P. A. 1986. Emergence, growth and development of dry bean seedlings in response to temperature, soil moisture and *Rhizoctonia solani*. Phytopathology 76:568-572.

Verdejo, S., Green, C. D., and Podder, A. K. 1988. Influence of *Meloidogyne incognita* on nodulation and growth of pea and black bean. Nematologica 34:88-97.

Walter, M. H., Kaiser, W. J., Klein, R. E., and Wyatt, S. D. 1992. Association between tobacco streak ilarvirus seed transmission and anther tissue infection in bean. Phytopathology 82:412-415.

Webster, D. M., Atkin, J. D., and Cross, J. E. 1983. Bacterial blights of snap beans and their control. Plant Dis. 67:935-940.

Webster, D. M., Temple, S. R., and Galvez, G. 1983. Expression of resistance to *Xanthomonas campestris* pv. *phaseoli* in *Phaseolus vulgaris* under tropical conditions. Plant Dis. 67:394-396.

Weller, D. M., and Saettler, A. W. 1980. Evaluation of seedborne *Xanthomonas phaseoli* and *X. phaseoli* var. *fuscans* as primary inocula in bean blights. Phytopathology 70:148-152.

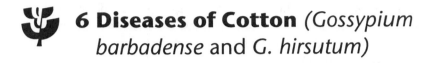

6 Diseases of Cotton (*Gossypium barbadense* and *G. hirsutum*)

Diseases Caused by Bacteria

Agrobacterium Root Rot and Wilt

Cause. *Agrobacterium* biovar 1.

Distribution. The United States (Arkansas, Louisiana, North Carolina, Tennessee, and Texas).

Symptoms. Symptoms are characterized by "copper top" and sudden wilt of diseased plants. Losses up to 50% have been reported.

Management. Not reported.

Bacterial Blight

Bacterial blight is also called angular leaf spot, and black arm.

Cause. *Xanthomonas campestris* pv. *malvacearum* (syn. *Bacterium malvacearum, Phytomonas malvacearum, Pseudomonas malvacearum, X. campestris,* and *X. malvacearum*) survives on seed lint, in seed, in residue left in the field after harvest, and in buds of symptomless plants. In arid and semiarid areas, survival occurs in soil residue consisting of stems and bolls. *Xanthomonas campestris* pv. *malvacearum* is disseminated by wind, water, insects, animals, and machinery, and on seed. Seed is the most important mode of long-distance spread. Volunteer plants that grow from infected seed often have infected cotyledons, which serve as a source of inoculum.

The optimum conditions for the entry of bacteria into plants through stomata and wounds are temperatures of 30° to 36°C and abundant moisture, or relative humidities of 85% or more. Boll rot frequently occurs during hot, humid weather when numerous insects may puncture or wound bolls. Seed becomes infected if excess moisture is present at harvest time.

Different physiologic races of *X. campestris* pv. *malvacearum* are present in the field.

Distribution. Cosmopolitan; bacterial blight occurs wherever cotton is grown.

Symptoms. Initially, green, round to elongate, water-soaked, translucent spots of different sizes appear on lower cotyledon surfaces 7 to 10 days after sowing. Eventually, infections penetrate to the upper sides of leaves as angular or irregular-shaped, brown to black lesions that are between veins and are along the main leaf veins and petioles. Diseased areas dry up, become sunken, and are reddish brown with a brownish to purplish margin. Leaf defoliation may occur.

Lesions on seedling hypocotyls are black, elongated cankers that may girdle and kill the plant. This phase of the disease is called "black arm." Similar lesions may also occur and girdle stems of older plants, causing death of the distal portion of the diseased plants.

Bolls have round to irregular-shaped or angular, black, sunken spots. Boll-rotting fungi may invade bacterial lesions and discolor or destroy bolls. A creamy, bacterial exudate occurs on leaf lesions and bolls during high humidity or after rain. Dried exudate will appear as a sugar-like coating or film on a dark lesion.

Management
1. Grow resistant cultivars.
2. Acid delint seed.
3. Rotate cotton for at least 1 year with a resistant crop.
4. Plow under residue immediately after harvest.
5. Sow seed from disease-free plants.

Bacterial Lint Discoloration

Cause. *Pseudomonas* sp. survives in soil and is introduced into bolls by insects.

Distribution. The United States.

Symptoms. Fiber is discolored and fluoresces yellow to green under ultraviolet light (366 nm).

Management. Not reported.

Crown Gall

Cause. *Agrobacterium radiobacter* var. *tumefaciens* and new biotype 2. Infection is enhanced by the root knot nematode, *Meloidogyne incognita*, but bacteria can penetrate root tissues without nematode involvement.

Distribution. Israel and the United States. Crown gall on cotton is rare.

Symptoms. Plants are generally stunted; the galls on stunted taproots are 1 to

4 cm long, irregular-shaped, and dark brown. Sometimes spherical-shaped galls approximately 1 cm in diameter are found on lateral roots.

Management. Not practiced because of the infrequency of crown gall.

Erwinia Internal Necrosis

Cause. *Erwinia herbicola.* Necrosis depends on development of a secondary internal boll that splits the placentae and causes an opening to the outside of the boll and intertwining of fibers of affected locules. This allows the bacteria to enter the boll, where it spreads from the interplacentae space into locules. *Erwinia herbicola* may have some association with the stinkbug *Euschistus impictiventris.*

Distribution. The United States (California).

Symptoms. Immature cotton bolls have reddish-brown necrotic tissue. Diseased locules are soft and slimy. Affected seed coats are discolored and seed contents are completely decayed. The lint of diseased mature open bolls is discolored tan, and the locules are compact.

Management. Not reported.

Pantoea agglomerans Cotton Lint Rot

This disease had been associated with abnormal boll morphology previously described as supernumerary carpel syndrome (SCS).

Cause. *Pantoea agglomerans* is a gram-negative bacterium with yellow pigmentation. Previously, *P. agglomerans* was reported to cause lint rot of cotton locules associated with SCS and injury caused by the stinkbug *Euschistus impictiventris.*

Distribution. The United States (California and Georgia).

Symptoms. Cotton bolls contain internal lint rot within single or multiple locules. Internal lint damage occurs in the tips of bolls that have formed abnormal fissures or openings nearest the sutures. Such bolls often contain an additional locule. Rotten lint within the locules, which is usually discolored a reddish brown, is mixed with diseased yellow lint and olive to olive-brown lint.

Management. Not reported.

Pseudomonas Leaf Spot

Cause. *Pseudomonas syringae.* Disease is favored by wet weather.

Distribution. The United States (Texas).

Symptoms. Symptoms occur on cotton seedlings. Circular spots (2–4 mm in diameter) consist of tan necrotic centers surrounded by dark brown to purple borders. Large, irregular-shaped, necrotic areas appear on severely diseased leaves. Symptoms are reported to be similar to those of Ascochyta blight.

Management. Not reported.

Diseases Caused by Fungi

Aerolate Mildew

Aerolate mildew is also called dahiya disease, false mildew, frosty mildew, gray mildew, moho blanco, and white mold.

Cause. *Mycosphaerella areola*, anamorph *Ramularia gossypii* (syn. *Cercosporella gossypii* and *R. areola*). *Mycosphaerella areola* survives as mycelia, conidia, or perithecia in infested residue. Maximum infection occurs at temperatures of 20° to 30°C, together with nightly wetting and daily drying of leaf tissues. Conidia are produced in lesions and are disseminated through the air to provide secondary inoculum. Disease is most severe under wet, humid conditions.

Distribution. Generally distributed, but gray mildew is most common in Africa and Asia.

Symptoms. Symptoms normally appear toward the end of the growing season. Cotyledons initially have circular, water-soaked, dark green spots, which eventually become chlorotic, wither, and turn reddish brown.

Typical symptoms on the upper leaf surfaces are small, light green to yellowish angular spots that are limited by the veins. During high humidity, the undersides of diseased leaves are covered with a white growth consisting of mycelia, conidia, and conidiophores of *M. areola*. Severely diseased leaves become chlorotic, dry, curly, and reddish brown, and defoliate prematurely. Bolls may open prematurely, resulting in loss of fiber quality.

Resistant cultivars have tiny, reddish-brown spots that are surrounded by a chlorotic area.

Management
1. Grow resistant cultivars.
2. Apply foliar fungicides.

Alternaria Leaf Spot

Cause. *Alternaria gossypina* survives as mycelium in infested residue. Conidia are produced from mycelium during moist conditions and are disseminated by wind to leaves. Conidia produced in lesions on diseased leaves and disseminated to other host plants during the growing season are secondary inoculum. *Alternaria gossypina* may infect cotton after other pathogens have caused disease, or plants are predisposed to infection by weak pathogens, such as *A. gossypina,* while under stress from low fertility or unfavorable environmental conditions.

Distribution. Wherever cotton is grown under warm, moist conditions.

Symptoms. Symptoms occur on young plant parts (such as cotyledons, seedlings, petioles, and stems) and on mature leaves. Spots are brown, have a "papery" consistency, and are roughly circular (average diameter of 13 mm) with concentric rings that give them a target-like appearance. During moist weather, spots become black because of the production of fungal conidia. Severely diseased leaves shrivel and defoliate prematurely, especially during periods of moist weather later in the growing season.

Management
1. Treat seed with a fungicide seed protectant.
2. Maintain proper soil fertility.
3. Avoid insect and mechanical injury.
4. Plow under residue.
5. Grow late-maturing cultivars. Early-maturing cultivars that fruit heavily are more susceptible to Alternaria leaf spot than cultivars that mature later in the growing season.

Alternaria Stem Blight and Leaf Spot

Alternaria stem blight and leaf spot is also called Alternaria leaf spot.

Cause. *Alternaria alternata* and *A. macrospora. Alternaria macrospora* survives as mycelium in alternative hosts, seed, and infested leaf residue. Primary inoculum consists of conidia produced from the overwintered mycelium and disseminated by wind to host plants.

Generally, cotton is relatively resistant to infection at the beginning of the growing season. At this stage of plant growth, the cotyledons are infected because they are more susceptible than true leaves. Once flowering begins, host plants become progressively more susceptible to infection as the canopy closes over to create a microclimate more favorable for disease development. Disease caused by *A. alternata* also increases with nutrient deficiency, presence of other diseases, moisture stress, and leaf senescence. However, in Israel, *A. alternata* occurred on young leaves of the cultivar 'Acala' grown in well-fertilized fields in which trickle irrigation was used to prevent moisture stress.

Exposure of both naturally infected and inoculated cotton *(Gossypium hirsutum* cv. Acala) to sunlight greatly increases expression of *A. alternata* symptoms. Exposure for 8 hours was sufficient to produce this effect in plants of different ages and under a range of temperatures and wetness, including marginal but recurrent dew periods interrupted by dryness.

Higher temperatures later in the growing season favor disease development. However, lesion formation may be influenced by extreme temperatures. In Arizona, a 40% to 100% reduction in the number of lesions occurred on plants grown at elevated temperatures compared to plants grown at 30°C.

Distribution. Africa, China, South America, and the southeastern United States.

Symptoms. Initially, tiny, dark brown, circular spots with deeply sunken centers appear on stems and leaf petioles of mature cotton. Spots develop into elliptical to oval-shaped cankers that cause stems and petioles to split longitudinally or to crack into small pieces. Eventually, the diseased stem or petiole breaks off at the canker. Leaves are frequently shed prematurely, resulting in yield losses of 20% to 40%.

Alternaria macrospora also infects bolls, bracts, and leaves of all but very young cotton plants under stress. At first, lesions on leaves are small, brown, and have reddish-purple borders. Eventually, lesions enlarge to approximately 1 cm in diameter and coalesce, resulting in irregular-shaped dead areas and premature defoliation. Mature spots have dead, gray centers in which cracks develop, causing the center to fall out.

Management
1. Grow resistant cultivars. Pima cotton *(G. barbadense)* is highly susceptible to *A. macrospora,* but *G. hirsutum* is considered highly resistant. However, *G. hirsutum* cv. Acala is susceptible to *A. alternata.*
2. Maintain cotton in a vigorous and well-fertilized condition.
3. In experimental plots, application of fungicides and removal of flowers reduced the effect of both pathogens. In Israel, the recommended spray program was to initiate fungicide application at flowering rather than when an average of one lesion per 10-m row of plants was detected on true leaves.

Anthracnose

Cause. *Glomerella gossypii,* anamorph *Colletotrichum gossypii,* is seedborne as mycelium in or on seed and as conidia contaminating the outside of seeds. *Glomerella gossypii* also survives as acervuli and perithecia in residue.

Colletotrichum capsici is also reported to be a pathogen of seedlings and bolls. *Colletotrichum capsici* reportedly survives on several weeds that serve as sources of primary inoculum. These weed species include cocklebur, *Xanthium strumarium*; johnsongrass, *Sorghum halepense*; smooth pigweed, *Amaranthus hybridus*; pitted morning glory, *Ipomoea lacunosa*; prickly sida, *Sida spinosa*; sicklepod, *Cassia obtusifolia*; smallflower morning glory, *Jacquemontia tamnifolia*; and spotted spurge, *Euphorbia maculata*.

Primary inoculum for *G. gossypii* and *C. gossypii* may originate from three different sources. The first source is from infected seed that gives rise to diseased cotyledons on which conidia are produced and are disseminated by wind to host plants. The second primary inoculum originates from infested residue on which perithecia produce ascospores that are disseminated by wind to host plants. A third source is from acervuli and mycelia that produce conidia, which are disseminated by splashing rain and by wind.

Disease is favored by moist or humid conditions. Seedling blight is most severe at temperatures of 20° to 26°C.

Distribution. Anthracnose occurs wherever cotton is grown; however, it is most prevalent under humid conditions.

Symptoms. Symptoms occur on all aboveground plant parts. Seedling cotyledons have small, reddish to light brown spots or a brown necrosis of the margin. Oblong, dark brown lesions, which occur on young hypocotyls and stems, become discolored pink in humid or wet weather because of the sporulation of conidia. Severe disease results in stem girdling, yellowing of leaves, and lack of plant thriftiness. The greatest injury to diseased plants occurs on stems at the soil surface. Following inoculation, isolates of *C. capsici* from weeds caused reddish-brown to black lesions on cotyledons, leaves, and stems of cotton seedlings.

Stems and, less frequently, leaves of mature or older plants become infected. Small, brown spots associated with injuries or angular leaf spots occur.

Diseased bolls initially have sunken, small, reddish or red-brown spots that become black to dark brown and have a reddish margin. During damp weather, the centers of spots become pink due to formation of slimy masses of conidia. Infection through a dead pistil results in internal rot of the boll.

Acervuli, which look like "pin cushions" under magnification, develop on necrotic lesions. Perithecia may also develop in dead tissue, but they are difficult to observe, even with magnification, as they appear only as tiny, dark bumps. These are the "beaks" of the perithecia that protrude above the plant epidermis; most of the perithecial structure is submerged in the plant tissue.

Symptoms caused by *C. capsici* on inoculated seedlings first appear 6 days after inoculation as pinpoint, reddish-brown lesions that are generally distributed over both surfaces of the cotyledons. Cotyledonary veins on the underside of cotyledons are conspicuously discolored. Two weeks after inoculation, cotyledonary symptoms had progressed to sunken, localized necrotic lesions 1 to several millimeters in size, but there was no associated chlorosis. A general chlorosis of cotyledons eventually occurs and is accompanied or followed by loss of turgidity. Seventeen days after inoculation, dark brown to black linear lesions up to several millimeters in length appear on cotyledonary petioles. Severely diseased cotyledons dehisce.

Management
1. Treat seed with a seed-protectant fungicide.
2. Grow recommended cultivars.
3. Rotate cotton with other crops.
4. Plow under infested residue.
5. Control insects.
6. Fertilize properly.
7. Practice timely defoliation.
8. Skip-row planting provides an open stand and prevents buildup of high humidity around lower bolls.

Ascochyta Blight

Ascochyta blight is also called wet weather blight, and ashen spot.

Cause. *Ascochyta gossypii* overwinters as pycnidia and mycelia in infested soil residue and as spores and mycelia on seed. Primary inoculum occurs during wet weather in the spring as pycnidiospores are produced in the pycnidia and are disseminated by splashing rain or by wind to host plants. Spores from seedborne inoculum also serve as an important source of primary inoculum. *Ascochyta gossypii* is most likely to infect young plants that are deficient in minerals and that are damaged by insects and hail or have other wounds. Pycnidia are formed in mature lesions, and pycnidiospores serve as secondary inoculum. Disease is most severe under cool, humid or moist conditions.

Distribution. Higher elevations in Mexico, South America, Africa, and Asia. In the United States, Ascochyta blight is most important along the northern fringe of the eastern and central cotton-growing areas.

Symptoms. Either preemergence or postemergence seedling blight occurs. Later in the growing season, symptoms occur on bolls, leaves, stems, and branches of diseased plants. Initially, round, brown spots approximately 2 mm in diameter occur on leaves. The spots rapidly enlarge during humid weather, become gray or light brown, and sometimes have reddish-brown borders. As spots enlarge and age, their centers fall out and give diseased leaves a ragged appearance. During wet, cool weather, spots coalesce and cause defoliation and blighting of younger plants.

At the bases of leaf petioles, stems and branches have dark brown, elongated lesions that enlarge rapidly and become sunken and light brown in the center. Lesion centers decay early and tend to shred, break up, and fall out of the diseased tissue. Lesions eventually encircle stems, killing plants above the infections. Disease symptoms on bolls are similar to those on stems and branches.

Management

1. Treat seed with a fungicide seed protectant.
2. Rotate cotton with resistant crops. Ascochyta blight is more severe when cotton follows cotton.
3. Plow under residue.
4. Balance soil fertility.

Ashbya Fiber Stain

Cause. *Ashbya gossypii* survives as mycelium in residue and enters the boll through feeding wounds of the cotton stainer bug. Asci are formed within mycelial strands.

Distribution. Not reported.

Symptoms. No external symptoms are visible on the boll. Fibers and seed have a tan to brown discoloration and eventually dry out.

Management. Not reported.

Aspergillus Boll Rot

Cause. *Aspergillus flavus* survives in soil and residue as spores and mycelia. *Aspergillus flavus* enters bolls through natural openings or wounds made by the pink bollworm *(Pectinophora gossypiella)* and other insects. Infection is favored by high humidity and temperatures of 32°C and above. *Aspergillus flavus* moves from infected locks to adjacent unwounded locks and contaminates seed produced there with low levels of aflatoxin. Certain pectolytic enzymes produced during host infection contribute to fungal aggressiveness as it grows in locular tissues, through the lint, and into the cottonseed, where the aflatoxin is produced.

Drying is an important factor in fungal entry into seeds. The nucellus membrane surrounding the embryo forms a possible physical barrier to fungal entry. Upon drying, the integrity of the nucellus membrane is affected, allowing the fungus to break through the membrane, enter the embryo, and form toxin. Diseased seed levels are also affected by water potentials on the day of anthesis and infection, with the highest disease levels generally occurring in seed from plants with water potentials of −1.6 to −1.9 bars. Aflatoxin is highest in seed from wounded locks and in seed harvested near the end, rather than at the beginning, of the harvest season.

Infection by *A. flavus* and subsequent aflatoxin production occurs primarily in green bolls close to maturity and has been reported to be most severe when carpels are infected and separated by *A. niger* and *Rhizopus* sp. However, *A. flavus* grows into unwounded locks of young bolls more extensively than in older bolls. Maturation of the intercarpellary membrane may hinder the interlock spread of *A. flavus.*

Distribution. Wherever cotton is grown.

Symptoms. Lint is stained yellow and otherwise is discolored. Lint fluoresces greenish yellow under ultraviolet light. In the United States (Arizona and California), relatively little aflatoxin is formed in bolls produced near the end of the season, and most toxin occurs in bolls borne near the soil.

Management. Control insects such as pink bollworm on cotton.

Black Boll Rot

Cause. *Aspergillus niger* survives saprophytically on several kinds of infested plant residue. Conidia are produced from mycelia and are disseminated by wind to host plants, where they germinate and enter bolls through wounds. Black boll rot is more severe in rank cotton growth and during frequent rains that occur at boll development and opening. In irrigated areas, boll rot is most severe when excessive moisture has occurred late in the growing season.

Distribution. Wherever cotton is grown.

Symptoms. Initially, a soft pinkish rot develops on the side or base of a boll. Eventually, the color of the rot changes from pink to brown and an abundant production of black spores in the diseased area gives the boll a black sooty appearance.

Management
1. Avoid practices that promote rank growth, especially excessive nitrogen.
2. Use correct timing and control of irrigation.
3. Control insects after boll development.
4. Shield equipment to reduce mechanical injuries.
5. Use bottom defoliation and two-stage harvesting to reduce exposure.
6. Keep cotton free of weeds.
7. Use skip-row plantings.
8. Rotate cotton with other crops.
9. Plow under residues to prevent rotting bolls from being a source of inoculum.

Black Root Rot

Black root rot is also called internal collar rot.

Cause. *Thielaviopsis basicola*, synamorph *Chalara elegans,* is a common soil inhabitant in both cultivated and noncultivated soils. However, Hood and Shew (1997) report that *T. basicola* does not display significant saprophytic ability and should be classified ecologically as an obligate parasite. *Thielaviopsis basicola* does survive as chlamydospores on and in roots; internally in the pith, phloem, and xylem; and in soil. Endoconidia are also produced, but their relevance to the survival of *T. basicola* is unknown.

Chlamydospores germinate to produce mycelia that infect roots through root hairs, then grow to the endodermis. Wet, alkaline soils at temperatures below 26°C increase disease severity. Some reports state that soil temperatures of 15° to 20°C favor disease. These relatively low soil temperatures may stress plants and favor *T. basicola,* although the optimum temperatures for pathogen growth in culture are 25° to 28°C. Internal colonization of root tissues is favored by high inoculum levels and soil temperatures of 18° to 20°C. Seedling blight also occurs in cool, wet weather.

Thrips are reported to predispose seedlings to infection.

Distribution. Egypt, Peru, Russia, and the United States.

Symptoms. Seedlings are stunted and have small, cupped, chlorotic leaves that have purplish borders with brown margins. Leaves are shed from severely diseased plants.

Roots are brown and decayed and have taproots that are smaller in diameter than those of healthy plants, which allows diseased plants to be easily pulled from the soil. There is a sharp line between diseased and healthy tissue. Cortical tissue at the stem base may appear healthy, but vascular tissue becomes black or purple. The stem base and taproot are swollen; a black crust on the roots consists of chlamydospores. Within a few weeks, diseased tissue is sloughed off and no symptoms remain.

An internal collar rot phase occurs on *Gossypium barbadense* from midseason to late season. Leaves and stems wilt and suddenly collapse. Reddish lenticels, swelling, and brown to purple-black discoloration of the stele occur in the crown.

Management
1. Sow into warm soils.
2. Rotate cotton with monocotyledons.

Botryodiplodia Seedling Blight and Boll Rot

Cause. *Botryodiplodia theobromae* (syn. *Lasiodiplodia theobromae)* is seedborne and is considered a weak pathogen.

Distribution. Brazil.

Symptoms. Seedlings are killed, and diseased bolls rot.

Management. Not reported.

Cercospora Leaf Spot

Cause. *Cercospora gossypina,* teleomorph *Mycosphaerella gossypina,* survives as conidia and mycelia on seed and in infested residue. *Mycosphaerella gossypina* may survive as perithecia in infested residue. Primary infection occurs by conidia or by ascospores disseminated by wind to host plants. Secondary inoculum is provided by conidia produced in spots on diseased leaves. Because *C. gossypina* is not a highly virulent pathogen, infection by *C. gossypina* may follow infections caused by other pathogens or may occur in cotton plants that are predisposed to disease by being under various stresses. Examples of stresses are cotton growing on sandy soils, lack of soil moisture at the time of maximum boll set, and lack of proper rates of nitrogen and potash throughout the growing season.

Distribution. Worldwide in warm and humid cotton-growing areas.

Symptoms. Symptoms are observed first on older upper leaves and eventually on younger leaves. Initially, leaf spots are barely visible reddish points, but eventually spots enlarge to approximately 2 cm in diameter, are round to irregular-shaped, and have white centers surrounded by a purple margin. Conidia growing on diseased areas give both the diseased lower and upper leaf surfaces a dark appearance. The centers of lesions may drop out as they enlarge and give a "shot-hole" effect to the lesion.

Diseased leaves turn brown and dry up, resulting in partial or complete defoliation. Young squares and bolls may fall from plants. The main terminal and upper fruiting branches may die back under severe disease conditions. Other branches may be unaffected.

Management
1. Treat seed with a fungicide seed protectant.
2. Maintain proper soil fertility.
3. Avoid insect and mechanical injury.
4. Plow under residue.
5. Grow later-maturing cultivars rather than the more susceptible early-maturing cultivars that fruit heavily.

Cercospora-Alternaria Leaf Blight Complex

This disease has been called black rust, and the cotton leaf blight disease complex.

Cause. *Cercospora gossypina* and *Alternaria alternata (syn. A. tenuis)*. The disease complex occurs on plants under stress from moisture and mineral deficiencies or following leaf infections caused by *Xanthomonas campestris* pv. *malvacearum (syn. X. malvacearum)*.

Distribution. The United States (Arizona, Arkansas, the upper San Joaquin Valley of California, Louisiana, Oklahoma, southern New Mexico, and southwestern Texas).

Symptoms. The disease complex appears at "cut-out," or approximately the time plants become dormant and cease setting bolls. Early symptoms somewhat resemble the beginning stages of potash deficiency or rust disease. Each pathogen causes circular, sometimes zonate, leaf spots late in the season. *Cercospora* lesions are scattered over the entire leaf surface, while *Alternaria* lesions occur on leaf margins. A bluish color occurs where the different lesions overlap. All lesions enlarge rapidly and coalesce; diseased leaves become necrotic and dry from the margins inward. Normally, the younger leaves, squares, and bolls die prematurely and drop, but when they are killed rapidly, dropping does not occur. As symptoms of the disease complex become more severe, all leaves may desiccate and defoliate prematurely.

Dieback of the main stem terminal and uppermost fruiting branches frequently occurs. Cut-out of the upper third to half of the plant is initiated prematurely, and the more mature bolls in this upper third of the plant open prematurely. Many plants have one or two unaffected lower vegetative branches whose leaves fail to defoliate or defoliate poorly when treated with defoliants.

Management. No disease management strategy is currently successful.

Charcoal Rot

Charcoal rot is also called ashy stem.

Cause. *Macrophomina phaseolina* (syn. *M. phaseoli, Macrophoma phaseolina, Macrophoma phaseoli, Rhizoctonia bataticola,* and *Sclerotium bataticola*). *Macrophomina phaseolina* survives as sclerotia in soil and infested residue and as saprophytic mycelia in soil residue. A pycnidial stage is formed; however, pycnidia are rarely formed in the diseased stem. Sclerotia, which likely constitute the source of primary inoculum, germinate on root surfaces to produce one or multiple germ tubes that directly penetrate roots. Wounds are not necessary for infection to occur. Infection occurs from the seedling to blossom stage.

Disease is most severe during high temperatures (35° to 39°C) and periods of moisture stress. It is generally thought that moisture is necessary for infection to occur but that later moisture stress disposes the plant to disease development.

Distribution. Throughout the world under hot, dry conditions.

Symptoms. Diseased plants wilt and die rapidly. Leaves wilt, droop, become chlorotic, and are prematurely shed. Grayish-white lesions develop on stems near the soil line and extend for several centimeters up the diseased stem. Diseased plants have a "dry" rot with numerous tiny, black sclerotia distributed throughout the infected stem, scattered through the lesion, and embedded in the cortex and wood that give tissue a grayish appearance, similar to charcoal.

Roots may be partially or totally decayed. Subsequently, diseased plants may be stunted or may die but remain standing.

Management
1. Rotate cotton with other crops, preferably small grains.
2. Do not oversow. Crowded seedlings are more subject to infection.
3. Ensure that soils are well fertilized.
4. Plow under infested residue.
5. Manage water supply for irrigated cotton to reduce water stress during periods of high temperatures.

Choanephora Mold

Cause. *Choanephora cucurbitarum* survives as mycelia, chlamydospores, and possibly spores in infested residue and soil. Blossoms become infected during wet, humid conditions.

Distribution. The United States.

Symptoms. The corolla becomes covered with a white, fuzzy, fungal growth consisting of young mycelia and conidiophores. Later, the fungal growth becomes sprinkled with black specks that are the conidial heads borne on the end of conidiophores.

Management. General controls for other boll rots are presumed to be of benefit.

Cochliobolus Leaf Spot

Cause. *Cochliobolus spicifer,* anamorph *Curvularia spicifer,* and *C. lunatus.*

Distribution. India and possibly Malawi. *Cochliobolus spicifer* is the more important pathogen.

Symptoms. Seed rot and preemergence damping off occurs. Leaf spots occur on seedling leaves as light yellow areas that eventually increase in size and become dark brown. During wet weather, discoloration spreads from the leaf to the leaf stalk. Defoliation of older plants may occur.

Management
1. Grow the most resistant cultivar.
2. Treat seed with a seed-treatment fungicide.
3. Apply a foliar fungicide.

Colletotrichum Boll Rot

Cause. *Colletotrichum capsici* and *C. indicum.* The two fungi are considered synonymous by some researchers. Boll rot is reported to occur during exceptionally wet conditions.

Distribution. The United States (Louisiana).

Symptoms. Bolls initially have circular, dark lesions that are associated with sutures. Eventually the entire boll turns black. In advanced stages, the carpel wall is destroyed, leaving the skeletonized remains of the wall. Severely rotted bolls often open, but the discolored lint usually does not fluff, leaving a "tight-locked" boll that dislodges and falls to the ground.

Management. Not reported.

Corynespora Leaf Spot

Cause. *Corynespora cassiicola* overwinters as mycelia and chlamydospores in residue or as chlamydospores in soil. *Corynespora cassiicola* can also grow saprophytically on different kinds of infested plant residue in the soil. Conidia are splashed or blown onto leaves and cause infection when free moisture is present or when the relative humidity is 80% and above.

Distribution. The United States (Mississippi) and possibly other cotton-growing areas of the world.

Symptoms. Initially, brick-red, pinpoint spots develop on diseased leaves. Later, the spots enlarge to become roughly circular (2–6 mm in diameter) with a dark brown margin and a light brown center that falls out. Alternating light and dark brown areas give mature spots a zonate appearance. On severely diseased leaves, spots coalesce to form large, irregular-shaped, necrotic areas that cause leaves to become chlorotic, wilt, and defoliate from the plant. Reddish spots frequently girdle petioles.

Management. Maintain plants in a vigorous growing condition.

Cylindrocladium Black Rot

Cause. *Cylindrocladium crotalariae* survives as microsclerotia on infested residue and as mycelia in residue. Microsclerotia are disseminated by several factors that transport soil. Microsclerotia also constitute the source of primary inoculum by germinating to form mycelia that penetrate roots.

Distribution. The southeastern United States.

Symptoms. Cylindrocladium black rot is considered a minor problem and ordinarily no aboveground symptoms are produced. Roots of susceptible cultivars are blackened and reduced.

Management. Not necessary. However, cotton maintains or increases levels of inoculum in soil that may harm other crops, such as peanut.

Diplodia Boll Rot

Cause. *Lasiodiplodia theobromae* (syn. *Diplodia gossypina)* survives as pycnidia in infested residue. Pycnidiospores are produced in pycnidia during wet weather and are disseminated by splashing rain and by wind to bolls, where infection occurs through weevil punctures and wounds. Lower bolls in contact with soil are frequently infected in this manner. Injured and un-injured bolls are infected at all stages of maturity but become the most severely diseased when they are infected at ages of 7 or fewer days and at 40 or more days. Disease development is optimal during moist conditions and a temperature of 30°C.

Distribution. Wherever cotton is grown.

Symptoms. Initially, small, brown, water-soaked, and sunken lesions occur on capsules or bracts. Pycnidia appear as tiny, black specks on the boll surface. Black, sooty masses of spores released from pycnidia under high humidity conditions cover the boll surface. Lesions may enlarge and cover the entire boll. Bolls eventually blacken completely, dry up, and split to expose the blackened and matted lint.

Management. In general, avoid practices that prevent plants from drying out.
1. Avoid practices that promote rank growth.
2. Use correct timing and control of irrigation.
3. Control insects after boll development.
4. Shield equipment to reduce mechanical injuries.
5. Keep cotton free of weeds.

Escobilla

Escobilla is also known as "little broom," ramulose, superbrotamento, super-sprouting, and witches' broom.

Cause. *Colletotrichum gossypii* (syn. *C. gossypii* var. *cephalosporioides),* teleomorph *Glomerella gossypii.* Disease is most severe during periods of prolonged rainfall. Plants of all ages may be diseased; however, those infected while young are most severely affected.

Distribution. Brazil and Venezuela.

Symptoms. Infected plants have both healthy and diseased branches and a dense shrub-like appearance from numerous lateral branches and "tufts" caused by terminal buds being killed. Swollen and twisted branches with shortened internodes and swollen nodes result in an excessive number of short twigs and leaves.

Leaves on diseased branches are small, dark green, wrinkled, and have reduced lobes and short petioles. Healthy-appearing leaves may occur on affected branches but have turned-up margins and swollen petioles with internal browning. Leaves and shoots have small, brown spots that later elongate along the veins, turn gray, and develop cracks in their centers. Gray mycelia in which black, dot-like acervuli occur eventually appear on lesions.

Bolls on diseased plants may remain green without opening. Seeds in unopened bolls may germinate abnormally.

Management. Not practiced.

Fusarium Boll Rot

Cause. *Fusarium moniliforme, F. oxysporum, F. roseum,* and *F. solani. Fusarium oxysporum* and *F. roseum* attack uninjured green bolls by initially infecting bracts, then invading the capsule base through the receptacle. Production of vast numbers of spores on plant debris, such as flowers, bolls, and squares, apparently provides the airborne inoculum necessary to initiate boll deterioration. Boll age is the most important factor in limiting rot. Bolls 33 or more days old are more susceptible than younger fruits. Boll rot proceeds rapidly during moist conditions.

Distribution. The southeastern United States.

Symptoms. Initially, small lesions occur along the bract margin, particularly at the apices of the toothed areas. Eventually, the entire bract becomes necrotic. The capsule becomes invaded through the receptacle; tissues show a dark blue-green discoloration. Decay is more rapid in dehiscing bolls.

Management. Not reported, but development of resistant cultivars may be possible.

Fusarium Damping Off

Cause. *Fusarium* spp., including *F. equiseti, F. graminearum, F. moniliforme, F. oxysporum,* and *F. solani.* Most *Fusarium* spp. survive as saprophytic mycelia in infested residue and as chlamydospores in the soil. Seed and roots are infected under cool, wet soil conditions. Thrips predispose seedlings to infection.

Distribution. Wherever cotton is grown.

Symptoms. Seed may be rotted or seedlings may die before emergence. Roots and hypocotyls are tan to dark brown and are rotted. Frequently, the tap-root dies.

Management
1. Sow high-quality seed treated with a seed-protectant fungicide.
2. Apply fungicide as a spray or dust in the seed furrow.
3. Delay planting until soil temperature has warmed to 20°C or more for at least three or four mornings before sowing.

4. Sow in bedded rows or slightly raised seedbeds, where the soil temperature normally warms up faster than in a furrow or level seedbed.

Fusarium nygami **Root Colonization**

Cause. *Fusarium nygami.*

Distribution. Widespread.

Symptoms. Fungus was isolated from roots, but "infected" plants were symptomless.

Management. Not necessary.

Fusarium Seed Infection

Cause. *Fusarium equiseti* and *F. semitectum* were isolated from embryos of weathered cottonseed. Several isolates produced equisetin, a phytotoxin.

Distribution. The United States (Texas).

Symptoms. Equisetin inhibits germination of various monocotyledonous and dicotyledonous seed when seed is germinated at 30°C under aqueous, shake conditions. Under certain conditions, equisetin is highly toxic to rapidly growing plant tissues; therefore, seed and seedlings infected by *F. equiseti* and *F. semitectum* may be affected. Necrotic lesions were produced on the cotyledons and radicles of young seedlings.

Management. Not reported.

Fusarium Wilt

Cause. *Fusarium oxysporum* f. sp. *vasinfectum* survives as chlamydospores and sclerotia in soil and as mycelia in infested residue. *Fusarium oxysporum* f. sp. *vasinfectum* has also been isolated from seed and gin trash, which may also serve as means of dissemination. Conidia, chlamydospores, and sclerotia germinate and infect roots through wounds made by nematodes, including *Belonolaimus* spp., *Conemoides* spp., *Helicotylenchus* spp., *Meloidogyne* spp., *Pratylenchus* spp., *Rotylenchulus* spp., *Trichodorous* spp., *Tylenchorhynchus* spp. and *Xiphinema* spp. Without nematode wounds, Fusarium wilt would not be a serious problem.

The mycelium then grows into xylem tissue and wilting occurs when water is prevented from moving up the plant. Sporulation occurs on the outside of dead plants, and conidia are disseminated by wind to other areas. Dissemination is also by any means, including furrow irrigation water, that moves soil. Fusarium wilt is most serious in an acid, sandy soil.

There are several races of *F. oxysporum* f. sp. *vasinfectum*.

Distribution. Wherever cotton is grown.

Symptoms. Seedlings may damp off, but Fusarium wilt is normally observed at approximately blossom time. The general appearance of a field will be spotty or uneven. Areas that range in size from a few scattered plants to several acres will be affected; the most severe symptoms occur on the sandiest soils.

Initially, diseased plants are stunted and unthrifty. Leaves turn yellow, beginning at the leaf margins and proceeding into interveinal areas, then brown, and eventually are shed from the bottom of the plant upward. The entire plant eventually becomes leafless, dies, and turns black due to growth of saprophytic fungi.

If a diseased plant does not die, it remains stunted and the lateral branches outgrow the main stalk, but it flowers and matures earlier than healthy plants. The interior of the main stem has a dark brown discoloration. Symptoms of nematode injury will be present on roots.

Management
1. Grow resistant cultivars.
2. Rotate cotton with crops, including millet, peanut, small grains, sorghum, and sudangrass, on which the cotton root knot nematode will not reproduce.
3. Mow and bury cotton stalks after harvest to reduce nematode populations. If erosion is a problem, plant a cover crop of rye.
4. Summer-fallow and control weeds to reduce hosts on which nematodes reproduce.
5. Ensure adequate levels of potash.
6. Use a soil nematicide if the cost is warranted.
7. Avoid growing upland cotton in infested sandy, acid soils.

Helminthosporium Leaf Spot

Cause. *Helminthosporium gossypii* overseasons in infested leaves. Leaves, flower bracts, and bolls are infected.

Distribution. Asia.

Symptoms. Spots on leaves and bracts are small (1–8 mm in diameter) and light red. In time, the spots turn dark purple and have a brown center that frequently falls out and gives the spot a "shot-hole" effect. Spots may be numerous, causing diseased leaves to die and defoliate prematurely. Spots on bolls are small and purplish but usually cause no damage to fiber.

Management. Not reported.

Myrothecium Leaf Spot

Cause. *Myrothecium roridum* is a common soil saprophyte. Primary inoculum is presumed to arise from infested soil and infected weeds. Optimum spore germination occurs at a temperature of 29°C.

Distribution. India and the United States.

Symptoms. Preemergence and postemergence damping off may occur. Initial symptoms appear on leaves of young plants 4 to 6 weeks old. The circular, tan leaf spots have violet-brown margins. Eventually, spots may enlarge to 3 cm in diameter and become surrounded by translucent areas that are concentrically zoned and have black, pinhead-sized dots that are the sporodochia of the fungus. Spots increase in size and number and ultimately coalesce, affecting large areas of the leaf. Eventually, the leaves defoliate.

Dark brown to black cankers on diseased stems occur relatively close to the soil line.

Management
1. Destroy infested residue.
2. Control susceptible weeds.
3. Fungicides have been used to manage disease.

Nematospora Boll Stain

Cause. *Nematospora coryli, N. gossypii,* and *N. nagpuri* survive in infested residue and, probably, as parasites on nearby wild plants. Infection occurs through insect wounds.

Distribution. *Nematospora coryli* is reported from the United States (California), *N. gossypii* from West India and Africa, and *N. nagpuri* from India.

Symptoms. Initially, diseased bolls do not display external symptoms except for tiny, inconspicuous wounds or scars. In time, lint becomes discolored tan, and seed is shriveled and killed. Leaf sheaths are discolored, and diseased bolls are shed.

Management
1. Avoid practices that promote rank growth, especially excessive nitrogen.
2. Use correct timing and control of irrigation.
3. Control insects after boll development.
4. Shield equipment to reduce mechanical injuries.
5. Use bottom defoliation and two-stage harvesting to reduce exposure.
6. Keep cotton free of weeds.
7. Use skip-row plantings.
8. Rotate cotton with other crops.
9. Plow under residues to prevent rotting bolls from being a source of inoculum.

Nigrospora Lint Rot

Cause. *Nigrospora oryzae* is disseminated and introduced into the boll by the mite *Siteroptes reniformis*. A mutualistic form of symbiosis exists between the two organisms.

Distribution. Africa, Russia, and the United States.

Symptoms. Lint is discolored by the presence of dark brown to black conidia on the lint surface and in the lumen. Lint also fails to fluff and the fibers are weakened.

Management. Not reported.

Phomopsis Leaf Spot

Cause. *Phomopsis* sp. survives as pycnidia in infested residue.

Distribution. The United States (Louisiana).

Symptoms. All aboveground plant parts become diseased, but the principal damage is to flower buds, which then are not capable of opening. Gray, water-soaked, slightly sunken areas occur on diseased leaves and stems. Later, the diseased areas are covered with numerous pycnidia, which resemble dark specks.

Management. Not known.

Phyllosticta Leaf Spot

Cause. *Phyllosticta gossypina* and *P. malkoffi*. Although *Phyllosticta* species have been reported from Africa as the causal organisms, some researchers question the identification of the fungi.

Distribution. Tanzania and Zimbabwe.

Symptom. Spotting occurs on diseased leaves.

Management. Not reported.

Phymatotrichum Root Rot

Phymatotrichum root rot is also called cotton root rot and Texas root rot.

Cause. *Phymatotrichopsis omnivora* (syn. *Phymatotrichum omnivorum)* survives in soil as sclerotia and hyphal strands. Survival may be facilitated by mycelial strands from other diseased dicotyledonous plants and, possibly, by monocotyledonous plants in a rotation. However, unless they are attached to sclerotia, mycelial strands will not infect taproots of cotton.

Early in the growing season, when soil temperatures have risen, sclerotia germinate to produce hyphal strands that contact roots of seedlings that have been emerged for 5 or more days. Strands grow ectotrophically on the taproot to the crown before penetrating and killing the plant. Root contact is not necessary for plant-to-plant spread because fungal growth occurs through soil in advance of diseased roots; however, disease is inhibited when plants are spaced 20 to 69 cm apart. Plant-to-plant spread of *P. omnivora* in rows is more important than spread across rows.

Sclerotia are abundantly produced on rotted roots. When the soil is moist, conidia are produced on mycelial mats that form on the soil surface near dead plants, but they are not known to germinate. Clay soils with high pH are especially conducive for disease caused by *P. omnivora*. High soil moisture during the early summer months is favorable for disease development, but moisture levels of −12 to −16 bars reduces disease. Because of the increased inoculum resulting from the saprophytic growth of *P. omnivora* on sorghum roots, disease incidence increases on ratoon sorghum that has preceded cotton in a rotation.

Almost 2000 species of dicotyledonous plants may be infected by *P. omnivora*.

Distribution. Mexico and the United States (Arizona and Texas).

Symptoms. Foliar symptoms do not develop until soil temperatures at the inoculum depth are 22°C and above. Leaves are chlorotic initially, then become a bronze color and wilt. Later leaf symptoms are associated with cortical senescence in the roots: diseased leaves become brown and dry up but remain attached to plants.

Roots become water-soaked and discolored, and the cortex is sloughed 18 to 25 days after plant emergence, or at approximately the same time as foliar symptoms occur. When taproots are girdled, plants rapidly wilt and die. Bark and cambial tissues of diseased roots turn brown and have light tan strands of mycelium growing on the root surface. The root bark easily peels off and reveals a reddish stain along the white woody tissue.

When soils are moist, mycelial mats form on the soil surface near dead plants. Mats are white initially but later become tan and "powdery" due to production of conidia on the mat surface. Death of diseased plants normally occurs 27 to 50 days after emergence, regardless of the age of the plants when they were first exposed to sclerotia.

The following atypical symptoms have been observed during periods of low soil moisture: The aboveground symptoms were a gradual wilting, followed by leaf chlorosis and defoliation. Mature bolls and young leaves remained attached to stems, and plants remained alive. Roots had discolored, sunken lesions 10 to 20 cm below the soil surface.

Management. There are no economic cultural practices that can be routinely used to manage cotton root rot; however, the following are aids to management of disease:

1. Plow under green manure crops. Residues can also be plowed under if nitrogen is applied to promote rapid decay.
2. Rotate cotton with cereals or grasses.
3. Plow deeply during hot, dry weather, thus exposing and killing the fungus.
4. Provide adequate fertility to ensure vigorous cotton growth.
5. Sow early and control insects and diseases to promote early maturation of bolls before *P. omnivora* becomes fully active.
6. When sorghum precedes cotton in a rotation, do not allow sorghum to regrow (ratoon) until killed by frost.

Phytophthora Boll Rot

Cause. *Phytophthora parasitica* (syn. *P. nicotianae* var. *parasitica)*. Infection occurs two ways: directly by sporangia forming a germ tube, or indirectly by the formation of zoospores within sporangia. Zoospores are then liberated and "swim" for a few minutes, become rounded, and germinate to form germ tubes that penetrate the bolls. Disease is most severe during conditions of high humidity.

Other species of *Phytophthora* reported to cause a boll rot are *P. cactorum, P. capsici,* and *P. palmivora.* Under conditions of high humidity, high rainfall, and irrigation, *Phytophthora capsici* is reported to cause boll rot in Arizona.

Distribution. The United States (Arizona and Louisiana).

Symptoms. Bolls are bluish black to black, and within a few days, the entire surface of an infected boll becomes black and spongy with a soft, watery rot. Eventually, a white, mealy fungal growth consisting of sporangia and mycelium occurs on the boll surface. Bolls, particularly those approaching maturity, rot in 2 to 3 days.

Management. Not reported.

Phytophthora boehmeriae Boll Rot

Cause. *Phytophthora boehmeriae* survives as oospores in the soil. Inoculum is splashed from soil onto bolls. Maximum temperature for fungal growth is 33°C.

Distribution. China and Greece.

Symptoms. Boll rot is restricted to the lower one-half to two-thirds of cotton plants. Diseased bolls are severely decayed; the subtending dried carpels are dark brown to black. White patches of fungal mycelia bearing numerous sporangia are present on the surface of diseased bolls. Oospores are present on the cotton lint and internal carpel surface.

Management. Not reported.

Powdery Mildew

Powdery mildew is also called manta blanca, or white mantle, when caused by *Salmonia malachrae.*

> **Cause.** *Leveillula taurica* (syn. *E. taurica* and *Oidiopsis taurica)* and *Salmonia malachrae* (syn. *Erysiphe malachrae).* The anamorph of *L. taurica* is *Oidiopsis sicula* (syn. *O. taurica),* but the anamorph of *S. malachrae* is unknown. *Oidiopsis gossypii* is also reported to be a causal agent.
>
> The fungi survive by continually infecting cotton or alternate hosts throughout the year. Infection is optimum at 85% to 100% relative humidity. The optimum temperatures for *L. taurica* are 25° to 30°C; however, *S. malachrae* requires cooler temperatures to cause disease.

> **Distribution.** *Leveillula taurica* is found in India, Peru, Russia, Sudan, the United States (California), and the West Indies. *Salmonia malachrae* is found in the Antilles and South America.

> **Symptoms.** Initial symptoms caused by *Leveillula taurica* depend on the diseased cultivar. Generally, symptoms are angular or rounded, powdery, white patches (1.5–2.0 mm in size) on the undersides of diseased leaves. Later, the patches coalesce to cover the entire underside of the leaf. Sometimes no external symptoms occur until leaves become chlorotic and defoliate. Cleistothecia are not present.
>
> Scattered, circular areas of white mycelia of *S. malachrae* initially appear on the upper leaf surfaces and later coalesce to cover the entire leaf area. White to tan cleistothecia are scattered throughout the mycelium. Eventually, diseased leaves become chlorotic, curl, and may defoliate. Later in the growing season, severe disease gives a field the appearance of being covered with snow.

Management. Not practiced.

Pythium Damping Off

> **Cause.** *Pythium ultimum* is the most prevalent species found on cotton. Other *Pythium* species associated with damping off include *P. aphanidermatum, P. debaryanum, P. heterothallicum, P. irregulare, P. polytylum, P. splendens,* and *P. sylvaticum. Pythium* spp. generally survive as oospores in soil or residue and saprophytically as mycelia in infested residue. Oospores germinate in wet soils at temperatures of 10° to 15°C and form sporangia in which zoospores are borne. Zoospores are then liberated from sporangia and "swim" to a plant surface, where they encyst and germinate to form germ tubes that penetrate the plant. At higher temperatures, oospores may germinate directly to form a single germ tube instead of a sporangium.

Thrips predispose seedlings to infection.

Distribution. Wherever cotton is grown.

Symptoms. Seed is rotted and soft. Initially, hypocotyls and roots become water-soaked and pale green. Lesions on hypocotyls range from small, discolored spots to large sunken, necrotic areas. Eventually, the diseased hypocotyls and roots become gray and dry up. The hypocotyl becomes constricted above the soil line, which causes the seedling to fall over.

Severe root rot has not been observed under field conditions, although small, light brown lesions occur on taproots and secondary roots. Root rot caused by *P. splendens* results in reduction of shoot growth.

Management
1. Sow high-quality seed treated with a seed-protectant fungicide.
2. Apply fungicide as a spray or dust in the seed furrow.
3. Delay sowing until soil temperature has warmed to 20°C or more for at least three or four mornings before sowing.
4. Sow in bedded rows or slightly raised seedbeds, where the soil temperature warms up faster than in a furrow or a level seedbed.

Ramulose

Ramulose is also called witches' broom.

Cause. *Colletotrichum gossypii* var. *cephalosporioides*.

Distribution. Brazil.

Symptoms. Necrotic spots occur on diseased leaves. Growth abnormalities, or the witches' brooms, occur on the stem.

Management. Resistance has been identified.

Rhizoctonia Leaf Spot

Cause. *Rhizoctonia solani* survives as sclerotia in soil or infested residue and as mycelia growing saprophytically on residue of many different plants. Sclerotia germinate to produce mycelia that infect hypocotyls near the soil surface, and in very humid weather, the fungus grows up the external surface of the stem.

The teleomorph *Thanatephorus cucumeris* produces basidiospores on a loose hyphal network growing on the stem; however, it is not certain what function basidiospores have in the life cycle of the fungus. Leaves may be infected either by basidiospores or other soilborne propagules.

Disease is favored, particularly in thick foliage, by high rainfall and warm temperatures.

Distribution. El Salvador and the United States (Louisiana).

Symptoms. The light brown, irregular-shaped leaf spots have dark purple borders. As spots enlarge, centers become chlorotic, then necrotic; cracks develop and centers of older spots drop out, giving a "shot-hole" effect and a general ragged appearance to diseased leaves. During periods of high humidity, the brownish growth of the fungus may appear on the underside of the diseased leaves.

Management. Not practiced.

Rhizopus Boll Rot

Cause. *Rhizopus nigricans* (syn. *R. stolonifer*) survives as mycelia and spores in infested residue and soil. Under moist and warm weather conditions, spores infect bolls.

Distribution. Wherever cotton is grown.

Symptoms. Diseased areas of the boll are olive green initially but become black as they dry up. Fungal growth consisting of mycelium, sporangiophores, and sporangia forms a dark gray mold over the boll.

Management
1. Avoid practices that promote rank growth, especially excessive nitrogen.
2. Use correct timing and control of irrigation.
3. Control insects after boll development.
4. Shield equipment to reduce mechanical injuries.
5. Use bottom defoliation and two-stage harvesting to reduce exposure.
6. Keep cotton free of weeds.
7. Use skip-row plantings.
8. Rotate cotton with other crops.
9. Plow under residues to prevent rotting bolls from being a source of inoculum.

Rust

Cause. *Puccinia schedonnardi* is a heteroecious long-cycled rust whose alternate host consists of several grasses, including *Muhlenbergia* spp. and *Sporobolus* spp. The pycnial and aecial stages occur on cotton, the uredinial and telial stages on grasses.

Distribution. Mexico and southwestern United States.

Symptoms. The inconspicuous pycnia are yellowish orange. Aecia develop around pycnia as large, orange, raised spots on leaves, bolls, and bracts. Uredinia are light cinnamon brown; telia are dark brown.

Management. Not reported.

Sclerotium Boll Rot

Cause. *Sclerotium rolfsii,* teleomorph *Athelia rolfsii,* overwinters as sclerotia in soil and as mycelia growing saprophytically in infested plant residue. Cool soil temperatures kill both sclerotia and mycelia. Sclerotia germinate during high soil temperatures and in dry soil to form mycelia that infect plants of all ages. Usually, plant parts in contact with soil or near the air-soil interface become infected. Disease is most common in neutral to acid sandy soils.

Distribution. Wherever cotton is grown under warm conditions.

Symptoms. Initially, light brown spots appear on carpels. Spots enlarge rapidly over whole carpels, peduncles, and leaf petioles, become brown, and cause all tissues to dry up. Eventually, white, feather-like mycelium is produced on surfaces of bolls. Later, numerous round to irregular-shaped, white to light brown sclerotia are produced in the mycelium. When a diseased boll splits, white cottony mycelial growth and sclerotia are seen in locules, lint, and seeds. Inner tissues of bolls dry up and become brown, and lint becomes discolored and eventually disintegrates. Frequently, seeds within a diseased boll decay.

Management
1. Grow the most resistant cultivar.
2. Rotate cotton with other crops, such as small grains.
3. Plow infected cotton residue 12 cm deep.

Sclerotium Stem and Root Rot

Sclerotium stem and root rot is also called Sclerotium rot and southern blight.

Cause. *Sclerotium rolfsii,* teleomorph *Athelia rolfsii,* overwinters as sclerotia in soil and as mycelia growing saprophytically in infested plant residue. Cool soil temperatures kill both sclerotia and mycelia. Sclerotia germinate during high soil temperatures and in dry soil to form mycelia that infect plants of all ages. Usually, plant parts in contact with soil or near the air-soil interface become infected. Disease is most common in neutral to acid sandy soils.

Distribution. Wherever cotton is grown.

Symptoms. Initially, leaves wilt over time until plants become defoliated and blossoms or bolls drop. Eventually, diseased plants die. Plant wilting is caused by a canker that develops on a stem at or just below the soil surface. The stem base and taproot may enlarge or swell. If soil moisture is present, a white, cottony growth composed of mycelia grows up the stem for 3 or more centimeters. Dark brown to black sclerotia will eventually develop in and on the mycelium. Dense white mycelium, in which successive concentric rings of sclerotia are formed, will grow out from the stem over moist soil for a distance up to 10 cm.

Management
1. Grow the most resistant cultivar.
2. Rotate cotton with other crops, such as small grains.
3. Plow infected cotton residue 12 cm deep.

Soreshin

Soreshin is also called damping off, postemergence damping off, and Rhizoctonia seedling blight.

Cause. *Rhizoctonia solani* survives as sclerotia in soil or residue and as mycelia growing saprophytically on infested residue of many different plant species. Sclerotia germinate to produce mycelia that infect hypocotyls near the soil line, and in very humid weather, the fungus grows up the external surface of the stem.

The teleomorph *Thanatephorus cucumeris* produces basidiospores on a loose hyphal network growing on the stem; however, it is not certain what function basidiospores have in the life cycle of the fungus. Leaves may be infected either by basidiospores or other soilborne propagules.

Disease is favored, particularly in thick foliage, by high rainfall and warm temperatures.

Distribution. Wherever cotton is grown.

Symptoms. During cool, wet weather, seedlings may damp off either preemergence or postemergence. Hypocotyls are infected near the soil line and become necrotic, which causes cotyledons to droop, then wilt. During warm, dry soil conditions, hypocotyl lesions become reddish-brown, sunken cankers that are superficial and grow slightly into cortical tissue, or that may completely girdle stems near the soil line. If soil remains warm and dry, plants partially recover by producing new roots above the diseased areas of the stem or hypocotyl. Infrequently, angular, brown leaf spots, whose centers may fall out and give leaves a shot-hole appearance, will occur during moist weather. *See* Rhizoctonia leaf spot.

Management
1. Sow after soils have "warmed up." Temperatures warm up faster on a bed than in a furrow or on level soil.
2. Rotate cotton with other crops; however, *R. solani* has a wide host and exists saprophytically for several years in the absence of a host crop.
3. Adjust soil pH to 6.0–6.5.
4. Ensure that soil fertility is adequate.
5. Avoid chemical injury.
6. Sow only healthy, high-quality seed treated with a seed-protectant fungicide.
7. Treat soil with an in-furrow granule or spray fungicide.

Southwestern Cotton Rust

Southwestern cotton rust is also called cotton rust.

Cause. *Puccinia cacabata* (syn. *P. stakmanii)* is a heteroecious long-cycle rust that has grama grass, *Bouteloua* spp., as the alternate host. Rust overwinters as teliospores that form on grama grass in late summer. During wet summer weather and night temperatures below 28.5°C, teliospores germinate to produce basidiospores, which are disseminated by wind to cotton. Pycnia are formed on the upper surface of diseased cotton leaves, and aecia are produced on the underside of cotton leaves, bracts, green bolls, and stems. Aeciospores are released from aecia and are disseminated by wind to the alternate host, grama grass, which becomes infected during wet conditions. Uredinia, in which urediniospores are produced, are formed on grama grass and are disseminated by wind to continually reinfect grama grass throughout the growing season. Telia, in which teliospores are produced, appear on grama grass in late summer as the grass matures. High humidity and abundant rainfall favor disease.

Distribution. Mexico and the southeastern United States.

Symptoms. Uredinia on grama grass leaves are small, pale brown, raised areas with the leaf epidermis turned back around the pustule in a ragged edge. Telia are formed either in the uredinium or separately as dark brown to black raised pustules. Pycnia initially appear as small, inconspicuous yellowish spots, usually on the upper side of leaves; however, they may occur on any aboveground plant part. Pycnia develop into bright yellow pustules that later turn brown. Aecia are large, slightly raised, circular, orange-yellow lesions on the underside of cotton leaves that later fade to light yellow. In severe rust infestations, leaves curl and plants are defoliated. Stems and branches become girdled and break, and bolls are smaller in size.

Management
1. Grow resistant cultivars.
2. Apply a foliar fungicide spray before rust becomes severe.
3. Destroy grama grass in the vicinity of cotton fields; however, this is of limited value since spores may be airborne a considerable distance.

Stemphylium Leaf Spot

Cause. *Stemphylium* sp.

Distribution. Not know for certain but thought to be widespread.

Symptoms. Small, roughly circular spots have concentric rings that give a target-like appearance to the spot. Spots tend to be discrete and usually do not spread or coalesce. The disease is of minor importance.

Management. Not necessary.

Tropical Cotton Rust

Cause. *Phakopsora gossypii* (syn. *Aecidium desmium, Cerotelium desmium, Doassansia gossypii, Kuehneola gossypii,* and *P. desmium).* Only the uredinial stage is known. Dissemination is by airborne urediniospores. High relative humidity favors disease.

Distribution. India, South America, the South Pacific, the southeastern United States, and the West Indies.

Symptoms. Leaves are primarily affected, but other aboveground plant parts may also become diseased. Initially, the upper leaf surface is marked by small, brown spots, which are uredinia immersed in host tissue. Yellow-brown, circular (0.5–3.0 mm in diameter) uredinia occur in purple lesions (1–5 mm in diameter) on both leaf surfaces. Lesions later become powdery. The spots coalesce into large patches and cause defoliation. Uredinia are elongated on pedicels and branches. Telia normally are not present, but if present, they are found on lower leaf surfaces.

Management
1. Grow resistant cultivars.
2. Do not sow seed from diseased plants.
3. Remove infected residue.

Verticillium Wilt

Cause. *Verticillium dahliae;* however, many researchers refer to the causal organism as *V. albo-atrum.* The causal fungus will be referred to as *V. dahliae* in this description because of the speciation criteria stated by Isaac (1967). *Verticillium dahliae* forms microsclerotia, whereas *V. albo-atrum* forms one- and two-celled conidia and sporophores with black bases. *Verticillium tricorpus* is also reported to colonize cotton.

Several strains of *V. dahliae* exist. Pathotypes of *V. dahliae* infecting cotton are differentiated mainly by whether or not they defoliate the diseased cotton plant. The two strains most commonly found in the United States are the P-1 and P-2 strains, formerly designated as T-1 and SS-4, respectively. The P-1 strain is considered the defoliating strain, while the P-2 strain is the non-defoliating strain.

Verticillium dahliae survives as microsclerotia, or dark resting hyphae, in soil or in infested residue. Dissemination is by seed lots contaminated with microsclerotia, and by wind, surface water, and equipment that carry microsclerotia, hyphae, or conidia to host plants. *Verticillium dahliae* has a large host range; therefore, inoculum may also originate from indigenous inoculum present on weeds in an area.

Infectious propagules are thought to originate from several sources. Microsclerotia may produce multiple germ tubes and conidia. Conidia and hyphae are also found either free in the soil or in residue. Although both conidia and hyphae form hyphal germ tubes, those from hyphae infect roots and grow into the vascular system, while it is not known for certain what the role of conidia are in infection. Infectious propagules, primarily in the form of microsclerotia, are stimulated to germinate by root exudates and establish colonies on the cotton root surfaces, usually 3 to 10 mm back from the root tip. Infectious hyphae from the root colonies grow through the root cortex into the xylem tissue. In the xylem, mycelium colonize vessels and produce conidia that move upward in the tissue. Eventually, hyphae penetrate from xylem cells into surrounding parenchyma cells and eventually throughout the necrotic tissue of dead leaves, stems, and roots, where new microsclerotia are formed after several weeks or months. When tissue decays, the microsclerotia are released into the soil.

Seasonal changes in soil moisture and temperature are reported to have no influence on the colonization of roots by *V. dahliae*. However, colonization of roots by *V. tricorpus* is affected by soil temperature: Colonization is greater at 20° to 23°C than at 28° to 31°C. After infection, disease development is most severe during cool, wet weather and ceases to occur during warm, dry weather. Foliar symptoms caused by strain P-2 of Verticillium wilt decline at air temperatures of 25°C and above, but vascular wilt symptoms continue to progress. Foliar symptoms caused by strain P-1 occur at temperatures of 26°C and above.

Disease is also favored by alkaline soils. *Verticillium dahliae* grows, sporulates, and forms microsclerotia under relatively dry conditions. The only significant saprophytic growth in soil occurs at the end of the growing season when infested dead plants are returned to soil. As a pioneer colonist, *V. dahliae* grows throughout infested debris and forms numerous microsclerotia.

Distribution. In most cotton-growing areas of the world except Egypt.

Symptoms. The major effect of Verticillium wilt is a reduction of growth and development of plants that results in reduced height, lateral branching, and dry matter accumulation in leaves, stems, roots, squares, and bolls. Cotton is susceptible at all stages of growth, but symptoms appear at a blossom time that follows rainy weather and vary according to the fungus strain and soil temperature.

Cotyledons of young plants infected with strain P-2 become chlorotic and dry up. Chlorotic areas that eventually become necrotic appear on leaf margins and between veins of lower leaves. Chlorosis and necrosis of leaves, which proceed upward on the plant, result in defoliation of diseased plants.

Normally, the number of plants with foliar symptoms is fewer than the number of plants with vascular discoloration. Vascular discoloration by P-2 is limited to the lower half of diseased plants. If infection occurs later in the season, plants may only be stunted and have dark green leaves. Epinasty also may occur, but this is usually a very mild symptom.

In susceptible plants infected by the P-1 strain, symptoms in both young and old plants progress rapidly. Initially, the terminal leaf curls downward, followed by severe epinasty and general chlorosis of upper leaves, which soon defoliate. Terminal dieback normally occurs. At temperatures of 26°C and above, leaf symptoms similar to those caused by P-2 occur except that they are more severe and progress to the top of the plant. Fruiting branches and bolls may be dropped. A high percentage of bolls on the top of diseased plants do not mature, and dry up. If plants are not killed, branches may grow on lower plant portions and form bushy plants. Vascular discoloration may occur the length of the stem to the plant top.

Another foliar symptom mimics potassium deficiency symptoms; it initially appears in younger leaves as a "bronzing" at the leaf margins. The leaves eventually develop a metallic sheen and become thick and brittle, which results in an extreme distortion of leaf tissues. Apparently, *V. dahliae* causes an impairment in the uptake and translocation of potassium that is often associated with the development of potassium deficiency symptoms in leaves of plants with large boll loads.

Management

1. Grow tolerant cultivars. However, increases of inoculum of more aggressive strains occur when tolerant varieties are sown successive years in the same infested soil.
2. Rotate cotton with resistant crops. Do not grow cotton in the same soil more than once every 3 years. Most short-term rotations are of little value. Rotation with perennial rye grass has been shown to reduce the population of propagules.
3. Avoid deep cultivation.
4. Use skip-row planting where possible.
5. Sow plant populations at higher than normal levels.
6. Do not apply gin trash to field in areas where Verticillium wilt is a problem.
7. Any cultural practice that increases soil temperature will partially aid in managing the disease.

Diseases Caused by Nematodes

Dagger

Cause. *Xiphinema americanum* is an ectotrophic nematode.

Distribution. Africa.

Symptoms. Galls occur on root tips. Diseased plants are stunted.

Management. Not reported.

Lance

Lance nematode is also called Columbia lance nematode.

Cause. *Hoplolaimus columbus* and *H. galeatus*. *Hoplolaimus columbus* feeds primarily in the cortex, but *H. galeatus* feeds as both an ectoparasite and an endoparasite. Eggs are deposited in the cortex.

Distribution. The United States. *Hoplolaimus columbus* is found mainly in Alabama, Georgia, and South Carolina, and *H. galeatus* in the southeastern Coastal Plains.

Symptoms. Both nematodes cause plants to be stunted and chlorotic. *Hoplolaimus columbus* has been reported to cause defoliation during moisture stress conditions.

Management
1. Rotate cotton with nonhost plants.
2. Apply nematicides.
3. Fumigate soil.

Lesion

Cause. *Pratylenchus* spp.

Distribution. Wherever cotton is grown.

Symptoms. Little injury occurs, but root growth may be reduced. Diseased roots may be coarse and stubby and have some decay caused by secondary fungi.

Management. Not necessary.

Needle

Cause. *Longidorus africans* is an ectoparasite.

Distribution. Africa.

Symptoms. Diseased plants are stunted. Galls occur at root tips.

Management. Not reported.

Pin

Cause. *Paratylenchus brachyurus, P. hamatus,* and *P. sudanensis* are endoparasitic nematodes.

Distribution. Africa and the United States.

Symptoms. Cell breakdown and necrosis occur in diseased roots. Yields have been reported reduced; however, the nematodes are thought to be part of a disease complex rather than a single causal agent.

Management. Not reported.

Reniform

Cause. *Rotylenchulus reniformis* is a nematode where the female either is completely in the root or has just the posterior end protruding. Fifty to 80 eggs are laid outside roots in a gelatinous matrix that encompasses the exposed portion of the female body. Eggs are resistant to drying and hatch under favorable conditions. Larvae molt four times and develop into male and females; however, only females have been observed feeding. The life cycle is completed in 17 to 23 days.

Distribution. The southeastern United States in localized areas in fields; however, reniform nematodes are a potential problem where cotton is grown in subtropical and tropical areas.

Symptoms. Diseased plants are severely stunted and chlorotic, and may wilt. Root growth is reduced; few large roots are produced, but numerous coarse and stubby lateral roots are present. Decay caused by secondary fungi may occur. Weedy areas occur within a field where cotton has been affected.

Management
1. Rotate cotton with rice, sorghum, oat, mustard, turnip, corn, pepper, and grasses.
2. Fumigate soil.

Ring

Cause. *Criconemella* spp. are ectoparasites.

Distribution. The southeastern United States.

Symptoms. Symptoms on cotton are not well known. The aboveground symptom is a mild stunting. Belowground symptoms are lesions on roots and some decay caused by secondary organisms.

Management. Not reported.

Root Knot

Cause. *Meloidogyne incognita* does not survive if the temperature averages below 3°C during the coldest month. Eggs in infested soil hatch to produce larvae that move through soil to penetrate root tips; however, entry is not restricted only to this area of the root. Larvae migrate through the root to vascular tissues, become sedentary, and commence feeding on undifferentiated provascular tissue. Excretions of larvae stimulate cell proliferation, which results in the development of syncythia from which the larvae derive their food. Females lay 500–1000 eggs in a gelatinous matrix that may be pushed out of roots into soil.

Distribution. Wherever cotton is grown in sandy or sandy loam soil.

Symptoms. Diseased plants are less vigorous and shorter than healthy plants. Severely diseased plants are lighter green and quickly wilt when under moisture stress during the day, but recover turgidity at night. Eventually, wilting becomes permanent.

The taproot is frequently missing or is greatly reduced in size. Lateral roots contain numerous spindle-shaped galls or knots that become noticeable 3 to 4 weeks after infection. Galls, which may be found in lesser numbers on the taproot, eventually grow up to 6 mm in diameter.

Management
1. Rotate cotton with alfalfa, oat, barley, and sorghum for 2 or more years.
2. Keep an infested field fallow for a "reasonable" time and control weeds.
3. Fumigate soil.
4. Grow resistant cultivars.
5. Apply nematicide.

Scutellonema spp.

Cause. *Scutellonema* spp. are endoparasites.

Distribution. Africa and the United States.

Symptoms. Symptoms are not well defined on cotton, but a yield loss has been reported with diseased plants.

Management. Not reported.

Spiral

Cause. *Helicotylenchus* spp. and *Scutellonema* spp. are endoparasites.

Distribution. Africa and the United States.

Symptoms. The aboveground symptom for diseased plants is stunting. Belowground symptoms are a reduction in the number of feeder roots and some root rot caused primarily by secondary fungi.

Management. Not reported.

Sting

Cause. *Belonolaimus longicaudatus* feeds mostly on the outside of root tips and along the sides of succulent roots without penetrating them or becoming attached to them.

Distribution. Sandy soils in the Coastal Plains of the southeastern United States.

Symptoms. Diseased plants are stunted and chlorotic; frequently, seedlings die. Roots are stubby and coarse and have necrotic lesions. Roots may end in enlargements caused by repeated forming and killing of new branches. Nematode injury may provide an entrance wound for the Fusarium wilt fungus, *Fusarium oxysporum* f. sp. *vasinfectum.*

Management
1. Rotate cotton with tobacco, watermelons, and crotalaria.
2. Fumigate soil.
3. Apply nematicides.

Stubby Root

Cause. *Paratrichodorus* spp., *P. minor (syn. Trichodorus christiei).*

Distribution. Wherever cotton is grown.

Symptoms. Little injury generally occurs to diseased plants, but root growth is reduced. Roots may be coarse and stubby, and decay caused by secondary fungi may occur.

Management. Not necessary.

Stunt

Cause. *Merlinius* spp. and *Tylenchorhynchus* spp. are ectoparasites.

Distribution. Thought to be widespread.

Symptoms. Aboveground symptoms for diseased plants are moderate plant stunting and chlorosis. The belowground symptom is a reduction in root growth.

Management. Not reported.

Diseases Caused by Phytoplasmas

Phyllody

Cause. A phytoplasma disseminated by the leafhopper *Orosius* sp.

Distribution. Ivory Coast, Mali, and Upper Volta.

Symptoms. Diseased leaves are string- or strap-like. Diseased plants are sterile and have virescent, or green, floral parts.

Management. Not reported.

Small Leaf

Cause. A phytoplasma.

Distribution. Cuba and India.

Symptoms. Diseased leaves are chlorotic, small, and malformed. Flowers tend to be small and may abort. Bolls are prematurely shed.

Management. Not reported.

Diseases Caused by Viruses

Abutilon Mosaic

Cause. Abutilon mosaic virus (AbMV) is in the geminivirus III group. AbMV survives in a wide number of host plants, but *Sida* spp. are the most important. AbMV is disseminated by the whitefly, *Bemisia tabaci,* and is graft-transmissible. AbMV is not transmitted by seed.

Distribution. Wherever cotton is grown.

Symptoms. Symptoms are most obvious on young diseased plants and become less pronounced on old plants. Diseased plants are dwarfed because of shortened internodes. Leaves are crinkled, blistered, and malformed, and have a conspicuous mosaic of chlorotic and green areas. Chlorotic areas, which may be limited by veins, become reddish and disappear in older leaves. Young leaves are smaller and wrinkled, and have fewer lobes than healthy leaves.

Management. Abutilon mosaic is not a serious disease and management is not necessary.

Anthocyanosis

Cause. Cotton anthocyanosis virus (CAV) survives in *Gossypium barbadense, Sida* spp., and ratoon cotton. CAV has been artificially inoculated into and recovered from several other plants. CAV is disseminated by the aphid *Aphis gossypii* but not by seed or sap. A feeding period of 12 hours or longer is necessary for aphids to become moderately infective. CAV is persistent within the aphid.

Distribution. Brazil.

Symptoms. Anthocyanosis is most severe on lower and middle leaves but may affect the upper leaves of older plants. Leaves develop chlorotic areas that turn reddish purple in sunlight. Such purple areas are limited by veins, but the entire leaf, except for veins and a narrow band adjacent to the veins, may become purple.

Management. Control insects on cotton throughout the growing season.

Blue Disease

Cause. An unknown virus that is transmitted by the aphid *Aphis gossypii*. Plants of various ages are affected.

Distribution. Africa.

Symptoms. Diseased plants are dwarfed, and stems have a "zigzag" pattern of growth. Apical leaves initially bulge between main veins and toward the base. Eventually, the entire leaf is involved. Affected areas are light green but later turn a blue-green.

Management
1. Grow resistant varieties.
2. Practice sanitation and destroy residue.

Cotton Leaf Crumple

Cause. The cotton leaf crumple virus (CLCV) is in the geminivirus group. CLCV is transmitted by the whitefly *Bemisia tabaci* at an optimum temperature of 32°C under experimental conditions, and by grafting techniques. In Arizona, the causal organism overwinters in cheeseweed, *Malva parviflora,* and in cultivated beans, *Phaseolus vulgaris.* Numerous plant species in the Malvaceae and Leguminosae are also hosts. Perennial cotton serves as an inoculum reservoir. Plants are infected at all stages of development.

Distribution. India, northern Mexico, and the southwestern United States.

Symptoms. Symptoms are most severe in perennial cotton. Leaves curl downward due to hypertrophy of interveinal tissue. There is frequent vein distortion and clearing and a mosaic appearance to leaves. Reddish spots sometimes occur in chlorotic, interveinal tissue of senescent leaves. Floral parts tend to be irregular-shaped. Stunting commonly occurs in perennial cotton. Although plants of all ages may become diseased, yield losses are most severe when plants are infected early in the growing season.

Management. Tolerance and resistance occurs in some cultivars; however, the following have been suggested as management measures:
1. Grow tolerant or resistant cultivars.
2. Control weed hosts.
3. Eliminate diseased perennial cotton.
4. Sow clean seed.

Cotton Small Leaf

Cotton small leaf is also called cotton stenosis, and smalling. Cotton small leaf may be the same disease as small leaf, which is caused by a phytoplasma. However, the descriptions of the symptoms in the literature are sufficiently different to discuss them here as two separate diseases.

Cause. Cotton small leaf virus (CSLV); however, CSLV may be a phytoplasma. It is not known how CSLV survives and is disseminated.

Distribution. India and Pakistan.

Symptoms. The aerial portions of diseased cotton plants are stunted. Leaves, which develop in clusters, are malformed, variously lobed, and of different sizes and shapes. Leaves and epicalyx are mottled, and enations are produced on the lower surfaces of veins. Flowers remain small, and bolls do not form.

The taproot ends abruptly and gives rise to a large number of adventitious roots. Diseased plants can be easily pulled out of the ground.

Management. Not reported.

Cotton Terminal Stunt

Cause. Cotton terminal stunt virus (CTSV). It is not known how CTSV survives. CTSV is graft-transmissible, and insects are suspected of being vectors.

Distribution. Mexico and the adjacent areas in the southern United States (Texas).

Symptoms. Terminal growth of stems and branches is stunted. Young leaves are small, misshapen, and mottled and cup either upward or downward. Tan to brown streaks occur in the xylem of the main stem. Immature bolls have a dark internal discoloration. Bolls, blooms, and squares are shed.

Management. Not reported.

Leaf Curl

Leaf curl is also called cotton leaf curl and cotton leaf crinkle.

Cause. Cotton leaf curl virus (CLCuV) is in the geminivirus group. CLCuV survives in several different hosts and is transmitted to cotton at all stages of growth by the whitefly, *Bemisia tabaci*. CLCuV is not transmitted by sap, seed, or soil but has been transmitted by grafting. Different strains of CLCuV exist.

Distribution. Africa and Pakistan.

Symptoms. Only immature leaves develop disease symptoms. The lower surface of smaller veins has an intermittent thickening that eventually becomes continuous. When backlighted, veins of a diseased leaf are darker green than the rest of the leaf.

Leaves formed after infection are shortened, small, and crinkled and curl either upward or downward at the edges. Enations are often formed on the thickened lower sides of primary veins. Internodes are lengthened, twisted, and flattened. All parts of a diseased plant are brittle. Yield may be greatly reduced.

Management
1. Grow resistant varieties.
2. Control whiteflies in the greenhouse.

Leaf Mottle

Cause. Leaf mottle virus (LMV). It is not known how LMV survives and is disseminated.

Distribution. Sudan.

Symptoms. Younger diseased leaves have a pronounced mottling near the veins. Leaf lobes are distorted and elongated, and severely diseased leaves are a light green. The main stem is stunted, and flowering is reduced.

Management. Not reported.

Leaf Roll

Cause. Leaf roll virus (LRV) is transmitted by aphids, including *Aphis gossypii, A. laburni, Myzus persicae,* and *Epitetranychus althaeae.*

Distribution. Russia.

Symptoms. Diseased plants are stunted, droop, and become prostrate and spreading. Leaves have chlorotic margins, are shiny and brittle, and curl upward at tips but downward at the edges. Stems and petioles become reddish and sticky. Fruiting is reduced. Roots are poorly developed.

Management. Some cultivars are less susceptible to infection by LRV in some geographic areas of Russia.

Mosaic

Cause. An unknown virus transmitted by the whitefly, *Bemisia tabaci.*

Distribution. Africa.

Symptoms. Diseased leaves have a mosaic and are mottled.

Management. Grow tolerant varieties.

Psylosis

Cause. Virus that is transmitted by psyllids.

Distribution. Africa.

Symptoms. Diseased leaves are purple and have enations and thickened veins. There is a twisting and curving of internodes and petioles.

Management. Not reported.

Terminal Stunt

Terminal stunt is also called terminal.

Cause. Possibly a virus, although the causal agent has been suspected to be a phytoplasma.

Distribution. The United States.

Symptoms. Terminal diseased leaves are puckered and twisted. Xylem tissue is frequently mottled and otherwise discolored.

Management. Not reported.

Tobacco Streak

Cause. Tobacco streak virus (TSV) is in the ilarvirus group. TSV survives in a wide number of host plants, but it is not known how it is disseminated to cotton, although insects are suspected. Disease usually occurs too late in the season to do much damage.

Distribution. TSV occurs on several other hosts around the world but is only known to infect cotton in Brazil.

Symptoms. Diseased plants are slightly stunted and produce more axillary twigs than healthy plants. Initial infection of leaves results in necrotic rings or spots. Later, the leaf size is reduced, and there are blisters along the veins and mosaic patterns between secondary veins.

Management. Not necessary.

Vein Clearing

Cause. Possibly a virus.

Distribution. The United States (Texas).

Symptoms. Diseased leaves have a vein clearing initially, followed by a vein banding and, eventually, mottling. Young leaves cup downward, and plants are stunted.

Management. Not reported.

Veinal Mosaic

Cause. Veinal mosaic virus (VMV). It is not known how VMV survives or is disseminated, but it is graft-transmissible in inoculation studies.

Distribution. Brazil.

Symptoms. Plants have shortened internodes and are slightly stunted, which gives them a compact and often bunchy appearance. Diseased leaves are dark green and have veinal mosaic, roughening, and a downward curling of margins. Veinal mosaic may also occur on bracts. Severely diseased leaves may be rolled. Necrotic lesions of veins sometimes occur.

Management. Not reported.

Selected References

Al-Beldawi, A. S., and Pinckard, J. A. 1970. Control of *Rhizoctonia solani* on cotton seedlings by means of benomyl. Plant Dis. Rptr. 54:76-80.

Alderman, S. C., and Hine, R. B. 1981. Pathogenicity and occurrence of strands of *Phymatotrichum omnivorum*. (Abstr.) Phytopathology 71:198.

Arndt, C. H. 1944. Infection of cotton seedlings by *Colletotrichum gossypii* as affected by temperature. Phytopathology 34:861-869.

Arndt, C. H. 1946. Effect of storage conditions on the survival of *Colletotrichum gossypii*. Phytopathology 36:24-29.

Arndt, C. H. 1953. Survival of *Colletotrichum gossypii* on cotton seed in storage. Phytopathology 43:220.

Ashworth, L. J., Jr., Hildebrand, D. C., and Schroth, M. N. 1970. *Erwinia*-induced internal necrosis of immature cotton bolls. Phytopathology 60:602-607.

Ashworth, L. J., Jr., Rice, R. E., McMeans, J. L., and Brown, C. M. 1971. The relationship of insects to infection of cotton bolls by *Aspergillus flavus*. Phytopathology 61:488-493.

Ashworth, L. J., Jr., Galanopoulos, N., and Galanopoulou, S. 1983. Selection of pathogenic strains of *Verticillium dahliae* and their influence on the useful life of cotton cultivars in the field. Phytopathology 74:1637-1639.

Baehr, L. F., and Pinckard, J. A. 1970. Histological studies on the mode of penetration of boll-rotting organisms into developing cotton bolls. (Abstr.) Phytopathology 60:581.

Bagga, H. S. 1970. Fungi associated with cotton boll rot in the Yazoo-Mississippi Delta, 1966-1968. Plant Dis. Rptr. 54:796-798.

Baird, R. E., Gitaitis, R. D., and Herzog, G. A. 1997. First report of cotton lint rot by *Pantoea agglomerans* in Georgia. Plant Dis. 81:551.

Brown, J. K., and Nelson, M. R. 1984. Geminate particles associated with cotton leaf crumple disease in Arizona. Phytopathology 74:987-990.

Brown, J. K., and Nelson, M. R. 1986. Cotton leaf crumple virus transmitted from naturally infected bean from Mexico. Plant Dis. 70:981.

Brown, J. K., and Nelson, M. R. 1987. Host range and vector relationships of cotton leaf crumple virus. Plant Dis. 71:522-524.

Brown, J. K., Mihail, J. D., and Nelson, M. R. 1987. Effects of cotton leaf crumple virus on cotton inoculated at different growth stages. Plant Dis. 71:699-703.

Calvert, O. H., Sappenfield, W. P., Hicks, R. D., and Wyllie, T. D. 1964. The Cercospora-Alternaria leaf blight complex of cotton in Missouri. Plant Dis. Rptr. 48:466-467.

Cleveland, T. E., and Cotty, P. J. 1991. Invasiveness of *Aspergillus flavus* isolates in wounded cotton bolls is associated with production of a specific fungal polygalacturonase. Phytopathology 81:155-158.

Colyer, P. D. 1988. Frequency and pathogenicity of *Fusarium* spp. associated with seedling diseases of cotton in Louisiana. Plant Dis. 72:400-402.

Colyer, P. D., Micinski, S., and Vernon, P. R. 1991. Effect of thrips infestation on the development of cotton seedling diseases. Plant Dis. 380-382.

Correll, J. C. 1986. Powdery mildew of cotton caused by *Oidiopsis taurica* in California. Plant Dis. 70:259.

Cotty, P. J. 1987. Evaluation of cotton cultivar susceptibility to Alternaria leaf spot. Plant Dis. 71:1082-1084.

Cotty, P. J. 1987. Temperature-induced suppression of Alternaria leaf spot of cotton in Arizona. Plant Dis. 71:1138-1140.

Cotty, P. J. 1989. Effects of cultivar and boll age on aflatoxin in cottonseed after inoculation with *Aspergillus flavus* at simulated exit holes of the pink bollworm. Plant Dis. 73:489-492.

Cotty, P. J. 1991. Effect of harvest date on aflatoxin contamination of cottonseed. Plant Dis. 75:312-314.

DeVay, J. E., Forester, L., Garber, R. H., and Butterfield, E. J. 1974. Characteristics and concentrations of propagules of *Verticillium dahliae* in air-dried field soils in relation to the prevalence of Verticillium wilt in cotton. Phytopathology 64:22-29.

DeVay, J. E., Weir, B. L., Wakeman, R. J., and Stapleton, J. J. 1997. Effects of *Verticillium dahliae* infection of cotton plants *(Gossypium hirsutum)* on potassium levels in leaf petioles. Plant Dis. 81:1089-1092.

Dizon, T. O., and Reyes, T. T. 1986. Phytopathological note: Occurrence of Sclerotium boll rot of cotton in the Philippines. Philipp. Phytopathol. 22:68-69.

Evans, G., Wilhelm, S., and Snyder, W. C. 1966. Dissemination of the Verticillium wilt fungus with cottonseed. Phytopathology 56:460-461.

Farr, D. F., Bills, G. R., Chamuris, G. P., and Rossman, A. Y. 1989. Fungi on Plants and Plant Products in the United States. American Phytopathological Society, St. Paul, MN. 1252 pp.

Garber, R. H., and Presley, J. T. 1971. Relation of air temperature to development of Verticillium wilt on cotton in the field. Phytopathology 61:204-207.

Gergon. E. B. 1982. Diplodia boll rot of cotton: Pathogenicity and histopathology. (Abstr.) Philipp. Phytopathol. 18:6.

Grinstein, A., Fishler, G., Katan, J., and Hakohen, D. 1983. Dispersal of the Fusarium wilt pathogen in furrow-irrigated cotton in Israel. Plant Dis. 67:742-743.

Hancock, J. G. 1972. Root rot of cotton caused by *Pythium splendens*. Plant Dis. Rptr. 5:973-975.

Hillocks, R. J. (Ed.). 1992. Cotton Diseases. C.A.B. International. Wallingford, United Kingdom. 415 pp.

Hood, M. E., and Shew, H. D. 1997. Reassessment of the role of saprophytic activity in the ecology of *Thielaviopsis basicola*. Phytopathology 87:1214-1219.

Hopkins, J. C. F. 1931. *Alternaria gossypina* (Thuem.) Comb. nov. causing a leaf spot and boll rot of cotton. Trans. Br. Mycol. Soc. 16:136-144.

Huisman, O. C. 1988. Seasonal colonization of roots of field-grown cotton by *Verticillium dahliae* and *V. tricorpus*. Phytopathology 78:708-716.

Huisman, O. C., and Ashworth, L. J., Jr. 1976. Influence of crop rotation on survival of *Verticillium albo-atrum* in soils. Phytopathology 66:978-981.

Isaac, I. 1967. Speciation in *Verticillium*. Ann. Rev. Plant Pathol. 5:201-222.

Jeffers, D.P., Smith, S. N., Garber, R. H., and DeVay, J. E. 1984. The potential spread of the cotton Fusarium wilt pathogen in gin trash and planting seed. (Abstr.) Phytopathology 74:1139.

Joanne, N.R. W., Groan, R. G., and, Stapleton, J. J. 1977. Effect of water potential and temperature on growth, sporulation, and production of microsclerotia by *Verticillium dahliae*. Phytopathology 67:637-644.

Joaquin, T. R., and Owe, R. C. 1990. Reassessment of vegetative compatibility relationships among strains of *Verticillium dahliae* using nitrate-nonutilizing mutants. Phytopathology 80:1160-1166.

Johnson, L. F., and Chambers, A. Y. 1973. Isolation and identity of three species of *Pythium* that cause cotton seedling blight. Plant Dis. Rptr. 57:848-852.

Johnson, L. F., Baird, D. D., Chambers, A. Y., and, Shamiyeh, N. B. 1978. Fungi associated with postemergence seedling disease of cotton in three soils. Phytopathology 68:917-920.

Johnson, W. M., Johnson, E. K., and, Brinkerhoff, L. A. 1980. Symptomatology and formation of microsclerotia in weeds inoculated with *Verticillium dahliae* from cotton. Phytopathology 70:31-35.

Jones, J. P. 1961. A leaf spot of cotton caused by *Corynespora cassiicola*. Phytopathology 51:305-308.

King, C. J., and Presley, J. T. 1942. A root rot of cotton used by *Thielaviopsis basicola*. Phytopathology 32:752-761.

Klich, M. A. 1987. Relation of plant water potential at flowering to subsequent cottonseed infection by *Aspergillus flavus*. Phytopathology 77:739-741.

Koch, D. O., Jeger, M. J., Gerik, T. J., and Kenerly, C. M. 1987. Effects of plant density on progress of Phymatotrichum root rot in cotton. Phytopathology 77:1657-1662.

Lee, L. S., Lacey, P. E., and Goynes, W. R. 1987. Aflatoxin in Arizona cottonseed: A model study of insect-vectored entry of cotton bolls by *Aspergillus flavus*. Plant Dis. 71:997-1001.

Ling, I., and Yang, J. Y., 1941. Stem blight of cotton caused by *Alternaria macrospora*. Phytopathology 32:752-761.

Mathre, D. E., Ravescroft, A. V., and Garber, R. H. 1966. The role of *Thielaviopsis basicola* as a primary cause of yield reduction in cotton in California. Phytopathology 56:1213-1216.

Mauk, P. A., and Hine, R. B. 1987. Internal root colonization and chlamydospore production by *Thielaviopsis basicola* in mature Pima cotton. (Abstr.) Phytopathology 77:1240.

Mauk, P. A., and Hine, R. B. 1988. Infection, colonization of *Gossypium hirsutum* and *G. Barbadense*, and development of black root rot caused by *Thielaviopsis basicola*. Phytopathology 78:1662-1667.

McCarter, S. M. 1972. Effect of temperature and boll injuries on development of Diplodia boll rot of cotton. Phytopathology 62:1223-1225.

McLean, K. S., and Roy, K. W. 1991. Weeds as a source of *Colletotrichum capsici* causing anthracnose on tomato fruit and cotton seedlings. Can. J. Plant Pathol. 13:131-134.

Mertely, J., Gannaway, J., and Kaufman, H. 1994. A new disease of seedling cotton caused by *Pseudomonas syringae*. (Abstr.). Phytopathology 84:1094.

Nelson, M. R., and Stowell, L. J. 1989. Dispersion characteristics of a whitefly-transmitted geminivirus in the field. (Abstr.). Phytopathology 79:1219.

Paplomatas, E. J., Elena, K., and Lascaris, D. 1995. First report of *Phytophthora boehmeriae* causing boll rot of cotton. Plant Dis. 79:860.

Pinckard, J. A., and Guidroz, G. F. 1973. A boll rot of cotton caused by *Phytophthora parasitica*. Phytopathology 63:896-899.

Pizzinatto, M. A., Soave, J., and Cia, E. 1983. Patogenicidade de *Botryodiplodia theobromae* Pat. a plantas de diferentes idades e macao de algodoeiro (*Gossypium hirsutum* L.). Fitopatologia Brasileira 8:223-228 (in Portuguese).

Pullman, G. S., and DeVay, J. E. 1982. Effect of soil flooding and paddy rice culture on the survival of *Verticillium dahliae* and incidence of Verticillium wilt in cotton. Phytopathology 72:1285-1289.

Pullman, G. S., and DeVay, J. E. 1982. Epidemiology of Verticillium wilt of cotton: A relationship between inoculum density and disease progression. Phytopathology 72:549-554.

Pullman, G. S., and DeVay, J. E. 1982. Epidemiology of Verticillium wilt of cotton: Effects of disease development of plant phenology and lint yield. Phytopathology 72:554-559.

Rane, M. S., and Patel, M. K. 1956. Diseases of cotton in Bombay. I. Alternaria leaf spot. Indian Phytopathol. 9:106-113.

Rane, M. S., and Patel, M. K. 1956. Diseases of cotton in Bombay. II. Helminthosporium leaf spot. Indian Phytopathol. 9:169-173.

Rathaiah, Y. 1976. Reaction of cotton species and cultivars to four isolates of *Ramularia areola*. Phytopathology 66:1007-1009.

Rathaiah, Y. 1977. Spore germination and mode of cotton infection by *Ramularia areola*. Phytopathology 67:351-357.

Rotem, J., Eidt, J., Wendt, U., and Kranz, J. 1988. Relative effects of *Alternaria alternata* and *A. macrospora* on cotton crops in Israel. Plant Pathol. 37:16-19.

Rotem, J., Wendt, U., and Kranz, J. 1988. The effect of sunlight on symptom expression of *Alternaria alternata* on cotton. Plant Pathol. 37:12-15.

Rush, C. M., and Gerik, T. J. 1989. Relationship between postharvest management of grain sorghum and Phymatotrichum root rot in the subsequent cotton crop. Plant Dis. 73: 304-305.

Rush, C. M., Gerik, T. J., and Kenerley, C. M. 1985. Atypical disease symptoms associated with Phymatotrichum root rot of cotton. Plant Dis. 69:534-537.

Rush, C. M., Gerik, T. J., and Lyda, S. D. 1984. Factors affecting symptom appearance and development of Phymatotrichum root rot of cotton. Phytopathology 74:1466-1469.

Rush, C. M., Lyda, S.D., and Gerik, T. J. 1984. The relationship between time of cortical senescence and foliar symptom development of Phymatotrichum root rot of cotton. Phytopathology 74:1464-1466.

Schnathorst, W. C. 1973. Nomenclature and physiology of *Verticillium* spp. with emphasis on the *V. albo-atrum* vs *V. dahliae* controversy, pp. 1-19. *In* Verticillium Wilt of Cotton. ARS-S19. Proceedings of a workshop conference held at the National Cotton Pathology Research Laboratory, College Station, Texas, 30 Aug.-1 Sept. 1971. Published by U.S. Dept. Agric., Agric. Res. Serv., Publication Div., Beltsville, MD. 398 pp.

Sciumbato, G. L., and Pinckard, J. A. 1972. *Alternaria macrospora* leaf spot of cotton in Louisiana in 1974. Plant Dis. Rptr. 58:201-202.

Shtienberg, D. 1992. Development and evaluation of guidelines for the initiation of chemical control of Alternaria leaf spot in Pima cotton in Israel. Plant Dis. 76:1164-1168.

Snow, J. P., and Mertley, J. C. 1979. A boll rot of cotton caused by *Colletotrichum capsici*. Plant Dis. Rptr. 63:626-627.

Snow, J. P., and Sanders, D. E. 1979. Role of abscised cotton flowers, bolls and squares in production of inoculum by boll-rotting *Fusarium* spp. Plant Dis. Rptr. 63:288-289.

Sobers, E. K., and Littrell, R. H. 1974. Pathogenicity of three species of *Cylindrocladium* to select hosts. Plant Dis. Rptr. 58:1017-1019.

Sparnicht, R. H., and Roncadori, R. W. 1972. Fusarium boll rot of cotton: Pathogenicity and histopathology. Phytopathology 62:1381-1386.

Tarr, S. A. J. 1964. Virus diseases of cotton. Commonw. Mycol. Inst. Misc. Pub. No. 18.

Watkins, G. M. (Ed.). 1981. Compendium of Cotton Diseases. The American Phytopathological Society, St. Paul, MN. 85 pp.

Wheeler, J. E., and Hine, R. B. 1972. Influence of soil temperature and moisture on survival and growth of strands of *Phymatotrichum omnivorum*. Phytopathology 62:828-832.

Wheeler, M. H., Stipanovic, R. D., and Puckhaber, L. S. 1996. Equisetin as a phytotoxin and its discovery in cultures of *Fusarium equiseti* and *F. semitectum* isolated from cottonseed. (Abstr.). Phytopathology 86:S46.

Wilhelm, S., Sagen, J. E., and Tietz, H. 1974. Resistance to Verticillium wilt in cotton: Sources, techniques of identification, inheritance trends, and the resistance potential of multiline cultivars. Phytopathology 64:924-931.

Wilhelm, S., Sagen, J. E., and Tietz, H. 1985. Phenotype modification in cotton for control of Verticillium wilt through dense plant population culture. Plant Dis. 69:283-288.

Wrather, J. A., Sappenfield, W. P., and Baldwin, C. H. 1986. Colonization of cotton buds by *Xanthomonas campestris* pv. *malvacearum*. Plant Dis. 70:551-552.

Zutra, D., and Orion, D. 1982. Crown gall bacteria *(Agrobacterium radiobacter* var. *tumefaciens)* on cotton roots in Israel. Plant Dis. 66:1200-1201.

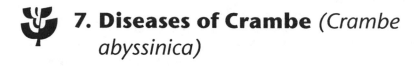

7. Diseases of Crambe *(Crambe abyssinica)*

Disease Caused by Bacteria

Bacterial Blight

Cause. *Xanthomonas campestris* is seedborne on the silicle and surface of the seed coat and possibly is systemic within the seed. The bacterium is yellow-pigmented and gram-negative.

Distribution. The United States (Missouri).

Symptoms. Diseased tissues are light tan, dry, and necrotic and often are bordered by a chlorotic halo. As the disease progresses, leaf symptoms are accompanied by stems with black streaks that affect the vascular tissue.

Management. Not reported.

Diseases Caused by Fungi

Alternaria Leaf Spot

Cause. *Alternaria circinans* (syn. *A. brassicicola*) is seedborne.

Distribution. Presumably wherever crambe is grown, but this is not known for certain.

Symptoms. Inoculated plants have small, black, linear lesions on the petioles, stems, veins of lower leaf surfaces, and seedpods. Stem lesions (1–3 mm long) are discolored brown to black and frequently coalesce or girdle diseased stems, which causes them to dry up. Seedpods are blackened and reduced in size, and seed maturation is prevented. Leaf spots (1–5 mm) are brown to black and oval-shaped. Diseased leaves quickly become chlorotic and fall from the plant. Seedlings may damp off after emergence through the soil.

Management

1. Because some crambe cultivars have a low level of resistance to *A. circinans,* it is possible to screen for resistance.
2. Treat seed with a seed-protectant fungicide.

Aphanomyces Root Rot

Cause. *Aphanomyces raphani* presumably survives as oospores in infested residue and soil. During the life cycle of *A. raphani* on other crops, zoosporangia, in which zoospores are produced, are formed from oospores. Zoospores are released from the zoosporangia and "swim" through soil moisture to the roots of host plants, where they germinate to form a germ tube. Oospores are eventually produced in diseased roots. Disease is most severe under wet soil conditions.

Distribution. Not known. Aphanomyces root rot has not been observed in the field but has been studied under greenhouse conditions.

Symptoms. In inoculation studies, diseased seedlings were stunted; however, inoculated seedlings did not damp off. Roots and hypocotyls were discolored black. *Aphanomyces raphani* moves up the hypocotyl into the cotyledons, causing a chlorosis and blackening of the diseased tissues. Under magnification, oospores are evident in the diseased root cortex.

Management. Not reported.

Fusarium Wilt

Cause. *Fusarium oxysporum* f. sp. *conglutinans* race 2. Seedlings are susceptible to damping off until they grow to a height of approximately 6.5 cm.

Distribution. Not known.

Symptoms. Diseased plants are stunted. An early symptom of diseased leaves is an interveinal yellowing that is frequently followed by leaf abscission.

Management. Not reported.

Sclerotinia Stem Rot

Cause. *Sclerotinia sclerotiorum.*

Distribution. The United States.

Symptoms. The first symptom is the presence of prematurely ripened plants that are scattered or are grouped together among green plants in a field. Diseased plants are easily pulled from the soil.

    Diseased stems have pale gray lesions, with faint concentric markings, that develop from the soil line or from the axils of branches or leaves. Initially, lesions are water-soaked, then expand to girdle the stems and kill the plant. Diseased stems become bleached, have a chalky white appearance, and tend to shred longitudinally. Sclerotia commonly appear as hard, black, grain-sized bodies inside the bottom stems of dead plants. Under moist conditions, sclerotia occur in all parts of diseased stems and pods.

Management. Not reported.

Diseases Caused by Viruses

Beet Western Yellows

Cause. Beet western yellows virus (BWYV) is in the luteovirus group. BWYV overwinters in several weed hosts and in sugar beet. BWYV is transmitted primarily by the green peach aphid, *Myzus persicae,* and infrequently by other aphid species. Known hosts for the virus are *Brassica napus* var. *napobrassica, Capsella bursa-pastoris, Cardamine oligosperma, Cardaria draba, Erodium cicutarium, Lactuca scariola, Matricaria maritima, Polygonum lapathifolium, Raphanus raphanistrum, Senecio vulgaris, Sisymbrium officinale,* and *Stellaria media.*

Myzus persicae can acquire BWYV from a diseased plant after a 5-minute feeding and can transmit the virus after a 10-minute feeding on a healthy host plant. An aphid can remain infective for several days. Different strains of BWYV exist.

An ST9-associated RNA is a newly discovered type of plant infectious agent that depends on BWYV for encapsidation but not for replication. The agent is not a satellite virus.

Distribution. BWYV has been reported on sugar beet from Asia, Europe, and North America. It is not known what the distribution of BWYV on crambe is.

Symptoms. Inoculated plants initially are generally a light green, followed by an interveinal reddening of diseased lower leaves. The reddening intensifies and involves more leaf tissue as the disease progresses. Older diseased leaves become thickened, brittle, and discolored red except for green areas adjacent to the veins.

Young leaves inoculated with a severe virus strain develop black, necrotic, pinpoint-sized spots. Later, necrotic areas develop along diseased stems. Severely diseased plants may be killed, and yield is reduced on plants that are not killed.

Management. Not reported.

Turnip Mosaic

Cause. Turnip mosaic virus (TuMV) is in the potyvirus group. TuMV is transmitted mechanically and by the aphids *Myzus persicae* and *Brevicoryne brassicae.* TuMV is not seedborne.

Distribution. TuMV is distributed in Europe, Japan, North America, and South Africa on other hosts, but distribution on crambe is not reported.

Symptoms. Inoculated plants have a systemic mottling that occurs in all diseased leaves. Stems are necrotic and inflorescences are stunted, distorted, and chlorotic. Seedpods do not mature.

Management. Not reported.

Selected References

Armstrong, G. M., and Armstrong, J. K. 1974. Wilt of *Brassica carinata, Crambe abyssinica,* and *C. hispanica* caused by *Fusarium oxysporum* f. sp. *conglutinans* race 1 or 2. Plant Dis. Rptr. 58:479-480.

Duffus, J. E. 1975. Effects of beet western yellows virus on crambe in the greenhouse. Plant Dis. Rptr. 59:886-888.

Holcomb, G. E., and Newman B. E. 1970. *Alternaria circinans* and other fungal pathogens on *Crambe abyssinica* in Louisiana. Plant Dis. Rptr. 54:28.

Horvath, J. 1972. Reaction of crambe (family: Cruciferae) to certain plant viruses. Plant Dis. Rptr. 56:665-666.

Humaydan, H. S., and Williams, P. H. 1975. Additional cruciferous hosts of *Aphanomyces raphani.* Plant Dis. Rptr. 59:113-116.

Kilpatrick, R. A. 1976. Fungal flora of crambe seeds and virulence of *Alternaria brassicicola.* Phytopathology 66:945-948.

Leppik, E. E. 1973. Diseases of crambe. Plant Dis. Rptr. 57:704-708.

Mihail, J. D., Taylor, S. J., Verslues, P. E., and Hodge, N. C. 1993. Bacterial blight of *Crambe abyssinica* in Missouri caused by *Xanthomonas campestris.* Plant Dis. 77:569-574.

Thornberry, H. H., and Phillippe, M. R. 1965. Crambe: Susceptibility to some plant viruses. Plant Dis. Rptr. 49:74-77.

White, G. A., and Higgins, J. J. 1966. Culture of crambe, a new industrial oilseed crop. USDA A. Res. Serv. Production Res. Rept. No. 95.

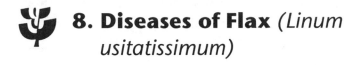

8. Diseases of Flax *(Linum usitatissimum)*

Diseases Caused by Fungi

Alternaria Flower and Stem Blight

Alternaria flower and stem blight is sometimes called Alternaria blight.

Cause. *Alternaria lini* is a good saprophyte that overwinters as mycelia and conidia in infested residue. During moist weather, conidia are produced from mycelia on residue and disseminated by wind to host plants. *Alternaria lini* infects either moribund flowers or flowers that have died. Using the colonized flowers as an energy source, the mycelium spreads to stems and leaves.

Distribution. India.

Symptoms. Initially, the flowers fail to open during the day. Minute, dark brown spots appear near the base of the calyx and gradually extend to the pedicel, causing the flower to decay. Capsules formed at the time of infection may be similarly diseased.

When the growing-point leaves are diseased, the fungus spreads from the bases of the leaves to the stem, which causes wilting and distortion. During severe disease, all diseased plant parts become black-green in color due to the growth of mycelia and conidia on their surfaces.

Management. Not reported.

Anthracnose

Cause. *Colletotrichum lini* and *C. linicola* commonly survive as mycelium within seed and as conidia on seed coats. *Colletotrichum lini* also overwinters as mycelia, acervuli, and possibly conidia in infested residue. Seedlings are infected primarily by conidia and mycelia during cool, wet weather.

After disease has progressed for a time, secondary inoculum, in the form of conidia, is produced in the following manner: Acervuli are formed subepidermally in cotyledon lesions caused by infections from the primary inoculum. The acervuli rupture the epidermis and release conidia that are disseminated by wind to the foliage of host plants or are washed to the soil to infect stems. Conidial sporulation occurs throughout the growing season on plants that have died or on stem cankers. Conidia from stem lesions infect flower sepals, which are a substrate, or energy source, from which *C. lini* infects bolls and seed.

Distribution. Wherever flax is grown, but anthracnose is most common in cool, humid flax-growing areas.

Symptoms. Symptoms are most noticeable when plants are 5 to 7 cm tall. Initially, small, circular, red-brown, zonate spots that occur on one or both cotyledons eventually cause both cotyledons to become brown and shriveled. Seedlings may be killed if the growing points become diseased. A reddish canker develops on the stem at or below the soil line, girdles the stem, and kills the plant. Diseased plants that are not killed may only be stunted.

Acervuli can be observed on leaf spots and stem cankers of diseased plants throughout the growing season. Acervuli appear in older lesions during moist weather as a pink spore mass with dark hair-like objects that project stiffly upward. Lodged plants are likely to have diseased seed.

Management
1. Sow healthy seed that is treated with a seed-protectant fungicide.
2. Grow resistant cultivars.
3. Rotate flax with other crops.

Basal Stem Blight

Basal stem blight is also called foot rot.

Cause. *Phoma exigua* var. *linicola* survives as pycnidia in infested residue, conidia on seed, and mycelium in seed coats. Primary infection occurs from conidia (produced either in pycnidia or from mycelium) that are washed into soil or disseminated by wind to host plants. Foot rot becomes more severe under wet conditions when numerous secondary infections are caused by conidia produced in pycnidia from lesions on recently killed plants.

Distribution. Europe.

Symptoms. The roots and stems at the soil level of diseased seedlings become discolored brown. The entire seedling then becomes chlorotic, wilts, and dies. Pycnidia, which look like small, black "pinpoints," are produced in dead seedling tissue.

Symptoms on older plants are the most conspicuous at flowering time, when diseased plants become chlorotic and wilt. Similar to the earlier symptoms on diseased seedlings, the lower part of the stem and the roots are light brown. Later, numerous, small, black pycnidia form in the discolored area. The epidermis is easily detached, causing the lower part of the diseased stem to appear "ragged." Eventually, plants are killed; their brown color contrasts with the normal green color of healthy plants.

Management
1. Sow only healthy seed that is treated with a fungicide seed protectant.
2. Rotate flax with other crops.
3. Do not place flax straw or residue on soil in which flax will grow the following year.
4. Grow resistant or less susceptible cultivars.

Brown Stem Blight

Brown stem blight is also called Alternaria seedling blight.

Cause. *Alternaria linicola* survives primarily as saprophytic mycelium in the residue of different plant species. *Alternaria linicola* may also be seedborne when conidia contaminate the outside of seed coats. Only seedlings weakened by other causes become infected.

Distribution. Canada and Europe.

Symptoms. Infested seeds either do not germinate or produce stunted seedlings after germination. The root tips of diseased seedlings may have a reddish discoloration, and dark red lesions on the hypocotyls and cotyledons develop into a moist, brown rot.

Dark spots and brown patches occur on diseased leaves, which soon become chlorotic and die. The larger lesions may have a margin or edge that is not typical of the concentric rings–symptom caused by *Alternaria* spp. on other crops.

Management
1. Apply a fungicide seed treatment to seed.
2. Foliar fungicide treatments applied toward the end of flowering have been effective in managing disease.

Browning and Stem Break

Cause. *Aureobasidium lini* (syn. *Polyspora lini*), teleomorph *Guignardia fulvida*, is primarily seedborne; it overwinters as conidia on and mycelium in seed coats. *Aureobasidium lini* also overwinters as conidia and mycelium in acervuli on infested residue. When seeds germinate, conidia are produced during warm, wet weather on seed coats that remain attached to cotyledons and are disseminated by wind to adjacent flax stems.

Distribution. Wherever flax is grown.

Symptoms. Initially, water-soaked spots are produced on diseased cotyledons. Soon, cankers form approximately 2.5 cm above the soil line on diseased stems. Diseased plants in the bud or flower stage may break over at the site of the stem canker but continue to live; however, any seed that is produced usually cannot be harvested.

Diseased areas on older plants appear brown, hence the name browning. The browning symptom occurs on upper stems as oval to elongated (6 mm long) brown spots that are surrounded by narrow purple borders. Spots do not coalesce unless they are very numerous. *Aureobasidium lini* grows into bolls and kills young seed but survives in the seed coats of mature seed that has not been killed.

Management
1. Sow healthy seed treated with a fungicide seed protectant.
2. Rotate flax with other crops.
3. Grow resistant cultivars.
4. If resistant cultivars are not grown, sow flax early in the growing season to attain sufficient growth before disease becomes severe.

Dieback

Cause. *Selenophoma linicola* overwinters as pycnidia on infested residue. Pycnidia are formed in the late summer, primarily on early-maturing varieties.

Distribution. Canada (Saskatchewan) and the United States (California). Dieback is of no economic significance on flax.

Symptoms. *Selenophoma linicola* was isolated from the upper third of plants that prematurely dried up or died back. In these cases it was observed that stems and bolls failed to develop properly. Since the fungus has also been found on plants apparently killed by other causes, researchers conjectured the fungus may be more saprophytic that pathogenic.

When fungus growing on agar disks was inoculated onto flaxseed that had germinated on filter paper, it caused a slight inhibition in the length of roots and a general curling and increase in branch roots.

Management. Not necessary.

Fusarium Secondary Rot

Cause. Several *Fusarium* spp., of which *F. roseum* is the main incitant. The *Fusarium* spp. overwinter as perithecia, chlamydospores, or saprophytic mycelia in infested residue. *Fusarium* spp. are secondary invaders of lesions caused by *Aureobasidium lini* or by the telia stage of *Melampsora lini*.

Distribution. Northern Ireland and Russia.

Symptoms. Diseased areas become light brown and extend up to 13 mm beyond a telial pustule of *M. lini*. Sporodochia, a mass of conidia supported on conidiophores and pseudoparenchyma, produced in the centers of diseased areas, give them a reddish to pinkish color. Under moist conditions, white mycelial strands grow on the outside of diseased stems.

Management. Not reported.

Gray Mold

Cause. *Botrytis cinerea* survives as sclerotia in soil, saprophytic mycelia in infested residue, and mycelia in seed coats. Conidia are produced on mycelia and disseminated by wind to plant hosts. Disease is favored by warm, wet conditions.

After the seedling stage, *B. cinerea* does not infect plants until the beginning of maturity. Fallen petals or pollen grains that adhere to the stem, leaves, and capsules after rain are colonized by *B. cinerea* and used as an energy source from which to infect mature plants. Airborne spores usually enter through damaged plant tissues.

Lodged flax is more subject to infection.

Distribution. Europe and the United States.

Symptoms. The first symptoms occur as cotyledonary leaves unfold. Initially, a barely visible, dark red neck rot occurs at the point the hypocotyl emerges from the seed coat. Bright red lesions subsequently develop on the hypocotyl and cotyledons. These lesions either develop directly into infection sites or serve as an inoculum source for disease of seedling stems at the soil level. Infections are observed as light brown spots that weaken the diseased stems and cause seedlings to wilt and fall over. The fallen stem becomes overgrown with gray mycelial growth of *B. cinerea*, particularly during warm, wet weather.

Later in the growing season as plants begin to mature, portions (1.25 to 7.5 cm long) of the stems of diseased older plants become soft and discolored a light brown with yellow margins. The stem, leaves, and petioles above the diseased stem become chlorotic and die. The dead plant residue also becomes overgrown with gray mycelial growth during warm, wet weather. Eventually, hard, black sclerotia that appear as relatively large, black objects form on the diseased stems.

Management
1. Sow healthy seed treated with a seed-protectant fungicide.
2. Plants that are too close together and excessive nitrogen may encourage lodging and more severe disease.

Pasmo

Cause. *Mycosphaerella linicola* (syn. *M. linorum*), anamorph *Septoria linicola*. Fungi overwinter as conidia produced within pycnidia on infested residue, and as mycelia in infested residue and seed. Most primary inoculum is conidia from pycnidia, but cotyledons may become infected from seed-borne inoculum. During warm, wet weather, conidia are exuded from pycnidia and disseminated by wind and rain. Secondary inoculum comes from conidia produced in pycnidia on lesions of diseased plants. Lodged plants are more susceptible to infection than standing flax plants. Low temperatures, which are conducive to flowering, are not favorable for disease development.

Distribution. Wherever flax is grown.

Symptoms. Symptoms are not prevalent on cultivars with long flowering periods. Initially, circular, green-yellow to dark brown lesions develop on diseased cotyledons and then, later, on lower leaves. Pycnidia develop in the older lesions on diseased cotyledons and leaves. Diseased leaves die and defoliate or cling tightly to stems.

As flax ripens later in the growing season, stem symptoms develop as small, brown, and elongated lesions. Stem lesions enlarge and coalesce, forming brown bands that encircle diseased stems and alternate with bands of healthy, green tissue to give the stem a mottled appearance, a typical pasmo symptom. Eventually, entire stems become brown and diseased plants defoliate. Pycnidia develop on stem lesions and appear as numerous, small, black "specks" scattered throughout the lesion.

Flowers and young bolls may also be blighted. Lesions form on older bolls, and seed inside the diseased boll may not develop properly or may be shriveled. The slender stems, or pedicels, supporting bolls may become weakened and cause ripe bolls to break off during wind or rain.

Management
1. Grow resistant or tolerant cultivars. Early-maturing cultivars tend to be more susceptible than later-maturing cultivars.
2. Sow only healthy seed treated with a seed-protectant fungicide.
3. Plow under infested residue to bury inoculum.

Phytophthora Root Rot (*Phytophthora aphanidermatum*)

Cause. *Phytophthora aphanidermatum.* Disease is favored by water-saturated soil.

Distribution. The United States (Arizona and Missouri).

Symptoms. Initial symptoms of a general chlorosis of plant foliage, stems, and petioles are followed by a necrosis of the leaves as the disease progresses. Taproots have a cortical rot, but the bright pink discoloration symptomatic of disease caused by *P. nicotianae* var. *parasitica* is lacking.

Management. Not reported.

Phytophthora Root Rot *(Phytophthora nicotianae* var. *parasitica)*

Cause. *Phytophthora nicotianae* var. *parasitica.* Disease is favored by water-saturated soil.

Distribution. The United States (Missouri).

Symptoms. Initial symptoms of a general chlorosis of plant foliage, stems, and petioles are followed by a necrosis of the leaves as the disease progresses. Taproots have a cortical rot, and lateral roots have a bright pink discoloration.

Management. Not reported.

Powdery Mildew

Cause. *Oidium lini,* teleomorph *Erysiphe polygoni,* generally overwinters as cleistothecia on infested residue. However, frequently only the conidial stage occurs and cleistothecia are not present. In such cases, it is not known for certain how *E. polygoni* overwinters but it is presumed overwintering occurs on volunteer flax or other hosts. Primary infection either may be by conidia from mycelium or ascospores from cleistothecia. Secondary infection is from conidia disseminated by wind from diseased plants to other hosts.

Distribution. Wherever flax is grown; however, powdery mildew is not considered an important disease.

Symptoms. A white, powdery growth consisting of mycelium, conidiophores, and conidia develops on diseased stems, both surfaces of leaves, and sepals. Cleistothecia appear as small, round, black structures intermingled within the white mycelial growth.

Management. No management is necessary in the field. In the greenhouse, apply a sulfur fungicide as a foliar spray.

Rhizoctonia Seedling Blight and Root Rot

Cause. *Rhizoctonia solani* survives as sclerotia and mycelium in soil or as saprophytic mycelium in residue. Disease is most severe in warm, moist soils that have been summer-fallowed.

Distribution. Wherever flax is grown.

Symptoms. Single plants and groups of seedlings in a row or within a circular area may become diseased. Diseased seedlings become chlorotic, wilt, and die. Root rot symptoms occur at flowering time when plants appear to ripen prematurely. Roots are stunted or rotted and have brown to red-brown, superficial lesions. Eventually, roots turn dark brown or black, shrivel, and dry up. Few or no seeds are formed.

Management
1. Sow healthy seed that is free of cracks in the seed coat.
2. Sow as early as feasible so plants are well developed when conditions conducive to root rot occur.
3. Do not sow into summer-fallowed soil.
4. Treat seed with a seed-protectant fungicide.

Rust

Rust is also called flax rust.

Cause. *Melampsora lini* is an autoecious long-cycled rust fungus that produces uredinia, telia, pycnia, and aecia on the flax plant. *Melampsora lini* overwinters as teliospores within telia on infested residue and seed. In the spring, teliospores germinate to produce basidiospores that infect young flax tissues. Pycnia then develop and each pycnium produces pycniospores and special mycelia called receptive hyphae. Pycniospores are exuded out of the pycnium in a thick, sticky, sweet liquid that is attractive to insects. Pycniospores are splashed by water or carried by insects from one pycnium to another, where they become attached to receptive hyphae. The pycniospore germinates and the nucleus from the spore enters into the receptive hypha. Aecia are usually initiated on the opposite side of the leaf or stem, and the resultant aeciospores are disseminated by wind to host tissue. Uredinia are initiated in aeciospore infections, and urediniospores are disseminated by wind to other flax plants. Secondary infections from urediniospores account for most of the spread of rust throughout the summer. Telia, which form around uredinia as flax matures, provide the overwintering stage.
Several physiologic races exist.

Distribution. Wherever flax is grown.

Symptoms. Telia are brown to black structures covered by epidermis that occur mostly on stems but also on leaves and capsules. Pycnia and aecia occur on leaves and stems early in the growing season as light orange-yellow sori. Uredinia are red-yellow and occur on leaves, stems, and bolls.

Management
1. Grow resistant varieties. There are at least 30 genes for resistance to rust.
2. Sow cleaned seed treated with a seed-protectant fungicide.
3. Sow as early as possible in the growing season to allow susceptible plants to escape early infection.
4. Rotate flax with other crops.

Scorch

Scorch is also called flax fire.

Cause. *Pythium megalacanthum* survives in infested soil as oospores. The life cycle is presumed to be typical of most *Pythium* spp. Slow growth of early-sown crops due to cool and wet soil conditions favors disease.

Distribution. Northern Europe.

Symptoms. Scorch symptoms occur within circular areas of a field. Symptoms start at the bottom of diseased plants and progress upward. Leaves become brown and shriveled approximately half way up the stem. Leaves just above this point are yellow with brown margins, and higher on the diseased plant, leaves are only partially yellow and flaccid. Leaves at the very top of a diseased plant may appear green and healthy. If soils dry and warm up, diseased plants may recover unless they are too severely diseased.

Management. Sow later in the growing season after soils have warmed up.

Seed Rot, Seedling Blight, and Root Rot

Cause. Several fungi, including *Aureobasidium lini; Fusarium* spp.; *Olpidium brassicae; Pythium aphanidermatum, P. debaryanum, P. intermedium, P. irregulare, P. mamillatum, P. megalacanthum, P. splendens, P. vexans; Rhizoctonia* spp.; and *Thielaviopsis basicola*. The most important fungi are the *Pythium* spp. Infection of susceptible plants usually occurs under wet soil conditions and when mechanically injured seed is sown. However, some fungi may cause more injury under cool soil temperatures, while other fungi grow better at higher soil temperatures.

Distribution. Generally distributed wherever flax is grown.

Symptoms. Seeds do not germinate and become soft, mushy, and overgrown with white mycelium that causes soil to adhere to the seeds. Seeds that do not germinate are difficult to find in soil because of the adhering soil. Seeds may germinate but seedlings may be killed before they emerge and will be brown and water-soaked. Again, such plants would be difficult to find in the soil. Postemergence infection is characterized by stunted plants whose lower leaves become brown and necrotic. Diseased roots are light brown and rotted. The entire root system is destroyed if soils remain wet.

Management
1. Sow healthy seed treated with a seed-protectant fungicide.
2. Rotate flax with other crops. However, this is a questionable practice as many causal fungi can also infect a large number of host plants.

Stem Mold and Rot

Stem mold and rot is also called Sclerotinia disease.

Cause. *Sclerotinia sclerotiorum* survives for several years in soil as sclerotia. Sclerotia produced on other hosts, such as canola, common bean, soybean, and sunflower, also may provide a source of primary inoculum. Sclerotia on or close to the soil surface germinate during wet conditions to form apothecia on which ascospores are produced in a layer of asci. Ascospores are then disseminated by wind to host plants. Initially, plant stems are infected at or just above the soil line. Sclerotia, eventually formed in mycelia growing on and in diseased stems, are returned to soil during harvest. *Sclerotinia sclerotiorum* has a wide host range.

Distribution. Canada (Alberta), England, the warmer flax-growing areas of the United States, and Russia. However, Sclerotinia disease is not considered serious.

Symptoms. Diseased plants are quickly killed, lodge, become pale brown in color, and dry up. White, fluffy mycelial growth occurs on diseased stems, which are easily shredded. Eventually, sclerotia of variable sizes and shapes are frequently formed in mycelium growing in the stem pith cavity and less frequently on stems. Sclerotia are easily seen as large, spherical-shaped bodies that initially are white but shortly become the characteristic black color.

Management
1. Rotate flax with resistant crops, such as small grains and maize.
2. Control weeds in and around a flax field.
3. Cultural practices that promote rank, thick growth should be avoided.

Thielaviopsis Root Rot

Cause. *Thielaviopsis basicola* survives in infested residue and soil as chlamydospores. Most primary inoculum, in addition to the inoculum that causes infections throughout the growing season, is presumed to be chlamydospores that germinate to produce infective hyphae. The infective hyphae penetrate only roots. Conidia are produced on diseased roots but their function in the etiology and epidemiology of Thielaviopsis root rot is not well understood. Chlamydospores are produced within diseased roots and released into the soil upon their decomposition. *Thielaviopsis basicola* is disseminated by any means that transports soil.

Distribution. Europe and the United States.

Symptoms. Diseased aboveground plant parts are stunted and chlorotic. Diseased roots, in which dark brown chlamydospores are borne in short chains, are black and necrotic.

Management
1. Crop rotation. Do not grow flax in the same soil for several years.
2. Control weeds in a field since they may be a host for *T. basicola*.

Wilt

Wilt is also called Fusarium wilt, soil sickness, and dead stalks.

Cause. *Fusarium oxysporum* f. sp. *lini* (syn. *F. lini)* survives as chlamydospores in soil and as chlamydospores and saprophytic mycelia in infested residue. Microconidia and macroconidia are seedborne on the outside of weathered seed. *Fusarium oxysporum* f. sp. *lini* is disseminated by seed, wind, and any means that moves soil. Infection hyphae enter host plants through the root hairs of young seedlings; the resulting mycelia grow into the xylem tissue, inhibiting water uptake and causing diseased plants to wilt.

Soil temperatures of 25°C and higher and low soil moisture favor wilt development. During moist weather, both microconidia and macroconidia are produced on diseased plant parts, particularly near the soil line. Chlamydospores, formed from macroconidia, microconidia, and mycelia, survive for several years in soil and intact infested residue.

Several physiologic races of *F. oxysporum* f. sp. *lini* exist.

Distribution. Widespread. Wilt is thought to occur wherever flax is grown.

Symptoms. Wilt symptoms are classified into four types: Early, or seedling, wilt; late wilt; partial wilt; and one-sided, or unilateral, wilt. Symptoms appear at any stage of plant growth and vary with plant age, plant cultivar, environmental conditions, and the physiologic race that caused the disease.

Early wilt is the most common type. Seedlings wilt and die from the cotyledonary stage until they are approximately 15 cm tall. Diseased plants wilt at their tops and become dry and brown. Stems become constricted at the soil level, and seedlings fall over. If the soil is moist, abundant mycelial growth and sporulation of macroconidia and microconidia occur on the outside of the dead seedling.

Late wilt occurs from flowering through boll set. Diseased internal and external stem tissues become discolored, necrotic, and brittle. Stem discoloration is a uniform light brown but is darker than the color symptomatic of premature ripening caused by root rot or other causes.

Partial wilt occurs when a seedling wilts and dies except for the roots and buds at the base of the stem. During cool weather, these buds at the stem base develop into new shoots after the death of the original shoot. With a return to high temperatures, lateral shoots wilt and the whole plant dies.

Unilateral wilt occurs when only one side of the stem is diseased. All branches on the diseased side become brown and necrotic.

Management. Grow resistant cultivars.

Disease Caused by Phytoplasmas

Aster Yellows

Cause. The aster yellows phytoplasma survives in several dicotyledonous plants and leafhoppers. The phytoplasma is transmitted primarily by the aster leafhopper, *Macrosteles fascifrons,* and less commonly by the leafhoppers *M. laevis* and *Endria inimica.*

Distribution. Central and western North America.

Symptoms. Symptoms are most conspicuous during and after flowering and appear only on the lateral branches of plants infected late in the growing season. Diseased plants are stunted and chlorotic. Flower parts are distorted and petals remain green and leaf-like. Sometimes diseased and normal flowers are both present on the same diseased plant. No seed is ordinarily set.

Management. Not reported.

Diseases Caused by Viruses

Crinkle

Cause. Oat blue dwarf virus (OBDV) is in the marafivirus group. OBDV overwinters in several species of plants and is transmitted by the adult aster leafhopper, *Macrosteles fascifrons.* Immature leafhoppers may occasionally transmit OBDV.

Distribution. Canada and the north central United States.

Symptoms. The only leaves that are affected and display symptoms are those formed after exposure of plants to viruliferous leafhoppers. A "crinkle" of diseased leaves occurs due to a swelling of lateral veins on the margins of leaves, small indentations along the upper surfaces of veins, and enations or pimples on the lower leaf surfaces. Diseased plants are also stunted and have reduced boll development and seed set.

Management. Not reported.

Curly Top

Cause. The curly top virus (CTV) overwinters in a large number of perennial plants. CTV is transmitted by the beet leafhopper, *Circulifer tenellus.*

Distribution. Western North America.

Symptoms. Diseased seedlings turn bronze, then yellow, with the leaves curled along the stem. Severely diseased plants die prematurely, while surviving plants are stunted and have a reduced number of tillers. Terminal leaves and flower parts are discolored, crinkled, and distorted, and there is little or no seed set.

Management
1. Grow resistant cultivars.
2. Sow at times during the growing season that avoid high populations of the beet leafhoppers.

Diseases Caused by Physiologic or Weather-oriented Problems

Boll Blight

Cause. Warm, dry weather following cool, moist weather.

Distribution. Not known.

Symptoms. Buds, flowers, and young bolls fail to develop.

Management. Not reported.

Heat Canker

Cause. High temperature at the soil line.

Distribution. Generally distributed in semihumid plains and at high altitudes throughout the world.

Symptoms. Cortical tissues collapse, resulting in seedling death or sunken brown lesions on affected stems. Plants that survive have an enlarged stem just above the canker. Cankers provide an entrance wound for secondary microorganisms that decay stems.

Management
1. Sow early to avoid high soil temperature.
2. Rows should be sown in a north and south direction to provide maximum shading.

Top Dieback

Cause. Physiologic. High temperatures at seed ripening.

Distribution. Not known. Affected plants may occur in a limited area or throughout an entire field.

Symptoms. The top portion of affected plants turns brown, which results in thin and lightweight seed.

Management. Not reported.

Selected References

Christensen, J. J. 1954. The present status of flax diseases other than rust. Adv. Agron. 6:161-168.

Farr, D. F., Bills, G. R., Chamuris, G. P., and Rossman, A. Y. 1989. Fungi on Plants and Plant Products in the United States. American Phytopathological Society, St. Paul, MN. 1252 pp.

Ferguson, M. W., Lay, C. L., and Evenson, P. D. 1987. Effect of pasmo disease on flower production and yield components of flax. Phytopathology 77:805-808.

Frederiksen, R. A., and Goth, R. W. 1959. Crinkle, a new virus disease of flax. (Abstr.) Phytopathology 49:538.

Hoes, J. A. 1975. Diseases of flax in western Canada. *In* Oilseed and Pulse Crops in Western Canada. Modern Press, Saskatoon, Sask.

Mederick, F. M., and Piening, L. J. 1982. *Sclerotinia sclerotiorum* on oil and fibre flax in Alberta. Can. Plant Dis. Surv. 62:11.

Mihail, J. D. 1993. Diseases of alternative crops in Missouri. Can. J. Plant Pathol. 15:119-122.

Muskett, A. E. 1947. The diseases of the flax plant. Northern Ireland Flax Development Committee, Belfast.

Turner, J. 1987. Linseed Law: A Handbook for Growers and Advisors. Alderman Printing CI, Ipswich, United Kingdom. 356 pp.

Vanterpool, T. C. 1947. *Selenophoma linicola* sp. nov. on flax in Saskatchewan. Mycologia 39:341-348.

9. Diseases of Maize *(Zea mays)*

Diseases Caused by Bacteria

Bacterial Leaf Blight and Stalk Rot

Cause. *Acidovorax avenae* subsp. *avenae* (syn. *Pseudomonas avenae* and *P. albo-precipitans*) does not survive well in infested residue or soil. Vaseygrass, *Paspalum urvillei*, is considered a primary source of inoculum in Florida. Contaminated farm equipment is a primary means of dissemination within fields and, likely, between fields.

Infection typically occurs through stomata of leaves that are in whorls and harbor large populations of *A. avenae* subsp. *avenae* and other epiphytic bacteria. Moisture in the whorl is considered necessary for infection to occur. Bacteria also are splashed and blown onto the plant and enter stalks through hail wounds and other injuries. Warm, rainy weather favors disease development but leaf blight will occur below 18°C.

Distribution. The central and southern United States. The disease is of relatively minor importance.

Symptoms. In Illinois, symptoms, which occur on diseased leaves from late April to early June, were severe following heavy rains. Leaves emerging from whorls have water-soaked, brown lesions that become elliptical and gray or white. Lesions may coalesce during wet weather and form large necrotic areas that cause leaves to shred when diseased plants are buffeted by the wind.

Plant tops become gray to brown and have shortened internodes. The outside of the stalk may be brown to black and appear water-soaked. The inside of the diseased stalk is brown and slimy and has a foul odor that somewhat resembles that of silage. Diseased plants may be stunted.

Management. Grow the most resistant hybrid. Full-season field maize hybrids are more susceptible than short-season hybrids and sweet maize. Bacterial leaf blight and stalk rot is not considered a serious disease.

247

Bacterial Leaf Spot

Cause. *Xanthomonas campestris* pv. *holcicola* (syn. *X. campestris* pv. *zeae*).

Distribution. Bacterial leaf spot occurs in most warm and humid maize-growing areas of the world.

Symptoms. Narrow, water-soaked streaks (33 × 20 mm long) occur on diseased leaves. Reddish-brown streaks covered by dried bacterial exudate coalesce and form large, irregular-shaped, necrotic areas that destroy most of the leaf.

Management. Differences in resistance and susceptibility occur in maize inbreds.

Bacterial Stalk and Top Rot

Bacterial stalk and top rot is also called bacterial stalk rot.

Cause. *Erwinia carotovora* subsp. *carotovora* and *E. chrysanthemi* pv. *zeae* (syn. *E. carotovora* var. *zeae* and *E. chrysanthemi*) live as saprophytes on infested residue in soil and are seedborne. Bacteria disseminated by wind or splashed onto host plants during wet conditions caused by rainfall or overhead irrigation enter the host plant through hydathodes, stomates, or wounds on leaves and stalks.

Disease development is aided by high temperatures (30° to 35°C) and poor air circulation. The enzyme xylanase is produced by *E. chrysanthemi* pv. *zeae* and has been shown to kill cells and macerate tissue in monocotyledonous plants.

Distribution. Generally wherever maize is grown.

Symptoms. Initially, water-soaked lesions in diseased leaf sheaths extend, as streaks, into leaf laminae and cause decay in the pith. Diseased nodes are tan to dark brown, water-soaked, soft, and slimy and have a foul odor that resembles spoiled silage. At midseason, plants fall to the ground as one to several diseased internodes twist and collapse. The fallen plant may remain green for several days because the vascular strands remain intact and do not decay.

Management. No disease management is practical, but chlorine in irrigation water has been reported to reduce disease incidence and severity.

Bacterial Stripe

Bacterial stripe is also called bacterial stripe and leaf spot.

Cause. *Pseudomonas andropogonis* overwinters in residue. During extended periods of warm and wet weather, bacteria enter into leaves through stomata and possibly leaf wounds caused by different means.

Distribution. The central and eastern United States.

Symptoms. Lesions occur first on diseased lower leaves and progress up the plant, but leaves above the ears are rarely infected. Lesions (sometimes described as stripes) initially are water-soaked, long and narrow with parallel sides, and olive to amber in color. Some investigators have described lesion color as pale yellow or greenish white. Lesions eventually enlarge, or elongate further, and coalesce with other lesions to involve a large area of the diseased leaf. As lesions enlarge and coalesce, their color becomes lighter. Severely diseased leaves are easily shred by wind.

Although upper leaves are normally not diseased, they become completely white, a secondary effect of the infection of the lower leaves. In susceptible inbred lines, most leaves below the ear may be killed.

Other foliar symptoms are slightly sunken, circular to ellipsoidal-shaped (1–4 mm in diameter) spots that are tan to brown with one or more darker brown rings and irregular margins. Some spots are surrounded by a chlorotic ring 1 mm wide. Sometimes spots coalesce into elongated blotches.

Management. Grow resistant hybrids and varieties. Bacterial stripe is not normally economically important but has been severe on a few susceptible inbred lines.

Chocolate Spot

Cause. *Pseudomonas syringae* pv. *coronafaciens* (syn. *P. coronafaciens* pv. *zeae*). Bacteria are disseminated by wind and enter through wounds or leaf stomates. Chocolate spot occurs only in fields with potassium-deficient soils.

Distribution. The United States (Minnesota and Wisconsin).

Symptoms. Lesions occur only on leaves. The translucent, dark brown, elongated and elliptical (2 × 5 mm) lesions are surrounded by a broad chlorotic halo. Lesions are most numerous along diseased leaf edges, where potassium-deficiency symptoms first appear. Eventually, lesions coalesce and cause large areas of the leaf to die.

Management. Maintain proper fertility levels.

Goss's Bacterial Wilt and Blight

Goss's bacterial wilt and blight is also called bacterial wilt, blight, freckles, Goss's leaf freckles, leaf freckles and wilt, and Nebraska bacterial wilt.

Cause. *Clavibacter michiganensis* subsp. *nebraskensis* (syn. *Corynebacterium michiganense* pv. *nebraskense*) survives in infested residue that consists of leaves, stalks, cobs, and ears on or near the soil surface; in or on seed; and in irrigation water. Bacteria are disseminated primarily in residue and less frequently by seed.

Infested maize residue is the major primary inoculum source for Goss's wilt. Seedborne inoculum is of minor concern as an inoculum source in an area where the disease is already established. However, infected or contaminated seeds could introduce the pathogen into new areas. Small abrasions and wounds caused by sand blasting, hail, severe rainstorms, and wind allow bacteria to enter plant tissue and become established in the vascular system. Infection occurs either during wet weather or under irrigation at temperatures of 27°C and higher.

Plants can be infected at all growth stages, but seedlings are more susceptible than older plants. Disease development is greatest when sweet maize hybrids are inoculated at the three- to five-leaf stage. Different strains of *C. michiganense* subsp. *nebraskense* have been classified into several groups.

Distribution. The United States (Colorado, Iowa, Kansas, Nebraska, and South Dakota).

Symptoms. Infection of seedlings causes diseased plants to wilt, wither, and die. Infection of older plants causes stunting, wilting, and various degrees of leaf blight.

Young lesions are yellow to gray-green streaks that are parallel to leaf veins. In certain hybrids, red streaks that are also parallel to veins occur. The streaks enlarge and appear "greasy," and their margins become wavy. Dark green to black, water-soaked, angular-shaped spots resembling freckles occur along the edge of the enlarged streak. Streaks eventually coalesce and form larger lesions that kill diseased leaves and cause them to dry up. Droplets of dried-up bacterial exudate leave a glistening crystalline residue on dead leaf surfaces.

Systemically diseased plants display either discolored water-conducting tissues from which orange bacterial exudate oozes when a stalk is cut in cross section, or leaf symptoms that resemble those of drought stress. A dry or water-soaked, slimy, brown rot of roots and lower stalks may occur. Seedlings and older plants may be killed following systemic infection.

Bacteria are in the chalazal region (the area between the scutellum and the endosperm) and in the vicinity of the embryo in heavily infected kernels.

Management
1. Grow resistant hybrids.
2. Rotate maize with soybean, small grains, and alfalfa.
3. Plow residue deep to reduce inoculum.

Holcus Spot

Cause. *Pseudomonas syringae* pv. *syringae* (syn. *P. holci*) overwinters in infested residue. During temperatures of 25° to 30°C and wet, windy weather early in the growing season, bacteria are splashed or blown onto host plants, where they invade the leaf either through stomates or small injuries.

Distribution. The eastern and midwestern United States.

Symptoms. Lesions are more numerous toward the tips of diseased lower leaves and initially are dark green with water-soaked margins. Later, lesions dry up and become tan to brown, round to elliptical-shaped (2–10 mm in diameter) areas with reddish to brown margins, depending on the maize hybrid. Larger lesions may be surrounded by a yellowish halo.

Holcus spot initially may appear to be a serious disease on young plants, but warm and dry weather conditions arrest bacterial spread on the plant and little injury usually occurs. Holcus spot has little or no effect on yield.

Management
1. Grow hybrids that are more resistant than others.
2. Rotate maize with resistant crops.

Pseudomonas fuscovaginae Disease

Cause. *Pseudomonas fuscovaginae* is an opportunistic pathogen favored by low temperatures that occur above 1300 m.

Distribution. Burundi.

Symptoms. Symptoms first occur at the silking stage of plant growth. Initially, glossy, brown-black, water-soaked spots occur on the adaxial side of diseased leaf sheaths. Spots are distinct and approximately 2 mm in diameter but later coalesce to form areas up to 20 cm long. Inside the diseased leaf sheath, light brown patches, which occur on the abaxial side, have withered centers and sharp, dark purplish-brown borders that fade into healthy tissue. Similar purple-brown lesions occur on husks.

Management. Not reported.

Purple Leaf Sheath

Cause. Hemiparasitic bacteria plus fungi. Some of the microorganisms may grow on exogenous substrate, such as pollen, then into the surrounding leaf sheath.

Distribution. Widespread.

Symptoms. A dark brown to purplish discoloration of the diseased leaf sheath occurs when excess moisture is present. Symptoms are more pronounced on inner leaf sheath than on the exterior. Although discoloration frequently is limited to discrete areas, the entire inner leaf sheath may be discolored in some plants. The integrity and strength of the plant is normally not harmed.

Management. Not necessary. Apparently no injury occurs to the plant; there simply is a discoloration of the leaf sheath tissue.

Seed Rot and Seedling Blight

Cause. *Bacillus subtilis.*

Distribution. Widespread.

Symptoms. The insides of seeds have a light to dark brown discoloration. Diseased seeds become soft and lose their integrity. If seeds germinate, the resulting seedlings have a soft rot with a light brown discoloration.

Management. Treat seed with a seed-treatment fungicide.

Stewart's Wilt

Stewart's wilt is also called bacterial leaf blight, bacterial wilt, maize bacteriosis, Stewart's bacterial wilt, and Stewart's leaf blight.

Cause. *Pantoea stewartii (Erwinia stewartii)* overwinters primarily in bodies of the corn flea beetle, *Chaetocnema pulicaria*. Warm winter weather favors beetle and bacterial survival, thereby favoring disease development the following growing season. If the sum of the mean monthly temperatures for December, January, and February in the Northern Hemisphere total 37.8°C or more, Stewart's wilt may be potentially severe. Little or no disease is likely to occur when the sum of the mean temperatures is below 32.2°C.

Some researchers have reported that *P. stewartii* is rarely spread through infected dent maize seed but is commonly spread by infected sweet maize seed. Bacteria may spread through the vascular system and pass into kernels. It has been reported that *P. stewartii* was isolated from the endosperm of seed produced on diseased plants and from the interior of seed 5 months after harvest. Khan et al. (1996) reported *P. stewartii* was detected in seed produced on hybrids that had systemic Stewart's wilt following leaf inoculation. The bacterium was not detected in seed from hybrids or inbreds with nonsystemic Stewart's wilt. They also reported no evidence of seed-to-seedling transmission of *P. stewartii* when seeds from diseased plants were sown in field and greenhouse trials. This may have been due, in part, to low rates of plant-to-seed transmission in seed produced on plants that were not systemically diseased.

Adult beetles feed on maize seedlings in late spring and early summer and place bacteria in wounds made by their feeding. Disease development is greatest when plants are infected at the three- to five-leaf stage. Beetles continue to spread bacteria throughout the growing season by feeding on diseased plants, then flying to host plants. The twelve-spotted cucumber beetle, *Diabrotica undecimpunctata howartii*; toothed flea beetle, *Chaetocnema denticulata*; larvae of the seed corn maggot, *Hylemya cilicrura*; wheat wireworm, *Agriotes mancus*; and May beetle, *Phyllophaga* sp., may infrequently spread *P. stewartii* but are not important in its overwintering.

High levels of ammonium and phosphorus increase plant susceptibility, whereas high calcium and potassium tend to decrease susceptibility of maize plants to *P. stewartii* infection.

Distribution. Central America, China, eastern and southern Europe, eastern United States, and Russia.

Symptoms. Except for a few susceptible inbreds, dent maize is not as susceptible as sweet maize. In sweet maize, bacteria may be found in every part of the plant, including the roots.

Conspicuous streaks occur on leaves after tasseling. Initially, streaks tend to follow leaf veins and are long, have irregular or wavy margins, and are light green to yellow but later turn tan as the diseased tissue dies. Streaks may coalesce and kill the whole leaf. When the leaf is held up to light, streaks will show beetle-feeding scars that appear as tiny scratches at right angles to the streak.

Tassels may die, and brown cavities form in stalk pith at the soil line of severely diseased plants. Bacteria may spread through the vascular system and pass into kernels. Bacteria may ooze as yellow, moist beads from cut ends of stalks or stream from cut edges of leaf tissue.

Management
1. Grow resistant cultivars or hybrids.
2. Apply insecticides early to control corn flea beetles.
3. Experimental results with insecticide seed treatments have been effective in managing disease in young maize plants.

Yellow Leaf Blotch

Cause. *Pseudomonas* sp. Plants are infected in the seedling stage.

Distribution. West Africa.

Symptoms. Scattered cream, yellow, or light tan, rectangular lesions (10–15 × 14–45 mm) occur on diseased leaves. There is a tendency for streaks, or runners, to follow veins. Entire leaves of some plants become water-soaked, wither, and die. Seedlings may outgrow the disease and produce healthy kernels.

Management. Not necessary.

Diseases Caused by Fungi

Alternaria Leaf Blight

Cause. *Alternaria alternata* (syn. *A. tenuis*) is a common saprophyte of infested residue. Infection occurs during heavy dew periods 16 to 20 hours after injury has occurred.

Distribution. The midwestern United States.

Symptoms. Chlorotic streaks that form on diseased leaves of all ages later become necrotic. Leaves may be killed in 1 week if heavy and prolonged dews occur.

Management. Not practiced.

Anthracnose Leaf Blight

Cause. *Colletotrichum graminicola*. No teleomorphic stage is known. *Colletotrichum graminicola* survives as mycelia and conidia in infested residue on top of the soil and as stroma or hyphae in endosperm. When exposed directly to soil, mycelia and conidia will lyse in approximately 16 and 14 days, respectively. Therefore, surface residue or any infested residue that is not buried is an important source of inoculum.

During warm, wet weather early in the growing season, conidia are produced in acervuli on residue and disseminated by splashing rain or wind to leaves. Disease development is optimum at 30°C and long periods of cloudy weather since low light intensity enhances lesion development.

Seedlings may be infected, but lower leaves of older plants are the most conducive for disease development. Leaf blight is most severe in continuous maize, and the spread of inoculum is more rapid within a row, in relation to inoculum sources, than across rows. Infection by the lesion nematode, *Pratylenchus hexincisus*, also increases disease severity.

Isolates of *C. graminicola* from sorghum, shatter cane, johnsongrass, and barnyard grass cause chlorotic flecks on juvenile maize leaves and susceptible-type lesions on senescing maize leaves. Isolates that infect small grains do not infect maize.

Distribution. France, Germany, India, the Philippines, Thailand, and the eastern United States.

Symptoms. Plants are susceptible as seedlings and, again, as mature plants, resulting in disease development early and late in the growing season. Chlorotic flecks occur on diseased leaves, enlarge, and become gray-green lesions that mature into brown, spindle-shaped lesions (5–15 mm long) with yellow to reddish-brown borders. Lesions sometimes have concentric rings or zones, and their centers become gray with numerous acervuli that, under magnification, appear as black spines. When lesions coalesce, leaves may rapidly dry up (top dieback), which can cause stalk lodging during warm, wet conditions later in the season. Symptoms may appear as a variety of more restricted lesion types in cultivars with different levels of resistance.

Management
1. Grow resistant hybrids. However, hybrids susceptible to leaf blight may be resistant to the stalk rot phase, and vice versa.

2. Rotate maize with nongrass crops.
3. Plow under infested residue where feasible.
4. Balance soil fertility.
5. Severity of stalk rot has been reported to decrease in artificially and naturally infected plants with increased nitrogen rates applied in either spring or autumn as anhydrous ammonia.

Anthracnose Stalk Rot

Cause. *Colletotrichum graminicola* survives as mycelia and conidia in infested residue on the soil surface and as stroma or hyphae in endosperm. When infested residue is buried, mycelia and conidia lyse in 16 and 14 days, respectively, because they are exposed directly to soil. During warm, wet weather early in the growing season, conidia are produced in acervuli on residue and disseminated by splashing rain or wind to host plants. Conidia may wash behind leaf sheaths and invade stalks or roots. The lower internodes may become infected as normal plant senescence progresses.

When inoculated with conidia prior to the rapid stalk elongation (late whorl) stage of plant growth, susceptible hybrids were resistant to stalk rot. Transition of the maize plant to susceptibility occurs during rapid stalk elongation just prior to tasseling and extends through anthesis. Therefore, plants are most susceptible to anthracnose stalk rot at tasseling or after tasseling. Subsequently, maximum stalk rot development occurs by the kernel dent stage. Susceptibility of young tissue to stalk rot is not consistent in all maize hybrids; moderately susceptible hybrids have delayed and reduced stalk rot development.

Disease is more severe in continuous maize and is associated, in New York, with early or midseason infestations by the European corn borer, *Ostrinia nubilalis*.

Distribution. France, Germany, India, the Philippines, Thailand, and the eastern United States. Anthracnose stalk rot may also occur where no leaf blight is present.

Symptoms. Symptoms may occur at different stages of plant growth, depending on susceptibility of the plant. Very susceptible plants may be killed before pollination. Initially, narrow, oval, water-soaked lesions that occur on the surfaces of diseased lower internodes after tasseling become tan to reddish and finally dark brown to black linear streaks. Frequently, large oval areas develop and cause internodes to be dark brown or shiny black. When such stalks are split, the pith, starting at the nodes, is a dark brown. Some genotypes may have a soft and watery pith. Sometimes internal discoloration may be present when there is little or no blackening of the surface. Severely diseased stalks are likely to lodge.

Although stalks above the ear sometimes become grayish green and die several weeks after pollination, the lower stalk remains green. Upper leaves may turn yellow or a reddish color and drop off, sometimes just before lower leaves begin to normally senesce.

Management
1. Grow resistant hybrids; however, resistance is not correlated with resistance to other stalk rot fungi.
2. Rotate maize with other crops.
3. Plow under infested residue where feasible.
4. Ensure that soils have a balanced fertility.

Ascochyta Leaf and Sheath Spots

Cause. *Ascochyta ischaemi* (syn. *A. zeae*), *A. maydis,* and *A. zeina* survive as pycnidia in infested residue. During wet weather, conidia are produced in pycnidia and disseminated by splashing rain or wind to host plants.

Distribution. The United States.

Symptoms. Lesions, which initially are ellipsoidal and reddish purple to brown, become elongated and irregular-shaped and have brown margins. Tiny black dots in the lesion are pycnidia, whose flask-like structures develop in rows below the leaf surface.

Management. Ascochyta leaf and sheath spots has not been considered important enough to warrant management.

Aspergillus Ear Rot

Cause. Several species of *Aspergillus,* including *A. flavus, A. glaucus, A. niger,* and *A. parasiticus. Aspergillus flavus* is generally considered to have a greater potential to naturally infect maize kernels than most other *Aspergillus* spp.

All *Aspergillus* spp. presumably survive as sclerotia on infested residue. However, *A. flavus* overwinters as sclerotia more successfully in the southern than in the northern United States maize-growing areas. This may partially explain the lower incidence of infection of maize by *A. flavus* in the northern maize-growing areas. Sclerotia of *A. flavus* var. *flavus* and *A. flavus* var. *parasiticus* survived burial in soil for 36 months in Georgia and Illinois. Apparently, sclerotia did not germinate sporogenically when placed on sand in moist chambers after 36 months burial in soil. However, sclerotia may have germinated sporogenically earlier in their soil burial and produced large numbers of fungal propagules in the soil. *Aspergillus flavus* and *A. parasiticus* also were isolated from soils in Iowa. Deposits of waste maize infested with *A. flavus* are frequently found in the vicinity of maize storage cribs and bins and may be potential sources of inoculum.

Growth and sporulation of *A. flavus* is favored by temperatures of 32° to 38°C and high humidity. Because airborne spores of *A. flavus* and other *Aspergillus* spp. are an important source of inoculum, open-pollinated maize is more susceptible to infection than hybrid maize. Spores enter kernels through injuries from growth cracks, insect damage from corn earworms, senescing silks, and cracks caused by drought stress. *Aspergillus niger* is observed most frequently in dry years; therefore, it is suspected that this fungus enters the plant through drought-stress cracks.

Aspergillus flavus readily infects wounded kernels at 32% or less moisture content or when physiological maturity occurs. The fungus does not normally infect uninjured kernels with higher moisture contents. However, *A. flavus* colonizes senescing silks at optimal temperatures of 32° to 38°C in the absence of injuries and grows on the senescent silk into the ear. *Aspergillus flavus* grows onto the adjacent pericarp at the silk attachment site on the kernel, then over the kernel surface and eventually penetrates the tip-cap region. Spread to other kernels is by mycelial growth on pericarps or glumes.

Aflatoxin accumulation in silk-inoculated ears follows a similar pattern to that produced in injured kernels except toxin levels are lower and decline as kernel moisture decreases to 14%.

Optimum moisture for toxin production may be higher than that for infection. Most aflatoxin is produced in wounded kernels that are not representative of the moisture content of intact kernels from the same ear. In mature maize kernels, *A. flavus* is reported to produce aflatoxin at kernel moisture contents of 18.0% to 18.5% but not at lower moisture contents. Other reports state maximum aflatoxin production occurred in maize at 41% to 58% moisture. Maize sown and harvested late in the growing season and produced under nitrogen stress is a better substrate for preharvest aflatoxin production than maize grown under good management practices and supplied with adequate nitrogen. When kernel moisture content is 16% to 21%, aflatoxin appears in wound-inoculated ears within 1 week and increases linearly until 7 to 9 weeks after inoculation. All *Aspergillus* spp. will grow on kernels stored at 15% and higher moisture content.

Common fungal colonists of maize kernels, such as *Fusarium moniliforme* and *Acremonium strictum*, interfere with the ability of *A. flavus* to infect maize before harvest. However, this interference is associated primarily with uninjured kernels.

Distribution. Generally wherever maize is grown.

Symptoms. *Aspergillus niger* is a black mold that is sometimes scattered over the whole ear. *Aspergillus flavus* appears as a green-yellow mold. *Aspergillus glaucus* is a greenish mold. Most other *Aspergillus* spp. are similar green-yellow-brown-colored molds. Frequently, *A. flavus* can be seen growing in the tracks of corn earworms or other insect damage and on injured kernels at the ear tip. However, kernels at any position on the ear may be diseased.

Management
1. Harvest maize as early as possible to limit aflatoxin contamination.
2. Dry maize to 15% or less moisture content soon after harvest.

3. Use tillage practices, such as irrigation, that reduce water stress.
4. Differences in aflatoxin levels exist between genotypes. However, any comparison of maize lines for resistance to aflatoxin accumulation should be done during the linear phase of aflatoxin accumulation.
5. Sources of resistance exist in F-1 maize hybrids to ear rot caused by *A. flavus*.

Banded Leaf and Sheath Spot

Banded leaf and sheath spot is also called banded leaf and sheath blight.

Cause. *Rhizoctonia solani* (syn. *R. microsclerotia*) infects plants under warm, humid conditions.

Distribution. Semitropics and tropics.

Symptoms. Twenty-five to 30 days after sowing, symptoms appear on the lowest leaf sheaths as water-soaked areas that later become dirty white bands 3 to 5 cm wide. Lesions increase in size and become discolored gray, tan, or brown areas that alternate with dark bands. Mycelium radiates and spreads upward to leaf blades, causing a similar band within 24 hours, then mycelium grows progressively to the diseased upper leaf sheaths, eventually killing the leaf blades. Infection continues until all leaves and ears have died and dried up. Mycelium on dried leaf sheaths and blades condenses to form sclerotia (1.4–2.7 mm in diameter) that are initially white but later become brown. Ears have a brownish rot under warm, humid conditions.

Management
1. Grow resistant genotypes.
2. Apply foliar fungicides.

Black Bundle Disease

Cause. *Acremonium strictum* (syn. *Cephalosporium acremonium*) is seedborne and soilborne, surviving on infested residue in the soil. Infection of the stalk may be directly through the epidermis or through wounds. There are reports of transmission from seeds to seedlings and eventually to adult plants.

Distribution. Generally wherever maize is grown.

Symptoms. Initially, diseased leaves become purple or red at the dough stage of plant development. Vascular bundles in the diseased stalk are blackened or reddened through several internodes. Diseased plants tiller excessively and are barren or have either small ears or multiple ears at one node.

Management
1. Resistance has been identified in some inbreds.
2. Treat seed with a seed-treatment fungicide.

Black Kernel Rot

Cause. *Lasiodiplodia theobromae* (syn. *Botryodiplodia theobromae*).

Distribution. Unknown.

Symptoms. Diseased kernels have a brown discoloration that later becomes black and has small eruptions containing pycnidia. A dry rot develops.

Management. Not reported.

Borde Blanco

Borde blanco is also called white border.

Cause. *Marasmiellus* sp. is primarily a pathogen of the leaf blade and sheath and is not associated with roots or the basal regions of the stalk.

Distribution. Costa Rica, Mexico, and Nicaragua. Borde blanco occurs in humid lowlands and highlands where "hanging mists" occur during the rainy season.

Symptoms. Leaf blight is characterized by blanched marginal lesions that always begin at the leaf margin but originate at any point along the diseased leaf from the sheath-blade juncture to the apex. Lesions (1–3 × 5–25 cm) on diseased leaves of all ages are characterized by concentric zones of slightly different discoloration. The white border lesions of severely infected plants glisten in sunlight.

Stalk lesions originate at the sheath margin and *Marasmiellus* sp. penetrates through several sheath layers into the stalk. Stalk lesions are numerous in fields with a high percentage of diseased leaves. Severe stalk rot has been observed in humid areas where foliage is dense.

Management. Not reported.

Brown Spot and Stalk Rot

Brown spot is also called black spot, Physoderma brown spot, and Physoderma disease.

Cause. *Physoderma maydis* (syn. *P. zeae-maydis*) overwinters as thick-walled, brown sporangia in infested residue and soil. Sporangia released from infection pustules, disintegrating infested residue, and soil are disseminated by wind, water, insects, and machinery.

A sporangium germinates by a lid, or cap, opening on one side of the sporangium to release a vesicle that, in turn, ruptures to release 20 to 50 zoospores. Zoospores swim for 1 to 2 hours then become amoeba-like and germinate to form infection hyphae that penetrate meristematic tissue of the host plant. The resulting mycelium enters mesophyll or parenchyma cells and forms vegetative structures from which sporangia develop in 16 to 20 days. Secondary spread is by zoospores produced within the thin-walled sporangia. Infection commonly occurs in a diurnal cycle that results in alternating bands of diseased and healthy tissue as the diseased leaf emerges from the whorl. Germination is optimum when moisture is present in whorls or behind leaf sheaths and temperatures are 22° to 32°C.

Plants become increasingly susceptible to infection until they are 45 to 50 days old. Disease development is favored by warm, wet weather and is more severe on plants sown into infested residue that has overwintered.

Distribution. Central America and the southeastern and midwestern United States.

Symptoms. Initially, lesions occur near the base of the leaf blade as small, round, yellowish spots. Lesions may occur in bands across the leaf blade, with diseased tissues turning chocolate brown to reddish brown and merging to form large, irregular to angular-shaped blotches. Cells of diseased maize tissue disintegrate and expose dusty pustules or brown blisters containing large numbers of microscopic light to dark brown sporangia.

Infections at the nodes beneath the leaf sheaths and premature death of diseased plants due to leaf blighting cause stalk rot and lodging. Symptoms of stalk infections are water-soaked lesions that occur initially on the stalk nodes beneath the leaf sheaths and later coalesce to form brown blotches. Small pockets of brown sporangia may form in the diseased stalk.

Management
1. Grow resistant hybrids.
2. Do not grow susceptible hybrids in river bottoms or other high-humidity locations.
3. Shred and bury infested maize residue.

Brown Stripe Downy Mildew

Cause. *Sclerophthora rayssiae* var. *zeae* survives as oospores in infested residue in the soil and as mycelia in different grasses. Oospores germinate in a film of free water and form sporangia in which zoospores are produced. Sometimes sporangia germinate directly and form a germ tube. Mycelia in other plant hosts give direct rise to sporangia. Zoospores liberated from sporangia germinate in free water on leaves at optimum temperatures of 22° to 25°C; thereafter, temperatures of 28° to 32°C favor disease development. Secondary spread is by the formation of sporangia, at temperatures of 20° to 22°C, that are disseminated by wind and water.

Distribution. India, Nepal, and Thailand.

Symptoms. Narrow (3–7 mm wide), yellow stripes with definite margins that are limited by veins develop on diseased leaves. Later, the stripes become reddish to purple. Further development of lesions causes a severe striping and blotching of leaves that may either kill plants or suppress seed development if the symptoms occur prior to plant flowering. Sporangia, spo-rangiophores, and mycelia develop on both sides of leaf lesions and appear as a downy growth. Floral and vegetative plant parts are not malformed nor are diseased leaves shredded.

Management
1. Grow resistant cultivars or hybrids.
2. Sow before the rainy season begins.
3. Apply fungicides as a foliar spray after symptoms appear or as a soil drench.

Cephalosporium Kernel Rot

Cephalosporium kernel rot is the kernel rot phase of black bundle disease.

Cause. *Acremonium strictum* (syn. *Cephalosporium acremonium*) is seedborne and survives within infested residue in the soil.

Distribution. Generally wherever maize is grown.

Symptoms. A white, cottony growth occurs on and between kernels on the diseased ear. A white streaking on the kernels has been associated with the disease, but this has been disputed by some researchers.

Management
1. Resistance has been identified in some inbreds.
2. Treat seed with a seed-treatment fungicide.

Cercospora Leaf Spot

Cause. *Cercospora* sp. The fungus presumably survives in infested residue in soil.

Distribution. The United States (Pennsylvania).

Symptoms. Greenhouse inoculations caused pale tan lesions on diseased leaves. Lesions are limited in lateral spread by the major leaf veins, and le-sion ends are more irregular-shaped than the rectangular lesions caused by *C. zeae-maydis*.

Management. Not reported.

Charcoal Rot

Cause. *Macrophomina phaseolina* (syn. *M. phaseoli*, *Botryodiplodia phaseoli*, and *Sclerotium bataticola*) survives as sclerotia in infested residue and soil. Strains of *M. phaseolina* that infect other host plants produce pycnidia, but strain(s) that infect maize do not. Sclerotia are disseminated by any means that moves residue and soil.

Sclerotia germinate in dry soil at optimum temperatures of 37° to 38°C and produce infection hyphae that infect the roots of seedlings, young plants, and mature plants. Hyphae eventually grow through the cortical tissues and into the stalks. Small sclerotia are eventually produced on diseased tissues.

Association with high populations of the Banks grass mite, *Oligonychus pratensis*, and the two-spotted spider mite, *Tetranychus urticae*, is reported to increase incidence and severity of stalk rot. Low soil temperatures or high soil moisture will hinder the development of charcoal rot.

Distribution. Europe, North America, and South Africa.

Symptoms. Plants nearing maturity are most likely to display disease symptoms. Brown, water-soaked lesions, which later become black, appear on roots. As plants mature, *M. phaseolina* grows into the lower stalk internodes and causes premature ripening and shredding. Diseased stalks frequently break over at the crown. The surface of rotted stalk internodes may turn gray due to the growth of numerous tiny, black sclerotia that grow through the epidermis of stalks and roots. The interiors of diseased stalks disintegrate and are covered with sclerotia, leaving only the vascular bundles intact. The numerous sclerotia give the diseased plant tissue the appearance of being covered with charcoal, hence the name charcoal rot. Kernels are also diseased and become completely black.

Management
1. Keep soils moist.
2. Rotate with resistant plants for 2 to 3 years.
3. Grow full-season, adapted hybrids.
4. Plow under infested residue.

Common Corn Rust

Common corn rust is also called common rust and common maize rust.

Cause. *Puccinia sorghi* is a heteroecious long-cycled rust fungus whose alternate host is wood sorrel, *Oxalis* spp. Commonly, urediniospores overwinter in warmer climates and are disseminated by wind into temperate maize-growing regions, thereby serving as primary inoculum and bypassing the alternate host. Temperatures of 16° to 23°C and high relative humidity favor disease development and spread.

The full cycle is as follows: Teliospores replace urediniospores in pustules as maize nears maturity. In the spring, teliospores germinate and form basidiospores that infect only *Oxalis* spp. The spermatial (pycnial) stage occurs on the diseased upper leaf surface at the points of infection by the basidiospores. Pycniospores (spermatia) and receptive hyphae are formed in pycnia (spermagonia) together with a sticky, sweet liquid that is attractive to insects. The pycniospores are extruded from the pycnia and disseminated by insects to the receptive hyphae of an opposite mating type. Nuclei of the pycniospores and receptive hyphae merge to initiate development of aecia on the diseased lower leaf surface. Aeciospores from aecia are disseminated by wind and infect only maize. These infections give rise to urediniospores.

Distribution. Uredinial and telial stages are generally distributed wherever maize is grown. The aecial stage occurs infrequently in temperate areas of Europe, India, Mexico, Nepal, Russia, South Africa, and the United States.

Symptoms. Symptoms appear soon after the silking stage of plant growth, but older tissue is usually resistant. Sweet maize is generally more susceptible than field maize. Oval to elongate, golden-brown to reddish-brown pustules, which consist of urediniospores breaking through the epidermis, are sparsely scattered over both surfaces of diseased leaves but may appear on any aboveground plant part. As maize matures, pustules become brownish black with the formation of teliospores that also break through the epidermis.

The rupture of the epidermis by urediniospores and teliospores differentiates common rust from southern corn rust. The epidermis over a southern corn rust uredinia remains intact for a long period of time but eventually does rupture.

During severe disease conditions, chlorosis and death of diseased leaf sheaths and leaves may occur.

Management
1. Grow resistant maize hybrids.
2. Apply foliar fungicides when pustules first appear, particularly in seed production fields.

Common Smut

Common smut is also called boil smut.

Cause. *Ustilago zeae* (syn. *U. maydis*) overwinters as teliospores (chlamydospores) in the soil. Teliospores germinate either directly or indirectly in the presence of moisture at a wide range of temperatures (10° to 35°C). Direct germination is by an infection hypha produced directly from a germinating teliospore. Indirect germination is by a promycelium first being formed by a teliospore and on which sporidia (basidiospores) are borne. Sporidia are then disseminated by wind to host leaves, where they germinate. The resultant mycelium fuses with the mycelium from an opposite mating type to form an infection hypha that infects meristematic tissues. Ears of maize are infected by infection hyphae growing through silk.

The parasitic mycelium stimulates host cells to increase in size and numbers, which results in the formation of a gall. Eventually, teliospores develop within the mycelium and gall, causing the galls to be converted to a black, powdery mass consisting of teliospores.

Disease development is favored by dry conditions that retard plant growth and temperatures of 26° to 34°C. Abnormally cool, wet weather, which retards growth of young maize plants, is also conducive to disease development. Smut incidence is higher when plants are grown in soil high in nitrogen, particularly where heavy applications of manure have been made. Injuries from hail, cultivation, herbicide, and detasseling increase the infection incidence and the resulting disease severity.

Distribution. Worldwide; however, smut rarely occurs in tropical areas.

Symptoms. All aboveground plant parts are susceptible, but meristematic tissue is particularly susceptible to infection and disease. Galls are first covered with a glistening white membrane that eventually ruptures within a few weeks to expose a black, powdery mass of teliospores. Galls on ears and stalks grow to 15 cm in diameter or even larger in some instances. However, galls on leaves rarely develop beyond the size of peas (0.6–1.2 cm in diameter), become hard and dry, and contain few spores. Large galls on the ear and on the stalk above the ear are more destructive than galls on the stalk below the ear.

Early infection may kill seedlings. Galls on the lower part of the stalk may cause plants to be barren or to produce several small ears. Such plants become a reddish color in the autumn.

Management
1. Grow resistant hybrids.
2. Maintain a well-balanced soil fertility.
3. Avoid mechanical injuries to plants.
4. Remove and burn galls from infected plants before they rupture.

Corticium Ear Rot

Cause. *Thanatephorus cucumeris* (syn. *Corticium sasakii*). The fungus presumably overseasons as sclerotia in infested residue. Ear rot occurs under extremely wet conditions.

Distribution. India.

Symptoms. Initially, water-soaked spots that vary in size and shape occur on diseased leaf sheaths. The spots gradually enlarge in moist and warm weather and profuse fungal growth sometimes appears on their surfaces. Brown mycelium grows upward from the base of the ear and is observed late in the season on husks, floral bracts, and pericarps of rotted kernels. Sclerotia are formed on diseased ears in the later stages of disease.

Management. Not reported.

Crazy Top Downy Mildew

Crazy top downy mildew is also called crazy top.

Cause. *Sclerophthora macrospora* (syn. *Sclerospora macrospora* and *Phytophthora macrospora*) survives in infested residue and wild grasses as oospores that germinate and form sporangia in which zoospores are produced. Zoospores are liberated in water-saturated soil and "swim" to seedlings, where they germinate to produce mycelia that penetrate the plant and become systemic.

Germination and infection occur over a wide range of temperatures and in soil saturated with water for 24 to 48 hours before plants are in the four- to five-leaf stage. Sporangia are rarely produced directly on sporangiophores projecting from stomates on leaves, but numerous oospores are produced within diseased leaves and leaf sheaths. Over 140 grass species are hosts of *S. macrospora*.

Distribution. Africa, Asia, Canada, Europe, Mexico, and the United States. Crazy top downy mildew occurs in localized wet areas within a field.

Symptoms. The most conspicuous symptom is the partial to complete replacement of the tassel by a large, bushy mass of small leaves. No pollen is produced since flower parts within the tassel are completely deformed. Ear shoots may be numerous, elongated, leafy, and barren. Diseased plants vary in height. Some plants are severely stunted and have narrow, yellow to brown streaks on leaves and numerous tillers. Other plants are taller than normal and have an increased number of nodes and leaves above the ear and in the shank.

Management
1. Provide adequate soil drainage.
2. Control grassy weeds.
3. Do not grow in low, wet spots.

Curvularia Leaf Spot

Cause. *Curvularia clavata, C. eragrostidis* (syn. *C. maculans*), *C. inaequalis, C. intermedia, C. lunata, C. pallenscens, C. senegalensis*, and *C. tuberculata*.

Distribution. Generally wherever maize is grown in warm or mild climates.

Symptoms. Initially, circular (1–2 mm in diameter), straw-colored lesions with reddish or dark brown margins develop on diseased leaves and may coalesce and form necrotic areas up to 1 cm long. The centers of some lesions have circular (1 mm in diameter), grayish spots that may have a chlorotic halo.

Management. Not reported.

Didymella Leaf Spot

Cause. *Didymella exitialis*. It is presumed *D. exitialis* overseasons as perithecia in infested residue.

Distribution. Not reported.

Symptoms. Small, light-colored, elongate to elliptical-shaped spots with brown margins form on diseased leaves. Later, spots enlarge and coalesce to form streaks, or irregular-shaped areas, in which perithecia may be formed.

Management. Not practiced.

Diplodia Ear Rot

Diplodia ear rot is also called white ear rot. Diplodia ear rot, Diplodia seedling blight, and Diplodia stalk rot caused by *Diplodia maydis* are considered as parts of a single disease complex.

Cause. *Diplodia maydis* (syn. *D. zeae*, *D. zeae-maydis* and *Stenocarpella maydis*). *Diplodia maydis* overwinters as conidia within pycnidia and mycelium in infested residue, and as conidia and mycelium on seed. During warm, moist weather in the late summer, conidia are exuded from pycnidia in long cirri and disseminated by splashing rain or, after cirri have dried up, by wind. Insects are reported to disseminate spores.

Infection occurs late in the development of the kernel. Some researchers state ears are most susceptible during a plant growth stage beginning at silking and continuing for 21 to 24 days afterward, when kernels have reached a moisture level of 75.4% to 90.0%. During this time, a decrease of 5% moisture in kernels is reported to be associated with a decrease of up to 22% in ear rot incidence. Resistance occurs 28 days after midsilk, when moisture is approximately 66%. As ears approach maturity, rot slows and ceases completely when kernels reach approximately 21% moisture content.

Infection commonly starts at the base of husks or on an exposed ear tip, but mycelium can advance from the stalk through the shank. *Diplodia maydis* enters the pedicel and penetrates through the hilar region. The hilar layer (black layer) of the hilum is impermeable to the fungus; therefore, entrance into the kernel occurs before the closing layer is formed or delayed, or if suberization of testa membrane is incomplete. As ears grow in length, husks become loose and provide less protection at the ear tip, allowing kernels to be more accessible to spores.

Pycnidia may be found on rotted ears, particularly on the inner husks. During wet weather, conidia are exuded and become disseminated by wind. Dry weather early in the growing season followed by wet conditions before and after silking favors ear infection.

Distribution. Africa, Australia, the Philippines, Romania, and the United States.

Symptoms. Disease on the kernels of maturing plants begins at the ear base and progresses toward the tip. In contrast to the green color of healthy ears, the husks of diseased ears that were infected early in the growing season appear to be bleached. When infection occurs 2 weeks after silking, the entire ear becomes grayish brown, shrunken, and light in weight by harvest time. Such ears remain upright and have tight husks that adhere to the ear due to the growth of the fungus between the ear and the husk. Black pycnidia are found at the base of husks and on the faces of kernels. Ears infected later in the growing season have no apparent symptoms until the ear is broken or kernels are removed. White mycelium of *D. maydis* will be found growing between the kernels on a diseased ear.

The germination of diseased kernels is reduced.

Management. Diseased stalks may produce inoculum that will infect ears. Therefore, any management practice that lessens stalk rot will likely lessen the incidence of ear rot.

1. Grow the most resistant hybrid.
2. Balance soil fertility. Excess levels of nitrogen may predispose plants to stalk rot. Stalk rot is lowest when the level of nitrogen is adequate enough to allow a hybrid to reach its yield potential. The severity of stalk rot in artificially inoculated plants has been reported to decrease with increasing rates of nitrogen applied as anhydrous ammonia in either spring or autumn.
3. Plant the recommended plant population. Higher than recommended plant populations cause plants to be stressed and have less resistance to infection.
4. Harvest early to prevent grain from being weathered.
5. Dry maize kernels to 15% or less moisture content.
6. In programs for breeding resistance to ear rot, ear inoculations should be made at or shortly after midsilk.

Diplodia Ear Rot and Stalk Rot

Cause. *Diplodia frumenti*, teleomorph *Botryosphaeria festucae*. *Diplodia frumenti* is seedborne, but it is not known if this is important in the life cycles of the fungi. The fungus presumably overseasons either as pycnidia or perithecia.

Distribution. Brazil and the United States (Florida).

Symptoms. Stalk pith is discolored black. Pycnidia occur in longitudinal cracks on the outer diseased stalk tissues. Diseased ears have a dark brown, felt-like growth at the butt end. Seeds are black and embedded with pycnidia.

Management. Not reported. The disease is of minor importance.

Diplodia Leaf Spot

Diplodia leaf spot is also called Diplodia leaf streak, southern leaf spot, and Stenocarpella leaf spot.

> **Cause.** *Stenocarpella macrospora* (syn. *Diplodia macrospora*) survives as mycelium within infested residue and seed. Ears, leaves, and leaf sheaths are generally considered to be most susceptible to infection. However, all plant tissue can be infected at any stage of growth. Disease occurs in the semitropics under warm, humid conditions with a minimum of 50% relative humidity during the day and 95% relative humidity at night.

> **Distribution.** Central America and the United States (North Carolina).

> **Symptoms.** Diseased seeds or seedlings may damp off. Initially, small, yellowish spots occur on leaves. Later, the spots become water-soaked lesions with chlorotic margins. The lesions enlarge and become long, narrow, chlorotic lesions or streaks (1.5 × 10.0 cm). Pycnidia, which appear as numerous black specks, are scattered throughout the lesion. Frequently, *S. macrospora* grows along the borders of leaf veins. Inoculated plants had lesions along veins the entire length of a diseased leaf blade. Toxin production was indicated.
>
> A leaf spot caused by *S. macrospora* has been described from North Carolina. However, these symptoms were described as circular to elongate lesions that were up to 25 cm long.

> **Management.** Differences in resistance exist between genotypes.

Diplodia Seedling Blight

Diplodia ear rot, Diplodia seedling blight, and Diplodia stalk rot caused by *Diplodia maydis* are considered as parts of a single disease complex.

> **Cause.** *Diplodia maydis* (syn. *D. zeae, D. zeae-maydis* and *Stenocarpella maydis*). *Diplodia maydis* overwinters as conidia within pycnidia and mycelium in infested residue, and as conidia and mycelium on seed. During warm, moist weather in late summer, conidia are exuded from pycnidia in long cirri and disseminated by splashing rain or, after cirri have dried up, by wind. Insects are also reported to disseminate spores.
>
> Infection occurs late in the development of the kernel. Some researchers state ears are most susceptible to infection beginning at the silking stage of plant growth and continuing for 21 to 24 days afterward, when kernels have reached a moisture level of 75.4 to 90.0%. A decrease of 5% in the moisture content of kernels is associated with a reported de-

crease of up to 22% in the incidence of ear rot. Resistance to infection occurs 28 days after midsilk, when the moisture content of kernels is approximately 66%. As ears approach maturity, rot slows and ceases completely when kernels reach approximately 21% moisture.

Infection commonly starts at the base of husks or on an exposed ear tip, but mycelium can advance from the stalk through the shank. *Diplodia maydis* enters the pedicel and penetrates through the hilar region. The hilar layer (black layer) of the hilum is impermeable to the fungus; therefore, entrance into the kernel occurs before the closing layer is formed or delayed, or if suberization of testa membrane is incomplete. As ears grow in length, husks become loose and provide less protection at the ear tip, allowing kernels to be more accessible to spores.

Pycnidia may be found on rotted ears, particularly on the inner husks. During wet weather, conidia are exuded and become windborne. Dry weather early in the season followed by wet conditions before and after the silking stage of plant growth favors ear infection.

Diplodia maydis causes diplodiosis in sheep and cattle.

Distribution. Africa, Australia, the Philippines, Romania, and the United States.

Symptoms. Diseased seeds fail to germinate and become soft or overgrown with white mycelium. Seedling mesocotyls become water-soaked and seedlings die in 2 to 3 days.

Management
1. Treat seed with a seed-treatment fungicide.
2. Harvest early to prevent grain from being weathered.
3. Dry maize to 15% or less moisture content.
4. In programs for breeding resistance to ear rot, ear inoculations should be made at or shortly after midsilk.

Diplodia Stalk Rot

Diplodia ear rot, Diplodia seedling blight, and Diplodia stalk rot caused by *Diplodia maydis* are considered as parts of a single disease complex.

Cause. *Diplodia maydis* (syn. *D. zeae*, *D. zeae-maydis* and *Stenocarpella maydis*). *Diplodia maydis* overwinters as conidia within pycnidia and mycelium in infested residue, and as conidia and mycelium on seed. During warm, moist weather in late summer, conidia are exuded from pycnidia in long cirri and disseminated by splashing rain or, after cirri have dried up, by wind. Insects are reported to disseminate spores.

Stalk rot infection occurs most frequently at the origin of leaf sheaths at any of the first three nodes above the soil line and infrequently at a higher node. Some infections also occur at the junction of the brace roots with stalks, on roots at the soil level, and through the mesocotyl from infected seed or infested soil. *Stenocarpella maydis* then grows up into the stalk and down into roots, but the entire plant is usually not invaded. Temperatures of 28° to 30°C and wet weather 2 to 3 weeks after silking favor spread and development of disease. Date of inoculation has little effect on rate of disease development in stalks.

Plants are predisposed to infection by a number of factors, such as excess nitrogen and low potassium, higher than recommended plant populations, and loss of leaf area by hail, insects, and other diseases. Early-maturing hybrids are more susceptible to stalk rot than longer-season hybrids. All of the factors are combined in the photosynthetic stress-translocation balance concept of predisposition hypothesis: Predisposition of maize to stalk rot is associated with a carbohydrate shortage in root tissue caused by a combination of the reduction in photosynthesis by leaves and intraplant competition for carbohydrates by developing kernels. The carbohydrates, formerly in root and stalk tissue, conferred a resistance to these tissues. When carbohydrates are transferred into the developing kernels, root tissue then has a weakened defense system, which allows soil microorganisms to invade and decompose roots and stalks. Parenchyma rot and crown rot have been reported to be more severe in plants with greater numbers of kernels, which is possibly related to the competition for carbohydrates.

Distribution. Africa, Australia, the Philippines, Romania, and the United States.

Symptoms. Stalk rot does not have a direct physiological effect on yield but is an important factor in stalk breakage. Symptoms appear several weeks after silking as grayish-green leaves that later turn brown, seemingly in a day or two, and give the impression of frost damage. Dark brown lesions extend in either direction from a diseased node. The stalk pith is discolored and disintegrated, with only vascular bundles remaining intact, which causes stalks to become weak. In the autumn, dark brown to black pycnidia form in clusters just beneath the epidermis of the lower diseased internodes. White fungal growth of *D. maydis* may be present on the surface of nodes.

Management
1. Grow the most resistant hybrid. No inbred line or hybrid is completely resistant to stalk rot.
2. Balance soil fertility. Excess levels of nitrogen may predispose plants to stalk rot. Stalk rot is lowest when the level of nitrogen is adequate enough to allow a hybrid to reach its yield potential. The severity of stalk rot in artificially inoculated plants has been reported to decrease

with increasing rates of nitrogen applied as anhydrous ammonia in either spring or autumn.
3. Grow the recommended plant population. Higher than recommended plant populations cause plants to be stressed and have less resistance to infection.

Dry Ear Rot

Dry ear rot is also called cob rot and Nigrospora ear rot.

Cause. *Nigrospora oryzae* overwinters as mycelium in infested soil residue. Conidia are formed on mycelium and are disseminated by wind. *Nigrospora oryzae* is a weak parasite that infects either dead plants or plants stressed by other factors, such as stalk rot, leaf blight, cold weather, and root injury. Maize grown on soil with poor fertility is more susceptible to infection than maize grown in fertile soil. Infection normally starts either at the butt end or at the tip of the ear.

Distribution. Generally wherever maize is grown.

Symptoms. Diseased ears are light in weight but do not have external symptoms until harvest time. The ear cobs are soft and easily broken and have kernels that are either bleached or have white streaks which start at the kernel tip and extend toward its crown. Kernels may be overgrown with gray mycelium and covered with groups of spores that resemble small, round, black dots. When viewed under magnification, individual spores appear to be the size of pinpoints.

Management
1. Grow full-season, adapted hybrids that are resistant to stalk rots and leaf blights.
2. Balance soil fertility.

Ergot

Ergot is also called horse's tooth and diente de caballo.

Cause. *Claviceps gigantea*, anamorph *Sphacelia* sp., survives as sclerotia either on the soil surface or mixed with seed. In the spring, sclerotia germinate to produce stroma on the ends of stipes. The stromatic head contains perithecia in which ascospores are produced, then are disseminated by wind to flowers. Macroconidia and microconidia are produced in a sweet, sticky substance called honeydew, which is attractive to insects. Insects inadvertently carry conidia from diseased to healthy flowers while feeding on honeydew. Disease development is optimum at temperatures of 13° to 15°C and high rainfall.

Distribution. At higher elevations in central Mexico.

Symptoms. Sclerotia replace kernels. Sclerotia initially are cream-colored but later become dark gray, enlarge to 5 × 8 cm in size, and become comma-shaped, superficially resembling a horse's tooth. Insects are present on honeydew, which becomes dark-colored due to dust and other material adhering to it.

Management. No disease management is practiced, but some lines are more susceptible than others.

Eyespot

Eyespot is also called brown spot.

Cause. *Aureobasidium zeae* (syn. *Kabatiella zeae*) overwinters as mycelium in infested residue on the soil surface and is seedborne. Conidia produced on residue are disseminated by wind to nearby plants. Secondary infection is by conidia produced on new lesions and disseminated by wind and splashed by rain to other host plants. Disease development is favored by cool, humid weather. Older leaves are more susceptible to infection than younger ones. Maize is the only plant that is known to be a host for *A. zeae*.

Distribution. Argentina, Austria, Brazil, Canada (Ontario), France, Germany, Japan, New Zealand, the United States (north central states and Pennsylvania), and Yugoslavia.

Symptoms. The first symptoms on leaves are small, translucent or water-soaked, ovoid to circular spots (2–5 mm in diameter). These symptoms may be confused with genetic spotting or Curvularia leaf spots. Later, the centers of the spots die, become tan-colored, and are surrounded by a narrow, brown to purple ring with a chlorotic halo. Spots develop in zones or patches on diseased leaves rather than occurring randomly scattered over leaf blades. Eventually, the spots coalesce and form large necrotic areas that are still distinguishable as spots after leaf death. Spots also develop on leaf sheaths, husks, and kernels.

Management
1. Grow resistant hybrids.
2. Plow under infested residue where feasible.
3. Rotate maize with resistant crops.

False Smut

Cause. *Ustilaginoidea virens*. Disease development is favored by hot, wet weather. Rice and other grasses in the Graminae are hosts.

Distribution. Generally wherever maize is grown, but false smut is rare.

Symptoms. Sclerotia, sometimes referred to as galls or smut balls, usually replace a few to several flowers on diseased tassels. Sclerotia are irregular and subspherical-shaped (4–15 mm in diameter), olive-green to black, and have a velvety texture. A sclerotium is composed of three layers: (1) an outer greenish layer of mature spores, (2) a middle layer of a light yellow-ish-green mass composed of mycelium, and (3) an inner yellowish layer consisting of host tissue, growing mycelium, and spores. As sclerotia mature, the membrane ruptures and exposes and liberates spores. The surface of the gall gradually becomes roughened.

Management. Not necessary.

Fusarium avenaceum Stalk Rot and Seedling Root Rot

Fusarium avenaceum stalk rot may be the same or similar to Gibberella stalk rot.

Cause. *Fusarium avenaceum*. The fungus is reported to form toxins harmful to animals.

Distribution. Widely distributed.

Symptoms. Diseased stalks are brown to gray-green and have a pinkish, disintegrated pith in which just the vascular bundles remain. Dark lesions may sometimes appear at nodes and extend into internodes. Roots of seedlings have a brown discoloration typical of symptoms of other root rots.

Management
1. Grow the least susceptible hybrid.
2. Balance soil fertility.
3. Grow the recommended plant population.

Fusarium globosum Kernel Infection

Cause. *Fusarium globosum*.

Distribution. South Africa.

Symptoms. *Fusarium globosum* was isolated from harvested maize kernels that were asymptomatic.

Management. Not reported.

Fusarium Kernel Rot

Fusarium kernel rot is also called Fusarium ear rot.

Cause. *Fusarium moniliforme*, teleomorph *Gibberella fujikuroi*. According to Nelson et al. (1983), *F. subglutinans* and *F. proliferatum* share many morphological characteristics with *F. moniliforme* and may be involved in the same disease syndrome.

Fusarium subglutinans is separated taxonomically from *F. moniliforme* on the following basis: *Fusarium subglutinans* bears microconidia on polyphialides only in false heads and not on microconidial chains, whereas *F. moniliforme* bears microconidia both in false heads and in long chains. *Fusarium proliferatum* is separated from *F. moniliforme* in that *F. proliferatum* also bears microconidia in chains of various lengths and in false heads but it forms microconidia in chains on polyphialides, whereas *F. moniliforme* forms microconidia in chains on monophialides. It is likely these fungi have been misidentified or confused with each other. Some researchers still consider the three fungi to be synonymous, presumably on the basis of their morphology.

Fusarium moniliforme survives in crop residue as thickened hyphae or chlamydospore-like structures and in kernels, but the fungus does not survive for long periods in soil. Macroconidia and abundant microconidia are thought to be produced on crop residue and disseminated by wind to host plants. The pathway of infection into kernels can occur by one of four ways:

1. Airborne conidia infect exposed kernels. This entry may be facilitated by injuries from cracks made by weather or other means and by insects, including corn earworms; corn borers, particularly the European corn borer, *Ostrinia nubilalis;* and rootworm beetles, such as the western corn rootworm beetle, *Diabrotica virgifera*. The insects also may be important in spreading fungi to ears and kernels. It has also been suggested that a relatively thin pericarp layer, which is found in susceptible hybrids, allows access of the fungus into the kernels, especially through insect wounds.
2. Spores germinate and mycelia grow down silks through the tip ends of ears, over the surface of kernels into bracts and pedicels, into the vascular cylinder of the cob, and, finally, into kernels. The location of *F. moniliforme* in asymptomatic kernels is the pedicel, or tip cap end, of kernels, suggesting the point of entry is the same. Sporulation of the fungus on the tassels may contribute to silk infection.
3. A systemic invasion of the shank, where fungal growth proceeds through the stalk from some entryway, or by entering the ruptured scar on the pericarp produced by the emerging coleorhiza after kernel germination. The fungus then proceeds to grow through the stalk into the cob, rachilla, and pedicel.
4. Symptomless infection can exist throughout an infected plant, and seed-transmitted strains of the fungus can develop systemically to infect kernels in an endophytic relationship.

Munkvold et al. (1997) concluded that systemic development of *F. moniliforme* from maize seed and stalk infections can contribute to kernel infection, but silk infection is a more important pathway for this fungus to reach the kernels.

Extensive colonization of kernel surfaces occurs early in the season, but little internal infection occurs until kernel moisture is lower than 34%. Disease development is generally favored by dry, warm weather. The presence of *F. moniliforme* increases as harvest is delayed beyond physiological maturity. *Fusarium moniliforme* does not grow in grain with a moisture content below 18 to 20%.

Generally fumonisin (a class of mycotoxins that have cancerous activity) concentrations are not thought to increase during storage as long as proper conditions of grain moisture and temperature are maintained. *Fusarium moniliforme* undergoes extensive growth and sporulation in the embryo and endosperm of kernels associated with toxicity to animals.

Gibberella fujikuroi consists of at least seven distinct mating populations, designated A through G. Populations of A, D, and E are the most common on maize and correspond to the particular anamorph species. Both A and F mating populations are considered to be *F. moniliforme*; population A contains many prolific fumonisin-producing strains, while members of the F population produce little or no fumonisin. Most strains of *F. moniliforme* found on maize are population A, and most found on sorghum are F. Population D is considered to be *F. proliferatum* and contains many strains that produce copious amounts of fumonisins. *Fusarium proliferatum* was demonstrated to produce beauvericin, fumonisin B1, and moniliformin in maize grown in Italy. Populations B and E are considered to be *F. subglutinans* and produce little or no fumonisin. In maize grown in temperate regions, *F. moniliforme* and *F. proliferatum* represent the greatest threat of fumonisin production.

Distribution. Generally wherever maize is grown.

Symptoms. Fusarium kernel rot does not involve the entire ear; it occurs on kernels at the tips of ears, scattered over the ear, or where there has been insect damage. The color of diseased kernels varies from faint pink to reddish brown. As disease progresses, kernels have an external powdery or cottony pink mold growth on them.

Symptomless infections can exist throughout some plants in an endophytic relationship. Such kernels are a source of inoculum for seedling blight and stalk rot. Asymptomatic kernels are as toxic as cracked kernels and plant parts obviously colonized by fungi. Hyphae are restricted to intercellular spaces during symptomless infection, while the hyphae of disease-causing strains are found in both intercellular and intracellular sites.

Seedlings grown from inoculated kernels had suppressed shoot diameter, plant height, leaf length, and plant weight 7 days after sowing. However, at 28 days, seedling growth from inoculated kernels was similar to or greater than that from noninoculated kernels. Histological modifications in seedlings grown from inoculated kernels included accelerated lignin deposition in shoots and modified chloroplast orientation in leaves.

Fumonisins are most common in maize screenings, which are the broken kernels and other fine material removed from grain passed over a wire screen. In Nebraska, fumonisins are responsible for the moldy corn disease of horses, cattle, mules, hogs, and chickens. The most dramatic manifestation of moldy corn disease is equine leucoencephalomalacia (ELEM), a fatal brain disease of horses, donkeys, mules, and rabbits. In horses, this disease results in death within a few hours to 1 week after fumonisin B1 is ingested at concentrations of 5 to 10 ppm. ELEM in horses is known by the popular term "blind staggers."

Other mycotoxins include fusarin C and moniliformin. Infested maize is carcinogenic, immunosuppressive, hepatotoxic, and nephrotoxic to animals. Fumonisin-contaminated maize has also been associated with porcine pulmonary edema in swine and is reported to be a mild carcinogen or tumor promoter in animals.

Management. There are no proven, practical methods for significantly reducing fumonisin concentrations in maize. The following are aids in disease management:

1. Avoid growing hybrids that are most susceptible. Hybrids with poor husk covers or weak seed coats, in which kernels pop or silk-cut, are susceptible to infection. Differences in resistance are conditioned by genotype of the pericarp. In mature stages, sweet maize and some high-lysine hybrids are very susceptible to Fusarium ear rot. Commercial hybrids differ in their tendency to accumulate fumonisins. Hybrids grown outside their adapted range tend to accumulate higher concentrations of fumonisins.
2. Harvest grain early to prevent deterioration from weather.
3. Dry maize to 15% or less moisture content.
4. Resistance has been identified in a sweet maize inbred.

Fusarium Leaf Spot

Cause. *Fusarium moniliforme* is a fungus that also causes kernel rot and stalk rot. Disease is favored by dry, warm weather.

Distribution. The Caribbean region.

Symptoms. Water-soaked spots that initially appear on whorl leaves eventually become white with brown borders and have a papery texture. The whorl does not open, and leaf tips turn down. Occasionally, necrotic spots develop high on a stalk.

Management. Not reported.

Fusarium scirpi in Grain

Cause. *Fusarium scirpi* (syn. *F. equiseti*).

Distribution. Australia, China, South Africa (Transkei).

Symptoms. *Fusarium scirpi* was isolated from grain and is toxic to animals when grown both in vivo and in vitro. Fungal metabolites induced chromosomal aberrations in animal cells.

Management. Not reported.

Fusarium Stalk Rot

Cause. *Fusarium moniliforme, F. proliferatum,* and *F. subglutinans,* teleomorph *Gibberella fujikuroi.* According to Nelson et al. (1983) *F. subglutinans* and *F. proliferatum* share many morphological characteristics with *F. moniliforme* and may be involved in the same disease syndrome.

Fusarium subglutinans is separated taxonomically from *F. moniliforme* on the following basis: *Fusarium subglutinans* bears microconidia on polyphialides only in false heads and not on microconidial chains, whereas *F. moniliforme* bears microconidia both in false heads and in long chains. *Fusarium proliferatum* is separated from *F. moniliforme* in that *F. proliferatum* also bears microconidia in chains of various lengths and in false heads but it forms microconidia in chains on polyphialides, whereas *F. moniliforme* forms microconidia in chains on monophialides. It is likely these fungi have been misidentified or confused with each other. Some researchers still consider the three fungi to be synonymous, presumably on the basis of their morphology.

Fusarium moniliforme is seedborne and survives as thickened hyphae or chlamydospore-like structures and mycelium in infested corn residue that was either precolonized parasitically or colonized when placed in soil. Residue of other plants may also be colonized in the soil by *F. moniliforme.* The teleomorph, *G. fujikuroi,* survives more than 1 year in residue that is buried in the soil. *Fusarium moniliforme* does not survive in soil outside of residue. Infected kernels lacking symptoms are a source of inoculum for seedling blight and stalk rot.

Before pollination, stalks are infected by conidia produced on residue and disseminated by wind to the bases of leaf sheaths and into corn borer wounds, where infective hyphae enter the plant. Roots are infected, either through direct penetration or insect wounds, by mycelia that grow through soil from a previously colonized substrate. *Fusarium moniliforme* also attracts the root lesion nematodes, *Hoplolaimus indicus, Meloidogyne incognita, M. javanica, Pratylenchus brachyurus, P. zeae,* and *Tylenchorhynchus vulgaris,* and aids their penetration into seedling roots.

Disease development is aided by warm, dry conditions, excess soil nitrogen, and higher than recommended plant populations, which stress plants, predisposing them to infection. Stalk rot is reported to be less in no-tillage systems than in either conventional tillage systems or plowing harvested fields in the autumn. Senescing tissues caused by water stress, high populations of the Banks grass mite, *Oligonychus pratensis*, and the two-spotted spider mite, *Tetranychus urticae*, are reported to increase incidence and severity of stalk rot. Parenchyma rot and crown rot are reported to be more severe in diseased plants with a greater numbers of kernels.

Gibberella fujikuroi consists of at least seven distinct mating populations, designated A through G. Populations of A, D, and E are the most common on maize and correspond to the particular anamorph species. Both A and F mating populations are considered to be *F. moniliforme*; population A contains many prolific fumonisin-producing strains, while members of the F population produce little or no fumonisin, a class of mycotoxins that have cancerous activity. Most strains of *F. moniliforme* found on maize are population A, and most found on sorghum are F. Population D is considered to be *F. proliferatum* and contains many strains that produce copious amounts of fumonisins. *Fusarium proliferatum* was demonstrated to produce beauvericin, fumonisin B1, and moniliformin in maize grown in Italy. Populations B and E are considered to be *F. subglutinans* and produce little or no fumonisin. In maize grown in temperate regions, *F. moniliforme* and *F. proliferatum* represent the greatest threat of fumonisin production.

Distribution. Generally wherever maize is grown.

Symptoms. Symptoms are generally similar to Gibberella stalk rot. Rotting affects diseased roots and lower internodes. Nodes have dark lesions without definite margins, and leaves become brown, resembling frost injury. Pith disintegration begins after pollination, although plants are infected earlier in the season. Pith tissue is shredded; only the vascular bundles remain intact. White to pale pink–colored mycelium grows on the surface of diseased nodes in wet weather.

Symptomless infections can exist throughout some infected plants in an endophytic relationship. Hyphae are restricted to intercellular spaces during symptomless infection, whereas hyphae of disease-causing strains are found in both intercellular and intracellular sites.

Management
1. Grow the most resistant hybrids.
2. Balance soil fertility.
3. Grow the recommended plant population.

Fusarium thapsinum Infection

Cause. *Fusarium thapsinum*, teleomorph *Gibberella thapsina*. *Fusarium thapsinum* and *F. moniliforme* are similar morphologically and may be confused in the literature.

Distinguishing *F. thapsinum* from *F. moniliforme* is accomplished by crosses with standard female-fertile tester strains. *Fusarium thapsinum* produces a yellow pigment in culture; *F. moniliforme* does not. *Fusarium thapsinum* does not produce fumonisins; *F. moniliforme* does. *Fusarium thapsinum* produces moniliformin; *F. moniliforme* does only rarely. *Fusarium thapsinum* is sensitive to hygromycin; *F. moniliforme* is not. Therefore, it is suggested that the numerous reports in the literature assigning the predominant *Fusarium* species associated with head blight of sorghum to *F. moniliforme* and consider this organism to be the same one responsible for stalk and ear rot in maize may be erroneous.

It is implied that the nonproduction of fumonisins or the production of large amounts of moniliformin by isolates of *F. moniliforme* from sorghum probably refer to strains of *F. thapsinum* rather than to strains of *F. moniliforme*. This difference also implies that sorghum is more likely to be naturally contaminated with moniliformin than with fumonisins, and maize is more likely to be naturally contaminated with fumonisins instead of moniliformin.

Distribution. Egypt, the Philippines, South Africa, Thailand, and the United States (Alabama, Arkansas, Georgia, Illinois, Kansas, Mississippi, Missouri, Ohio, and Texas).

Symptom. *Fusarium thapsinum* has been isolated from diseased stalks and seeds. The fungus has also been isolated from asymptomatic seed, stems, roots, and crowns of sorghum, suggesting *F. thapsinum* may be endophytic in this plant. Pathogenicity of the fungus has not been tested under field conditions on maize, but under greenhouse conditions some strains can cause lesions in sorghum stalks following toothpick inoculations.

Management. Not reported.

Gibberella Ear Rot

Gibberella ear rot is also called red ear rot.

Cause. *Gibberella zeae*, anamorph *Fusarium graminearum*, overwinters as chlamydospores in soil, perithecia and mycelia in residue, and mycelia and conidia on seed. The fungus *F. crookwellense* has been isolated from ears that bear symptoms similar to those caused by *F. graminearum*. To date, no teleomorphic stage is known for *F. crookwellense*.

Perithecia are produced on residue in the autumn or spring. Rainfall may be needed for perithecial and ascospore formation and maturity on crop residue but not to trigger the actual release of ascospores. Perithecial drying during the day followed by sharp increases in relative humidity may provide the stimulus for release of ascospores. In eastern Canada, ascospores were released during the first 3 weeks of July in a diurnal pattern that began around 1600 to 1800 hours and reached a peak before midnight, then declined to low levels by 0900 the following morning. The beginning of ascospore release was correlated with a rise in relative humidity during the early evening hours.

During warm, moist weather early in the season, conidia of *F. gramin-*

earum are produced from mycelia in 1-year-old infested residue. Shortly after pollination, during similar weather conditions, ascospores are exuded from perithecia. Both conidia and ascospores are disseminated by wind to host plants. Sap beetles, *Glischrochilus quadrisignatus*, also disseminate conidia and ascospores to ears, primarily as secondary inoculum. In some areas, Group 2 isolates, which are mostly windborne and cause disease of aboveground plant parts, dominate. These eventually form perithecia. On wheat, *F. graminearum* Group 2 readily produces perithecia in culture, but *F. graminearum* Group 1, which is associated with crown and root rot of wheat, does not.

Entry of *F. graminearum* into the ear often occurs via the silk and/or the silk channel and through bird or insect wounds. The pathogen spreads down the ear from the ear tip, with the silks possibly providing a pathway to the kernels and rachis. Silk is most susceptible to infection when it has some brown tips after elongation and pollination.

Cool, wet weather at silking favors development of disease. Ears with loose, open husks are more susceptible than those with good husk coverage. Inbreds and hybrids differ in susceptibility.

Distribution. Generally wherever maize is grown.

Symptoms. A reddish mold starts at the ear tip. Ears infected early in the growing season may rot completely. Husks adhere tightly to the ear because of mycelial growth between husks and ears. Blue-black perithecia, which can be easily scraped off, are sometimes found on husks and ear shanks.

Fusarium graminearum produces the mycotoxins zearalenone, deoxynivalenol (vomitoxin), 3-acetyldeoxynivalenol, and 15-acetyldeoxynivalenol. *Fusarium crookwellense* is known to produce the toxins nivalenol and fusarenone. The toxins deoxynivalenol and zearalenone are toxic to swine and may be produced in corn affected by *G. zeae*. Zearalenone concentration may vary according to corn line and fungus isolate; however, it is interesting to note that less-virulent isolates have been reported to produce large quantities of zearalenone in culture and in diseased ears. Swine may refuse to eat maize containing vomitoxin, which causes vomiting, dizziness, and, in severe cases, death. Infected maize is also toxic to humans, dogs, and other animals with similar digestive systems. Zearalenone produces estrogenic disturbances that cause young gilts to come in heat and mammary glands of boar pigs to enlarge.

The highest concentration of deoxynivalenol is in the cob, followed by symptomatic kernels, with the lowest concentrations in symptomless kernels.

Management
1. Harvest early.
2. Dry maize to 15% or less moisture content.

3. Several inbreds display resistance. High concentrations of kernel (E)-ferulic acid confer resistance.
4. Complete burial prevents inoculum production on infested residues.
5. Avoid successive crops of wheat or maize.

Gibberella Stalk Rot

Cause. *Gibberella zeae*, anamorph *Fusarium graminearum*, overwinters as chlamydospores in soil, perithecia and mycelia in infested residue, and mycelia and conidia on seed. Perithecia are produced on infested residue in the autumn or spring.

During warm, moist weather early in the season, conidia of *F. graminearum* are produced from mycelia in 1-year-old residue. Shortly after pollination, ascospores are exuded from perithecia. Rainfall may be needed for perithecial and ascospore formation and maturity on crop residue but not to trigger the actual release of ascospores. Perithecial drying during the day followed by sharp increases in relative humidity may provide the stimulus for release of ascospores. In eastern Canada, ascospores were released during the first 3 weeks of July in a diurnal pattern that began around 1600 to 1800 hours and reached a peak before midnight, then declined to low levels by 0900 the following morning. The beginning of ascospore release was correlated with a rise in relative humidity during the early evening hours.

Both spore types are disseminated by wind and infect nodes just above the soil line. *Fusarium graminearum* grows into pith tissue late in the growing season as tissues senesce. In some areas, Group 2 isolates, which are mostly windborne and cause disease of aboveground plant parts by infecting at the leaf sheaths, dominate. These eventually form perithecia. On wheat, *F. graminearum* Group 2 readily produces perithecia in culture, but *F. graminearum* Group 1, which is associated with crown and root rot of wheat, does not.

Another mode of infection common in some areas is the infection of roots by chlamydospores or mycelia. Eventually, the fungus grows into the crown and stalk. Seedborne infection is not thought to result in stalk rot.

Gibberella stalk rot is prevalent in plants subjected to stresses that resulted in early senescence and a reduction of sugar to roots and stalks. Such plants are readily invaded by fungi. Stresses are caused by a variety of factors, such as excessively high levels of nitrogen, low levels of potassium, and high plant populations.

The incidence and severity of stalk rot is lowest when the nitrogen level is adequate enough to allow a hybrid to reach its yield potential. Severity of stalk rot is reported to decrease with increasing rates of nitrogen applied as anhydrous ammonia in spring or autumn. Stalk rot is less in no-tillage systems than in either conventional tillage or fall plowing. There is evidence that Gibberella stalk rot is prevalent in plants growing under favorable conditions early in the season but subjected to unfavorable conditions after silking.

Parenchyma rot and crown rot are reported to be more severe in plants with greater numbers of kernels.

Distribution. Wherever maize is grown. In the United States, Gibberella stalk rot is most common in the eastern and northern maize-growing areas.

Symptoms. Symptoms appear several weeks after pollination. Leaves become grayish green, then quickly turn brown, which resembles damage caused by low temperatures. Stalks are brown to gray-green and have a pinkish, disintegrated pith in which just the vascular bundles remain intact. Dark lesions appear at nodes and extend into internodes. Infrequently, concentric rings appear in lesions on the outside of the stalk. Small, round, black perithecia, which are easy to scrape off, may occur on diseased tissue.

Management
1. Grow the least-susceptible hybrid.
2. Balance soil fertility.
3. Grow the recommended plant population.

Gray Ear Rot

Cause. *Botryosphaeria zeae* (syn. *Physalospora zeae*), anamorph *Macrophoma zeae*, overwinters as sclerotia in infested ear and kernel residue and as pycnidia and perithecia in infested leaf residue. In the spring, ascospores from perithecia and conidia from pycnidia are disseminated by wind or splashing rain to infection sites on leaves and ears. Occasionally, infection occurs on the tassel neck or under the sheath of the uppermost leaf. Perithecia and pycnidia are eventually formed only in leaf lesions, and sclerotia only in ears and kernels. Disease development is favored by long periods of warm to hot, wet weather for several weeks after silking.

Distribution. The eastern United States.

Symptoms. Gray ear rot resembles Diplodia ear rot in its early stages. A gray mycelial growth develops on and between kernels near the base of the ear. Early infection results in bleached husks that adhere tightly to ears. At harvest, ears are slate gray, light in weight, and remain upright. Cobs will have sclerotia, evident as black specks, scattered within them. Kernels may have black streaks beneath the seed coat.

Management. Grow hybrids resistant to Diplodia ear rot since these genotypes may also be resistant to gray ear rot.

Gray Leaf Spot

Gray leaf spot is also called Cercospora leaf spot.

Cause. *Cercospora zeae-maydis* and *C. sorghi* (syn. *C. sorghi* var. *maydis*). A *Mycosphaerella* sp. is thought to be the teleomorph of *C. zeae-maydis*.

Cercospora zeae-maydis overwinters as mycelium in infested residue on the soil surface. Conidia are then produced from mycelia during wet weather and disseminated by wind to host plants. In the north central United States, sporulation does not occur on overwintered tissue buried in soil.

Infested residue has a significant effect on disease severity. Any tillage method leaving residue on the soil surface favors gray leaf spot development. Disease incidence and severity increases proportionately to the amount of infested residue on the soil surface. Tillage systems leaving greater than 35% residue cover may result in high disease levels, especially under environmental conditions favorable for disease development.

Plants are infected at all stages of growth. Although early infection favors disease development, symptoms develop best later in the growing season, possibly due to a thick plant canopy creating a favorable microclimate. Research in Maryland found low populations of maize sown early in the growing season had more severe disease than later-sown maize. No differences were observed between low and high plant populations with late-sown maize.

Disease development is favored by temperatures of 22° to 30°C and long daily periods of high relative humidity and leaf wetness. Maize dwarf mosaic virus is reported to predispose plants to gray leaf spot.

Genetic differences exist between isolates of *C. zeae-maydis*. Barnyard grass, johnsongrass, and *Sorghum* spp. are susceptible to *C. zeae-maydis*. *Zeae diploperennis*, *Z. mays* subsp. *luxurians*, and *Z. mays* subsp. *mexicana* develop typical eyespot lesions when inoculated.

Distribution. Africa, Central America, China, Europe, India, Mexico, the Philippines, northern South America, Southeast Asia, and the eastern and midwestern United States.

Symptoms. General symptoms appear on the lower leaves of mature plants that are at or near anthesis and on the higher leaves of diseased plants later in the season. Initially, reddish, water-soaked spots occur that later become narrow, rectangular-shaped (5 cm long), tan or pale brown lesions restricted by leaf veins. Lesions may have a grayish cast when *C. zeae-maydis* sporulates. Lesions may eventually coalesce and kill the leaf. During severe disease conditions, all leaves may be killed. Severe stalk rot and breakage may occur.

Lesion type influences disease progress. Many susceptible inbreds display necrotic lesions, moderately resistant hybrids display chlorotic-type lesions, and resistant inbreds display fleck-type lesions.

Management
1. Grow resistant hybrids.
2. Plow under infested residue.

Green Ear Downy Mildew

Green ear downy mildew is also called Graminicola downy mildew.

Cause. *Sclerospora graminicola* overwinters as oospores in soil and on seed. Primary infection likely occurs by oospores germinating to form germ tubes that penetrate corn leaves. Infection is favored by temperatures of 24° to 32°C for 2 days after planting. Secondary spread is by formation on diseased leaves of sporangia, which are disseminated by wind to host leaves. Sporangia commonly germinate indirectly by releasing three or more zoospores. Each zoospore then germinates to form a germ tube. Infrequently, sporangia germinate directly to form only a single germ tube. Oospores are eventually formed in diseased tissue later in the growing season. Disease development is favored by high humidity, light rains or heavy dews, and a temperature of 17°C (optimum).

Distribution. Israel and the United States. Green ear downy mildew occurs on various grasses throughout the world, but it is not an economically important disease on corn.

Symptoms. Symptoms appear about 10 days after the plumule emerges. Diseased plants are stunted, and chlorotic streaking, together with gray blotching and mottling, occurs on leaves. A white, downy growth consisting of mycelium and sporangia occurs on discolored areas. Leaves may be thick, corrugated, and fragile. Sometimes diseased plants will outgrow the disease.

Management. Not necessary.

Head Smut

Head smut is also called ear and tassel smut.

Cause. *Sporisorium holci-sorghi* (syn. *Sphacelotheca reiliana*, *Sorosporium reilianum*, and *Ustilago reiliana*) overwinters as teliospores in soil and on seed. Survival is better in dry than in moist soil. Teliospores germinate to infect seedlings either directly by producing a germ tube or indirectly by producing a basidium on which sporidia (basidiospores) are formed. Sporidia of opposite mating types fuse to form a germ tube or hyphae. Soil temperatures of 23° to 30°C and moderate to low soil moistures (−1.5 bar is the maximum) are optimum for seedling infection. Mycelia develop systemically and invade undifferentiated floral tissues that partly or wholly develop into smut sori in which teliospores are produced. Disease is accentuated by nitrogen deficiency.

Other hosts include pitscale grass, sorghum, and sudangrass. Pathogenic specialization occurs. It has been determined that a single race infects maize; another race infects grain sorghum, forage-sorghum hybrids, and some sudangrass varieties. A "hybrid" of *S. reiliana* has been identified that infects both maize and sorghum.

Distribution. Drier soils of Australia, southeastern Europe, India, western Mexico, New Zealand, Russia, South Africa, the United States, and Yugoslavia.

Symptoms. The most obvious symptom is the development of sori (galls) on tassels and infrequently on "normal-appearing" vegetative leaves. Occasionally, sori develop as long, thin stripes in leaves. Galls are covered by a membrane that breaks open to expose a powdery spore mass and the host vascular bundles. The appearance of vascular bundles and leafy proliferations on tassels or ears when they are only partially converted to galls distinguishes head smut from common smut. Plants with smutted tassels do not produce pollen, and they may be dwarfed and have numerous tillers. When the tassel is not diseased, most ears except for a few "nubbins" will be smutted, rounded, and lack silks. Sometimes multiple ears appear at nodes. Less obvious chlorotic spots begin to develop on the fourth or fifth emerged leaf of seedlings and, thereafter, on successive leaves.

Management
1. Grow resistant hybrids.
2. Rotate maize with other crops.
3. Fungicides applied to soil at or before sowing may be useful in reducing inoculum.
4. Apply a fungicide seed protectant.

Hormodendrum Ear Rot

Hormodendrum ear rot is also called Cladosporium rot.

Cause. *Cladosporium herbarum* and *C. cladosporioides* (syn. *Hormodendrum cladosporioides*) infect kernels through growth cracks or cracks caused by early frost.

Distribution. Generally wherever maize is grown.

Symptoms. Symptoms are not obvious until harvest, when greenish-black streaks occur on kernels. The discoloration starts where kernels are attached to the cob and grows upward but seldom reaches the crown. After harvest, deterioration may occur if maize is stored when its moisture content is too high.

Management. Not necessary.

Hyalothyridium Leaf Spot

Cause. *Hyalothyridium maydis* causes disease under humid conditions.

Distribution. Colombia, Costa Rica, and Mexico.

Symptoms. Two types of symptoms occur on diseased leaves. One symptom is small, tan, elliptical spots with brown borders. The other is large (1.5 cm in diameter), tan, circular blotches that cover much of the leaf surface. Pycnidia are found in both types of lesions. In some areas only the smaller leaf spots are found.

Management. Not reported.

Java Downy Mildew

Cause. *Peronosclerospora maydis* (syn. *Sclerospora maydis*) overseasons in diseased maize plants. The fungus is seedborne as internal mycelium, but this is not thought to be an important source of inoculum. *Peronosclerospora maydis* is not known to form oospores. Conidia are produced at 18° to 23°C in the presence of dew for 5 to 6 hours.

All Java downy mildew outbreaks on maize and sweet maize in northern Australia since 1970 have been in areas where plume sorghum, *Sorghum plumosum*, occurs, suggesting it is an alternate host. It is not known for certain if resting spores are produced in plume sorghum.

Conidia are disseminated by wind to young plants and germinate under conditions of darkness, free moisture, and temperatures below 24°C. However, some researchers state the temperature range for optimum germination is 10° to 30°C and for optimum germ tube growth, 18° to 30°C. Germ tubes enter plants through stomata. High levels of systemic infection occur at temperatures of 8° to 36°C. Conidia are produced on wet leaves in the dark. Maize sown late in the growing season is most vulnerable to infection early in the rainy season, particularly if maize follows maize or sugarcane and is overfertilized with nitrogen.

Distribution. Australia and Indonesia.

Symptoms. Diseased leaves of plants younger than 4 weeks initially have white to chlorotic streaking that later becomes necrotic. A white, downy growth consisting of mycelium, conidiophores, and conidia commonly occurs in chlorotic streaks. Systemic infection results in chlorosis of the upper leaves. Diseased plants may be stunted and sterile, and lodge.

Management
1. Grow resistant hybrids and varieties.
2. Treat seed with a systemic fungicide.

Late Wilt

Cause. *Cephalosporium maydis* is soil and seedborne. Seedlings are more susceptible to infection than older plants. Infection occurs through the roots and mesocotyls of host plants, then develops slowly in the xylem tissue until rapid plant growth occurs at tassel emergence. Disease development is favored by light, sandy and heavy, clay soils.

Distribution. Egypt and India.

Symptoms. Diseased leaves wilt at the tasseling stage of plant growth, turn dull green, then dry up. Vascular bundles are discolored, and the lower portions of stalks dry up and become hollow and shrunken. Secondary organisms may cause a wet rot.

Management
1. Grow resistant hybrids.
2. Rotate maize with other crops.
3. Balance soil fertility.

Leptosphaeria Leaf Spot

Cause. *Leptosphaeria maydis*.

Distribution. Humid areas at high altitudes in the Himalayas.

Symptoms. Leptosphaeria leaf spot is most obvious on lower diseased leaves at silking and tasseling stages of plant development. Small, light tan lesions form that enlarge to become either concentric-shaped or streaked. These latter symptoms suggest two separate diseases.

Management. Not reported.

Nigrospora Stalk Rot

Nigrospora stalk rot is associated with dry ear rot.

Cause. *Nigrospora oryzae* overwinters as mycelium in infested soil residue. Conidia are formed on mycelia and are disseminated by wind to host plants. Plants nearing maturity and weakened by poor fertility, diseases, or injury are infected through stomates in the stalks.

Distribution. Generally wherever maize is grown, but Nigrospora stalk rot is not a serious disease.

Symptoms. Superficial, dark gray to black lesions with a slight bluish cast form on diseased lower internodes. Discoloration may appear as several tiny lesions or larger, irregular-shaped blotches.

Management. Maintain well-fertilized soil.

Northern Leaf Blight

Northern leaf blight is also called crown stalk rot, northern blight, northern corn leaf blight, stripe, Turcicum leaf blight, and white blast.

Cause. *Exserohilum turcicum* (syn. *Bipolaris turcica, Dreschlera turcica*, and *Helminthosporium turcicum),* teleomorph *Setosphaeria turcica. Exserohilum turcicum* is a heterothallic ascomycete that overwinters as chlamydospores (converted from conidia), conidia, and mycelia in leaf midribs and sheaths. Successive nights of temperatures at approximately 10°C induce formation of chlamydospores. In Israel, *E. turcicum* also overwinters on sorghum plants.

In spring, conidia are formed from mycelia in residue and are disseminated for long distances by wind or splashed by rain to lower leaves. Infection occurs when leaf surfaces are wet during temperatures of 18° to 27°C. Successive spores are produced following warm nights with 10 to 12 hours of 90% or more relative humidity and are spread to higher leaves. The proximal portions of leaf blades are less susceptible than distal portions. Disease is more severe under no-tillage conditions and when no tillage has occurred before or after sowing than with mulch-tillage and with autumn and spring cultivation without the use of a moldboard plow.

Five physiologic races were known to exist and several new races have now been reported from maize grown in China, Mexico, Uganda, and Zambia. Race 0 is widely distributed in many parts of the world. Race 1 was first detected in 1974 in Hawaii but is now widespread in the maize belt of the United States. Race 23 and race 23N are restricted to the continental United States. Race 2N was reported in Hawaii. Other races have been reported from matings of isolates in the laboratory.

Sweet maize is more susceptible than field maize. Sorghum, sudangrass, johnsongrass, garnagrass, and teosinte also become diseased.

Distribution. Wherever maize is grown in a humid area. Northern corn leaf blight is common in the eastern United States.

Symptoms. Lesions first appear on diseased lower leaves. As the season progresses, lesions develop on upper leaves. Susceptible plants that become severely diseased may be killed and their leaves covered with lesions that give the appearance of low temperature or frost injury. Because lesions compete for resources within a leaf, heavy disease pressure often results in smaller, numerous lesions, rather than fewer, larger ones. Typically, lesions on susceptible plants are gray-green to tan and elliptical-shaped. Lesions usually are 1.3×5.0–10.0 cm but occasionally grow as large as 3.8×15.0 cm. During damp weather, dark olive to black conidia produced on leaf surfaces often form concentric zones that give lesions a target-like appearance. Husks can be infected, but ears and kernels are not.

Symptoms on plants with monogenic resistance conferred by genes Ht1, Ht2, and Ht3 are characterized by chlorotic lesions with little or no sporulation. When the Ht2 gene is present, lesions on some hybrids will be long, narrow, chlorotic streaks extending the length of leaves and may be confused with symptoms of Stewart's wilt. The HtN gene delays lesion development until shortly after pollination. Plants with polygenic resistance have fewer and smaller lesions.

A chlorotic halo-type symptom is governed by a single recessive gene. The symptom is characterized by a distinct dark orange-brown pigment that later becomes surrounded by a circular chlorotic halo about 1 cm in diameter. Most infection points in this genetic stock retain the chlorotic halo phenotype until plant senescence, but some develop into typical elongated, necrotic northern leaf blight lesions.

Management
1. Grow resistant hybrids. Resistance is of two types: monogenic and polygenic. Monogenic resistance genes are Ht1, Ht2, Ht3, and HtN. Polygenic resistance is partial resistance effective against all biotypes.
2. Apply a foliar fungicide before disease becomes severe.
3. Plow under infested residue to reduce inoculum.

Northern Leaf Spot

Northern leaf spot is also called Carbonum corn leaf spot, Helminthosporium leaf spot, and northern corn leaf spot.

Cause. *Bipolaris zeicola* (syn. *Drechslera carbonum*, *D. zeicola*, *Helminthosporium carbonum*, and *H. zeicola*) teleomorph *Cochliobolus carbonum*. *Bipolaris zeicola* overwinters as mycelia and conidia in infested residue. Conidia are produced from mycelia in the spring during moderate temperatures and wet weather and are disseminated by wind or splashed to the lower leaves of maize. Wet weather also favors infection and subsequent disease development. Five physiologic races of *B. zeicola* are known. These are designated 0 through 4. Northern corn leaf spot does not occur on commercial single- or double-cross hybrids.

Distribution. In the temperate maize-growing areas of the world.

Symptoms. Race 0 is avirulent on maize genotypes susceptible to races 1, 2, and 3.

Race 1 produces the host-specific HC toxin, is highly virulent, and infects only a few inbred lines. Symptoms on diseased leaves are tan, oval to circular lesions that vary from small spots to larger lesions (1 × 3 cm). A pattern of concentric zones often occurs in the lesions. During optimum disease conditions, lesions become numerous. A black mold grows over kernels to give the diseased ear a charred appearance.

Race 2 isolates do not produce HC toxin, are not considered pathogenic, and show no distinct host specialization. However, Race 2 will infrequently penetrate leaves and elicit small necrotic lesions. These Race 2 lesions are small, round to oblong or oval-shaped, chocolate-colored or necrotic spots that range in size from barely visible to 0.5 × 2.5 cm. The ear rot caused by Race 2 is indistinguishable from that caused by Race 1.

Race 3 causes symptoms on most maize inbred lines and hybrids and is frequently referred to as northern leaf spot or Helminthosporium leaf blight. Lesions are on diseased leaf blades, sheaths, husks, and ears. Lesions on leaves are narrow and linear (0.5–2.0 × 15.0–20.0 mm), grayish tan, and surrounded by a light to dark border. Ears are infected from the tip, butt, or through sides of the husk. A black mold develops over kernels.

Race 4 causes early and sudden leaf death. Leaves become reddish brown, which is suggestive of a toxin being produced by the fungus. Lesions within dead tissue are oval to circular-shaped (5–10 mm in diameter) and often contain concentric circles. Most susceptibility was expressed in inbreds with B73 background. Lesions on some hybrid genotypes are smaller, nearly cream-colored, and without concentric circles. There is no seed infection.

A similar pathotype from Wisconsin caused foliage damage, which reduced yields of inbreds having a B73 background. Seed size, but not germination, was also affected. A dry, brown to black internal and external discoloration of crowns occurred, but roots were unaffected.

Bipolaris zeicola also causes stalk rot characterized by rapid and severe necrosis of the nodes and internode pith.

Management
1. Grow resistant hybrids.
2. Treat seed with a fungicide seed treatment.
3. Differences in resistance exist among inbreds to Race 3. Some are highly resistant, others are susceptible, and many are intermediate in reaction. Inbreds are usually more susceptible than hybrids.

Penicillium Rot

Penicillium rot is also called blue-eye.

Cause. *Penicillium chrysogenum*, *P. expansum*, and P. *oxalicum* occur on ears that have been injured by corn earworms or other causes. *Penicillium* spp. are soil-inhabiting fungi that enter kernels through uninjured pericarps (even at moisture contents as low as 14%), grow in the germ, and sporulate beneath the pericarp. *Penicillium* spp. may invade seed embryos and cause seedling blight at low temperatures of 15° to 20°C. *Penicillium oxalicum* loses its viability in seeds after storage at room temperatures of 20° to 25°C for 12 months.

Distribution. Wherever maize is grown.

Symptoms. Rot is most likely to occur on tips but may be found on other parts of ears. Kernels infected with *P. oxalicum* will have a grayish blue-green mold, while other *Penicillium* spp. cause a brighter bluish-green mold.

Management. Grain should be dried to 14% or less moisture content during storage.

Phaeocytostroma Stalk Infection

Cause. *Phaeocytostroma ambiquum* (syn. *Phaeocytosporella zeae*) overwinters as pycnidia in infested residue.

Distribution. France and the United States (Illinois).

Symptoms. Light tan, oblong blotches appear near the soil line on maturing stalks. Elongated pycnidia appear within lesions and have a short to long slit as an opening.

Management. Not necessary.

Phaeosphaeria Leaf Spot

Cause. *Phaeosphaeria maydis* (syn. *Leptosphaeria zea-maydis*, *Metasphaeria maydis*, and *Sphaerulina maydis*), anamorph *Phoma* sp. *Phaeosphaeria maydis* survives in infested residue, presumably as pseudothecia and less commonly as pycnidia. Disease is favored by high rainfall and low night temperatures.

Distribution. Brazil, Colombia, Costa Rica, India, and the United States (Florida).

Symptoms. Initially, round to oblong, light green or chlorotic lesions (0.3–2.0 cm in diameter) occur randomly over diseased leaves. Later, lesions become white to light tan and have distinct dark brown margins. Lesions may coalesce and form larger irregular-shaped diseased areas in which perithecia and, infrequently, pycnidia are formed.

Management. Most germplasm in the United States is not susceptible.

Philippine Downy Mildew

Cause. *Peronosclerospora philippinensis* (syn. *Sclerospora philippinensis*). Some researchers consider *P. philippinensis* to be conspecific with *P. sacchari*.

Conidia are disseminated by wind and splashed by water to young plants. *Peronosclerospora philippinensis* is also seedborne. Leaves are penetrated by germ tubes growing into stomata and then into the stem, where *P. philippinensis* becomes established in the shoot apex.

Conidia are produced at temperatures of 18° to 23°C in the presence of moisture for 5 to 6 hours. Germination occurs at temperatures of 10° to 30°C, and germ tube growth occurs at 18° to 30°C.

Other hosts include johnsongrass, sorghum, sugarcane, and teosinte.

Distribution. Africa, India, Indonesia, Nepal, and the Philippines.

Symptoms. The first diseased leaf may have systemic symptoms of either a general chlorosis or chlorotic streaks. The chlorotic streaks will commonly develop a downy growth consisting of mycelium, conidia, and conidiophores. Ears may abort, which results in partial or complete sterility, and tassels may be malformed, resulting in less pollen. Infection of plants early in the growing season results in stunted and dead plants.

Infection by maize streak virus often masks symptoms of Philippine downy mildew.

Management
1. Grow resistant hybrids and varieties.
2. Rogue and destroy diseased plants.
3. Apply systemic fungicide seed treatment and foliar spray.

Phomopsis Seed Rot

Cause. *Phomopsis* sp.

Distribution. The United States.

Symptoms. Dark, ostiolate, globose, and erumpent pycnidia are present on diseased seed. Thin, smooth, creamy-white cirri emerge from pycnidia. A loss in germination occurs.

Management. Not reported.

Physalospora Ear Rot

Physalospora ear rot is also called Botryosphaeria ear rot.

Cause. *Botryosphaeria festucae* (syn. *Physalospora zeicola*), anamorph *Diplodia frumenti*, overseasons as perithecia and pycnidia in infested residue. Disease is favored by warm, humid weather.

Distribution. In tropical maize-growing areas.

Symptoms. Partially diseased ears have a few blackened kernels near the base. Severely diseased ears are covered by a dark brown to black, "felt-like" mold.

Management. Not reported.

Pyrenochaeta Stalk Rot and Root Rot

Pyrenochaeta stalk rot and root rot is also called red root rot.

Cause. *Phoma terrestris* (syn. *Pyrenochaeta terrestris*) survives as pycnidia in infested soil residue. Plants are usually infected before pollination.

Distribution. The United States (Delaware and Maryland), but Pyrenochaeta stalk rot and root rot is likely more widely spread.

Symptoms. Only stalks below the soil surface are infected. Dark brown, shallow blotches occur on stalks. Later, a pink discoloration that is evident in the outer rind blends into the dark brown areas.

Management. Maize hybrids vary in susceptibility. Pyrenochaeta stalk infection is not a serious disease.

Pythium Root Rot

Cause. *Pythium arrhenomanes* and *P. graminicola*. *Pythium torulosum* and *P. dissotocum* are also reported to be associated to a lesser degree with Pythium root rot. *Pythium graminicola* is reported to be the dominant causal fungus.

Pythium spp. survive in infested residue and soil as oospores and sporangia; however, some isolates do not form oospores. Infection is either by direct germination of oospores and sporangia to form a germ tube or indirectly by zoospores formed in either structure and released into water-saturated soil. Zoospores "swim" to root surfaces, where they encyst and germinate to form a germ tube.

Disease severity is greatest in poorly drained, wet soils. Pythium root rot is prevalent in the early spring and late summer.

Distribution. Wherever maize is grown.

Symptoms. Secondary roots of diseased young plants have scattered yellowish-brown lesions that eventually coalesce and become dark brown to black, resulting in dark streaks and girdling of roots. Diseased primary roots become black and necrotic. Similar lesions will be formed on new roots under wet soil conditions later in the growing season. Sometimes new roots form above a lesion.

Stunting and chlorosis of lower leaves occurs on young plants. Diseased older plants may be generally stunted and also display chlorosis, rolling, and wilting of leaves.

Management. Sow seeds into well-drained soil.

Pythium Seed and Seedling Blight

Cause. Several *Pythium* species, including *P. debaryanum*, *P. irregulare*, *P. paroecandrum*, *P. rostratum*, *P. splendens*, *P. ultimum*, and *P. vexans*. The fungi survive either as oospores, sporangia, or mycelia in infested residue.

Disease is severest in wet soils at temperatures of 10° to 13°C. Under these conditions, seed germination and growth is slow, causing plants to become more susceptible to infection. Sporangia either germinate directly to form a germ tube or germinate indirectly to form a vesicle in which zoospores are formed. Similarly, oospores either germinate directly to form a germ tube or germinate indirectly to form a zoosporangium in which zoospores are formed. Zoospores "swim" by means of flagella through soil water to host seeds and roots. Upon contact with a plant, a zoospore will encyst and germinate to produce a germ tube that directly penetrates the seed or root. As mycelium grows through a plant, sporangia and oospores are produced both on and in the diseased plant.

Generally sweet maize is more susceptible than dent maize. Popcorn is the most resistant.

Distribution. Wherever maize is grown.

Symptoms. Seeds may be rotted before germination and seedlings may be infected before or after emergence. Seeds are brown, soft, and overgrown with mycelia. It is difficult to find diseased seeds in soil because of their rapid decomposition after infection and because of adhering soil that is bound to the seed by mycelium, making a diseased seed indistinguishable from the surrounding soil. Roots are brown and water-soaked and have a "sloughed" appearance.

Aboveground symptoms are chlorosis, wilting, and death of leaves. Lesions on mesocotyls are initially brown and sunken. Eventually, mesocotyls become soft and water-soaked.

Management
1. Treat seed with a seed-protectant fungicide.
2. Do not sow seed stored for 2 or more years at temperatures between 15.5° and 35.0°C.
3. Do not sow seed with cracked or split seed coats because cracks are an entrance for fungi.
4. Delay sowing until the soil temperature is 13°C and above.

Pythium Stalk Rot

Cause. *Pythium aphanidermatum* (syn. *P. butleri*) survives as oospores, sporangia, or mycelia in infested soil residue. During humid weather at temperatures of 32°C and higher or where air and soil drainage is poor, sporangia germinate either directly to form a germ tube or indirectly by releasing zoospores into the soil. Oospores likewise may germinate either directly to form a germ tube or indirectly to form a zoosporangium in which zoospores are formed. Zoospores "swim" through soil water to a host plant, where it germinates to form a germ tube that penetrates the plant.

Distribution. Wherever maize is grown under hot, humid conditions.

Symptoms. Pythium stalk rot can occur any time during the growing season but is most frequent at the tasseling stage of plant growth. The disease is recognized when plants twist and fall over due to a loss of strength at a brown, water-soaked, and softened internode(s) on the diseased stalk immediately above the soil line. Vascular bundles within the fallen stalk remain intact, which causes the fallen plants to remain green and turgid for several weeks.

Management. Grow resistant hybrids.

Red Kernel Disease

Cause. *Epicoccum nigrum* is a weak parasite that infects injured or weakened kernels.

Distribution. The United States.

Symptoms. A red discoloration occurs in the endosperm of sweet maize.

Management. Not reported.

Rhizoctonia Ear Rot

Rhizoctonia ear rot is also called sclerotial rot.

Cause. *Rhizoctonia zeae*, teleomorph *Waitea circinata*, overseasons as mycelia and sclerotia on or in infested residue, in kernels, and in soil. Sclerotia are the main means of dissemination. Disease is favored by warm, humid weather.

Distribution. Asia, Europe, North America, and South America.

Symptoms. Initially, pink mycelial growth occurs on diseased ears. Later, the mycelium becomes dull gray. Numerous white, salmon-colored, dark brown or black sclerotia develop on the outer husks.

Management. Not reported.

Rhizoctonia Root Rot and Stalk Rot

Cause. Several anastomosis groups of *Rhizoctonia solani* and *R. zeae*.

Distribution. The United States (Georgia).

Symptoms. AG-2 causes the most severe symptoms. *Rhizoctonia solani* causes postemergence damping off, stunting, leaf chlorosis, and necrosis of diseased seedlings. Isolates from other plant species cause brown to black lesions on the seminal, crown, and brace roots. Roots frequently decay and disintegrate 2 to 5 cm from the crown.

    Rhizoctonia zeae, which causes buff to tan lesions with dark brown borders on diseased corn roots, is less virulent than *R. solani.*

Management. Not practiced.

Rostratum Leaf Spot

Rostratum leaf spot is also called Helminthosporium leaf disease, and Helminthosporium leaf disease, ear, and stalk rot.

Cause. *Setosphaeria rostrata*, anamorph *Exserohilum rostratum* (syn. *Bipolaris rostrata, Drechslera rostrata, D. halodes, E. halodes,* and *Helminthosporium ro-*

stratum), survives as mycelia and conidia in infested residue. Conidia are produced from mycelia and disseminated by wind or splashing rain to maize. Mature plants or older leaves are more susceptible to infection than younger leaves. A temperature of 30°C and a water-saturated atmosphere are conditions for optimum infection. Sudangrass, pearl millet, and several other grasses are also hosts.

Distribution. Where maize is grown under hot, humid conditions. Rostratum leaf spot is more severe in India and parts of Asia.

Symptoms. Initially, small, pale yellow, elongated spots (1–2 × 2–5 mm) on diseased leaves gradually elongate into longitudinal stripes between the leaf veins. In cases of severe disease, the lesions will coalesce and extend across the veins. Older lesions become light tan to cream with a light brown margin and enlarge to 2–3 × 2–40 mm.

A stalk rot with typical symptoms in the nodal and internodal tissue has been reported from Florida.

Management. Grow resistant hybrids.

Sclerotium Ear Rot

Sclerotium ear rot is also called southern blight.

Cause. *Sclerotium rolfsii*. Disease occurs under extremely wet conditions.

Distribution. India.

Symptoms. Initially, water-soaked spots on diseased leaf sheaths vary in size and shape, but they gradually enlarge during moist and warm weather. Profuse fungal growth sometimes appears on the surface of diseased spots. White mycelium grows upward from the base of a diseased ear and is observed late in the growing season on husks, floral bracts, and pericarps of rotted kernels. In the later stages of disease, sclerotia form on ears.

Management. Not reported.

Seed and Seedling Blight Other than Pythium

Cause. *Stenocarpella maydis, Gibberella zeae, Fusarium moniliforme, Rhizoctonia solani*, and possibly other fungi. *Aspergillus flavus* var. *columnaris* is reported from South Africa to be a seed storage fungus that can colonize seedlings grown from spore-contaminated seed. Most of these fungi are discussed under the stalk rot or ear and kernel blight caused by each. Seedling blight generally occurs under wet, cool, soil conditions that do not vary appreciably from those favoring Pythium blight.

Distribution. Wherever maize is grown.

Symptoms. Symptoms are nondescript and may be entirely absent. Failure of seeds to germinate, the failure of seedlings to emerge, the killing of seedlings after emergence, or short, stunted plants are general symptoms. Diseased seed, which may be rotted and overgrown with mycelium, eventually disintegrates. The mesocotyl may be rotted and discolored, cutting off water and nutrients from the seed and primary root. If the adventitious roots from the first node above the mesocotyl are not well developed, leaves wilt and plants die. Roots have brown to black, water-soaked, flaccid lesions.

Management. Treat seed with a fungicide seed protectant.

Selenophoma Leaf Spot

Cause. *Selenophoma* sp. Disease occurs at low temperatures (10° to 18°C).

Distribution. Colombia, at an altitude of 2700 m.

Symptoms. Lesions on diseased leaves are typical "eyespots" with concentric zonations bordered by a yellow halo. When relative humidity is 70% or greater, lesions grow to 2 cm long and eventually coalesce and cause severe leaf necrosis.

Management. Not reported.

Septoria Leaf Blotch

Cause. *Septoria* sp. causes disease under cool, humid conditions.

Distribution. Septoria leaf blotch is presumed to be common in cool, humid climates.

Symptoms. Small, light green, chlorotic or brown lesions form on diseased leaves. Lesions coalesce and form large necrotic areas in which numerous pycnidia may be produced.

Management. Not necessary.

Sheath Blight

Cause. *Rhizoctonia solani*. The causal fungus is probably multinucleate and has been tentatively identified as a sasakii form. Disease occurs in irrigated fields with over 100 units of N fertilizer applied and a history of rice production in the last 1 to 6 years.

Distribution. The United States (east central Arkansas).

Symptoms. Reddish eyespot lesions with dark red to purple margins occur on diseased stalks near the soil line and develop upward to approximately 2 m above the soil line. The fungus penetrates leaf sheaths to the rind and appears as a yellow-green discoloration with a thin black border. Prolific sclerotial development occurs in the older lesions on diseased leaf sheaths, ear husks, and silks.

Management. Not reported.

Sheath Rot

Sheath rot is also called *Gaeumannomyces graminis* root rot.

Cause. *Gaeumannomyces graminis.* Inoculum likely originates as mycelia and perithecia in the infested residue of cereals, especially wheat. Maize was reported to be severely infected during the early seedling stages but not later in the growing season. Maize may act as a "carrier" crop from season to season for the fungus to infect small grains.

Distribution. Europe. However, disease is likely widespread wherever maize and small grains are grown in rotation.

Symptoms. Primarily the cortical tissue of roots becomes diseased. Diseased roots have a dark brown to black discoloration. Apparently the disease on corn is not severe and little damage occurs.

Management. Not necessary.

Shuck Rot

Cause. *Myrothecium gramineum.* The *Myrothecium* spp., in general, are soil inhabitants and widespread facultative parasites. Disease development is normally favored by wounds and hot, humid weather. Infection of the shucks is presumably by inoculum within soil that is disseminated by wind to host plants.

Distribution. Not known.

Symptoms. Small, dark, water-soaked lesions occur that expand into larger necrotic areas on diseased shucks. Sporodochia may develop in lesions. It is not known if the fungus grows into the kernels from diseased shucks.

Management. Not necessary.

Silage Mold

Cause. *Monascus purpureus* and *M. ruber.*

Distribution. Widespread.

Symptoms. Silage is stained a reddish-purple to dark purple color.

Management. Not reported.

Sordaria fimicola Stalk Infection

Cause. *Sordaria fimicola* is an opportunistic inhabitant of maize stalks.

Distribution. South Africa.

Symptoms. *Sordaria fimicola* was isolated only from the nodes of diseased stalks. Plants grown in infested soil showed a significant reduction in dry matter. Root length was reduced by 33.3% and plant height by 34.0%. A reduction in seedling vigor is possibly due to phytotoxic by-products released by the fungus into the soil.

Management. Not reported.

Sorghum Downy Mildew

Sorghum downy mildew is also called downy mildew.

Cause. *Peronosclerospora sorghi* (syn. *Sclerospora sorghi*) survives for several years as oospores in infested residue and in soil after the infested residue has decomposed. Isolates from Java or Thailand do not form oospores. *Peronosclerospora sorghi* is not seedborne but does survive for a short time in immature seeds that are sown shortly after harvest. Oospores germinate and infect young plants that are emerging in warm soil. Conidia are also reported to provide primary inoculum. Mycelium grows systemically in infected plants and, eventually, conidia are produced on leaf surfaces. Conidia are disseminated by wind but infect only the leaves of host plants that are younger than 3 weeks old.

Reports of temperatures for optimum disease development differ. Some researchers report optimum germination occurs at high humidity and temperatures of 12° to 20°C for some isolates and 12° to 32°C for other isolates. Others report temperature ranges are 10° to 30°C for optimum germination and 18° to 30°C for optimum germ tube growth. Conidia are produced at temperatures of 18° to 23°C in the presence of dew for 5 to 6 hours. High levels of systemic infection occur at temperatures of 8° to 36°C. Eventually, oospores are formed in diseased tissues. Johnsongrass, sorghum, and teosinte are also hosts.

Distribution. Africa, Asia, Central America, Israel, Italy, South America, and the United States (primarily in the south central and midwestern states).

Symptoms. Systemic infection occurs only in young plants. Initially, chlorotic to white streaks develop parallel to the leaf midrib. Sometimes half of an entire leaf will be chlorotic. A downy growth consisting of conidia and conidiophores appears in chlorotic tissue on both surfaces of the diseased leaf. Severely diseased plants are stunted and chlorotic and have narrow, erect leaves. Tassels are phyllodied, with floral parts converted to small leaves, which results in an abnormal seed set. Symptoms of secondary infections are small lesions that contain downy growth scattered over the leaves of older plants.

A remission of symptoms, which is possibly a resistance mechanism, has been reported from Nigeria. Cobs from such plants formed normally but weighed significantly less than those from uninfected plants. When seeds that were produced on maize plants with symptom remission were sown, seedlings did not develop downy mildew symptoms. Infection by maize streak virus will mask symptoms of downy mildew.

Management
1. Grow resistant hybrids.
2. Rogue out and destroy diseased plants as they appear in the field.
3. Do not rotate maize with sorghum, and vice versa.
4. Treat seed with a fungicide seed treatment.
5. Sow in cool soil at the onset of the rainy season.

Southern Corn Leaf Blight

Southern corn leaf blight is also called Maydis leaf blight, southern blight, and southern leaf blight.

Cause. *Cochliobolus heterostrophus*, anamorph *Bipolaris maydis* (syn. *Drechslera maydis* and *Helminthosporium maydis*), overwinters as conidia and mycelia in infested residue. Race T also survives on kernels in storage. In the spring, conidia are disseminated by wind or splashed by rain to host plants, where infection occurs in the presence of free moisture and temperatures of 15.5° to 26.5°C. Lesions form and expand more rapidly at 30°C than at lower temperatures. Secondary sporulation occurs in lesions in 60 to 72 hours. Race T produces five times more spores than Race O at 22.5°C and twice as many at 30°C.

Races O and T are the most common physiologic races, but other races are present. Race O infects only leaves, while Race T infects leaves, stalks, leaf sheaths, ear husks, and kernels. Race C has been designated from China, where isolates of *Bipolaris maydis* induced significantly larger lesions on leaves of inbred lines of maize with cms-C than on leaves of the same lines with cms-T, cms-S, or normal (N) cytoplasm. Maize infected with maize dwarf mosaic virus is more susceptible to infection by *B. maydis* than are healthy maize plants. Disease is most severe when susceptible cultivars are grown in continuous maize culture utilizing a minimum-tillage system and overhead irrigation.

Distribution. Generally wherever maize is grown, but Race O is most likely to be present in the southeastern and midwestern United States.

Symptoms. Lesions on leaves caused by Race O are tan with buff to brown borders and parallel sides (2–6 × 3–22 mm). Race T lesions are larger (0.6–1.2 × 0.6–2.7 cm), elliptical or spindle-shaped, and surrounded by yellow-green or chlorotic halos. Lesions often may have dark reddish-brown borders. Lesions may coalesce and kill leaves.

Race T produces similar lesions on other plant parts. Such lesions are usually very large (several centimeters in length) and tan with dark purple-brown borders. Mycelium grows through husks to the kernels, where a gray-black color from the production of conidia gives ears and kernels a moldy or charcoal appearance. Seedlings from infected kernels wilt and die a few days after sowing.

During temperatures of 20° to 30°C and a relative humidity of 100%, race T produces large numbers of spores, which give lesions a black "velvety" appearance. Severely diseased plants are predisposed to stalk rot.

Management

1. Grow resistant hybrids.
2. Apply foliar fungicides before disease becomes too severe and when weather conditions are conducive to disease development.
3. Plow under infested residue.

Southern Corn Rust

Southern corn rust is also called Polysora rust and southern rust.

Cause. *Puccinia polysora* is a rust fungus that produces the urediniospore stage and rarely the teliospore stage. No alternate host is known.

Urediniospores constitute both primary and secondary inoculum. Optimum infection occurs at a temperature of 26°C and a minimum of 16 hours of dew. Subsequent disease development occurs at temperatures of 21° to 27°C and high relative humidity, with temperature being the limiting factor to disease development in the northern United States. Disease becomes progressively more severe as plants develop and is greatest on maize sown later in the growing season.

Epiphytotics coincide with an increase in double cropping of maize in the lower Mississippi River Valley and the Coastal Plain of the southeastern United States. This cropping practice may provide conditions for development of large quantities of inoculum to infect maize in the northern corn belt at an early growth stage.

Nine physiologic races of *P. polysora* have been described.

Distribution. Africa, Central and South America, Southeast Asia, the United States (primarily the southeastern states but occasionally the Midwest), and the West Indies.

Symptoms. Pustules occur on both leaf surfaces and resemble common rust pustules except they are smaller (0.2–2.0 mm). Pustules are a light cinnamon brown, circular to oval-shaped, and densely scattered over the diseased upper leaf surface. Pustule development on lower leaf surfaces is slower and less abundant than on the upper leaf surfaces. The epidermis remains intact over the pustules for a longer time than common rust but eventually ruptures. Telia, which are chocolate brown to black and circular to elongate-shaped (0.2 to 0.5 mm in diameter), often form a ring around the initial uredinium. The epidermis over the telia will eventually rupture, but it also remains intact longer than common rust. Severely diseased leaves may become chlorotic and eventually dry up. During severe disease conditions, pustules may occur on leaf sheaths and stalks.

Management
1. Grow resistant hybrids.
2. Apply foliar fungicides.

Spontaneum Downy Mildew

Cause. *Peronosclerospora spontanea* (syn. *Sclerospora spontanea*).

Distribution. The Philippines and Thailand.

Symptoms. The first leaf may have systemic symptoms of either general chlorosis or chlorotic streaks. The most common symptom is long, chlorotic streaks with downy growth consisting of mycelium, conidia, and conidiophores that occurs on diseased leaves. Diseased plants are generally stunted and chlorotic. During humid conditions at night, diseased leaves are covered with a white "fuzz" consisting of mycelium, conidia, and conidiophores.

Ears may abort, resulting in partial or complete sterility, and tassels may be malformed, thus producing less pollen. Early infection results in stunted and dead plants.

Management
1. Grow resistant hybrids and varieties.
2. Rogue and destroy diseased plants.
3. Apply systemic fungicide seed treatment and foliar spray.

Sugarcane Downy Mildew

Cause. *Peronosclerospora sacchari* (syn. *Sclerospora sacchari*) survives as mycelium in sugarcane. Sporangia produced from mycelia at nightly temperatures of 20° to 25°C and in the presence of free water are disseminated by wind to maize. Sporangia germinate in the presence of free water and form mycelia that penetrate through the stomata of plants that are less than 1 month old. Mycelia then develop systemically in the infected plant. Sporangia are eventually produced in diseased tissue and serve as secondary inoculum. Oospores are also produced, but their function is not known. Older plants are resistant.

Other hosts are broomcorn, gamagrass, grain sorghum, sugarcane, and teosinte.

Distribution. Southeast Asia.

Symptoms. Initially, small, round, chlorotic spots appear on diseased leaves. Eventually, systemic infection occurs as yellow to white streaks, or stripes, starting at the base of the third to sixth oldest leaf and extending the length of the leaf. Streaks may be discontinuous in some hybrids or varieties and may disappear on plants that are infected later in the growing season or are less severely diseased. Downy masses of sporangia appear at night on both leaf surfaces, leaf sheaths, and husks during periods of high humidity and an optimum temperature of 25°C. Plants, especially tassels, are distorted and have poorly filled ears and elongated ear shanks.

Management
1. Do not grow maize near sugarcane.
2. Grow resistant hybrids and varieties.

Tar Spot

Cause. *Phyllachora maydis;* however, at least two other organisms, *Monographella maydis* and *Coniothyrium phyllachorae*, can be found in tar spot lesions. The fungus *M. maydis* is commonly found on the surface of leaves, generally without any apparent reaction by the plant to the presence of the fungus. It is only in association with *P. maydis* that *M. maydis* becomes pathogenic and highly virulent. The effect of *C. phyllachorae* is not completely known but it seems to be a hyperparasite of both *P. maydis* and *M. maydis.*

    Phyllachora maydis is an obligate parasite and *M. maydis* requires its presence to become pathogenic. Ascospores of *P. maydis* are disseminated by wind to host plants. The presence of a toxin produced by *P. maydis* has been associated with the rapid killing of plant tissue.

    Disease development is favored by cool, humid weather.

Distribution. Central America and South America in moderately cool and humid, tropical or subtropical, mountainous areas.

Symptoms. Different disease symptoms have been reported. Disease development generally starts on the lower leaves at flowering time and moves rapidly up the diseased plant. Disease becomes most severe after pollination and may cause premature desiccation of the spike.

    Phyllachora maydis initially induces small (0.5–2.0 mm in diameter), oval to circular-shaped, light brown lesions that may be surrounded by a dark brown border. This border is caused by *M. maydis* and has been described as a brown, elliptic, necrotic halo that surrounds each *P. maydis* lesion. The development of *P. maydis* lesions is frequently followed by the development of *M. maydis* lesions.

    Lesions coalesce and form longer lesions (up to 10 mm in length).

Glossy, black, sunken spots frequently occur in the middle of a lesion but are also abundant outside a lesion, particularly when plants are severely diseased.

Management. Grow the most resistant genotype.

Trichoderma Ear Rot and Root Rot

Cause. *Trichoderma viride* (syn. *T. lignorum*) is a common soilborne fungus that is considered to be a secondary invader of ears and roots.

Distribution. Widespread.

Symptoms. A greenish, cottony fungal growth occurs on and between seeds. Roots are generally discolored, and diseased plants are reported to be stunted.

Management. Treat seed with a fungicide seed treatment.

Tropical Corn Rust

Tropical corn rust is also called tropical rust.

Cause. *Physopella pallescens* and *P. zeae* (syn. *Angiospora zeae*) are rust fungi that produce the urediniospore and teliospore stages on maize. No alternate host is known. Urediniospores constitute both the primary and secondary inoculum. Disease development is favored by warm to hot, and humid weather.

At least two races of *P. zeae* are known. Teosinte is also a host.

Distribution. Central America, South America, and the Caribbean.

Symptoms. Uredinial pustules occur in small groups (0.3-1.0 mm long) on diseased upper leaf surfaces. Pustules are yellow or cream-colored and covered by an epidermis except for a small pore or split. Later, pustules develop into purple blotches that are circular to oblong-shaped (0.5 cm in diameter) with cream-colored centers. Dark brown to black telia develop in groups around uredinia.

Management. Grow resistant hybrids.

Yellow Leaf Blight

Yellow leaf blight is also called Phyllosticta leaf spot.

Cause. *Ascochyta ischaemi* and *Phyllosticta maydis*, teleomorph *Mycosphaerella zeae-maydis*. *Phyllosticta maydis* overwinters as pycnidia or pseudothecia in infested maize or grass residue. Under cool, wet conditions, primarily in the spring, conidia ooze from pycnidia and are disseminated to host plants by diseased leaves rubbing against healthy leaves, by wind for a short distance, and by splashing rain. Ascospores are produced that are also windborne. Young plants growing through residue on the soil surface are especially vulnerable to infection; however, plants may be infected at any stage of their growth. Conidia from pycnidia on older lesions provide secondary inoculum. Sudangrass and foxtail are also susceptible.

Distribution. Africa, Argentina, Asia, Brazil, Canada, Romania, Taiwan, and the United States (primarily in the northern corn-growing areas but also in the midwestern and southern states).

Symptoms. *Phyllosticta maydis* does not spread rapidly over a large area; it is usually restricted to isolated fields. Plants growing in fields containing infested residue from last year's crop may be most vulnerable to disease since residue harbors the pathogen.

Lesions develop first on lower leaves. The gray, tan, or brown lesions surrounded by a narrow red or purple margin with a wide yellowish-green halo are oval to elliptical-shaped (0.3×1.3 cm). The yellow discoloration surrounding lesions gives the disease its name. Lesions on older leaves contain pycnidia that look like tiny, brownish-black specks but, under magnification, resemble small "flask-shaped" structures.

Severe disease results in yellowed leaves that resemble symptoms of nitrogen deficiency. Diseased leaves eventually die, and plants are stunted and more susceptible to stalk rot. Leaf sheaths and outer husks are also diseased.

Management
1. Grow the most resistant hybrids. Hybrids containing T cytoplasm are more susceptible than hybrids containing other cytoplasms.
2. Apply a foliar fungicide if economic and disease conditions warrant it.
3. Plow under infested residue.
4. Rotate maize with other crops.

Zonate Leaf Spot

Cause. *Gloeocercospora sorghi* survives as sclerotia and mycelia in infested residue. Conidia are produced in sporodochia, exuded in a slimy matrix, and disseminated by splashing rain. Sclerotia develop in diseased tissue later in the growing season. Bentgrass, johnsongrass, sudangrass, and sugarcane are also hosts.

Distribution. Africa, Central America, South America, and in the southern maize-growing areas of the United States.

Symptoms. Lesions on leaves initially are water-soaked, then become reddish brown and enlarge to 5 cm in diameter, forming a target-like or zonate pattern. Small (0.1–0.2 mm diameter), black sclerotia develop in dead tissue.

Management. Not reported. Zonate leaf spot is not known to cause severe damage.

Diseases Caused by Nematodes

Several kinds of nematodes attack maize, but in most cases the distribution is not well known. Disease management tends to be general for all nematode-caused diseases. Following are practices that tend to reduce effects of nematode injury. Specific management practices are mentioned, where applicable, to control a disease caused by nematodes.

1. Fertilize soils according to soil tests. Plants suffering from nutrient deficiency are more susceptible to injury.
2. Maintain good weed control. Weeds act as reservoirs of nematodes to infect next year's crop.
3. Rotate maize with other crops.
4. Apply nematicides to severely affected areas.

Awl

Cause. *Dolichodorus heterocephalus* and likely other *Dolichodorus* spp. Awl nematodes are ectoparasites that are limited to wet soils. Some dissemination of the nematode occurs by water. Feeding is generally restricted to the root tips.

Distribution. The southeastern United States.

Symptoms. The aboveground symptom is plant stunting. The belowground symptom is stubby, thick roots that have some lesions present.

Bulb and Stem

Cause. *Ditylenchus dipsaci* occurs in association with heavy soils, high rainfall, and cool temperatures.

Distribution. Widespread, but usually occurring in small areas within a field. This is not a serious disease.

Symptoms. *Ditylenchus dipsaci* feeds on leaves and stems, where it may cause cell hypertrophy and hyperplasia. Stalks may be slightly stunted and distorted. Infrequently, a growing point may be destroyed.

Burrowing

Cause. *Radopholus similis* is an endoparasite that reproduces within roots and is capable of surviving for only a few months outside of host tissue. Feeding occurs in the cortex. Some dissemination occurs by water.

Distribution. Tropical and subtropical areas around the world.

Symptoms. The aboveground symptom is a slight stunting of diseased plants. Belowground symptoms are root lesions and decay.

Columbia Lance

Cause. *Hoplolaimus columbus* is an ectoparasitic, semi-endoparasitic, and endoparasitic nematode.

Distribution. Widespread.

Symptoms. Lesions occur on diseased roots. Foliage is chlorotic and a severely diseased plant may be stunted.

Corn Cyst

Cause. *Heterodera avenae*, H. *zeae*, and *Punctodera chalcoensis*.

Distribution. Egypt, India, Pakistan, and the United States (Maryland and Virginia).

Symptoms. The aboveground symptom is a general stunting. The belowground symptom is white to tan cysts on the roots of diseased plants.

Dagger

Cause. *Xiphinema americanum* and *X. mediterraneum* are ectoparasites.

Distribution. Not known.

Symptoms. The aboveground symptom is a slight to moderate stunting of diseased plants. The belowground symptom is a reduction in feeder roots accompanied by necrosis of a portion of the root system.

False Root Knot

Cause. *Naccobus dorsalis*. Maize grown in a rotation with susceptible crops becomes infrequently infected.

Distribution. The United States.

Symptoms. Epidermal and cortical cells of the diseased root hypertrophy and give the impression of a swelling along portions of the root. The "swellings" bear a superficial resemblance to root knots caused by *Meloidogyne* spp.

Lance

Cause. *Hoplolaimus galeatus* and *H. indicus* are primarily endoparasites but sometimes feed as semi-endoparasites and ectoparasites. *Hoplolaimus indicus* does more damage to maize when it is in association with *Fusarium moniliforme* than it does alone. Other lance nematodes may also be parasitic to maize.

Distribution. Not known.

Symptoms. Aboveground symptoms are stunting and chlorosis of diseased plants. The belowground symptom is lesions that reduce root growth.

Lesion

Cause. *Pratylenchus brachyurus*, *P. crenatus*, *P. hexincisus*, *P. neglectus*, *P. penetrans*, *P. scribneri*, *P. thornei*, and *P. zeae* are endoparasites that feed and reproduce in the root cortex. Some species thrive in warm, sandy soils, while others grow and reproduce better in heavy textured soils. Maize is tolerant to populations up to 500 nematodes per gram of dry root tissue if plant growth conditions are favorable. *Pratylenchus brachyurus* and *P. zeae* cause more injury to maize when they are in association with *Fusarium moniliforme* than they do alone. Disease is most severe during the seedling stage.

Distribution. Temperate climates of the world.

Symptoms. Aboveground symptoms include plant stunting. Leaves are chlorotic and a purple to reddish discoloration may occur. Diseased roots have reduced growth with few fibrous roots. Brown lesions may be present and a general brown to blackish root decay may occur.

Management. A seaweed concentrate from *Ecklonia maxima* suppressed reproduction of *P. zeae* on excised maize roots in vitro; however, in greenhouse experiments, reproduction of *P. zeae* was not affected.

Needle

Cause. *Longidorus breviannulatus*.

Distribution. Sandy soils in temperate regions. *Longidorus breviannulatus* can potentially cause severe disease in the United States (Illinois, Indiana, and Iowa).

Symptoms. The aboveground symptom is thin stands of severely stunted and chlorotic plants. Occasionally plants are killed in irregular patches during the first 6 to 8 weeks after sowing. Later, damaged plants may grow as tall as healthy plants, but stalks remain slender. If ears are present, they are reduced in size.

Belowground symptoms include yellow discoloration, slightly swollen root tips, stubby roots, pruning of lateral roots, and fewer small feeder roots than healthy plants. When soil moisture is high, seminal roots are destroyed and bush-like crown roots proliferate near the soil surface. The prop root system is unaffected.

Ring

Cause. *Criconemoides ornata* and *C. citui* are ectoparasites.

Distribution. Southeastern United States.

Symptoms. The aboveground symptom is a mild stunting. Belowground symptoms are lesions on roots and some decay.

Root Knot

Cause. *Meloidogyne arenaria*, *M. chitwoodi*, *M. incognita*, and *M. javanica* are endoparasitic nematodes. Populations of *M. incognita* are highest under minimum-tillage systems and irrigation. *Meloidogyne incognita* and *M. javanica* are more restrictive of maize growth when they are in association with *Fusarium moniliforme* than they are alone. *Meloidogyne arenaria*, *M. incognita*, and *M. javanica* do not survive if the temperature averages below 3°C during the coldest month.

Distribution. Warmer regions of the world.

Symptoms. Aboveground symptoms are stunting and chlorosis. Belowground symptoms are galls on roots and stunted and abnormally branched roots.

Management. No commercial hybrids have been identified as resistant to *M. incognita*, but some are resistant to *M. arenaria*.

Spiral

Cause. *Helicotylenchus* spp. are either ectoparasites, semi-endoparasites, or endoparasites.

Distribution. Generally wherever maize is grown.

Symptoms. The aboveground symptom is plant stunting. Belowground symptoms are a reduction in number of feeder roots and some root rot.

Sting

Cause. *Belonolaimus longicaudatus* is an ectoparasite, found only in sandy soils, that produces a phytotoxic enzyme.

Distribution. Generally wherever maize is grown.

Symptoms. Aboveground symptoms are stunting and chlorosis. The belowground symptom is deep necrotic lesions on roots. The frequent destruction of root tips results in thick, stubby roots.

Stubby Root

Cause. *Paratrichodorus christiei*, *P. minor*, *Quinisulcius acutus*, and *Trichodorus christiei* are ectoparasites found primarily in sandy soils. Most nematode feeding is around root tips.

Distribution. Not known.

Symptoms. Aboveground symptoms include severe stunting and chlorosis. The belowground symptom is a "devitalizing" of the root tips that causes stubby lateral roots. Roots sometimes appear thicker than normal.

Stunt

Cause. *Tylenchorhynchus dubius* and *T. vulgaris* are ectoparasites that are more restrictive of maize growth when they are in association with *Fusarium moniliforme* than when they are alone.

Distribution. Wherever maize is grown.

Symptoms. Aboveground symptoms are moderate plant stunting and chlorosis. The belowground symptom is a reduction in root growth.

Diseases Caused by Phytoplasmas

Corn Stunt

Corn stunt is also called achaparramiento, maize stunt, Rio Grande–type maize stunt, and Rio Grande corn stunt.

Cause. *Spiroplasma kunkelii,* or corn stunt spiroplasma (CSS), is a helical, motile phytoplasma. CSS is not seedborne or mechanically transmitted and it is not known how it survives in the absence of maize. In areas where maize is grown continuously, CSS is transmitted from maturing to newly sown fields by the leafhoppers *Dalbulus maidis*, *D. elimatus*, *Graminella nigrifrons*, and *Baldulus tripsaci*. The mollicute is introduced into the sieve tubes of host plants by phloem-feeding leafhoppers. Spiroplasmas move into tassels as they develop.

The age and/or size of the plant at the time of infection affects pathogen movement. Spiroplasmas are detected earlier in roots of plants inoculated at the one-leaf seedling stage than in roots of plants inoculated at the four-leaf stage.

CSS tolerates wide temperature regimes (18° to 31°C).

Distribution. Argentina, Central America, Mexico, and the United States (Texas and possibly throughout the southern and southwestern United States and California). Infected plants have been found at 940 m above sea level in Mexico, but corn stunt is more prevalent in tropical lowlands than at higher elevations.

Symptoms. The following symptoms are observed in the United States. A chlorosis that initially occurs on the margin of the whorl leaf is followed by a reddening of the tops of older leaves. Small, circular to elongated, chlorotic spots develop at the bases of leaves on young plants and often coalesce to become elongated stripes that may or may not have a distinct margin. Plants are stunted and bear numerous small ear shoots and suckers. When disease is severe, there is a proliferation of roots.

Three types of symptoms are described from Mexico. First, leaves of stunted plants observed 60 to 940 m above sea level have well-defined, broad, chlorotic streaking. Second, diseased plants are not always stunted, but leaf margins have red to purple streaks. Third, diseased plants usually are not stunted, but their leaves have either a diffuse yellow or a chlorotic stripe with or without red margins. Both the second and third symptom types are observed at all elevations in Mexico and appear 7 days before or after anthesis.

Symptoms of diseased plants from Argentina include chlorotic streaks that originate at the bases of young leaf blades. Older leaf margins are reddened, and there is a progressive shortening of upper stalk internodes and a proliferation of ears.

Management. Resistant germplasm is being sought.

Maize Bushy Stunt

Maize bushy stunt is also known as Mesa Central corn (maize) stunt.

Cause. Maize bushy stunt phytoplasma (MBSP) is a nonhelical or pleomorphic phytoplasma found only in the phloem of diseased plants that is, to date, not culturable. It is not known how MBSM survives in the absence of maize. MBSM is transmitted by several leafhoppers but only *Dalbulus maidis* and *D. elimatus* are efficient vectors. Maize and annual teosinte are the only known hosts for MBSM.

Distribution. Central America, Mexico, South America, and the southern United States.

Symptoms. An initial chlorosis of the whorl leaf margin occurs and is followed by a reddening of older leaf tips. Reddening may not occur in certain hybrids. Subsequent leaves develop a chlorosis of the margins, turn yellow or red, tear, twist, and are shortened. Plants are stunted, have numerous small ears, and have a bushy appearance due to numerous tillers that develop at the leaf axils and base.

Management. Not reported.

Unnamed Disease

Cause. Spiroplasma. Disease occurs in late-planted corn.

Distribution. The United States (California).

Symptoms. The leaf blades of diseased plants are bright red at harvest time.

Management. Not reported.

Diseases Caused by Viruses

American Wheat Striate Mosaic

American wheat striate mosaic is also called wheat striate mosaic.

Cause. American wheat striate mosaic virus (AWSMV) is in the rhabdovirus group. AWSMV is transmitted mainly by the painted leafhopper, *Endria inimica,* and less frequently by the leafhopper *Elymana virenscens.*

Distribution. Canada and the north central United States.

Symptoms. Diseased plants are stunted and have slight to moderate leaf striation that consists of thin yellow to white parallel streaks.

Management. Not reported.

Australian Maize Stripe

Cause. The Australian maize stripe agent is thought to be a virus. The agent is vectored by the planthopper *Peregrinus maidis.* The major reservoir host is *Sorghum verticilliflorum.* Acquisition and inoculation threshold periods were approximately 1 hour each.

Distribution. Australia (Queensland). Incidence may be high in sweet maize grown in coastal districts where the vector and plant reservoir is common; however, the disease is rare in the major inland agricultural areas.

Symptom. Broad yellow stripes occur on leaves of diseased plants.

Management. Not reported.

Barley Stripe Mosaic

Cause. Barley stripe mosaic virus (BSMV) is in the hordeivirus group. BSMV is transmitted by seed. Maize is more susceptible to systemic infection at 25°C than at higher temperatures.

Distribution. Generally wherever maize is grown.

Symptoms. Chlorotic stripes occur on leaves of diseased plants.

Management. Not practiced.

Barley Yellow Dwarf

Cause. Barley yellow dwarf virus (BYDV) is in the luteovirus group. BYDV is transmitted by several aphids. Irrigated maize is a reservoir of BYDV and its aphid vectors between summer harvest and autumn sowing of winter grains in eastern Washington. Several grasses are hosts. *See* Barley Yellow Dwarf in Chapter 2, "Diseases of Barley."

Distribution. Africa, Asia, Australia, Europe, New Zealand, and North America.

Symptoms. Leaves of diseased plants become chlorotic at the tips and margins. Later, beginning with the oldest leaves, all leaves become reddish to purple in color. Little or no stunting occurs.

Management. Disease is normally not severe enough to warrant management; however, avoid growing maize adjacent to or following diseased grass hosts.

Brome Mosaic

Cause. Brome mosaic virus (BMV) is in the bromovirus group. BMV is transmitted by the corn rootworms *Diabrotica undecimpunctata* and *D. virgifera* and the nematodes *Longidorus macrosoma, Xiphinema coxi,* and *X. diversicaudatum.* Several grasses and some dicotyledons are hosts.

Distribution. Worldwide in other hosts.

Symptoms. Chlorotic stripes of varying widths occur on young diseased leaves. Occasionally, young plants may wilt and die.

Management. Avoid growing maize adjacent to or following diseased grasses in a rotation.

Cereal Chlorotic Mottle

Cause. Cereal chlorotic mottle virus (CeCMV) is in the rhabdovirus group. CeCMV is transmitted by the leafhopper *Nesoclutha pallida.* Several other grasses are hosts.

Distribution. Australia (Queensland).

Symptoms. The initial chlorotic stripes on diseased leaves are followed by a slight chlorotic mottling.

Management. Not reported.

Cereal Tillering Disease

Cause. The causal virus is unknown. It is disseminated by the planthoppers *Laodelphax striatellus* and *Dicranotropis hamata.* Barley, oat, and several grasses are also hosts.

Distribution. Sweden.

Symptoms. Diseased plants are stunted and have numerous shoots. Leaves have vein enations.

Management. Not reported.

Corn Chlorotic Vein Banding

Corn chlorotic vein banding is also called Brazilian maize mosaic.

Cause. Corn chlorotic vein banding virus (CCVBV) is transmitted by the corn planthopper, *Peregrinus maidis*. CCVBV may be related to maize mosaic virus.

Distribution. Brazil.

Symptoms. The disease occurs sporadically in a field. Chlorotic vein banding occurs along the leaf veins. Diseased plants are generally stunted. The disease is generally not considered serious, and significant losses have not occurred.

Management. Not necessary.

Corn Lethal Necrosis

Cause. Corn lethal necrosis is caused by the synergistic action of maize chlorotic mottle virus (MCMV) in combination with either maize dwarf mosaic virus (MDMV) strain B or wheat streak mosaic virus (WSMV). (*See* "maize chlorotic mottle virus" (MCMV), "maize dwarf mosaic virus" (MDMV), and "wheat streak mosaic virus" (MSMV) in this chapter for specific information.)

It is presumed that maize plants are susceptible at all stages of development.

Distribution. Mexico in winter seed maize nurseries, the United States (Hawaii in winter seed maize nurseries, Kansas, and Nebraska).

Symptoms. An initial bright yellow mottling in diseased leaves is followed by a necrosis inward from the leaf margins and, eventually, premature death of plants. In diseased maturing plants, necrosis normally begins at the tassel and progresses downward, causing plants to die from the top down. Ears may be small and distorted and have little or no kernel development. Developed kernels are often wrinkled and shriveled. Losses of 75% to 80% may occur in susceptible hybrids.

Management
1. Grow a nonhost crop, such as alfalfa, a small grain, sorghum, or soybean, in fields where corn lethal necrosis has occurred.
2. Grow a tolerant hybrid.

Cucumber Mosaic

Cause. Cucumber mosaic virus (CMV) is in the cucumovirus group. CMV is disseminated by several aphid species. Other hosts include numerous dicotyledons and monocotyledons.

Distribution. Worldwide in several other hosts.

Symptoms. A mild symptom is chlorotic, elliptical spots that eventually form stripes on diseased leaves. Severe symptoms are the death of seedlings or the stunting of diseased plants that have necrotic leaf spots.

Management. Not reported.

Cynodon Chlorotic Streak

Cause. Cynodon chlorotic streak virus (CCSV) is in the rhabdovirus group. CCSV is transmitted by the planthopper *Toya propinqua*, but it is not mechanically transmitted. CCSV is endemic in Bermuda grass, *Cynodon dactylon*.

Distribution. Morocco.

Symptoms. Diseased plants are stunted and have narrow, longitudinal, chlorotic streaks on leaves.

Management. Not reported.

Elephant Grass Mosaic

Cause. Elephant grass mosaic virus (EGMV). EGMV has been tentatively identified as a potyvirus that causes a mosaic of elephant grass but has a very limited host range. In greenhouse tests, EGMV was transmitted mechanically, but not by aphids.

Distribution. Brazil.

Symptoms. Chlorotic spots and streaks developed on diseased leaves 1 to 2 weeks after inoculation.

Management. Not reported.

High Plains Virus

Cause. High plains virus (HPV) is transmitted by the wheat leaf curl mite, *Aceria tosichella*, which also transmits wheat streak mosaic virus, resulting in mixed infections by these two viruses. HPV is also transmitted at a low rate in the infected seed of sweet maize. The virus is associated with a 32-kDa protein that resembles tenuiviruses; however, it is also suggested there is some resemblance to tospoviruses.

HPV is also transmitted to barley and wheat.

Distribution. The midwestern United States (Colorado, Idaho, Kansas, Nebraska, Texas, and Utah). Dent maize, sweet maize, and blue maize are susceptible.

Symptoms. Disease is often worse on the side of the maize field that is near wheat. It is speculated that, generally, higher temperatures or fluctuating day/night temperatures may exacerbate symptom expression. Symptoms are often most severe on diseased lower leaves and develop more slowly on upper leaves. Many other viruses are typically more noticeable on the upper leaves.

The first symptoms appear when the plants are small, approximately at the 30- to 45-cm stage. The incidence and severity increase for several weeks. In some reports, the incidence appears to decline because the smaller, more severely diseased plants die, while the healthy and less severely diseased plants survive and improve the appearance of the field.

The first symptoms on diseased plants infected early in their growth are plant stunting and a pronounced chlorosis in the form of an interveinal general mosaic and flecking, or streaking, in the youngest leaves. The chlorosis is discrete, small, white spots in linear clusters or rows that roughly follow the veins. There is no sharp delimitation between symptomatic and nonsymptomatic areas of a diseased leaf.

As diseased plants develop, new growth continues to be chlorotic, while older tissue reddens at the leaf margin, beginning at the leaf tip. The reddening progresses down the leaf and is followed by necrosis that begins at the tip and moves down the leaf. In the most severe cases, the plants die. In mild cases, some plants eventually recover from the shock phase and grow to sexual maturity. However, in sweet maize, these plants produce unacceptable ears; in field maize, the ear matures but is often stunted or deformed and ear and kernel size are reduced. At maturity, the infected dent maize plants are stunted (2 m tall versus a normal 2.7 m tall) and display a striking chlorosis with red stripes, sectors, and flecks. Mature plants approaching harvest have clearly defined red sectors and stripes on the leaves.

Symptoms of the disease are very severe in susceptible genotypes, killing maize plants in as little as 2 weeks following infection. Dent maize yields are reduced by 25% to 75%. Severely infected sweet maize is a total loss.

Management. Genetic variability for resistance to HPV exists among maize inbred lines.

Johnsongrass Mosaic

Cause. Johnsongrass mosaic virus (JGMV) is in the potyvirus group. JGMV now consists of former strain JG of sugarcane mosaic virus from Australia and strain O of maize dwarf mosaic virus from the United States. JGMV is transmitted mechanically and nonpersistently by the corn leaf aphid, *Rhopalosiphum maidis.*

Distribution. Likely wherever maize is grown near johnsongrass.

Symptoms. Initially, diseased plants have a stippled mottle or mosaic of light and dark green that develops into narrow streaks on the youngest leaves. Symptoms may occur on all diseased leaves, leaf sheaths, and husks that develop after the plant is infected by JGMV. As diseased plants mature, the mosaic disappears and leaves become yellowish green and frequently show blotches or streaks of red.

Management
1. Grow resistant hybrids.
2. Destroy johnsongrass or other overwintering hosts.

Maize Chlorotic Dwarf

Cause. Maize chlorotic dwarf (MCDV) is composed of at least two strains: MCDV-T, the type isolate; and MCDV-M1. Although the two strains are similar in host plant range, vector range, and retention of inoculativity by the vector, they can be distinguished from one another by symptomatology as well as by transmission rate by some vector species. MCDV-M1 can also be differentiated from MCDV-T on the basis of serology and the electrophoretic patterns of the coat proteins.

MCDV overwinters primarily in johnsongrass, but other plants, such as beard grass, *Andropogon virginicus*; muhly, *Muhlenbergia sobolifera*; indian grass, *Sorghastrum nutans*; and blue stem, *Schizachyrium scoparium,* are also susceptible. MCDV is semipersistently transmitted by several leafhoppers, including *Amblysellus grex, Dalbulus maidis, Graminella nigrifrons, G. sonora, Planicephalus flavicostatus*, and *Stirellus bicolor. Graminella nigrifrons* is the most efficient vector. *Amblysellus grex, P. flavicostatus*, and *S. bicolor* transmit MCDV-T at higher rates than they transmit MCDV-M1. A helper component produced in MCDV-infected plants is necessary for the transmission of MCDV by leafhopper vectors, but MCDV is not transmitted mechanically.

In single infections, each isolate causes no stunting or only mild stunting and vein banding, but in infections together, a synergistic interaction between MCDV-T and MCDV-M1 causes severe stunting.

Distribution. The southeastern United States.

Symptoms. There is a more severe chlorosis and stunting of plants when MCDV-T infects maize in combination with MCDV-M1 than in infections by either virus alone. Singly, each isolate causes only mild symptoms, with isolate M1 causing the mildest symptoms.

Initially, chlorosis occurs on young leaves in the whorl. As diseased leaves unfurl, a fine chlorotic striping associated with the smallest visible leaf veins (tertiary leaf veins) runs for some length parallel to the veins. This diagnostic symptom is more pronounced when induced by MCDV-T than by MCDV-M1. The early striping may become obscured as diseased plants mature and their leaves become yellowish and reddish. Additional symptoms are plant stunting and a reddening, yellowing, and eventual necrosis of leaf margins that are horizontally split. Diseased leaves are duller than bright, shiny healthy leaves.

Athough leaf discoloration and plant stunting are correlated with MCDV infection, these symptoms are not diagnostic since not all plants infected with MCDV are stunted and discolored.

Management
1. Grow tolerant hybrids.
2. Sow early in the growing season to avoid large leafhopper populations.
3. Control johnsongrass to eliminate overwintering hosts.

Maize Chlorotic Mottle

Cause. Maize chlorotic mottle virus (MCMV) is in the sobemovirus group. MCMV survives in corn residue and is transmitted by six species of beetles: the cereal leaf beetle, *Oulema melanopa*; the corn flea beetle, *Chaetocnema pulicaria*; the flea beetle, *Systena frontalis*; the southern corn rootworm, *Diabrotica undecimpunctata*; the northern corn rootworm, *D. longicornis*; and the western corn rootworm, *D. virgifera*. MCMV is also transmitted by seed, by thrips *(Frankliniella williamsi)*, mechanically, and possibly by soil. In the absence of fresh maize roots, newly hatched larvae of vectors forage on infested crop residues and acquire MCMV, which is transmitted later by larvae feeding on developing maize roots.

MCMV survival in maize tissues, and perhaps viability of beetle eggs in soil, may occur only in soils with high water-holding capacities. Such soils can maintain infected crop residues in a "proper" state of hydration and preserve virus particles. Continuous maize production greatly increases the incidence of MCMV. Other hosts of MCMV include barley, johnsongrass, rye, sorghum, and wheat.

MCMV, together with either maize dwarf mosaic virus strain B or wheat streak mosaic virus, is one of the viruses that causes corn lethal necrosis. At least two strains of MCMV exist: the Peru strain, MCMV-P; and the Kansas strain, MCMV-K.

Distribution. Brazil, Peru, and the United States (Hawaii, Kansas, Nebraska, and Texas).

Symptoms. Initially, fine, chlorotic stripes are parallel to the veins of the youngest diseased leaves. Stripes coalesce and produce elongated, chlorotic blotches that eventually become necrotic. Eventually, leaves curl downward and plant death follows. Diseased plants are stunted, have distorted tassels, and form fewer ears. Significant yield losses may occur from infection by MCMV alone.

Management
1. Grow tolerant hybrids.
2. Rotate maize with soybean.

Maize Dwarf Mosaic

Cause. Maize dwarf mosaic virus (MDMV) now consists of former MDMV strains A, D, E, and F; former sugarcane mosaic virus strain J infective to johnsongrass in the United States; and maize mosaic virus, European type, from Yugoslavia.

MDMV is transmitted mechanically from the mouthparts of several aphids, and is seedborne in sweet maize. Disease appearance and spread are likely related to aphid numbers. Plants infected when young are more severely diseased than plants infected when they are older; however, late sowings of sweet maize are more severely diseased than early sowings. Soil moisture and nitrogen availability are associated with virus infection.

Several wild and cultivated grasses are susceptible to MDMV. In Mississippi, 70% of all grass species are susceptible in various degrees to MDMV.

Distribution. Wherever maize is grown.

Symptoms. Symptoms are variable and most severe on plants infected when young; those infected at the pollination stage of plant growth or later may appear normal. Initially, plants have a stippled mottle or mosaic of light and dark green that develops into narrow streaks on the youngest diseased leaves. Sometimes the mosaic appears as dark green "islands" on a chlorotic background. Symptoms may occur on all leaves, leaf sheaths, and husks that develop after infection. As diseased plants mature, the mosaic disappears and leaves become yellowish green and frequently show blotches or streaks of red that are generally observed after periods of cool night temperatures of 15.5°C and below.

Pollen germ tube lengths are shortened and virus may be present in silks of some sweet maize cultivars. Severely diseased plants are barren. Plants are predisposed to root rot, and stalk strength is reduced because diseased stalks become smaller in diameter than those of healthy plants. Upper internodes may be shortened, giving a "feather duster" appearance to diseased plants.

Infection of sweet maize by MDMV and drought stress are additive in effect on ear and plant characteristics. Weight of ears is reduced by 16%, butt blanking of ears is increased, leaf area is reduced, and plant height is reduced.

Management
1. Grow resistant hybrids.
2. Destroy johnsongrass or other overwintering hosts.
3. Plant sweet maize early.

Maize Leaf Fleck

Cause. Maize leaf fleck virus is disseminated in a persistent manner by the aphids *Myzus persicae, Rhopalosiphum padi,* and *R. maidis.* Another host is harding grass, *Phalaris stenoptera.*

Distribution. The United States in the San Francisco Bay area of California.

Symptoms. Older diseased leaves and the tips of other leaves have small, round, yellowish to orange spots and tip and marginal burning. Eventually, diseased leaves become necrotic.

Management. Not reported.

Maize Line

Cause. Maize line virus (MLV) is disseminated by the corn planthopper, *Peregrinus maidis.*

Distribution. East Africa.

Symptoms. Wide, chlorotic bands occur on diseased leaves. Veins on leaf undersurfaces become thickened and give a rough texture to the surface. Diseased plants are slightly stunted.

Management. Not reported.

Maize Mosaic

Maize mosaic is also called corn leaf stripe, corn stripe, corn mosaic, enanismo rayado, and sweet corn mosaic stripe. In some literature, maize mosaic has been confused with maize dwarf mosaic.

Cause. Maize mosaic virus (MMV) is in the rhabdovirus group. MMV is transmitted by the corn planthopper, *Peregrinus maidis*, and may be seedborne. MMV infects sorghum and wild grasses, which provide another means for MMV to overseason.

Distribution. In tropical areas of Africa, Australia, Caribbean, India, northern South America, and the United States (Hawaii and the mainland).

Symptoms. Symptoms vary. Initially, small, white flecks associated with whitening of leaf veins occur on one side of the midrib near the base of a young leaf. The specks elongate and form fine, discontinuous stripes parallel to the midrib. Sometimes stripes coalesce and cause yellow bands to form on leaves. Severely diseased tissue becomes red to purple, with a necrosis of the formerly chlorotic areas of diseased leaves. Stripes also occur on sheaths, husks, and stalks. Sometimes only broad, chlorotic bands occur on leaves that have both veins and interveinal tissue affected. Bands, which may appear on one or both sides of the midrib, are not continuous but fade into spots or yellow areas at different lengths on the leaf. Moderate to severe dwarfing of new growth can occur.

Management. Grow resistant hybrids and inbreds.

Maize Mottle

Maize mottle and chlorotic stunt have been viewed as the same disease complex (Rossel 1984).

Cause. Mottle strain of maize streak virus (MSV-MS) is in the geminivirus group. MSV-MS is transmitted mainly by the leafhopper *Cicadulina mbila*, but other leafhoppers also are vectors. MSV is not seedborne or mechanically transmitted.

Distribution. East Africa.

Symptoms. Symptoms are not described well but generally have been characterized as a chlorotic mottling of leaves and a stunting of diseased plants.

Management. Some maize genotypes have been described as having levels of resistance.

Maize Necrotic Lesion

Cause. Maize necrotic lesion virus infects roots of maize. Isometric virus-like particles (17 and 29 nm in diameter) have been found in crude extracts from lesions and mesophyll cells, suggesting a satellite and/or helper virus may be involved in the disease syndrome.

Distribution. Not known.

Symptoms. Symptoms on rub-inoculated leaves of seedlings first appear as chlorotic local lesions, then become necrotic after 24 to 36 hours. Masses of amorphous inclusions and fibril-containing vesicles occur in the cytoplasm.

Management. Not known.

Maize Pellucid Ringspot

Cause. Maize pellucid ringspot virus is thought to be disseminated by an aphid.

Distribution. New Guinea and western Africa.

Symptoms. Diseased leaves have round, water-soaked spots.

Management. Not reported.

Maize Raya Gruesa

Cause. Maize raya gruesa virus (MRGV) is transmitted in a persistent manner by the corn planthopper, *Peregrinus maidis*. MRGV is acquired by a feeding acquisition period of 48 to 72 hours and an incubation period in the insect vector of 4 to 22 days.

Distribution. Colombia.

Symptoms. Chlorotic stripes 1 mm wide occur on diseased leaves.

Management. Not reported.

Maize Rayado Fino

Maize rayado fino is also called fine striping disease and rayado fino. Maize Colombian stripe (del rayado Colombiana del maize) and Brazilian corn streak are diseases of maize caused by strains of MRFV.

Cause. Maize rayado fino virus (MRFV) is in the marafivirus group. MRFV virions are isometric (approximately 30 nm in diameter) and contain a single-stranded RNA genome. The virus is able to replicate in both maize and in the leafhopper *Dalbulus maidis*. MRFV often occurs in field infections associated with mollicutes (corn stunt spiroplasma and maize bushy stunt phytoplasma) that are transmitted by the same vector.

MRFV is transmitted by nymphs and adults of the leafhoppers *Baldulus tripsaci*, *D. maidis*, *D. elimatus*, *Graminella nigrifrons*, and *Stirellus bicolor*. The incubation period is 7 to 21 days, with leafhoppers retaining MRFV for prolonged periods, sometimes for life.

Distribution. Central America, South America, and the United States (Florida and Texas). MRFV is the only maize-infecting virus existing solely in the center of origin of maize.

Symptoms. Initially, a few small, chlorotic dots or short stripes develop at the base and along veins of young diseased leaves. Dots become numerous and begin to fuse; however, long continuous stripes are seldom found and a characteristic chlorotic stipple striping of veins prevails. As diseased plants become older, symptoms become less conspicuous and may disappear when plants reach maturity. Diseased plants may be partially stunted and chlorotic.

Management. Genotypes have been identified that display mild, delayed symptoms and low virus concentrations in diseased plants.

Maize Red Stripe

Cause. Maize red stripe virus (MRSV). It is not known how MRSV is spread.

Distribution. Bulgaria.

Symptoms. Red streaks occur on leaves. Diseased plants are dwarfed and have small ears.

Management. Not reported.

Maize Ring Mottle

Cause. Maize ring mottle virus is transmitted by seed.

Distribution. Bulgaria.

Symptoms. Chlorotic spots and rings occur on diseased leaves.

Management. Not reported.

Maize Rio Cuarto Disease

Cause. A reo-like virus.

Distribution. Argentina.

Symptoms. Diseased plants are severely stunted, malformed, and dark green and have stiff or brittle stalks and leaves. Enations on leaf veins may be present.

Management. Grow resistant inbreds or hybrids.

Maize Rough Dwarf

Maize rough dwarf is also called nanismo ruvido.

Cause. Maize rough dwarf virus (MRDV) is in the fijivirus group. MRDV is disseminated only by the planthopper *Laodelphax striatellus*.

Distribution. Southern and southeastern Europe.

Symptoms. Early infection results in severely stunted plants that have numerous enations on the undersides of diseased leaves and leaf sheaths. Root systems are reduced in size and discolored. Ears are not formed or if they are, tend to be small and malformed. Later infection does not result in noticeable stunting but enations are still formed.

Management. Not reported.

Maize Sterile Stunt

Cause. Maize sterile stunt virus (MSSV) is in the rhabdovirus group. MSSV is present in several grassy weeds and is transmitted by planthoppers.

Distribution. Australia, on some inbreds.

Symptoms. Mild chlorosis occurs on emerging leaves; older diseased leaves become red to purple. Tassels will not form and ears become malformed.

Management. Resistance has been identified.

Maize Streak

Maize streak is also called corn streak, maize streak disease, and streak disease. Bajra streak (Chapter 10) and maize mottle are caused by strains of the same virus.

Cause. Maize streak virus (MSV) is in the geminivirus group. MSV is transmitted mainly by the leafhopper *Cicadulina mbila,* but other leafhoppers are also vectors. MSV is not seedborne or mechanically transmitted.

Several specialized strains of MSV from maize, sugarcane, and grasses exist. Fifty-four grass species are hosts, including African millet, barley, broomcorn, millet, oat, rice, rye, sugarcane, wheat, and several wild grasses. In South Africa, isolates MSV-CT and MSV-Pe caused different symptoms in maize and also differed antigenically. In Nigeria, MSV isolated from many grasses was not readily transmissible to susceptible field maize, but the grasses *Axonopus compressus*, *Brachiaria lata*, and *Setaria barbata* were likely to be involved in perpetuating strains of MSV pathogenic to maize.

Infection by MSV will mask symptoms of *Peronosclerospora philippinensis* and *P. sorghi.*

Distribution. Kenya, Nigeria, and South Africa.

Symptoms. Initially, circular (0.5–2.0 mm in diameter), colorless spots occur on the lowest exposed portions of the youngest diseased leaves. With incident light, lesion color varies from white to yellow. In transmitted light, white lesions appear translucent. Spots are scattered but eventually become more numerous and grouped together. Eventually, narrow, broken, chlorotic stripes that may consist of only a few flecks or small dots occur along veins. However, symptoms may become more severe and stripes may coalesce and form wider stripes that are evenly distributed on leaves formed after infection. Eventually, the broad chlorotic stripes diffuse together, producing both yellow and white lesions that cause an entire leaf to become chlorotic.

Insect toxin is responsible for stunting, leaf curling, and vein enations often associated with symptoms caused by MSV. Plants infected early are more severely diseased than plants infected when older; consequently, yield losses are directly proportional to the age of the plant at infection. Virus strains from maize cause severe stunting if plants are infected before the 4- to 5-leaf stage.

Management
1. Grow resistant hybrids. Some lines of *Zea mays* are resistant but mildly susceptible at the 1- to 2-leaf stage. Perennial species of maize, *Z. perennis* and *Z. diploperennis,* are resistant to MSV.
2. Apply insecticides.
3. Do not grow maize adjacent to other hosts.

Maize Stripe

Maize stripe is also called maize chlorotic stripe, maize hoja blanca (white leaf of corn), and maize white leaf and may be the same as Tsai's disease.

Cause. The maize stripe virus (MStpV) is a member of the tenuivirus group. MStpV persists and multiplies with a minimum latent period of more than 1 week in the planthoppers *Toya catilina* and *Peregrinus maidis*. MStpV is generally transovarially transmitted, but an isolate from Venezuela was mechanically transmitted to the sweet maize cultivar 'Iochief' but not to field maize. Other hosts include barley, sorghum, and teosinte.

Distribution. In all tropical areas where *P. maidis* occurs.

Symptoms. Initially, numerous chlorotic spots and streaks appear at the bases and extend outward on the youngest diseased leaves. As leaves expand, spots and stripes coalesce and form broad chlorotic bands that later cause leaves to be completely yellow. Diseased plants may have acute bending at the apical tip, causing them to become necrotic and die. Severe infection results in reduced yields.

Management. Not reported.

Maize Tassel Abortion

Cause. Maize tassel abortion virus is disseminated by the insect *Malaxodes farinosus*.

Distribution. East Africa.

Symptoms. The tassels are aborted. Small leaves grow at right angles to the stalk.

Management. Not reported.

Maize Vein Enation

Cause. Maize vein enation virus (MVEV) is disseminated by the leafhopper *Cicadulina mbila*. Other hosts of MVEV include numerous grasses, oat, rice, rye, sorghum, sugarcane, and wheat.

Distribution. India.

Symptoms. Galls form on the lower surfaces of diseased leaves. Leaves become dark green. Diseased plants are stunted.

Management. Not reported.

Maize Wallaby Ear

Cause. Maize wallaby ear virus (MWEV) is transmitted by the leafhoppers *Cicadulina bipunctella bimaculata, C. bimaculata,* and *Nesoclutha pallida.* Other hosts for MWEV are rice, rye, sorghum, sugarcane, wheat, and several other grasses.

Distribution. Australia.

Symptoms. There are two types of symptoms. Mild symptoms consist of galls and enations from which diseased plants recover. Severe symptoms consist of numerous galls on most veins and upright dark green leaves with edges that roll upward and inward and stand stiffly at right angles to the stalk. Diseased plants may be severely stunted and have reduced yields.

Management. Not reported.

Maize White Line Mosaic

Maize white line mosaic is also called white line mosaic and stunt.

Cause. Maize white line mosaic virus (MWLMV) is not mechanically transmitted but has been transferred from diseased roots placed in sterile soil with zoospores of an *Olpidium*-like fungus considered to be the likely vector. Infection is associated with time of season rather than plant age. Maize white line mosaic often occurs locally in poorly drained areas of a field and along edge rows. A satellite virus (SV-MWLMV), which is serologically related to a satellite-like particle associated with maize dwarf ringspot virus, is associated with MWLMV.

Distribution. The United States (Michigan, New York, Ohio, Vermont, and Wisconsin).

Symptoms. An interveinal mosaic pattern that varies in intensity from a mild mottle to severe necrosis develops on leaves. Mild mottling consisting of small, discrete, chlorotic lines that are seen in the early growth stages of diseased plants develops into chlorotic, white lines ($1–2 \times 1–4$ cm) near the veins. In the latter stages of growth, diseased plants may be severely stunted, develop a severe mosaic and mottled appearance, and may develop a crozier-like hooking of the terminal portions. Plants may be barren or produce small ears with reduced kernel numbers. Diseased plants may be symptomless.

Management. Grow the most resistant genotype. There is no difference in susceptibility to MWLMV among 11 tested sweet maize cultivars.

Maize Yellow Stripe

Cause. Maize yellow stripe virus (MYSV) is associated with tenuivirus-like filaments. MYSV is transmitted in a persistent manner by both nymphs and adults of the leafhopper *Cicadulina chinai*. Acquisition and inoculation threshold times are 30 minutes each, with a latent period of 4.5 to 8.0 days, depending on temperature (14°C minimum and 25°C maximum). The maximum retention period is 27 days.

Wheat, barley, and graminaceous weeds are winter hosts. Different strains of MYSV exist.

Distribution. Egypt.

Symptoms. Three symptom types exist: fine stripe, coarse stripe, and chlorotic stunt. Each type may represent different MYSV strains. Experimentally, fine stripes appear on the first leaves of diseased plants, followed by coarse stripes on younger leaves; however, some leaves have both symptom types.

Management. Not reported.

Mal de Rio Cuarto

Cause. Mal de rio cuarto virus (MRCV) is in the fijivirus group. MRCV is transmitted in a persistent, propagative way by the planthopper *Delphacodes kuscheli*. The insects develop mainly on oat and wheat, then migrate to maize during their first stages of growth. Disease is most severe when infection takes place during the early stages of plant development and during periods of high rainfall. MRCV has been detected in 12 species of weeds in the families Poaceae and Cyperaceae.

Distribution. Argentina. Mal de Rio Cuarto is the most important disease of maize in Argentina.

Symptoms. Diseased plants have shortened internodes, thickened and flattened stalks, degenerated leaves or leaves reduced to sheaths, malformed cobs, and proliferating grainless ears. Enations protrude from the veins in the backs of leaves.

Management. Not reported.

Millet Red Leaf

Cause. Millet red leaf virus (MRLV) is disseminated by the aphids *Macrosiphum granarium*, *Rhopalosiphum maidis*, and the greenbug, *Schizaphis graminum*. Other hosts of MRLV include species of *Panicum* and *Setaria*.

Distribution. China.

Symptoms. Chlorosis to a reddening discoloration occurs on diseased leaf blades.

Management. Not reported.

Northern Cereal Mosaic

Cause. Northern cereal mosaic virus (NCMV) is in the rhabdovirus group. NCMV is disseminated by the planthoppers *Delphacodes albifascia, Laodelphax striatellus, Meuterianella fairmairei,* and *Unkanodes sapporonus.* Other hosts include barley, grasses, oat, rye, and wheat.

Distribution. Japan and Korea.

Symptoms. Chlorotic striping occurs along one side of diseased leaf blades.

Management. Not reported.

Oat Pseudorosette

Oat pseudorosette is also known as zakuklivanie.

Cause. Oat pseudorosette virus (OPRV) is disseminated by the planthopper *Laodelphax striatellus.* Other hosts of OPRV include barley, oat, rice, wheat, and several species of grassy weeds.

Distribution. Russia (western Siberia).

Symptoms. Diseased plants have numerous tillers and malformed tassels. Leaves have a reddish discoloration.

Management. Not reported.

Oat Sterile Dwarf

Cause. Oat sterile dwarf virus (OSDV) is in the fijivirus group. OSDV is disseminated by the planthoppers *Dicranotropis hamata, Javesella discolor, J. obscurella,* and *J. pellucida.*

Distribution. Sweden.

Symptoms. Diseased plants have narrow leaves with enations that last a short time.

Management. Not reported.

Rice Black-Streaked Dwarf

Cause. Rice black-streaked dwarf virus (RBSDV) is in the fijivirus group. RBSDV is disseminated by the planthoppers *Laodelphax striatellus, Ribautodel-*

phax albifascia, and *Unkanodes Sapporonus.* Other hosts include barley, several grasses, rice, and wheat.

Distribution. Japan.

Symptoms. Diseased plants are stunted and have chlorotic streaks on leaves.

Management. Not reported.

Rice Stripe

Cause. Rice stripe virus (RSV) is in the tenuivirus group. RSV is disseminated by the planthoppers *Laodelphax striatellus, Ribautodelphax albifascia,* and *Unkanodes sapporonus.* Other hosts of RSV include barley, millet, rice, and wheat.

Distribution. Japan.

Symptoms. Diseased plants are stunted and have chlorotic streaks on the leaves.

Management. Not reported.

Sorghum Stunt Mosaic

Cause. Sorghum stunt mosaic virus (SSMV) is in the rhabdovirus group. SSMV is disseminated by the leafhopper *Graminella sonora.*

Distribution. The United States (Arizona and California).

Symptoms. Diseased plants are stunted and leaves are chlorotic and have a mosaic pattern. Seed may not set.

Management. Do not grow maize in the vicinity of sorghum.

Sugarcane Fiji Disease

Cause. Sugarcane Fiji disease virus (FDV) is in the fijivirus group. FDV is disseminated by the leafhoppers *Perkinsiella saccharicida* and *P. vastatrix.* Other hosts include sorghum and sugarcane.

Distribution. Australia (New South Wales), Java, New Guinea, and the Philippines.

Symptoms. Galls occur on veins of diseased leaves.

Management. Not reported.

Sugarcane Mosaic

Cause. Sugarcane mosaic virus (SCMV) is in the potyvirus group. SCMV is now grouped as follows and consists of the former SCMV strains A, B, D, E, and MB; the former SCMV strains SC, BC, and Sabi from Australia; and the former maize dwarf mosaic virus (MDMV) strain B in the United States.

A new strain of SCMV in Yugoslavia occurs on maize mixed with MDMV.

SCMV is transmitted by several aphid species in a nonpersistent manner and by sap inoculation. SCMV is seedborne at a low rate in sweet maize seed but not in seed of other types of maize. SCMV does not normally cause severe disease in maize except for sweet maize.

Distribution. Tropical and subtropical areas.

Symptoms. Initially, a mild mottle occurs in interveinal areas of diseased leaves. Chlorotic patterns soon occur and elongate to produce streaks or stripes with irregular margins. Plants may become severely stunted, are lighter in appearance than healthy plants, and produce ears with few or no kernels.

Management
1. Do not grow maize near sugarcane.
2. Control grassy weeds.
3. Grow resistant hybrids, especially for sweet maize.

Vein Enation

Vein enation is also called leaf gall.

Cause. Unknown virus that is spread by the leafhopper *Cicadulina bipunctella*.

Distribution. The Philippines.

Symptoms. Diseased plants are stunted, and leaves are reduced in length and width. Vein enation, or galling, makes leaves appear corrugated or creased. The young leaves enveloping the tassel often will fail to unfurl.

Management
1. Rogue diseased plants.
2. Application of insecticides has been effective.

Wheat Spot Mosaic

Cause. Wheat spot mosaic virus (WSMV) is disseminated by the wheat curl mite, *Aceria tulipae* (syn. *Eriophyes tulipae*). Other hosts of WSMV include barley, numerous grasses, rye, and wheat.

Distribution. Canada and the United States (Ohio).

Symptoms. Chlorotic spots, streaks, and mottling occur on diseased leaves.

Management. Not reported.

Wheat Streak Mosaic

Cause. Wheat streak mosaic virus (WSMV) is in the potyvirus group. WSMV overwinters in winter wheat, other winter small grains, and a number of wild grasses. WSMV is transmitted in the field by the wheat curl mite, *Aceria tulipae* (syn. *Eriophyes tulipae*), and mechanically, but WSMV is not seedborne, even though the virus can be isolated from kernels before maturity.

Young plants are most susceptible to infection. Some maize lines are more susceptible to infection at 35°C than at lower temperatures; however, many lines are susceptible regardless of temperature. Other hosts include cereals and numerous grasses.

Distribution. Europe, northern Africa, and North America.

Symptoms. Small, oval to elliptical-shaped spots occur at the tips of younger diseased leaves and later elongate parallel to leaf veins. Older diseased leaves may become chlorotic at the tips. Ears may be poorly developed, and there may be a general yellowing and stunting of diseased plants.

Management. Not necessary, but differences in resistance exist among inbreds.

Wheat Striate Mosaic

Cause. American wheat striate mosaic virus is in the rhabdovirus group. *See* "American Wheat Striate Mosaic" in Chapter 23, Diseases of Wheat.

Distribution. South central Canada and north central United States.

Symptoms. Upper diseased leaves have distinct, long, thin, white to chlorotic streaks.

Management. Not reported.

Unknown Virus

Cause. Unknown virus that is possibly in the tenuivirus group.

Distribution. The United States (Kansas).

Symptoms. Diseased plants are severely stunted and have a pronounced chlorosis of the youngest leaves. The chlorosis is a general mosaic with flecking or streaking. As plants mature, new growth becomes chlorotic, while older tissue reddens along the leaf margins, beginning at the tip. Reddening progresses down the leaf and is followed by a necrosis that also begins at the tip. Severely diseased plants die. Ear and kernel sizes are reduced. Blue maize and popcorn are also affected.

Management. Not reported.

Disease Caused by Unknown Factors

North Queensland Stripe

Cause. Unknown. Virus particles have not been found.

Distribution. Northern Australia.

Symptoms. Symptoms closely resemble leaf stripe symptoms of Java downy mildew caused by *Peronosclerospora maydis*.

Management. Not known.

Selected References

Adipala, E., Lipps, P. E., and Madden, L. V. 1993. Occurrence of *Exserohilum turcicum* on maize in Uganda. Plant Dis. 77:202-205.

Ammar, E. D., Elnagar, S., Abul-Ata, A. E., and Sewify, G. H. 1989. Vector and host-plant relationships of the leafhopper-borne maize yellow stripe virus. J. Phytopathology 126:246-252.

Arny, D. C., et al. 1971. Eyespot of maize, a disease new to North America. Phytopathology 61:54-57.

Assabgui, R. A., Reid, L. M., Hamilton, R. I., and Arnason, J. T. 1993. Correlation of kernel (E)-ferulic acid content of maize with resistance to *Fusarium graminearum*. Phytopathology 83:949-953.

Attwater, W. A., and Busch, L. V. 1982. The role of sap beetles (*Glischrochilus quadrisignatus* (Say)) in the epidemiology of Gibberella corn ear rot. (Abstr.) Can. J. Plant Pathol. 4:303.

Bacon, C. W., Bennett, R. M., Hinton, D. M., and Voss, K. A. 1992. Scanning electron microscopy of *Fusarium moniliforme* within asymptomatic corn kernels and kernels associated with equine leukoencephalomalacia. Plant Dis. 76:144-148.

Bains, S. S., Jhooty, J. S., Sokhi, S. S., and Rewal, H. S. 1978. Role of *Digitaria sanquinalis* in outbreaks of brown stripe downy mildew of maize. Plant Dis. Rptr. 62:143.

Bajet, N. B., and Renfro, B. L. 1989. Occurrence of corn stunt spiroplasma at different elevations in Mexico. Plant Dis. 73:926-930.

Biddle, J. A., McGee, D. C., and Braun, E. J. 1990. Seed transmission of *Clavibacter michiganense* subsp. *nebraskense* in corn. Plant Dis. 74:908-911.

Bonde, M. R., Peterson, G. L., and Duck, N. B. 1985. Effects of temperature on sporulation, conidial germination, and infection of maize by *Peronosclerospora sorghi* from different geographical areas. Phytopathology 75:122-126.

Bonde, M. R., Peterson, G. L., Kenneth, R. G., Vermeulen, H. D., Sumartini, and Bustaman, M. 1992. Effect of temperature on conidial germination and systemic infection of maize by *Peronosclerospora* species. Phytopathology 82:104-109.

Boosalis, M. G., Sumner, D. R., and Rao, A. S. 1967. Overwintering of conidia of *Helminthosporium turcicum* on corn residue and in soil in Nebraska. Phytopathology 57:990-996.

Boothroyd, C. W. 1981. Virus diseases of sweet corn, pp. 103-109. *In* D. T. Gordon, J. A.

Knoke, and G. E. Scott (eds.). Virus and viruslike diseases of maize in the United States. Southern Cooperative Series Bull. 247. 218 pp.

Boothroyd, C. W., and Israel, H. W. 1980. A new mosaic disease of corn. Plant Dis. 64:218-219.

Bowden, R. L., and Stromberg, E. L. 1982. Chocolate spot of corn in Minnesota. Plant Dis. 66:744.

Bradfute, O. E., and Louie, R. 1988. Maize necrotic lesion virus (MNLV) and satellite-like particles. (Abstr.) Phytopathology 78:1585.

Bradfute, O. E., Teyssandier, E., Marino, E., and Dodd, J. L. 1981. Reolike virus associated with maize rio cuarto disease in Argentina. (Abstr.) Phytopathology 71:205.

Burns E. E., and Shurtleff, M. C. 1973. Observations of *Physoderma maydis* in Illinois: Effects of tillage practices in field corn. Plant Dis. Rptr. 57:630-633.

Bustamante, P. I., Hammond, R., and Ramirez, P. 1998. Evaluation of maize germplasm for resistance to maize rayado fino virus. Plant Dis. 82:50-56.

Campbell, K. W., and Carroll, R. B. 1991. Red root rot, a new disease of maize in the Delmarva Peninsula. Plant Dis. 75:1186.

Campbell, K. W., White, D. G., and Toman, J. 1993. Sources of resistance in F 1 corn hybrids to ear rot caused by *Aspergillus flavus*. Plant Dis. 77:1169.

Carrera, L. M., and Grybauskas, A. 1992. Effect of planting dates and plant density on the development of gray leaf spot of corn. (Abstr.). Phytopathology 82:718.

Carson, M. L. 1995. A new gene in maize conferring the "chlorotic halo" reaction to infection by *Exserohilum turcicum*. Plant Dis. 79:717-720.

Carson, M. L., and Goodman, M. M. 1991. Phaeosphaeria leaf spot of maize in Florida. Plant Dis. 75:968.

Castillo, J., and Herbert, T. T. 1974. Nueva enfermedat virosa afectano al maiz en al Peru (A new virus disease of maize in Peru). Fitopatologia 9:79-84.

Ceballos, H., and Deutsch, J. A. 1992. Inheritance of resistance to tar spot complex in maize. Phytopathology 82:505-512.

Chambers, K. R. 1987. Isolation of *Sordaria fimicola* from maize stalks. J. Phytopathology 120:369.

Chambers, K. R. 1988. Effect of time of inoculation on Diplodia stalk and ear rot of maize in South Africa. Plant Dis. 72:529-531.

Christensen, J. J., and Wilcoxson, R. D. 1966. Stalk Rot of Corn. Monograph No. 3. The American Phytopathological Society, St. Paul, MN.

Cohen, Y., and Sherman, Y. 1977. The role of airborne conidia in epiphytotics of *Sclerospora sorghi* on sweet corn. Phytopathology 67:515-521.

Cook, G. E., Boosalis, M. G., Dunkle, L. D., and Odvody, G. N. 1973. Survival of *Macrophomina phaseoli* in corn and sorghum stalk residue. Plant Dis. Rptr. 57:873-875.

Cullen, D., Caldwell, R. W., and Smalley, E. B. 1983. Susceptibility of maize to *Gibberella zeae* ear rot: Relationship to host genotype, pathogen virulence, and zearalenone contamination. Plant Dis. 67:89-91.

Damsteegt, V. D., Bonde, M. R., and Hewings, A. D. 1993. Interactions between maize streak virus and downy mildew fungi in susceptible maize cultivars. Plant Dis. 77:390-392.

Davis, R. M., Farrar. J. J., Peters, T. L., Kegel, F. R., and Mauk, P. A. 1991. Relationship between ear husk morphology and Fusarium ear rot of corn. (Abstr.) Phytopathology 81:1135.

De Agudelo, F. V., and Martinez-Lopez, G. 1983. Maize raya gruesa: A rhabdovirus transmitted by *Peregrinus maidis*. (Abstr.) Phytopathology 73:125.

de Nazareno, N. R. X., Lipps, P. E., and Madden, L. V. 1991. Gray leaf spot of corn as influenced by the amount of surface corn residue. (Abstr.) Phytopathology 81:1143.

de Nazareno, N. R. X., Lipps, P. E., and Madden, L. V. 1992. Survival of *Cercospora zeaemaydis* in corn residue in Ohio. Plant Dis. 76:560-563.

de Nazareno, N. R. X., Lipps, P. E., and Madden, L. V. 1993. Effect of levels of corn residue on the epidemiology of gray leaf spot of corn in Ohio. Plant Dis.77:67-70.

De Waele, D., McDonald, A. H., and De Waele, E. 1988. Influence of seaweed concentrate on the reproduction of *Pratylenchus zeae* (Nematoda) on maize. Nematologica 34:71-77

Dodd, J. L. 1980. Grain sink size and predisposition of *Zea mays* to stalk rot. Phytopathology 70:534-535.

Dodd, J. L., and Hooker, A. L. 1990. Previously undescribed pathotype of *Bipolaris zeicola* on corn. Plant Dis. 74:530.

Doupnik Jr., B., and Jensen, S. G. 1991. Suppression of maize chlorotic mottle virus by soybean rotation in first-year corn and fumigation in third-year corn. (Abstr.) Phytopathology 81:1158.

Duveiller, E., Snacken, F., Maraite, H., and Autrique, A. 1989. First detection of *Pseudomonas fuscovaginae* on maize and sorghum in Burundi. Plant Dis. 73:514-517.

Farr, D. F., Bills, G. R., Chamuris, G. P., and Rossman, A. Y. 1989. Fungi on Plants and Plant Products in the United States. The American Phytopathological Society, St. Paul, MN. 1252 pp.

Forster, R. L., Strausbaugh, C. A., Jensen, S. G. Ball, E. M., Harvey, T., and Seifers, D. L. 1997. Seed transmission of the High Plains virus in sweet corn. (Abstr.). Phytopathology 87:S31.

Gamez, R. 1969. A new leafhopper-borne virus of corn in Central America. Plant Dis. Rptr. 53:929-932.

Garcia, A. S., Reloba, R. G., and Castro, F. B. 1985. Field symptoms, spread and signs of the banded leaf and sheath blight disease of corn hybrids in Mindanao. (Abstr.) Philipp. Phytopathol. 21:4.

Gendloff, E. H., Rossman E. C., Casale, W. L., Isleib, T. G., and Hart, L. P. 1986. Components of resistance to Fusarium ear rot in field corn. Phytopathology 76:684-688.

Gilbertson, R. L., Brown, W. M., Jr., and Ruppel, E.G. 1985. Effect of tillage and herbicides on Fusarium stalk rot of corn. (Abstr.) Phytopathology 75:1296.

Gilbertson, R. L., Brown, W. M., Jr., Ruppel, E. G., and Capinera, J. L. 1986. Association of corn stalk rot *Fusarium* spp. and western corn rootworm beetles in Colorado. Phytopathology 76:1309-1314.

Gingery, R. E., and Nault, L. R. 1990. Severe maize chlorotic dwarf disease caused by double infection of mild virus strains. Phytopathology 80:687-691.

Gingery, R. E., Nault, L. R., Tsai, J. H., and Lastra, R. J. 1979. Occurrence of maize stripe virus in the United States and Venezuela. Plant Dis. Rptr. 63:341-343.

Gordon, D. T., Bradfute, O. E., Gingery, R. E., Knoke, J. K., Louie, R., Nault, L. R., and Scott, G. E. 1981. Introduction: History, geographical distribution, pathogen characteristics, and economic importance, pp. 1-12. *In* D. D. Gordon, J. K. Knoke, and G. E. Scott (eds.), Virus and viruslike diseases of maize in the United States. Southern Cooperative Series Bull. 247. 218 pp.

Greber, R. S. 1982. Maize sterile stunt: A delphacid-transmitted rhabdovirus affecting some maize genotypes in Australia. Aust. J. Agric. Res. 33:13-23.

Guthrie, E. J. 1978. Measurement of yield losses caused by maize streak disease. Plant Dis. Rptr. 62:839-841.

Halfon-Meiri, A, and Solel, Z. 1990. Factors affecting seedling blight of sweet corn caused by seedborne *Penicillium oxalicum*. Plant Dis. 74:36-39.

Headrick, J. M., and Pataky, J. K. 1989. Resistance to kernel infection by *Fusarium moniliforme* in inbred lines of sweet corn and the effect of infection on emergence. Plant Dis. 73:887-892.

Herold, F. 1972. Maize mosaic virus. CMI/A.A.B. Descriptions of plant viruses. Set 5. No. 94.

Hirrel, M. C., Lee, F. N., Dale, J. L., and Plunkett, D. E. 1988. First report of sheath blight *(Rhizoctonia solani)* on field corn in Arkansas. Plant Dis. 72:644.

Hoenisch, R. W., and Davis, R. M. 1994. Relationship between kernel pericarp thickness and susceptibility to Fusarium ear rot in field corn. Plant Dis. 78:517-519.

Hollier, C.A., and King, S. B. 1985. Effect of dew period and temperature on infection of seedling maize plants by *Puccinia polysora*. Plant Dis. 69:219-220.

Hsia, C. C., Kommedahl, T., Tziang, B. L., and Wu, J. L. 1988. Toxigenic *Fusarium scirpi* in maize grain from midnorthern China. Phytopathology 78:978-980.

Huff, C. A., Ayers, J. E., and Hill, R. R., Jr. 1988. Inheritance of resistance in corn *(Zea mays)* to gray leaf spot. Phytopathology 78:790-794.

Hunt, R. E., Nault, L. R., and Gingery, R. E. 1988. Evidence for infectivity of maize chlorotic dwarf virus and for a helper component in its leafhopper transmission. Phytopathology 78:499-504.

Jamil, F. F., and Nicholson, R. L. 1987. Susceptibility of corn to isolates of *Colletotrichum graminicola* pathogenic to other grasses. Plant Dis. 71:809-810.

Jardine, D. J., Bowden, R. L., Jensen, S. G., and Seifers, D. L. 1994. A new virus of corn and wheat in western Kansas. (Abstr.). Phytopathology 84:1117.

Jensen, S. G., Lane, L. C., and Seifers, D. L. 1996. A new disease of maize and wheat in the High Plains. Plant Dis. 80:1387-1390.

Jiang, X. Q., Meinke, L. J., Wright, R. J., Wilkinson, D. R., Campbell, J. E., and Berry, J. A. 1991. Corn thrips *(Frankliniella williamsi* Hood) a major vector associated with a 1990 maize chlorotic mottle virus epiphytotic in Hawaii. (Abstr.) Phytopathology 81:1243.

Jones R. K., and Duncan, H. E. 1981. Effect of nitrogen fertilizer, planting date, and harvest date on aflatoxin production in corn inoculated with *Aspergillus flavus*. Plant Dis. 65:741-744.

Jones, R. K., Duncan H. E., Payne, G A., and Leonard, K. J. 1980. Factors influencing infection by *Aspergillus flavus* in silk-inoculated corn. Plant Dis. 64:859-863.

Jons, V. L., Timian, R. G., Gardner, W. S., Stromberg, E. L., and Berger, P. 1981. Wheat striate mosaic virus in the Dakotas and Minnesota. Plant Dis. 65:447-448.

Jordann, E. M., Loots, G. C., Jooste, W. J., and De Waele, D. 1987. Effects of root-lesion nematodes *(Pratylenchus brachyurus* Godfrey and *P. zeae* Graham) and *Fusarium moniliforme* Sheldon alone or in combination, on maize. Nematologica 33:213-219.

Keller, N. P., and Bergstrom, G. C. 1988. Developmental predisposition of maize to anthracnose stalk rot. Plant Dis. 72:977-980.

Keller N. P., Bergstrom, G. C., and Carruthers, R .I. 1986. Potential yield reductions in maize associated with an anthracnose/European corn borer pest complex in New York. Phytopathology 76:586-589.

Khan, A., Ries, S. M., and Pataky, J. K. 1996. Transmission of *Erwinia stewartii* through seed of resistant and susceptible field and sweet corn. Plant Dis. 80:398-403.

Khonga, E. B., and Sutton, J. C. 1988. Inoculum production and survival of *Gibberella zeae* in maize and wheat residues. Can. J. Plant Pathol. 10:232-239.

Kingsland, G. C. 1980. Effect of maize dwarf mosaic virus infection on yield and stalk strength of corn in the field in South Carolina. Plant Dis. 64:271-273.

Kitajima, E. W., and Costa, A. S. 1982. The ultrastructure of the corn chlorotic vein banding (Brazilian maize mosaic) virus-infected corn leaf tissues and viruliferous vector. Fitopatologia Brasileria 7:247-259.

Klittich, C. J. R., Leslie, J. F., Nelson, P. E., and Marasas, W. F. O. 1997. *Fusarium thapsinum (Gibberella thapsina):* A new species in section Liseola from sorghum. Mycologia 89:643-652.

Kloepper, J. W., Garrott, D. G., and Kirkpatrick, B. C. 1982. Association of spiroplasmas with a new disease of corn. (Abstr.) Phytopathology 72:1004.

Kommedahl, T., Sabet, K. K., Burnes, P. M., and Windels, C. E. 1987. Occurrence and pathogenicity of *Fusarium proliferatum* on corn in Minnesota. Plant Dis. 71:281.

Kucharek, T. A. 1973. Stalk rot of corn caused by *Helminthosporium rostratum.* Phytopathology 63:1336-1338.

Kumar, V., Vasanthi, H. J., and Shetty, H. S. 1987. Detection, location and transmission of *Nigrospora oryzae* in maize. Int. J. Tropical Plant Diseases 5:153-163.

Lastra, R., and Carballo, O. 1985. Mechanical transmission, purification and properties of an isolate of maize stripe virus from Venezuela. Phytopathol. Z. 114:168-179.

Latterell, F. M., and Rossi, A. E. 1983. *Stenocarpella macrospora (Diplodia macrospora)* and *S. maydis (D. maydis)* compared as pathogens of corn. Plant Dis. 67:725-729.

Latterell, F. M., and Rossi, A. E. 1984. A *Marasmiellus* disease of maize in Latin America. Plant Dis. 68:728-731.

Latterell, F. M., and Rossi, A. E. 1984. An unidentified species of *Cercospora* pathogenic to corn. (Abstr.) Phytopathology 74:852.

Latterell, F. M., Rossi, A. E., and Trujillo, E. E. 1986. A previously undescribed Selenophoma leaf spot of maize in Colombia. Plant Dis. 70:472-474.

Leach, C. M., Fullerton, R. A., and Young, K. 1977. Northern leaf blight of maize in New Zealand: Relationship of *Drechslera turcica* airspora to factors influencing sporulation, conidium development, and chlamydospore formation. Phytopathology 67:629-636.

Leonard, K. J., and Thompson, D. L. 1976. Effects of temperature and host maturity on lesion development of *Colletotrichum graminicola* on corn. Phytopathology 66:635-639.

Leonard, K. J., Levy, Y., and Smith, D. R. 1989. Proposed nomenclature for pathogenic races of *Exserohilum turcicum* on corn. Plant Dis. 73:776-777.

Levy, Y. 1984. The overwintering of *Exserohilum turcicum* in Israel. Phytoparasitica 12:177-182.

Levy, Y., and Leonard, K. J. 1987. Development of northern corn leaf blight lesions as affected by position and density of lesions on leaves. (Abstr.) Phytopathology 77:409.

Lipps, P. E. 1988. Spread of corn anthracnose from surface residues in continuous corn and corn-soybean rotation plots. Phytopathology 78:756-761.

Lland, A., and Schieber, E. 1980. *Diplodia macrospora* of corn in Nicaragua. Plant Dis. 64:797.

Lockhart, B. E. L., Khaless, N., El Maataoui, M., and Lastra, R. 1985. Cynodon chlorotic streak virus, a previously undescribed plant rhabdovirus infecting Bermuda grass and maize in the Mediterranean area. Phytopathology 75:1094-1098.

Logrieco, A., Moretti, A., Ritieni, A., Bottalico, A., and Corda, P. 1995. Occurrence and tox- igenicity of *Fusarium proliferatum* from preharvest maize ear rot, and associated my- cotoxins, in Italy. Plant Dis. 79:727-731.

Lopes, J. R. S., Nault, L. R., and Gingery, R. E. 1994. Leafhopper transmission and host plant range of maize chlorotic dwarf waikavirus strains. Phytopathology 84:876-882.

Louie, R., Gordon, D. T., and Lipps, P. E. 1981. Transmission of maize white line mosaic virus. (Abstr.) Phytopathology 71:1116.

Louie, R., Gordon D. T., Madden, L. V., and Knoke, J. K. 1983. Symptomless infection and incidence of maize white line mosaic. Plant Dis. 67:371-373.

Maiti, S. 1978. Two new ear rots of maize from India. Plant Dis. Rptr. 62:1074-1076.

Malek, R. B., Norton, D. C., Jacobsen, B. J., and Acosta, N. 1980. A new corn disease caused by *Longidorus breviannulatus* in the Midwest. Plant Dis. 64:1110-1113.

March, G. J., Balzarini, M., Ornaghi, J. A., Beviacqua, J. E., and Marinelli, A. 1995. Predic- tive model for "Mal de Rio Cuarto" disease intensity. Plant Dis. 79:1051-1053.

Marcon, A., Kaeppler, S. M., and Jensen, S. G. 1997. Genetic variability among maize in- bred lines for resistance to the High Plains virus–wheat streak mosaic virus com- plex. Plant Dis. 81:195-198.

Martins, C. R. F., and Kitajima, E. W. 1993. A unique virus isolated from elephant grass. Plant Dis. 77: 726-729.

Martinson, C. A., Foley, D. C., and Marton, C. 1988. The effect of plant density and kernel sink level on stalk rot and strength of maize. (Abstr.) Phytopathology 78:1525.

Matyac, C. A., and Kommedahl, T. 1985. Factors affecting the development of head smut caused by *Sphacelotheca* on corn. Phytopathology 75:577-581.

Matyac, C. A., and Kommedahl, T. 1985. Occurrence of chlorotic spots on corn seedlings infected with *Sphacelotheca reiliana* and their use in evaluation of head smut resis- tance. Plant Dis. 69:251-254.

McDaniel, L. L., and Gordon, D. T. 1985. Identification of a new strain of maize dwarf mo- saic virus. Plant Dis. 69:602-607.

McKee, D. C. 1988. Maize Diseases: A Reference Source for Seed Technologists. The Amer- ican Phytopathological Society. St. Paul, MN. 150 pp.

McLennan, S. R., and Rijkenberg, F. H. J. 1987. *Stenocarpella* (*Diplodia*) *macrospora* on maize (*Zea mays*). (Abstr.) Phytophylactica 19:122.

Mesfin, T., Bosque-Perez, N. A., Budenhage, I. W., Thottappilly, G., and Olojede, S. O. 1992. Studies on maize streak virus isolates from grass and cereal hosts in Nigeria. Plant Dis. 76:789-795.

Misra, A. P. 1959. Diseases of millets and maize. Indian Agriculturist 3:75-89.

Muller, G. J. et al. 1973. A Compendium of Corn Diseases. The American Phytopatholog- ical Society. St. Paul, MN.

Munkvold, G. P., and Desjardins. A. E. 1997. Fumonisins in maize. Can we reduce their oc- currence? Plant Dis. 81:556-565.

Munkvold, G. P., McGee, D. C., and Carlton, W. M. 1997. Importance of different path- ways for maize kernel infection by *Fusarium moniliforme*. Phytopathology 87:209-217.

Mycock, D. J., Lloyd, H. L., and Berjak, P. 1987. Infection of *Zea mays* seedlings by *As- pergillus flavus* var. *columnaris*. (Abstr.) Phytophylactica 19:121.

Nankam, C., and Pataky, J. K. 1996. Resistance to kernel infection by *Fusarium moniliforme* in the sweet corn inbred IL125b. Plant Dis. 80:593-598.

Nault, L. R., and Gordon, D. T. 1988. Multiplication of maize stripe virus in *Peregrinus maidis*. Phytopathology 78:991-995.

Nelson, P. A., Toussoun, T. A., and Marasas, W. F. O.1983. *Fusarium* Species: An Illustrated Manual for Identification. The Pennsylvania State University Press, University Park. 193 pp.

Niblett, C. L., and Claflin, L. E. 1978. Corn lethal necrosis, a new virus disease of corn in Kansas. Plant Dis. Rptr. 62:15-19.

Nicholson, R. L., Bergeson, G. B., Degennaro, F. P., and Viveiros, D. M. 1985. Single and combined effects of the lesion nematode and *Colletotrichum graminicola* on growth and anthracnose leaf blight of corn. Phytopathology 75:654-661.

Norton, D. C., and De Agudelo, F. V. 1984. Plant-parasitic nematodes associated with maize in Cauca and Valle del Cauca, Colombia. Plant Dis. 68:950-952.

Nwigwe, C. 1974. Occurrence of *Phomopsis* on maize *(Zea mays)*. Plant Dis. Rptr. 58:416-417.

Ochor, T. E., Trevathan, L. E., and King, S. B. 1987. Relationship of harvest date and host genotype to infection of maize kernels by *Fusarium moniliforme*. Plant Dis. 71:311-313.

Olanya, O. M., and Fajemisin, J. M. 1992. Remission of symptoms on maize plants infected with downy mildew in northern Nigeria. Plant Dis. 76:753.

Olanya, O. M., Hoyos, G. H., Tiffany, L. H., and McGee, D. C. 1997. Waste corn as a point source of inoculum for *Aspergillus flavus* in the corn agroecosystem. Plant Dis. 81:576-581.

Paulitz, T. C. 1996. Diurnal release of ascospores by *Gibberella zeae* in inoculated wheat plots. Plant Dis. 80:674-678.

Payne, G. A., and Leonard, K. J. 1985. *Stenocarpella macrospora* on corn in North Carolina. Plant Dis. 69:613.

Payne, G. A., and Waldron, J. K. 1983. Overwintering and spore release of *Cercospora zeae-maydis* in corn debris in North Carolina. Plant Dis. 67:87-89.

Payne, G. A., Cassel, D. K., and Adkins, C. R. 1985. Reduction of aflatoxin levels in maize due to irrigation and tillage. (Abstr.) Phytopathology 75:1283.

Payne, G. A., Cassel, D. K., and Adkins, C. R. 1986. Reduction of aflatoxin contamination in corn by irrigation and tillage. Phytopathology 76:679-684.

Payne, G. A., Duncan, H. E., and Adkins, C. R. 1987. Influence of tillage on development of gray leaf spot and number of airborne conidia of *Cercospora zeae-maydis*. Plant Dis. 71:329-332.

Payne, G. A., Hagler, W. M., Jr., and Adkins, C. R. 1988. Aflatoxin accumulation in inoculated ears of field-grown maize. Plant Dis. 72:422-424.

Payne, G. A., Thompson, D. L., Lillehoj, E. B., Zuber, M. S., and Adkins, C. R. 1988. Effect of temperature on the preharvest infection of maize kernels by *Aspergillus flavus*. Phytopathology 78:1378-1380.

Payne, G. A., Kamprath, E. J., and Adkins, C. R. 1989. Increased aflatoxin contamination in nitrogen-stressed corn. Plant Dis. 73:556-559.

Pedersen, W. L., and Brandenburf, L. J. 1986. Mating types, virulence, and cultural characteristics of *Exserohilum turcicum* race 2. Plant Dis. 70:290-292.

Pedersen, W. L., and Oldham, M. G. 1992. Effect of three tillage practices on development of northern corn leaf blight *(Exserohilum turcicum)* under continuous corn. Plant Dis. 76:1161-1164.

Pinner, M. S., Markham, P. G., Markham, R. H., and Dekker, E. L. 1988. Characterization of maize streak virus: Description of strains, symptoms. Plant Pathol. 37:74-87.

Pordesimo, A. N., and Aday, B. A. 1984. Vein enation or leaf gall of corn. (Abstr.) Philipp. Phytopathol. 20:15.

Raid, R. N., Pennypacker, S. P., and Stevenson, R. E. 1988. Characterization of *Puccinia polysora* epidemics in Pennsylvania and Maryland. Phytopathology 78:579-585.

Ramsey, M. D., and Jones, D. R. 1988. *Peronosclerospora maydis* found on maize, sweet corn and plume sorghum in Far North Queensland. Plant Pathol. 37:581-587.

Rao, B., Schmitthenner, A. F., Caldwell, R., and Ellett, C. W. 1978. Prevalence and virulence of *Pythium* species associated with root rot of corn in poorly drained soil. Phytopathology 68:1557-1563.

Reid, L. M., Bolton, A. T., Hamilton, R. I., Woldemariam, T., and Mather, D. E. 1992. Effect of silk age on resistance of maize to *Fusarium graminearum*. Can. J. Plant Pathol. 14:293-298.

Reid, L. M., Mather, D. E., and Hamilton, R. I. 1996. Distribution of deoxynivalenol in *Fusarium graminearum*–infected maize ears. Phytopathology 86:110-114.

Reifschneider, F. J. B., and Arny, D. C. 1980. Host range of *Kabatiella zeae*, causal agent of eyespot of maize. Phytopathology 70:485-487.

Reifschneider, F. J. B., and Lopes, C. A. 1982. Bacterial top and stalk rot of maize in Brazil. Plant Dis. 66:519-520.

Rheeder, J. P., Marasas, W. F. O., and Nelson, P. E. 1996. *Fusarium globosum*, a new species from corn in southern Africa. Mycologia 88:509-513.

Rich, J. R., and Schenck, N. C. 1981. Seasonal variations in populations of plant-parasitic nematodes and vesicular-arbuscular mycorrhizae in Florida field corn. Plant Dis. 65:804-807.

Roane, M. K., and Roane, C. W. 1983. New grass hosts of *Polymyxa graminis* in Virginia. (Abstr.) Phytopathology 73:968.

Robinson, R. K., and Lucas, R. L. 1967. Observations on the infection of *Zea mays* by *Ophiobolus graminus*. Plant Pathol. 16:75-77.

Rossel, H. W. 1984. On geographical distribution and control of maize mottle/chlorotic stunt in Africa. Maize Virus Dis. Newsl. 1, 17-19.

Rupe, J. C., Siegel, M. R., and Hartman, J. R. 1982. Influence of environment and plant maturity on gray leaf spot of corn caused by *Cercospora zeae-maydis*. Phytopathology 72:1587-1591.

Sardanelli, S., Krusberg, L. R., and Golden, A. M. 1981. Corn cyst nematode, *Heterodera zeae*, in the United States. Plant Dis. 65:622.

Schneider, R. W., and Pendery, W. E. 1983. Stalk rot of corn: Mechanism of predisposition by an early season water stress. Phytopathology 73:863-871.

Schurtleff, M. C. (ed.). 1980. Compendium of Corn Diseases. The American Phytopathological Society, St. Paul, MN. 105 pp.

Seifers, D. L., Harvey, T. L., Martin, T. J., and Jensen, S. G. 1997. Identification of the wheat curl mite as the vector of the High Plains virus of corn and wheat. Plant Dis. 81:1161-1166.

Sharma, H. S. S., and Verma, R. N. 1979. False smut of maize in India. Plant Dis. Rptr. 63:996-997.

Sinha, R. C., and Benki, R. M. 1972. American wheat striate mosaic virus. CMI/A.A.B. Descriptions of Plant Viruses. Set 6, No. 99.

Smidt, M. L., and Vidaver, A. K. 1986. Population dynamics of *Clavibacter michiganense* subsp. *nebraskense* in field-grown dent corn and popcorn. Plant Dis. 70:1031-1036.

Smidt, M. L., and Vidaver, A. K. 1987. Variation among strains of *Clavibacter michiganense* subsp. *nebraskense* isolated from a single popcorn field. Phytopathology 77:388-392.

Stevens, C., and Gudauskas, R. T. 1982. Relation of maize dwarf mosaic virus infection to *Helminthosporium maydis* race O. Phytopathology 72:1500-1502.

Sumner, D. R., and Bell, D. K. 1980. Root diseases of corn caused by *Rhizoctonia solani* and *Rhizoctonia zeae*. (Abstr.) Phytopathology 70:572.

Sumner, D. R., and Bell, D. K. 1982. Root diseases induced in corn by *Rhizoctonia solani* and *Rhizoctonia zeae*. Phytopathology 72:86-91.

Sumner, D. R., and Schaad, N. M. 1977. Epidemiology and control of bacterial leaf blight of corn. Phytopathology 67:1113-1118.

Suparyono, and Pataky, J. K. 1989. Influence of host resistance and growth stage at the time of inoculation on Stewart's wilt and Goss's wilt development and sweet corn hybrid yield. Plant Dis. 73:339-345.

Tosic, M., Ford, R. E., Shukla, D. D., and Jilka, J. 1990. Differentiation of sugarcane, maize dwarf, johnsongrass, and sorghum mosaic viruses based on reactions of oat and some sorghum cultivars. Plant Dis. 74:549-552.

Traut, E. J., and Warren, H. L. 1991. Reactions of corn inbreds to *Bipolaris zeicola* races 1, 2 and 3, and the new pathotype. (Abstr.) Phytopathology 81:1189.

Trujillo, G. E., Acosta, J. M., and Pinero, A. 1974. A new corn virus disease found in Venezuela. Plant Dis. Rptr. 58:122-126.

Tsai, J. H. 1979. Occurrence of a corn disease in Florida transmitted by *Peregrinus maidis*. Plant Dis. Rptr. 59:830-833.

Ullstrup, A. J. 1970. A comparison of monogenic and polygenic resistance to *Helminthosporium turcicum* in corn. Phytopathology 60:1597-1599.

Ullstrup, A. J. 1978. Corn Diseases in the United States and Their Control. USDA Agric. Res. Serv. Agric. Hdbk. No. 199 (Rev.).

Uyemoto, J. K. 1983. Biology and control of maize chlorotic mottle virus. Plant Dis. 67:7-10.

Uyemoto, J. K., Phillips, N. J., and Wilson, D. L. 1981. Control of maize chlorotic mottle virus by crop rotation. (Abstr.) Phytopathology 71:910.

Vakili, N. G., and Booth, G. D. 1981. *Helminthosporium carbonum*, a cause of stalk rot of corn in Iowa. (Abstr.) Phytopathology 71:910.

VanGessel, M. J., and Coble, H. D. 1993. Effect of nitrogen and moisture stress on severity of maize dwarf mosaic virus infection in corn seedlings. Plant Dis. 77:489-491.

Vidaver, A. K., and Carlson, R. R. 1978. Leaf spot of field corn caused by *Pseudomonas andropogonis*. Plant Dis. Rptr. 62:213-216.

Vidaver, A. K., Gross, D. C., Wysong, D. S., and Doupnik, B. L., Jr. 1981. Diversity of *Corynebacterium nebraskense* strains causing Goss's bacterial wilt and blight of corn. Plant Dis. 65:480-483.

Visconti, A., Chelkowski, J., Solfrizzo, M., and Bottalico, A. 1990. Mycotoxins in corn ears naturally infected with *Fusarium graminearum* and *F. crookwellense*. Can. J. Plant Pathol. 12:187-189.

Wallin J. R. 1986. Production of aflatoxin in wounded and whole maize kernels by *Aspergillus flavus*. Plant Dis. 70:429-430.

Warren, H. L. 1975. Temperature effects on lesion development and sporulation after infection by races O and T of *Bipolaris maydis*. Phytopathology 65:623-626.

Warren, H. L. 1977. Survival of *Colletotrichum graminicola* in corn kernels. Phytopathology 67:160-162.

Warren, H. L., Huber, D. M., and Tsai, C. Y. 1987. Nutrient interactions with stalk rot of maize. (Abstr.) Phytopathology 77:1745.

Waudo, S. W., and Norton, D. C. 1986. Pathogenic effects of *Pratylenchus scribneri* in maize inbreds and related cultivars. Plant Dis. 70:636-638.

Weston, W. J., Jr. 1921. Another conidial *Sclerospora* of Philippine maize. J. Agric. Res. 20:669-685.

White, D. G., Hoeft, R. G., and Touchton, J. T. 1978. Effects of nitrogen and nitrapyrin on stalk rot, stalk diameter and yield of corn. Phytopathology 68:811-814.

Wicklow, D. T., Horn, B. W., Shotwell, O. L., Hesseltine, C. W., and Caldwell, R. S. 1988. Fungal interference with *Aspergillus flavus* infection and aflatoxin contamination of maize grown in a controlled environment. Phytopathology 78:68-74.

Wicklow, D. T., Wilson, D. M., and Nelsen, T. C. 1993. Survival of *Aspergillus flavus* sclerotia and conidia buried in soil in Illinois or Georgia. Phytopathology 83:1141-1147.

Widham, G. L., and Williams, W. P. 1988. Resistance of maize inbreds to *Meloidogyne incognita* and *M. arenaria*. Plant Dis. 72:67-69.

Windels, C. E., and Kommedahl, T. 1984. Late-season colonization and survival of *Fusarium graminearum* Group II in cornstalk in Minnesota. Plant Dis. 68:791-793.

Wei, J.-K; Lui, K.-M; Chen, J.-P, Luo, P.-C., and Lee-Stadelmann, O. Y. 1988. Pathological and physiological identification of race C of *Bipolaris maydis* in China. Phytopathology 78:550-554.

Welz, H. G., and Leonard, K. J. 1993. Phenotypic variation and parasitic fitness of races of *Cochliobolus carbonum* on corn in North Carolina. Phytopathology 83:593-601.

Williams, R. J., Frederiksen, R. A., and Mughogho, L. K. (eds.). 1978. Proceedings of the International Workshop on Sorghum Diseases. ICRISAT, Patancheru, India. 469 pp.

Worf, G. L., and Delahaut, K. A. 1991. Effects and control of northern corn leaf spot disease in Wisconsin. (Abstr.) Phytopathology 81: 1201.

Wright, W. R., and Billeter, B. A. 1974. "Red kernel" disease of sweet corn on the retail market. Plant Dis. Rptr. 58:1065-1066.

Yates, I. E., Bacon, C. W., and Hinton, D. M. 1997. Effects of endophytic infection by *Fusarium moniliforme* on corn growth and cellular morphology. Plant Dis. 81:723-728.

Young, G. V., Lefebvre, C. L., and Johnson, A. G. 1947. *Helminthosporium rostratum* on corn, sorghum and pearl millet. Phytopathology 47:180-183.

Zeyen, R. J., and Morrison, R. H. 1975. Rhabdoviruslike particles associated with stunting of maize in Alabama. Plant Dis. Rptr. 59:169-171.

Zhang, L., Zitter, T. A., and Lulkin, E. J. 1991. Artificial inoculation of maize white line mosaic virus into corn and wheat. Phytopathology 81:397-400.

Zuber, M. S., Darrah, L. L., Lillehoj, E. B., Josephson, L. M., Manwiller, A., Scott, G. E., Gudauskas, R. T., Horner, E. S., Widstrom, N. W., Thompson, D. L., Bockholt, A. J., and Brewbaker, J. L. 1983. Comparison of open-pollinated maize varieties and hybrids for preharvest aflatoxin contamination in the southern United States. Plant Dis. 67:185-187.

Zummo, N. 1976. Yellow leaf blotch: A new bacterial disease of sorghum, maize, and millet in West Africa. Plant Dis. Rptr. 60:798-799.

10. Diseases of Millet

The species included are foxtail millet, or Italian millet, *Setaria italica* (syn. *Chaeotchloa italica*); proso, or browntop, millet, *Panicum miliaceum*; Japanese millet, or sawan millet, *Echinochloa crus-qalli* var. *frumentacea*; jungle rice millet, *E. colonum*; pearl millet, or bajra, *Pennisetum glaucum;* and finger millet, or ragi or African millet, *Eleusine coracana*. Certain diseases are reported to affect some millets. Any known differences between millets in susceptibility to pathogens are noted in the text.

Diseases Caused by Bacteria

Bacterial Blight

Cause. *Xanthomonas campestris* pv. *coracanae*.

Distribution. India on ragi, or African millet.

Symptoms. Initially, water-soaked, translucent, pale yellow, linear to elongate spots (5–10 mm long) develop parallel to the midribs of diseased leaves. Eventually, spots become brown and coalesce to cover the entire leaf, causing it to wither.

Management. Not reported.

Bacterial Leaf Streak

Cause. *Xanthomonas campestris* pv. *pennamericanum*. A *Xanthomonas* spp. causing a leaf streak has been described in Colorado; however, this may be different than bacterial leaf streak caused by *X. campestris*.

Distribution. Northern Nigeria and the United States (northeastern Colorado).

Symptoms. Symptoms are not fully reported, but they are presumed to be streaks on the diseased foliage. The bacterial streak from Colorado is described as a brown-red, water-soaked streak on diseased leaves.

Management. Not reported.

Bacterial Leaf Stripe

Cause. *Acidovorax avenae* subsp. *avenae* (syn. *Pseudomonas avenae*) is not known to be seedborne; however, the high incidence of disease on pearl millet in Nigeria could be due to infected seed. Hydathodes are the possible portals of entry into host plants by the bacteria.

Distribution. Nigeria, but it is thought to be present in most areas of Africa where pearl millet is grown.

Symptoms. Water-soaking normally occurs at the tips of diseased leaves, to where the ends of interveinal lesions (several centimeters to more than 25 cm long) have elongated. Older lesions are a uniform light brown. Lesions are similar to those of yellow leaf blotch, but lesions of yellow leaf blotch are reported to not be limited by veins.

Management. Not reported.

Bacterial Spot

Cause. *Pseudomonas syringae.* The gram-negative bacterium produces fluorescing, opaque, colorless, smooth-margined, domed colonies. Disease appears after hard rains.

Distribution. The United States.

Symptoms. Water-soaked spots on diseased leaves expand to form oval to elongated, tan, necrotic lesions with a thin dark brown margin.

Management. Not reported.

Bacterial Stripe

Cause. *Pseudomonas syringae* pv. *panici.*

Distribution. The United States on proso millet.

Symptoms. Stripes on diseased leaves initially are water-soaked and later become brown. Dried scales of bacterial exudate may be present on the lesion surface.

Management. Not reported.

Pantoea agglomerans Leaf Spot

Cause. *Pantoea agglomerans* (syn. *Erwinia herbicola*).

Distribution. India and Zimbabwe.

Symptoms. Initially, water-soaking occurs at the leaf tips and margins. About a day later, necrotic lesions that are surrounded by chlorotic halos occur at the leaf tips and margins. Straw-colored lesions with a chlorotic edge often extend the length of the diseased leaf.

Management. Not reported.

Yellow Leaf Blotch

Cause. *Pseudomonas* sp. is seed-transmitted.

Distribution. West Africa.

Symptoms. Symptoms occur at the five-leaf stage of plant growth as large (30 × 150 mm), cream yellow or light tan lesions that elongate to the tips of diseased leaves. Young plants that become infected in the two- to three-leaf stage are sometimes severely stunted or killed, but plants that become infected later in the growing season are not severely stunted. Maturing diseased plants have large lesions on young leaves but outgrow the disease.

Management. Not reported.

Diseases Caused by Fungi

Bipolaris Leaf Spot

Cause. *Bipolaris setariae. Drechslera setariae* is considered a synonym by Farr et al. (1989), but since the two fungi are reported to cause separate disease syndromes by some researchers, *D. setariae* is considered as a separate pathogen.

    Disease becomes severe during humid conditions with intermittent rains and temperatures of 20° to 25°C. Disease symptoms in the form of lesions are fewer and smaller during dry, hot weather.

Distribution. India.

Symptoms. *Bipolaris setariae* causes elliptical, brown spots with dark brown centers that are often surrounded by a pale yellow halo.

Management. Grow resistant cultivars.

Blast

It has been suggested "Pyricularia leaf spot" is a more appropriate name, since millet is not "blasted" in the same way that foliage is severely diseased on rice.

Cause. *Pyricularia setariae.* Some researchers consider *P. grisea* to be the proper name for this organism.

Distribution. India on finger millet, or ragi.

Symptoms. Small, brown, circular to elongated spots, which occur on diseased leaves and leaf sheaths, eventually develop into large elongated or spindle-shaped areas that coalesce to cover most of the leaf blade. The centers of spots are grayish and have irregular red-brown margins. During humid conditions, olive gray fungal growth occurs in the middle of spots on the upper diseased leaf surfaces.

Elongated, irregular-shaped patches occur just below the ear head, resulting in a condition known as rotten neck. The main stalk below the lowest spike becomes discolored brown to black and sunken for 25 to 51 mm. An olive gray fungal growth may be seen in the center of these spots. Kernels may not develop or may be light in weight and "chaffy." A seedling blight phase also occurs and causes heavy losses.

Management. Apply foliar fungicides before disease becomes severe.

Brown Leaf Spot

Cause. *Bipolaris urochloae* (syn. *Drechslera urochloae* and *Helminthosporium urochloae*).

Distribution. Zimbabwe.

Symptoms. Leaf spots are extremely variable in shape and size. Initially, spots develop in various ways: as small dots, brown flecks, fine streaks, or small oval to large rectangular or spindle-shaped lesions (2–15 × 2–25 mm). Under humid conditions, spots develop brown margins and coalesce to cover large areas of the leaf and leaf sheath. The centers of the spots are covered with a dense, grayish growth of mycelium, conidia, and conidiophores of the fungus. At the latter stages of disease, the affected tissues die and severely diseased plants fail to produce heads.

Management. Some genotypes may have degrees of resistance.

Cercospora Leaf Spot

Cause. *Cercospora penniseti.* This fungus may be synonymous with *C. apii.*

Distribution. The United States (Georgia).

Symptoms. Symptoms develop late in the growing season, usually in September or October. Spots on diseased leaves are usually oval (0.8–2.5 × 1.0–8.0 mm), but occasionally may be oblong to rectangular-shaped, and have pale tan to gray or chalky white centers that are dotted with rows of black conidiophore tufts and surrounded by a dark brown margin. In moist weather, spots are covered with silvery layers of conidia.

Spots similar to those on leaves sometimes occur on stems (0.5–2.0 × 1.0–5.0 mm). Sometimes dark brown runners extend longitudinally for 3 mm from the ends of these spots.

Management. Some cultivars appear to have more resistance than others. Management is usually not necessary since the disease occurs too late in the season to cause much damage to foliage. Disease management may be desirable if millet is grown for seed.

Curvularia Leaf Spot

Curvularia leaf spot is also called leaf mold.

Cause. *Curvularia penniseti*. Other Curvularia species reported to cause leaf spots are *C. affinis, C. geniculata, C. lunata,* and *C. maculans.* It is not certain if fungi are primary pathogens or secondary organisms. While no pathogenicity tests have been made, the organisms are often the only fungi associated with various spots and blotches.

Distribution. India and the United States.

Symptoms. Oval to oblong, brown spots with yellow margins occur on diseased leaves. Heads are also diseased.

Management. Not necessary.

Dactuliophora Leaf Spot

Dactuliophora leaf spot is also called circular leaf spot.

Cause. *Dactuliophora elongata* presumably overseasons as sclerotia in infested residue.

Distribution. Higher rainfall areas of western Africa.

Symptoms. Diseased leaves have circular lesions, sometimes with zonate circles. The inside of the spot is frequently speckled with black sclerotia.

Management. Not reported.

Downy Mildew (*Plasmopara penniseti*)

Cause. *Plasmopara penniseti* survives as oospores in soil and as mycelia and zoosporangia in seed. During wet soil conditions in the spring, oospores germinate to produce zoosporangia in which zoospores are formed. Eventually, zoospores are released from the zoosporangia and "swim" to roots, where they encyst and germinate to form a germ tube that infects the plant host.

Distribution. Presumed to be worldwide but most common in the temperate millet-growing regions.

Symptoms. During wet weather, downy, whitish, fungal growths consisting of zoosporangia and mycelium occur on the undersides of diseased leaves, opposite yellow areas on the topsides of these leaves. As diseased plants continue to grow, leaves become wrinkled and distorted and the entire plant is stunted.

Management. Not reported.

Downy Mildew (*Sclerospora graminicola*)

This disease is also called green ear disease. Symptoms are similar to those caused by *Sclerospora macrospora* and confusion regarding these two diseases appears in the literature. Singh (1995) describes *S. graminicola* as causing two types of symptoms, downy mildew and green ear, and implies the downy mildews caused by *S. macrospora* and *S. graminicola* are the same disease. However, Farr et al. (1989) do not list *S. graminicola*. Both symptoms, as described by Singh, will be discussed here.

> **Cause.** *Sclerospora graminicola* produces asexual spores in the form of sporangia, in which zoospores are formed, and sexual spores called oospores. Oospores are thick-walled resting spores produced in diseased leaves. *Sclerospora graminicola* has been reported to survive for several years in soil as oospores and as oospores on seed. However, Singh and King (1988) reported isolates from different hosts are highly host-specific and considerable variability occurs in *S. graminicola*. Therefore, when a host is withdrawn, the population of the pathogen dies out.
>
> *Sclerospora graminicola* is disseminated by seedborne oospores and by any means that moves oospore-infested residue or soil. Oospores germinate to produce one to many germ tubes that directly penetrate meristematic tissue of young plants.
>
> Sporangia (conidia) are produced from 0100 to 0400 hours during the night on the ends of conidiophores that have emerged through stomatal openings at temperatures of 20° to 25°C and a relative humidity of 95% to 100%. Sporangia are disseminated by wind to other hosts and germinate either directly by germ tubes or, more commonly, by releasing one to 12 zoospores that first encyst, then germinate to produce a germ tube that infects meristematic tissue. Zoospores retain their infectivity for about 4 hours at 30°C and for a longer time at lower temperatures.
>
> Downy mildew is most severe under moist conditions.

> **Distribution.** Wherever millet is grown. Although downy mildew s a serious disease in Africa and India, it is found only on the weed *Setaria viridis* in the United States. Downy mildew is considered the most important disease of pearl millet in India and in the Sahelian and sub-Sahelian zones of Africa.

> **Symptoms.** Downy mildew symptoms may appear on the first leaf but normally occur on the second and third leaves in the form of chlorosis of the leaf lamina, beginning at the base of the diseased leaf. Chlorosis progresses to successively higher leaves on the plant and covers the entire lamina on the third or fourth leaf. A downy, grayish growth consisting of sporangia

and conidiophores develops on diseased tissue during wet or humid weather, generally on the abaxial leaf surface. As diseased plants approach maturity, leaves become brown, necrotic, and either split or shred.

Another characteristic symptom is the half-leaf symptom, which is a distinct margin between the diseased area toward the basal portion of the leaf and the nondiseased area toward the leaf tip. Because of the systemic nature of the disease, after leaf symptoms develop, all the subsequent leaves and the panicle have symptoms except in the case called recover resistance, where plants outgrow the disease.

Green ear symptoms appear on panicles due to the transformation of floral parts into leaf structures, which have no kernel development. Sometimes symptoms may appear as local lesions on the leaves.

Symptomless shoots have been reported to develop from plants that showed typical systemic symptoms of downy mildew. Injection of the growing points of the symptomless shoots with a sporangial suspension or clipping the stems and shoots of recovered plants growing under high disease pressure generally does not result in subsequent disease development. Lines produced from recovered plants maintained their resistance at three locations in India. This phenomenon was also observed in Mali and Niger.

Severely diseased plants remain stunted and do not produce panicles. Plants are dwarfed, have excessive tillering from the crown, and develop axillary buds along the culm.

Management
1. Treat seed with a seed-protectant fungicide.
2. Control weeds that may serve as alternate hosts.
3. Grow the most resistant cultivar. Recovery resistance may be an effective defense mechanism that could be exploited in breeding pearl millet for resistance to downy mildew.
4. Practice good fertilizer management. Phosphorous as superphosphate (16% P) decreased severity of downy mildew in field trials.

Downy Mildew (*Sclerospora macrospora*)

Downy mildew is sometimes called green ear.

Cause. *Sclerospora macrospora* (syn. *Sclerophthora macrospora*) survives as oospores in soil and infested residue. Oospores germinate to produce mycelia, which may directly infect the host plant. More commonly, sporangia are produced in which zoospores are formed. Zoospores are liberated from sporangia and "swim" to meristematic tissue, where they encyst and germinate to form a germ tube that penetrates the host plant. Secondary infection occurs by production of sporangia borne on the ends of sporangiophores that have emerged through stomates. Sporangia are disseminated by wind to host plants, where zoospores are again released and germinate to form germ tubes that penetrate meristematic tissue.

Oospores are eventually formed in diseased tissue. The disease is most severe under wet conditions.

Distribution. Africa and India.

Symptoms. Diseased plants are stunted, have shortened internodes, and are chlorotic or pale green. There is a proliferation of the first glumes of the lower spikelets. Ultimately, the whole inflorescence becomes involved, giving the diseased plant a bush-like appearance.

Management. Grow resistant cultivars.

Drechslera Leaf Spot

Cause. *Drechslera dematioidea* (syn. *Helminthosporium dematioidea*). Overwintering is likely as mycelia in infested residue. Conidia are disseminated by wind from residue to plant hosts.

Distribution. Africa, temperate Europe, and southern North America.

Symptoms. Symptoms on diseased leaves are primarily lesions that are elliptical, yellowish to grayish or tan, and several centimeters long. The centers of lesions are usually grayish to tan with a tan margin. In warm, humid weather, sporulation gives lesions a dark gray appearance. During humid conditions, lesions enlarge and kill large parts of diseased leaves.

Management. Not reported.

Ergot

Cause. *Claviceps fusiformis* overseasons as sclerotia in the soil and in stored grain. Sclerotia germinate, coinciding with flowering, and produce one to 16 stipes, which terminate in a globular capitulum in which perithecia are produced. Ascospores are produced in perithecia and infect inflorescences through young, fresh stigmas. Infection is prevented or reduced after the stigmas wither either subsequent to pollination or during aging. Honeydew containing macroconidia is produced in 4 to 6 days. The macroconidia germinate and produce microconidia in 2 to 3 days.

Several insects function in the secondary spread of the fungus by being attracted to the honeydew and becoming contaminated with conidia. Included are *Aphis indica, Dysdercus ungulatus, Monomorium salomonis, Musca domestica, Syrphus confractor, Tabanus rubidus, Vespa orientalis,* and *V. tropica. Aphis indica* and *M. domestica* are the most efficient disseminators.

Infection is favored by 16 to 24 hours of panicle wetness and moderate temperatures. Minimum temperature is more critical for infection than maximum temperature. In susceptible genotypes, temperatures in the range of 14° to 35°C with 8 hours/day at less than 20°C and 4.6 hours/day at greater than 30°C are favorable for infection.

Cytoplasmic male sterility is associated with increased susceptibility.

Distribution. India on pearl millet. Ergot also occurs throughout Africa, but it is not considered a serious disease.

Symptoms. Symptoms are typical of ergot found on other plants. Hard, black sclerotia, or ergot bodies, replace grain kernels. Sclerotia vary in shape from elongated to round (1.3–1.8 × 3.6–6.1 mm), are light to dark brown, and are hard to brittle. Sweet, sticky honeydew containing conidia is produced on diseased inflorescences and becomes black from adhering dust and attracts insects.

Management. Grow resistant cultivars.

Exserohilum Leaf Blight

Exserohilum leaf blight is also called zonate eyespot.

Cause. *Exserohilum rostratum* (syn. *Helminthosporium rostratum*) survives as mycelium in infested residue and is seedborne. Conidia production from mycelium is optimum during wet conditions at a temperature of 30°C; conidia are disseminated by wind to plant hosts. Conidia produced in leaf spots during warm, wet conditions provide secondary inoculum.

Distribution. The United States (Georgia) on pearl millet. Exserohilum leaf blight is considered to be of minor importance.

Symptoms. Lesions on diseased leaves are small (1–2 × 2–5 mm) and limited laterally by veins. Lesions may later coalesce and form large, necrotic areas that extend across veins. Lesions are dark brown at first but later become light brown, particularly on older leaves. The necrotic center of all lesions may bleach to a straw color with a slight browning at the edges and some yellowing at the lesion ends. *Exserohilum rostratum* has also been found associated with head mold.

Management. Not necessary, as Exserohilum leaf blight occurs infrequently.

False Mildew

Cause. *Beniowskia sphaeroidea.* Disease is favored by humid weather.

Distribution. Malawi and Uganda.

Symptoms. Typical white growth consisting of conidia, conidiophores, and mycelium occurs on diseased leaves.

Management. Not reported.

Fusarium nygamai Disease

Cause. *Fusarium nygamai. Fusarium nygamai* is considered very similar to *F. moniliforme.* Polyphialides and chlamydospores are produced in cultures that are 21 days or older.

Distribution. Southern Africa. However, *F. nygamai* likely occurs in other areas where millet is grown.

Symptoms. The mycotoxin moniliformin is produced. *Fusarium moniliforme* produces moniliformin in small amounts or not at all.

Management. Not reported.

Head Mold

Cause. Several fungi, including *Alternaria* spp., *Aspergillus flavus*, *Bipolaris setariae*, *B. stenospila* (syn. *Helminthosporium stenospilum*), *Curvularia lunata*, *Drechslera dematioidea*, *Epicoccum* spp., *Exserohilum rostratum* (sometimes incorrectly spelled rostrata), *Fusarium moniliforme*, *Gloeocercospora sorghi*, *Olpitrichum tenellum*, and *Phyllosticta penicillariae*. Other *Fusarium* species reported to cause head mold are *F. chlamydosporum*, *F. equiseti*, *F. napiforme*, *F. nygamai*, and *F. pallidoroseum* (syn. *F. semitectum*). Infection is favored by high rainfall and low maximum temperatures.

    Fusarium moniliforme infects mature millet; the disease severity is increased by dew or other moisture. Other pathogens infect in a similar manner, with disease incidence increasing more on late-sown millet than on early-sown millet. Head mold is most severe following insect injury and under humid, wet weather conditions.

Distribution. Africa, India, and the United States.

Symptoms. *Oidium tenellum* causes a mealy, white coating on diseased heads. Under magnification, the coating appears as white hairs with white beads at their tips.

    Helminthosporium spp. and *C. lunata* cause a black woolly mat to grow over heads. *Fusarium moniliforme* initially causes a whitish mycelial growth on spikelets of immature heads that later turns pink or causes individual grains to be covered with an orange crust. Moldy spikelets are easily shaken off. Seed quality and germination are greatly reduced.

    Aspergillus flavus causes a typical brown-green discoloration of seed. The aflatoxins deoxynivalenol, nivalenol, zearalenone, and 15-acetylscirpentriol have been detected in pearl millet.

    Incidence of *F. chlamydosporum* has been positively correlated with concentrations of trichothecenes and zearalenone.

Management. No acceptable management is known except timely harvest of millet grown for grain. It may be possible to avoid disease in wet millet-growing areas by timing sowing so heads mature after heavy rains have stopped.

Head Smuts

Cause. *Sporisorium destruens* (syn. *Sphacelotheca destruens*) is a pathogen of proso millet. *Ustilago trichophora* (syn. *U. crus-galli*) is a pathogen of Japanese barnyard millet. Both *S. destruens* and *U. trichophora* are smut fungi that survive in soil as chlamydospores.

Distribution. Wherever millet is grown.

Symptoms. *Sporisorium destruens* completely destroys the inflorescence except for the vascular strands. The sorus is covered with a gray mycelium that does not persist long before exposing red-brown chlamydospores that become windborne.

Ustilago trichophora forms sori in gall-like swellings on nodes and flowers that remain covered by host tissue. When exposed, chlamydospores are yellow to olive brown.

Management
1. Treat seed with a seed-protectant fungicide.
2. Rotate millet with other crops.

Helminthosporiosis

Helminthosporiosis is also called blight.

Cause. *Bipolaris nodulsa* (syn. *Helminthosporium nodulosum*) survives in infested soil residue and is seedborne. The fungus is present in the pericarp and endosperm but not in the embryo. Helminthosporiosis is most severe at temperatures of 30° to 32°C and high moisture.

Distribution. Africa and India on finger millet, or ragi.

Symptoms. Roots, culms, leaves, heads, and inflorescences are diseased. Numerous small, oval, brown spots commonly occur on upper leaf surfaces and gradually elongate to an approximate size of 1×10 mm. Other descriptions note the spots are pale yellow to black.

Spots on leaf sheaths are larger than the spots on diseased leaves, are dark brown to nearly black, and usually occur at the junction of the leaf blade and sheath. Lesions on culms and internodes are similar to those on leaf sheaths. Diseased heads dry up, especially at the rachis.

Diseased seed is discolored gray or black. Preemergence damping off may occur with seeds and seedlings. Most infected seed does not germinate, and the fungus sporulates heavily on ungerminated seeds and dead seedlings. In less severe disease instances, rot develops as a browning of the coleoptile and slowly progresses upward. The main roots of such seedlings are discolored and there is a tinge of brown discoloration on root hairs. Emerged seedlings have discolored, dull brown coleoptiles with some etiolation at the tip. With time, browning and etiolation of the coleoptile increases and seedlings collapse within 9 to 12 days. Yield losses as high as 50% have been reported.

Management
1. Grow the most resistant cultivar.
2. Treat seed with a seed-protectant fungicide.

Helminthosporium Leaf Spot

Helminthosporium leaf spot is also called Bipolaris leaf spot.

Cause. *Bipolaris stenospila* (syn. *Helminthosporium stenospilum*), *B. sacchari* (syn. *H. saachari*), and *B. setariae* (syn. *H. setariae*). Optimum disease development caused by *B. setariae* is at temperatures of 15° to 21°C.

Distribution. The United States (Georgia) on pearl millet.

Symptoms. Disease symptoms usually develop extensively during August through October after plants head. Lesions on diseased leaves will vary in size and shape from brown flecks, fine linear streaks, and small oval spots to large, irregular-shaped, oval, oblong, or almost rectangular spots (0.5–3.0 ×1.0–10.0 mm). Sometimes large, elongated lesions are produced. Spots may expand and coalesce and form long streaks (2.5–5.0 × 20.0–90.0 mm). Lesions may be dark brown initially but later become tan or gray-brown with a less distinct dark brown border. During moist weather, a dense gray-brown growth consisting of mycelium, conidiophores, and conidia grows in lesions. A black conidial and mycelial growth may also occur over individual grains in the head.

Management. Some cultivars are more resistance than others; however, management is usually not necessary since disease occurs too late in the season to cause much damage to foliage production. Management may be desirable if millet is grown for seed.

Kernel Smuts

Cause. *Ustilago crameri* and *Sporisorium neglectum* (syn. *U. neglecta*). Both *U. crameri* and *S. neglectum* survive in dry soil as chlamydospores. Chlamydospores germinate and produce a basidium on which either lateral branches or sporidia, both of which infect seedlings, are formed.

Distribution. Africa, Asia, and the United States. *Ustilago crameri* is a pathogen of foxtail millet.

Symptoms. Sori resemble enlarged kernels and are enclosed in floral bracts. The floral bracts rupture during the growing season or at harvest time and release yellow-brown to olive green chlamydospores of *U. crameri* or purple-brown chlamydospores of *S. neglectum*.

Management
1. Treat seed with a seed-protectant fungicide.
2. Rotate millet with other crops.

Leaf Spot (*Drechslera setariae*)

Cause. *Drechslera setariae* infects only areas of leaves infected by *Sclerospora graminicola*.

Distribution. India.

Symptoms. Lesions vary in size and have ash-colored centers with dark margins. Sometimes under severe disease conditions, diseased leaves are completely blighted.

Management. Not reported.

Leaf Spot (*Helminthosporium frumentacei*)

Cause. *Helminthosporium frumentacei*.

Distribution. India on Japanese millet.

Symptoms. Initially, numerous, yellowish spots (1–2 mm in diameter) occur on both sides of diseased leaves and on leaf sheaths. Spots gradually increase in size and have a light brown center with a yellow margin that becomes dark brown as the spots mature. Spots elongate or coalesce and form long stripes.

Management. Not reported.

Long Smut

Long smut is also called pearl millet smut and pearl millet kernel smut. Much confusion surrounds the etiology of this disease and the proper name(s) of the causal organism.

Cause. *Moesziomyces penicillariae* (syn. *Tolyposporium penicillariae*). *Ustilago penniseti* has also been reported to be a synonym. *Moesziomyces penicillariae* survives as chlamydospores in soil. Chlamydospores are the only source of inoculum and germinate during humid conditions and produce a basidium on which sporidia bud, germinate, and infect host plants through the flowers. Flowers are most susceptible at an early stage of development, before the stigma or anthers are visible. After pollination, flowers are virtually immune from infection. Later in the growing season, chlamydospores are formed on grain instead. The chlamydospores are released and fall to the soil when their enclosing membrane ruptures.

Distribution. Africa and Asia on pearl millet.

Symptoms. The membrane enclosing the chlamydospores and the chlamydospores themselves is collectively called a sorus. The sorus is brown to black, is somewhat pear-shaped, and protrudes from the floral bracts of the diseased plant. Upon rupturing of the sori membranes, the exposed chlamydospores are green-brown.

Management. In experimental plots, applying foliar fungicides shows promise of partial control.

Myrothecium Leaf Spot

Cause. *Myrothecium roridum.*

Distribution. Worldwide, but most common in warmer areas.

Symptoms. Necrotic spots occur on diseased leaves. Sometimes the centers of the spots fall out and give leaves a shot-hole appearance.

Management. Not reported.

Phyllosticta Leaf Blight

Cause. *Phyllosticta penicillariae* survives as pycnidia in infested residue and is seedborne. Dissemination is by any method that moves residue and by airborne conidia.

Distribution. Africa and the United States (Georgia).

Symptoms. Diseased plants are stunted and chlorotic and have numerous leaf lesions that are approximately 2.5–5.0 mm in size. Lesions are parallel-sided and have light brown necrotic tissue in the center and dark brown margins. Coalesced lesions usually result in tattered leaf tissue. Leaf margins are frequently necrotic. Pycnidia, which average 75.8 µm in diameter, frequently can be observed in the necrotic tissue. Husks also become diseased.

Management. Not reported.

Pyricularia Leaf Spot

Pyricularia leaf spot is also called blast and may be the same disease as blast. Some researchers consider *Pyricularia grisea* and *P. setariae* to be synonymous.

Cause. *Pyricularia grisea* is seedborne in finger millet. The fungus is present in the pericarp and endosperm but not in the embryo. Disease is favored by wet conditions or extended periods of high relative humidity.

Distribution. Australia, India, Kenya, Russia, Uganda, and the United States (Georgia).

Symptoms. Numerous small, brown flecks enlarge and become brown, spindle-shaped spots on diseased leaves. Later, ash gray, water-soaked spots with brown margins develop. Spots increase in size (0.5–1.5 × 3.0–8.0 mm), become necrotic and eventually coalesce and give leaves a blasted appearance. Severe disease is usually accompanied by extensive chlorosis.

Culm nodes blacken, become fragile, and may lodge at these weakened areas. Brown to black spots appear at the panicle base, enlarge, and often girdle the neck below the panicle.

The diseased neck becomes shriveled and covered with mycelium, conidiophores, and conidia. Abundant conidia and conidiophores are also formed on nodes and necks in wet weather. When neck blast is severe, panicles fail to emerge and spikelets become diseased and blackened. When neck infection occurs early in the growth of the host plant, the grains do not fill and panicles remain erect. Later infection in the life of the plant results in lodging of partially filled panicles.

Diseased seed is discolored gray or black. Most diseased seed does not germinate and the fungus sporulates heavily on ungerminated seeds and dead seedlings. Less severe disease results in a rot that develops as a browning of the coleoptile and slowly progresses upward. The main roots of such seedlings are discolored and there is a light brown discoloration of the root hairs.

Emerged seedlings have discolored, dull brown coleoptiles with some etiolation at the tip. With time, browning and etiolation of the coleoptile increases and seedlings collapse within 9 to 12 days. Yield losses as high as 50% have been reported.

Management
1. Apply foliar fungicide before disease becomes severe.
2. Treat seed with fungicide seed treatments.

Rhizoctonia Blight

Cause. *Rhizoctonia solani* and *R. zeae* survive in soil and infested residue as sclerotia. Sclerotia are disseminated by wind, splashing water, or any means that transports soil or residue. When sclerotia are deposited on a host plant, they germinate and form infective hyphae that penetrate the plant. Disease is more severe on millet types with a low, dense growth habit, which creates high humidity conditions that are more conducive to disease development.

Distribution. The United States (Georgia).

Symptoms. Lesions are irregular-shaped blotches (4–7 cm long) that occur mostly on diseased leaf sheaths and the bases of leaf blades. Lesions are water-soaked at first but eventually become light tan and have an indistinct brown border. Diffuse white to tan mycelia grow over the surfaces of lesions and adjacent green tissue. Older lesions may be covered with tiny, brown sclerotia that are usually associated with *R. solani*. *Rhizoctonia zeae* is distinguished by tiny, globular, red sclerotia that may occur on the same lesions and be mixed together with *R. solani*.

Apparently little damage is caused except that an occasional tiller may be killed. Rarely, an entire plant may be so severely blighted that it has the appearance of being scalded.

Management. Not necessary.

Rust (*Puccinia substriata* var. *indica*)

Cause. *Puccinia substriata* var. *indica* (syn. *P. penniseti*) is a macrocyclic heteroecious rust that overwinters as teliospores in the United States and has eggplant, *Solanum melongena*, as an alternate host. Other aecial hosts include *S. anguivi, S. ferox, S. gilo, S. incanum, S. linaeanum, S. nodiflorum,* and *S. rostratum. Solanum americanum* and *S. aviculare* are resistant. Uredinia and telia are produced on millet; pycnia are produced on the upper leaf surfaces of eggplant and aecia on the lower leaf surfaces of eggplant.

Optimal temperatures for urediniospore germination on artificial media is 19° to 22°C for incubation periods of 6 to 7 hours. Continuous light for 2 hours incubation delayed germination, but exposure to light during the first hour of incubation followed by 1 hour of dark is stimulatory.

At least 11 races of *P. substriata* var. *indica* have been identified.

Distribution. India and the United States on pearl millet. Rust is a late-season disease in the southeastern United States.

Symptoms. Uredinia occur as orange-brown pustules that rupture the leaf epidermis. Uredinia occur on both leaf surfaces but are more abundant on the upper surfaces of diseased leaves.

Management
1. Grow resistant cultivars.
2. Remove eggplants from vicinity of millet.

Rust (*Puccinia substriata* var. *pennicillariae*)

Cause. *Puccinia substriata* var. *pennicillariae* (syn. *P. penniseti*). The aecial stage is not known.

Distribution. Africa.

Symptoms. Typical brownish uredinia occur on both leaf surfaces but are most abundant on the upper surfaces of diseased leaves.

Management. Not reported.

Smut

Cause. *Melanopsichium eleusinis* (syn. *Ustilago eleusinis*). Chlamydospores are disseminated by wind to flowers, where infection occurs. *Melanopsichium eleusinis* is not seedborne.

Distribution. India.

Symptoms. After flowering, symptoms are evident on scattered diseased kernels throughout the ear. Diseased grains are transformed into galls that are five to six times the size of a normal grain. Initially, diseased grains, which are slightly greenish and 2 to 3 mm in diameter, project slightly beyond the glumes. Eventually, galls reach a diameter up to 16 mm and turn pinkish green and have cracks in the outer wall of the sorus, which is the membrane plus the enclosed chlamydospores. The sorus eventually ruptures and the chlamydospores are disseminated by wind to developing flowers or fall to the soil surface.

Management

1. Some cultivars are apparently more resistant than others.
2. Rotate millet with other crops.

Other Smuts Reported on Millet

Sorosporium paspali var. *verrucosum*
Tilletia barclayana (syn. *Neovossia barclayana*)
Tilletia ajrekari
Tolyposporium senegalense
Ustilago paradoxa

Southern Blight

Southern blight is also called foot rot, and southern blight and wilt.

Cause. *Sclerotium rolfsii* overwinters as sclerotia in soil and as mycelia in infested residue and seeds. Mycelia from germinated sclerotia, residue, or seed serve as the primary inoculum. High soil temperatures and a low soil pH of 3 to 6 favor disease development. High relative humidity stimulates aerial growth of mycelia on stems, branches, and leaves. Cool temperatures in winter kill sclerotia and mycelia. The disease is reported to be most severe under wet soil conditions, although on other crops southern blight is more severe under relatively dry soil conditions.

Distribution. Africa and India on finger millet, or ragi.

Symptoms. Symptoms develop when diseased plants are 6 to 8 weeks old. Single tillers and occasionally an entire plant may be killed. Diseased plants are stunted and chlorotic and have a poorly developed root system. A brown to black discoloration is present on crowns, and the root cortex is rotted. Later, the basal portion of stems becomes discolored brown and the cortical tissue disintegrates and rots, causing plant death. White mycelia, in which black sclerotia eventually develop, grow on crowns and lower stems.

Aboveground symptoms are white mycelial mats that grow upward inside leaf sheaths for as much as 30 cm above the soil line. Tissue beneath mats becomes discolored and is eventually killed and shredded. Sclerotia form inside the leaf sheaths and around the base of the diseased plant.

Management. Grow the most resistant cultivar.

Top Rot

Top rot is also called pokkah boeng.

Cause. *Fusarium subglutinans* (syn. *F. moniliforme* var. *subglutinans*). *Fusarium subglutinans* overwinters as mycelia and perithecia in infested residue and is seedborne as mycelia and conidia. Conidia are produced from mycelia growing on residue during wet weather and disseminated by wind or splashed by rain to plant hosts, where they lodge in the leaf sheaths or whorls and germinate. During wet weather, *F. subglutinans* also grows upward on the outside of stalks. Metabolites produced by *F. subglutinans* cause distortions in plants.

Distribution. Tropical or semitropical millet-growing areas. Top rot is not a common disease of millet.

Symptoms. Mild symptoms on diseased leaves may appear as a mosaic that somewhat resembles a virus symptom. Leaves near the top of diseased plants are deformed and discolored and have wrinkled bases and numerous transverse cuts in the leaf margin. Leaves may be so wrinkled that they do not unfold properly. In severe cases, *F. subglutinans* may grow from whorls and sheaths into stalks, causing the tops of plants to die.

Management. No management measures are necessary.

Zonate Leaf Spot

Cause. *Gloeocercospora sorghi* overwinters as sclerotia in infested soil residue. Most primary inoculum presumably comes from residue, However, *G. sorghi* is also seedborne as mycelia and possibly sclerotia that contaminate seed, thus affording another potential source of primary inoculum. In the spring, sclerotia germinate and produce conidia, which are disseminated by splashing rain or wind to millet. During warm, wet weather, sporulation occurs within lesions in the form of pink sporodochia that form above the leaf stomates. Conidia produced in sporodochia are borne in a pinkish, gelatinous mass and serve as secondary inoculum by being splashed by rain to nearby plant hosts. Sclerotia are eventually formed in lesions.

Distribution. The United States (Georgia). Zonate leaf spot is of minor importance.

Symptoms. Lesions on diseased leaves initially appear as water-soaked spots (2.0–3.5 × 3.5–5.0 mm) that eventually develop tan centers with dark brown borders. Spots enlarge to form roughly semicircular blotches that may cover half or more of the leaf width. Blotches are different shades of dark brown mottled with light tan spots that tend to arrange in circles. This gives the appearance of incomplete tan rings alternating with dark brown rings. This "zonation" pattern is often absent on narrow-leafed strains of millet. During moist weather, tiny, salmon-colored globules consisting of masses of spores are visible under magnification on both surfaces of spots. As leaves die, lenticular black sclerotia appear in the dead tissue.

Management. Not necessary.

Diseases Caused by Nematodes

Burrowing

Cause. *Radopholus similis* is an endoparasite that survives for a few months outside of host tissue. Some dissemination occurs by water. Feeding occurs in the cortex and reproduction occurs in the roots.

Distribution. Tropical and subtropical areas around the world.

Symptoms. The aboveground symptom is a slight stunting of diseased plants. Belowground symptoms are root lesions and decay.

Management. Not reported.

Cyst

Cause. *Heterodera gambiensis.*

Distribution. Gambia and Niger.

Symptoms. Small, whitish to light yellow sacs of immature females can be observed without magnification on the outside of diseased roots. Plant growth is stunted and unthrifty.

Management. Not practiced.

Dagger

Cause. *Xiphinema americanum* is an ectoparasite.

Distribution. Not known.

Symptoms. The aboveground symptom is a slight to moderate stunting of the entire diseased plant. Belowground symptoms are a reduction in feeder roots accompanied by necrosis of portions of the root system.

Management. Not reported.

Lance

Cause. *Hoplolaimus indicus* is an endoparasite that sometimes feeds as a semi-endoparasite and an ectoparasite.

Distribution. Not known.

Symptoms. Aboveground symptoms are stunting and chlorosis of the entire diseased plant. The belowground symptom is lesions that occur on roots and reduce root growth.

Management. Not reported.

Ring

Cause. *Criconemella ornata* is an ectoparasite.

Distribution. The southeastern United States.

Symptoms. The aboveground symptom is a mild stunting of the whole diseased plant. Belowground symptoms are lesions on roots and some root decay.

Management. Not reported.

Root Knot

Cause. *Meloidogyne incognita, M. incognita,* and *M. javonica* do not survive if the temperature averages below 3°C during the coldest month.

Distribution. Not known for certain.

Symptoms. Diseased plants are stunted and chlorotic and have reduced yields. Roots have galls, elongated swellings, and discrete knots or swellings. Root proliferation is common.

Management. Not reported.

Root Lesion

Cause. *Pratylenchus brachyurus, P. mulchandi,* and *P. zeae* are endoparasites.

Distribution. Worldwide.

Symptoms. Plants are stunted and chlorotic. Roots are poorly developed and have necrotic lesions.

Management. Not reported.

Sting

Cause. *Belonolaimus longicaudatus* is an ectoparasite found only in sandy soils. It produces a phytotoxic enzyme.

Distribution. Not known for certain.

Symptoms. Aboveground symptoms are stunting and chlorosis of the entire diseased plant. Belowground symptoms include deep necrotic lesions on the roots. Root tips are frequently destroyed, resulting in thick, stubby roots.

Management. Not reported.

Stubby Root

Cause. *Paratrichodorus minor* is an ectoparasite found primarily in sandy soils. Most nematode feeding is around the root tips.

Distribution. Not known.

Symptoms. Aboveground symptoms include severe stunting and chlorosis of the entire diseased plant. Roots are stubby and sometimes appear thicker than healthy roots.

Management. Not reported.

Stunt

Cause. *Tylenchorhynchus phaseoli, T. vulgaris,* and *T. zeae* are ectoparasites.

Distribution. Not reported.

Symptoms. Diseased plants may be stunted and chlorotic. The root system is poorly developed and has fewer feeder roots than healthy plants. Numerous root tips may be short and thickened.

Management. Not reported.

Diseases Caused by Viruses

Bajra (Pearl Millet) Streak

Cause. Bajra streak virus (BSV) is transmitted by the leafhopper *Cicadulina mbila*. BSV is not mechanically transmitted. BSV is considered to be a strain of maize streak virus. Plants may become infected at all stages of growth. Maize, sorghum, finger millet, barley, oat, and wheat are also infected.

Distribution. India.

Symptoms. Long, parallel, chlorotic streaks develop the entire length of diseased leaves. Plants infected early in their growth are stunted and produce empty heads.

Management. Not reported.

Guinea Grass Mosaic

Cause. Guinea grass mosaic virus is transmitted by several species of aphids. Millet was inoculated.

Distribution. Ivory Coast.

Symptoms. Leaves had a slight green mosaic, and diseased plants were slightly stunted.

Management. Not reported.

Maize Dwarf Mosaic

Cause. Maize dwarf mosaic virus (MDMV) now consists of former MDMV strains A, D, E, and F; former sugarcane mosaic virus strain J infective to johnsongrass in the United States; and maize mosaic virus, European type (Et), from Yugoslavia.

MDMV persists in several annual and perennial grasses. In the spring, aphids, particularly the corn leaf aphid, *Rhopalosiphus maidis,* and the greenbug *Schizaphis graminum,* feed on overwintering hosts and acquire the virus.

Distribution. Potentially, wherever millet is grown.

Symptoms. Symptoms of MDMV infection can vary according to virus strain and temperature. Symptoms are most evident on the upper two or three leaves as an irregular mottling of dark and light green areas interspersed with longitudinal white or light yellow streaks.

Management. Control grassy weeds, particularly johnsongrass, in a field.

Maize Streak

Cause. Maize streak virus (MSV) is in the geminivirus group. MSV is transmitted mainly by the leafhopper *Cicadulina mbila*, but other leafhoppers are also vectors. MSV is not seedborne or mechanically transmitted in maize.

Distribution. Kenya, Nigeria, and South Africa.

Symptoms. Chlorotic streaks and stripes occur on diseased leaves.

Management. Not reported.

Panicum Mosaic

Cause. Panicum mosaic virus (PMV) is in the sobemovirus group. PMV is mechanically transmitted. The grasses *Panicum* spp.; crabgrass, *Digitaria sanguinalis*; foxtail millet, *Setaria italica*; and barnyard grass, *Echinochloa crusgalli* are also susceptible. Different strains of PMV are present.

Distribution. The United States.

Symptoms. Diseased plants may be stunted and chlorotic. Other symptoms include a yellow-green mosaic and mottling of leaves and stunted, deformed, or blasted heads. In severe disease cases, plants may die. Some cultivars of millet may be symptomless carriers.

Management. Grow resistant cultivars. Generally, vigorous-growing types are more susceptible and slow-growing plants with smaller leaves are more resistant.

Wheat Streak Mosaic

Cause. Wheat streak mosaic virus (WSMV). WSMV is vectored by the wheat curl mite, *Aceria tosichella* (syn. *A. tulipae*).

Distribution. The United States (Kansas).

Symptoms. Symptoms on diseased leaves range from a mosaic with interveinal reddening to a mosaic with interveinal reddening and necrosis within the red areas.

Management. Not reported.

Diseases Caused by Unknown Factors

Brown Leaf Mottle

Cause. Unknown.

Distribution. Unknown.

Symptoms. Initially, a faint brown, irregular mottle of the leaf and sheath appears at about the time of rapid stem elongation of the affected plant. Symptoms become more pronounced as plants mature. Affected tissue eventually becomes necrotic before normal senescence occurs, and leaves on affected plants senesce sooner than healthy plants.

Management. Not reported.

Red Leaf Spot

Cause. Unknown.

Distribution. Unknown.

Symptoms. Symptoms first appear about the time of flower initiation. Irregular or zonate, water-soaked blotches appearing on leaves soon become a mahogany color. As affected plants mature, the discolored areas coalesce and become necrotic.

Management. Not reported.

Selected References

Burton, G. W., and Wells, H. D. 1981. Use of near-isogenic host populations to estimate the effect of three foliage diseases on pearl millet forage yield. Phytopathology 71:331-333.

Claflin, L. E., Ramundo, B. A., Leach, J. E., and Erinle, I. D. 1989. *Pseudomonas avenae,* causal agent of bacterial leaf stripe of pearl millet. Plant Dis. 73:1010-1014.

Desaum S. G., Thirumalachar, M. J., and Patel, M. K. 1965. Bacterial blight disease of *Eleusine coracana* Gaertn. Indian Phytopathol. 18:384-386.

Deshmukh, S. S., Mayee, C. D., and Kulkarni, B. S. 1978. Reduction of downy mildew of pearl millet with fertilizer management. Phytopathology 68:1350-1353.

Farr, D. F., Bills, G. R., Chamuris, G. P., and Rossman, A. Y. 1989. Fungi on Plants and Plant Products in the United States. The American Phytopathological Society, St. Paul, MN. 1252 pp.

Frederickson, D. E., Monyo, E. S., King, S. B., and Odvody, G. N. 1997. A disease of pearl millet in Zimbabwe caused by *Pantoea agglomerans.* Plant Dis. 81:959.

Govindu, H. C., Shivanandappa, N., and Renfro, B. L. 1970. Observations on diseases of *Eleusine coracana* with special reference to resistance to the Helminthosporium disease, pp. 415-424. *In* Plant Disease Problems. International Symposium on Plant Pathology. Indian Phytopathology Society, New Delhi.

Jensen, S. G., Lambrecht, P., Odvody, G. N., and Vidaver, A. K. 1991. A leaf spot of pearl millet caused by *Pseudomonas syringae.* (Abstr.) Phytopathology 81:1193.

Keshi, K. C., and Mohanty, N. N. 1970. Efficacy of different fungicides and antibiotics on control of blast of ragi, pp. 425-429. *In* Plant Disease Problems. International Symposium on Plant Pathology. Indian Phytopathology Society, New Delhi.

Lee, T. A., Jr., and Toler, R. W. 1977. Resistance and susceptibility to Panicum mosaic virus–St. Augustine decline strain in millets. Plant Dis. Rptr. 61:60-62.

Luttrell, E. S. 1954. Diseases of pearl millet in Georgia. Plant Dis. Reptr. 38:507-514

Marasas, W. F. O., Rabie, C. J., Lubben, A., Nelson, P. E., Toussoun, T. A., and Van Wyk, P. S. 1987. *Fusarium napiforme,* a new species from millet and sorghum in southern Africa. Mycologia 79:910-914.

Marasas, W. F. O., Rabie, C. J., Lubben, A., Nelson, P. E., Toussoun, T. A., and Van Wyk, P. S. 1988. *Fusarium nygamai* from millet in southern Africa. Mycologia 80:263-266.

Mehta, P. R., Singh, B., and Mathur, S. C. 1952. A new leaf spot of bajra (*Pennisetum typhoides* Stapf. & Hubbard) caused by a species of *Piricularia.* Indian Phytopathology 5:140-143.

Misra, A. P. 1959. Diseases of millets and maize. Indian Agriculturist 3:75-89.

Mohan, L., Sivaprakasam, K., and Jeyarajan, R. 1988. A new leaf blight disease of pearl millet caused by *Exserohilum rostratum* Leonard and Suggs. Madras Agric. J. 75:414-415.

Ohobela, M., and Claflin, L. E. 1987. Identification of the causal agent of bacterial leaf streak of pearl millet (*Pennisetum americanum* (L.) Leeke). (Abstr.) Phytopathology 77:1767.

Onesirosan, P. T. 1975. Head mold of pearl millet in southern Nigeria. Plant Dis. Rptr. 59:336-337.

Onyike, N. B., and Nelson, P. E. 1987. Distribution of *Fusarium* species on millet from Nigeria, Lesotho and Zimbabwe. (Abstr.) Phytopathology 77:1617.

Pande, S., Mukuru, S. Z., Odhiambo, R. O., and Karunakar, R. I. 1994. Seedborne infection of *Eleusine coracana* by *Bipolaris nodulosa* and *Pyricularia grisea* in Uganda and Kenya. Plant Dis. 78:60-63.

Pathak, V. N., and Gaur, S. C. 1975. Chemical control of pearl millet smut. Plant Dis. Rptr. 59:537-538.

Ramakrishnan, R. S., and Soumini, C. K. 1948. Studies in cereal rust. I. *Puccinia penniseti* Zimm and its alternate host. Indian Phytopathol. 1:97-103.

Rangawami, G., Prasad, N. N., and Eswaran, K. S. S. 1961. Bacterial leaf spot diseases of *Eleusine coracana* and *Setaria italica* in Madras State. Indian Phytopathol. 14:105-107.

Safeeulla, K. M. 1970. Studies on the downy mildews of bajra, sorghum and ragi, pp. 405-414. *In* Plant Disease Problems. International Symposium on Plant Pathology. Indian Phytopathology Society, New Delhi.

Seifers, D. L., Harvey, T. L., Kofoid, K. D., and Stegmeier, W. D. 1996. Natural infection of pearl millet and sorghum by wheat streak mosaic virus in Kansas. Plant Dis. 80:179-185.

Seth, M. L., Raychaudhuri, S. P., and Singh, D. V. 1972. Bajra (pearl millet) streak: A leafhopper-borne cereal virus in India. Plant Dis. Rptr. 56:424-428.

Sharma, S. B., Waliyar, F., and Ndunguru, B. J. 1990. First report of *Heterodera gambiensis* on pearl millet in Niger. Plant Dis. 74:938.

Singh, S. D. 1990. Sources of resistance to downy mildew and rust in pearl millet. Plant Dis. 74:871-874.

Singh, S. D. 1995. Downy mildew of pearl millet. Plant Dis. 79:545-550

Singh, S. D., and King, S. B. 1988. Recovery resistance to downy mildew in pearl millet. Plant Dis. 72:425-428.

Singh, S. D, and Williams, R. J. 1980. The role of sporangia in the epidemiology of pearl millet downy mildew. Phytopathology 70:1187-1190.

Singh, S. D., de Milliano, W. A. J., Mtisi, E., and Chingombe, P. 1990. Brown leaf spot of pearl millet caused by *Bipolaris urochloae* in Zimbabwe. Plant Dis. 74:931-932.

Sundaram, N. V., Palmer, L. T., Nagarajan, K., and Prescott, J. M. 1972. Disease survey of sorghum and millets in India. Plant Dis. Rptr. 56:740-743.

Swift, C. E., Ishimaru, C. A., and Brown, W. M., Jr. 1991. Bacterial streak of millet caused by *Xanthomonas* spp. in Colorado. (Abstr.) Phytopathology 81:1159.

Tapsoba, H., and Wilson, J. P. 1997. Effects of temperature and light on germination of urediniospores of the pearl millet rust pathogen, *Puccinia substriata* var. *indica*. Plant Dis. 81:1049-1052.

Thakur, R. P., and Williams, R. J. 1980. Pollination effects on pearl millet ergot. Phytopathology 70:80-84.

Thakur, R. P., Ras, V. P., and Williams, R. J. 1984. The morphology and disease cycle of ergot caused by *Claviceps fusiformis* in pearl millet. Phytopathology 70:201-205.

Thakur, R. P., Rao, V. P., and King, S. B. 1989. Ergot susceptibility in relation to cytoplasmic male sterility in pearl millet. Plant Dis. 73:676-678.

Thakur, R. P., Rao, V. P., and King, S. B. 1991. Influence of temperature and wetness duration on infection of pearl millet by *Claviceps fusiformis*. Phytopathology 81:835-838.

Thirumalachar, M. J., and Mundkur, B. B. 1947. Morphology and the mode of transmission of ragi smut. Phytopathology 37:481-486.

Verma, O. P., and Pathak, V. N. 1984. Role of insects in secondary spread of pearl millet ergot. Phytophylactica 16:257-258.

Weber, G. F. 1973. Bacterial and Fungal Diseases of Plants in the Tropics. University of Florida Press, Gainesville, FL.

Wells, H. D. 1967. Effects of temperature on *Helminthosporium setariae* on seedlings of pearl millet, *Pennisetum typhoides*. Phytopathology 57:1002.

Wells, H. D. 1978. Eggplant may provide primary inoculum for rust of pearl millet caused by *Puccinia substriata* var. *indica*. Plant Dis. Rptr. 62:469-470.

Wells, H. D., Burton, G. W., and Hennen, J. F. 1973. *Puccinia substriata* var. *indica* on pearl millet in the Southeast. Plant Dis. Rptr. 57:262.

Wilson, J. P. 1994. Field and greenhouse evaluations of pearl millet for partial resistance to *Puccinia substriata* var. *indica*. Plant Dis 78:1202-1205.

Wilson, J. P., and Burton, G. W. 1990. *Phyllosticta penicillariae* on pearl millet in the United States. Plant Dis. 74:331.

Wilson, J. P., and Hanna, W. W. 1991. Fungal flora of pearl millet grain as affected by date of planting. (Abstr.) Phytopathology 81:1200.

Wilson, J. P., Hanna, W. W., Wilson, D. M., Beaver, R. W., and Casper, H. H. 1993. Fungal and mycotoxin contamination of pearl millet grain in response to environmental conditions in Georgia. Plant Dis 77:121-124.

Wilson, J. P., Phatak, S. C., and Lovell, G. 1996. Aecial host range of *Puccinia substriata* var. *indica*. Plant Dis. 80:806-808.

Young, G. V., Lefebvre, C. L., and Johnson, A. G. 1947. *Helminthosporium rostratum* on corn, sorghum, and pearl millet. Phytopathology 47:180-183.

Zummo, N. 1976. Yellow leaf blotch: A new bacterial disease of sorghum, maize, and millet in West Africa. Plant Dis. Rptr. 60:798-799.

11. Diseases of Oat *(Avena sativa)*

Diseases Caused by Bacteria

Bacterial Blight (*Acidovorax avenae* subsp. *avenae*)

Cause. *Acidovorax avenae* subsp. *avenae* (syn. *Pseudomonas avenae*). Infection and disease development is most severe at high temperatures.

Distribution. The southeastern United States.

Symptoms. Lesion shape varies from circular to linear. Lesions are water-soaked, then dry up and become brown to black.

Management. Sow later in the autumn when temperatures are cool.

Bacterial Blight (*Pseudomonas syringae* pv. *coronafaciens*)

Bacterial blight is also called halo blight and blade blight.

Cause. *Pseudomonas syringae* pv. *coronafaciens* (syn. *P. coronafaciens*) can survive for 2 or more years on seed and infested soil residue. During wet weather in the spring, bacteria are splashed by rain onto the first leaves, where they enter the plant through wounds or natural openings, such as stomates. The bacteria are spread in cool, wet weather by plants rubbing against each other during a stormy or windy period. Bacteria are also spread by insects, particularly aphids that feed on diseased tissue and inadvertently carry bacteria to a plant host. Warm, dry weather stops spread of the bacteria and subsequent disease development.

Distribution. Australia, Europe, North America, and South America.

Symptoms. Symptoms of halo blight occur earlier in the season than any diseases caused by leaf-spotting fungi. Symptoms are first noticeable on leaves as small, light green, oval to oblong, water-soaked spots with slightly sunken centers. The spots eventually change color from light green to yellow to light brown and the tissue around the spots forms a water-soaked and light yellow halo. Due to an increased number of spots that eventually coalesce, severely diseased leaves turn brown and die back from the tip. Portions of leaves of highly susceptible cultivars may die when lesions are not present.

Management
1. Grow resistant cultivars.
2. Treat seed with a seed-protectant fungicide.
3. Plow under residue to reduce inoculum.

Bacterial Stripe Blight

Bacterial stripe blight is also called stripe blight.

Cause. *Pseudomonas syringae* pv. *striafaciens* (syn. *P. striafaciens*) survives on seed and in infested residue for at least 2 years. During cool, wet weather, bacteria are disseminated by splashing rain and by wind to leaves. Warm, dry weather stops the spread of bacteria and subsequent disease development.

Distribution. Australia, Europe, North America, and South America.

Symptoms. Bacterial stripe blight is seldom a serious disease. Symptoms first appear on leaves as sunken, water-soaked dots. If dots are abundant, the leaf may die. Dots enlarge into water-soaked stripes or blotches, often with narrow yellowish margins, that may extend the length of the diseased leaf blade. As stripes age, they become a translucent rusty brown. In moist weather, bacteria are exuded in droplets from the stripes; the droplets of bacteria later dry and form white scales on diseased leaves.

Management
1. Grow resistant cultivars if they are available.
2. Plow under residue.
3. Treat seed with a seed-protectant fungicide.

Black Chaff

Cause. *Xanthomonas campestris* pv. *translucens* (syn. *X. campestris* and *X. translucens* pv. *undulosa*). The taxonomy of the causal organism(s) continues to be in flux. Black chaff is considered to be caused by five of the former *X. campestris* pathovars (pv. *cerealis*, pv. *hordei*, pv. *secalis*, pv. *translucens*, and pv. *undulosa)* that are often grouped together under the name "*translucens* group" or are considered to be a single species, *X. translucens*, or a pathovar named *X. campestris* pv. *translucens*. Bragard et al. (1997) agreed that *X. translucens* pv. *translucens* and *X. translucens* pv. *hordei* are true synonyms. They also reported that when using restriction fragment length polymorphism and fatty acid methyl esters analysis, the pathovars *X. translucens* pv. *cerealis, X. translucens* pv. *translucens,* and *X. translucens* pv. *undulosa* cluster in different groups and correspond to true biological entities.

The bacteria overwinter in soil residue, on seed, and directly in soil. Primary infection occurs by bacteria being disseminated from soil, host plants, and infested residue by splashing water and insects, particularly aphids. Primary infection may also occur by seedborne bacteria, but seed stored 6 or more months is not considered an important source of inoculum. Bacteria enter into plants through natural openings such as stomata and wounds. Since free water is necessary for bacteria to enter plants, small spaces and grooves that contain water may act as a reservoir for bacteria. Secondary spread of bacteria occurs by plant-to-plant contact, splashing rain, and sucking and chewing insects.

On other small grains, black chaff is called bacterial blight, bacterial streak, bacterial leaf streak, or black chaff.

Distribution. Wherever oat is grown.

Symptoms. Symptoms of black chaff, which occur after several days of damp or rainy weather, are more obvious on leaves than on heads. The inner portions of the glumes have brown or black spots. In moist weather, tiny, yellow beads of bacteria ooze to the surface of these discolored areas.

Small water-soaked spots occur on tender green leaves, on leaf sheaths of older plants, and sometimes on seedlings. The spots enlarge and coalesce, becoming glossy, olive green, translucent stripes, or streaks, of various lengths that later turn yellow-brown. Stripes may extend the length of a diseased leaf and are usually narrow, being limited by leaf veins. Occasionally, a spot becomes large and blotch-like, causing the leaf to shrivel, turn light brown, and die. Severely diseased leaves die back from the tips. Under humid conditions early in the morning, droplets of bacterial exudate may be seen on the surface of spots. Droplets dry into hard, yellowish granules that may be easily removed as dry flakes from leaf surfaces. Grain is not destroyed, but it may be brown and shrunken and may carry bacteria to infect next year's crop.

Management
1. Do not sow seed from diseased plants. Seed should be cleaned to remove lightweight, infected kernels.
2. Treat seed with a fungicide seed protectant.
3. Rotate oat with other crops, preferably noncereals.

Diseases Caused by Fungi

Anthracnose

Cause. *Colletotrichum graminicola,* teleomorph *Glomerella graminicola,* overwinters as mycelium and spores on infested residue, but the fungus is not seedborne. Optimum production of conidia from mycelia occurs during wet weather at a temperature of 25°C; conidia are disseminated by wind to plant hosts.

Distribution. Wherever oat is grown, but anthracnose is most severe when oat is grown on sandy soils that are low in fertility.

Symptoms. Red-brown, elongated, lens-shaped spots are produced on diseased leaves, especially where they are attached to the leaf sheath. After the death of leaf tissue, acervuli appear as dark brown objects scattered throughout the spots. During severe disease, most of the diseased leaf tissue is covered with acervuli. Diseased basal and crown tissue becomes bleached, then turns brown and is covered with acervuli. Under magnification, acervuli appear "hairy" or have a pincushion appearance. Roots and stems are also infected. Infection of panicles results in lightweight, shriveled grain. Later in the season, severely diseased plants are small and have few tillers.

Management
1. Ensure soils have adequate fertility.
2. Rotate oat with legumes.

Cephalosporium Stripe

Cause. *Hymenula cerealis* (syn. *Cephalosporium gramineum*) survives for up to 5 years as conidia and mycelia within infested residue in the top 8 cm of soil. Conidia serve as primary inoculum and infect roots through mechanical injuries caused by soil heaving and insects during the winter and early spring. Infection is most severe in wet, acid soils (pH 5.0) due to a combination of increased fungal sporulation and root growth. Subcrown internodes can also be infected as seed germinates. After infection, conidia enter xylem vessels and are carried upward, where they lodge and multiply at nodes and leaves. No further damage is done to roots. *Hymenula cerealis* prevents water movement up plants and also produces metabolites that are harmful to plants. At harvest time, *H. cerealis* is returned to the soil in infested residue, where it is a successful saprophytic competitor with other soilborne microorganisms.

Distribution. Great Britain, Japan, and most oat-growing areas of North America.

Symptoms. Oat is ordinarily not severely diseased. Diseased plants are scattered throughout a field but are most numerous in the lower and wetter areas of the field.

Diseased plants are dwarfed. At jointing and heading time, one to four—but normally one or two—distinct yellow stripes develop the length of the plant on leaf blades, sheaths, and stems. Thin brown lines consisting of infected veins occur in the middle of the stripes. The stripes eventually become brown and are highly visible on green leaves. The stripes continue to remain noticeable, even on yellow straw. Toward harvest, culms at or below nodes become darkly discolored due, in part, to fungal sporulation and discoloration of diseased tissue. Heads of diseased plants are white and contain shriveled seed or no seed at all.

Management
1. Rotate oat with a noncereal for at least 2 years.
2. Plow residue deeper than 8 cm.

Covered Smut

Cause. *Ustilago segetum* (syn. *U. kolleri*) is seedborne and seed is the only source of inoculum. After infected seed is sown, mycelium from the seed grows into the developing oat shoots. Plants cease to be susceptible to infection when the first seedling leaves have emerged more than 1 cm beyond the leaf sheaths. Once inside the seedlings, mycelium grows systemically, keeping pace with the growing tip of the plant and finally entering the young, developing kernels. By heading time, the kernels and hulls are completely replaced with chlamydospores.

Chlamydospores are scattered by wind and harvesting operations. Some chlamydospores lodge either outside or inside the hulls of healthy-appearing seed, where they remain dormant until the seed is sown. Other chlamydospores germinate immediately after being disseminated by wind to healthy kernels. The infective mycelia then grow into hulls or seed coats, where they remain inactive until the infected seed is sown.

Soil temperature and moisture at the time of seed germination have an influence on infection. Infection occurs at 5% to 60% soil moisture (35%–40% is optimum) and at soil temperatures of 5° to 30°C (15°–25°C is optimum).

Distribution. Wherever oat is grown.

Symptoms. All spikelets and panicles on a diseased plant are usually smutted. Smutted panicles do not spread as much as healthy panicles. As a smutted oat panicle emerges from its enclosing sheath, a dark brown to black, powdery mass of chlamydospores enclosed within a white-gray membrane has replaced the grains and, often, the awns and glumes. The white-gray membrane is more persistent than that of loose smut and remains intact until the chaff dries or until grain is harvested. The persistence and the extent of damage to panicles vary with the oat cultivar. Smutted plants are generally shorter than healthy plants.

Management
1. Grow resistant cultivars.
2. Treat seed with a systemic fungicide seed protectant.

Crown Rust

Crown rust is also called oat leaf rust.

Cause. *Puccinia coronata* is a heteroecious long-cycled rust fungus that has the buckthorn species *Rhamnus carthartica, R. dahurica,* and *R. lanceolata* as alternate hosts to oat and other grasses.

Puccinia coronata causes crown rust of oat by following either one of two distinct life cycles. Depending on weather conditions, urediniospores are produced in uredinia on oat every 8 to 10 days and are disseminated by wind to other oat plants. The urediniospores germinate in the presence of moisture and infect leaves and less frequently leaf sheaths, stems, and panicles. This cycle will continue indefinitely as long as susceptible oat plants are available.

Urediniospores produced in northern oat-producing areas are blown south during the summer and autumn and infect oat and grasses in Mexico and the southern United States. Urediniospores continue to recycle on oat in these southern oat-producing areas until they are disseminated north and infect oat during late winter and spring.

A second, but complete, life cycle that utilizes both plant hosts occurs where species of buckthorn are present in climates with low temperatures. As oat ripens and during periods of drought, excessive moisture, or high temperatures, telia containing teliospores form in or around uredinia. Teliospores, which function as the means of survival through the winter, germinate in the spring and form basidiospores that are disseminated by wind to buckthorn, where the leaves are infected. At the point of basidiospore infection, a pycnium forms on the upper leaf surface.

By exchanging genetic material between different fungal strains, pycnia are the means by which new races of crown rust arise. Each pycnium produces pycniospores and special mycelia called receptive hyphae. Pycniospores are exuded from pycnia in a thick, sticky, sweet liquid that is attractive to insects. Pycniospores are then splashed by water or carried by insects from one pycnium to another, where they become attached to receptive hyphae. The nucleus from the pycniospore enters the receptive hypha, which results in the formation of an aecium on the lower leaf surface.

Aeciospores differ genetically from both pycniospores and receptive hyphae. The aeciospores are disseminated by wind to oat, where infection occurs. Each infection gives rise to a uredinium in which urediniospores are again formed. The urediniospores serve as the repeating stage, reinfecting only oat.

The production of urediniospores is optimum at a temperature of 21°C in the presence of dew, fog, or some other form of moisture. The higher relative survival ability of some races on *Avena sterilis* is due to a shorter time between inoculation and uredinia appearance, higher infection density, higher yield of urediniospores per uredinium, and extended longevity of urediniospores. Telia are again produced as oats mature, thus completing the life cycle. Numerous physiologic races exist.

Experimentally, oat becomes resistant when incubated at 35°C for 8 hours at the vegetative stage of fungal development. The effect of this temperature stress disappears by cutting the stem prior to the heat treatment, which suggests the effect of heat treatment on disease is through its effect on the host.

Distribution. Wherever oat is grown.

Symptoms. Uredinia appear primarily on leaves as bright orange-yellow, round to oblong pustules. There is no loose epidermis around pustules, which distinguishes crown rust from stem rust. Similar pustules may occur on sheaths, stems, and panicles. Pustules burst open and release urediniospores that look like orange powder. Under conditions favorable for infection, uredinia become numerous and may coalesce. Stems may be so weakened that severe lodging results.

As plants begin to ripen, telia form either in a ring around old uredinia or develop independently of uredinia. Telia remain covered indefinitely by epidermis, giving them a gray-black, oblong, and slightly raised appearance.

Pycnia on buckthorn leaves are bright orange to yellow spots that sometimes have a small drop of liquid consisting of pycniospores in the middle. Opposite pycnial spots on the underleaf surface are aecia that resemble raised, orange cluster cups.

Management
1. Grow resistant cultivars.
2. Plant oat early in spring to escape rust.
3. Eradicate buckthorn shrubs within 1.5 km of oat fields.

Downy Mildew

Cause. *Sclerophthora macrospora* (syn. *Sclerospora macrospora*) survives as oospores in infested leaf and stem residue in soil. When residue decays, oospores are liberated into soil, where they may persist for long periods of time. Oospores are disseminated by seed and within soil and residue by wind or water. Oospores germinate in saturated soil and form sporangia in which zoospores are formed. Zoospores are released from sporangia and "swim" for a short distance to a host, where they encyst, then germinate and form a germ tube that penetrates the host plant. Disease is most severe during wet, cool environmental conditions.

Distribution. Wherever oat is grown in wet areas.

Symptoms. Symptoms are most noticeable within a field at heading time in areas that are low or liable to flood. Diseased plants are stiff, upright, and remain green for several days after healthy plants have ripened. Severely diseased plants are dwarfed to less than a third the height of healthy plants and tiller excessively. Leaves are leathery, and the upper leaves are curled around panicles which, in turn, are also curled and twisted into a cluster of frayed and tangled spikelets.

Less severely diseased plants may be slightly dwarfed and stiff and may produce distorted heads that stand above those of healthy plants. The flag leaf is curled and twisted about a poorly developed head. Heads are deformed and spikelets are not recognizable since viable seed is not produced.

Management. Drain areas in fields that are likely to be flooded.

Ergot

Cause. *Claviceps purpurea* survives as sclerotia buried in soil or on the soil surface. Sclerotia germinate before flowering time and produce stromatic heads in which perithecia are produced at the end of a stipe or stalk. Ascospores form in perithecia and are disseminated by wind to young flowers, where they germinate and form infective hyphae that penetrate ovaries. Mycelium in an ovary produces conidia in a sweet, sticky, liquid called honeydew on the outside of infected flowers. Insects are attracted to honeydew and inadvertently disseminate conidia from diseased to healthy flowers. Eventually, sclerotia replace oat kernels.

Distribution. Wherever oat is grown, but ergot is relatively uncommon on oat.

Symptoms. Ergot is easily recognized by the purple-black, horn-shaped sclerotia produced in place of one or several kernels on a head. Sticky honeydew covered with dust and foreign matter gives diseased flowers a grayish to black appearance.

Management
1. Place sclerotia at a soil depth where ascospores cannot be released into the air. Sclerotia are decomposed in a year or so by soil microorganisms.
2. Control grassy weeds around fields.
3. Rotate with crops, such as legumes, that are not susceptible to *C. purpurea.*
4. Clean seed to remove ergot bodies.

Eyespot

Eyespot is also called Cercosporella foot rot, stem break, foot rot, strawbreaker, and culm rot.

Cause. *Pseudocercosporella herpotrichoides* survives as mycelium in infested soil residue for several years in the absence of grain crops. Conidia produced during cool, damp weather in autumn or spring infect crown and basal culm tissue. Roots are not infected.

Distribution. Ireland. Reports from Europe and the United States indicate that oat is resistant.

Symptoms. Characteristic eyespot lesions are found on oat but are not as common as they are on barley, rye, and wheat. Eyespot is most conspicuous near the end of the growing season when diseased plants lodge. Lodging caused by eyespot causes straw to fall in all directions. In contrast, lodging caused by wind or rain causes straw to fall in one direction. White heads, similar to symptoms produced in take-all, are also present at maturity.

Eye-shaped or ovate lesions with white to tan centers and brown margins develop first on the basal leaf sheaths. Similar spots form on stems adjacent to sheaths and eventually cause stems to lodge. Roots are not infected, but necrosis occurs around roots in the upper crown nodes. Under moist conditions, lesions enlarge and the black stroma-like mycelium that develops over the surface of diseased crowns and the base of culms gives them a "charred" appearance. Stems then shrivel and collapse. Plants that do not collapse are yellowish or pale green and heads are reduced in size and number.

Oat often has an indentation on one or both sides of culms where infection has occurred that extends up from the plant base for 2.5 to 5.0 cm and sometimes causes a split in the culm. *Pseudocercosporella herpotrichoides* is often visible when infected stems are cut longitudinally. Oat is usually not stunted but plants that are pulled up break off above the soil line.

Management. Rotate with noncereals.

Fusarium Foot Rot

Cause. *Fusarium culmorum* survives in soil as chlamydospores or as mycelia in infested residue.

Distribution. Wherever oat is grown.

Symptoms. Seedlings are infected below or at the soil surface. Light brown to red-brown lesions may be noticed at the base of the seedling leaf or just below the soil surface on the stem. Some seedlings may also have these spots on the mesocotyl near the seed. Seedlings may damp off either before or after emergence.

Management
1. Treat seed with a seed-protectant fungicide.
2. Plow under infested residue.

Fusarium Head Blight

Fusarium head blight is also known as scab, Fusarium blight, and head blight.

Cause. *Gibberella zeae,* anamorph *Fusarium graminearum.* Although not well documented, the *Fusarium* spp. that cause Fusarium head blight of oat are probably the same as those that cause Fusarium head blight of wheat.

Gibberella zeae is seedborne and overwinters as mycelia in infested residue and as perithecia on oat, barley, maize, wheat, and other grass residue that was diseased the previous season. Ascospores produced in perithecia during warm, moist weather are disseminated by wind to plant hosts. Conidia of *Fusarium* spp. are produced from mycelia and also are disseminated by wind to hosts. Spores, primarily conidia, are also produced on diseased heads and serve as a source of secondary inoculum to cause new infections. Wet autumn weather favors disease development on lodged grain and on grain in swaths.

Distribution. Wherever oat is grown.

Symptoms. Fusarium head blight on oat is not as readily detected as it is on wheat or barley because a hull covers the oat kernel. Hulls of infected spikelets are ashen gray and may be partially covered by a pink growth consisting of mycelium and conidia. In severe cases, hulls are shriveled and rough but do not turn brown. Small, black perithecia may be produced on hulls as grain nears maturity.

If seed is heavily diseased, seedlings may be killed before emergence. Seedlings frequently are only stunted, but many become chlorotic and die after emergence. Roots of infected seedlings are red-brown and rotted. Later in the season, roots may be covered by pink mycelium. If a joint on the stem becomes infected, the plant portion above it becomes white and perithecia are produced on the affected joint.

Management
1. Treat seed with a seed-protectant fungicide.
2. Plow under maize residue when oat follows maize. Fusarium head blight is more severe when maize stalks are left on the soil surface than when they are plowed into the soil.
3. Rotate oat with legumes rather than cereals. Oat is the least susceptible of the cereals.

Head Blight

Head blight is also called Helminthosporium foot rot and Helminthosporium culm rot.

Cause. *Bipolaris sorokiniana* (syn. *Drechslera sorokiniana, Helminthosporium sativum,* and *H. sorokinianum*) overwinters as mycelia and conidia in infested residue and seed.

Distribution. Wherever oat is grown.

Symptoms. Roots and basal portion of young diseased seedlings are rotted. Leaves droop and wilt or whole leaves turn chlorotic and die. Disease of older plants appears as a blackening and weakening of the lower stems. Frequently, diseased stems bend over and lodge on the soil surface but the upper portion of the stems stays alive for some time after falling over. Often the live tips of fallen stems turn and grow upward.

Management
1. Treat seed with a seed-protectant fungicide.
2. Rotate oat with legumes.

Helminthosporium Seedling Blight, Crown, and Lower Stem Rot

Cause. *Drechslera avenacea* (syn. *Helminthosporium avenaceum*) and *D. avenae* (syn. *H. avenae*), teleomorph *Pyrenophora avenae*. *Drechslera* spp. overwinter as mycelium under the seed coat, spores on seed, and mycelia or perithecia in infested residue. Seedborne infection results in spots on seedling leaves in which conidia are produced during wet weather and disseminated by wind to other leaves. Infections of older leaves may originate either from conidia produced on residue or primary spots, or from ascospores produced in perithecia on residue during wet weather later in the growing season. Disease development is favored by cool, wet weather.

Distribution. Wherever oat is grown.

Symptoms. Oat seedlings are stunted. Diseased leaves initially turn yellow, then reddish, and eventually brown as severely diseased plants die. Dark brown to black lesions develop on lower stems and leaf sheaths. Disease initiated as seedling infections can continue as plants develop, or infections of roots, crowns, and lower stems of older plants may occur that are characterized by dark brown areas of decay near the oat crown. Severely diseased plants are unproductive, are more subject to lodging than healthy plants, and may die.

Management
1. Sow high-quality seed that was properly harvested. Delayed harvest can result in increased *D. avenacea* damage to kernels.
2. Treat seed with a seed-protectant fungicide.
3. Rotate oat with other crops.

Leaf Blotch and Crown Rot

Leaf blotch and crown rot is also called blotch, Helminthosporium leaf blotch, Helminthosporium leaf spot, leaf blotch, leaf streak, oat leaf blotch, oat leaf spot, and Pyrenophora leaf blotch.

Cause. *Drechslera avenacea* (syn. *Helminthosporium avenaceum*) and *D. avenae* (syn. *H. avenae*), teleomorph *Pyrenophora avenae. Drechslera* spp. overwinter as mycelia under the seed coat, as spores on seed, and as mycelia or perithecia in residue. Seedborne infection results in spots on seedling leaves in which conidia are produced during wet weather and disseminated by wind to other leaves. Infections of older leaves may originate either from conidia produced on residue or primary spots, or from ascospores produced in perithecia on residue during wet weather later in the growing season. Disease development is favored by cool, wet weather.

Distribution. Africa, Asia, Australia, Europe, and North America.

Symptoms. In the seedborne phase of the disease, oblong to elongate, light red-brown spots appear on seedling leaves soon after emergence. Seedling leaves may also be twisted or contorted.

On older leaves, lesions start as small, brown flecks that develop into longitudinal strips of dead tissue. The brown to yellow or reddish outer edges of spots merge and frequently spread over the greater part of a diseased leaf blade. Sometimes, well-defined spots are not produced but leaves will have a withered appearance as if they had been injured by adverse weather. As leaves die, they change from green to yellow or gray and the brown spots fade. When kernels are diseased, they will turn brown at the basal end.

Management
1. Treat seed with a seed-protectant fungicide.
2. Rotate oat with other crops.
3. Plow under residue to reduce inoculum originating from the field.
4. Grow the least susceptible cultivars. Several lines of spring oat are identified as having a degree of resistance.

Loose Smut

Loose smut is also called black loose smut, black head, false loose smut, and naked smut.

Cause. *Ustilago avenae* is seedborne and seed is the only source of inoculum. After seed is sown, mycelia grow into and infect young oat shoots. Infection occurs when seeds sprout at soil moistures of 5% to 60% (35% to 40% is optimum) and soil temperatures of 5° to 30°C (15° to 25°C is optimum).

Plants cease to be susceptible to infection when the first leaves have emerged more than 1 cm beyond the sheath. Once inside a seedling, mycelium grows systemically, keeping pace with the growing tip of the infected plant and finally entering the young, developing kernels. By heading time, kernels and hulls are completely replaced by chlamydospores, which are released into the air and disseminated by wind to healthy heads before harvest. Some spores lodge outside or inside hulls of healthy seed, where they remain dormant until the seed is sown. Other spores germinate immediately and grow into the hulls or seed coats of kernels but remain inactive until the seed is sown.

Distribution. Wherever oat is grown.

Symptoms. Smutted panicles do not spread as much as healthy ones. As a smutted oat panicle emerges from its enclosing sheath, a dark brown to black powdery mass of smut spores (chlamydospores) contained within a delicate, white-gray membrane has replaced the grains and, often, the awns and glumes. All spikelets and panicles on a diseased plant are usually smutted. The thin membrane breaks and disintegrates soon after oat panicles emerge. Smutted plants are shorter than healthy ones and often are overlooked at harvest because the naked mass of chlamydospores quickly scatter by wind and rain and leave a bare panicle, which is difficult to see.

Management
1. Grow resistant cultivars.
2. Treat seed with a systemic seed-protectant fungicide.

Microdochium Root Rot

Cause. *Microdochium bolleyi* (syn. *Aureobasidium bolleyi* and *Gloeosporium bolleyi*). *Microdochium bolleyi* is primarily a saprophyte.

Distribution. The United States (California).

Symptoms. Disease symptoms are mild and consist of discoloration and necrosis on the subcrown internode.

Management. Not necessary.

Pink Snow Mold

Pink snow mold is also called snow mold.

Cause. *Microdochium nivale* (syn. *Fusarium nivale*), teleomorph *Monographella nivalis* (syn. *Calonectria nivalis*). The causal fungus oversummers as perithecia or mycelia in infested residue. Plants are infected in the autumn by ascospores disseminated by wind to host plants or by mycelia growing from previously infested residue to nearby plants. Primary infections occur on leaf sheaths and blades near the soil level. Disease is spread by mycelial growth during cool, wet weather, or by mycelial growth underneath a snow cover. Secondary infection in the spring occurs in two ways: by ascospores produced in perithecia that developed during cool, humid weather and by conidia produced on mycelia and sporodochia that grew on oat plants under the winter snow cover. Both types of spores are then disseminated by wind to other host plants.

Distribution. Canada, central and northern Europe, and the United States.

Symptoms. Diseased plants have chlorotic and dry necrotic leaves. A good diagnostic characteristic is the pink color of mycelia and sporodochia on leaf and crown tissues as snow melts in the spring.

Management. Grow resistant cultivars.

Powdery Mildew

Cause. *Blumeria graminis* f. sp. *avenae* (syn. *Erysiphe graminis* f. sp. *avenae*) overwinters as cleistothecia on wild grasses and oat residue. Ascospores form within cleistothecia in spring and are disseminated by wind to oat plants. Conidia are produced as soon as mycelia become established on leaf surfaces, especially during cool, humid weather but not in the presence of free water. Conidia account for most of the secondary inoculum and are disseminated by wind from plant to plant. Cleistothecia are formed on leaf surfaces as diseased plants approach maturity.

Distribution. Wherever oat is grown.

Symptoms. Powdery mildew first appears as patches of white, fluffy mycelium, conidia, and conidiophores on diseased lower leaves and leaf sheaths. As disease progresses, the patches become powdery and turn gray, then brown. The "powder" is a combination of mycelium, conidia, and conidiophores that has continued to grow on the leaf surface. Eventually, when large areas of the leaves are diseased, the leaves become yellow and shrivel up. As plants near maturity, cleistothecia are seen as small, brown to black, round objects interspersed within the mycelium and spores. Yield may be reduced.

Management. Management usually is not necessary; however, the following are aids for managing the disease.
1. Apply a sulfur fungicide to foliage.
2. Grow cultivars that have adult resistance.

Pythium Seed and Seedling Rot

Cause. *Pythium debaryanum, P. irregulare,* and *P. ultimum* survive as oospores in residue or soil. In moist soil, oospores germinate and form sporangia in which zoospores are produced. Zoospores then "swim" in moist soil to root tips or seeds, where they germinate and form germ tubes that infect plants.

Distribution. Wherever oat is grown.

Symptoms. Seed and seedlings may be killed before or after emergence. Plants that are not killed are chlorotic and stunted but may recover and become as green as healthy plants. However, diseased plants usually remain less vigorous than healthy plants. Water-soaked, translucent areas that occur on roots later turn red-brown.

Management. Treat seed with a seed-protectant fungicide.

Scoliocotrichum Leaf Blotch

Cause. *Cercosporidium graminis* (syn. *Scoliocotrichum graminis* var. *avenae*) presumably survives as mycelium in infested residue. During wet weather in the spring, conidia are produced and disseminated by wind to plant hosts. Sporulation ceases during dry weather.

Distribution. Generally distributed in most oat-growing areas, but Scoliocotrichum leaf blotch is uncommon.

Symptoms. Oblong to linear, red-brown to purple-brown blotches with definite margins develop on leaves. The necrotic area is dry and sunken. A good diagnostic characteristic that can be seen with the aid of magnification is the rows or tufts of conidiophores that emerge through stomata.

Management. Management is not practiced ordinarily, but sanitation will reduce inoculum.

Sharp Eyespot

Sharp eyespot is also called Rhizoctonia root rot.

Cause. *Rhizoctonia cerealis.* Survival is as sclerotia in soil and infested residue or as mycelia in infested residue of several susceptible plant species. Plants may be infected any time during the growing season. Roots and culms are infected by mycelia that grow either from germinating sclerotia or from a precolonized substrate, particularly in dry, cool soils at less than 20% moisture-holding capacity.

Distribution. Wherever oat is grown.

Symptoms. Oat is one of the most susceptible cereals to *R. cerealis.* Diseased seedlings may be killed. Diamond-shaped lesions with tan centers and dark brown margins occur on lower leaf sheaths. These lesions somewhat resemble the symptoms of eyespot disease. Dark sclerotia and mycelium may develop in lesions; sclerotia also may develop between culms and leaf sheaths.

When roots are diseased, plants may produce new roots to compensate for those that have rotted off, but plants may still lodge and produce white heads. Diseased plants generally appear stiff, have a grayish cast, and are delayed in maturity.

Management. No management measure is effective. Vigorous plants growing in well-fertilized soil are not as likely to become as severely diseased as unthrifty plants.

Speckled Blotch

Speckled blotch is also called dark stem of oat, leaf spot of oat, Septoria black stem, Septoria blight, Septoria disease, Septoria leaf blotch, and speckled leaf blotch. There is a great deal of confusion in the literature concerning the disease name. Farr et al. (1989) list a speckled blotch as being caused by both *Stagnospora avenae* and *Septoria tritici* f. *avenae,* whereas the list of diseases by the American Phytopathological Society lists only *S. avenae* as the causal fungi.

Cause. *Stagnospora avenae* (syn. *Septoria avenae*) and *Septoria tritici* f. *avenae.* Some researchers consider *Leptosphaeria avenaria* to be the teleomorph. The causal fungi overwinter as mycelia and pycnidia (micropycnidia) in infested residue; *S. tritici* f. *avenae* is also seedborne, but only in very susceptible cultivars. Conidia (microspores) are disseminated by wind to plant hosts during cool, wet weather in the spring. Perithecia are not thought to be an important means of overwintering nor are ascospores important as a means of primary inoculum or in the secondary spread of disease.

Pycnidia form in mature lesions if weather continues to be cool and wet after the initial infections. Conidia, which are disseminated mostly by rain and water, are primarily responsible for stem infections. Perithecia and ascospores form primarily in spring, but some perithecia form in the summer and overwinter on infested residue.

Distribution. Africa, Australia, Europe, and North America.

Symptoms. Blotches first appear on lower leaves and then spread up a diseased plant. Round to elongate or diamond-shaped, yellow to light or dark brown blotches are surrounded by a band of dull brown that changes to yellow as it blends into the green of healthy plant tissue. As blotches enlarge, the diseased leaf tissue dies. Pycnidia appear as black dots scattered throughout the centers of older blotches, giving them a speckled appearance. After heading, infection at a leaf base spreads into an adjoining leaf sheath, which becomes discolored chocolate or red-brown.

Above the upper two joints, gray-brown to shiny black lesions develop on the stems or on culms under diseased leaf sheaths. Dark gray mycelium fills the hollow areas of diseased culms. During cool, wet weather, black or dark brown lesions may extend to lemma and palea and eventually to the groat of the kernel. Diseased plants frequently lodge as they near maturity.

Management
1. Grow cultivars that mature late; early cultivars are more susceptible.
2. Sow certified seed.
3. Treat seed with a seed-protectant fungicide.
4. Rotate oat every 3 or 4 years.
5. Plow under infested residue.

Speckled Snow Mold

Speckled snow mold is also called gray snow mold and Typhula blight.

Cause. *Typhula idahoensis, T. incarnata* (syn. *T. itoana*), and *T. ishikariensis* overseason as sclerotia in infested residue and soil or as parasitic growth on live plants. Saprophytic growth on dead plant tissue provides an inoculum source for increased disease in the spring; however, the fungi are poor saprophytes and depend primarily on parasitism for existence. Sclerotia germinate during wet weather in autumn and form either a basidiocarp or mycelium. Optimum infection occurs at temperatures of 1° to 5°C, when basidiospores form on basidiocarps and are disseminated by wind to host plants or when mycelia directly infect plants by growing over the surface of the soil. Further infection occurs by mycelial growth under snow cover. Sclerotia are formed within necrotic tissue and in mycelia.

Distribution. Canada, central and northern Europe, Japan, and the northwestern United States.

Symptoms. At snowmelt, gray-white mycelia are present on leaves and crowns of diseased plants and on the soil. The numerous sclerotia in diseased tissues or scattered in mycelia growing over diseased plant surfaces give the plants a speckled appearance. Dead leaves are common, but unless the crown is diseased, plants recover during warm, dry weather although they will not be as vigorous as healthy plants. Dead leaves crumple easily and are covered with gray to white mycelium.

Management
1. Rotate oat with a legume to reduce the inoculum in soil since the fungi are not good saprophytes.
2. In wheat experiments disease incidence was decreased by fungicide seed treatments.

Stem Rust

Cause. *Puccinia graminis* (syn. *P. graminis* f. sp. *avenae*) is a heteroecious long-cycle rust that has the common barberry, *Berberis vulgaris,* as the alternate host to oat.

Rust is caused by *P. graminis* in one of two ways. Urediniospores, called summer or repeating spores, are produced in uredinia on oat several weeks before it ripens. The urediniospores are disseminated by wind to other oat plants, where infection and production of secondary urediniospores in new uredinia occur under moist conditions and moderate temperatures. This cycle will continue indefinitely as long as green oat plants are available. Urediniospores produced in northern oat-producing areas are blown south during the summer and autumn and infect oat in Mexico and the southern United States. Urediniospores continue to recycle on oat during the winter in southern oat-producing areas until they are disseminated by wind northward during late winter and spring to infect oat in the north.

In the second method of infection, *P. graminis* goes through a complete life cycle. Telia containing teliospores form on maturing oat. The teliospores are the means of survival through the winter. In spring, teliospores germinate and form basidiospores (sometimes called sporidia) that are disseminated by wind to barberry bushes, where infection of the foliage occurs. At the point of infection by the basidiospore, a pycnium (sometimes called a spermogonium) forms on the upper leaf surface.

Pycnia exchange genetic material between different isolates of *P. graminis,* thereby creating new races that are genetically different from the parent isolates. Each pycnium produces pycniospores and special mycelia called receptive hyphae. Pycniospores are exuded from pycnia in a thick, sticky, sweet liquid that is attractive to insects. Pycniospores are splashed by water or carried by insects from one pycnium to another, where they become attached to receptive hyphae and germinate, with the nucleus from the spore from one pycnium entering into the receptive hypha of another pycnium. This results in formation of an aecium on the lower leaf surface.

Aeciospores, which differ genetically from both pycniospores and receptive hyphae, are produced in aecia and are disseminated by wind to oat, where infection occurs. Each infection gives rise to a uredinium in which urediniospores are formed. Urediniospores then serve as the repeating stage, becoming windborne to oat, where new generations of urediniospores are formed. Telia are once again produced as oat matures, thus completing the life cycle.

Distribution. Wherever oat is grown.

Symptoms. Uredinia and telia occur on diseased stems, leaf sheaths, leaf blades, and panicles. Uredinial pustules are large, oblong, and dark red-brown. The epidermis of leaves and culms is ruptured and pushed back around the pustule, giving it a jagged appearance and exposing the urediniospores. The dark color and jagged appearance of the pustule distinguish stem rust from the relatively smooth margin and light orange-yellow color of crown rust. As plants approach maturity, black, oblong telia form in and around uredinia, particularly on stems and sheaths. Teliospores are exposed when the epidermis ruptures.

Pycnia on barberry leaves appear as bright orange to yellow spots with a small drop of moisture in the middle. Opposite pycnial spots on the under side of leaves are aecia that resemble raised, orange clusters.

Management
1. Grow resistant cultivars.
2. Eradicate common barberry shrubs from the vicinity of oat fields.

Take-All

Take-all is also called white head.

Cause. *Gaeumannomyces graminis* var. *avenae* survives as mycelia in plants or as mycelia and perithecia in infested soil residue. However, ascospores produced in perithecia are not considered important in the dissemination of *G. graminis* var. *avenae,* which is a poor saprophyte and does not survive long in soil in the absence of a host.

In the autumn or spring, seedlings are infected by roots growing into the vicinity of previously colonized residue and coming in contact with mycelia growing from the residue. Plant-to-plant spread of *G. graminis* var. *avenae* occurs by hyphae growing through soil from a diseased plant to a healthy plant or by a healthy root coming in contact with a diseased root. Ascospores are produced in perithecia during wet weather but are not disseminated a great distance by splashing water or by wind. Take-all is usually more severe on oat grown in alkaline soils.

Distribution. Wherever oat is grown.

Symptoms. The first symptoms are light to dark brown necrotic lesions on roots. By the time a diseased plant reaches the jointing stage, most of its roots are brown and dead. At this point, many plants die, but if plants are still alive, they are stunted and their leaves are chlorotic.

Disease becomes most obvious as plants approach heading. The plant stand is uneven in height and plants appear to be in several stages of maturity. At heading, roots are sparse, blackened, and brittle, and diseased plants have few tillers, ripen prematurely, and have bleached and sterile heads. When pulled from the soil, diseased plants easily break free from their crowns. A dark discoloration occurs on stems just above the soil line. Mats of dark brown mycelia grow under the lower sheaths between the stems and inner leaf sheaths.

Management. Rotate oat with a nongrass host crop.

Victoria Blight

Victoria blight is also called Helminthosporium blight.

Cause. *Bipolaris victoriae* (syn. *Helminthosporium victoriae*) overwinters as conidia on seeds and as mycelia and spores on infested residue. *Bipolaris victoriae* produces a toxin that spreads throughout a plant.

Distribution. Wherever oat cultivars with the Victoria type of crown rust resistance are grown. Victoria blight is now rare.

Symptoms. Diseased seed may rot in soil before germination. Usually, plants are infected in the seedling stage as they emerge from the soil. Diseased seedling leaves are a dull blue-gray and have red-brown stripes that may cover half a leaf blade and extend the length of a leaf. Later, seedling leaves develop a reddish color and the entire plant may die. Still later, as the lower joints and leaf sheaths become covered with conidia and conidiophores, the basal parts of stems turn a brownish color and have a dark, velvety appearance. Severe disease causes root rot and the death of stems, which results in lodging and premature ripening of plants.

Management
1. Grow oat cultivars that do not have the Victoria type of crown rust resistance.
2. Treat seed with a seed-protectant fungicide.
3. Rotate oat with other crops.

Diseases Caused by Nematodes

Bulb and Stem

Cause. *Ditylenchus dipsaci.* Nematodes are associated with heavy soils, high rainfall, cool growing seasons, and winter grains. Free moisture permits the nematode to migrate to and feed on aerial plant parts. The nematode penetrates leaves and stems.

Distribution. Worldwide. Damage is confined to small areas within fields.

Symptoms. Cell hypertrophy and hyperplasia occur. Stunting distortions and swollen stems are common symptoms. In severely infested soils, the growing points of diseased plants are destroyed and the plants may die. Other diseased plants have reduced spike growth and reduced grain yields.

Management
1. Rotate with noncereal crops.
2. Grow winter cereals to reduce damage.

Cyst

Cause. *Heterodera hordecalis, H. latipons,* and *Punctodera chalcoensis.*

Distribution. *Heterodera latipons* has been reported in Israel.

Symptoms. Diseased plants are stunted and chlorotic. Cysts are not described on roots.

Management. Rotate oat with a nongrass crop.

Dagger

Cause. *Xiphinema americanum* nematodes are external parasites that prefer to feed on young, succulent roots.

Distribution. Widespread. Damage on oat is likely not very severe.

Symptoms. Moderate swelling occurs on young roots. Severely diseased plants have clusters of short, stubby branches and a shriveling of small roots at the points of attachment. The most obvious symptom is one or several light brown to dark brown or black necrotic flecks on roots.

Management. Not necessary.

Lesion

Lesion is also referred to as the pin nematode.

Cause. *Pratylenchus thornei* but other *Pratylenchus* spp. are also thought to infect oat. Nematodes live free in soil as migratory endoparasites and overwinter as eggs, larvae, or adults in host tissue or soil. Both larvae and adult nematodes penetrate roots, where they move through cortical cells and the females deposit eggs as they migrate. Older roots are abandoned and new roots are sought as sites for penetration and feeding.

Distribution. Wherever oat is grown.

Symptoms. Plants in limited areas within a field are chlorotic and under moisture stress. Roots and crowns are rotted when *Rhizoctonia solani* infects plants through nematode wounds. New roots become dark and stunted, and the yield of diseased plants is reduced.

Management
1. Sow in autumn when soil temperatures are below 13°C.
2. Use soil fumigants where the high costs warrant them.
3. Application of nitrogen and phosphorous reduced populations of *P. thornei* on roots of wheat grown in a rotation.

Needle Nematode

Cause. *Longidorus cohni* builds to high populations during the winter, but populations decrease during higher soil temperatures in the spring.

Distribution. Israel.

Symptoms. Terminal swellings occur on roots. Diseased plants are stunted and chlorotic and can cause large bare patches in a field.

Management. Rotate oat with a nongrass crop.

Oat Cyst Nematode

Oat cyst nematode is also called the cereal cyst nematode.

Cause. *Heterodera avenae.* The life cycle from egg to larva to adult is completed in 9 to 14 weeks. Females enter roots and begin feeding. Shortly after entering roots, females swell up and break through the root epidermis but remain attached to the plant roots by a thin neck. Females are inseminated by males and several hundred eggs develop within a female body. Eventually, the female body forms a hard body or cyst that is resistant to physical adversities. The cysts detach or break off from the roots and can survive in soil for several years. The following spring, second-stage larvae hatch from an average of 75% of the eggs.

Heterodera avenae is also a pathogen of barley, rye, and wheat but is most prevalent on oat.

Distribution. Africa, Australia, southeastern Canada, Europe, Japan, Russia, and the United States (Oregon).

Symptoms. The lemon-shaped cysts, which are barely visible to the naked eye, are initially white, then become dark brown as they harden. Eggs and larvae are white and can only be observed with magnification.

The first symptom on oat is poor growth in one or more localized areas within a field. Leaf tips of young, severely diseased plants are red or purple. The discolored leaves die and the diseased plant becomes chlorotic. Roots are thickened and branched more than healthy roots. Large infestations of nematodes in soils cause diseased plants to wilt, particularly during times of water stress, followed by stunted growth, poor root development, and early plant death.

Management
1. Grow resistant oat cultivars. However, resistant cultivars will vary from area to area because the oat cyst nematode is known to have up to 20 races. Normally, not all races are present in one locality.
2. Rotate oat with a legume crop. The number of nematodes will decline the longer oat is not grown in infested soil.
3. Apply a nematicide before planting.

Ring

Cause. *Criconemella* spp. and *Nothocriconemella mutabilis*. The nematodes are ectoparasites.

Distribution. Not reported.

Symptoms. Symptoms on oat are not well known. Feeding on roots of other plant species does not result in necrosis.

Management. Not reported.

Root Gall

Cause. *Subanguina radicicola* survives in host roots. Larvae penetrate roots and develop in the cortical tissue, forming a root gall in 2 weeks. Mature females begin egg production within a gall. Eventually, galls weaken and release larvae that establish secondary infections. Each generation is completed in about 60 days.

Distribution. Canada and northern Europe.

Symptoms. Diseased seedlings have reduced and chlorotic top growth. Galls on roots are inconspicuous and vary in diameter from 0.5 to 6.0 mm. Roots may be bent at the gall site. At the center of larger galls is a cavity filled with nematode larvae.

Management. Rotate oat with a noncereal crop.

Root Knot

Cause. *Meloidogyne chitwoodi* and *M. naasi* are endoparasitic nematodes that overwinter as eggs in the soil. In spring, larvae hatch from eggs but enter roots at any time during the growing season. By the middle of summer, females inside root knot tissue release eggs into soil.

Distribution. Wherever oat is grown.

Symptoms. Root knots are found on diseased roots in the spring and summer. These root swellings or thickenings are comprised of swollen root cortical cells and the bodies of nematodes containing egg masses. When root knots are cut open, the inner egg masses turn dark.

Management
1. Sow oat in the autumn.
2. Rotate oat with root crops.

Sheath

Cause. *Hemicycliophora* spp. feed near plant root tips as ectoparasites.

Distribution. Widespread, but probably most common under warmer climate conditions.

Symptoms. Symptoms on oat are not well known, but galls have been reported to form near the root tips of other grass hosts. Other plant species did not display any obvious symptoms.

Management. Not known.

Spiral

Cause. *Helicotylenchus* spp. are ectoparasites.

Distribution. Not reported.

Symptoms. Symptoms on oat are not well known. Small necrotic flecks on roots and a general overall decline in plant vigor have been suggested.

Management. Not reported.

Sting

Cause. *Belonolaimus longicaudatus.*

Distribution. Presumably in warmer climates.

Symptoms. Root tips are injured, which results in stubby roots or a reduced root system. Lesions may occur along diseased roots.

Management. Not known.

Stubby Root

Cause. *Paratrichodorus minor* survives in soil or on roots as eggs, larvae or adults. *Paratrichodorus minor* feeds only on the outside of oat roots and moves relatively rapidly through soil at a speed of 5 cm/hour, especially in fine, sandy soils.

Distribution. Widely distributed in most agricultural soils.

Symptoms. Diseased roots are thickened, short, and stubby; the root tips may have brown lesions. The tops of diseased plants decline in growth, and the entire plant is easily pulled from soil due to a lack of a fibrous root system.

Management. Not reported.

Stunt

Cause. *Quinisulcius capitatus, Tylenchorhynchus* spp., and *Merlinius* spp. *Merlinius* spp. survive as different morphological stages in soil and in association with host tissue. They are ectoparasitic nematodes that feed on the outside of wheat roots, often in association with the fungus *Olpidium brassicae.*

Distribution. Indigenous to most soils but rarely found in high populations.

Symptoms. The most severe damage occurs on winter cereals growing in wet soils. The lower leaves die and a few short tillers with small seed form on diseased plants. The diseased roots of seedlings are shortened, dark in color, and shriveled.

Management. Not known.

Disease Caused by Phytoplasmas

Aster Yellows

Cause. The aster yellows phytoplasma survives in several dicotyledonous plants and leafhoppers. Aster yellows phytoplasma is transmitted primarily by the aster leafhopper, *Macrosteles fascifrons,* and less commonly by the painted leafhopper, *Endria inimica,* and the leafhopper *M. laevis.* Leafhoppers acquire the phytoplasma by feeding on diseased plants, then transmit the phytoplasma by flying to healthy plants and feeding on them. Different strains vary in their ability to infect oat. Canadian workers have designated a western strain of the aster yellows phytoplasma as CAYA and three eastern strains as DAYA, MAYA, and NAYA.

Distribution. Eastern Europe, Japan, and North America.

Symptoms. Aster yellows is rarely severe on oat. Symptoms become most obvious at temperatures of 25° to 30°C. Seedlings either die 2 to 3 weeks after infection or, if seedlings live, they become stunted and have chlorotic blotches on leaves or have completely chlorotic leaves and distorted and sterile heads. Normal-appearing floral parts may still be nonfunctional.

Infection of older plants causes leaves to be stiff and discolored in shades of yellow, red, or purple, which begins from the leaf tip or margin and moves inward to include the entire leaf blade. Diseased root systems may not be well developed.

Management. No management is practical.

Diseases Caused by Viruses

African Cereal Streak

Cause. African cereal streak virus (ACSV) is limited to the phloem, where it induces a necrosis. ACSV is transmitted by the planthopper *Toya catilina* but is not seed- or mechanically transmitted. The natural virus reservoir is native grasses. Disease development is aided by high temperatures.

Distribution. East Africa.

Symptoms. Initially, faint, broken, chlorotic streaks begin near leaf bases and extend upward. The broken nature of young streaks is clearly defined. Later, definite alternate yellow and green streaks develop along entire leaf blades. Eventually, leaves become almost completely chlorotic. New leaves tend to develop a shoestring habit and die.

Plants infected while young become chlorotic and severely stunted, and die. Seed yield is almost completely suppressed. Plants become soft, flaccid, and velvety to the touch.

Management. Not reported.

American Wheat Striate Mosaic

Cause. American wheat striate mosaic virus (AWSMV) is in the rhabdovirus group. AWSMV is transmitted mainly by the leafhopper *Endria inimica* and occasionally by *Elymana virescens.*

Distribution. Canada and the central United States.

Symptoms. Leaves have obvious striations consisting of yellow to white parallel streaks. Older leaves are stunted and chlorotic, then become necrotic.

Management. Not reported.

Barley Yellow Dwarf

The disease caused by the barley yellow dwarf virus is sometimes called oat red leaf, red leaf, and yellow dwarf.

Cause. Barley yellow dwarf virus (BYDV) is a phloem-restricted virus in the luteovirus group. Several luteoviruses are grouped under the name BYDV and share the characteristic of infecting gramineous plants but differ in various other properties, such as virulence, host range, serological behavior, and transmissibility by aphid vectors.

BYDV survives or is reservoired in grains sown in the autumn and in several perennial grasses. Perennial grasses are a large reservoir of BYDV but may not serve as the most important source of inoculum. Inoculum carried by aphids from elsewhere in Canada was considered to be the main source of infection, not nearby grasses, winter wheat, or maize. In areas where oversummering is a factor, maize is a reservoir for both aphids and some isolates of BYDV. A major source of BYDV in Indiana is exogenous aphid populations moving from distant plants in wind currents, especially in the spring. Transmission from local grasses was sporadic and less common.

BYDV is transmitted by several species of aphids, including the corn leaf aphid, *Rhopalosiphum maidis*; oat bird-cherry aphid, *R. padi*; rice root aphid, *R. rufiabdominelis*; Russian wheat aphid, *Diuraphis noxia*; early instars of greenbug, *Schizaphis graminum*; and the English grain aphid, *Sitobion avenae*. Once BYDV is acquired, aphids are capable of transmitting it for the rest of their lives. One very active aphid feeding for short periods on different plants is a more important vector than several stationary aphids. In the autumn, aphids migrate to autumn-planted small grains and perennial grasses.

Based on transmission by different insects, five types of BYDV are recognized. MAV is transmitted specifically by *S. avenae*; RMV by *R. maidis*; RPV by *R. padi*; SGV by *S. graminum*; and PAV nonspecifically by *R. padi* and *M. avenae* and infrequently by *D. noxia*. However, some BYDV isolates obtained from cereal plants in Victoria, Australia, were serologically similar to MAV but distinct in being readily transmissible by the aphid *R. padi*. Some RMV isolates from Montana were transmitted efficiently by *R. maidis* and *S. graminum*, and two RMV isolates were occasionally transmitted by *R. padi*. An aphid may transmit BYDV for the rest of its life.

BYDV is not transmitted through eggs, newborn aphids, seed, soil, or mechanical means. Disease development occurs during temperatures of

10° to 18°C and moist conditions that favor grass and cereal growth and aphid reproduction and migration. Storm fronts aid aphid flights, which may cover hundreds of kilometers from southern areas, where most over-wintering occurs, to the upper Midwest in the United States.

Distribution. Africa, Asia, Australia, Europe, New Zealand, North America, and South America.

Symptoms. BYDV on oat is first seen in plants along the edges of fields. Areas of diseased plants varying in size from a few to several kilometers in diameter give fields a patchy and uneven appearance.

The plant discoloration symptoms that lend the often-used name "red leaf" to the disease vary considerably. Initially, faint yellowish-green blotches occur near the leaf tip, on the margin, or on the leaf blade. The blotches enlarge, coalesce at the leaf base, and turn various shades of yellow-red, orange, red, or red-brown; however, tissue next to the midrib remains green longer than the rest of the diseased leaf. The discoloration eventually involves the entire leaf blade progressively from the leaf tips to the bases. The entire flag leaf ultimately becomes a yellow to reddish-orange discoloration. Margins of symptomatic leaves may show inward curling. The still-green areas of infected plants may be a darker green than that of healthy plants. It has been reported that the bright yellow discoloration occurs on older but not the youngest leaves except in late infections when only flag leaves show symptoms.

Plants infected early are severely dwarfed and may die prematurely. Eventually, most diseased plants are dwarfed to some extent and leaf color becomes bright yellow or occasionally red or purple. Root elongation is restricted on spring oat.

Infection of young plants has the most significant influence on yield. Depending on the cultivar, heads of diseased plants have many blasted spikelets. Test weight of grain is lowered. The biggest losses have been reported on spring oat.

Management
1. Grow resistant cultivars. Resistance has been identified in breeding lines.
2. Properly fertilize soil. Vigorously growing plants are more tolerant of BYDV than weaker ones.

Cereal Chlorotic Mottle

Cause. Cereal chlorotic mottle virus (CeCMV) is in the rhabdovirus group. CeCMV is transmitted by cicadellids.

Distribution. Australia and northern Africa.

Symptoms. Severe necrotic and chlorotic streaks occur on leaves.

Management. Not reported.

Flame Chlorosis

Cause. Double-stranded RNAs ranging in size from 900 to 2800 base pairs present in vesicles rather than virions is likely the disease agent or its replicative intermediate. The causal agent is soil-transmitted and has been transmitted by sowing seed or by placing seedlings in soil where diseased plants had previously grown.

The grassy weeds green foxtail, *Setaria viridis,* and barnyard grass, *Echinochloa crus-galli,* are also hosts.

Distribution. Canada (Manitoba).

Symptoms. A striking flame-like pattern of leaf chlorosis and severe stunting occur in spring barley. Chloroplasts and mitochondria of affected cells are hypertrophied and contain an extensive proliferation of fibril-containing vesicles that form within the organellar envelope. Diseased plants continue to produce leaves with symptoms after being transferred to sterile potting medium.

Management. Not reported.

Maize Dwarf Mosaic

Cause. Maize dwarf mosaic virus strain O (MDMV-O) is in the potyvirus group. MDMV-O is transmitted from maize to oat by several aphid species. Infections occur on plants grown near MDMV-infected maize.

Distribution. The United States, but infection of oat is rare.

Symptoms. Leaves have a mild mottling.

Management. Not necessary.

Oat Blue Dwarf

Cause. Oat blue dwarf virus (OBDV) is in the marafivirus group. OBDV is transmitted by adult aster leafhoppers, *Macrosteles fascifrons,* and occasionally by immature leafhoppers. OBDV overwinters in wild grasses in southern oat-growing areas and is transmitted into northern oat-growing areas by leafhoppers. OBDV is limited to the phloem. At least 18 plant species are infected by OBDV.

Distribution. Canada and the north central United States.

Symptoms. Diseased plants are dwarfed, uniformly dark blue-green, and mature later than healthy plants. The dwarfed condition causes diseased plants to be overlooked. Leaves, especially flag leaves, and stems are stiffer, shorter, and stand out at a greater angle from the stem than those of healthy plants. Tillers appear in larger numbers and form above the crown. Severely blasted heads produce little or no seed.

Management. Not reported.

Oat Golden Stripe

Cause. Oat golden stripe virus (OGSV) is in the furovirus group. OGSV is transmitted by the soilborne fungus *Polymyxa betae*. *Polymyxa betae* survives in field soil as cystosori. Cysts give rise to zoospores, which swim through free soil water until they contact a host root and encyst. Encysted zoospores produce a structure called a stachel through which zoosporic cytoplasm enters the host cell and becomes a plasmodium. Only primary root tissue of young roots is infected; the optimum temperature for infection is about 25°C. The host cell then becomes infected with OGSV if *P. betae* is viruliferous. After a period of time the plasmodium develops into a zoosporangium, which releases additional zoospores that repeat the infection cycle. However, some plasmodia develop into cysts, and frequently nearly every cell in the small feeder roots will contain a cyst. As root cells senesce, cysts are eventually released into the soil, where they can remain viable for years without loss of virulence.

Distribution. France, United Kingdom, and the United States (North Carolina).

Symptoms. When plants are systemically infected, a yellow striping occurs on the tips of leaves and on the two youngest leaves. Many plants have infected roots but show no systemic symptoms.

Management. Not reported.

Oat Mosaic

Some researchers have reported oat mosaic virus and soilborne mosaic virus to be the same organism causing the same disease. However, because the symptoms are described somewhat differently in the literature, they are reported as two different diseases here.

Cause. Oat mosaic virus (OMV) is in the bymovirus group. OMV is transmitted by the soilborne fungus *Polymyxa graminis* and by mechanical means.

Distribution. The United Kingdom and the United States.

Symptoms. Diseased leaves have a general mottling, especially the first leaves formed on the plant.

Management. Not reported.

Oat Necrotic Mottle

Cause. Oat necrotic mottle virus (ONMV) is in the rymovirus group. ONMV is a flexuous filament that is transmitted mechanically but not by aphids. Symptoms in the growth chamber developed optimally at 20°C.

Distribution. Canada (Manitoba).

Symptoms. Initially, fine chlorotic lines develop on leaves. As leaves mature, the chlorotic lines enlarge and merge to form a general chlorotic mottle. Later, pale orange-brown spots, some with pale brown borders and green centers, develop at random on leaf blades and sheaths and are accompanied by a reddish pigmentation that varies with oat variety. Eventually, a necrosis develops at random on leaves and sheaths.

Management. Not reported.

Oat Soilborne Mosaic

Cause. Oat soilborne mosaic virus (OSBMV) is soilborne and survives in soil for up to 5 years, possibly in the soil fungus *Polymyxa graminis*. Circumstantial evidence suggests that OSBMV is transmitted to roots by soil fungi that previously invaded roots for at least 2 weeks before the virus was transmitted. OSBMV is composed of at least two different strains.

Distribution. Europe and North America.

Symptoms. The leaves of less severely diseased plants have light green to yellow dashes and streaks and sometimes a necrotic mottling that parallels the axis of leaves. Other strains of OSBMV cause eyespot lesions that are spindle-shaped with light green to ash gray borders and green centers. One strain of OSBMV causes severely diseased plants to grow in small rosettes.

Management. Grow resistant cultivars.

Oat Sterile Dwarf

Cause. Oat sterile dwarf virus (OSDV) is in the fijivirus group. OSDV is transmitted mainly by the planthopper *Javesella pellucida* but also by *J. dubia, J. discolor, J. obscura,* and *Dicranotropis hamata*.

Distribution. Czechoslovakia, Finland, Germany, Norway, Poland, Sweden, and the United Kingdom.

Symptoms. Plants have a general bushy, dwarfed, dark green, grass-like appearance. Leaves are malformed and have yellow or white enations that are centered on veins or vein swellings on the back of leaves.

Management. Not reported.

Oat Striate Mosaic

Cause. Oat striate mosaic virus (OSMV) is in the rhabdovirus group. OSMV is transmitted by the leafhopper *Graminella nigrifrons*. The highest incidence of disease was reported in the autumn on voluntary spring oats.

Distribution. The United States (Illinois). All laboratory isolates are thought to be lost.

Symptoms. Diseased leaves have a striation pattern parallel to the veins, similar to wheat striate. A necrosis of leaves and stems eventually occurs.

Management. Not reported.

Orchardgrass Mosaic

Cause. Orchardgrass mosaic virus (OGMV) is mechanically transmitted. Orchardgrass is a source of inoculum.

Distribution. Canada (Quebec).

Symptoms. Typical mosaic symptoms on leaves, considerable stunting, and reduced or delayed heading.

Management. Not reported.

Wheat Streak Mosaic

Cause. Wheat streak mosaic virus (WSMV) is in the potyvirus group. WSMV survives in infected plants and is transmitted from plant to plant only by the feeding for one or more minutes of young wheat curl mites, *Aceria tosichella* (syn. *A. tulipae* and *Eriophyes tulipae*). Once a mite has picked up WSMV from a diseased plant, it carries the virus internally for several weeks. It was thought that neither mites nor WSMV could survive longer than 1 to 2 days in the absence of a living plant, including seed. However, active virus reportedly has been detected in dead plants.

As plants mature during late summer or early autumn, mites migrate or are windborne for at least 2.4 km to nearby volunteer cereals, grasses, or maize and infect them with WSMV. Consequently, WSMV is disseminated from volunteer plants to cereals sown in autumn. However, some plants are hosts for mites but not WSMV, and vice versa. Other plants are susceptible to both mites and WSMV, and some are resistant to both.

WSMV has a wide host range that includes barley, maize, oat, rye, wheat, and several annual and perennial grasses.

Distribution. Eastern Europe, western and central North America, and Russia.

Symptoms. The greatest disease severity occurs in autumn-sown oat. Oat sown in the spring is generally not severely diseased but may serve as a reservoir for WSMV to be carried to other autumn-sown cereals. The first symptoms on leaves consist of light green to light yellow blotches, dashes, or streaks parallel to veins. Plants become stunted and develop a large number of tillers that vary in height and an overall yellow mottling. As diseased plants mature, the yellow-striped or yellow-mottled leaves turn brown and die. After harvest, stunted plants with sterile heads remain standing at the same height or shorter than stubble.

Management

1. Destroy all volunteer cereals and grasses in adjoining fields 2 weeks before sowing and 3 to 4 weeks in the field to be sown.
2. Sow winter oat as late as practical after the Hessian fly–free date.

Disease Caused by Physiological Disorders

Blast

Blast is also called blight, blindness, or white star.

Cause. Any factor that interferes with the normal development of the oat plant when heads are forming. Late sowing, lack of moisture, high temperatures, nutrient imbalance, crowding, disease, insect attacks, or a combination of factors can cause blast.

Distribution. Wherever oat is grown.

Symptoms. Spikelets, recognized by light color, delicate texture, and lack of grain as heads emerge from boots, are blighted and fail to develop. Ordinarily a few spikelets on the lower half of heads are blasted, but occasionally half and rarely entire heads are affected.

Management. Not reported.

Selected References

Boewe, G. H. 1960. Disease of wheat, oats, barley and rye. Illinois Nat. Hist. Surv. Circ. 48.

Bragard, C., Singer, E., Alizadeh, A., Vauterin, L., Maraite, H., and Swings, J. 1997. *Xanthomonas translucens* from small grains: Diversity and phytopathological relevance. Phytopathology 87:1111-1117.

Brodney, V., Wahl, I., and Rotem, J. 1983. Factors affecting the survival of physiologic races of *Puccinia coronata avenae* on *Avena sterilis* in Israel. (Abstr.) Phytopathology 73:363.

Clement, D. L., Lister, R. M., and Foster, J. E. 1986. ELISA-based studies on the ecology and epidemiology of barley yellow dwarf virus in Indiana. Phytopathology 76:86-92.

Cohn, E., and Ausher, R. 1973. *Longidorus cohni* and *Heterodera latipons,* economic nematode pests of oats in Israel. Plant Dis. Rptr. 57:53-54.

Elliott, C. 1927. Bacterial stripe of oats. J. Agric. Res. 35:811-824.

Farr, D. F., Bills, G. R., Chamuris, G. P., and Rossman, A. Y. 1989. Fungi on Plants and Plant Products in the United States. American Phytopathological Society, St. Paul, MN. 1252 pp.

Frank, J. A., and Christ, B. J. 1988. Rate-limiting resistance to Pyrenophora leaf blotch in spring oats. Phytopathology 78:957-960.

Gildow, F. E., and Frank, J. A. 1988. Barley yellow dwarf virus in Pennsylvania: Effect of the PAV isolate on yield components of Noble spring oats. Plant Dis. 72:254-256.

Gill, C. C. 1967. Oat necrotic mottle, a new virus disease in Manitoba. Phytopathology 57:302-307.

Gough, F. J., and McDaniel, M. E. 1974. Occurrence of oat leaf blotch in Texas in 1973. Plant Dis. Rptr. 58:80-81.

Gray, S. M., Smith, D. M., and Sorrells, M. E. 1991. Isolate-specific resistance to barley yellow dwarf virus in spring oats. (Abstr.). Phytopathology 81:122.

Haber, S., and Chong, J. 1993. Flame chlorosis induces vesiculations in chloroplasts and mitochondria: What does it mean? (Abstr.) Can J. Plant Pathol. 15:57.

Haber, S., and Hardener, D. E. 1992. Green foxtail (*Setaria viridis*) and barnyard grass (*Echinochloa crus-galli*), new hosts of the virus-like agent causing flame chlorosis in cereals. Can. J. Plant Pathol. 14:278-280.

Harder, D. E., and Bakker, W. 1973. African cereal streak, a new disease of cereals in east Africa. Phytopathology 63:1407-1411.

Huffman, M. D. 1955. Disease cycle of Septoria disease of oats. Phytopathology 45:278-280.

Jedlinski, H. 1976. Oat striate mosaic, a new virus disease in Illinois spread by the leafhopper *Graminella nigrifrons* (Forbes). (Abstr.) Am. Phytopathol. Soc. 3:208.

Jons, V. L. 1986. Downy mildew *(Sclerophthora macrospora)* of wheat, barley, and oats in North Dakota. Phytopathology 70:892.

Kolb, F. L., Cooper, N. K., Hewings, A. D., Bauskey, E. M., and Teyker, R. H. 1991. Effects of barley yellow dwarf virus on root growth in spring oat. Plant Dis. 75:143-145.

Lockhart, B. E. L. 1986. Occurrence of cereal chlorotic mottle virus in Northern Africa. Plant Dis. 70:912-915.

McKay, R. 1957. Cereal Disease in Ireland. Arthur Guiness, Son & Co. Ltd., Dublin.

Peterson, J. F. 1989. A cereal-infecting virus from orchardgrass. Can. Plant Dis. Surv. 69:13-16.

Poole, D. D., and Murphy, H. C. 1953. Field reaction of oat varieties to Septoria black stem. Agron. J. 45:369-370.

Richardson, M. J., and Noble, M. 1970. *Septoria* species on cereals: A note to aid their identification. Plant Pathol. 19:159-163.

Roane, M. K., and Roane, C. W. 1983. New grass hosts of *Polymyxa graminis* in Virginia. (Abstr.) Phytopathology 73:68.

Scardaci, S. C., and Webster, R. K. 1982. Common root rot of cereals in California. Plant Dis. 66:31-34.

Schaad, N. W., Sumner, D. R., and Wate, G. O. 1980. Influence of temperature and light on severity of bacterial blight of corn, oats and wheat. Plant Dis. 64:481-483.

Simmonds, P. M. 1955. Root disease of cereals. Can. Dept. Agric. Pub. 952.

Simons, M. D., and Murphy, H. C. 1952. Kernel blight phase of Septoria black stem of oats. Plant Dis. Rptr. 36:448-449.

Simons, M. D., and Murphy, H. C. 1968. Oat Diseases and Their Control. USDA. Agric. Res. Serv. Agric. Hdbk. No. 343.

Sinha, R. C., and Benk, R. M. 1972. American Wheat Striate Mosaic Virus. Commonwealth Mycological Institute. Descriptions of Plant Viruses, Set 6., No. 99. Kew, England.

Timian, R. G. 1985. Oat blue dwarf virus in its plant host and insect vectors. Plant Dis. 69:706-708.

Walker, J. 1975. Take-all disease of Graminae: A review of recent work. Rev. Plant Pathol. 54:113-144.

Yamamoto, H., and Tani, T. 1987. Induction of crown rust resistance in oat by high temperature stress. Ann. Phytopathol. Soc. Japan 53:616-621.

12. Diseases of Peanut *(Arachis hypogaea)*

Disease Caused by Bacteria

Bacterial Wilt

Bacterial wilt is also called brown rot and slime disease.

Cause. *Pseudomonas solanacearum* survives in soil and may be seedborne. Infection occurs through wounds or lenticels on host roots. Disease development is favored by heavy clay soils that tend to stay moist and soils in which peanut has been grown for several successive years. *Pseudomonas solanacearum* has a wide host range.

Distribution. Wherever peanut is grown. Bacterial wilt is of minor importance in the United States.

Symptoms. Different symptoms of the disease occur in different geographical peanut-growing areas. A symptom from the East Indies is sudden wilting with leaves remaining attached to dead plants. Symptoms from India are a general loss of turgidity, drying and drooping of leaves, and the ultimate death of the plant. Internal symptoms include browning of the vascular system of the roots and stem.

Symptoms of disease in the United States are not as severe. There are a large number of dead roots. Brown or black streaks occur throughout diseased roots, main stems, and lower branches and can be observed as dark brown spots in the xylem and pith when stems and roots are cut in cross section. Eventually, diseased tissue is blackened and has extensive plugging and necrosis. If plants are infected when young, pods are invaded and remain small or become wrinkled and develop a spongy or soft decay. When mature plants are infected, there is no invasion of pods.

Management
1. Grow resistant cultivars.
2. Treat seed with a seed-protectant fungicide.

403

3. Grow plants on light, well-drained soils.
4. Rotate peanut with resistant crops, such as sweet potatoes, small grains, and certain legumes.

Diseases Caused by Fungi

Aerial Blight

Cause. *Rhizoctonia solani*. Fungal propagules are splashed onto leaves during rainstorms.

Distribution. India, Malaya, and the United States.

Symptoms. Leaves of diseased plants rot or blight.

Management. Not reported.

Alternaria Leaf Blight

Cause. *Alternaria tenuissima*.

Distribution. India.

Symptoms. Diseased leaves have light to dark brown spots and frequently are totally blighted.

Management. Not reported.

Alternaria Leaf Spot

Cause. *Alternaria arachidis*.

Distribution. India.

Symptoms. Orange-brown lesions in interveinal areas of diseased leaves often extend into veins and veinlets.

Management. Not Reported.

Alternaria Spot and Veinal Necrosis

Cause. *Alternaria alternata*.

Distribution. India and the United States.

Symptoms. Orange-brown necrotic spots that occur in the interveinal areas of diseased leaves often extend into veins and veinlets.

Management. Not reported.

Anthracnose

Anthracnose is also called Colletotrichum leaf spot.

Cause. *Colletotrichum arachidis, C. dematium,* and *C. mangenoti* survive as mycelia in infested residue. Conidia are produced in acervuli and disseminated by wind and splashing rain to host plants.

Distribution. Africa, Argentina, India, Niger, Taiwan, Uganda, and the United States.

Symptoms. Different symptoms have been described. When plants are infected by *C. mangenoti,* brown-gray, elongate to circular lesions occur on both leaflet surfaces but rarely on petioles or stems. Lesions may be large and involve up to half the leaflet.

Other symptom descriptions are scattered circular to irregular-shaped lesions with gray-white centers surrounded by dark brown borders. Some researchers describe lesions as initially small, water-soaked, yellow spots that enlarge to 1–3 mm in diameter and become dark brown. These spots grow rapidly, become irregular-shaped, and spread over the entire leaflet. Spots may extend into petioles and branches and cause the death of the entire plant.

Management. Management measures usually are not warranted.

Aspergillus Crown Rot

Cause. *Aspergillus niger* and *A. pulverulentus* survive in several different ways: As saprophytes in infested residue and as conidia and mycelia in soil, on seed surfaces, and in or under tissues of the testae. The abundant spores that are produced are disseminated by wind for long distances and by any agent that moves soil.

Most infection occurs within 10 days after seed germination as elongating hypocotyls come in contact with soilborne inoculum. Hyphae penetrate directly into hypocotyls or cotyledons. Disease is most severe when plant emergence is delayed, plants are predisposed by high soil and air temperatures, and plants grow in soils low in organic matter. The causal fungi grow best in warm, moist conditions.

Distribution. Wherever peanut is grown.

Symptoms. The most common symptom is that, after seed germinates, the elongating hypocotyl is infected, becomes water-soaked and light brown, and is covered by black conidia and mycelium. Eventually, the diseased hypocotyl tissue becomes dark brown, then lighter in color, and shreds. Necrosis and shredding eventually extend into the branches.

The aboveground symptom is a rapid wilting of the entire plant during dry weather. Recovery of the turgidity of diseased plants, which may occur during high soil moisture, is related to the growth of adventitious roots above the infection site. However, most plants succumb and die in less than 30 days after infection.

Seed is infected when placed in a moist environment, where it becomes covered with sooty black masses of spores.

Management
1. Treat seed with a seed-protectant fungicide.
2. Grow bunch-type cultivars.

Blackhull

Cause. *Thielaviopsis basicola* overwinters most commonly as chlamydospores in the endocarp of unharvested peanuts, as saprophytic mycelium in infested residue, and as endoconidia in soil. As pods deteriorate the following season, chlamydospores germinate and produce mycelia that infect new pods. Dissemination is by any means that transports soil. Optimum growth of *T. basicola* in vitro is at 22° to 28°C. Blackhull is more severe in wet soils that are neutral or slightly alkaline.

Distribution. Occurs in most countries where 'Spanish' and 'Valencia' peanuts are grown.

Symptoms. Infections occur on external schlerenchymatous shell tissue during development of the fruit and are first seen as tiny black spots. Eventually, shells become black due to an aggregation of chlamydospores in the developing shells. Dark, crusty patches develop where great numbers of lesions have coalesced. *Thielaviopsis basicola* grows throughout shell tissue and produces masses of chlamydospores. Internal shell tissue and testae of kernels are discolored brown.

Management
1. Sow only healthy seed treated with a seed-protectant fungicide.
2. Rotate peanut with grain sorghum, maize, or small grains.
3. Leave soil fallow.
4. Sow as late as feasible in the growing season.

Black Root Rot

Cause. *Thielaviopsis basicola* overwinters most commonly as chlamydospores in the endocarp of unharvested peanuts, as saprophytic mycelium in infested residue, and as endoconidia in soil.

Distribution. The United States.

Symptoms. Diseased roots become dark, disintegrate, and die. Leaves become stunted and chlorotic and the entire plant wilts, especially during dry soil conditions. Diseased plants die slowly.

Management
1. Sow only healthy seed treated with a seed-protectant fungicide.
2. Rotate peanut with grain sorghum, maize, or small grains.
3. Leave soil fallow.
4. Sow as late as feasible in the growing season.

Botrytis Blight

Botrytis blight is also called gray leaf mold and stem rot.

Cause. *Botrytis cinerea* survives as sclerotia in soil and infested residue. Sclerotia germinate and form either mycelia or support conidiophores and conidia on sclerotial surfaces. Conidia are abundantly produced on infested tissue and disseminated by wind and water. Sclerotia are disseminated by any means that moves soil.

Disease is most severe during damp, cool weather (20°C or below) in late autumn. *Botrytis cinerea* requires no wounds or necrotic tissue to infect; however, senescing frost-injured and mechanically injured tissues are most prone to colonization. Leaves, stems, and subterranean organs are infected. Organic debris on the soil surface aids in the infection process by serving as an energy source for *B. cinerea*.

Distribution. Japan, Nyasaland, Rhodesia, Tanganyika, the United States, and Venezuela. However, Botrytis blight is seldom a severe problem since climatic conditions that favor disease are rarely present.

Symptoms. Diseased tissues decay rapidly and are sparsely covered with dark gray mycelium, conidiophores, and conidia. Disease progresses rapidly from leaves and stems down into pegs and fruit. Flattened, black, irregular-shaped sclerotia develop on decayed stems and pods.

Management
1. Regulate sowing dates to prevent plants from growing or maturing during cool, wet weather.
2. Apply foliar fungicides.

Charcoal Rot and Macrophomina Leaf Spot

Charcoal rot and Macrophomina leaf spot is also called root rot disease.

Cause. *Macrophomina phaseolina* (syn. *Rhizoctonia bataticola*) survives primarily as sclerotia or microsclerotia in infested residue and soil. Other types of survival are as mycelium in seeds and residue and, presumably, as pycnidia in infested residue. Sclerotia remain viable in dry soil for several years but rapidly lose their viability in wet soil. Dissemination is by any means that moves soil residue and seed. Disease is most severe at a high soil temperature (35°C) and low soil moisture.

Distribution. Worldwide. Charcoal rot is of minor importance.

Symptoms. Water-soaked necrotic areas that develop on stems at the soil line become dull brown and extend up stems into the branches and down into the roots. Plants wilt if stems become girdled by lesions or if roots are decayed. Athough infection of leaflets is rare, symptoms consist of large, marginal, zonate spots in which pycnidia that resemble small, black bumps are found. Partial defoliation may occur. When diseased plants die, a blackening or sooty appearance caused by numerous sclerotia occurs over plant parts. A few pycnidia that resemble small, black pimples may also develop.

Root infection can occur independently of stem rot. At first, roots are blackened; later, the taproot is completely rotted.

When peanuts are physically damaged before or after harvest, *M. phaseoli* grows through shells into the kernels. Sclerotia are evident as a black or sooty growth that occurs on internal and external shell surfaces.

Management
1. Treat seed with a seed-protectant fungicide.
2. Apply soil fungicides to reduce fruit infection.
3. Reduce row spacing. Incidence of root rot is lower at 30 cm than at either 45- or 60-cm row spacings.
4. Grow varieties that are the most resistant.
5. In experimental conditions, gypsum applied at the rate of 150 kg/ha resulted in significant reduction of disease incidence regardless of row spacings or varieties used.

Choanephora Leaf Spot

Cause. *Choanephora* spp.

Distribution. The Philippines, Senegal, Thailand, and Uganda.

Symptoms. Brown lesions containing faint concentric circles originate at leaflet margins and eventually spread over the entire leaflet. Abundant sporulation occurs on both leaflet surfaces. Defoliation of diseased plants may occur.

Management. Not reported.

Collar Rot

Collar rot is also called Diplodia collar rot.

Cause. *Lasiodiplodia theobromae* (syn. *Diplodia gossypina*) survives as mycelium and pycnidia in infested residue. During wet weather, conidia are produced in pycnidia and disseminated by splashing water or by wind to host plants. Mycelia in residue are also disseminated by running water, cultivation, or any means that moves soil. When predisposed by heat injuries, plants of all ages are infected. *Lasiodiplodia theobromae* rapidly colonizes heat-injured tissue and grows intercellularly through cortical parenchyma into adjacent nonwounded tissue. Hot, dry weather favors infection.

Distribution. Worldwide, but collar rot is not considered an important disease.

Symptoms. Initially, there is a rapid wilting of branches or of entire diseased plants. During warm weather, plants die within a few days of infection and symptom expression. Stem lesions become gray-brown to black and extend toward the taproot. Necrotic stems become shredded. Infection of a branch usually results in death of that branch. Numerous pycnidia, which appear as tiny, black pimple-like dots, develop in necrotic tissue.

Management
1. Heat canker may be more prevalent in 'Runner' than in 'Spanish' cultivars because the greater leaf development in 'Spanish' cultivars provides shade.
2. Do not rotate peanut with cotton and soybean.
3. Plow under infested residue.
4. Sow peanut rows so plants shade each other.
5. A finely clodded soil surface is most favorable for reducing reflective sunlight energy.

Colletotrichum Leaf Spot

Cause. *Colletotrichum gloeosporioides.* Overseasoning is likely as mycelium in seeds, seed coats, infested residue, and on weeds.

Distribution. Not certain, but presumably occurs in humid tropics and subtropics where peanut is grown.

Symptoms. Foliar symptoms develop after prolonged periods of high humidity and include dark brown to black spots on leaves, necrosis of laminar veins, petiole cankering, and premature defoliation. Diseased lower branches and leaves on a plant may be killed.

Management. Not reported.

Cylindrocladium Black Rot

Cause. *Cylindrocladium parasiticum* (syn. *C. crotalariae*), teleomorph *Calonectria ilicicola,* survives as small (35–425 μm in diameter), irregular-shaped microsclerotia in infested residue, soil, and possibly seed. Microsclerotia survive better buried in soil than on the soil surface. Primary infection originates from germination of microsclerotia and from mycelia growing from infested residue. Microsclerotia 150 μm and larger induce root rot more efficiently than do smaller microsclerotia. Seeds are infected at a higher incidence shortly before storage than after 7 months storage in an unheated building. Hyphae grow intracellularly and intercellularly

throughout the testae of discolored seed. Hyphae grow in the cotyledonary tissues from seed with discolored testae.

Long-distance dissemination occurs as propagules on seed surfaces and
windborne plant fragments that contain microsclerotia. Other means of
dissemination are birds, farm implements, movement of hay, and dispersal in roots.

During moist conditions, ascospores are formed in newly developed
perithecia and disseminated by splashing rain or insects. However, the epidemiological importance of ascospores is thought to be limited to short-
range spread.

A soil temperature of 25°C and a moisture content near field capacity
are most conducive for infection and subsequent root rot development.
Disease is most severe in heavy clay soils that stay wet and waterlogged for
long periods of time after a rainfall.

Cylindrocladium crotalariae does not survive well at soil temperatures of
5°C and below. However, at these low temperatures survival is better in
moist than in dry soil.

Distribution. Australia, India, Japan, and the United States.

Symptoms. Symptoms typically appear midway through the growing season
and may increase rapidly as the season progresses. Some researchers state
aboveground symptoms are optimum early in the growing season after
moisture stress follows a period of high rainfall that accentuates root rot
and causes loss of functional roots. Diseased plants are stunted and
chlorotic, wilt on warm days, eventually collapse, and remain debilitated
although they normally do not die. Taproots, immature pegs, and pods are
blackened and taproots are shredded and have few or no lateral roots. Microsclerotia form in roots and appear as small black structures the size and
appearance of ground pepper.

Perithecia look like small red-orange spherical objects in dense clusters
on stems, pegs, and pods just above or beneath the soil surface. Perithecia
are especially abundant in moist areas under dense foliage. During wet
weather, ascospores are exuded from perithecia as a viscous yellow ooze in
2 to 3 weeks.

Management
1. Grow the most resistant cultivars.
2. Disk under all diseased plants prior to digging and combining operations. Avoid passing through these areas with equipment during harvest.
3. Dig and remove all diseased plants and either burn them completely or
   bury them in a sanitary landfill.
4. Practice sanitation on all equipment since microsclerotia can be moved
   in soil left on equipment. Wash equipment with a strong stream of water immediately before equipment is moved from infested areas.

5. Cropping history has been reported to not affect disease development. However, some reports suggest rotating peanut every 3 to 5 years with maize, small grains, or perennial grasses but not with soybean since peanut and soybean have many pathogens in common.
6. Remove all diseased plants prior to harvest in small, localized areas.
7. Perform cultural practices that increase soil temperature. Disease development is slowed when soil temperatures exceed 25°C and stops when soil temperatures exceed 35°C.

Cylindrocladium Leaf Spot

Cause. *Cylindrocladium scoparium.*

Distribution. Not known for certain.

Symptoms. Spots on diseased leaves vary from light to dark brown to black.

Management. Not reported.

Diplodia Infection

Cause. *Diplodia* sp. Disease was favored when stems were wounded with a shallow cut and inoculated plants were placed in a mist chamber during artificial inoculation.

Distribution. The United States (Florida).

Symptoms. Black lesions greater than 5 cm in length occur on diseased upper stems. Following wounding, internal necrosis was greater in upper stems than in lower stems.

Management. Not reported.

Drechslera Leaf Spot

Cause. *Bipolaris spicifera* (syn. *Drechslera spicifera*).

Distribution. Asia.

Symptoms. Reddish to brown elongated specks occur on diseased leaves.

Management. Not reported.

Early Leaf Spot

Early leaf spot is also called tikka, viruela, peanut cercosporosis, Mycosphaerella leaf spot, brown leaf spot, and early Cercospora leaf spot.

Cause. *Cercospora arachidicola,* teleomorph *Mycosphaerella arachidis,* survives as conidia, asci, and mycelia either in infested residue or soil. Conidia produced from mycelium and ascospores produced in asci are splashed or blown to host leaves. Penetration by germ tubes is through open stomata or directly through lateral faces of epidermal cells. Secondary infection occurs from conidia produced in leaf spots and disseminated by wind, rain, insects, and machinery. Infection is favored by temperatures of 26° to 31°C and long periods of high relative humidity. Perithecia are produced in diseased tissue later in the growing season.

Disease severity is greatest where peanut follows peanut in a rotation. The incidence and severity is usually greater in sprinkle-irrigated plants and in plants receiving sodic water. Prior infection with peanut green mosaic virus increases susceptibility to infection by *C. arachidicola*. Different races of *C. arachidicola* are present.

Distribution. Wherever peanut is grown.

Symptoms. Symptoms occur earlier in the season than those of late leaf spot. *Cercospora arachidicola* is more frequent on domestic cultivars of *Arachis hypogaea*, while *Phaeoisariopsis personata* is more common on wild species; therefore, differences in the occurrence of symptoms may be related to host differences rather than the time of the growing season. The terms "early" and "late" leaf spot may not be significant in peanut-growing areas where both pathogens occur.

Initially, small chlorotic spots occur, then enlarge (1–10 mm or more in diameter) and become brown to black and subcircular. A chlorotic halo around a spot or lesion on the upper leaflet surfaces is common but is not always present. The spots will coalesce during wet weather. Sporulation on upper lesion surfaces gives lesions a sooty appearance. Defoliation may occur. Petioles and stems have dark, elongated, superficial lesions with indistinct margins. Both species of *Cercospora* may be present in late-season infections.

Management
1. Rotate peanut with other crops.
2. Plow under or remove infested residue. Destroy volunteer peanut plants.
3. Apply a foliar fungicide.

Fusarium Peg and Root Rot

Cause. Several species of *Fusarium,* including *F. roseum, F. solani, F. tricinctum,* and *F. moniliforme.* These fungi survive as chlamydospores in soil and as saprophytic mycelia in infested residue. Primary inoculum is either chlamydospores that germinate and produce hyphae or conidia produced from mycelia. Conidia are disseminated by wind, water, or any means that moves soil or residue. Several *Fusarium* spp. are apparently only parasitic on belowground plant parts and do not become pathogenic. Injury from *Fusarium* spp. is ordinarily not serious if host plants are in a vigorous growing condition.

Distribution. The disease occurs wherever peanut is grown.

Symptoms. Seedlings that haven't emerged have gray, water-soaked tissues that are overgrown with mycelia.

Fusarium solani causes a dry root rot. The lower taproot is brown to red-brown, withers, and often curls. Secondary roots become brown and slough off. Eventually, disease progresses up the hypocotyl, causing plants to wilt. Sometimes adventitious roots that develop above the diseased area allow plants to survive.

Older plants infected by *F. solani* have small, elongate, slightly sunken, brown lesions on roots just below the crowns. Eventually, lesions girdle the roots and cause cortical tissue to become shredded and diseased plants to be chlorotic, wilted, and moribund.

Fusarium-infected pods do not have typical symptoms. However, a violet-white discoloration of pods is reported to be characteristic.

Management
1. Increase organic matter in the soil.
2. Avoid growing peanut in acid soils.

Fusarium Stem Canker

Cause. *Fusarium oxysporum.* Wounding of the central stem by mechanical or environmental means is necessary for infection to occur. Prolonged wet weather enhances disease development.

Distribution. The United States (Alabama).

Symptoms. Reddish to dark brown, sunken, elongated (0.5–3.0 cm), girdling lesions occur at varying distances above the soil line along the central stems of diseased plants. Foliage distal to the cankers becomes chlorotic and wilts. Sporodochia of *F. oxysporum* may be observed on the surface of some lesions.

Management. Not reported.

Fusarium Wilt (*Fusarium martii* var. *phaseoli*)

Although the disease is called Fusarium wilt in some literature, the fungus is not systemic in the xylem tissue and the disease is more of a stem or root rot that results in wilt symptoms. *Fusarium martii* is not listed in Farr et al. (1989) or in Nelson et al. (1983).

Cause. *Fusarium martii* var. *phaseoli.*

Distribution. The United States.

Symptoms. Symptoms first occur 8 to 10 weeks after sowing. The youngest leaves become chlorotic, degenerate rapidly through a drooping or wilt stage, and finally die because of stem lesions at or near the soil line. Roots may also be girdled by lesions. The entire diseased stem may eventually be destroyed.

Management. Not reported.

Fusarium Wilt *(Fusarium oxysporum)*

Cause. *Fusarium oxysporum* survives as chlamydospores in soil and as mycelia in infested residue. Primary inoculum is either chlamydospores that germinate and produce hyphae or conidia produced from mycelia. Conidia are disseminated by wind, water, or any means that moves soil or residue.

Distribution. Fusarium wilt occurs wherever peanut is grown.

Symptoms. Fusarium wilt occurs sporadically from year to year and within a field. Wilt is characterized by gray-green leaves that quickly permanently wilt. In dry weather, wilted leaves become dry and brittle and appear to be bleached. In slower wilting, the leaves may first become chlorotic and the plant defoliates before death occurs. Taproots have vascular discoloration but secondary roots appear healthy.

Management. Not reported.

Late Leaf Spot

Late leaf spot is also called brown leaf spot, leaf spot, late Cercospora leaf spot, Mycosphaerella leaf spot, peanut cercosporosis, tikka, and viruela.

Cause. *Phaeoisariopsis personata* (syn. *Cercospora personata* and *Cercosporidium personatum*), teleomorph *Mycosphaerella berkeleyi*. *Phaeoisariopsis personata* survives as conidia, perithecia, and mycelia in infested residue, and as mycelia in soil. Primary inoculum is conidia produced directly from mycelia growing in infested residue and disseminated by wind to host plants, and ascospores produced in perithecia and splashed by rain to peanut leaves. Penetration of leaves is by germ tubes that grow through open stomata or directly through the lateral faces of epidermal cells.

Some report temperatures of 16° to 20°C to be very favorable for germination of conidia. Others report infection is favored by temperatures of 26° to 31°C and long periods of high relative humidity. However, in experiments using detached leaves, maximum infection occurred at 20°C if leaves were exposed to 12 hours/day of more than 93% relative humidity. Using this method, infection decreased with increasing temperature, regardless of genotype resistance. Daily periods of high relative humidity for less than 12 hours also reduced the number of infections on all genotypes, regardless of temperature.

After leaves are infected, conidia that have sporulated in lesions are readily disseminated by wind, rain, insects, and machinery throughout the growing season and function as secondary inoculum. A minimum of 4 hours of relative humidity at or greater than 95% per day is required for conidial production. The highest numbers of conidia are produced when lesions are subjected to daily periods of 16 or more hours of relative humidity at or above 95% and a temperature of approximately 20°C. Perithecia are produced in diseased tissue later in the growing season.

Disease severity is greater where peanut follows peanut in a rotation and in fields receiving tractor traffic. Prior infection with peanut green mosaic virus increases susceptibility to infection by *P. personata*. Different races of *P. personata* are present.

Distribution. Late leaf spot occurs wherever peanut is grown. However, the disease predominates in the tropics and subtropics and is reported to be the major peanut leaf spot in the southernmost United States.

Symptoms. Symptoms occur later in the season than those of early leaf spot. However, *Cercospora arachidicola* is more frequent on domestic cultivars of *Arachis hypogaea*, while *P. personata* is more common on wild species; therefore, differences in occurrence of symptoms may be related to host differences rather than the time of the growing season. The terms "early" and "late" leaf spot may not be significant in peanut-growing areas where both pathogens occur.

Initially, small chlorotic spots occur, then enlarge (1–10 mm in diameter), become dark brown to black, are subcircular, and do not have yellow halos. If halos are present, they are normally on the upper leaf surfaces. The color of spots on the lower leaf surfaces tends to be black and is a good diagnostic characteristic to separate symptoms from those of early Cercospora leaf spot.

Sporulation first occurs on the lower leaf surfaces and later will occur sparsely on the upper leaf surfaces. Stroma on which spores are produced are often arranged in concentric circles and are visibly raised above the lesion surface. Defoliation may occur. Diseased petioles and stems have dark, elongated and superficial lesions with indistinct margins.

Management
1. Rotate peanut with other crops.
2. Plow under or remove infested residue. Destroy volunteer peanut plants.
3. Apply a foliar fungicide.

Macrophoma Leaf Spot

Cause. *Macrophoma* sp. persists as saprophytic mycelia and pycnidia in infested residue.

Distribution. Europe.

Symptoms. Lesions on diseased leaves are dark and composed of firm necrotic tissue. Marginal necrosis occurs along the apical portions of diseased leaflets. Pycnidia appear as black pimple-like objects scattered throughout necrotic tissue.

Management. Since Macrophoma leaf spot is not a serious disease, management is not necessary.

Melanosis

Melanosis also is called Stemphylium leaf spot in the United States. Some researchers consider symptoms similar to those described here as a melanosis disease caused by *Macrosporium* sp. and *Alternaria* sp.

Cause. *Stemphylium botryosum.*

Distribution. Argentina. Melanosis is considered a minor disease.

Symptoms. Symptoms are most prominent on the diseased lower leaf surfaces. Spots are irregular, circular to oval (0.5–1.0 mm in diameter), or elongated (1.5 mm long). The numerous dark brown spots give the impression leaflets are covered with fly specks. Initially, spots are slightly submerged but later become elevated and "crust-like." Defoliation does not occur even in severe disease conditions.

Management. Some cultivars are more susceptible than others, but management is usually not necessary.

Myrothecium Leaf Blight

Myrothecium leaf blight is also called groundnut leaf blight.

Cause. *Myrothecium roridum.*

Distribution. India.

Symptoms. Lesions on diseased leaves are gray, round to irregular-shaped (5–10 mm in diameter), and surrounded by chlorotic halos. Lesions eventually coalesce and give diseased leaves a blighted appearance. Black fruiting bodies, which are frequently arranged in concentric rings, form on diseased lower and upper leaf surfaces.

Management. Not reported.

Neocosmospora Root Rot

Cause. *Neocosmospora vasinfecta.*

Distribution. South Africa.

Symptoms. Diseased taproots and lateral roots are discolored and split. External discoloration of stems occurs about 1 cm above the soil surface and internal discoloration of vascular bundles and pith occurs about 3 cm above the soil surface. Small, dark lesions and larger dark cracks occur in the surface tissues of diseased pods and the underlying parenchymatous tissue is brown. Perithecia are produced on diseased tissue.

Management. Not reported.

Net Blotch

The common name of this disease may be confused with web blotch (sometimes called net blotch) caused by *Phoma arachidicola*. Net blotch may be the same disease syndrome as pod wart.

Cause. The reported cause is *Streptomyces* sp. in Israel and *S. scabies* in South Africa. However, disease may be caused by the same fungus at both locations. Disease may be accentuated when peanut follows potatoes in a rotation.

Distribution. Israel and South Africa.

Symptoms. Disease symptoms are accentuated by the herbicide ethalfluralin. The dark brown to black blotches that initially occur on diseased pods become net-like lesions that eventually may cover the entire pod. Dark brown, necrotic, wart-like lesions were reported on the cultivar 'Sellie' in South Africa.

Management
1. Methyl bromide reduces disease incidence and severity.
2. Do not grow peanut following potatoes.

Olpidium Root Rot

Cause. *Olpidium brassicae.*

Distribution. India and the United States (Texas). Olpidium root rot is a minor disease of peanut.

Symptoms. Diseased root cortex is brown to black.

Management. Not necessary.

Pepper Spot and Scorch

Cause. *Leptosphaerulina crassiasca* overwinters as pseudothecia and saprophytic mycelia in infested residue. Ascospores are produced in pseudothecia and are forcibly ejected from the pseudothecia and disseminated a considerable distance by wind. Ascospores become closely attached to leaflet surfaces and germinate when free water is available. Optimum conditions for germination are a temperature of 28°C and 100% relative humidity. Eventually, numerous pseudothecia are produced in detached dead leaves.

Distribution. Argentina, India, Madagascar, Malawi, Mauritius, Senegal, South Africa, Taiwan, the United States, and Zambia.

Symptoms. Symptoms are of two types and occur only on diseased leaves. Pepper spots are less than 1 mm in diameter, dark brown to black, irregular-shaped to circular, and occasionally depressed. Lesions do not rapidly enlarge with age. Spots are more common on the upper leaf surfaces but also occur on lower leaf surfaces. When spots are sparse, there is not an obvious deleterious effect to the diseased leaf. Conversely, numerous spots tend to coalesce, which gives leaflet surfaces a "netted" appearance and causes leaflets to die and defoliate. Perithecia are abundantly produced in detached diseased leaflets.

Leaflets with scorch symptoms become chlorotic, then necrotic at discrete points along the margins. Necrotic tissue becomes dark brown and has a chlorotic zone along the edges of the discolored tissue. Lesions commonly develop from the tips of diseased leaflets along a wedge-shaped front toward the petiole. Pseudothecia form abundantly in necrotic tissue. Necrotic tissue tends to fragment along the leaflet margins, which gives a tattered appearance to the leaf.

Management
1. Apply foliar fungicides.
2. Grow resistant cultivars.

Pestalotia Leaf Spot

Cause. *Pestalotia arachidicola.*

Distribution. Brazil.

Symptoms. Circular (up to 12 mm in diameter) necrotic spots occur on diseased leaves.

Management. Not reported.

Pestalotiopsis Leaf Spot

Cause. *Pestalotiopsis arachidis* persists as saprophytic mycelia in infested residue. Conidia are produced in acervuli.

Distribution. India.

Symptoms. Diseased leaves have dark brown circular spots that are surrounded by yellow halos. Spots on the diseased lower leaf surfaces are marked by concentric rings and have prominent black acervuli scattered throughout the spots.

Management. Not necessary.

Phanerochaete omnivorum Association

Cause. *Phanerochaete omnivorum.*

Distribution. The United States (Texas).

Symptoms. An orange resupinate fruiting body is produced on the lower stems and prostrate branches of diseased peanut plants.

Management. Not reported.

Phomopsis Foliar Blight

Phomopsis foliar blight is also called Phomopsis blight.

Cause. *Phomopsis phaseoli* (syn. *P. sojae*) survives as pycnidia or mycelia in infested residue.

Distribution. India and the United States.

Symptoms. Diseased leaflet margins have necrotic lesions that are brown to black and have a chlorotic zone between the healthy and necrotic tissue. Lesions start at leaflet tips and grow to the petioles in a V-shaped pattern. Pycnidia appear as small, black, pimple-like structures in rows parallel to midribs and smaller veins.

Discrete lesions have occurred in the center of leaflets. Such lesions are small (1–10 mm in diameter), circular to irregular-shaped, and surrounded by a red-brown margin. The centers of the lesions are white to tan, have a papery texture, and contain pycnidia.

Diseased stems become blackened.

Management. Since Phomopsis foliar blight is not considered to be a serious disease, disease management is not necessary.

Phyllosticta Leaf Spot

Cause. *Phyllosticta arachidis-hypogaea* and *P. sojicola* survive as pycnidia and saprophytic mycelia in infested residue. Phyllosticta leaf spot is most prevalent early in the growing season. Other causal organisms may also be involved.

Distribution. Wherever peanut is grown. Phyllosticta leaf spot is not considered to be an important disease.

Symptoms. Different types of symptoms may occur. Lesions are circular to oval (1.5–5.0 mm in diameter) with gray to tan centers and definite borders surrounded by red-brown margins. Sometimes the centers of lesions fall out, which gives diseased leaflets a "shot-hole" appearance.

Other symptom descriptions note that lesions are tan to red-brown with a dark brown margin surrounded by a chlorotic halo. Such lesions are found predominantly near the tips of diseased leaflets and extend along the leaflet midribs.

Management. Management measures are generally not considered necessary.

Phymatotrichum Root Rot

Phymatotrichum root rot is also called cotton root rot, Ozonium root rot, and Texas root rot.

Cause. *Phymatotrichopsis omnivora* (syn. *Phymatotrichum omnivorum*) survives for several years as sclerotia at depths of 30 to 75 cm in soil and on weeds; however, *P. omnivora* is very susceptible to freezing temperatures. Dissemination of *P. omnivora* is primarily by machinery moving infested soil within a field and from field to field. Phymatotrichum root rot is most severe in alkaline, poorly aerated black clay soils.

Distribution. The southwestern United States. Phymatotrichum root rot is not an important disease of peanut.

Symptoms. Diseased plants are stunted, chlorotic, and die. Diseased roots become tan and have strands of whitish to tan mycelia on the outside of the root.

Management. Disease management is generally not necessary, the following are aids to reduce disease:
1. Plow under green manure crops. Crop residues can also be plowed under if nitrogen is applied to promote rapid decay.
2. Rotate peanut with cereals or grasses.
3. Plow deeply during hot, dry weather to expose and kill the fungus.
4. Provide adequate fertility to ensure vigorous peanut growth.

Pod Wart

Pod wart may be the same disease syndrome as net blotch.

Cause. *Streptomyces* spp. that occur in warted tissue, soil, and the rhizosphere.

Distribution. Israel.

Symptoms. Dark brown, necrotic, wart-like lesions occur on diseased pods.

Management. Not reported.

Powdery Mildew

Cause. *Oidium arachidis.* Development of powdery mildew is optimum at a temperature of 25°C.

Distribution. Israel, Mauritius, Portugal, and Tanganyika.

Symptoms. Symptoms first occur in midsummer. Portions of the diseased upper leaflet surfaces are covered with a white, powdery growth consisting of the mycelium, conidia, and conidiophores of the fungus. As the disease progresses, upper leaflet surfaces become covered with large spots that have brown, necrotic centers.

Management. Grow the most resistant cultivar.

Pythium Damping Off

Cause. *Pythium aphanidermatum, P. debaryanum, P. irregulare, P. myriotylum,* and *P. ultimum. Pythium* spp. survive as oospores and mycelia in soil and infested residue. Oospores germinate and produce mycelia that may either infect plants directly by producing a germ tube or indirectly by forming sporangia. Zoospores form within sporangia, are released, and "swim" through moist soil to a host root, where they encyst. The encysted zoospores germinate and produce hyphae that infect host plants. Dissemination is by any means that moves soil, such as water and cultivation equipment.

In Oklahoma, greater populations of *Pythium* spp. were found in field soil containing peanut roots and pods than in soil without roots and pods.

Peanut diseases caused by *Pythium* spp. are more severe under high soil moisture and temperature conditions.

Distribution. Pythium damping off has been reported from several countries and likely occurs wherever peanut is grown.

Symptoms. Initially, diseased plants wilt. Water-soaked, necrotic tissue can be observed on hypocotyls and lateral branches near the soil line. Elongate, sunken, tan lesions may partially or completely encircle stems and extend upward for 2 to 4 cm above the soil line. Diseased seedlings frequently topple over at the soil line.

Management
1. Reduce irrigation water and allow topsoil to dry.
2. Treat seed with a seed-protectant fungicide.
3. Treat soil with a combined fungicide/nematicide.
4. Gypsum has been applied to soil with varying success. Deep plow to bury infested organic matter along with a gypsum application. The disparity in control may be due to differences in soils, with some soils retaining more calcium under irrigation. Another factor may be the inoculum potential of different soils.

Pythium Pod Rot

Cause. *Pythium myriotylum* is considered the major pod-rotting organism in the United States, but other *Pythium* spp. are also likely involved, often in combination with *Rhizoctonia solani* and *Fusarium* spp. *Pythium* spp. survive as oospores and mycelia in soil and residue. Oospores germinate and produce mycelia that may infect plants directly by forming a germ tube or indirectly by forming sporangia. Zoospores form within sporangia, are released, and "swim" through soil water to a plant host, where they encyst. The encysted zoospores germinate and produce hyphae that infect plants. Dissemination is by any means that moves soil, such as water and cultivation equipment.

The mite *Caloglyphus rodionovi* is associated with pod rot by enhancing the spread of the pathogen to adjacent healthy pods, and by introducing fungal propagules to the pod surface or the pod interior. However, it is thought the primary role of mites is as a disseminating agent, not a wounding agent. Colonization of pods may be regulated by pod development. Most commonly, *Pythium* populations peak several weeks after pegging, then rapidly decline by harvest time.

Distribution. Most peanut-growing areas.

Symptoms. Initially, diseased pods have a light brown discoloration and extensive water-soaking of the tissue. Eventually, entire pods appear watery and have a brown to black necrosis. Immature pods are usually completely destroyed. Seeds in mature fruit show various degrees of water-soaking and brown to black necrosis. Pegs may become infected as they contact wet soil, causing peg tips to become rotted and blackened.

Management
1. Reduce irrigation water and allow topsoil to dry.
2. Treat seed with a seed-protectant fungicide.
3. Treat soil with a combined fungicide/nematicide.
4. Gypsum has been applied to soil with varying success. Deep plow to bury infested organic matter along with the gypsum application. The disparity in control may be due to differences in soils, with some soils retaining more calcium under irrigation. Another factor may be the inoculum potential of different soils.

Pythium Wilt

Cause. *Pythium myriotylum* survives as oospores and mycelia in soil and residue. Oospores germinate and produce mycelia that may infect plants either directly by producing a germ tube or indirectly by forming sporangia. Zoospores form within sporangia, are released, and "swim" through moist soil to a host root, where they encyst. The encysted zoospores germinate and produce hyphae that infect plants. Dissemination is by any means that moves soil, such as water and cultivation equipment.

Distribution. Widespread.

Symptoms. Initially, one or more branches may wilt, but the entire diseased plant seldom wilts. Soon, foliage on wilted branches becomes chlorotic and scorched, beginning at the leaflet margins and extending inward until entire leaflets and, eventually, entire leaves are dry and crinkled. Petioles often remain green even if petiolules become dry. Vascular systems in the taproot-hypocotyl region show a brown to black discoloration.

Management. Not specifically reported but likely the same as for Pythium damping off.

Rhizoctonia Diseases

A portion of the disease syndrome is also called Rhizoctonia limb rot.

Cause. *Rhizoctonia solani,* teleomorph *Thanatephorus cucumeris,* is seedborne and survives as sclerotia in soil or infested residue and saprophytically as mycelia in infested residue. In a greenhouse study, peanut pods left in soil for 22 weeks after harvest contained *R. solani* AG-4, which infected and killed soybean and peanut seedlings. However, recovery of *R. solani* AG-4 was greater from infested shells placed on the soil surface than from infested shells buried in the soil. Most *R. solani* isolates that infect peanut are in anastomosis group AG-4 and rarely in AG-2.

Dissemination is by any means that moves soil containing sclerotia and mycelia. Basidiospores are disseminated by water and wind to host plants and infect either through wounds or directly through the epidermis. Disease development is favored by temperatures of 19° to 36°C and moderate soil moisture. Disease incidence is increased by tractor traffic.

Distribution. Wherever peanut is grown.

Symptoms. Seedlings may damp off before emergence. Lesions on the hypocotyls of emerged plants are sunken, elongated (2–3 cm), dark brown areas. A rapid browning of diseased hypocotyls sometimes occurs. Similar lesions develop on taproots and lateral roots. Infection of hypocotyls also results in an increase in total and individual lipid components.

Pods are subject to infection during all stages of development. Pegs and small pods first become discolored brown or black at the tips, then rot, and wither. Older pods have slight superficial russeting to browning of entire pods and a decay of their contents. Lesions on older pegs are discolored brown to black and vary from having a slight to an extensive sunken area. Diseased seed have discolored, faded, or stained seed coats. Cultivars with pink testae have a light brown to gray discoloration.

Older diseased plants have sunken, dark brown cankers on primary roots and browning of secondary roots. Stems have dry, sunken, dark brown lesions several centimeters long near the soil line. Stems are eventually girdled and diseased areas are thinner than "healthy" areas. With excess moisture, branches may become infected and appear brown and shredded. Lower diseased leaves have temporary brown speckled or blotchy areas that disappear as the leaves die. Diseased plants become wilted and have one to several branches that die.

Management
1. Plow under infested residue. Burial of peanut pods 20 to 25 cm deep with a moldboard plow could effectively reduce the inoculum from the root zone of the following crop.
2. Direct soil away from the plants during cultivation.
3. Do not rotate peanut with common bean, soybean, cotton, or southern pea.
4. Treat seed with a seed-protectant fungicide.
5. Apply fungicide to soil where disease potential is great.
6. Trace elements, especially copper, reduce root rot.

Rhizopus Seed and Preemergence Seedling Rot

Cause. *Rhizopus arrhizus* (syn. *R. oryzae*) and *R. stolonifer* are seedborne and survive in infested residue and soil, with the highest populations in the upper 15 cm of the soil profile. The fungi are heterothallic and may go through one of two life cycles. One cycle entails sexual reproduction and requires compatible and physiologically distinct mycelia. When compatible mycelia fuse, zygospores are formed that are capable of survival in soil for lengthy periods of time. Each zygospore germinates and forms a sporangium in which sporangiospores (chlamydospores) are formed.

The second life cycle involves sporangiospores that either survive for periods of time in soil or each one germinates and forms a sporangium in which more sporangiospores are formed. Sporangiospores may be disseminated a great distance by wind. Peanut seed and seedlings are infected by mycelia from germinating sporangiospores. Physiologic strains of *R. arrhizus* and *R. stolonifer* vary considerably in their responses to different temperatures.

Distribution. Wherever peanut is grown.

Symptoms. When infected seeds are sown, decay of seeds is very rapid. In 36 to 96 hours after sowing, the cotyledons are invaded first, followed by destruction of primary roots. A partial destruction of plumules and cotyledonary laterals results in stunted seedlings. Seed and emerged seedlings are reduced to a discolored dark brown to black, rotted, pulpy mass. Frequently, a mat of mycelium, with adhering soil particles, envelops each seed. In about 5 days, seeds are indistinguishable from the surrounding soil.

Management
1. Sow only healthy seed treated with a seed-protectant fungicide.
2. Plow deeply to bury infested organic litter 7.5 to 15.0 cm in the soil.
3. Control soilborne insects to prevent pod damage.

4. Harvest before pods and fruits become overly mature.
5. Cure peanuts as rapidly as possible after harvest and store in cool, dry storage facilities.

Rust

Rust is also known as peanut leaf rust.

Cause. *Puccinia arachidis* is a rust fungus in which only the uredinial and telia stages are known. However, some researchers do not consider the fungus to be a *Puccinia* sp. Hennen and Buritica (1993) propose the name *Peridipes arachidis* for the uredinia (conidiomata) of the peanut rust fungus. The telial stage is uncommon in most areas but is common on wild *Arachis* spp. in Brazil. Urediniospores do not survive long in infested residue; therefore, overseasoning occurs on plants in continuous cultivation and on volunteer plants. In North America, *P. arachidis* overwinters in the West Indies and tropics and is disseminated as windborne urediniospores to the southern United States. Other long-range dissemination includes movement of residue, pods, and seeds that are externally contaminated with urediniospores. Short-range dissemination is by wind, splashing water, and insects.

Plants of all ages are susceptible. Infection occurs at temperatures of 20° to 30°C and in the presence of moisture. Dew provides sufficient moisture to promote infection.

Distribution. Most peanut-growing areas of the world.

Symptoms. Pustules develop on all aerial plant parts except flowers. Uredinia are most common on diseased lower leaflet surfaces. Initially, whitish flecks appear on diseased lower leaflet surfaces 8 to 10 days after inoculation. A few hours later, yellow-green flecks appear on upper leaflet surfaces opposite the lower leaflet pustules. Uredinial pustules become visible in the whitish flecks on the lower leaf surface. At first, the pustules are yellow, then change to orange, then to tan, then to brown, and enlarge and rupture within 2 days. Uredinia are 0.5–1.4 mm in diameter, circular, and often surrounded by a light green to tan margin. Infection sites may coalesce and cause irregular-shaped patches of uredinia. Eventually, tissue surrounding infection sites becomes necrotic and dries up in irregular patches. Diseased leaflets may curl but tend to remain attached to plants. On highly susceptible genotypes, the original pustules may be surrounded by secondary pustules.

Management
1. Grow the most resistant cultivars.
2. Apply foliar fungicides.
3. Do not grow successive peanut crops in fields where viable urediniospores are present.

4. Eradicate volunteer peanut plants in a field.
5. Sow at times that avoid infection from outside sources and environmental conditions conducive to rust buildup.

Scab

Cause. *Sphaceloma arachidis* persists as mycelia in infested residue. Conidia are produced in acervuli and presumably function as primary inoculum. Scab may occur under both dry and humid conditions.

Distribution. Argentina and Brazil.

Symptoms. Small, round to irregular-shaped spots (approximately 1 mm in diameter) with sunken centers and raised margins occur beside the principal vein on both diseased leaf surfaces. On the upper leaf surface, spots are tan with narrow brown margins. During humid conditions, lesions are covered with a gray, velvety growth consisting of mycelium, conidia, and conidiophores. After conidia fall away, acervuli become evident as small brown to black objects within lesions. Spots on lower leaf surfaces are pink-brown to red and may have brown margins.

Petioles and branches have numerous spots that are a maximum of 3 mm long. Lesions coalesce and cover extensive areas and appear as cankerous growths that cause distortion of branches and petioles, making them appear wavy or sinuous. Diseased plants have a general "burned" appearance and look as if they are covered with "scabs."

Management. Grow resistant cultivars.

Sclerotinia Blight

Cause. *Sclerotinia minor* and rarely *S. sclerotiorum*. When *S. sclerotiorum* is found, it is in conjunction with *S. minor*. *Sclerotinia minor* overwinters as sclerotia in soil and is occasionally seedborne. Dissemination from field to field and from one geographic area to another is by several means, including infested residue, hay, irrigation, water, manure, sclerotia on surfaces of seed, windborne ascospores, and soil adhering to farm equipment, animals, or humans. Sclerotia in infested debris fed to animals remain viable after passing through their digestive tracts. Approximately 2% of the sclerotia of *S. sclerotiorum* remain viable after passing through the digestive tract of sheep.

During periods of cool temperatures and high relative humidity, sclerotia at or near the soil surface initiate infection directly. Sclerotia germinate myceliogenically and form fast-growing white mycelia that infect plant tissue near or in contact with soil. The reported optimum temperature for germination varies. Some researchers report the optimum temperatures to be from 20° to 25°C. Others report a temperature range of 6° to 30°C for germination, with 18°C being the optimum temperature. Relative humidity for germination is from 95% to 100% for more than 12 hours; however, some report relative humidity need only be greater than 94%.

Germination of sclerotia is favored by a soil pH 6.0 to 6.5 and the presence of infested residue in the canopy. At 6.0 to 6.5 pH in the presence of remoistened leaves, mycelia from sclerotia may be capable of direct infection without an exogenous food base.

Apothecia of *S. minor* are rarely observed during the growing season but are common on the soil surface in the spring and autumn. Sclerotia germinate and form mushroom-like structures called apothecia. Asci are formed in a layer on the apothecial surface. The ascospores liberated from the asci are disseminated by wind to nearby host plants and germinate under moist conditions. All parts of the plant, including branches, leaves, roots, pegs, and pods, can be infected, but infection of aerial parts is uncommon. Senescing or mechanically injured leaflets are easily colonized by *S. minor,* but they are not prerequisites for infection. Disease development is low when plants are small and are without a dense canopy or complete ground cover.

Distribution. Sclerotinia blight occurs in most peanut-growing countries of the world.

Symptoms. The initial symptom is a rapid wilting of branch tips. Cankers approximately 2.5 cm wide may girdle stems at the soil line. Diseased areas become sunken and dry and have a white, cotton-like, mycelial growth on them. Large, black sclerotia develop in mycelium on stem surfaces or inside stems. Diseased branch and peg tissue becomes shredded, which results in pod losses.

Management
1. Grow the least susceptible cultivars.
2. Apply foliar fungicides to give partial control.
3. Pruning peanut canopies to alter microclimate or enhance fungicide penetration may reduce disease and increase yield when *S. minor* damage is yield-limiting.

Sclerotinia Leaf Spot

Cause. *Sclerotinia homoeocarpa.*

Distribution. The United States (Florida) on perennial peanut, *Arachis glabrata.*

Symptoms. Circular, semicircular, or elliptical (1.3–1.9 cm) leaf spots are tan in the center and surrounded by a brown to purple ring. Leaf spots were reported to be associated with weak and dying plants in a thin stand.

Management. Not necessary because the disease is of little economic consequence when forage is harvested at recommended intervals.

Southern Blight or Stem Rot

Southern blight is also called foot rot, root rot, Sclerotium blight, Sclerotium rot, Sclerotium wilt, and white mold.

Cause. *Sclerotium rolfsii* survives as sclerotia in soil and saprophytically as mycelia in infested residue. *Sclerotium rolfsii* is most active near the soil surface because of an adequate food base and absence of competitive or antagonistic fungi. Antagonism may be increased by high soil moisture.

Infection of host plants occurs either by sclerotia germinating under low relative humidity and producing infection hyphae or by mycelia growing through soil from a precolonized substrate. Sclerotia may not have sufficient reserve energy to establish mycelia in a living host. A precolonized substrate is utilized as an energy source or food base from which mycelia can grow to and infect a healthy host plant. Growth of mycelia from these food bases may be an important mode of infection. Eventually, sclerotia form in mycelia on diseased tissue. Temperatures of −2°C to −10°C kill mycelia and germinating sclerotia, but do not kill dormant mature sclerotia. Disease is favored by warm and moist environmental conditions, which may be accentuated under a dense foliar canopy.

Distribution. Wherever peanut is grown.

Symptoms. Symptoms caused by infection of stems at the soil line are most common. Stem bases are covered with elongated, eroded lesions that are tan to reddish in color. During dry weather, brown lenticular lesions may occur on stems just below the soil surface. Diseased stems are overgrown with white mycelium near the soil line, particularly if a leaf canopy is present to maintain high humidity. Diseased areas of stems become shredded and numerous sclerotia are produced in the mycelium growing over the stems. Excessive moisture may prevent mycelium from growing on the outside of diseased stems. Sclerotia are spherical, velvety-appearing, soft, and initially white but eventually become light brown to brown. Mycelium also grows over the soil surface and organic debris.

Initially, branches suddenly wilt and leaves become chlorotic, then brown as they dry up. Branches adjacent to the first diseased branches then wilt and die. Eventually, all branches on a plant will wilt, but dead plants tend to remain upright.

Pegs have light to dark brown lesions (0.5 to 2.0 cm long) that eventually cause tissue shredding and pod loss. Lesions on young pods of 'Spanish' peanuts are orange-yellow to light tan. Older pods have light brown to black discolored zonate lesions. Severely decayed kernels are shriveled and lightly covered with mycelium.

Management
1. Grow resistant cultivars.
2. Bury surface residue by deep plowing.

3. Sow seed in a flat or slightly raised bed area.
4. Move soil away from the row during cultivation to prevent accumulation of debris around the bases of plants.
5. Manage early leaf spot and late leaf spot to prevent leaf drop and subsequent accumulation of dead leaves at the bases of plants.
6. Apply fungicides to soil before planting.
7. Rotate peanut with cotton, small grains, or maize.
8. Apply transparent polyethylene sheets to soil in the off-season to raise the soil temperature, where practical.

Verticillium Wilt

Verticillium wilt is also called Verticillium pod rot.

Cause. *Verticillium dahliae* and *V. albo-atrum*. *Verticillium dahliae* is considered the primary causal agent and survives as microsclerotia on infested residue and possibly on weeds. When residue decomposes, microsclerotia are released into the soil or remain imbedded in bits of residue in the upper soil profile for several years. Root exudates stimulate microsclerotia to germinate and produce mycelia that infect the root system of host plants. Mycelium then grows into the xylem tissue of the infected plant and blocks water from moving up the plant. Sclerotia are disseminated by being moved in soil and residue by machinery, water, and wind and possibly by seed.

Distribution. Verticillium wilt occurs in all peanut-growing countries, but the disease is not considered a serious problem.

Symptoms. Symptoms occur at flowering time as a dull green or chlorotic discoloration of the lower leaflets. Eventually, many leaflets over an entire diseased plant become withered, brown, and defoliate. If adequate moisture is present, diseased plants remain alive but are stunted, have sparse foliage, and become relatively unproductive. In the advanced stages of the disease, the vascular system of roots, stems, and petioles has a brown to black discoloration. Diseased pods are black, rotted, and sprinkled with white powdery patches that consist primarily of conidia.

Management
1. Grow resistant cultivars in areas where they are adapted. Bunch-type peanuts are more resistant than 'Valencia' and 'Spanish' types.
2. Rotate infested soil to grass, grain, sorghum, or alfalfa.
3. Practice clean fallow with occasional plowing during dry periods to deplete the amount of soilborne inoculum.
4. Practice field sanitation, such as burning or removing infested residue, to reduce inoculum.
5. Irrigate infested fields frequently to reduce stress.
6. Practice good weed control.

Web Blotch

Web blotch is also called Ascochyta, leaf spot, muddy spot, net blotch, Phoma leaf spot, and spatselviek.

Cause. *Phoma arachidicola*, teleomorph *Didymosphaeria arachidicola,* overwinters as pycnidia, pseudothecia, chlamydospores, and clusters of chlamydospores called microsclerotia. Ascospores and pycnidiospores serve as primary inoculum. Under experimental conditions, chlamydospores have also been reported to serve as primary inoculum. Disease is most severe during cool temperatures and high relative humidity.

Distribution. Africa, Asia, Australia, North America, and South America.

Symptoms. Several different symptoms may be present. The most distinct symptom is a webbing or netting pattern caused by brown fungal strands that radiate beneath the waxy cuticle of leaf surfaces from different points of infection. Webbing may occur without blotch symptoms, but this symptom occurs more commonly when blotches are present.

Circular, dark brown blotches have a lighter-colored, irregular margin. Under moist conditions, blotches are surrounded by a grayish margin. Varying degrees of webbing may surround blotches.

Another symptom develops during conditions favorable for rapid disease development. Tan spots are surrounded by gray margins made up of fungal strands growing from the dark centers of spots.

Management
1. Apply a foliar fungicide during the growing season.
2. Plow under crop residue.
3. Rotate peanut with other crops.

Yellow Mold

This disease may be similar to "Aflaroot" as described by Chohan and Gupta (1968).

Cause. *Aspergillus flavus* and *A. parasiticus* survive as conidia and hyphal fragments in soil or infested residue. The fungi are seedborne and also are disseminated by wind, water, and any means that moves soil. The fungi grow on most plant material below or above the soil surface and in most soil types.

Infection by *Aspergillus flavus* and *A. parasiticus* is optimum over a wide range of soil moistures at a temperature of 32°C. *Aspergillus flavus* and *A. parasiticus* infect through wounds or other injuries made during harvest when the moisture content of peanuts is 12% to 35%. The fungi produce the toxin called aflatoxin. When peanut follows maize in a rotation, fungal populations increase in the lower half of the plow layer.

Distribution. Wherever peanut is grown.

Symptoms. Cotyledons of germinating seed are invaded first, then the emerging radicle and hypocotyl are infected and rapidly decay. Seed and seedlings that have not emerged are shriveled and dried and become a brown to black "mass" 4 to 8 days after sowing. Diseased plant parts may be covered with a yellow-green growth consisting of fungal conidia, conidiophores, and mycelium.

Management
1. Use an inverter during the digging operation. Inverting allows peanuts to dry faster and minimizes fruit contact with the soil.
2. Dry peanuts to a 12% or less moisture content after harvest. Do not allow a container full of peanuts to stand without proper drying.
3. Minimize injury to pods during harvest.
4. Store peanuts in cool, dry facilities.

Zonate Leaf Spot

Cause. *Cristulariella moricola* (syn. *C. pyramidalis*). The sclerotial stage is *Sclerotium cinnamomi*. *Cristulariella moricola* has been associated with leaf spots on several deciduous trees.

Distribution. The United States (Georgia). Zonate leaf spot is considered to be of minor importance.

Symptoms. Symptoms occur late in the season. Necrotic spots (1–13 mm in diameter) occur on diseased leaves. Small spots have a light brown center surrounded by a darker brown ring of necrotic tissue. Large spots have a zonate appearance on both diseased leaf surfaces.

Management. Not reported.

Diseases Caused by Nematodes

Aphasmatylenchus straturatus

Cause. *Aphasmatylenchus straturatus* is a migratory endoparasite and ectoparasite. The nematode survives the dry season adjacent to roots of the karite tree, *Butyrospermum parkii*.

Distribution. Burkina Faso.

Symptoms. Diseased plants are generally stunted and have chlorotic foliage and reduced root development. *Rhizobium* nodulation and yields are reduced.

Management. Not reported.

Dagger

Cause. *Xiphinema americanum* and *X. diversicaudatum.*

Distribution. Widespread, but populations in peanut fields tend to be low.

Symptoms. Galls and curly tips are produced on diseased roots.

Management. Not necessary.

Pod Lesion or Kalahasti Malady

Cause. *Tylenchorhynchus brevilineatus* (syn. *T. brevicadatus*).

Distribution. India.

Symptoms. Small brown-yellow lesions occur on diseased pegs, pod stalks, and young developing pods. The margins of the lesions are slightly elevated. Pod stalks are greatly reduced in length, and in advanced stages of disease, the pod surface is completely discolored. Diseased plants are stunted and root growth is reduced at maturity. Pods are reduced in size, but kernels from such pods are apparently healthy.

Management. Application of a nematicide in research plots has been effective.

Ring

Cause. *Criconemella ornata* is thought to be ectoparasitic.

Distribution. The United States.

Symptoms. Diseased plants are chlorotic. Roots, pods, and pegs are discolored and have brown necrotic lesions that are superficial if small but extend deep into tissue if they are large. Young roots and root primordia may be killed. Pod yields are reduced.

Management
1. Rotate peanut with nonhost crops.
2. Apply nematicides.

Root Knot

Cause. The peanut root knot nematode, *Meloidogyne arenaria,* and the northern root knot nematode, *M. hapla.* The Javanese root knot nematode, *M. javanica,* also has been reported to attack peanut. *Meloidogyne arenaria* and *M. javanica* do not survive if the temperature averages below 3°C during the coldest month. *Meloidogyne hapla* is limited by temperatures at or greater than 24° to 27°C.

   Larvae are produced from eggs in infested soil and move through the soil to penetrate root tips. Larvae migrate through the root to vascular tissues, become sedentary, and commence feeding. Excretions of larvae stimulate cell proliferation, which results in development of syncythia from which food is derived. Females lay eggs that are pushed out of roots into the soil.

Distribution. The United States.

Symptoms. Mild symptoms caused by *M. arenaria* are a slight yellowing of foliage and the imperceptible stunting of diseased plants. Severe disease symptoms consist of stunted plants that are chlorotic, wilt, and die during dry weather. Along their length, roots and pegs have different-sized galls that may be several times the diameter of a normal root or peg. Pods develop galls that appear as knobs, protuberances, or small warts. Galls on roots, pegs, and pods sometimes begin to deteriorate by maturity and the resulting necrotic tissue is colonized by a number of fungi.

The northern root knot nematode causes similar aboveground symptoms, but galls are usually smaller than those caused by *M. arenaria*. Diseased roots tend to branch near the point of invasion, resulting in a dense, bushy root system. Total crop loss may occur when disease caused by both nematodes is severe.

Management
1. Rotate peanut with a nonlegume crop for 2 to 3 years.
2. Apply nematicides either banded at time of planting or as a fumigant, using plow-sole method during deep plowing of soil prior to sowing.

Root Lesion

The root lesion is also called the meadow nematode.

Cause. *Pratylenchus brachyurus, P. coffeae,* and *P. scribneri.* Both adults and larvae infect roots, pegs, and pods, but higher nematode populations are present on the pegs and pods. The nematodes directly penetrate plant tissues and feed on the parenchyma tissue.

Nematode numbers may increase on roots of other plants, especially rye, during winter months.

Distribution. The United States.

Symptoms. Symptoms appear on plants in circular areas within fields. Diseased plants are stunted and chlorotic because of small, discolored roots. Pegs have brown lesions, and pods initially have small, tan spots with dark centers that eventually become discolored brown to black, angular-shaped lesions. Lesions may become so numerous that the entire pod is discolored. Damage caused by nematodes allows secondary fungi to enter and cause further discoloration and decomposition of roots, pods, and pegs.

Management
1. Apply nematicides that are either banded at time of sowing or as a fumigant, using plow-sole method during deep plowing of soil prior to planting.
2. Fallowing reduces the nematode population.

Seed and Pod

Cause. *Ditylenchus africanus* is a migratory endoparasite that completes its life cycle from egg to adult in 8 days at 30°C. The nematode enters the pods soon after pegging. Nematodes have been observed in the exocarp, endocarp, testa, and embryo and on cotyledons. Eggs have been observed in all invaded tissues.

Distribution. South Africa (Transvaal).

Symptoms. Diseased hulls have a black discoloration that first appears along longitudinal veins. Kernels are shrunken, testae are brown to black, and embryos are brown. Early germination may occur and seed weight may be greatly reduced.

Management. Not reported.

Spiral

Cause. *Scutellonema cavenessi* is active during the rainy season. The nematode goes into anhydrobiosis when soil moisture drops to 0.2%.

Distribution. Mali, Nigeria, and Senegal.

Symptoms. Foliage of diseased plants is chlorotic. Roots have reduced growth and *Rhizobium* nodulation.

Management
1. Apply nematicides.
2. Leave fields fallow between crops.

Sting

Cause. *Belonolaimus longicaudatus* and *B. gracilis* are nematodes with a wide host range. Nematodes feed at root tips and along sides of succulent roots and other belowground parts without becoming attached to the roots. The nematodes are restricted to soils with 84% to 94% sand and occur in the top 30 cm of the soil profile. The nematodes are most active at temperatures of 20° to 34°C.

Distribution. The United States.

Symptoms. Diseased plants are stunted and chlorotic and have stubby, sparse roots. Roots and pods may have small, dark, necrotic spots.

Management
1. Rotate peanut with a nonlegume crop for 2 to 3 years.
2. Apply nematicides that are either banded at the time of sowing or a fumigant, using the plow-sole method during deep plowing of soil prior to sowing.

Testa

Cause. *Aphelenchoides arachidis* is a facultative endoparasite found primarily in the parenchymatous tissue and around the tracheids of the seed, testae, shells, roots, and hypocotyls. *Aphelenchoides arachidis* is disseminated in infected seed and can survive desiccation in stored pods for up to 12 months. Volunteer plants may also serve as a source of inoculum. Seed is predisposed to invasion by soilborne fungi.

Distribution. Nigeria.

Symptoms. Seed coats are discolored and translucent and have dark vascular strands. Seed coats may become unevenly thickened.

Management
1. Immerse seed in water at a temperature of 60°C.
2. Sun-dry pods after harvest.

Diseases Caused by Phytoplasmas

Peanut Yellows

Peanut yellows is sometimes referred as peanut yellows diseases.

Cause. Aster yellows phytoplasma-like organism is transmitted by the leafhopper *Macrosteles quadrilineatus* and possibly other leafhoppers.

Distribution. Asia.

Symptoms. Diseased plants are chlorotic.

Management. Not reported.

Witches' Broom

Cause. A phytoplasma-like organism.

Distribution. China, India, Indonesia, Japan, Taiwan, and Thailand.

Symptoms. Diseased plants are bushy due to the excessive proliferation of axillary shoots. Leaflets are chlorotic and smaller than normal. Pegs tend to grow upward. Yields may be greatly reduced.

Management. Not reported.

Disease Caused by Rickettsia

Rugose Leaf Curl

Cause. A rickettsia-like organism that is transmitted by leafhoppers.

Distribution. Australia.

Symptoms. Leaflets are distorted and puckered.

Management. Not reported.

Diseases Caused by Viruses

Bean Yellow Mosaic

Cause. Bean yellow mosaic virus (BYMV) is in the potyvirus group. BYMV is transmitted mechanically and by the cowpea aphid, *Aphis craccivora*, in a nonpersistent manner. Arrowleaf clover, *Trifolium vesiculosum*, is a reservoir for BYMV.

Distribution. The United States (Georgia and Texas).

Symptoms. Initial symptoms are chlorotic rings and spots on diseased leaves. After 2 to 3 weeks, the symptoms disappear.

Management. Not reported.

Cowpea Chlorotic Mottle

Cause. Cowpea chlorotic mottle virus (CCMV) is in the bromovirus group.

Distribution. The United States (Georgia).

Symptoms. Infected plants are symptomless. However, CCMV is reported to have significant disease potential when it is in mixed infections with other viruses.

Management. Not reported.

Cowpea Mild Mottle

Cause. Cowpea mild mottle virus (CPMMV) is in the carlavirus group. CPMMV is transmitted by sap and by the whitefly, *Bemisia tabaci*.

Distribution. India. Cowpea mild mottle is not considered to be a serious disease.

Symptoms. The youngest leaves show a vein clearing followed by a downward rolling. Later, diseased leaves and petioles become necrotic and defoliation occurs. Diseased plants are severely stunted and rarely produce pods.

Management. Avoid growing peanut near soybean or cowpea fields.

Cucumber Mosaic

Cause. A strain of cucumber mosaic virus (CMV-CA). CMV is in the cucumovirus group and is transmitted by seed and by the aphid *Macrosiphum euphorbiae.* CMV infects 31 species of plants in six families.

Distribution. China.

Symptoms. Chlorotic spots occur on young emerging leaves. Other symptoms include young expanded leaves that are small, rolled, and generally chlorotic. A mosaic or mottling occurs on some leaves. Diseased plants are stunted to one-half to two-thirds the size of a healthy plant.

Management. Not reported.

Eyespot

Eyespot may be similar to or the same disease as groundnut eyespot, although the symptoms reported in the literature appear to be different.

Cause. Peanut eyespot virus (PEV) is in the potyvirus group. PEV is transmitted by sap and by the cowpea aphid, *Aphis craccivora,* in a nonpersistent manner.

Distribution. Ivory Coast.

Symptoms. Dark green spots on diseased leaves are surrounded by chlorotic spots, which give an eyespot symptom consisting of concentric colored rings.

Management. Not reported.

Groundnut Crinkle

Cause. Groundnut crinkle virus is transmitted mechanically and possibly by whiteflies.

Distribution. Ivory Coast.

Symptoms. Plants are slightly stunted and have a slight reduction in leaf size. Leaves are faintly mottled and display a crinkling of the lamia.

Management. Not reported.

Groundnut Eyespot

Groundnut eyespot may be similar to or the same disease as eyespot, although the symptoms reported in the literature appear to be different.

Cause. Groundnut eyespot virus.

Distribution. West Africa. Groundnut eyespot is not considered a serious disease.

Symptoms. Yellow eyespots occur on diseased leaves. Banding along the main veins may occur.

Management. Not reported.

Groundnut Rosette

Groundnut rosette is also called rosette.

Cause. Groundnut rosette virus (GRV) is an umbravirus and is transmitted mechanically. GRV depends on the groundnut rosette assistor virus (GRAV), a luteovirus, for transmission by the black cowpea aphid, *Aphis craccivora,* in a persistent or circulative manner. However, GRAV causes no symptoms in peanut on its own. Neither virus is seedborne. Variants of the satellite RNA, which differ in different parts of Africa, are responsible for the different symptoms.

Two forms of GRV exist: chlorotic rosette (GRV-C) and green rosette (GRV-G). Aphids acquire GRV-C within 4 hours and GRV-G within 8 hours. Median latent periods are 26.4 and 38.4 hours for GRV-C and GRV-G, respectively. After a 24-hour latent period, both viruses are transmitted within 10 minutes. Multiple plant infections occur more frequently with GRV-C than with GRV-G. Maximum retention time for both viruses is the lifetime of aphids or about 14 days.

Peanut is the main source of inoculum. Diseased plants survive longer than healthy plants and are not normally harvested. Aphids colonize diseased plants, then rainy fronts are responsible for the dissemination of the aphids.

Distribution. Gambia, Java, Madagascar, Nigeria, Senegal, Sierra Leone, Tanganyika, and Uganda. Chlorotic rosette occurs throughout Africa south of the Sahara Desert. Green rosette disease occurs in western Africa and Uganda, and mosaic rosette occurs only in eastern and central Africa. Groundnut rosette has also been reported from Argentina, India, Indonesia, and the Philippines.

Symptoms. There are three basic types of symptoms: chlorotic rosette, green rosette, and mosaic rosette.

Chlorotic rosette symptoms, which are caused by GRV-C, initially occur on young leaflets as a faint mottling with a few green islands. Later, leaflets are pale yellow with green veins. Diseased plants infected when they are young produce progressively smaller, chlorotic, curled, and distorted leaflets. Symptoms on older infected plants may be restricted to a few branches or to the apical portion of the plants. Plants infected when they are young are severely stunted and have thickened stems and a severe reduction in the number and size of pods.

Green rosette symptoms, which are caused by GRV-G, occur on young leaflets as a mild chlorotic mottling and isolated flecks. Although symptoms are masked in older leaflets, the older leaflets are reduced in size and display outward rolling but are not distorted. Plants infected while young are severely stunted and are a darker green than healthy plants.

Mosaic rosette symptoms occur on young leaflets as conspicuous mosaic symptoms that resemble those of chlorotic rosette except stunting is less severe.

Management

1. Destroy volunteer and unharvested diseased plants.
2. Sow early in the season at a high seeding rate.
3. Apply insecticides.
4. Sources of resistance are available in peanut germplasm.

Groundnut Streak

Cause. Groundnut streak virus (GSV).

Distribution. Ivory Coast.

Symptoms. Necrotic streaks along leaflet veins in seedlings disappear as diseased plants mature.

Management. Not reported.

Indian Peanut Clump

Indian peanut clump is also called peanut clump.

Cause. Indian peanut clump virus (IPCV) is in the furovirus group. Two different isolates have been identified: the West African and the Indian. However, the Indian virus is serologically unrelated to the West African virus and is considered by some researchers to cause a separate disease. The soil fungus *Polymyxa graminis* transmits the West African IPCV and is thought to transmit the Indian isolate also. Both isolates are seed-transmitted.

Polymyxa graminis survives in field soil as cystosori. Cysts give rise to zoospores, which "swim" through free soil water until they contact a host root and encyst. Encysted zoospores produce a structure called a stachel through which zoosporic cytoplasm enters the host cell and becomes a plasmodium. Only primary root tissue of young roots is infected; optimum infection occurs at a temperature of about 25°C. The host cell then becomes infected with IPCV if *P. graminis* is viruliferous. After a period of time, the plasmodium develops into a zoosporangium, which releases additional zoospores that repeat the infection cycle. However, some plasmodia develop into cysts and often nearly every cell in the small feeder roots will contain a cyst. As root cells senesce, cysts are eventually released into the soil, where they can remain viable for years without loss of virulence.

IPCV is considered by some researchers to cause a different disease than peanut clump virus.

Distribution. India and West Africa.

Symptoms. Symptoms are areas of diseased plants within fields that enlarge in succeeding years. Diseased plants are stunted and quadrifoliates are small initially and have mosaic mottling and chlorotic rings but subsequently become dark green and have faint mottling. Plants eventually become bushy and have small, dark green leaves. Several flowers are produced on erect petioles. The number and size of pods is reduced, resulting in smaller-sized seeds. Root systems are small and diseased roots darken with an epidermal layer that easily peels off. However, isolates from different locations in India caused different symptoms on five different host range plants.

Management
1. Sow virus-free seed.
2. Grow resistant cultivars in India.

Marginal Necrosis

Cause. Possibly a virus. The disease agent is seed-transmitted.

Distribution. Papua New Guinea.

Symptoms. Diseased leaflets have marginal necrosis and crinkling.

Management. Not reported.

Peanut Bud Necrosis

Peanut bud necrosis is also called peanut ringspot, spotted wilt, tomato spotted wilt, and tomato ringspot.

Cause. Peanut bud necrosis is caused by both the peanut bud necrosis virus (PBNV) and the tomato spotted wilt virus (TSWV). PBNV and TSWV are in the tospovirus group. In Louisiana, virus inoculum originates from noncultivated areas. TSWV overwinters in several species of perennial plants in areas that have cold winter temperatures and in annual plants where winters are not cold enough to kill plants.

The presence of TSWV has been reported in spiny amaranthus, *Amaranthus spinosus*; blackseed plaintain, *Plantago rugelii*; buttercup, *Ranunculus* spp., with *R. sardous* being the species most often associated with natural TSWV infection; coneflower, *Rudbeckia amplexicaulis*; horsenettle, *Solanum carolinense*; dandelion, *Taraxacum officinale*; spiny sow-thistle, *Sonchus asper*; blue vervain, *Verben brasiliensis*; and *Lactucua* spp., with *L. floridana* as an important overwintering host.

TSWV is transmitted mechanically and by several species of thrips, including potato or onion thrips, *Thrips tabaci*; blossom or cotton thrips, *Frankliniella schultzei*; western flower thrips, *F. occidentalis*; and tobacco thrips, *F. fusca*. The chili thrip, *T. setosus*, has been shown to transmit the virus in Japan. TSWV must first be acquired by the larval stage; the subsequent adults are able to transmit TSWV. TSWV persists in insects as long as the insect lives, but the virus is not transmitted through insect eggs. Disease is more severe during high temperatures and abundant moisture.

Distribution. Australia, Brazil, India, Nigeria, South Africa, Sri Lanka, and the United States (Alabama, Florida, and Georgia).

Symptoms. Symptoms caused by TSWV and PBNV are similar. Young diseased leaflets have chlorotic spots or a mild mottle that develops into necrotic and chlorotic rings and streaks. The bud necrosis symptom occurs when temperatures are above 30°C during the day, when petioles bearing fully expanded leaflets with initial symptoms usually become flaccid and droop. This symptom is followed by a necrosis of the terminal bud. When a young plant is infected, the necrosis spreads toward the base of the plant, resulting in plant death.

Other symptoms include stunting and proliferation of axillary shoots. Leaflets produced on the axillary shoots are reduced in size and show puckering, mosaic mottling, and general chlorosis. These symptoms are most common on early-infected plants and give them a stunted and bushy appearance. Plants infected later may also be stunted but the symptoms may be restricted to a few branches or to the apical parts of the plants.

Similar symptoms are reported for the bud necrosis disease from India. The quadrifoliate leaf below the terminal bud displays distinct chlorotic ring spots or chlorotic speckling and often becomes flaccid. Terminal buds become necrotic, followed by a proliferation of small shoots with mottled leaves of a much reduced size. In advanced stages of the disease, the whole plant becomes bushy and stunted and often dies. Pods are distorted and reduced in size. Early infection results in a greater yield loss than that caused by a later infection.

Seed from early-infected plants are small and shriveled, and the seed coats are discolored red, brown, or purple and are mottled. Late-infected plants produce seed of normal size that has mottled and cracked seed coats.

Management
1. Good sources of resistance have been identified.
2. Sow early in the growing season to avoid thrips.
3. Use high-quality seed treated with a fungicide seed protectant.
4. Sow at the recommended rate and spacing to give optimum plant populations.

Peanut Chlorotic Ringspot

Peanut chlorotic ringspot is also called chlorotic spot or peanut chlorotic spot.

Cause. Peanut chlorotic ringspot virus is transmitted mechanically.

Distribution. India.

Symptoms. Yellow spots (0.5–1.0 mm in diameter) first occur on young terminal leaves of the side branches. Spots enlarge and become irregular-shaped. Some of the spots develop rings around them, giving the appearance of ring spots.

Management. Not reported.

Peanut Chlorotic Streak

Cause. Peanut chlorotic streak virus (PClSV) is in the caulimovirus group. PClSV is mechanically transmissible to several plants in the Leguminosae and Solanaceae. Disease is favored by the very high temperatures of the semiarid tropics.

Distribution. India.

Symptoms. Young diseased leaflets have oval, chlorotic streaks along the veins 3 to 4 weeks after inoculation. Streaks are not distinct in older leaflets and can be seen only when viewed against light. In field infections, diseased plants are stunted and have chlorotic streaks on the younger leaflets.

Management. Not reported.

Peanut Green Mosaic

Cause. Peanut green mosaic virus (PGMV) is in the potyvirus group. PGMV can be mechanically transmitted.

Distribution. India.

Symptoms. Initially, chlorotic spots and vein clearing occur on young quadrifoliates, but later a severe mosaic occurs. The number of pegs, pods, and leaves, the amount of leaf area, and the dry weight of shoots, roots, and pods are reduced. Premature leaf drop also occurs.

Management. Bavistin (50% w/w Carbendazim) applied as a foliar spray reduced severity of symptoms in research trials.

Peanut Mild Mottle

Cause. Virus producing mild mottle (VPMM); the virus sometimes is referred to as peanut mild mottle virus. It has been proposed that this virus should be considered the same as the peanut stripe virus. VPMM is transmitted mechanically and in a nonpersistent manner by the black cowpea aphid, *Aphis craccivora,* and the green peach aphid, *Myzus persicae.*

Distribution. China.

Symptoms. Leaves are mildly mottled. Top leaflets frequently have distinct chlorotic spots or ring spots that disappear after several days and become mottled with dark green islands on a light green background. Most leaf symptoms disappear during high temperatures. Diseased plants are not stunted but yields are reduced.

Management. Not reported.

Peanut Mottle

Cause. Peanut mottle virus (PMoV) is a potyvirus that is disseminated mechanically and in a nonpersistent manner by several aphid species. Another important means of dissemination is by seed. PMoV is found at a low rate in embryos but not in seed coats or cotyledons. Wild peanut, *Arachis chacoense,* serves as a reservoir for PMoV. Different strains of PMoV exist.

Distribution. Worldwide.

Symptoms. Diseased plants have a leaf mottling, leaflets that curl upward, and depressed interveinal tissue. A mild mottle that occurs on the youngest leaves is best observed by transmitted light. Symptoms tend to become obscure during hot, dry weather as diseased plants mature. Seeds are discolored, and diseased pods are smaller than healthy pods and have gray to brown patches. Yield and nodulation is reduced.

Symptom expression varies with the strains of *Rhizobium* that produce nodules on the diseased plant. Plants harboring an ineffective *Rhizobium* strain show more severe symptoms than plants harboring an effective strain.

Management. Grow the most tolerant cultivar.

Peanut Stripe

Cause. Peanut stripe virus (PStV) is in the potyvirus group. Two symptom variants of peanut stripe virus occur: stripe (PStV-S) and blotch (PStV-B). It has been proposed that the virus producing mild mottle, peanut mild mottle virus, and peanut chlorotic ring mottle virus, a virus found on peanut in Thailand, should also be considered as PStV. Two different strains of PStV have also been identified in Taiwan and identified as PStV-Ts and PStV-Tc. The two strains are indistinguishable serologically.

Peptide profile data suggest that PStV is a strain of bean common mosaic virus.

PStV is transmitted mechanically, in a nonpersistent manner by aphids, and is seedborne in cotyledons and embryos.

Distribution. China, India, Indonesia, Malaysia, the Philippines, Taiwan, Thailand, and the United States, primarily in research areas of Florida, Georgia, North Carolina, Oklahoma, Texas, and Virginia.

Symptoms. PStV-S causes dark green stripes along the lateral leaf veins. Initially, there is a discontinuous dark striping or banding along lateral veins of diseased leaves that resembles a sergeant's stripes. Later, on older leaves, the striping fades and a mild oak-leaf pattern occurs and becomes the predominant symptom expression.

On diseased leaves, PStV-B infrequently causes large, dark green, circular areas or blotches that are not associated with veins. This is called peanut blotch.

PStV-Ts induces severe mosaic and systemic necrotic symptoms. PStV-Tc causes stripe symptoms.

Management. The University of Georgia has issued the following guidelines for managing PStV in Georgia:

1. Only Georgia-certified seed shall be sown in research plots with the exception of breeding and variety performance tests.
2. Seed from research plots shall not be processed through commercial cleaning and shelling operations.
3. Peanuts from experimental plots shall be processed or sold for processing only and are not to be used for seed except in breeding tests.
4. All residual peanut seed and debris shall be removed from harvesting, handling, and transporting equipment before such items are removed from areas contaminated with PStV to clean areas.
5. All research plots containing plants inoculated with virus shall be grown under screen cages with proper precautions to prevent spread and shall be planted only on experiment station land.
6. Breeders shall release only seed that is free from PStV.
7. Virus-free seed to be retained for future seed production shall not be sown in areas where PStV-infected peanuts have been grown previously.
8. Virus-free seed shall not be sown in proximity to leguminous crops or other hosts, and rigid weed control shall be practiced in and around virus-free plots.
9. Researchers shall not import peanuts unless they are either tested for PStV before sowing or grown in screened isolation for one growing season to allow for visual inspection and serological assay if necessary.
10. Peanut breeding plots shall be isolated from other research plots.
11. The Uniform Peanut Performance Trial shall be sown exclusively at one university farm.

12. Other legume crop-breeding nurseries shall be separated from peanut research plots.
13. All seed shall be removed from PStV-infected research plots and plots shall be fumigated.
14. Infected seed, except breeders' seed, shall be destroyed.

Other:

15. Some wild peanut accessions are resistant.

Peanut Stunt

Cause. Peanut stunt virus (PSV) is in the cucumovirus group. PSV overwinters primarily in white clover, *Trifolium repens;* crown vetch, *Coronella varia;* and other legumes. PSV is transmitted by mechanical means, grafting, dodder, and three species of aphids in a nonpersistent manner. PSV is rarely seedborne and this is not a factor in transmission. PSV has a wide host range.

Distribution. Europe, Japan, Morocco, Sudan, and the United States.

Symptoms. Diseased plants are severely dwarfed, but stunting may involve the entire vine or only a portion of the plant. Foliage is malformed and only a few to half of the leaves on a stem are normal. Apical to normal leaves are stiff and erect, and pointed leaves have light green leaflets that are less than half normal size. Discoloration is variable; the entire leaf or plant part may be slightly chlorotic, severely chlorotic, or mildly mottled or may be chlorotic and have dark green vein banding. Affected leaves also may be curled upward. Fruits are small and poorly shaped and have split pericarps.

Management
1. Do not grow peanut in the vicinity of legumes.
2. Rogue out infected plants.

Peanut Yellow Mottle

Cause. Peanut yellow mottle virus.

Distribution. Nigeria.

Symptoms. Young diseased plants have a bright yellow mottle. Yields are reduced.

Management. Not reported.

Unnamed Virus Disease

Cause. Virus isolated from an Erectoides peanut hybrid growing in a greenhouse in Stillwater, Oklahoma. The virus is transmitted mechanically and by grafting.

Distribution. The United States (Oklahoma).

Symptoms. A wide chlorotic ringspot is associated with a chlorotic line pattern and mottling on diseased leaves. The virus causes a mosaic in peas, a severe malformation and reduction in the leaf area of the lupine cultivar 'Tiftwhite,' and necrotic local lesions and chlorotic ringspot of the bean cultivars 'Topcrop' and 'Chenopodium,' respectively.

Management. Not reported.

Selected References

Adams, D. B., and Kuhn, C. W. 1977. Seed transmission of peanut mottle virus in peanuts. Phytopathology 67:1126-1129.

Alderman, S. C., and Nutter, F. W., Jr. 1994. Effect of temperature and relative humidity on development of *Cercosporidium personatum* on peanut in Georgia. Plant Dis. 78:690-694.

Baard, S. W., and Laubscher, C. 1985. Histopathology of blackhull incited by *Thielaviopsis basicola* in groundnuts. Phytophylactica 17:85-88.

Baard, S. W., and Van Wyk, P. S. 1985. *Neocosmospora vasinfecta* pathogenic to groundnuts in South Africa. Phytophylactica 17:49-50.

Bailey, J. E., and Brune, P. D. 1997. Effect of crop pruning on Sclerotinia blight of peanut. Plant Dis. 81:990-995.

Baird, R. E., Bell, D. K., Sumner, D. R., Mullinix, B. G., and Culbreath, A. K. 1993. Survival of *Rhizoctonia solani* AG-4 in residual peanut shells in soil. Plant Dis. 77:973-975.

Balasubramanian, R. 1979. A new type of alternariosis in *Arachis hypogaea* L. Curr. Sci. 48:76-77.

Bays, D. C., and Demski, J. W. 1986. Bean yellow mosaic virus isolate that infects peanut (*Arachis hypogaea*). Plant Dis. 70:667-669.

Behncken, G. M. 1970. The occurrence of peanut mottle virus in Queensland. Aust. J. Agric. Res. 21:465-492.

Bell, D. K., and Sobers, E. K. 1966. A peg pod and root necrosis of peanuts caused by a species of *Calonectria*. Phytopathology 56:1361-1364.

Bell, D. K., Locke, B. J., and Thompson, S. S. 1973. The status of Cylindrocladium black rot of peanut in Georgia since its discovery in 1965. Plant Dis. Rptr. 57:90-94.

Ben-Yephet, Y., Mhameed, S., Frank, Z. R., and Katan, J. 1991. Effect of the herbicide ethalfluralin on net blotch disease of peanut pods. Plant Dis. 75:1123-1126.

Bhowmik, T. P., Sharma, R. C., and Singh, A. 1985. Effect of gypsum, row spacing, and groundnut varieties on the incidence of root rot disease caused by *Macrophomina phaseolina*. Int. J. Tropical Plant Dis. 3:69-72.

Black, M. C., and Beute, M. K. 1984. Relationships among inoculum density, microsclerotium size, and inoculum efficiency of *Cylindrocladium crotalariae* causing root rot of peanuts. Phytopathology 74:1128-1132.

Brenneman, T. B., and Sumner, D. R. 1989. Effects of chemigated and conventionally sprayed tebuconazole and tractor traffic on peanut diseases and pod yields. Plant Dis. 73:843-846.

Chang, C. A., Purcefull, D. E., and Zettler, F. W. 1990. Comparison of two strains of peanut stripe virus in Taiwan. Plant Dis. 74:593-596.

Chohan, H. S., and Gupta, V. K. 1968. Aflaroot, a new disease of groundnut caused by *Aspergillus flavus*. Indian J. Agric. Sci. 38:568-570.

Clinton, P. K. S. 1962. The control of soilborne pests and diseases of groundnuts in the Sudan central rainlands. Emp. J. Exp. Agric. 30:145-154.

Culbreath, A. K., Beute, M. K., and Campbell, C. L. 1991. Spatial and temporal aspects of epidemics of Cylindrocladium black rot in resistant and susceptible peanut genotypes. Phytopathology 81:144-150.

Culver, J. N., Sherwood, J. L., and Melouk, H. A. 1987. Resistance to peanut stripe virus in *Arachis* germplasm. Plant Dis. 71:1080-1082.

de Klerk, A., McLeod, A., and Faurie, R. 1997. Net blotch and necrotic warts caused by *Streptomyces scabies* on pods of peanut *(Arachis hypogaea)*. Plant Dis. 81:958.

Demski, J. W., and Lovell, G. R. 1985. Peanut stripe virus and the distribution of peanut seed. Phytopathology 69:734-738.

Demski, J. W., Reddy, D. V. R., and Lowell, G., Jr. 1984. Peanut stripe, a new virus disease of peanut. (Abstr.) Phytopathology 74:627.

Demski, J. W., Reddy, D. V. R., Wongkaew, S., Kameya-Iwaki, M., Saleh, N., and Xu Z. 1988. Naming of peanut stripe virus. Phytopathology 78:631-632.

Diener, U. L. et al. 1965. Invasion of peanut pods in the soil by *Aspergillus flavus*. Plant Dis. Rptr. 49:931-935.

Dow, R. L., Porter, D. M., and Powell, N. L. 1988. Effect of environmental factors on *Sclerotinia minor* and Sclerotinia blight of peanut. Phytopathology 78:672-676.

Dubern, J., and Dollett, M. 1979. Groundnut crinkle, a new virus disease observed in Ivory Coast. Phytopathol. Zeitschr. 95:279-283.

Farr, D. F., Bills, G. R., Chamuris, G. P., and Rossman, A. Y. 1989. Fungi on Plants and Plant Products in the United States. The American Phytopathological Society, St. Paul, MN. 1252 pp.

Garren, K. H., and Jackson, C. R. 1973. Peanut diseases. *in* Peanuts: Culture and Uses, pp. 429-494. Peanut Research and Education Association, Stillwater, OK.

Germani, G. 1981. Pathogenicity of the nematode *Scutellonema cavenessi* on peanut and soybean. Rev. Nematol. 4:203-208.

Gopal, K., and Upadhyaya, H. D. 1991. Effect of bud necrosis disease on yield of groundnut under field conditions. Indian Phytopathol. 44:221-223.

Griffin, G. J., Garren, K. H., and Taylor, J. D. 1981. Influence of crop rotation and minimum tillage on the population of *Aspergillus flavus* group in peanut field soil. Plant Dis. 65:898-900.

Grinstein, A., Katan, J., Abdul Razik, A., Zeydan, O., and Elad, Y. 1979. Control of *Sclerotium rolfsii* and weeds in peanuts by solar heating of soil. Plant Dis. Rptr. 63:1056-1059.

Halliwell, R. S., and Philley, G. 1974. Spotted wilt of peanut in Texas. Plant Dis. Rptr. 58:23-25.

Hau, F. C., Beute, M. K., and Smith, T. 1982. Effect of soil pH and volatile stimulants from remoistened peanut leaves on germination of sclerotia of *Sclerotinia minor*. Plant Dis. 66:223-224.

Hennen, J. F., and Buritica, P. 1993. *Peridipes arachidis,* a conidial anamorph of the peanut rust fungus (Uredinales). (Abstr.) 1993 Annual Meeting of the Mycological Society of America.

Hoover, R. J., and Kucharek, T. A. 1995. First report of a leaf spot on perennial peanut caused by *Sclerotinia homoeocarpa*. Plant Dis. 79:1249.

Hull, R., and Adams, A. N. 1968. Groundnut rosette and its assistor virus. Ann. Appl. Biol. 62:139-145.

Jackson, C. R. 1962. Aspergillus crown rot in Georgia. Plant Dis. Reptr. 46:888-892.

Johnson, G. I. 1985. Occurrence of *Cylindrocladium crotalariae* on peanut (*Arachis hypogaea*) seed. Plant Dis. 69:434-436.

Jones, B. L., and De Waele, D. 1988. First report of *Ditylenchus destructor* in pods and seeds of peanut. Plant Dis. 72:453.

Kokalis-Burelle, N., Porter, D. M., Rodriguez-Kabana, R., Smith, D. H., and Subrahmanyam, P. (eds.). 1997. Compendium of Peanut Diseases, Second Edition. American Phytopathological Society, St. Paul, MN. 94 pp.

Kolte, S. J. 1984. Diseases of Annual Edible Oilseed Crops. Volume I: Peanut Diseases. CRC Press, Inc., Boca Raton, FL. 143 pp.

Kucharek, T. A., 1975. Reduction of Cercospora leaf spots of peanut with crop rotation. Plant Dis. Rptr. 59:822-823.

Kucharek, T. A. 1991. Unusual occurrence of lesions caused by *Diplodia* sp. in upper stems of peanuts in Florida. (Abstr.). Phytopathology 81:812.

Kuhn, C. W., and Demski, J. W. 1987. Latent virus in peanut in Georgia identified as cowpea chlorotic mottle virus. Plant Dis. 71:101.

Lana, A. F. 1980. Properties of a virus occurring in *Arachis hypogaea* in Nigeria. Phytopathol. Zeitschr. 97:169-178.

Lutterell, E. S. 1981. Peanut web blotch: Symptoms and field production of inoculum. (Abstr.) Phytopathology 71:892.

McDonald, D., Bos, W. W., and Gumel, M. H. 1979. Effects of infestation of peanut (groundnut) seed by the testa nematode *Aphelenchoides arachidis* on seed infection by fungi and on seedling emergence. Plant Dis. Rptr. 63:464-467.

Minton, N. A., and Baujard, P. 1990. Nematode parasites of peanut, pp. 285-320. *In* Plant Parasitic Nematodes in Tropical and Subtropical Agriculture, M. Luc, R. A. Sikora, and J. Bridge (eds.). C.A.B. International, Wallingford, England.

Misari, S. M., Abraham, J. M., Demski, J. W., Ansa, O. A., Kuhn, C. W., Casper, R., and Breyel, E. 1988. Aphid transmission of the viruses causing chlorotic rosette and green rosette diseases of peanut in Nigeria. Plant Dis. 72:250-253.

Mullen, J. M., Hagan, A. K., Nelson, P. E. 1996. A new stem canker of peanut in Alabama caused by *Fusarium oxysporum*: A wound-dependent disease. Plant Dis. 80: 1301.

Nelson, P. E., Toussoun, T. A., and Marasas, W. F. O. 1983. *Fusarium* Species: An Illustrated Manual for Identification. The Pennsylvania State Univ. Press, University Park, PA. 193 pp.

Nolt, B. L., Rajeshwari, R., Reddy, D. V. R., Bharathan, N., and Manohar, S. K. 1988. Indian peanut clump virus isolates: Host range, symptomatology, serological relationships, and some physical properties. Phytopathology 78:310-313.

Olorunju, P. E., Kuhn, C. W., Demski, J. W., Misari, S. M., and Ansa, O. A. 1992. Inheritance or resistance in peanut to mixed infections of groundnut rosette virus (GRV) and groundnut rosette assistor virus and a single infection of GRV. Plant Dis. 76:95-100.

Pataky, J. K., and Beute, M. K. 1983. Effects of inoculum burial, temperature, and soil moisture on survival of *Cylindrocladium crotalariae* microsclerotia in North Carolina. Plant Dis. 67:1379-1382.

Phipps, P. M., and Beute, M. K. 1977. Influence of soil temperature and moisture on severity of Cylindrocladium black rot in peanut. Phytopathology 67:1104-1107.

Porter, D. M. 1970. Peanut wilt caused by *Pythium myriotylum*. Phytopathology 60:393-394.

Porter, D. M., and Adamsen, F. J. 1993. Effect of sodic water and irrigation on sodium levels and the development of early leaf spot in peanuts. Plant Dis. 77:480-483.

Porter, D. M., Garren, K. H., Mozingo, R. W., and Van Schaik, P. H. 1971. Susceptibility of peanuts to *Leptosphaerulina crassiasca* under field conditions. Plant Dis. Rptr. 55:530-532.

Porter, D. M., Smith, D. H., and Rodriguez-Kabana, R. (eds.). 1984. Compendium of Peanut Diseases. American Phytopathological Society, St. Paul, MN. 73 pp.

Porter, D. M., Wright, F. S., Taber, R. A., and Smith, D. H. 1991. Colonization of peanut seed by *Cylindrocladium crotalariae*. Phytopathology 81:896-900.

Purss, G. S. 1962. Peanut diseases in Queensland. Queensland Agric. J. 88:540-543.

Recheigl, N. A., Tolin, S. A., Grayson, R. L., and Hooper, G. R. 1989. Ultrastructural comparison of peanut infected with stripe and blotch variants of peanut stripe virus. Phytopathology 79:156-161.

Reddy, D. V. R., Wongkaew, S., and Santos, R. 1985. Peanut mottle and peanut stripe virus diseases in Thailand and the Philippines. Plant Dis. 69:1101.

Reddy, D. V. R., Subrahmanyam, P., Sankara Reddy, G. H., Raja Reddy, C., and Siva Rao, D. V. 1984. A nematode disease of peanut caused by *Tylenchorhynchus brevilineatus*. Plant Dis. 68:526-529.

Reddy, D. V. R., Richins, R. D., Rajeshwari, R., Iizuka, N., Manohar, S. K., and Shepherd, R. J. 1993. Peanut chlorotic streak virus, a new caulimovirus infecting peanuts (*Arachis hypogaea*) in India. Phytopathology 83:129-133.

Rothwell, A. 1962. Diseases of groundnuts in southern Rhodesia. Rhodesia Agric. J. 59:199-201.

Rowe, R. C., and Beute, M. K. 1973. Susceptibility of peanut rotational crops (tobacco, cotton, and corn) to *Cylindrocladium crotalariae*. Plant Dis. Rptr. 57:1035-1039.

Rowe, R. C., Johnston, S. A., and Beute, M. K. 1974. Formation and dispersal of *Cylindrocladium crotalariae* microsclerotia in infected peanut roots. Phytopathology 64:1294-1297.

Sai Gopal, D. V. R., Siva Prasad, V., Gopinath, K., Satyanarayana, T., Sreenivasulu, P., and Nayudu, M. V. 1987. Chemical suppression of peanut green mosaic disease in peanut. (Abstr.) Phytopathology 77:1728.

Sanborn, M. R., and Melouk, H. A. 1983. Isolation and characterization of mottle virus from wild peanut. Plant Dis. 67:819-821.

Sharma, N. D. 1974. A previously unrecorded leaf spot of peanut caused by *Phomopsis* sp. Plant Dis. Rptr. 58:640.

Shew, B. B., Beute, M. K., and Wynne, J. C. 1988. Effects of temperature and relative humidity on expression of resistance to *Cercosporidium personatum* in peanut. Phytopathology 78:493-498.

Shew, H. D., and Beute, M. K. 1979. Evidence for the involvement of soilborne mites in Pythium pod rot of peanut. Phytopathology 69:204-207.

Sidebottom, J. R., and Beute, M. K. 1989. Control of Cylindrocladium black rot of peanut with cultural practices that modify soil temperature. Plant Dis. 73:672-676.

Sidebottom, J. R., and Beute, M. K. 1989. Inducing soil suppression to Cylindrocladium black rot of peanut through crop rotations with soybean. Plant Dis. 73:679-685.

Singh, B., and Hussain, S. M. 1991. Bacterial wilt of groundnut: A new record for India. Indian Phytopathol. 44:369-370.

Smith, D. H. 1972. *Arachis hypogaea,* a new host of *Cristulariella pyramidalis.* Plant Dis. Rptr. 56:796-797.

Smith, D. H., Burdsall, H. H., Jr., Black, M. C., and Crumley, C. R. 1990. Association of *Phanerochaete* with peanut in Texas. (Abstr.). Phytopathology 80:438.

Sobers, E. K., and Littrell, R. H. 1974. Pathogenicity of three species of *Cylindrocladium* to select hosts. Plant Dis. Rptr. 58:1017-1019.

Soufi, R. K., and Filonow, A. B. 1992. Population dynamics of *Pythium* spp. in soil planted with peanut. Plant Dis. 1203-1209.

Sreenivasulu, P., and Demski, J. W. 1988. Transmission of peanut mottle and peanut stripe viruses by *Aphis craccivora* and *Myzus persicae.* Plant Dis. 72:722-723.

Subrahmanyam, P., Reddy, L. J., Gibbons, R. W., and McDonald, D. 1985. Peanut rust: A major threat to peanut production in the semiarid tropics. Phytopathology 69:813-819.

Wadsworth, D. F., and Melouk, H. A. 1985. Potential for transmission and spread of *Sclerotinia minor* by infected peanut seed and debris. Plant Dis. 69:379-381.

Wagih, E. E., and Melouk, H. A. 1987. A virus causing chlorotic ringspot of peanut. (Abstr.) Phytopathology 77: 1731.

Warwick, D., and Demski, J. W. 1988. Susceptibility and resistance of soybean to peanut stripe virus. Plant Dis. 72:19-21.

Wills, W. H., and Moore, L. D. 1973. Pathogenicity of *Rhizoctonia solani* and *Pythium myriotylum* from rotted pods to peanut seedlings. Plant Dis. Rptr. 57:578-582.

Wongkaew, S., and Peterson, J. F. 1983. Peanut mottle virus symptoms in peanuts inoculated with different *Rhizobium* strains. Plant Dis. 67:601-603.

Woodard, K. E., and Jones, B. L. 1983. Soil populations and anastomosis groups of *Rhizoctonia solani* associated with peanut in Texas and New Mexico. Plant Dis. 67:385-387.

Xu, Z., and Barnett, O. W. 1984. Identification of a cucumber mosaic virus strain from naturally infected peanuts in China. Plant Dis. 68:386-389.

Xu, Z., Yu, Z., Liu, J., and Barnett, O. W. 1983. A virus causing peanut mild mottle in Hubei Province, China. Plant Dis. 67:1029-1032.

Yung, C. H., Wernsman, E. A., and Gooding, G. V., Jr. 1991. Characterization of potato virus Y resistance from gametoclonal variation in flue-cured tobacco. Phytopathology 81:887-891.

13. Diseases of Red Clover (*Trifolium pratense*)

Disease Caused by Bacteria

Bacterial Leaf Spot

Bacterial leaf spot is also called bacterial blight and bacterial leaf blight.

Cause. *Pseudomonas syringae* survives in infested residue. During cool, wet weather at any time of the growing season, bacteria are disseminated by splashing rain and enter leaves through wounds or natural openings.

Distribution. Europe and North America.

Symptoms. Bacterial leaf spot affects stems, petioles, petiolules, stipules, and flower pedicels, but symptoms are most conspicuous on diseased leaflets. At first, tiny, translucent dots appear on the lower leaf surfaces. The spots become black and have water-soaked margins and chlorotic halos that enlarge to fill the area between the leaf veins. During wet weather, milky, white bacterial exudate develops in spots and appears as drops or a thin film on the spot. When the exudate dries, it forms a thin, crusty film that glistens in the light. Leaves may become tattered and frayed by wind tearing out the dead portions. Lesions on petioles and stems are dark, elongated, and slightly sunken.

Management. None is recommended since bacterial leaf spot is not considered an important disease.

Diseases Caused by Fungi

Anther Mold

Cause. *Botrytis anthophila,* teleomorph *Sclerotinia spermophila.*

Distribution. Not known.

Symptoms. A gray to black fungal growth occurs on diseased flowers.

Management. Not reported.

Aphanomyces Root Rot

Cause. *Aphanomyces euteiches.* One-week-old red clover is most susceptible to infection. Seedlings that are 2 and 3 weeks old are significantly less susceptible to infection.

Distribution. The United States, but it is likely common wherever red clover is grown.

Symptoms. A necrosis of diseased roots, hypocotyls, and cotyledons occurs. Surviving plants are stunted.

Management. Low levels of resistance occur in some cultivars.

Black Patch Disease

Cause. *Rhizoctonia leguminicola* is seedborne and survives as mycelia in infested residue.

Distribution. The southern United States.

Symptoms. Black patch occurs on scattered plants and in scattered groups of plants within a field. Seedlings are often blighted and overgrown with black mycelium. Leaves, stems, flowers, and seeds become diseased. Leaf symptoms are lesions that vary in color from brown to gray-black and have concentric rings. Often all the bottom leaves of diseased plants are killed. Black, coarse, aerial mycelial growth occurs on petioles and on stems, which frequently are girdled. Dark lesions and dark, coarse mycelia occur on flowers and seeds.

Management
1. Treat seed with a seed-protectant fungicide.
2. Cut hay earlier than normal to reduce foliage loss.
3. Rotate red clover with resistant crops.

Black Root Rot

Cause. *Thielaviopsis basicola,* synamorph *Chalara elegans,* survives as chlamydospores in infested residue and soil. Chlamydospores germinate during moist soil conditions and produce mycelia that directly infect host roots. Chlamydospores are formed in diseased tissues and returned to the soil when tissues decay.

Distribution. Not known.

Symptoms. The cortex of the hypocotyl below the soil line and taproots and fibrous roots become dark brown and necrotic. A severe root rot may develop, with subsequent wilting of foliage or plant death.

Management. Not reported.

Brown Root Rot

Cause. *Phoma sclerotioides* (syn. *Plenodomus meliloti*) overwinters in infested residue as pycnidia and saprophytic mycelia. It is not known for certain what the source of primary inoculum is, but it is presumed to be conidia produced in pycnidia.

Distribution. Canada.

Symptoms. Diseased plants are observed in the spring as chlorotic, stunted plants that frequently die after beginning to grow. Circular, brown, necrotic areas occur on diseased roots and eventually spread into the crown. Numerous pycnidia occur on diseased areas.

Management. Rotate red clover with nonlegumes.

Common Leaf Spot

Common leaf spot is also called Pseudopeziza leaf spot.

Cause. *Pseudopeziza trifolii* survives either as apothecia within leaf spots or as saprophytic mycelia in infested residue. During cool, wet weather, ascospores produced in apothecia are disseminated by wind to host leaves.

Distribution. Wherever red clover is grown.

Symptoms. Very small, dark spots that vary in color from olive to red-brown, purple, or black develop on both diseased leaf surfaces. The angular to round spots with portions of their margins jutting out like a finger or branch give spots a star-like appearance. Under magnification, an apothecium, which resembles a raised cushion, is observed in the center of spots on the undersides of leaves during wet weather. Later, apothecia dry up and become dark, making them difficult to observe. While common leaf spot is confined mostly to leaves, small, elongated, dark streaks may occasionally occur on petioles.

Management. Not reported.

Curvularia Leaf Blight

Cause. *Curvularia trifolii.* Curvularia leaf blight is most severe during warm, moist weather.

Distribution. Worldwide. Curvularia leaf blight is considered to be of minor importance.

Symptoms. Water-soaked, angular, yellow to brown lesions develop on diseased leaflets. Lesions often develop on petioles, resulting in necrotic leaves that remain attached to the plant.

Management. Not necessary since Curvularia leaf blight is considered to be of minor importance.

Cylindrocarpon Root Rot

Cause. *Cylindrocarpon magnusianum* (syn. *C. ehrenbergii*).

Distribution. The United States.

Symptoms. Similar to symptoms caused by *C. crotalariae*. Diseased roots have black lesions.

Management. Not reported.

Cylindrocladium Root and Crown Rot

Cause. *Cylindrocladium crotalariae.*

Distribution. The United States (Florida).

Symptoms. Black lesions occur in crowns and roots of diseased plants. Diseased plants normally die.

Management. Not reported.

Downy Mildew

Cause. *Peronospora trifoliorum* survives in crown buds and infested leaves. Disease occurs during wet or humid weather.

Distribution. Distribution is not known for certain, but downy mildew likely occurs wherever red clover is grown.

Symptoms. The leaves at the tops of diseased plants are light green. A gray "fuzzy" mycelial growth consisting of mycelium, conidia, and conidiophores is present on the underside of diseased leaves during wet or humid weather. Leaves of severely diseased plants are twisted, and stem growth is retarded.

Management
1. Harvest early.
2. Rotate red clover with nonlegumes.

Fusarium Root Rot

Fusarium root rot is also called common root rot.

Cause. Several *Fusarium* spp., including *F. acuminatum, F. avenaceum, F. oxysporum, F. roseum,* and *F. solani,* are the fungi most commonly isolated from the diseased roots and crowns of red clover. However, several other fungi are also associated with root rot of red clover. Fusarium are usually present in soil and persist as chlamydospores and as saprophytic mycelia in infested residue.

Fusarium are generally considered to be weak pathogens and infect a plant after it has been weakened by some other cause, such as drought, low fertility, insects, other diseases, improper management, or winter injury. Wounding of roots results in localized increases in incidence and severity of necrosis caused by *F. acuminatum* and *F. avenaceum*. The main effect of wounding is not to provide an opening through which the fungus enters the host plant but to accelerate fungal penetration and increased disease development by altering the host–pathogen interaction. This is done, following fungal penetration into the plant, by favoring the formation of distributive hyphae throughout roots rather than the formation of chlamydospores.

Distribution. Wherever red clover is grown.

Symptoms. Diseased red clover plants may be killed in all stages of plant development, but stand loss is most conspicuous during the second year of stand establishment. Diseased plants appear unthrifty, stunted, and yellowish and will wilt during hot, dry weather. Roots and crowns are spongy or soft and are discolored from light brown to red-brown to dark brown. Lesions develop on the surface of larger roots and subsequently cause feeder roots to be pruned off. The inner core of the taproot may be decomposed.

Management. In general, practice management that promotes vigorous plant growth. This includes the following practices:
1. Sow certified disease-free seed of adapted cultivars. Some cultivars tend to be less affected by root rot than others.
2. Avoid overgrazing, particularly late in the growing season.
3. Get a soil test to determine the deficiencies or excesses of nutrients.
4. Soil should have a pH of 6.2 to 7.0.

Gray Stem Canker

Cause. *Ascochyta lethalis* (syn. *A. caulicola*).

Distribution. Canada. Gray stem canker is not considered a serious disease.

Symptoms. Silver-white cankers, which occur on midribs, leaf stalks, and diseased stems, sometimes girdle the lower stems. Fruiting bodies of *A. lethalis* are in the center of cankers and resemble small, black dots. Severely diseased stems may be stunted, swollen, and twisted and have a few small leaves.

Management. Not reported.

Leaf Gall

Leaf gall is also called crown wart.

Cause. *Physoderma trifolii* (syn. *Urophlyctis trifolii*) survives in soil and infested residue as resting sporangia. During wet soil conditions, sporangia liberate zoospores that "swim" through moist soil to the developing buds of crown branches on host plants. Zoospores germinate and form infection hyphae that penetrate the host plants.

Distribution. A minor disease found primarily in warmer red clover–growing areas of Europe, North America, and South America.

Symptoms. Irregular-shaped galls form around the crowns of diseased plants just below the soil level. Galls are first produced in late spring and continue to increase in size during the summer. Diseased plants wilt in hot weather.

Management. Not reported.

Midvein Spot

Cause. *Mycosphaerella carinthiaca* overseasons as perithecia and mycelia or sporodochia. It is not known for certain, but primary infection is presumed to be either ascospores liberated from perithecia or conidia produced on sporodochia.

Distribution. In warmer red clover–growing areas of the world.

Symptoms. Small, chlorotic spots occur on the veins of diseased leaves. The spots enlarge slightly and become dark brown. Infrequently, spots have gray centers.

Management. Not reported.

Mycoleptodiscus Crown and Root Rot

Cause. *Mycoleptodiscus terrestris* (syn. *Leptodiscus terrestris*) overwinters as sclerotia in infested residue and soil, and presumably as mycelia and acervuli in infested residue. Most primary inoculum likely originates from sclerotia that germinate and directly infect roots and crowns. Conidia are formed in summer and disseminated by being forcibly ejected from the acervulus by setae that unfold as the acervulus mucilage dries out. Disease is favored by warm, humid weather.

Distribution. The central and eastern United States. Mycoleptodiscus crown and root rot is not considered a serious disease of red clover.

Symptoms. A brown to black rot occurs in diseased lateral roots, taproots, and crowns of older plants. Numerous sclerotia form in decayed tissue. Small spots develop on diseased leaves, and reddish-brown lesions occur on stems. Spots on the lower stems expand and develop into crown rot.

Management. Not reported.

Myrothecium Leaf Spot

Cause. *Myrothecium roridum* and *M. verrucaria* are soil inhabitants and widespread facultative parasites. Myrothecium leaf spot is favored by wounds and hot, humid weather.

Distribution. The United States (Pennsylvania).

Symptoms. Initially, small, dark, water-soaked lesions occur on both leaf surfaces and become dark brown to black. Sporodochia develop in lesions, often in concentric circles in alternating light and dark brown rings that give the spot a "target-like" effect. A slight chlorosis sometimes occurs around lesions. At first, lesions are circular, but as they enlarge, they become irregular-shaped and coalesce, which causes a necrosis of petioles and subsequent wilting of the leaves. If humid conditions persist for several days, diseased leaves curl and die. Some lesions may be restricted by leaf veins. Symptoms appear most rapidly at wound sites but also may develop on nonwounded tissue.

Management. Not reported.

Myrothecium Root Rot

Cause. *Myrothecium roridum* and *M. verrucaria* are soil inhabitants and widespread facultative parasites. Both fungi are thought to be primary pathogens rather than secondary wound parasites.

Distribution. The United States (Pennsylvania and Wisconsin).

Symptoms. In slant-board evaluations, brown, water-soaked rots with poorly delineated margins occurred 3 days after inoculation. Foliage symptoms were chlorosis, purpling of leaflet margins, and, finally, the death of diseased leaves and petioles. Foliar symptoms also included stunting.

In greenhouse evaluations, dark brown rots occurred across the entire root and had occasional streaks that extended upward in the vascular system. Decayed portions of diseased roots remained firm and were not water-soaked.

Management. Not reported.

Northern Anthracnose

Northern anthracnose is also called clover scorch.

Cause. *Aureobasidium caulivorum* (syn. *Kabatiella caulivora*) overwinters as mycelia and acervuli in infested residue and may be seedborne. Temperatures of 20° to 25°C and wet weather are optimum for the production of conidia in acervuli and from mycelia; conidia are disseminated by wind or splashing water to host plants.

Distribution. The cooler red clover–growing areas of Asia, Europe, and North America.

Symptoms. From a distance, a field of diseased plants will appear "scorched" as if by fire. The first symptom is the presence of elongated, dark brown to black lesions that split in the middle and become light colored with dark margins on diseased leaf petioles and stems. Lesions may girdle and kill stems, causing leaves to become brown, wilt, and die. Leaves and flower heads droop like shepherd's crooks. Diseased plant parts dry out rapidly and become so brittle that leaflets are easily broken off.

Small, irregular-shaped, colorless acervuli that lack setae and white masses of conidia and conidiophores occur in deeper cracks and depressions.

Management. Grow resistant cultivars.

Powdery Mildew

Cause. *Erysiphe polygoni* overwinters as cleistothecia scattered on the surfaces of infested residue. Primary inoculum is provided by ascospores that are produced in asci within cleistothecia and disseminated by wind to host leaves. Infection may occur any time during the growing season but is most common during cool nights and warm days in the late summer and early autumn. Long periods of dry weather favor disease development. Conidia are subsequently produced on diseased leaf surfaces and are the primary source of secondary inoculum throughout the remainder of the growing season. In the autumn, cleistothecia form on the surfaces of diseased plant tissues.

Erysiphe polygoni is possibly an aggregate species consisting of several physiologic races. *Erysiphe polygoni* infects at least 359 species of plants in 154 different genera.

Distribution. Wherever red clover is grown in the temperate zones of the world.

Symptoms. Powdery mildew is not serious early in the growing season but can cause reductions in yield and hay quality later in the growing season. Initially, patches of fine, barely visible white mycelium develop on diseased upper leaf surfaces. The patches enlarge and merge, which gives the appearance of white flour or powder covering the diseased leaf surfaces. If disease is severe, an entire field may look white. Diseased leaves remain green but a few turn yellow and wither prematurely.

Management. Grow resistant cultivars.

Pseudoplea Leaf Spot

Pseudoplea leaf spot is also called pepper spot and burn.

Cause. *Pseudoplea trifolii* (syn. *Leptosphaerulina trifolii*) overwinters as perithecia or mycelia in infested leaf residue. Ascospores are released during cool, wet weather and disseminated by wind to host plants.

Distribution. Wherever red clover is grown, but Pseudoplea leaf spot is most common in temperate, humid areas.

Symptoms. Pseudoplea leaf spot occurs throughout the growing season but is most common during cool, wet weather in the spring or autumn. Tiny, sunken, black spots that develop on both leaf surfaces and petioles eventually become gray and have a red-brown border. Although spots do not enlarge (they grow to a diameter of only a few millimeters), they may become very numerous. Severely diseased leaves and petioles are chlorotic, then become brown, and wither, forming a mass of dead leaves. Parts of flowers and flower stalks become diseased and die.

Management. Cut hay before Pseudoplea leaf spot becomes too severe.

Rust

Cause. *Uromyces trifolii-repentis* var. *fallens* (syn. *U. trifolii* var. *fallens*) is an autoecious long-cycled rust fungus that overwinters as teliospores on red clover in the southern United States and northern Mexico. Each summer, urediniospores are disseminated by wind to the northern United States and Canada. Although infection occurs in the spring in warm red clover–growing areas, infection does not occur until late summer or autumn in the northern United States and Canada. Rust development is optimum during moist conditions at temperatures of 16° to 22°C.

Distribution. Wherever red clover is grown in humid or semihumid areas.

Symptoms. Rust is most important in new seedings and in stands left for seed instead of cut for hay purposes. When rust becomes severe, it is usually in fields where the cutting of plants has been delayed.

The uredinial stage is the most obvious symptom and appears as small, red-brown pustules on diseased petioles, stems, and leaves, especially on the undersides of leaves. When a pustule is rubbed, a red to brown dust or powder consisting of urediniospores is evident. Seriously rusted leaves may become chlorotic and defoliate. Telia occur either in old uredinia or independently on the diseased leaf and are darker than uredinia. The aecial stage develops as swollen, white to yellow pustules on stems, petioles, and leaves that appear under magnification to be clusters of tiny cup-like structures. Aecia may cause distortion of diseased leaves and petioles.

Management. Grow the most resistant cultivars.

Sclerotinia Crown and Root Rot

Sclerotinia root rot and crown rot is also called Sclerotinia crown and stem rot.

Cause. *Sclerotinia trifoliorum* survives for several years in the soil as sclerotia. Sclerotia germinate in the autumn and form apothecia in which ascospores are produced and disseminated by wind to host leaves. Disease development is optimum during wet weather at temperatures of 13° to 18°C.

Distribution. In regions with mild winters or heavy snow cover. Sclerotinia crown and root rot is most important in the southern red clover–growing areas of the United States.

Symptoms. Sclerotinia root rot and crown rot occurs in patches within a field that may merge and form large areas of dead plants. Symptoms first appear in the autumn as small, brown spots on diseased leaves and petioles. Leaves turn light brown, wither, and become overgrown with white mycelium that spreads to the crowns and roots. By late winter or early spring, crowns and basal portions of young stems have a brown, soft rot that extends into the roots. As stems and petioles are killed, they are also overgrown with white mycelium. Black sclerotia of various sizes form in mycelium and are imbedded or attached to the surface of diseased red clover tissue. When infected tissue decomposes, the sclerotia are released into the soil.

Management. Rotate red clover with a nonsusceptible host. The sclerotia will eventually decompose or the causal fungus will die out in the absence of a susceptible host.

Seed Mold

Cause. *Alternaria alternata.*

Distribution. Widespread.

Symptoms. Diseased seed is shriveled and discolored and, when moistened, gray mycelium will grow from the seed. Diseased seed will damp off when sown.

Management
1. Do not sow diseased seed.
2. Treat seed with a seed-protectant fungicide.

Seed Rot and Damping Off

Seed rot and damping off is also called Pythium blight.

Cause. *Pythium* spp., particularly *P. debaryanum.* Most *Pythium* spp. survive in soil as oospores. Under moist soil conditions, oospores germinate and form sporangia in which zoospores are produced. Zoospores are released and "swim" through free water in the soil to the root tips or seeds of host plants and germinate. Usually, only young plants are infected. Seedlings more than 5 days old apparently are almost immune, and older plants are rarely infected.

Distribution. Wherever red clover is grown.

Symptoms. Symptoms are not obvious because preemergence damping off is the most common form of Pythium blight. Seeds and seedlings become soft and are overgrown with white mycelia when they are placed in a moist chamber. Emerged seedlings that become diseased have a soft, brown rot on the hypocotyl and roots.

Management. Treat seed with a seed-protectant fungicide.

Sooty Blotch

Sooty blotch is also called black blotch.

Cause. *Cymadothea trifolii,* anamorph *Polythrincium trifolii,* overwinters as a stroma. Sometime during the growing season, perithecia form within the overwintered stroma and ascospores are produced, liberated, and disseminated by wind to host leaves. Infection occurs in spring in the southern United States and during late summer and autumn in the northern United States and Canada.

Distribution. The temperate zones of the world.

Symptoms. The first symptom occurs on lower leaf surfaces as tiny, olive-green dots. The dots enlarge, become thicker and darker, and eventually resemble a velvety, black, elevated cushion. In the autumn, the cushions are replaced by other black areas that have a shiny surface. Chlorotic spots that become necrotic appear on the upper leaf surfaces opposite a cushion. Severe disease causes the entire leaf to turn brown, die, and defoliate.

Management. Not known.

Southern Anthracnose

Cause. *Colletotrichum trifolii* overwinters as acervuli on diseased stems, crowns, and roots. In some literature, *Colletotrichum destructivum* has been reported to be the causal fungus of southern anthracnose. Conidia production is optimum during wet weather at a temperature of 28°C; conidia are splashed by rain or blown by wind to host plants.

Distribution. Africa, southern Europe, and the southern United States.

Symptoms. Symptoms closely resemble those of northern anthracnose except southern anthracnose is frequently found on the upper part of taproots and crowns, while northern anthracnose is not. Southern anthracnose also is more likely than northern anthracnose to be present on the new growth of the second crop in the northern and midwestern United States.

In the southern United States, southern anthracnose is present in the spring on the young, succulent parts of stems and petioles of the first crop. Dark brown, irregular-shaped spots develop on diseased leaves. These vary in size from barely visible dots to a general necrosis of entire leaves. Disease commonly develops on young tissue but may also occur on older plant tissue.

The first symptoms on petioles and stems are water-soaked spots that become elongated and dark brown to black with a gray or light brown center. When viewed under magnification, small, black setae that resemble tiny black spines can be seen in the centers of older, sunken lesions. Diseased petioles cause the attached leaflets to droop. Stems may be killed when girdled by a lesion.

Diseased crowns are blue-black and attached stems are brittle, causing plants to break off at the ground level. Normally, the entire plant wilts and dies when the crown is diseased.

Management
1. Grow adapted, resistant cultivars.
2. Use clean, certified seed from disease-free plants.

Southern Blight

Cause. *Sclerotium rolfsii,* teleomorph *Athelia rolfsii,* survives for several years in the soil or in plant residue as small, brown sclerotia that are disseminated by water and wind. Sclerotia germinate under hot, humid conditions and form mycelial strands that infect plants.

Distribution. In the warmer red clover–growing areas of the United States and possibly Europe.

Symptoms. Diseased plants are "bleached" to light tan and have a white, cotton-like mycelial growth on stems near the soil surface. Numerous small, tan to brown sclerotia that resemble "seeds" form in mycelia growing on the stems, crowns, and residue on the soil surface.

Management
1. Plow under infested residue.
2. Practice proper mowing or grazing. The causal fungus does not grow well when exposed to direct sunlight and air movement.

3. Grow a resistant plant host in the crop rotation. Avoid growing crops such as alfalfa and soybean in a rotation with red clover.

Spring Black Stem and Leaf Spot

Cause. *Phoma pinodella* (syn. *P. trifolii*) overwinters as pycnidia in infested residue and may be seedborne. Spores are produced in pycnidia and are disseminated by wind to host plants. Plants are infected during cool, wet weather in the autumn and spring.

Distribution. Europe, North America, and South America.

Symptoms. The most conspicuous symptom is on diseased stems. Initially, small, brown to black spots that vary in size and shape occur on stems. Spots merge and form large black areas, especially on the lower stems. Young shoots may be girdled and killed. The black discoloration increases when clover is not cut at the proper time or is left for seed.

Leaves have dark brown to black spots that are irregular in size and shape. Some spots may have gray centers. Diseased leaves, together with diseased stems, may result in severe defoliation.

Management
1. Sow disease-free seed.
2. Plow under infested residue.
3. Cut hay before defoliation becomes too severe.

Stagonospora Leaf Spot

Stagonospora leaf spot is also called gray leaf spot.

Cause. *Stagonospora recedens*. However, *S. meliloti,* teleomorph *Leptosphaeria pratensis,* has also been reported to be a causal fungus. *Stagonospora recedens* overwinters as mycelia and pycnidia in infested residue. *Leptosphaeria pratensis* overwinters as perithecia in infested residue. The fungi are poor soil saprophytes and do not survive in the absence of residue colonized during their parasitic phase. During wet weather in the spring, both conidia and ascospores are produced. The conidia are primarily disseminated to plant hosts by splashing rain and irrigation, and the ascospores by wind. Secondary inoculum is conidia produced in pycnidia during wet conditions and disseminated by splashing rain or irrigation to plant hosts.

Distribution. Generally distributed but most common in warm, humid areas or where red clover is grown under irrigation.

Symptoms. Spots are circular to irregular (3–6 mm in diameter). The centers of spots are pale buff to almost white and have light to dark brown margins. In some instances, spots have faint concentric zones that are the same color as the margins. Pycnidia appear as small, dark dots scattered throughout the older spots.

Management. Rotate red clover with a resistant crop for 2 to 3 years. Many legumes, including alfalfa and other clovers, are susceptible.

Summer Black Stem

Summer black stem is also called angular leaf spot and Cercospora leaf and stem spot.

Cause. *Cercospora zebrina* is seedborne and overwinters as mycelia in infested residue.

Distribution. Europe, central and eastern North America.

Symptoms. Summer black stem occurs in late summer and early autumn. Leaf spots are dark brown, angular, and somewhat limited by leaf veins. Older spots may appear ash gray due to the sporulation of *C. zebrina* in the spot. Sunken, dark brown lesions occur on stems and petioles. Lesions may merge and form extensive dark areas on the lower parts of stems. Seeds are also diseased.

Management
1. Treat seeds with a seed-protectant fungicide.
2. Harvest before defoliation becomes severe. This also helps to reduce inoculum.

Target Spot

Target spot is also called ring spot, Stemphylium leaf spot, and zonate leaf spot.

Cause. *Stemphylium sarciniforme* and *S. botryosum* overwinter as mycelia in infested residue. The teleomorph of *S. botryosum, Pleospora tarda,* may survive as perithecia in which ascospores are produced. At any time during the growing season, conidia are produced and disseminated by wind to host plant leaves. Diseased leaves fall to the soil surface at harvest and provide inoculum for infection the following year. Disease development is optimum during moist weather and temperatures of 20° to 24°C.

Distribution. Wherever red clover is grown.

Symptoms. Symptoms are confined to leaflets of plants growing in dense stands and are most severe in the late summer and autumn. The initial small, dark brown spots enlarge and develop into target-like spots with alternate light and dark brown rings. Entire leaves become wrinkled and dark brown and have a sooty appearance but remain attached to diseased plants. Sunken brown lesions infrequently appear on the stems, petioles, and pods.

Management. No satisfactory disease management is known, but selection of resistant cultivars may be possible. Harvest early to reduce losses.

Violet Root Rot

Cause. *Helicobasidium brebissonii*, anamorph *Rhizoctonia crocorum* (syn. *R. violaceae*), survives in soil as sclerotia or as saprophytic mycelia in residue. Disease is most severe under wet soil conditions.

Distribution. Europe and the United States.

Symptoms. Plants in a diseased area (which usually corresponds to a wet area of a field) are brown in contrast to surrounding green plants. Diseased roots have a brown to dark brown or black discoloration and are covered with thick mats of violet mycelia that extend below the soil surface. Later, roots are rotted and shredded and have a brown to dark violet discoloration. Barely visible tiny, black sclerotia are seen on diseased roots.

Management
1. Grow red clover in well-drained soil.
2. Harvest hay in the prebloom stage.
3. Rotate red clover with resistant crops, such as small grains and maize.

Winter Crown Rot

Winter crown rot is also called snow mold.

Cause. *Coprinus psychromorbidus*, *Phoma sclerotioides* (syn. *Plenodomus meliloti*) and *Typhula* spp. *Coprinus psychromorbidus* was previously called the low-temperature basidiomycete (LTB). Disease occurs at temperatures near 0°C at the soil surface and under a snow cover.

Distribution. Canada.

Symptoms. Young stands have irregular-shaped areas of dead plants. Older stands have scattered plants with rotted crowns. White fungal threads are present at the edge of melting snow.

Management. Rotate with nonlegume plants.

Diseases Caused by Nematodes

Bulb and Stem

Cause. *Ditylenchus dipsaci* survives for years as larvae in dry stems, plant crowns, soil, and other plant hosts. Larvae feed and reproduce in shoots near the crown, moving over the plant surface in a film of water and infecting plants through stomates. A female nematode lays 75 to 100 eggs with optimum reproduction and infection in heavy, wet soils at soil temperatures of 15° to 21°C. Dissemination is by machinery, rain, and irrigation water.

Distribution. Worldwide, but it is not considered a serious disease of red clover.

Symptoms. Diseased plants are stunted and have a bushy appearance due to swollen nodes and shortened internodes, which result in abnormal proliferation and swollen, short stems. White to light brown stems are scattered throughout a field. Shoots from infected buds are severely dwarfed and have swollen and spongy buds that are easily detached from the plant.

Management
1. Plow under diseased stands.
2. Do not grow red clover in a rotation with alfalfa or other clovers.

Clover Cyst

Cause. Clover cyst nematode, *Heterodera trifolii*.

Distribution. Europe and probably other red clover–growing areas.

Symptoms. Diseased plants are stunted.

Management. Follow general nematode management practices.

Dagger

Cause. *Xiphinema americanum* is an ectoparasite.

Distribution. Widely distributed.

Symptoms. Root growth of diseased plants is reduced. Necrotic areas form where feeding has occurred. Cells next to necrotic tissue may enlarge, causing a gall-like growth.

Management. Not known.

Lance

Cause. *Hoplolaimus* spp. are ectoparasites that feed in the cortex area.

Distribution. The southern United States.

Symptoms. Diseased plants are stunted and chlorotic. The taproot becomes necrotic and large numbers of secondary and tertiary roots are produced.

Management. Not reported.

Lesion

Cause. *Pratylenchus penetrans*. Not much information is known about *P. penetrans* on red clover, but the life cycle is assumed to be similar to that on other legumes. The preferred area of infection is root hairs on feeder roots, where nematodes force their way through or between epidermal and cortical cells. Gravid females deposit eggs in infected root tissue or soil. The second-stage larvae emerge. Injury is most severe in sandy and sandy loam soils.

Distribution. Widespread.

Symptoms. Initially, a lesion appears on the affected root as a water-soaked area that becomes yellow and elliptical-shaped. Dark brown cells later appear in the center of the lesion. Secondary infection by saprophytic fungi is thought to occur.

Management. Nematicides are effective but impractical.

Pin

Cause. *Paratylenchus* spp.

Distribution. Not know for certain but assumed to be widespread.

Symptoms. Necrotic areas occur on diseased roots.

Management. Not known.

Ring

Cause. *Criconemella* spp. are thought to be ectoparasitic.

Distribution. The United States.

Symptoms. Diseased plants are chlorotic. Roots are discolored with brown, necrotic lesions.

Management. Not reported.

Root Knot

Cause. *Meloidogyne arenaria, M. hapla, M. incognita,* and *M. Javanica. Meloidogyne arenaria, M. incognita,* and *M. javanica* do not survive if the temperature averages below 3°C during the coldest month. *Meloidogyne hapla* is limited by temperatures above 26°C.

Distribution. Wherever red clover is grown.

Symptoms. Diseased plants are stunted and wilt during moisture stress. Leaves frequently have purple undersides. Roots have galls, which causes them to be excessively branched and brittle.

Management
1. Rotate red clover with nonhosts.
2. Apply nematicides if practical.

Spiral

Cause. *Helicotylenchus* spp. are ectoparasites.

Distribution. Widely distributed.

Symptoms. Necrotic spots occur at the feeding sites on the roots. Roots may be severely damaged when high nematode populations are present.

Management. Nematicides control nematodes on some crops.

Diseases Caused by Phytoplasmas

Phyllody

Cause. Phytoplasma-like organism that survives in biennial and perennial plants. The organism is disseminated by leafhoppers.

Distribution. Canada and Europe.

Symptoms. Floral parts are turned into leaf-like organs. Leaves are reduced in size and have a mild chlorosis. Newly emerged leaves are slightly deformed and old leaves may have a bronze color.

Management. Not reported.

Proliferation

Cause. Phytoplasma-like organism that probably overwinters in biennial and perennial plants. Vectors are not known for certain, but leafhoppers may be involved.

Distribution. Canada (Alberta).

Symptoms. Foliar growth is profuse, giving the plant a witches' broom appearance. Flowers may be transformed into leaf-like appendages.

Management. Not reported.

Witches' Broom

Cause. Phytoplasma-like organism.

Distribution. Western Canada.

Symptoms. Diseased plants are severely dwarfed and the numerous short, spindly shoots that originate from crown and axillary buds along the stem give plants an overall bunchy appearance. Leaves are small and chlorotic.

Management. Not reported.

Yellow Edge

Cause. Phytoplasma-like organism that overwinters in biennial and perennial plants. Dissemination is by leafhoppers.

Distribution. Canada (Ontario and Quebec).

Symptoms. New growth has a mild chlorosis that later becomes more pronounced along the leaf margins. Chlorosis may eventually become a red-brown discoloration. Clusters of small leaves with short petioles and chlorotic margins develop at the nodes of creeping stems. Some plants do not have creeping stems; their numerous small leaves on upright stems give the plant a witches' broom appearance. Flower size and production may be reduced. The entire diseased plant is stunted.

Management. Not reported.

Diseases Caused by Viruses

Alfalfa Mosaic

Cause. Alfalfa mosaic virus is disseminated by several species of aphids.

Distribution. Canada and the United States. Alfalfa mosaic is possibly distributed wherever alfalfa and red clover are grown.

Symptoms. Diseased leaves are distorted and have a green and yellow mosaic pattern.

Management
1. Sow virus-free seed.
2. Control aphids where possible.
3. Grow the most resistant cultivar.

Alsike Clover Mosaic

Alsike clover mosaic virus is possibly a strain of bean yellow mosaic virus. However, because their symptoms are different in the literature, both viruses are described here.

Cause. Alsike clover mosaic virus (ACMV) is in the potyvirus group. ACMV overwinters in perennial plants, such as wild sweet clover, and is transmitted mechanically and by several species of aphids. ACMV is not thought to be seedborne.

Distribution. Widespread.

Symptoms. Symptoms are more prevalent along the borders of fields. The initial drooping of leaflets at the point of attachment to the stem is followed by development of angular, pale yellow spots. Diseased leaves become malformed and distorted.

Management. Not reported.

Bean Yellow Mosaic

Alsike clover mosaic virus may be a strain of bean yellow mosaic virus. However, because their symptoms are different in the literature, both viruses are described here.

Cause. Bean Yellow mosaic virus (BYMV) is in the potyvirus group. BYMV overwinters in diseased legumes and is transmitted by aphids, especially the pea aphid, *Macrosiphum pisi*. Arrowleaf clover, *Trifolium vesiculosum,* is reported to be a reservoir for BYMV. BYMV probably interacts with *Fusarium* spp. to increase root rot and cause premature stand decline.

Distribution. Africa, Asia, Europe, North America, and South America.

Symptoms. A systemic mottling of diseased leaves is the most common symptom. Other symptoms include vein yellowing, distortion, vein necrosis, and systemic necrosis.

Management
1. Grow the least susceptible cultivar. There is some evidence that resistant cultivars can be developed.
2. Avoid growing red clover near garden peas and beans.

Clover Yellow Mosaic

Cause. Clover yellow mosaic virus (ClYMV) is in the potexvirus group. ClYMV is seedborne.

Distribution. Canada and probably other red clover–growing areas.

Symptoms. Diseased plants are frequently stunted and bushy. Leaves have a mosaic appearance and the veins are yellowed.

Management. Sow disease-free seed. Other general management measures may be useful.

Common Pea Mosaic

Common pea mosaic is also called Pea common mosaic, red clover mosaic, and pea mosaic virus.

Cause. Common pea mosaic virus (CPMV) is reservoired in several legumes and transmitted by aphids, particularly the pea aphid, *Macrosiphum pisi,* but also by the aphids *Myzus persicae* and *Aphis rumicis.* The aphids acquire CPMV after feeding 5 or more minutes on diseased plants. CPMV is not seedborne.

Distribution. Canada, Europe, and the United States.

Symptoms. A yellow mottling consisting of various kinds of chlorotic streaks and spots is located on and between the veins of diseased leaves.

Management
1. Grow the least susceptible cultivar. There is some evidence that resistant cultivars can be developed.
2. Avoid growing red clover near garden peas and beans.

Cowpea Mosaic

Cause. Cowpea mosaic virus (CPMV) is in the comovirus group. CPMV is transmitted by sap and the bean leaf beetle, *Cerotoma ruficornis.*

Distribution. Central America and the United States (including Puerto Rico).

Symptoms. Diseased leaves have chlorotic and necrotic lesions. A systemic light green mosaic occurs on some diseased leaves.

Management. Not reported.

Cucumber Mosaic

Cause. Cucumber mosaic virus (CMV) is in the cucumovirus group. CMV is seedborne.

Distribution. Europe and the United States.

Symptoms. Diseased leaves have a mosaic appearance.

Management. Not reported.

Dwarf

Cause. Soybean dwarf virus (SDV) is in the luteovirus group. SDV is transmitted by grafting and in a persistent manner by the aphids *Aulacorthum solani* and *Acyrthosiphon pisum.* SDV is not seed- or sap-transmitted. The optimum temperature for transmission is 20°C.

Distribution. Japan. SDV may also occur in Australia and New Zealand.

Symptoms. Diseased plants are generally stunted. Leaves are rugose, curl down, and may have interveinal yellowing.

Management. Not reported.

Red Clover Vein Mosaic

Cause. Red clover vein mosaic virus (RCVMV) is in the carlavirus group. RCVMV overwinters in red clover and other legumes and is transmitted by the pea aphid, *Macrosiphum pisi,* and other aphids. The pea aphid can acquire the virus in 1 hour of feeding but loses infectivity in 24 hours. Seed transmission is suspected but has not been demonstrated. Diseased red clover plants serve as reservoirs for RCVMV, which also causes pea stunt.

Distribution. United States, particularly the eastern and north central states.

Symptoms. Symptoms are most conspicuous on leaves of new growth. The first symptom is a faint yellowing of the leaf veins. Gradually the yellowing becomes more chlorotic until veins and adjacent tissue become white. Yields are reduced and plants become more susceptible to root rot organisms, particularly *Fusarium* spp. Symptoms become masked during warm weather.

Management
1. Grow the least susceptible cultivar. There is some evidence that resistant cultivars can be developed.
2. Avoid growing red clover near garden peas and beans.

Ringspot

Cause. Tobacco ringspot virus is in the nepovirus group.

Distribution. Not known.

Symptoms. Symptom description is obscure but is thought to be a mild mottling of leaves that occasionally results in necrotic spots on diseased leaves.

Management. Not known.

Streak

Cause. Tobacco streak virus is in the ilarvirus group.

Distribution. Brazil and the United States.

Symptoms. Necrotic areas occur in the stem pith, especially near nodes. Axillary buds develop and leaves are dwarfed.

Management. Not reported.

Wilt

Cause. Broadbean wilt virus is in the fabavirus group.

Distribution. Not known.

Symptoms. Diseased plants are distorted and have a yellow mosaic. Plants eventually wilt during severe disease conditions.

Management. Not known.

Selected References

Blain, F., Bernstein, M., Khanizadeh, S., and Sparace, S. A. 1991. Phytotoxicity and pathogenicity of *Fusarium roseum* to red clover. Phytopathology 81:105-108.

Cunfer, B. M., Grahan, J. H., and Lukezic, F. L. 1969. Studies on the biology of *Myrothecium roridum* and *M. verrucaria* pathogenic on red clover. Phytopathology 59:1306-1309.

Dickson, J. G. 1956. Diseases of Field Crops, Second Edition. McGraw-Hill Book Co. Inc., New York.

Engelke, M. C., Smith, R. R., and Maxwell, D. P. 1975. Evaluation of red clover germplasm for resistance to leaf rust. Plant Dis. Rptr. 59:959-963.

Farr, D. F., Bills, G. R., Chamuris, G. P., and Rossman, A. Y. 1989. Fungi on Plants and Plant Products in the United States. American Phytopathological Society, St. Paul, MN. 1252 pp.

Hanson, E. W. 1959. Relative susceptibility of seven varieties of red clover to diseases common in Wisconsin. Plant Dis. Rptr. 43:782-786.

Hanson, E. W., and Kreitlow, K. W. 1953. The many ailments of clover, pp. 217-228. USDA Yearbook, Washington, D. C.

Kahn, M. A., Maxwell, D. P., and Smith, R. R, 1978. Inheritance of resistance to red clover vein mosaic virus in red clover. Phytopathology 68:1084-1086.

Leath, K. T., and Barnett, O. W. 1981. Viruses infecting red clover in Pennsylvania. Plant Dis. 65:1016-1017.

Leath, K. T., and Kendall, W. A. 1983. *Myrothecium roridum* and *M. verrucaria* pathogenic to roots of red clover and alfalfa. Plant Dis. 67:1154-1155.

Leath, K. T., Bloom, J. R., Hill, R. R., Kaufman, T. D., and Byers, R. A. 1983. Clover cyst nematode on red clover in Pennsylvania. Phytopathology 73:369.

Martens, J. W., Seaman, W. L., and Atkinson, T. G. (ed.). 1984. Diseases of Field Crops in Canada. Canadian Phytopathological Society, Ottawa, Canada. 160 pp.

McVey, D. V., and Gerdemann, J. W. 1960. Host-parasite relations of *Leptodiscus terrestris* on alfalfa, red clover and birdsfoot trefoil. Phytopathology 50:416-421.

Murray, G. M., Maxwell, D. P., and Smith, R. R. 1976. Screening *Trifolium* species for resistance to *Stemphylium sarcinaeforme*. Plant Dis. Rptr. 60:35-37.

Rao, A. L. N., and Hiruki, C. 1985. Clover primary leaf necrosis virus, a strain of red clover necrotic mosaic virus. Plant Dis. 69:959-961.

Roberts, D. A., and Kucharek, T. A. 1983. *Cylindrocladium crotalariae* associated with crown and root rots of alfalfa and red clover in Florida. (Abstr.) Phytopathology 73:505.

Smith, O. F. 1940. Stemphylium leaf spot of red clover and alfalfa. J. Agric. Res. 61:831-846.

Stutz, J. C., Leath, K. T., and Kendall, W. A. 1985. Wound-related modifications of penetration, development, and root rot by *Fusarium roseum* in forage legumes. Phytopathology 75:920-924.

Tofte, J. E., Smith, R. R., and Grau, C. R. 1992. Reaction of red clover to *Aphanomyces euteiches*. Plant Dis. 76:39-42.

14. Diseases of Rice *(Oryzae sativa)*

Diseases Caused by Bacteria

Bacterial Blight

Bacterial blight is also called kresek.

Cause. *Xanthomonas oryzae* pv. *oryzae* (syn. *X. campestris* pv. *oryzae*) survives in the temperate regions of the world in the rhizospheres of weeds, in straw piles, and at the bases of stems and on roots of rice stubble. In the tropics, *X. oryzae* pv. *oryzae* survives on weeds, on secondary growth from rice stubble, and in irrigation water in canals and rice fields.

Xanthomonas oryzae pv. *oryzae* is seedborne, although the extent to which it is transmitted through emerging rice seedlings is questionable. Glumes are readily infected, but recovery from infected seed is difficult because of the slow growth of *X. oryzae* pv. *oryzae* on media. Some evidence suggests that infected seed harbors viable bacteria for only 2 months, indicating seeds may not be an important source of inoculum.

Bacteria are disseminated locally by wind and water, particularly during a storm. Long-range dissemination is by irrigation water. Bacteria enter plants through hydathodes, growth cracks caused by the emergence of roots at the bases of leaf sheaths, and leaf wounds made by wind during storms. Bacteria do not enter through stomates. Seedlings are more susceptible to infection than are older plants. Infection at the tillering stage can lead to a total crop loss; however, plants are most commonly infected at the maximum tillering stage.

Disease is also associated with root and leaf wounds made when seedlings are pulled from seedbeds. This form of bacterial blight is called kresek, and plants less than 21 days old are the most susceptible. Kresek is favored by high temperatures (28° to 34°C).

Once inside a plant, bacteria affect the vascular tissue by cutting off water and causing infected plants to wilt. Bacteria ooze from lesions whenever moisture is present. The bacterial ooze dries and falls into the paddy water.

Bacterial blight became increasingly important when semidwarf, high-yielding cultivars were adopted. Six races of *X. campestris* pv. *oryzae* have been characterized in the Philippines. The races 1, 2, 3, and 5 are commonly detected in the Philippines, with race 2 being the most common. Race 1 is detected in lowlands where traditional cultivars are sown or in areas where cultivars with gene Xa-4 have been extensively cultivated. Race 3 is infrequently found. Race 4 is found only in Palawan. Race 5 is in the highlands and dominates the bacterial population in Banaue, the mountain terraces more 1500 m above sea level. Race 6 is a minute population. Bacterial blight is particularly destructive in Asia during the heavy rains of the monsoon season.

Distribution. Africa, Asia, Caribbean, Central and South America, and the United States (Louisiana and Texas).

Symptoms. Disease incidence increases with plant growth and is the highest at the flowering stage. Diseased seed may be discolored and poorly filled. However, symptomless seed may harbor low populations of *X. oryzae* pv. *oryzae* that later serve as an inoculum source for introducing the pathogen into disease-free fields.

Foliar symptoms are usually observed at the tillering stage. Water-soaked stripes, which occur along one or both margins of the upper portions of diseased leaf blades, enlarge and become yellow. Eventually, the lesions cover entire leaf blades and may extend to the lower ends of leaf sheaths, turning them white, then gray due to the growth of saprophytic fungi. Lesion development increases as the temperature increases and the relative humidity is greater than 70%. Discolored spots surrounded by water-soaked areas appear on the glumes of green grain; spots on mature grain are gray or light yellow.

The most severe form of bacterial blight is kresek, or wilting. Symptoms appear 2 to 3 weeks after transplanting but only in the tropics due to the practice of cutting off leaf tips before transplanting seedlings. Diseased leaves become gray-green, roll along the midrib, and eventually completely wither. Bacteria reach the growing point through the vascular system and infect the bases of other leaves and eventually kill the plant.

Another symptom in the tropics is the production of pale yellow leaves on older plants. The oldest leaves on these older plants remain green, but the youngest leaves are a uniformly pale yellow or whitish. Sometimes a broad green-yellow stripe may occur on leaves, and diseased tillers do not grow.

Epiphytotics that begin before flowering significantly reduce yield. Epiphytotics that occur after flowering have no effect on grain yield or yield components.

Management. Grow resistant cultivars. Resistance varies according to age of the plant and race of the causal bacterium. Genes Xa-3, Xa-4, Xa-5, and Xa-10 have been identified as resistant genes. Reactions caused by these genes have been divided into seedling and adult plant resistances. Seedling resistance is stable over the entire growth cycle and results, in adult plants, in lesions up to 15% the length of the lesions on the most susceptible cultivar. Cultivars with adult plant resistance based on the gene Xa-3 are similarly resistant in the adult plant stage but are susceptible in the seedling stage. Both resistances are race specific.

Seedling resistance increases with plant age; the fastest increase in resistance occurs 30 to 50 days after sowing. This differs from that of adult plant resistance, where immature leaves that are still extending are more susceptible than mature, fully extended leaves.

Bacterial Glume Blotch

Bacterial glume blotch is also called bacterial sheath rot. In several Asian countries and in Hungary, *Pseudomonas syringae* pv. *syringae* also has been reported as the cause of bacterial sheath brown rot.

Cause. *Pseudomonas syringae* pv. *syringae* (syn. *P. oryzicola*).

Distribution. Asia and Australia.

Symptoms. Symptoms vary from dark brown areas (1–2 mm in diameter) surrounded by green to light brown tissue on scattered florets in a panicle to a dark brown discoloration of complete florets. In severe disease, 75% of the panicle can be affected, but florets at the base of the diseased panicles rarely are completely discolored. Lesions also occur on florets just emerging from the boot.

Light green to dark brown lesions without definite margins occur on flag leaf sheaths. Veins of flag leaf sheaths also are darker than interveinal tissue. The stems of severely diseased plants may collapse at this point, but disease generally does not extend to the stem tissue. Bacterial ooze occurs on diseased tissue during warm, moist conditions.

Management. Bacterial glume blotch is not an economically important disease; therefore, management is not warranted.

Bacterial Halo Blight

Cause. *Pseudomonas syringae* pv. *oryzae*. The causal bacterium is also pathogenic to barley, oat, and common bean.

Distribution. Japan.

Symptoms. Disease symptoms develop in the tillering stage and disappear in the boot stage. Diseased leaves have small brown lesions that are surrounded by large yellow halos.

Management. Not reported.

Bacterial Leaf Streak

Cause. *Xanthomonas oryzae* pv. *oryzicola* (syn. *X. campestris* pv. *oryzicola*). The bacterium is a gram-negative, non-spore-forming rod (1.2 × 0.3–0.5 μm) with a single flagellum. The bacterium survives on infested residue, seed, and possibly in irrigation water. Dissemination is by splashing and wind-borne rain, irrigation water, and wind moving diseased leaves back and forth against healthy leaves. Bacteria enter plants through stomates but do not enter the xylem vessels until the later stages of infection, when bacterial multiplication becomes limited.

After lesions develop, bacterial exudate develops on the surface of lesions during moist conditions at night but dries up and falls into the water when dry conditions occur during the day. Disease development is favored by rain, high humidity, and temperatures of 28° to 30°C.

Distribution. Tropical Asia.

Symptoms. Initially, watery, dark green, translucent streaks (3–5 mm long) are confined between larger veins. The streaks enlarge into lesions and, under moist conditions, yellow beads of bacterial exudate appear on lesion surfaces. When conditions become dry, the bacterial exudate dries and forms numerous small yellow beads on the lesion surface. Susceptible cultivars may have yellow halos around lesions. Older lesions become light brown and translucent, and entire leaves die, become brown, then gray-white and are colonized by saprophytic organisms.

Bacterial streak can be distinguished from bacterial blight by their thinner, translucent lesions with the yellow bacterial ooze.

Management. Grow resistant cultivars.

Bacterial Sheath Brown Rot

Bacterial sheath brown rot is also called dirty panicle disease and manchado de grano.

Cause. *Pseudomonas fuscovaginae* survives in infested straw, on grain crops, and on wild grasses. The bacterium is also seedborne and seed-transmitted. Disease is most severe following early infection of the boots. Infection and disease development is favored by relatively cool (17°C and below), wet weather or high relative humidity at the booting and heading stages.

Distribution. Widespread. Bacterial sheath brown rot is an important disease in the high-altitude, high-rainfall environments of Nepal. The disease is also economically important in the mid- to high-altitude regions of Burundi, Japan, Latin American, Rwanda, and Zaire.

Symptoms. Symptoms of severe infections first appear at the heading stage and cause total spikelet sterility by preventing normal emergence of the panicle from the boot. Together with poor emergence, panicles and grain are partially to totally discolored. Grain discoloration may also be due to virus infection, especially rice hoja blanca virus, before flowering, or severe soil stress. Highly acid upland soils or flooded acid soils with high iron content cause grain discoloration if the rice lines are not adapted to these conditions. Several pathogenic fungi may also cause grain discoloration.

The flag leaf becomes rotted and has a brown discoloration. The disease can eventually lead to the death of the entire plant.

In mild infections, adult plants have characteristic gray-brown to brown lesions on the flag leaf sheaths from the booting to heading stages of plant growth. Although panicles still emerge, a reduced grain yield is produced.

Management. Resistance has been reported.

Bacterial Sheath Stripe

Bacterial sheath stripe is sometimes called bacterial brown stripe and bacterial stripe. There appear to be some similarities between bacterial sheath stripe and bacterial stripe in the literature and the two diseases may be the same. However, because symptoms appear to be dissimilar and the causal organisms are identified differently in the literature, bacterial sheath stripe and bacterial stripe are treated as two separate diseases here.

Cause. *Pseudomonas syringae* pv. *panici* (syn. *P. panici*). Disease is more severe under wet or humid conditions. Only immature plants become infected.

Distribution. Japan, the Philippines, and Taiwan.

Symptoms. Initially, water-soaked, longitudinal stripes form on the lower portions of leaf sheaths. Lesions enlarge (1–100 mm), coalesce, and cover entire diseased leaf sheaths. Bacterial exudate may form on lesion surfaces during humid conditions. Disease may progress into the unfolding leaves and cause a bud rot. Diseased plants may be stunted or killed.

Management. Not reported.

Bacterial Stripe

Bacterial stripe is also called bacterial brown stripe and brown stripe.

Cause. *Acidovorax avenae* subsp. *avenae* (syn. *Pseudomonas avenae, P. alboprecipitans* and *P. setariae*) is seedborne and survives up to 8 years in seed.

Distribution. Africa, Asia, Central America, Portugal, and South America.

Symptoms. Symptoms occur only on seedlings grown in nursery boxes. Leaves and leaf sheaths develop water-soaked, brown, longitudinal stripes. Growth of seedlings with only brown stripes is not affected in nursery boxes. However, seedlings with more severe symptoms of curving of leaf sheaths and abnormal elongation of mesocotyls, which cause leaves not to open, are generally stunted or curved.

Severely diseased seedlings are dwarfed and often turn yellow, wilt, and die. A heavy bacterial ooze occurs at cuts across brown lesions. Diseased seedlings die shortly after transplanting. Grain is discolored.

Management. Not reported.

Foot Rot

Foot rot is also called bacterial foot rot.

Cause. A bacterium similar to *Erwinia chrysanthemi*. Iris plants commonly found along canals, ponds, or irrigated rice fields, particularly in Japan, serve as overwintering hosts and sources of infection. Bacteria from diseased irises are liberated into the water and spread to rice plants by irrigation water. Other annual or perennial plants may also serve as sources of inoculum.

Distribution. Asia and the United States (Louisiana).

Symptoms. Diseased tillers display symptoms of either sheath rot or a wilting of the younger leaves. Sometimes decay is found on plants without detectable sheath rot symptoms. A dark brown rot occurs in diseased culms, which are soft and easily pulled apart, and bacterial ooze is present in the internodes. A black discoloration indicative of decay occurs at the nodes, but when the decay is restricted to the lower nodes, new roots that may develop from noninfected nodes above the decayed area allow new leaves to develop and panicles to keep growing.

Systemic crown infections occur at early tillering stages; sheath rot and restricted culm decay result from infections later than the maximum tillering stage.

Management. Not reported.

Grain Rot

This disease is also known as glume blight.

Cause. *Burkholderia glumae* (syn. *Pseudomonas glumae*). Bacteria are 1.0×1.5 μm and have multiple polar flagella. Emerging rice panicles were inoculated at 100% relative humidity for 24 hours.

Distribution. Asia, Central America, and South America.

Symptoms. A brown discoloration occurs on the diseased flag leaf sheath. Grain has a grayish discoloration.

Management. A dry heat treatment of 65°C for 6 days eradicated bacteria from small seed samples.

Seedling Blight

Cause. *Burkholderia plantarii* (syn. *Pseudomonas plantarii*).

Distribution. Japan.

Symptoms. Diseased seedlings damp off.

Management. Not reported.

Sheath Brown Rot

Sheath brown rot is also called bacterial sheath brown rot, bacteria brown rot, and sheath rot. The causal bacterium may be the principal causal agent of dirty panicle disease. The following symptoms may also be similar to those of dirty panicle disease.

Cause. *Pseudomonas fuscovaginae* is present on several grass hosts and is seed-borne. Grain discoloration is more severe under high humidity and rainfall and in fields with soil nutrient imbalances.

Pseudomonas syringae pv. *syringae* (syn. *P. oryzicola*) has been found to be associated with bacterial sheath rot in Hungary and Australia.

Distribution. Africa, Asia, Australia, Europe, Central America, and South America. The disease is found in highland swamps at 1450 to 1600 m in Burundi, Latin America, Madagascar, Rwanda, and Zaire.

Symptoms. Longitudinal, red-brown, necrotic stripes extend the full length of diseased sheaths and often the entire length of leaf laminae. Other sheath symptoms are small (1–5 mm), brown, elongated spots that expand and coalesce to form indistinct blotches on sheaths and, sometimes, on culms.

Symptoms on seedlings are limited to a water-soaked, brown necrosis of sheaths and occasional brown necrotic stripes on leaves. However, black stripes have been reported as a symptom on seedlings in Burundi. Seedlings often die.

When older sheaths are diseased, flag leaf sheaths enclosing emerging panicles sometimes have water-soaking, blotching, and brown necrosis. Severely diseased flag leaf sheaths and collars are necrotic and dry. Developing florets are brown, and panicles partially emerge from the boots. Florets that do emerge frequently do not fill with grain, are spotted or totally discolored, and are sterile. Seed has been described as "rusty-appearing." Roots may also have a brown discoloration.

Inoculated seedlings initially have water-soaked spots that evolve into dark brown streaks (5–10 cm long) on sheaths and laminae. Three to 5 days after inoculation, symptoms appear on flag leaf sheaths as water-soaked spots that enlarge to form dark- to grayish-brown blotches. Florets are dark- to grayish brown.

Management. Not reported.

Diseases Caused by Fungi

Aggregate Sheath Spot

Aggregate sheath spot is also called brown sclerotial disease. Aggregate sheath spot and sheath spot may be the same disease, but descriptions of the symptoms appear to be sufficiently different to consider these as two different diseases.

Cause. *Ceratobasidium oryzae-sativae,* anamorph *Rhizoctonia oryzae-sativae.* Some researchers consider *R. oryzae* and *R. oryzae-sativae* to be synonymous. The causal fungi survive as sclerotia in residue and soil. Initial infections occur by sclerotia floating on paddy water to plants. *Ceratobasidium oryzae-sativae* causes disease under warm weather conditions. Generally, disease is more severe at low soil nitrogen levels.

Distribution. China, Japan, India, Iran, Thailand, the United States (California), and Vietnam.

Symptoms. Lesions first appear at the waterline on lower leaf sheaths during the tillering stage. Lesions (0.5 to 4.0 cm long) are circular to elliptical-shaped and have gray-green to straw-colored centers that are surrounded by distinct brown margins. A strip of necrotic cells runs down the middle of the lesion center. Frequently additional margins that form around the initial lesion produce a series of concentric bands.

Disease progresses to upper leaf sheaths and sometimes to the lower portions of leaf blades. Lesions may coalesce and cover entire leaf sheaths, causing leaves to turn bright yellow, then die. During favorable disease conditions, culms, flag leaves, and panicle rachises may become diseased and cause the death of entire tillers.

As leaf sheaths become rotted, abundant brown sclerotia are produced in and on diseased tissue. Sclerotia inside sheaths are cylindrical to rectangular and visible through the diseased tissue. Sclerotia produced on the plant surface are irregularly globose.

Management. Grow tall cultivars. These are more resistant than semidwarf cultivars.

Bakanae Disease

Bakanae disease is also called white stalk and palay lalake (man rice).

Cause. *Fusarium moniliforme* survives between growing seasons as mycelia and conidia on or in seed. *Fusarium moniliforme* is also soilborne, but this is not considered to be an important source of inoculum.

When seed is sown, *F. moniliforme* infects seedlings and grows systemically within the xylem tissue as microconidia and mycelium. Diseased plants will either be stunted or elongated, depending on soil moisture conditions. *Fusarium moniliforme* produces fusaric acid, which stunts plants in dry soil conditions, and gibberellin, which causes plants to elongate abnormally under moist soil conditions. At flowering time, conidia are produced on diseased or dead culms and disseminated by wind to flowers of plant hosts or to developing grain, which becomes either infected or contaminated.

Distribution. Wherever rice is grown.

Symptoms. Plants may be either elongated or stunted, but the most common symptom is abnormal elongation of diseased plants in seedbeds or fields. Diseased seedlings are thin, chlorotic, and several centimeters taller than healthy plants. Severely diseased seedlings may die before or after transplanting.

In the field, diseased plants are taller than normal and have few tillers and chlorotic flag leaves. Leaves on a diseased plant die in a relatively short time. Diseased plants that live have empty panicles. A white or pink fungus growth may appear on the lower parts of diseased plants.

Management
1. Treat seed with a systemic seed-protectant fungicide.
2. Grow a resistant cultivar.
3. Experimentally, disease was reduced 1.9% to 77.0% when seeds were soaked in suspensions of antagonistic bacteria.

Basal Node Rot

Cause. *Fusarium oxysporum.* Basal node rot is prevalent under two conditions: in fields where rice is grown in rotation with pasture grass, and in the second and third year of successive rice cultivation.

Distribution. Brazil.

Symptoms. Areas of stunted plants with dull green leaves occur in fields. A black discoloration occurs at the point on underground basal nodes where adventitious roots are initiated and often extends to mesocotyls and adventitious roots. Severely diseased plants have poorly developed roots and few tillers. Plant death is rare.

Management. Not reported.

Black Kernel

Cause. *Curvularia lunata,* teleomorph *Cochliobolus lunatus.* Disease is favored by excess moisture in the form of rain, dew, or high humidity and temperatures of 24° to 30°C.

Distribution. Not reported, but likely common in warm, humid rice-growing areas.

Symptoms. At maturity, diseased kernels are covered with a dense, velvety, dark fungal growth consisting of conidia and conidiophores.

Management. Not reported.

Blast

Blast is also called Pyricularia blight and rotten neck.

Cause. *Pyricularia grisea* (syn. *P. oryzae*), teleomorph *Magnaporthe grisea,* overwinters for at least 4 years as mycelia in infested residue, and for 1 or more years as conidia on residue and seed in the temperate rice-growing regions and, likely, in the tropical rice-growing regions of the world. Although Rossman (1990) considered *Pyricularia grisea* and *P. oryzae* to be synonymous, past literature reported that *P. oryzae* also survives in the temperate regions as chlamydospores, perithecia, and sclerotia in soil, on winter cereals, diseased rice plants, and weed hosts. Survival is not important for *P. oryzae* in the tropics since airborne conidia are present throughout the year.

Conidia are produced from mycelium and become windborne at night in the presence of dew or rain. High humidity favors subsequent disease development. At temperatures of 18° to 20°C, rice cultivars adapted to temperate climates generally are more predisposed to infection than are tropical cultivars. Secondary inoculum is conidia that are produced on lesions when there has been more than 93% relative humidity for 6 days and are disseminated by wind, water, infested residue, or seed to other host plants.

Although plants of all ages are susceptible, blast is essentially a disease of young host tissues. For example, fully extended leaves are more resistant than partially extended leaves, leaves on older plants are more resistant to infection, and panicle resistance is reported to increase with age. Rice plants also tend to be most susceptible when they are grown in dry soil and are relatively resistant when they are grown under flooded conditions. Physical conditions tend to affect resistance; plant susceptibility increases when plants are grown too close together and excessive nitrogen is applied to soil. Several physiologic races of *P. oryzae* are thought to exist.

Distribution. Generally distributed wherever rice is grown.

Symptoms. Blast generally is more destructive in temperate environments than in the tropics. Symptoms of blast can be divided into two general phases, depending on when the plant was infected. The leaf blast phase, which occurs between the seedling and late tillering stage, frequently causes the death of plants. The panicle blast phase, or neck blast phase, occurs after flowering. Spots are produced on leaves, nodes, panicles, and grains, but seldom on leaf sheaths.

Spots on leaves begin as small, water-soaked, whitish, grayish, or bluish dots but quickly enlarge (0.3–0.5 × 10–15 mm) under moist conditions.

Mature leaf spots are elliptical and pointed at both ends and have a gray to whitish center surrounded by a brown to red-brown margin. Spots vary in shape and color depending on cultivar, environment, and age. Resistant cultivars have only brown specks, while moderately resistant cultivars have small, roundish lesions with necrotic centers and brown margins.

Later in the growing season, some cultivars are diseased where the flag leaf attaches to the sheath. The lesion enlarges downward on the sheath and upward on the leaf, becoming gray at the point of attachment. The flag leaf may break off and become detached from the plant.

Nodes at or near flood level, which generally are the second or third node from the soil line may be infected. The base of the sheath turns black due to production of conidia and the node eventually breaks apart and remains connected by a few vascular strands. The plant above the infected node dies.

Any portion of the panicle, panicle branches, and glumes may become diseased and have brown lesions. If the base of the panicle is diseased, rotten neck (or neck rot) symptoms occur, causing the stem below the panicle to snap.

Management
1. Grow resistant cultivars; however, a cultivar may not be resistant to all races of *P. oryzae*. Some cultivars display adult-plant resistance to leaf blast and panicle blast. It is uncertain if vegetative-stage resistance to leaf blast is correlated with resistance to neck blast.
2. Sow only clean, healthy seed treated with a seed-protectant fungicide.
3. Apply small increments of nitrogen at any one time.
4. Manage floodwater to maintain adequate but not excessive depth. Flooding is an effective way of managing blast.
5. Avoid excessive plant populations.
6. Apply a foliar fungicide if disease conditions warrant its use.
7. Destroy previous crop residue as soon after harvest as feasible.
8. Rotate rice with other crops.
9. Control grasses and other weeds.
10. Experimentally, silicon has been shown to lower disease intensity.

Brown Blotch

Cause. *Cercospora oryzae*. Symptom development is aided by the secretion of a red toxin, cercosporin, by the mycelium of some pathovars. Cercosporin is excited upon illumination and generates free radicals that damage cell membranes and cause cell death.

Distribution. The United States (Louisiana and Texas).

Symptoms. Initially, small, irregular-shaped, brown blotches occur on the diseased leaf sheath approximately 3 to 4 cm below the ligule. Eventually, the blotch spreads to cover the entire leaf sheath. Leaves are prematurely desiccated and yields are reduced.

Management
1. Grow resistant cultivars. Resistant cells have a mechanism for excluding, exporting, or destroying cercosporin.
2. Sow early-maturing cultivars early in the growing season to escape the disease.

Brown Spot

Brown spot is also called brown leaf spot, rice brown spot, and sesame leaf spot.

Cause. *Cochliobolus miyabeanus,* anamorph *Bipolaris oryzae,* (syn. *Drechslera oryzae* and *Helminthosporium oryzae*). Some researchers consider the causal fungus to be *Exserohilum oryzae,* but this name is not listed in Farr et al. (1989) as a synonym for *B. oryzae. Cochliobolus miyabeanus* overwinters primarily as mycelium in seed. The following spring, infected seeds give rise to seedlings with spots on roots and coleoptiles. Leaves are usually not infected by seedborne inoculum because of their rapid growth at this time. Most leaf spots originate from secondary infections that occur when conidia produced on spots that developed on coleoptiles are disseminated by wind to leaves. Plants that grow in soils with nutritional deficiencies or poorly drained soils where nutrient uptake is hindered are more susceptible to infection. Disease development is favored by relative humidities of 86% to 100% and temperatures of 20° to 25°C. Leaves must be wet for 8 to 24 hours for infection and disease to occur. *Bipolaris oryzae* produces a toxin that induces disease symptoms.

Distribution. Generally distributed in rice-growing areas of Africa, Asia, and the Western Hemisphere.

Symptoms. The first symptoms on coleoptiles are pale, yellow-brown spots or streaks. Young roots have blackish lesions. Young spots on leaves are small, circular, and dark brown or purple-brown. Typical spots are oval, about the size and shape of sesame seeds, and evenly distributed over leaf surfaces. Spots may become numerous enough to kill leaves. When fully developed, spots are brown with a gray or whitish center surrounded by a bright yellow halo. On very susceptible cultivars, spots may be larger (1 cm long).

Black or dark brown spots appear on glumes and may cover entire surfaces of glumes. During humid weather, conidia are produced in spots, giving them a velvet-like appearance. Grain has dark spots on endosperms.

Management
1. Balance soil fertility. Plants grown in silicon-deficient soils are more prone to disease than plants grown in soils with adequate silicon. In de-

ficient soils, the addition of silicon decreases disease severity and increases yields.
2. Rotate rice with other crops.
3. Sow healthy seed treated with a seed-protectant fungicide.
4. Apply foliar fungicides.
5. Grow the most resistant and adapted cultivar.

Collar Rot *(Ascochyta oryzae)*

Collar rot is also called rice collar rot.

Cause. *Ascochyta oryzae.*

Distribution. India.

Symptoms. Small, brown lesions expand to cover the whole collar region. Lesions extend down the diseased leaf sheath and blade, turning them dark brown to black. Leaf blades separate from leaf sheaths and defoliate.

Management. Not reported.

Collar Rot *(Pestalotiopsis versicolor)*

Collar rot is also called rice collar rot.

Cause. *Pestalotiopsis versicolor.*

Distribution. India.

Symptoms. Symptoms first appear at the collar joining the leaf blade and leaf sheath and consist of small, brown lesions that expand to cover the whole collar region. Lesions sometimes extend down the diseased leaf sheath and the blade, turning them dark brown to black. Subsequent rotting causes the leaf blade to separate from the sheath and drop off. In severe cases, several leaves on the same plant dry up and defoliate.

Management. Not reported.

Crown Sheath Rot

Crown sheath rot is also called Arkansas foot rot, black sheath, black sheath rot, and brown sheath rot.

Cause. *Gaeumannomyces graminis* var. *graminis* likely overseasons as perithecia and mycelia in infested residue and is seedborne to a slight extent. Ascospores are disseminated by wind to host plants and initiate primary infection when moisture is present later in the growing season. *Gaeumannomyces graminis* var. *graminis* also survives on St. Augustine grass and bermudagrass in Florida.

Distribution. Africa, India, Japan, the Philippines, and the United States.

Symptoms. Symptoms appear late in the growing season. A brown discoloration occurs on diseased sheaths from the crown to the waterline or above. Leaf blades die when sheaths are severely diseased. Numerous black perithecia are produced in the discolored areas but are not easily seen because only the perithecial peaks protrude above the surface of the epidermis. Red-brown mycelial mats may be found on the inner surface of the diseased sheaths. At maturity, straw has a dull brownish cast. Culms may also be infected, but perithecia are not produced. Few panicles—frequently only one—are produced on diseased plants. Lodging usually occurs at one of the nodes rather than from the internode area.

Management
1. Grow resistant varieties.
2. Treat seed with a seed-protectant fungicide.
3. Sow early in the growing season.
4. Delay permanent flood.

Dead Tiller Syndrome

Cause. An unidentified fungus is thought to be the cause.

Distribution. The United States (Arkansas).

Symptoms. Symptoms first appear 6 to 10 days after the permanent flood and progress from 8 to 14 days until the floodwater equilibrates to ambient soil and air temperatures. The initial symptom is a slight discoloration of plants and some wilting, but advanced symptoms include severe wilting. Occasional plants have yellow to orange chlorosis along older leaf tips or leaf margins. Diseased plants have a rotting culm (usually the main culm in a hill), which frequently produces a distinct odor when crushed. Tillers or nearby plants are not affected. Decay begins internally at a node and can be well advanced before this symptom can be observed; plants then die in 24 to 48 hours. Dead tiller syndrome continues to develop throughout the growing season in areas where cold floodwater is added to the field.

Management. Not reported.

Downy Mildew

Downy mildew is also called yellow wilt.

Cause. *Sclerophthora macrospora* survives as oospores in infested leaf residue. Oospores germinate in the spring by forming sporangia. Sporangia, in turn, germinate by their cell contents dividing into zoospores that are liberated into water and "swim" to a nearby rice seed that is germinating. Infection is optimum at temperatures of 18° to 20°C. Oospores and sporangia are produced on diseased plants throughout the growing season and provide secondary inoculum.

Distribution. Australia, China, India, Italy, Japan, and the United States (Arkansas). Downy mildew is not considered a serious disease.

Symptoms. Symptoms are most obvious at flowering time. Initially, the youngest leaves have chlorotic or whitish spots, and leaf sheaths may be distorted and twisted. Panicles are also distorted due to their failure to emerge from the distorted and twisted sheaths. These panicles also are smaller than normal and remain green, while the floral parts are either bare or have a "hairy" appearance.

Management. Treat seed with a seed-protectant fungicide.

Eyespot

Eyespot has also been called brown spot–type disease.

Cause. *Drechslera gigantea* (syn. *Helminthosporium giganteum*). Grasses such as *Cynodon dacytlon* and *Eleusine indica* may be alternate hosts. Prolonged leaf wetness favors disease.

Distribution. Colombia, Guatemala, Guyana, Honduras, Mexico, Panama, and Peru.

Symptoms. Initially, spots on leaves are small, water-soaked, olivaceous dots or rings. A yellow halo frequently appears around young spots but later disappears. Later, spots develop into minute, longitudinally elongated, oval lesions (0.5–1.0 × 1.0–4.0 mm) with white to straw-colored necrotic centers and surrounded by narrow, dark brown margins.

Under favorable disease conditions, successive development and coalescence of lesions produces a characteristic zonation, which is more irregular-shaped than the regular wavy pattern characteristic of leaf scald.

Management. Grow resistant cultivars.

False Smut

False smut is also called green smut.

Cause. *Ustilaginoidea virens* overwinters in temperate regions as sclerotia and chlamydospores. Most primary inoculum comes from sclerotia germinating and producing stromata on tips of stalks. Ascospores are produced in perithecia within stromata and disseminated by wind to host flowers. Rice is infected either at very early flowering or when grain is mature.

Individual grains are transformed into spore balls in which one to four sclerotia are produced. Secondary infection is by chlamydospores formed on spore balls that are disseminated by wind to rice plants, where they germinate and form germ tubes on which conidia are borne. Chlamydospores do not free easily from smut balls because of a sticky material that holds them in place. Optimum germination occurs at 28°C and, later, disease development is favored by high moisture.

Distribution. Most rice-growing areas, but false smut causes little damage.

Symptoms. A few to many grains of the panicle are transformed into ball-like objects that have a velvety appearance. Balls are small and visible between glumes at first but continue to grow and enclose the floral parts, reaching a diameter of 1.0 cm or more. Young balls are slightly flattened, smooth, yellow, and covered by a membrane. With continued growth, ball membranes burst and expose mycelium and spores, and the color of the ball changes to orange and still later to yellow-green or green-black. At this stage of disease development, the surface of the balls crack. When the balls are cut open, their white interior consists of mycelium together with floral parts and other host parts.

Management. Apply a foliar fungicide a few days before heading.

Fusarium Diseases

Cause. *Fusarium equiseti, F. semitectum* (syn. *F. pallidoroseum*), *F. nivale* (syn. *Microdochium nivale*), and *F. culmorum.*

Distribution. Wherever rice is grown.

Symptoms. *Fusarium culmorum* causes brown, necrotic leaf tips and spots on seedling leaves ranging from minute, irregular-shaped spots to rectangular spots (1–2 × 4–6 mm) that cover 3% to 5% of the leaf blade. Adult plants are usually not diseased.

    Fusarium equiseti causes discoloration of vascular tissues of culms that extends from the seedling base to the third node. Diseased culms are initially reddish but later become brown to black.

    Fusarium semitectum causes disease of panicles and culms. Panicles in the developing sheath at boot stage have irregular-shaped lesions that are yellowish but later become dark brown. Diseased culms have a reddish to brownish vascular discoloration that extends from the base of the plant to the third or fourth internode. A necrosis develops during flowering and continues until ripening to partially or totally cover diseased kernel surfaces.

    Fusarium nivale causes a necrosis of panicles and leaves.

Management. Not reported.

Fusarium Leaf Blotch

Cause. *Fusarium solani.*

Distribution. India.

Symptoms. Symptoms on diseased leaves are irregular-shaped, oblong, brown necrotic areas (5–6 × 14–20 cm) with water-soaked margins that eventually coalesce and result in a blight of the leaf blade. Disease usually appears a few centimeters from the leaf tip during the boot stage of plant development or later.

Management. Not reported.

Fusarium Sheath Rot

Cause. *Fusarium moniliforme* and *F. proliferatum.*

Distribution. *Fusarium moniliforme* occurs in India and *F. proliferatum* occurs in the United States (Arkansas and Texas).

Symptoms. Dark brown lesions occur on the flag leaf sheath, and there is a blanking/discoloration of the panicle. Young tillers of inoculated plants are sometimes killed. The mycotoxins beauvericin, fumonisin, and moniliformin are produced by cultures of *F. proliferatum.* The fumonsins FB 1, FB 2, and FB 3 have been detected in naturally contaminated rice in the United States.

Management. Not reported.

Helminthosporium Leaf Spot

The original name of the disease has been retained although the nomenclature of the causal organism has changed.

Cause. *Exserohilum rostratum* (syn. *Helminthosporium rostratum*).

Distribution. India.

Symptoms. Symptoms are most prevalent in August. Small, light-colored, oval to linear spots (1.0–1.5 × 2.0–3.0 mm) occur on diseased leaves.

Management. Not reported.

Kernel Smut

Kernel smut is also called black smut and bunt.

Cause. *Tilletia barclayana* (syn. *Neovossia barclayana, N. horrida,* and *T. horrida*). *Tilletia barclayana* survives as chlamydospores in endosperms, on the surface of seeds, and in soil. Chlamydospores germinate in the presence of moisture and produce a promycelium on which several sporidia are borne in a whorl. Secondary sporidia are produced either directly by budding from primary sporidia, or from mycelium produced by primary sporidia. The secondary sporidia are disseminated by wind to host plants and infect flowers that are just opening. After infection occurs, only the endosperms, not the embryos, are destroyed. *Tilletia barclayana* apparently does not become systemic in diseased seedlings.

Abundant moisture during flowering increases disease incidence and severity. Moisture in the form of rain or dew at harvest time causes chlamydospores to swell and break through the hull of the diseased seed and be released to infest soil and healthy seed. Diseased seed may also remain intact until sown the following spring.

Distribution. Wherever rice is grown.

Symptoms. Symptoms are obvious only at maturity. Generally only a few grains in a panicle and often only a part of a grain—usually the endosperm—are diseased. After a rain or dew, the mass of chlamydospores swells and bursts through a glume, appearing as tiny, black pustules or streaks. Severe disease causes a short, beak-like outgrowth to be produced by ruptured glumes. Chlamydospores look like a black powder has been scattered onto the leaves and grains of surrounding plants. Chlamydospores that do not break out of a glume can be seen through the wet hulls. Seeds that are not severely diseased may germinate, but the resulting seedlings are stunted. Smut spores give milled rice a grayish color.

Management
1. Grow the most resistant cultivars.
2. Sow early-maturing cultivars as early as feasible in the growing season.
3. Avoid high rates of nitrogen fertilizer.

Leaf Scald

Cause. *Microdochium oryzae* (syn. *Rhynchosporium oryzae*) overseasons in and on seeds. Although *Gerlachia oryzae* is reported as the causal organism in some literature, it is possibly a synonym of *M. oryzae* even though it is not listed as such by Farr et al. (1989). *Microdochium oryzae* is a weak pathogen that infects plants either predisposed by other factors or through wounds. Seeds are the primary source of inoculum. Disease development is favored by wet weather.

Distribution. West Africa, Central America, Southeast Asia, and the United States.

Symptoms. Symptoms most commonly occur near tips of mature leaves but may also occur on leaves, leaf sheaths, culms, and grains in all stages of development. Oblong or diamond-shaped, water-soaked blotches that are somewhat restricted by veins develop into large olive-colored areas encircled by alternate dark and light brown halos. Later, blotches become gray-olive and have a characteristic zonation of light and dark brown rings. When large portions of leaves are diseased, they are killed and become tan.

Management
1. Grow the most resistant cultivar.
2. Apply a foliar fungicide.
3. Apply a seed-treatment fungicide.

Leaf Smut

Cause. *Entyloma oryzae* overwinters as chlamydospores in the sori on leaves. The following summer the presence of moisture and temperatures of 28° to 30°C is optimum for chlamydospores to germinate and produce a promycelium on which sporidia are formed. The sporidia bud and form secondary sporidia that are disseminated by wind to leaves, where they germinate in the presence of water.

Distribution. Wherever rice is grown, but leaf smut is considered a minor disease.

Symptoms. Leaf smut symptoms occur late in the growing season. On both sides of diseased leaves, sori occur as small, black, slightly raised, rectangular spots that may become numerous but still remain distinct from each other. Sori have been mistaken for sclerotia. The chlamydospores in sori are held tightly together and are covered by the leaf epidermis. After soaking in water for a few minutes, the epidermis ruptures and reveals a black mass of chlamydospores beneath the epidermis surface. Severely diseased leaves may turn yellow, split, and sometimes die.

Management. No management is necessary since leaf smut is not considered a serious disease.

Narrow Brown Leaf Spot

Narrow brown leaf spot is also called Cercospora leaf spot.

Cause. *Cercospora janseana* (syn. *C. oryzae*), teleomorph *Sphaerulina oryzina*. Little is known of the life cycle of *C. janseana* as it relates to causing narrow brown leaf spot on rice. Several physiologic races exist.

Distribution. Generally distributed wherever rice is grown except in Europe.

Symptoms. Disease usually does not become severe until rice approaches maturity. Symptoms are most common on leaves but may occur on leaf sheaths, glumes, and pedicels. Tan or brown leaf spots (1.5–3.0 × 3.0–13.0 mm) occur only in interveinal areas of the diseased leaf.

Spots on glumes and pedicels are smaller and darker than the spots on leaves. Spots on leaf sheaths (7–50 mm long) are light to dark brown and have indefinite borders that may encircle the plant.

Management
1. Grow resistant cultivars.
2. Sow early-maturing cultivars early in the growing season to escape the disease.

Pecky Rice

Pecky rice is also called kernel spotting.

Cause. *Cochliobolus miyabeanus, Curvularia* spp., *Fusarium* spp., *Microdochium oryzae, Sarocladium oryzae,* and other fungi.

Distribution. Widespread.

Symptoms. Several or all florets per panicle are shriveled and discolored a light to dark brown, reddish-brown, purple, or white.

Management. Normally not practiced.

Phoma Seedling Blight

Cause. *Phoma glomerata* is seedborne.

Distribution. Ghana.

Symptoms. Diseased seed is covered with a pale brown to brown mycelium. Pycnidia are produced singly or in groups. Seedlings that rise from diseased seed damp off within 7 days after germination. Brown lesions occur on diseased coleoptiles, leaves, and culms of emerged seedlings. Roots show different degrees of root rot; pycnida develop in the diseased areas of the root. Eventually, seedlings are killed.

Management. Not reported.

Pithomyces Glume Blotch

Cause. *Pithomyces chartarum.*

Distribution. India.

Symptoms. Diseased plants occur in patches within a field. Although all glumes on a head may be diseased, the terminal seeds are diseased most frequently. Initially, light brown lesions occur on glumes of individual grains. In most cases, lesions develop toward the tip of the grain rather than the base. Lesions become gray in the center and are surrounded by a dark brown ring. Lesions enlarge and sometimes cover the whole surface of grains. Diseased glumes are grayer than healthy glumes. Grains enclosed by infected glumes remain small, have an undeveloped endosperm, and germinate poorly. Diseased grains also are shriveled, discolored, and light in weight. Short, dark conidiophores bearing conidia may be observed on diseased plant parts.

Management. Not reported.

Pythium Damping Off

Cause. *Pythium irregulare, P. spinosum,* and *P. sylvaticum* are the main causal agents of damping off of rice seedlings in nursery flats. Other *Pythium* spp. that contribute to early damping off are *P. aphanidermatum, P. myriotylum, P. splendens, P. ultimum,* and *P. vexans.* Disease is favored by low temperatures and a soil pH of 6.0. The main causal agents are pathogenic to rice seedlings at emergence but are no longer pathogenic to seedlings older than the first-leaf stage of plant development. *Pythium graminicola*, however, is reported to cause damping off of rice seedlings at the two- to three-leaf stage.

Distribution. Damping off tends to be more of a problem in the temperate rice-growing areas of the world than in the tropics.

Symptoms. Diseased seedlings at the four- to five-leaf stage are chlorotic and stunted. Roots are necrotic and undeveloped. Severely diseased plants are killed.

Management. Treat seed with a fungicide.

Rust

Cause. *Puccinia graminis* f. sp. *oryzae*. The life cycle and symptoms are similar to those caused by *P. graminis* on other cereals. Rust is a minor disease on rice and causes no economic loss.

Sheath Blight

Cause. *Rhizoctonia solani* AG-1 IA, teleomorph *Thanatephorus cucumeris*. *Rhizoctonia solani* survives as sclerotia in the top 1 cm of soil for up to 21 months in dry soil but for less than 7 months in water-saturated soils. As soil depth increases, the number of viable sclerotia decreases. Survival also occurs as mycelia in infested residue, and as sclerotia and the basidial stage of *T. cucumeris* on weeds. Thus weeds are hosts of the causal fungi and a potential source of inoculum to infect rice.

Primary inoculum likely consists of buoyant sclerotia; disease incidence is directly related to the number of these sclerotia. Sclerotia float on top of the water and tend to accumulate around plants at the water-plant interface. There, sclerotia germinate and form infection hyphae that penetrate plants at the waterline.

The fungus then forms infection cushions and/or lobate appressoria on the plant surface. Mycelium grows rapidly on or inside the diseased plant tissues, spreading by means of runner hyphae to the upper plant parts and initiating secondary infections that become spots. Eventually, sclerotia formed in the center of the spots are easily detached and reinfest soil and water.

Infection and disease development are optimum at temperatures of 28° to 32°C and a 95% or higher relative humidity. Sheath blight is associated with intensive and high-input production systems. Disease has been reported to be more severe under high nitrogen than low nitrogen conditions. Close sowing and heavy fertilization increases the crop canopy, thereby raising humidity and increasing disease incidence and severity. In

the United States, disease develops more extensively in semidwarf cultivars than in closely related standard height cultivars because of the shorter
distance between the waterline and panicles.

Distribution. Wherever rice is grown. Sheath blight is considered by many researchers to be the most important disease of rice in the southern United
States.

Symptoms. Symptoms usually occur as lesions on lower leaf sheaths after diseased plants have reached the late tillering or early internode elongation
stage of plant growth, but symptoms may also occur on young plants.
Spots or lesions appear mainly on diseased leaf sheaths but may extend
into the leaf blades if environmental conditions are favorable for disease
development.

    Symptoms appear near the waterline as green-gray, elliptical spots that
are about 1 cm long. Eventually, spots enlarge to 2 to 3 cm long, become
gray, and have irregular purple-brown margins. Lesions on the upper plant
coalesce and encompass the entire diseased leaf sheaths and stems. Brown
sclerotia in the center of spots are easily detached. During humid conditions, mycelium may grow upward to infect higher leaf sheaths on the
plant. With severe disease, all leaves of a plant may be blighted, allowing
increased sunlight penetration and decreased humidity. Lesions then dry
up and become white, tan, or gray with brown borders.

    Yields are reduced primarily from reduction in mean seed weight and a
lower percentage of filled spikelets. Additional losses result from increased
lodging or reduced ratoon production.

Management
1. Apply fungicides and other chemicals that are effective in managing
   sheath blight.
2. Sow the recommended population. In the Philippines, a wider plant
   spacing generally decreased disease incidence and increased yield in
   BR1 rice.
3. Grow tolerant cultivars. Indica types are more tolerant than Japonica
   types. In the United States, semidwarf lines are more susceptible than
   standard height lines. Most long-grain cultivars of rice in the United
   States are susceptible or very susceptible to sheath blight, whereas
   medium- and short-grain cultivars tend to be moderately susceptible to
   moderately resistant. Resistance in plants is associated with two factors:
   a reduction in numbers of infection cushions and the production of ox-
   idized phenolic compounds, which slow the spread of *R. solani* within
   a plant. The phenolic compound factor is manifested as a dark zone
   around the lesion.
4. Apply proper fertility rates. High nitrogen applications increase disease
   incidence.
5. A suspension of *Bacillus subtilis* or an antibiotic substance from *B. subtilis* has been reported to limit size of lesions.

Sheath Blotch

Cause. *Pyrenochaeta oryzae.* Survival occurs as pycnidia in infested residue. Conidia are the primary inoculum.

Distribution. Worldwide in all rice-growing areas. In the United States, sheath blotch has been reported from Arkansas, Florida, and Texas.

Symptoms. Oval, reddish-brown, blotch-like symptoms occur near the middle of the outer leaf sheaths of rice tillers. Lesions start from the margin of a sheath as dark reddish-brown, oblong blotches that enlarge, become bluish gray, and may cover the entire sheath. Lesions may grow to 10 cm and often appear at the junction of the sheath and leaf blade, just below the leaf collar, without causing the collar to break. When dry, the affected part of the sheath turns gray-brown and may or may not have distinct red-brown to brown margins. As diseased plants mature, lesions enlarge to eventually cover the whole sheath and kill the leaf. Occasionally, *P. oryzae* infects the leaf blade and glumes. Pycnidia may be seen embedded in diseased tissue.

Management. Not reported.

Sheath Rot

Cause. *Sarocladium oryzae* (syn. *Acrocylindrium oryzae*) is seedborne and survives as mycelia in infested residue. Conidia are produced from mycelia and disseminated by wind to leaf sheaths, where entry is gained through stomata and injuries. Insect wounding is especially important in the infection process. High insect populations injure the leaf sheaths enclosing panicles, thereby retarding emergence of panicles and facilitating development and spread of disease. Mycelium then grows intercellularly in vascular bundles and sheath mesophyll tissues. Midribs of leaf blades and grains are also infected.

In the Philippines, rice is infected at all stages of growth but sheath rot is reported to be most destructive when plants are infected after booting. Maximum disease occurs when minimum temperatures are 17° to 20°C and minimum relative humidity is 40% to 56% at flowering time. However, other reports from the Philippines state the optimum temperature and relative humidity for infection and disease development of seedlings are 25° to 35°C and 70%, respectively. Heavy nitrogen application and other fertility programs that result in poor growth favor disease development.

Distribution. Southeast Asia and the United States.

Symptoms. Symptoms vary with cultivar and the diseased plant part. Generally, sheath rot occurs on the top leaf sheaths, especially the flag leaf sheath enclosing the young panicles. Initially, lesions start as oblong or somewhat irregular-shaped spots (5–15 mm long) that either have gray centers with brown margins or are entirely gray-brown. Lesions enlarge, coalesce, and cover most of the leaf sheath. Panicles fail to emerge or partially emerge, then rot. A whitish, powdery growth consisting of mycelium, conidiophores, and conidia may be found inside diseased sheaths.

It has been reported that a large reduction in yield is due to a synergistic effect between *S. oryzae* and rice tungro virus.

Management
1. Grow resistant cultivars.
2. Balance soil fertility. Excessive nitrogen increases disease incidence and severity and increased potassium decreases it.

Sheath Spot

Sheath spot is also called bordered leaf spot, brown bordered leaf spot, red sclerotial disease, Rhizoctonia sheath spot, and sheath blight. Aggregate sheath spot and sheath spot may be the same disease. However, descriptions of the symptoms appear to be sufficiently different to consider these as two separate diseases.

Cause. *Rhizoctonia oryzae.* It is also reported that *R. oryzae-sativae* (syn. *Sclerotium oryzae-sativae*) is a causal organism. Although it is not known for certain how the fungi survive, it is presumed that *R. oryzae* overwinters as mycelia or sclerotia in infested residue and possibly in soil. Wild grasses have also been observed to act as overwintering hosts.

Dissemination is likely by wind or waterborne soil and residue. Sheath spot is most severe along levees and other places where rice growth is dense, and high humidity and temperatures of 29° to 32°C commonly occur.

Distribution. Japan, the United States, and Vietnam.

Symptoms. Symptoms, in the form of spots, first appear during the late tillering or early heading stage of plant growth. Spots are confined almost entirely to leaf sheaths and do occasionally occur on leaves but not on stems. Spots first appear on the lower leaf sheaths near the waterline but may occur on other sheaths of the diseased plant.

Spots initially are elliptical and have a red-brown discoloration. As spots mature, they have a bleached, straw-colored center with a wide, red-brown margin and average 1 to 3 cm in length but may become as long as 10 cm. Lesions may enlarge and coalesce to form large discolored areas that kill diseased leaves and cause plant lodging.

Management
1. Grow resistant cultivars.
2. Avoid excessive seeding rates.
3. Use proper timing of nitrogen top-dress applications.
4. Control grasses and other weeds.

Stackburn Disease

Cause. *Alternaria padwickii* is seedborne and overwinters in soil and infested residue as sclerotia and saprophytic mycelia. Conidia produced on mycelia are disseminated by wind to leaves, where infection is most likely to occur through a wound because *A. padwickii* is a weak pathogen.

Distribution. Wherever rice is grown.

Symptoms. A few oval or circular spots occur on leaves and vary in size from 3 to 10 mm long. Initially, the centers of spots are tan, then become white and encircled by one or two narrow, dark brown rings. The white spot is speckled with sclerotia that look like black dots scattered over the spot.

Glumes have tan to whitish spots that are surrounded by a relatively wide, dark brown margin. These spots also have sclerotia scattered over them. Diseased kernels are discolored, shriveled, and brittle.

Roots and coleoptiles of germinating seedlings may have dark brown to black lesions with sclerotia scattered over them. Seedlings may either outgrow the disease or wither and die.

Management. Not reported.

Stem Rot

Cause. *Magnaporthe salvinii,* synamorph *Sclerotium oryzae* (the sclerotial stage), anamorph *Nakataea sigmoidea* (syn. *Helminthosporium sigmoideum* and *Vakrabeeja sigmoidea*). A similar fungus, *H. sigmoideum* var. *irregulare,* also was reported to cause the same disease, but it may be the same as *N. sigmoidea.*

The fungi overwinter for up to 6 years as sclerotia in the top 5 to 10 cm of soil and on infested residue. Sclerotia are the primary inoculum. During tillage operations, sclerotia float on the water surface, germinate, and infect plants through wounds. Conidia and ascospores are produced approximately 60 and 100 days after sowing, respectively, but apparently do not cause severe enough disease to reduce yield. However, they may initiate secondary infections that, in turn, may increase the level of overwintering sclerotia.

Sclerotia develop between the leaf sheath and culm. After harvest, *S. oryzae* continues to grow saprophytically on infested residue and produce large numbers of sclerotia. Rice tissue parasitized by *S. oryzae* remains the most important source of sclerotia. Colonization of uninfected rice residue by mycelia from sclerotia is not an important source of new sclerotia. Inoculum levels are greatest where large amounts of infested residues are left on or near the soil surface. High nitrogen levels increase the incidence and severity of stem rot.

Distribution. Burma, India, Japan, the Philippines, Sri Lanka, the United States (Arkansas, California, Louisiana, and Texas), and Vietnam.

Symptoms. Symptoms appear late in the growing season, generally within a week after a field has been drained of water. Symptom development occurs rapidly only after plants are in the reproductive stage and is the most rapid at physiologic maturity.

Small, blackish, irregular-shaped lesions occur on outer leaf sheaths near the waterline. Lesions enlarge as the disease progresses. Eventually, leaf sheaths are rotted and numerous sclerotia form as *S. oryzae* grows between the diseased inner leaf sheaths and culms. One or two internodes of diseased stems rot and collapse, with only the epidermis remaining intact. Dark gray mycelium may be seen in the hollowed stems and numerous, small, black sclerotia occur on inner stem surfaces. The next internodes may appear to be healthy. Diseased stems lodge.

Occasionally in the tropics, leaf sheaths of transplanted seedlings rot at the waterline and cause the leaves above them to die.

Management
1. Sow early-maturing cultivars early to escape disease injury.
2. Application of potassium fertilizer reduces stem rot severity.
3. Drain water at the early stages of infection and keep soil saturated but not covered with water until rice has almost matured.
4. Rotate a field out of rice for 6 years or more.
5. Apply foliar fungicides.
6. Destroy or remove infested rice straw.

Water Mold Disease

Cause. Several aquatic Phycomycetes, including *Achlya conspicua, A. klebsiana, Pythium dissotocum,* and *P. spinosum. Fusarium* spp. also have been implicated as causal organisms. The disease occurs in water-seeded rice.

Distribution. Widely distributed.

Symptoms. Germinating seed is rotted and diseased seedlings are weakened or killed, resulting in losses through reduced density, irregular stands, and poor growth of diseased seedlings.

Management. Apply a seed-treatment fungicide.

White Leaf Streak

Cause. *Ramularia oryzae.*

Distribution. Africa and Asia.

Symptoms. White to gray lesions (0.5 × 1.0–2.5 mm) with narrow, brown borders develop on both leaf surfaces. However, lesions on diseased upper leaf surfaces may be entirely white and have no borders, while lesions on lower leaf surfaces may be entirely brown.

Management. Not reported.

Diseases Caused by Nematodes

Lesion

Cause. *Pratylenchus zeae* and probably other *Pratylenchus* spp. *Pratylenchus zeae* is an endoparasite that feeds and reproduces in the root cortex.

Distribution. Temperate climates of the world.

Symptoms. Aboveground symptoms include plant stunting. Diseased leaves are chlorotic and a purple to reddish discoloration may occur. Diseased roots have reduced growth and few fibrous roots. Brown lesions may be present on roots and a general brown to blackish root decay may occur.

Management. *Pratylenchus zeae* on upland rice can be controlled by crop rotation with at least two crops of cowpea or mung bean.

Rice Root

Cause. *Hirschmanniella oryzae* is an endoparasite found in aquatic environments. The nematode is associated with lowland rice and is adapted to flooded paddies and marshes. *Hirschmanniella oryzae* infects and reproduces on sedges and grasses but remains dormant when paddies are dry. It occurs in parenchyma tissue but not in thin lateral roots where there are no air channels. *Hirschmanniella oryzae* moves out of roots into soil soon after seed forms or plant growth ceases.

A total of seven species of *Hirschmanniella* are reported to infect rice.

Distribution. Wherever rice is grown.

Symptoms. Diseased roots become necrotic and discolored. Plants are stunted and the number of tillers, panicles, top growth, and yield are reduced. There is also a delay in tillering and flowering.

Management. Not reported.

Root Knot

Root knot is sometimes called the rice root-knot nematode.

Cause. *Meloidogyne* spp., including *M. arenaria, M. incognita, M. javanica, M. graminicola, M. oryzae,* and *M. salasi. Meloidogyne arenaria, M. incognita,* and *M. javanica* do not survive if the air temperature averages below 3°C during the coldest month and do not survive continuous flooding. *Meloidogyne graminicola* does not invade roots in flooded soils but does survive inside roots for 5 weeks and in flooded soil for 5 months or more. Nematodes survive as eggs in soil or reproduce on other plants. Sandy or coarse-textured soils favor reproduction.

The nematodes are obligate endoparasites. Second-stage juveniles infect the apical meristem of young seedling roots before flooding. Nematode secretions stimulate development of giant cells and root galls in 4 days. Females lay eggs inside the galls and juveniles hatch and reinfect the same root. A life cycle is completed in 19 days in well-drained soils at temperatures of 22° to 29°C.

Distribution. *Meloidogyne arenaria, M. incognita, M. javanica* are found throughout Asia. *Meloidogyne graminicola* is found in Bangladesh, India, Laos, Thailand, and the United States (Georgia, Louisiana, and Mississippi). *Meloidogyne oryzae* is found in Surinam, and *M. salasi* is found in Costa Rica and Panama.

Symptoms. Aboveground symptoms on diseased plants are chlorosis, wilting, late maturation, and reduced plant growth and tillering. The characteristic root galls that are formed prevent root elongation. Yield of upland rice is reduced. Deepwater rice infected by *M. graminicola* may not grow above the water.

Management
1. Continuous flooding of fields controls *M. arenaria, M. incognita,* and *M. javanica.*
2. Leave soil fallow or rotate to a nonhost plant, such as peanut, sweet potato, maize, and soybean.
3. Grow resistant cultivars where appropriate.
4. Apply nematicides in nurseries.

Stem

This disease is also called ufra or ufra disease.

Cause. *Ditylenchus augustus* survives as adults in dry soil and infested residue. Nematodes migrate from diseased to healthy plants through water and by stem and leaf contact at more than 75% relative humidity. Nematodes can also survive on and be dispersed in freshly harvested rice grain.

During periods of high humidity, active nematodes climb up seedlings, enter young growing tissue, and in 1 hour burrow between leaves and leaf sheaths, where they feed as ectoparasites. As diseased plants grow, the nematodes move up to new tissue. All reproduction occurs only on rice plants. Younger plants are more likely to become infected than older plants.

Distribution. Bangladesh, Egypt, India, and Southeast Asia.

Symptoms. Chlorosis and malformation of leaves occur 2 months after sowing. A few brown spots on diseased leaves and sheaths later become a darker brown, as do the upper stem internodes.

The most obvious symptoms are panicles that have few or no filled spikelets toward the base and, together with leaves, are crinkled, twisted, corrugated, or otherwise distorted. In some instances, two or three distorted ears may be surrounded by one sheath. There may also be more than the normal number of branches on a diseased stem.

Diseased plants may or may not be killed; however, panicles from surviving plants either emerge or remain enclosed within the flag leaf sheaths. In both cases, kernels usually are not produced.

Management
1. Grow the most resistant cultivars.
2. Dry rice grains to a moisture content of 14% or below before storage to eliminate the nematodes from grain.

Summer Crimp

Summer crimp is also called white tip.

Cause. *Aphelenchoides besseyi* survives on seed for up to 2 years. Infested seed provides the major source of inoculum for nematode infection. *Aphelenchoides besseyi* does not overwinter in soil but is disseminated by water through the soil. The nematode is mostly an ectoparasite, occurring within folded leaves in the early stages of plant growth, then entering spikelets and finally the glumes. The optimum conditions for nematode development are a temperature of 28°C and a relative humidity of 70% or higher.

Distribution. Africa, Australia, Cuba, Japan, Southeast Asia, and the United States.

Symptoms. Diseased leaves are darker green than leaves of healthy plants. However, the most obvious symptom is chlorotic or white leaf tips (up to 5 cm long) that are evident at the beginning of elongation and later become brown and tattered. Diseased plants are stunted, lack vigor, and give rise to tillers at the higher nodes of the plants. Panicles of severely diseased tillers mature late. The length and number of spikelets are reduced and there is a high percentage of sterile, small, distorted glumes and grains. Upper leaves, particularly flag leaves, are twisted so panicle emergence from boots is often incomplete.

Management
1. Grow resistant cultivars.
2. Sow in water and keep fields flooded.

Diseases Caused by Phytoplasmas

Rice Orange Leaf Disease

Cause. Possibly phytoplasma-like organisms (PLO) that are disseminated by the leafhoppers *Recilia dorsalis* and *Inazuma dorsalis*. The incubation period in *R. dorsalis* nymphs is 15 to 33 days but has been reported to be as short as 2 to 15 days. The causal organism(s), which is persistent in the vectors, has an average incubation period of 22 days in *R. dorsalis* and 10 to 28 days in rice plants. Some leafhoppers transmit the agent continuously and others transmit intermittently until they die. PLO have been reported to be bounded by a unit membrane and to contain ribosome particles and DNA-like fibrils. Shapes of PLO vary and range from 50 to 1100 nm in diameter.

Distribution. Malaysia, the Philippines, Sri Lanka, Thailand, and probably other countries in Southeast Asia.

Symptoms. Diseased plants are generally distributed sporadically within a field. Leaves become orange, roll inward, and dry up. Plants display symptoms 10 to 28 days after inoculation and die 2 to 4 weeks after symptoms appear. Inoculated seedlings are stunted and leaf emergence is delayed, leaves are short and often have chlorotic stripes, tip twisting, and ragged blades similar to symptoms of rice ragged stunt virus–infected leaves. Rice orange leaf disease does not cause serious yield loss.

Tungro symptoms are similar in some aspects to those of orange leaf.

Management. Not reported.

Rice Yellow Dwarf

Rice yellow dwarf is also called yellow dwarf.

Cause. A phytoplasma-like organism. The causal agent may overwinter in leafhoppers and the wild grass *Alopecurus aequalis*. The causal agent is disseminated primarily by the leafhopper *Nephotettix cincticeps* and also by *N. apicalis* and *N. impicticeps*. Leafhoppers acquire the causal agent by feeding on diseased plants for 1 to 3 hours. There is a 20 to 39 day incubation period, after which the inoculation feeding time is usually less than 1 hour. The latent period in rice is about 1 month in "warm" weather and 3 months in "cool" weather.

Distribution. Tropical Asia.

Symptoms. Diseased plants are generally chlorotic and stunted and produce numerous tillers. The uniformly pale green or pale yellow chlorosis first appears on the newest leaves and continues on the successive leaves. Sometimes only a faint mottling occurs. Plants infected early normally do not die but produce abnormal heads or no heads.

Management
1. Grow a resistant cultivar.
2. Apply an insecticide to kill overwintering insects.

Diseases Caused by Viruses

African Cereal Streak

Cause. African cereal streak virus (ACSV) is transmitted only by the plant-hopper *Toya catilina*. The natural virus reservoir probably is native grasses. Disease development is aided by high temperatures.

Distribution. East Africa.

Symptoms. ACSV is limited to the phloem of diseased plants, where it causes a necrosis. Initially, faint broken chlorotic streaks begin near the leaf base and extend upward. Later, definite alternate yellow and green streaks develop along the entire leaf blade but are more broken than on other cereals, resulting in a mottled appearance. Eventually, diseased leaves become almost completely yellow. Leaves formed after infection tend to develop a shoestring habit and die.

　　Young diseased plants become chlorotic and severely stunted, and die. Little or no seed is produced. Plants become soft, flaccid, and almost velvety to the touch.

Management. Not reported.

Giallume Disease

Giallume disease is also called barley yellow dwarf.

Cause. Barley yellow dwarf virus (BYDV).

Distribution. Italy and Spain.

Symptoms. Diseased plants are stunted and have fewer than normal tillers. Foliage is yellowed.

Management. Delaying the sowing date eliminated the occurrence of BYDV in Spain. The effect was attributed to the timing of insect vector flights and to temperature.

Grassy Stunt B

Cause. There is a question if the causal organism grassy stunt B (GSB) is a virus (perhaps a strain of rice grassy stunt virus) or a phytoplasma-like organism. GSB is transmitted in a persistent manner by the rice brown planthopper *Nilaparvata lugens*.

Distribution. Taiwan.

Symptoms. Symptoms vary somewhat depending on cultivar. Characteristic symptoms are excessive tillering with or without stunting. Initially, young diseased leaves are narrower than healthy leaves and are green but have faint to conspicuous chlorotic stripes on both sides of the midribs. Later, diseased leaves become pale green and have a conspicuous mottling. Still later, leaves become yellow and the stripes disappear. Mildly stunted plants live to maturity but produce few filled panicles.

Management. Not reported.

Grassy Stunt Y

Cause. There is question if the causal organism (GSY) is a virus (perhaps a strain of rice grassy stunt virus) or a phytoplasma-like organism. GSY is transmitted in a persistent manner by the rice brown planthopper *Nilaparvata lugens*.

Distribution. Taiwan.

Symptoms. Symptoms vary somewhat depending on cultivar. Tillering is reduced in the winter but increases in the summer. Plants are slightly to markedly stunted. Leaves initially may be somewhat chlorotic, narrow, and stiff and have an erect growth habit. Small, dark brown spots may occur on leaves of some cultivars and conspicuous yellow-white stripes may occur along leaf veins of other cultivars. Diseased plants are not killed but produce empty panicles.

Management. Not reported.

Rice Black Streak Dwarf

Rice black streak dwarf is also called black-streaked dwarf disease.

Cause. Rice black streak dwarf virus (RBSDV) is transmitted in a persistent manner by the planthoppers *Laodelphax striatellus*, *Unkanodes sapporonus*, and *Ribautodelphax albifascia*. Several grasses are hosts for RBSDV.

Distribution. Japan.

Symptoms. Galls appear as waxy, elongated, irregularly shaped protuberances extending along major veins on diseased lower leaf surfaces, the outer surface of leaf sheaths, and culms. Protuberances form gray to dark brown streaks of various lengths on older leaves. The proximal portion of leaf blades is often twisted.

Plants infected early in their growth are severely stunted and have more tillers than normal. Foliage is darker than normal and few or no panicles are formed. Grain often has dark brown blotches.

Management. Grow resistant cultivars.

Rice Dwarf

Rice dwarf is also called dwarf disease.

Cause. Rice dwarf virus (RDV) is in the reovirus group. RDV is transmitted by the planthopper *Nephotettix nigropictus.*

Distribution. Japan and Nepal.

Symptoms. Diseased plants are stunted to various degrees. Leaves are dark green and have large numbers of discrete translucent spots that often coalesce and form longitudinal lines parallel to the midrib. Spots are also present on leaf sheaths. Plants infected at an early growth stage are the most severely stunted, while those infected at a later growth stage are less stunted. Both root and shoot systems are reduced, and panicles are either absent or reduced in size.

Management. Not reported.

Rice Gall Dwarf

Cause. Rice gall dwarf virus (RGDV) is in the reovirus group. RGDV is transmitted in a persistent manner by the leafhopper *Recilia dorsalis* and the green rice leafhoppers *Nephotettix cincticeps, N. malayanus, N. nigropictus,* and *N. virenscens. Nephotettix* spp. have about a 2-week incubation period. RGDV is not known to be seed-transmitted.

Distribution. Thailand.

Symptoms. Diseased plants are severely stunted. Galls are most abundant on early-infected plants. Leaves are dark green and have galls or vein swellings on the under surfaces of diseased leaf blades and on the outer sides of leaf sheaths. Galls initially are light green and translucent but eventually become white. Most galls are less than 2 mm long, but they can vary in size (0.4–0.5 × 0.4–8.0 mm). Leaf tips may be twisted.

Management. Not reported.

Rice Grassy Stunt

Rice grassy stunt is also called grassy stunt.

Cause. Rice grassy stunt virus (RGSV) is in the tenuivirus group, but there is some question as to whether the causal organism is a phytoplasma-like organism. RGSV survives year-round in rice, *Oryzae sativa,* and several other species of rice. RGSV also persists in and is transmitted only by the rice brown planthopper *Nilaparvata lugens.* At least two different strains of RGSV, designated as RGSV-1 and RGSV-2, are known to occur.

Distribution. Ceylon, India, Malaysia, the Philippines, and Thailand.

Symptoms. RGSV-1 symptoms are severely stunted plants that have excessive tillers and an erect growth habit. Diseased leaves are short and narrow, chlorotic, and covered with numerous rusty spots. Young leaves of some cultivars may be mottled or may have narrow yellow and white stripes with diffuse margins parallel to the midribs. Stripes may be located only at the basal portion of leaves or extend the entire length of diseased leaves. Numerous small tillers give a diseased plant a grassy appearance. Diseased plants normally survive but produce few or no panicles.

RGSV-2 symptoms on diseased plants are stunting, a yellow to orange discoloration, rusty spotting of lower leaves, and a narrowing of leaf blades. Young emerging leaves are pale green to almost entirely chlorotic and tend to have striping and mottling symptoms. Leaves remain yellow even when the plants are adequately fertilized, but mature leaves of plants infected with RGSV-1 turn dark green when nitrogen fertilizer is applied. Severely diseased seedlings produce few tillers and eventually die.

Management. Some cultivars show a degree of resistance. Resistance is governed by a single dominant gene. Experimentally. Neem (*Azadirachta indica)* seed oil has been found to reduce survival of planthoppers and suppress transmission of grassy stunt.

Rice Hoja Blanca

Rice hoja blanca is also called chlorosis, cinta blanca (white band), hoja blanca, raya (stripe), raya blanca (white stripe), rayadilla (striped), and white leaf.

Cause. Rice hoja blanca virus (RHBV) is in the tenuivirus group. RHBV is transmitted by the planthopper *Sogatodes oryzicola,* and once RHBV is acquired, it is transmitted through eggs from infective females to their progenies for several generations. Little is known of how the virus overwinters, but since hoja blanca is most prevalent in the tropics, it is likely diseased rice plants grow year-round. Jungle rice, *Echinochloa colonum,* also is a host.

Distribution. Western Hemisphere. In the United States, rice hoja blanca has been reported from Florida, Louisiana, and Mississippi.

Symptoms. Initially, one or more white stripes or white mottling may occur on a diseased leaf, or the entire leaf may turn white. Diseased and normal tillers may be produced by the same plant. Often the second crop of tillers produced by a diseased plant displays no symptoms.

Diseased plants are stunted, and panicles may not reach their normal size or may not emerge from sheaths. Hulls, which often are distorted, turn brown and dry out. Flower parts are sterile or absent; therefore, few seeds are produced and heads remain upright instead of bending over.

Management. Grow resistant cultivars.

Rice Necrotic Mosaic

Rice necrotic mosaic is also called necrotic mosaic disease, yaika-sho, and eso mosaic.

Cause. Rice necrotic mosaic virus (RNMV) is in the potyvirus group. RNMV is sap-transmissible and is soilborne. Transmission is by soilborne fungi, possibly *Polymyxa graminis*. The weeds *Ludwigia perennis* and *Brachiaria ramosa* are hosts and potential reservoirs for the virus. Diseased plants occur mostly in upland seedbeds.

Distribution. India and Japan.

Symptoms. Initially, a mosaic mottling occurs on diseased lower leaves at about the maximum tillering stage, after seeds have been transplanted in paddy fields. Mottling consists of light green to yellow streaks that are oval to oblong (1 × 1–100 mm). Later, streaks on diseased leaves coalesce and form irregular-shaped patches that spread to the upper leaves. When several patches are present, a leaf becomes yellow. Mottling may also occur on the culm.

Plants are slightly reduced in size, but this is not very noticeable. The number of tillers is reduced and tillers that are present tend to lie flat, resulting in a spreading growth habit. Later, a few elongated, necrotic lesions appear on the surfaces of leaf sheaths and the basal portions of culms. There is also a reduction in the number of panicles, which produce only a few light grains.

Management. Not reported.

Rice Ragged Stunt

Cause. Rice ragged stunt virus (RRSV) is in the reovirus group. RRSV is transmitted by the brown leafhopper, *Nilaparvata lugens*, with an acquisition period of 24 hours or longer. RRSV is not mechanically or seed-transmitted. Vectors are most efficient at transmitting RRSV 5 days after hatching and become less efficient as they grow older. RRSV is systemic. Rice ragged stunt is more severe on deepwater rice cultivars when plants are infected at the seedling to early stem elongation growth stages than at the late stem elongation and flowering growth stages.

Distribution. The Philippines, India, Indonesia, and Sri Lanka.

Symptoms. Diseased plants are variously stunted at all growth stages. Severe stunting of deepwater rice varieties may lead to plant death because of drowning.

During the early growth of diseased plants, leaves are ragged; irregular edges consisting of one to 17 notches or indentations occur before leaf blades unroll. The ragged edges vary in length and usually occur only on a portion of one side of a leaf blade but occasionally occur on both sides of a leaf blade and the leaf sheaths. The ragged area is chlorotic, becomes yellow to brown-yellow, and eventually disintegrates. However, dark green leaves have also been reported as a symptom.

The top portion of leaf blades is often twisted, resulting in a spiral shape of one or more turns. Vein swellings caused by a proliferation of phloem cells occur on the upper portions of the outer surfaces of leaf sheaths near the collar and, less often, on the lower surfaces of leaf blades and culms. Vein swellings (1 × 1–100 mm) are pale yellow or, infrequently, brown.

At the boot stage of plant growth, the flag leaves are shortened and often are twisted, malformed, or ragged. Tillers often generate nodal branches, which are secondary or tertiary tillers produced at the upper nodes. Nodal branches often bear small panicles that are more numerous than on healthy plants.

Diseased plants flower later than healthy plants, which results in empty or partially filled panicles. As a consequence, yield losses of 50% to 100% have been reported.

Management. No acceptable management measure is reported. However, neem *(Azadirachta indica)* seed oil has been found to experimentally reduce the survival of planthoppers and to suppress the transmission of virus.

Rice Stripe

Rice stripe is also called stripe disease.

Cause. Rice stripe virus (RStV) is in the tenuivirus group. Transmission in a persistent and a transovarial manner is by the planthoppers *Laodelphax striatellus, Ribautodelphax albifascia,* and *Unkanodes sapporonus*. RStV multiplies and persists in vectors; therefore, disease incidence is correlated to the percentage of viruliferous insects. Several other plants are virus hosts.

Distribution. Japan, Korea, and Taiwan. RStV has been reported to cause the most serious disease of rice in Japan.

Symptoms. Plants infected while young will have more pronounced symptoms than plants infected at a later growth stage. Early infection may cause severe plant stunting or death, but plants infected later may be only slightly stunted.

Diseased leaves will emerge without unfolding, then elongate, become twisted, and droop. Leaves are chlorotic and often have a wide chlorotic stripe with diffuse margins. Frequently, a gray necrotic streak that appears in the chlorotic area enlarges and eventually kills the leaf.

Leaves that unfold properly have an irregular chlorotic mottling that appears as a stripe the length of the diseased leaf blade and on the leaf sheath. The number of tillers and panicles is reduced, and diseased panicles have malformed spikelets that do not emerge properly from leaf sheaths. Grain may be malformed.

Management. Grow resistant cultivars.

Rice Stripe Necrosis

Rice stripe necrosis is called rice "entorchamiento" in Colombia.

Cause. Rice stripe necrosis virus (RSNV) is in the furovirus group. RSNV is transmitted by the soilborne fungus *Polymyxa graminis*. *Polymyxa graminis* survives in field soil as cystosori. Cysts give rise to zoospores that "swim" through free soil water until they contact a host root and encyst. Encysted zoospores produce a structure called a stachel through which zoosporic cytoplasm enters the host cell and becomes a plasmodium. Only primary root tissue of young roots is infected; optimum infection occurs at a temperature of about 25°C. The host cell then becomes infected with RSNV if *P. graminis* is viruliferous. After a period of time, the plasmodium develops into a zoosporangium that releases additional zoospores, which repeats the infection cycle. However, some plasmodia develop into cysts and, often, nearly every cell in the small feeder roots will contain a cyst. As root cells senesce, cysts are eventually released into the soil, where they can remain viable for years without loss of virulence.

Distribution. Colombia and the Ivory Coast.

Symptoms. Marked mosaic-like discolorations occur on diseased leaves. Entire leaves may become twisted or severely malformed and exhibit chlorotic stripes and necrosis.

Management. Not reported.

Rice Transitory Yellowing

Rice transitory yellowing is also called brown wilt, rice transitory yellowing disease, and yellow stunt.

Cause. Rice transitory yellowing virus (RTYV) is in the rhabdovirus group. RTYV is transmitted by the leafhoppers *Nephotettix apicalis*, *N. cincticeps*, and *N. impicticeps*. RTYV persists and multiplies in vectors; however, virus is not infective when the temperature is below 16°C or above 38°C.

Distribution. China, Japan, Taiwan, and Thailand.

Symptoms. Diseased leaves become yellow, beginning at the distal portion of lower leaves; therefore, discoloration is initially more intense in the lower diseased leaves than in the upper ones. Brown, rusty flecks or patches may appear on discolored leaves. The yellow discoloration eventually becomes indistinct in lower leaves, which soon roll and wither, leaving only a few live upper leaves. Diseased plants have a reduced root system and produce few or no panicles.

Diseased plants often recover under greenhouse conditions. Following approximately a month of yellow leaves, normal leaves may be produced at a later plant growth stage. Consequently, the appearance of diseased plants may become normal after yellow leaves have fallen off.

Management. Not reported.

Rice Yellow Mottle

Rice yellow mottle is also called rice yellow mottle disease.

Cause. Rice yellow mottle virus (RYMV) is in the sobemovirus group. Survival of RYMV is in volunteer rice and ratoons, or rice regrowths, from previously harvested crops.

RYMV is mechanically transmitted through sap and is spread by adult leaf-feeding beetles in the Chrysomelidae, especially *Sesselia pusilla*. In Madagascar, *Dicladispa gestroi* is considered the most important vector. RYMV is also possibly spread by irrigation water. RYMV is not known to be seedborne. Disease development is more severe in plants infected during an early growth stage than in plants infected at a later growth stage. Different strains of RYMV exist.

Distribution. Burkina Faso, Ivory Coast, Guinea, Kenya, Liberia, Madagascar, Mali, Niger, Nigeria, and Sierra Leone.

Symptoms. Initially, scattered, orange-yellow spots occur either on the youngest leaves of recently transplanted rice or on mature rice. Eventually, the older leaves at the bottom of diseased plants are yellow and the younger leaves at the top of plants are orange. Young leaves have a thin mottle ranging from yellow to dark green. Both artificially and naturally infected plants may be dwarfed, and seedlings may be killed. Depending on the age of plants at infection, there is a reduced tillering and panicles are sterile, malformed, and only partially emerge, which results in reduced yields. Losses of up to 90% have been reported from Madagascar.

Management. Grow resistant cultivars. Resistant cultivars also possess horizontal resistance to blast.

Tungro

The following diseases are also thought to be tungro: accepta no pula (red disease), cadang-cadang (yellowing), leaf yellowing, mentek disease, panyakit merah (red disease), penyakit habang, rice dwarf, stunt disease, and yellow orange leaf. Tungro is also called rice tungro.

Cause. Two types of tungro virus particles occur: spherical particles (RTSV) and bacilliform particles (RTBV). RTSV contains single-stranded RNA and two major proteins of 24 and 23 kDa. RTBV contains circular double-stranded DNA and a single major protein of 32 kDa.

RTBV is transmitted only with the help of RTSV. RTBV causes tungro symptoms of plant stunting and yellow-orange discoloration, and RTSV enhances them. Infection by RTSV alone causes only very mild stunting. RTSV is known as a latent virus and acts as a helper or an RTSV-related helper factor for the transmission of RTBV by leafhoppers. Hence, RTBV is transmitted only when the virus source is infected with both RTBV and RTSV, or when leafhoppers first acquire RTSV, then RTBV.

Other evidence suggests RTSV may not be the helper or the only helper for RTBV transmission. When leafhoppers were placed on RTBV-infected plants, RTBV transmission occurred 7 days after acquisition. The helper factor for RTBV is lost after molting. RTSV causes disease as an independent pathogen, is efficiently transmitted from plants infected with RTSV alone, and is retained by adult leafhoppers 2 to 3 days after acquisition.

RTSV and RTBV survive year-round in diseased rice plants or when one rice crop only is grown a year: survival is in stubble of diseased plants that ratoon and in numerous grassy weeds.

Both viruses are transmitted in a semipersistent manner by the leafhoppers *Nephotettix virenscens, N. malayanus, N. nigropictus, N. parvus,* and *Recilia dorsalis.* Greater disease spread occurs in wet seasons due to the greater numbers of vectors. Insects feed at least 30 minutes to acquire the virus from diseased plants and transmit the virus by feeding for 15 minutes on healthy plants. Viruses are retained for 2 to 3 days in leafhoppers, after which they must be reacquired. *Nephotettix virenscens* is reported to be the most efficient vector of both viruses, with retention for 4 days after acquisition. *Nephotettix nigropictus* is the least efficient vector but retained both viruses for 5 days after acquisition. Leafhoppers feed predominantly on the phloem of leafhopper-susceptible cultivars but feed on the xylem of leafhopper-resistant cultivars.

Tungro is severe only in areas where host plants and vectors multiply year-round. Disease incidence generally increases with an increase in plant spacing. A wide spacing results in larger plants with a larger surface area for vectors to feed on and more tissue for virus multiplication than smaller plants grown in a closer spacing. High tungro incidence is associated with the vector population and the proportion of viruliferous vectors.

Several strains, differentiated on the basis of their characteristic symptoms on rice cultivars, are reported to exist. At least three strains of tungro virus have been identified and designated as S, M, and T.

Distribution. Bangladesh, India, Indonesia, Malaysia, Thailand, and the Philippines. Tungro is considered the most important viral disease of rice and the major constraint to rice production in south and southeastern Asia.

Symptoms. Plants infected with both viruses show severe tungro symptoms, while plants infected only with RTBV display mild stunting and yellowing. Plants infected only with RTSV have indistinct symptoms, including mild stunting.

The most obvious symptoms are stunting, reduction in the number of tillers, and a leaf discoloration that varies from shades of yellow to orange starting at the leaf tips and extending to the lower parts of diseased leaves. Discoloration of leaves in moderately resistant cultivars may gradually disappear at the later stages of plant growth. Young leaves may be mottled and older leaves have rusty-colored specks of various sizes.

Susceptible cultivars may be severely stunted and discolored, and die. The three strains of tungro virus—S, M, and T—each produce distinct symptoms on certain rice cultivars. On diseased leaves, the S strain causes an interveinal chlorosis that looks like yellow stripes. The M strain produces mottling. The T strain causes narrow leaf blades in some cultivars and a striping of diseased leaves in other cultivars.

Management
1. Grow resistant cultivars. Resistance to RTSV infection in some cultivars may not be due to resistance to leafhoppers.
2. Treat seed with a systemic insecticide.
3. Apply foliar insecticides to reduce vector populations.
4. Adjust planting date to correspond to low numbers of vectors.
5. Neem, *Azadirachta indica,* applied as a seed oil and cake at 150 and 250 kg/ha prevented tungro virus transmission by *N. virescens*. The feeding behavior was disrupted as it shifted from phloem to xylem feeding.

Unknown Virus–Caused Disease

This disease may be rice wilted stunt disease.

Cause. Unknown, but transmitted by a brown planthopper. The causal agent is possibly a virus and persists in the vector for a lengthy time even after molting.

Distribution. The Philippines, particularly where brown planthopper populations are high.

Symptoms. Initial symptoms occur 7 to 14 days after inoculation. Diseased plants are stunted. Leaves are yellow, abnormally narrow, and erect, which gives diseased plants an erect growth habit. Irregular-shaped, brownish blotches and chlorotic streaks that resemble a mosaic-type of mottling may be observed on the discolored leaves. Diseased plants die prematurely.

Management. Not reported.

Waika Disease

Cause. Rice waika virus (RWV) is transmitted by the green rice leafhoppers *Nephotettix cincticeps, N. malayanus, N. migropictus,* and *N. virenscens.* Strains designated C and S are present in nature. RWV may be related to or the same as rice tungro spherical virus. RWV overwinters in ratoons.

Distribution. Japan.

Symptoms. Symptoms caused by RWV somewhat resemble those caused by tungro virus. Symptoms of strain C occur after the booting stage of plant growth. Diseased plants have a slight stunting, slight discoloration and drooping of leaf blades, poor root systems, and a delay in the flowering time.

Symptoms of strain S are leaf discoloration and irregular-shaped brown blotches on diseased leaves. Plants are severely stunted and have few tillers, delayed flowering, and a reduction in yield.

Management. Grow resistant cultivars.

Wilted Stunt Disease

Cause. There is question if the causal organism is a strain of rice grassy stunt virus or a phytoplasma-like organism. The causal organism is transmitted in a persistent manner by the rice brown planthopper, *Nilaparvata lugens*.

Distribution. Taiwan.

Symptoms. Symptoms vary, depending on cultivar. Initially, rusty spotting and yellowing that later intensify occur on lower basal leaves. Diseased plants eventually wilt. The central, newly unfolded leaves are pale. Diseased plants produce few tillers in winter and slightly more tillers in summer.

Leaf twisting and trapping of unfolded leaves also occur in some cultivars. In other cultivars, diseased leaves are pale green, shortened, narrow, curled inward, and brittle. Tillering may increase during the summertime. In still other cultivars, young leaves are pale green to pale yellow and exhibit conspicuous mottling. Sometimes vague, chlorotic stripes that eventually disappear form along veins. Plants may be severely stunted and have few tillers. Eventually, leaves dry up, beginning at the basal portion.

Severely diseased plants die, and surviving plants produce no panicles or a few poorly developed panicles.

Management. Not reported.

Diseases Caused by Unknown or Abiotic Factors

Panicle Blight

Cause. Unknown.

Distribution. Not reported.

Symptoms. Aborted florets and poor yields occur on diseased plants.

Management. Not reported.

Sekiguchi Lesion

Cause. Sekiguchi lesion is conditioned by a single recessive gene, sl, and can be induced by a number of biotic and abiotic agents. Sekiguchi lesion is likely a manifestation of a flaw in the biological mechanisms that regulate the hypersensitive response of plants homozygous for the sl gene to pathogenic agents.

Distribution. Japan and the United States.

Symptoms. Initially, lesions (1–2 mm in diameter) are gray, water-soaked spots that enlarge rapidly into diurnally zonate orange-brown areas. Gradually, lesions coalesce until the whole plant is affected.

Management. Not reported.

Straighthead

Cause. The exact cause is not known, but it may be due to unfavorable soil conditions aggravated by prolonged flooding. It is more severe on sandy loam soils and seldom occurs on clay soils. It has often occurred when soil containing too much undecayed plant material had been plowed under. In some limited geographical areas, it has been caused by an accumulation of arsenic as a result of repeated applications of cotton insecticides.

Distribution. Colombia, Japan, Portugal, and the United States.

Symptoms. Rice heads remain upright at maturity because the few grains that form are too light to bend the heads over. Diseased heads often contain no fertile seed and hulls are distorted into a crescent or parrot-beak form, especially on long-grain cultivars. One or both hulls and other flower parts are missing. In severe cases, heads are smaller than normal and emerge slowly or incompletely from the boot. Diseased plants remain green and may produce shoots from the lower nodes. Diseased seeds may have a low or abnormal germination.

Management
1. Grow resistant cultivars. No cultivar is immune or highly resistant.
2. Drain fields just prior to the stem elongation stage.

Selected References

Abbas, H. K., Cartwright, R. D., Shier, W. T., Abouzied, M. M., Bird, C. B., Rice, L. G., Ross, P. F., Sciumbato, G. L., and Meredith, F. I. 1998. Natural occurrence of fumonisins in rice with Fusarium sheath rot disease. Plant Dis. 82:22-25.

Ahn, S. W. 1980. Eyespot of rice in Colombia, Panama, and Peru. Plant Dis. 64:878-880.

Anjaneyulu, A., Shukla, V. D., Rao, G. M., and Singh, S. K. 1982. Experimental host range of rice tungro virus and its vectors. Plant Dis. 66:54-56.

Atkins, J. G. 1972. Rice Diseases. USDA Farmers Bull. No. 2120.

Attere, A. F., and Fatokum, C. A. 1983. Reaction of *Oryza glaberrima* accessions to rice yellow mottle virus. Plant Dis. 67:420-421.

Bajet, N. B., Aguiero, V. M., Daquioag, R. D., Jonson, G. B., Cabunagan, R. C., Mesina, E. M., and Hibino, H. 1986. Occurrence and spread of rice tungro spherical virus in the Philippines. Plant Dis. 70:971-973.

Bakr, M. A., and Miah, S. A. 1975. Leaf scald of rice: A new disease in Bangladesh. Plant Dis. Rptr. 59:909.

Baravidan, M. R., Mew, T. W., and Aballa, T. 1982. Some studies on bacterial stripe of rice. (Abstr.) Philipp. Phytopathol. 18:9.

Batchvarova, R. B., Reddy, V. S., and Bennett, J. 1992. Cellular resistance in rice to cercosporin, a toxin of *Cercospora*. Phytopathology 82:642-646.

Bockus, W. W., Webster, R. K., and Kosuge, T. 1978. The competitive saprophytic ability of *Sclerotium oryzae* derived from sclerotia. Phytopathology 68:417-421.

Bockus, W. W., Webster, R. K., Wick, C. M., and Jackson, L. F. 1979. Rice residue disposal influences overwintering inoculum level of *Sclerotium oryzae* and stem rot severity. Phytopathology 69:862-865.

Bonman, J. M., Estrada, B. A., and Bandong, J. M. 1987. Correlation between leaf and neck blast resistance in rice. (Abstr.) Phytopathology 77:1723.

Bradbury, J. F. 1986. Guide to Plant Pathogenic Bacteria. C.A.B. International, Farnham Royal, United Kingdom. 332 pp.

Cabauatan, P. Q. 1982. An unknown disease of rice transmitted by the brown planthopper *Nilaparvata lugens* in the Philippines. (Abstr.) Philipp. Phytopathol. 18:13.

Cabauatan, P. Q., and Hibino, H. 1985. Transmission of rice tungro bacilliform and spherical viruses by *Nephotettix virenscens* Distant. Philipp. Phytopathol. 21:103-109.

Cabauatan, P. Q., Hibino, H., Labis, D. B., Omura, T., and Tsuchizaki, T. 1984. Rice grassy stunt virus 2: A new strain of rice grassy stunt in the Philippines. (Abstr.) Philipp. Phytopathol. 20:9.

Cabauatan, P. Q., Cabunagan, R. C., and Koganezawa, H. 1995. Biological variants of rice tungro viruses in the Philippines. Phytopathology 85:77-81.

Cartwright, R. D., Correl, J. C., and Crippen, D. L. 1995. Fusarium sheath rot of rice in Arkansas. (Abstr.). Phytopathology 85:1199.

Chattopadhyay, S. B., and Dasgyotam, C. 1959. *Helminthosporium rostratum* Drechs. on rice in India. Plant Dis. Rptr. 43:1241-1244.

Chen, C. C., and Chiu, R. J. 1982. Three symptomatologic types of rice virus diseases related to grassy stunt in Taiwan. Plant Dis. 66:15-18.

Choong-Hoe, K., Rush, M. C., and Mackenzie, D. R. 1985. Effect of water management on the epidemic development of rice blast. (Abstr.) Phytopathology 75:501.

Chuke, K. C. and Lapis, D. B. 1985. Pathogenicity of *Sarocladium oryzae* and factors affecting sheath rot development on rice. Philipp. Phytopath. 21:28-33.

Correa-Victoria, F. J., and Carrillo, D. 1997. Confirmation of the transmission of rice "entorchamiento" (stripe necrosis) by the fungus *Polymyxa graminis* in Colombia. (Abstr.) Phytopathology 87:S21.

Cother, E. J. 1974. Bacterial glume blotch of rice. Plant Dis. Rptr. 58:1126-1129.

Cottyn, B., Cerez, M. T., Van Outryve, M. F., Barroga, J., Swings, J., and Mew, T. W. 1996. Bacterial diseases of rice. I. Pathogenic bacteria associated with sheath rot complex and grain discoloration of rice in the Philippines. Plant Dis. 80:429-437.

Cottyn, B., Van Outryve, M. F., Cerez, M. T., De Cleene, M., Swings, J., and Mew, T. W. 1996. Bacteria diseases of rice. II. Characterization of pathogenic bacteria associated with sheath rot complex and grain discoloration of rice in the Philippines. Plant Dis. 80:438-445.

Damicone, J. P., Patel, M. V., and Moore, W. F. 1993. Density of sclerotia of *Rhizoctonia solani* and incidence of sheath blight in rice fields in Mississippi. Plant Dis. 77:257-260.

Danquah, O. A. 1975. Occurrence of *Phoma glomerata* on rice (*Oryza sativa*): A first record in Ghana. Plant Dis. Rptr. 59:844-845.

Datnoff, L. E., and Jones, D. B. 1992. Sheath blotch of rice: A disease new to the Americas. Plant Dis. 76:1182-1184.

Datnoff, L. E., Snyder, G. H., and Deren, C. W. 1991. Effect of silicon fertilizer grades on brown spot development and yield of rice. (Abstr.) Phytopathology 81:1201.

Datnoff, L. E., Elliott, M. L., and Jones, D. B. 1993. Black sheath rot caused by *Gaeumannomyces graminis* var. *graminis* on rice in Florida. Plant Dis. 77:210.

Datnoff, L. E., Elliott, M. L., and Krausz, J. P. 1997. Cross pathogenicity of *Gaeumannomyces graminis* var. *graminis* from bermudagrass, St. Augustinegrass, and rice in Florida and Texas. Plant Dis. 81:1127-1131.

Deighton, F. C., and Shaw, D. 1960. White leaf streak of rice caused by *Ramularia oryzae* sp. nov. Brit. Mycol. Soc. Trans. 43:516-518.

Disthaporn, S., Catling, H. D., Chettanachit, D., and Putta, M. 1985. Effect of time of infection of rice ragged stunt virus on deepwater rice. Int. J. Tropical Plant Dis. 3:19-25.

Duveiller, E., Miyajima, K., Snacken, F., Autrique, A., and Maraite, H. 1988. Characterization of *Pseudomonas fuscovaginae* and differentiation from other fluorescent pseudomonads occurring on rice in Burundi. J. Phytopathol. 122:97-107.

Farr, D. F., Bills, G. R., Chamuris, G. P., and Rossman, A. Y. 1989. Fungi on Plants and Plant Products in the United States. American Phytopathological Society, St. Paul, MN. 1252 pp.

Fauqet, C., and Thouvenel, J. C. 1977. Isolation of the rice yellow mottle virus in Ivory Coast. Plant Dis. Rptr. 61:443-446.

Febellar, N. G., and Mew, T. W. 1985. Crown sheath rot of rice. (Abstr.) Philipp. Phytopathol. 21:16.

Fomba, S. N. 1988. Screening for seedling resistance to rice yellow mottle virus in some rice cultivars in Sierra Leona. Plant Dis. 72:641-642.

Gangopadhyay, S., and Row, K. V. S. R. K. 1986. Perennation of *Pyricularia oryzae* Briosi et Cav. in sclerotial state. Int. J. Tropical Plant Dis. 4:187-192.

Ghosh, S. K. 1981. Weed hosts of rice necrosis mosaic virus. Plant Dis. 65:602-603.

Gnanamanickam, S. S., Shigaki, T., Medalla, E. S., Mew, T. W., and Alvarez, A. M. 1994. Problems in detection of *Xanthomonas oryzae* pv. *oryzae* in rice seed and potential for improvement using monoclonal antibodies. Plant Dis. 78:173-178.

Groth, D. E., and Nowick, E. M. 1992. Selection for resistance to rice sheath blight through number of infection cushions and lesion type. Plant Dis. 76:721-723.

Goto, M. 1979. Bacterial foot rot of rice caused by a strain of *Erwinia chrysanthemi*. Phytopathology 69:213-216.

Goto, M. 1979. Dissemination of *Erwinia chrysanthemi*, the causal organism of bacterial foot rot of rice. Plant Dis. Rptr. 63:100-103.

Gunnell, P. S., and Webster, R. K. 1983. Sheath blight of rice in California. (Abstr.) Phytopathology 73:796.

Gunnell, P. S., and Webster, R. K. 1984. Aggregate sheath spot of rice in California. Plant Dis. 68:529-531.

Gunnell, P. S., and Webster, R. K. 1985. The effect of cultural practices on aggregate sheath spot of rice in California. (Abstr.) Phytopathology 75:1340.

Gunnell, P. S., and Webster, R. K. 1985. The perfect state of *Rhizoctonia oryzae-sativae*, causal organism of aggregate sheath spot of rice. (Abstr.) Phytopathology 75:1383.

Harder, D. E., and Bakker, W. 1973. African cereal streak, a new disease of cereals in East Africa. Phytopathology 63:1407-1411.

Hibino, H. 1983. Transmission of two rice tungro-associated viruses and rice waika virus from doubly or singly infected source plants by leafhopper vectors. Plant Dis. 67:774-777.

Hibino, H., Roechan, M., and Sudarisman, S. 1978. Association of two types of virus particles with penyakit habang (tungro disease) of rice in Indonesia. Phytopathology 68:1412-1416.

Hibino, H., Cabauatan, P. Q., Omura, T., and Tsuchizaki, T. 1985. Rice grassy stunt virus strain causing tungro-like symptoms in the Philippines. Plant Dis. 69:538-541.

Hibino, H., Jonson, G. B., and Sta. Cruz, F. C. 1987. Association of mycoplasma-like organisms with rice orange leaf in the Philippines. Plant Dis. 7:792-794.

Hibino, H., Daquioag, R. D., Cabauatan, P. Q., and Dahal, G. 1988. Resistance to rice tungro spherical virus in rice. Plant Dis. 72:843-847.

Hibino, H., Ishikawa, K., Omura, T., Cabauatan, P. Q., and Koganezawa, H. 1991. Characterization of rice tungro bacilliform and rice tungro spherical viruses. Phytopathology 81:1130-1132.

Hwang, B. K., Koh, Y. J., and Chung, H. S. 1987. Effects of adult-plant resistance on blast severity and yield of rice. Plant Dis. 71:1035-1038.

Iboton Singh, N., and Tombisana, R. K. 1995. First report of Fusarium leaf blotch of rice. Plant Dis. 79:1186.

Iboton Singh, N., and Tombisana, R. K. 1995. *Pestalotiopsis versicolor*: A causal agent of rice collar rot. Plant Dis. 79:1186.

Imolehin, E. D. 1983. Rice seedborne fungi and their effect on seed germination. Plant Dis. 67:1334-1336.

Inoue, H. 1978. Strain S: A new strain of leafhopper-borne rice waika virus. Plant Dis. Rptr. 62:867-871.

Inoue, H., and Omura, T. 1982. Transmission of rice gall dwarf virus by the green rice leafhopper. Plant Dis. 66:57-59.

John, V. T., Heu, M. H., Freeman, W. H., and Manandhar, D. N. 1979. A note on dwarf disease of rice in Nepal. Plant Dis. Rptr. 63:784-785.

Jonson, G. B., Sta. Cruz, F. C., and Hibino, H. 1985. Etiology of orange leaf. (Abstr.) Philipp. Phytopathol. 21:6.

Kato, S., Nakanishi, T., Takahi, Y., and Nakagami, K. 1985. Studies on Pythium damping-off of rice seedlings: (1) *Pythium* species associated with damping-off in the early growth stage of rice seedlings in nursery flats. Phytopathol. Soc. Japan Ann. 51:159-167.

Kato, S., Nakanishi, T., Takahi, Y., Nakagami, K., and Ogawa, M. 1985. Studies on Pythium damping-off of rice seedlings: (2) *Pythium* species associated with damping-off at the middle and latter growth stages of rice seedlings in nursery flats. Phytopathol. Soc. Japan Ann. 51:168-175.

Kauffman, H. E., and Reddy, A. P. K. 1975. Seed transmission studies of *Xanthomonas oryzae* in rice. Phytopathology 65:663-666.

Koch, M. F., and Mew, T. W. 1991. Effect of plant age and leaf maturity on the quantitative resistance of rice cultivars to *Xanthomonas campestris* pv. *oryzae*. Plant Dis. 75:901-904.

Kondaiah, A., Rao, A. V., and Srinivasan, T. E. 1976. Factors favoring spread of rice "tungro" disease under field conditions. Plant Dis. Rptr. 60:803-806.

Krishna, P. G., and Rush, M. C. 1983. Role of *Pythium spinosum* in the fungal complex causing the water-mold disease of water-seeded rice. (Abstr.) Phytopathology 73:502.

Kuwata, H. 1985. *Pseudomonas syringae* pv. *oryzae* pv. nov., causal agent of bacterial halo blight of rice. Phytopathol. Soc. Japan Ann. 51:212-218.

Lee, F. N. 1990. Rice dead tiller syndrome in Arkansas. (Abstr.). Phytopathology 80:1005.

Lee, F. N., and Rush, M. C. 1983. Rice sheath blight: A major rice disease. Plant Disease 67:829-832.

Leu, L. S., and Yang, H. C. 1985. Distribution and survival of sclerotia of rice sheath blight fungus, *Thanatephorus cucumeris,* in Taiwan. Phytopathol. Soc. Japan Ann. 51:1-7.

Lindberg, G. D. 1970. Loss of rice yield caused by stem rot. (Abstr.) Phytopathology 60:1300.

Ling, K. C. 1975. Rice Virus Diseases. International Rice Research Institute, Los Banos, Philippines.

Ling, K. C., Tiongco, E. R., and Aguiero, V. M. 1978. Rice ragged stunt: A new virus disease. Plant Dis. Rptr. 62:701-705.

Lozano, J. C. 1977. Identification of bacterial leaf blight in rice caused by *Xanthomonas oryzae* in America. Plant Dis. Rptr. 61:644-648.

Marchetti, M. A. 1983. Potential impact of sheath blight on yield and milling quality of short-statured rice lines in the southern United States. Plant Dis. 67:162-165.

Marchetti, M. A., Bollich, C. N., and Vecker, F. A. 1983. Spontaneous occurrence of the Sekiguchi lesion in two American rice lines: Its induction, inheritance and utilization. Phytopathology 73:603-606.

Marin-Sanchez, J. P., and Jimenez-Diaz, R. M. 1982. Two new *Fusarium* species infecting rice in southern Spain. Plant Dis. 66:332-334.

Mew, T. W., and Rosario, M. B. 1980. Bacterial foot rot of rice. (Abstr.) Philipp. Phytopathol. 16:5.

Mew, T. W., Vera Cruz, C. M., and Medalla, E. S. 1992. Changes in race frequency of *Xanthomonas oryzae* pv. *oryzae* in response to rice cultivars planted in the Philippines. Plant Dis. 76:1029-1032.

Mew, T. W., Alvarez, A. M., Leach, J. E., and Swings, J. 1993. Focus on bacterial blight of rice. Plant Dis. 77:5-12.

Morinaka, T., Putta, M., Chettanachit, D., Parejarearn, A., Disthaporn, S., Omura, T., and Inoue, H. 1982. Transmission of rice gall dwarf virus by cicadellid leafhoppers *Recilia dorsalis* and *Nephotettix nigropictus* in Thailand. Plant Dis. 66:703-704.

Mueller, K. E. 1974. Field Problems of Tropical Rice. International Rice Research Institute, Los Banos, Philippines.

Nuque, F. L., Aguiero, V. M., and Ou, S. H. 1982. Inheritance of resistance to grassy stunt virus in rice. Plant Dis. 66:63-64.

Omura, T., Inoue, H., Morinaka, T., Saito, Y., Chettanachit, D., Putta, M., Parejarearn, A., and Disthaporn, S. 1980. Rice gall dwarf: A new virus disease. Plant Dis. 64:795-797.

Oster, J. J. 1992. Reaction of a resistant breeding line and susceptible California rice cultivars to *Sclerotium oryzae*. Plant Dis. 76:740-744.

Ou, S. H. 1973. A Handbook of Rice Diseases in the Tropics. International Rice Research Institute, Los Banos, Philippines.

Ou, S. H. 1986. Rice Diseases, Second Edition. Commonwealth Mycological Institute,

Kew, Surrey, England. Eastern Press Ltd., London. 391 pp.

Ou, S. H., Nuque, F. L., and Vergel de Dios, T. I. 1978. Perfect stage of *Rhynchosporium oryzae* and the symptoms of rice leaf scald disease. Plant Dis. Rptr. 62:524-528.

Padwick, W. 1950. Manual of Rice Diseases. Commonwealth Mycological Institute, Kew, Surrey, England. Oxford University Press, England.

Prabhu, A. A., and Bedendo, I. P. 1983. Basal node of rice caused by *Fusarium oxysporum* in Brazil. Plant Dis. 67:228-229.

Rahman, M. L., and Evans, A. A. F. 1987. Studies on host-parasite relationships of rice stem nematode, *Ditylenchus angustus* (Nematoda: Tylenchida), on rice, *Oryzae sativa* L. Nematologica 33:451-459.

Reddy, P. R., and Mohanty, S. K. 1981. Epidemiology of the kresek phase of bacterial blight of rice. Plant Dis. 65:578-580.

Reddy, A. P. K., Mackenzie, D. R., Rouse, D. I., and Rao, A. V. 1979. Relationship of bacterial leaf blight severity to grain yield of rice. Phytopathology 79:967-969.

Rivera, C. T., Ou, S. H., and Pathak, M. D. 1963. Transmission studies on the orange-leaf disease of rice. Plant Dis. Rptr. 47:1045-1048.

Rosales, A. M., Nuque, F. L., and Mew, T. W. 1986. Biological control of bakanae disease of rice with antagonistic bacteria. Philipp. Phytopathol. 22:29-35.

Rossman, A. Y. 1990. *Pyricularia grisea:* The correct name for the rice blast disease fungus. Mycologia 82:509-512.

Sahu, R. K., and Khush, G. S. 1989. Inheritance of resistance to bacterial blight in seven cultivars of rice. Plant Dis. 73:688-691.

Saito, Y., Chaimongkol, U., Singh, K. G., and Hino, T. 1976. Mycoplasma-like bodies associated with rice orange leaf disease. Plant Dis. Rptr. 60:649-651.

Saxena, R. C., and Khan, Z. R. 1985. Effect of neem oil on survival of the rice brown planthopper, *Nilaparvata lugens* (Stal) (Homoptera: Delphacidae), and on grassy stunt and ragged stunt virus transmission. Philipp. Phytopathol. 21:80-87.

Shahjahan, A. K. M., and Mew, T. W. 1986. Sheath blotch of rice in the Philippines. (Abstr.) Philipp. Phytopath. 22:5.

Shahjahan, A. K. M., Harahap, Z., and Rush, M. C. 1977. Sheath rot of rice caused by *Acrocylindrium oryzae* in Louisiana. Plant Dis. Rptr. 61:307-310.

Shahjahan, A. K. M., Ahmed, H. V., Sharma, N. R., and Miah, S. A. 1985. Spacings × rate of nitrogen effect on sheath blight disease severity in BRI rice. (Abstr.) Philipp. Phytopathol. 21:3.

Shakya, D. D., Vinther, F., and Mathur, S. B. 1985. Worldwide distribution of a bacterial stripe pathogen of rice identified as *Pseudomonas avenae*. Phytopathol. Zeitschr. 114:256-259.

Shukla, V. D., and Anjaneyulu, A. 1981. Adjustment of planting date to reduce rice tungro disease. Plant Dis. 65:409-411.

Shukla, V. D., and Anjaneyulu, A. 1981. Plant spacing to reduce rice tungro incidence. Plant Dis. 65:584-586.

Singh, R. A., and Raju, C. A. 1981. Some observations on sheath rot of rice. (Abstr.) Philipp. Phytopathol. 17:5.

Sthapit, B. R., Pradhanang, P. M., and Witcombe, J. R. 1995. Inheritance and selection of field resistance to sheath brown rot disease in rice. Plant Dis. 79:1140-1144.

Thomas, M. D., Mayango, D., and Oberly, W. 1985. Suppressions and elimination of *Rhynchosporium oryzae* by benomyl in rice foliage and seed in Liberia. Plant Dis. 69:884-886.

Tschen, J. S. M. 1987. Control of *Rhizoctonia solani* by *Bacillus subtilis*. Trans. Mycol. Soc. Japan 28:483-493.

Venkataraman, S., Ghosh, A., and Mahajan, R. K. 1987. Synergistic effect of rice tungro virus and *Sarocladium oryzae* on sheath rot disease of rice. Int. J. Tropical Plant Dis. 5:141-145.

Walawala, J. J., and David, R. G. 1984. Pathogenicity, damage assessment, and field population pattern of *Hirschmanniella oryzae* in rice. Philipp. Phytopathol. 20:39-44.

Webster, R. K., and Gunnell, P.S. 1992. Compendium of Rice Diseases. American Phytopathological Society, St. Paul, MN. 62 pp.

Whitney, N. G. 1980. Brown blotch: A new disease of rice. (Abstr.) Phytopathology 70:572.

Zaragoza, B. A., and Mew, T. W. 1979. Relationship of root injury to the "Kresek" phase of bacterial blight of rice. Plant Dis. Rptr. 63:1007-1011.

Zeigler, R. S., and Alvarez, E. 1987. Bacterial sheath brown rot of rice caused by *Pseudomonas fuscovaginae* in Latin America. Plant Dis. 71:592-597.

Zeigler, R. S. and Alvarez, E. 1989. Grain discoloration of rice caused by *Pseudomonas glumae* in Latin America. Plant Dis. 73:368.

Zeigler, R. S., Aricapa, G, and Hoyos, E. 1987. Distribution of fluorescent *Pseudomonas* spp. causing grain and sheath discoloration of rice in Latin America. Plant Dis. 71:896-900.

15. Diseases of Rye *(Secale cereale)*

Diseases Caused by Bacteria

Bacterial Streak

Bacterial streak is called bacterial blight, black chaff, and bacterial leaf blight.

Cause. *Xanthomonas campestris* pv. *translucens* (syn. *X. translucens* f. sp. *secalis*). The taxonomy of the causal organism(s) continues to be in flux. Bacterial streak (which is called bacterial blight, bacterial leaf streak, or black chaff on other small grains) is considered to be caused by five of the former *X. campestris* pathovars: pv. *cerealis*, pv. *hordei*, pv. *secalis*, pv. *translucens*, and pv. *undulosa*. These pathovars are often grouped together under the name "*translucens* group" or are considered to be a single species (*X. translucens)* or pathovar (*X. campestris* pv. *translucens*). Bragard et al. (1997) agreed that *X. translucens* pv. *translucens* and *X. translucens* pv. *hordei* are true synonyms. They also reported that when using restriction fragment-length polymorphism and fatty acid methyl esters analysis, the pathovars *X. translucens* pv. *cerealis, X. translucens* pv. *translucens,* and *X. translucens* pv. *undulosa* clustered in different groups and corresponded to true biological entities.

Bacteria are seedborne and overwinter in infested residue. Initial infection comes from seedborne bacteria that are disseminated by leaf-to-leaf and plant-to-plant contact, splashing rain, and possibly insects. Bacteria enter into plants through natural openings, such as stomates and wounds.

Distribution. Australia and North America.

Symptoms. Symptoms occur after several days of damp or rainy weather. Diseased plants may be stunted, but this is not usually noticed until plants mature. Small water-soaked spots occur on the youngest leaves and sheaths of older plants and sometimes on seedlings. Spots enlarge and may coalesce, become glossy and translucent, and eventually turn yellow or brown and have an irregular margin. Spots, when coalesced, do not become as stripe-like as bacterial blight on barley. Severely diseased leaves die back from the tip. Under humid morning conditions, droplets of milky bacterial exudate may be seen on surfaces of spots. Droplets dry into hard, yellowish flakes that are easily removed from leaf surfaces.

Water-soaked areas occur on heads but bacterial exudate is not common on heads. Grain is not destroyed but may be brown and shrunken and carry bacteria to infect next year's crop. If flag leaves are infected, heads may not emerge from the boot but may break through the side of the sheath and be destroyed and blighted. Bacterial blight on rye can be diagnosed in the field in the same manner as bacterial blight on barley. When a newly developed spot is cut crosswise, a bead of milky ooze will exude from the cut edge.

Management
1. Do not sow seed from diseased plants.
2. Treat seed with a fungicide as a precaution even though bacterial blight is not known to be present.
3. Rotate rye with other crops. The causal bacterium that infects rye does not infect barley.

Halo Blight

Cause. *Pseudomonas syringae* pv. *coronafaciens* (syn. *P. coronafaciens*) may be seedborne. Development of halo blight is most severe under cool, moist conditions.

Distribution. Wherever rye is grown; however, halo blight is rarely severe enough to be economically important.

Symptoms. Initially, minute, brown spots occur on diseased leaf sheaths, leaf blades, and along leaf margins, often in association with frost injury. A few days later, yellow halos up to 15 mm in diameter surround the spots. Eventually, the centers of spots become necrotic and have a narrow chlorotic halo. Lesions may coalesce and cause entire leaves to turn brown and have a scalded appearance that is a common symptom from flowering to maturity.

Normally, entire florets are blighted, but during less severe disease conditions, only small halos may occur on glumes. On early tillers, individual florets on heads are sterile and turn white. On later tillers, all florets of some heads are diseased and produce shriveled grain or no seed. As temperatures increase later in the season, few or no new lesions are formed.

Atypical symptoms reported from Virginia are as follows: Plants were slightly to severely stunted, often with chlorotic veins in the diseased leaves. Diseased blades on some plants had chlorotic and necrotic lesions but no halo blight was observed. On other diseased plants, entire blades were yellowed.

Management. Not reported.

Diseases Caused by Fungi

Anthracnose

Cause. *Colletotrichum graminicola,* teleomorph *Glomerella graminicola,* is an excellent saprophyte and overwinters as mycelia and conidia on winter cereals and infested residue. Conidia are produced during wet weather and disseminated by wind to host plants. Disease development is most severe when weather is wet and at a temperature of 25°C. Anthracnose is most likely to occur when rye is grown on coarse soils with low fertility in rotation with another cereal. Rye is one of the most susceptible cereals to infection by *C. graminicola.*

Distribution. Wherever rye is grown.

Symptoms. Symptoms become apparent toward plant maturity as a premature ripening or whitening of seed, a general reduction in plant vigor, and the eventual death of diseased plants. Specific symptoms occur most often on the lower parts of diseased plants because primary infections are initiated there. The crown and bases of some stems become bleached, then turn brown.

Later, acervuli, which appear as small, black, raised spots, develop on the surfaces of the diseased lower leaf sheaths and culms. Under magnification, acervuli will look like a "clump of spines." When moisture is plentiful, acervuli may develop on the leaves of dead plants. Round to oblong lesions bearing acervuli may occur on green leaves. If infection occurs early in the life of the diseased plant, diseased kernels may be shriveled. A seedling and crown infection may occur under severe disease conditions.

Management
1. Rotate rye with a noncereal or grass because *C. graminicola* can infect a large number of grass and cereal species. Legumes are not susceptible and would be a suitable crop in a rotation.
2. Improve soil fertility.

Bunt

Bunt is also called common bunt, covered smut, European bunt, hill bunt, and stinking smut.

Cause. *Tilletia caries* (syn. *T. tritici*) and *T. laevis* (syn. *T. foetida* and *T. foetens*). The species name "*laevis*" is sometimes spelled "*levis*" in the literature. The two fungi are closely related and have similar life cycles that may occur together in the same diseased plant.

Both fungi overwinter as teliospores (chlamydospores) on seed and in soil. When rye is sown, teliospores on or near seed in the soil germinate in the presence of moisture and cool temperatures of 5° to 15°C. A promycelium (basidium) is formed on which eight to 16 basidiospores (sporidia) are produced. Basidiospores fuse in the middle with a compatible basidiospore and form an H-shaped structure that germinates and produces secondary sporidia. Secondary sporidia then germinate and produce mycelia that infect seedlings. Mycelium grows behind the growing point or meristematic tissue of the infected plant, invades the developing head, and displaces the grain. Eventually, teliospores are formed in seed. At harvest time, infected seed is broken and teliospores are released, become windborne, and eventually contaminate host kernels or soil.

Distribution. *Tilletia caries* is generally distributed wherever rye is grown but is most prevalent in the northwestern United States. *Tilletia laevis* is limited to areas of Europe and North America, where it has been reported in the Midwest and the northwestern United States.

Symptoms. Common bunt is more severe when soil temperatures become cool following sowing. Rye sown in the spring may be less severely diseased. Diseased plants are somewhat stunted but cannot be readily distinguished from healthy ones until the heading stage of plant growth. Bunted heads are more slender than healthy ones and glumes of spikelets may be spread apart. A bunted head will often stay green for a longer time than normal and have a bluish cast. Bunted kernels are about the same size as healthy seed but are light brown and more round in shape. The pericarp ruptures at harvest and teliospores contaminate healthy seed, giving them a "fish-like" odor; hence, the name stinking smut. When crushed, the smut balls and diseased kernels have an oily feeling.

Management
1. Treat seed with a systemic seed-treatment fungicide.
2. Grow resistant cultivars.
3. Sow early in the autumn. Seedlings may be far enough advanced and less susceptible to infection by the time secondary sporidia develop.

Cephalosporium Stripe

Cause. *Hymenula cerealis* (syn. *Cephalosporium gramineum*) survives for up to 5 years as conidia and mycelia in residue and in the top 8 cm of soil. Conidia serve as primary inoculum and infect roots during winter and early spring through mechanical injuries caused by soil heaving and insects. Infection is most severe when increased fungal sporulation and root growth occur in wet and acid soils that have a pH of 5.0. The subcrown internode can also be infected just as seed is germinating.

After infection, conidia enter xylem vessels and are carried upward in the plant, where they lodge and multiply at the nodes and leaves. Normally, there is no further injury to roots. The fungus prevents water movement up the plant and also produces metabolites that are harmful to the plant. At harvest time, *C. gramineum* is returned to the soil in infested residue, where it is a successful saprophytic competitor with other soil organisms.

Winter rye is most severely diseased. Spring rye is susceptible but apparently escapes disease and usually does not show symptoms.

Distribution. Great Britain, Japan, and in most winter rye–growing areas of North America.

Symptoms. Diseased stunted plants are scattered throughout a field but are most numerous in the lower and wetter areas. During jointing and heading stages of plant growth, one to two, but sometimes up to four, distinct yellow stripes occur the length of the diseased plant on leaves, sheaths, and stems. Thin, brown lines consisting of diseased veins occur in the middle of the stripes. Stripes eventually become brown, are highly visible on green leaves, and remain noticeable on yellow straw. Toward harvest, culms at or below nodes may become dark due to fungus sporulation. Heads of diseased plants are white and contain shriveled seed or no seed.

Management
1. Grow tolerant cultivars. No cultivars are resistant.
2. Rotate rye with a noncereal for at least 2 years.
3. Sow later in the autumn or when the soil temperature 10 cm below the surface is less than 13°C. Such plants apparently have limited root growth, which reduces the number of infection sites.
4. Residue should be plowed deeper than 8 cm.

Common Root Rot

Cause. *Bipolaris sorokiniana,* teleomorph *Cochliobolus sativus,* and *Fusarium* spp. Both fungi survive in soil as saprophytes on infested residue. Conidia of *B. sorokiniana* and chlamydospores of *Fusarium* spp. survive for several months in soil and are seedborne.

Primary infections occur on coleoptiles, primary roots, and subcrown internodes. Infection of subcrown internodes is usually caused by *B. sorokiniana. Fusarium* spp. infect secondary roots as they emerge from the crown. Diseased plants normally do not die because new roots are continually being produced. Conidia are produced when disease progresses above the soil line on the diseased plants.

Most secondary sporulation is conidia produced on diseased crowns, with the remainder occurring on crown roots, subcrown internodes, seed pieces, and seminal roots. Sporulation is greatest on mature rye plants; the highest numbers of conidia are present on necrotic or senescent tissue, such as coleoptiles and the lower outside leaf sheaths.

Plants under stress due to drought, warm temperatures, lack of nutrition, and insect injury are most subject to infection. Moisture is initially required for infection, but once infection is initiated, disease development requires warm temperatures and moisture stress on the infected plant.

Distribution. Wherever rye is grown.

Symptoms. Diseased seedlings may be killed before or after emergence, particularly in dry soil when inoculum is seedborne. Surviving seedlings have brown lesions on coleoptiles, roots, and culms. Infections of crowns usually kill plants. Diseased older plants, which occur in random patches throughout a field, are stunted and lighter green and mature earlier than healthy plants. Diseased plants also have few tillers, and their heads are bronzed and bleached or white-headed and contain shriveled seed. There may be a browning of root systems that is observable only when roots are washed. *Fusarium* spp. in particular cause roots, culm bases, and lower nodes to become dry and dark brown or black. Diseased plants then become brittle and break off easily near the soil line.

Management
1. Treat seed with a seed-protectant fungicide.
2. Soil should have proper fertility to ensure growth of vigorous plants that are able to produce new roots and overcome root rot.
3. Sow rye in late autumn.

Cottony Snow Mold

Cottony snow mold is also called winter crown rot.

Cause. *Coprinus psychromorbidus.*

Distribution. Canada.

Symptoms. After snowmelt, patches of dead rye plants occur where the snow was deepest in the field. Sometimes a sparse mycelial growth gives diseased plants a gray sheen, but often, little mycelium can be found. Irregular-shaped, gray to black sclerotia up to 1 mm in length are loosely attached to the diseased host tissues inside leaf sheaths and, infrequently, on roots and subcrown internodes.

Management. Not reported.

Dilophospora Leaf Spot

Dilophospora leaf spot is also called twist.

Cause. *Dilophospora alopecuri* (syn. *D. graminis*) survives as mycelia and pycnidia in infested residue and as conidia on seed. Primary infection is by conidia produced in pycnidia and disseminated by wind and splashing rain to seedlings during wet weather. Secondary infection is by conidia produced in pycnidia within the spots on seedling leaves during moist weather. The twist phase of the disease is caused by secondary conidia disseminated into the whorls by larvae of the seed gall nematode, *Anguina tritici,* as it moves up the plant in a film of water.

Distribution. Canada, Europe, India, and the United States. Twist has not been observed for several years and is considered to be a rare disease on rye.

Symptoms. Initially, small, elongated, yellow spots appear only on diseased leaves. Spots become light brown with black centers that consist of stromata and pycnidia of *D. alopecuri*. Leaves may be killed if the spots become numerous. When the whorl is colonized, leaves do not emerge or emerge twisted, distorted, and covered with gray mycelium. This phase of the disease rarely occurs in the absence of *A. tritici*. *Dilophospora alopecuri* dries as stromata that occur as dark streaks on leaves and in which pycnidia are produced.

Management
1. Rotate rye with other crops. Both rye and wheat are susceptible.
2. Plow under infected residue.
3. Treat seed with a seed-protectant fungicide.
4. Sow certified rye seed.

Downy Mildew

Cause. *Sclerophthora macrospora* survives for years as oospores in infested residue and soil. *Sclerophthora macrospora* is an obligate parasite and cannot grow saprophytically on dead plant tissue. Oospores are disseminated on seed and in infested residue carried by wind and water. Oospores germinate in wet soil and produce sporangia in which zoospores are formed. Zoospores are then released, "swim" through soil water to seedlings, and infect host plants by forming a germ tube that directly penetrates the plant. Oospores may survive for months in dry soil, then germinate and form only a single germ tube that directly penetrates a host plant. Infection occurs at temperatures from 7° to 31°C. Following infection, *S. macrospora* develops systemically within infected plants, particularly in xylem tissue.

Distribution. Wherever rye is grown. The disease is most severe in the wetter areas of a field.

Symptoms. Downy mildew occurs only in localized areas of fields where seedlings have been growing in flooded or waterlogged soil for 24 hours or longer. Diseased plants are dwarfed, deformed, and twisted and have leathery, stiff, thickened leaves, stems, and heads. Diseased plants tiller excessively, and severely diseased plants form no seeds. In less severely diseased plants, dwarfing may be slight, one or more of the upper leaves may be stiff, upright, or variously curled and twisted, and heads and stems may not be deformed. Under magnification, numerous round, yellow-brown oospores may be seen in diseased tissue.

Management
1. Provide proper soil drainage, where possible.
2. Control grassy weeds that may serve as collateral hosts.
3. Sow cleaned seed from disease-free plants to ensure no infested residue is disseminated with the seed.

Dwarf Bunt

Cause. *Tilletia controversa* survives as teliospores in soil for up to 10 years or on the surfaces of infested seed. When rye is sown in moist soil, teliospores on or close to the seed germinate under warm, dry conditions and form promycelia on which eight to 16 basidiospores are borne. Basidiospores fuse in the middle with a compatible basidiospore and form an H-shaped structure that, in turn, germinates and forms a structure called a secondary sporidium. The secondary sporidia germinate and produce infective mycelia.

Optimum germination of *T. controversa* occurs at temperatures of 3° to 8°C; the incubation period is 3 to 10 weeks. Infection generally requires a heavy snow cover over unfrozen ground, which provides suitable conditions of moisture and temperature at the soil surface for teliospore germination and infection of plants.

After infection, mycelium grows behind the meristematic tissue of growing points, invades developing heads, and displaces grain. Eventually, teliospores form in the seed. At harvest time, diseased seed is broken and teliospores disseminated by wind contaminate healthy kernels and soil.

Distribution. In Canada, Europe, and the United States where a heavy snow cover is likely to occur over unfrozen ground.

Symptoms. Diseased plants are generally one-fourth to one-half normal size. Viable pollen does not develop, and the ovaries of diseased florets are larger than those in healthy florets. There are more bunt-infested kernels per spikelet of a diseased plant than seeds per spikelet on a healthy plant. Smut balls are smaller and rounder than common bunt and the spore mass feels dry. Spores smell like rotten fish.

Management
1. Grow resistant cultivars in adapted areas.
2. Treat seed with a fungicide seed treatment.

Ergot

Cause. *Claviceps purpurea,* anamorph *Sphacelia segetum,* survives as sclerotia on and in soil. Immediately before blossoming of host plants, a sclerotium germinates and produces one or many stipes (stalks), each bearing a stromatic head in which perithecia form. Eventually, ascospores produced in the perithecia are disseminated by wind to flowers. The ascospore germinates and produces a germ tube that grows into the young ovary. Conidia are then produced from mycelium growing in the ovary within a sweet, sticky liquid called honeydew. Secondary dissemination is by insects that are attracted to the honeydew and inadvertently carry conidia to healthy flowers. Sclerotia usually occur after the conidial stage, but under some conditions, particularly in tropical areas, no sclerotia are produced. As sclerotia enlarge, floral bracts spread apart and dark sclerotial bodies protrude beyond the floral bracts. Sclerotia eventually replace kernels in diseased flowers. Several grasses and cereals are susceptible to *C. purpurea,* but ergot is usually most severe on rye. Rye is predisposed to ergot by barley yellow dwarf virus infection.

Distribution. Wherever rye is grown.

Symptoms. Few to many sclerotia (ergot bodies) are produced on a spike or panicle and appear as hard, purple-black, horn-shaped structures in place of seed. Prior to sclerotial formation, honeydew and conidia accumulate in liquid droplets or adhere to the surfaces of floral structures. Insects feeding on honeydew are conspicuous around diseased flowers. Saprophytic fungi grow on honeydew and give infected flowers a black or sooty appearance.

Management
1. Plow infested residue deep in the soil to place sclerotia where they will not effectively germinate but will eventually decompose.
2. Control grassy weeds since they may be infected by *C. purpurea* and serve as a source of inoculum.
3. Rotate rye with crops, such as legumes or maize, that are not susceptible to *C. purpurea.*
4. Clean seed to remove sclerotia.

Eyespot

Eyespot is also called Cercosporella foot rot, culm rot, foot rot, stem break, strawbreaker, and strawbreaker foot rot.

Cause. *Tapesia yallundae,* anamorph *Pseudocercosporella herpotrichoides* (syn. *Cercosporella herpotrichoides*). At least two distinct morphological types that correspond to the two pathotypes of *P. herpotrichoides* are present. Daniels et al. (1991) reported the two pathotypes designated wheat (W) and rye (R) are distinguished on the bases of pathogenicity to wheat seedlings, spore morphology and colony morphology, pigmentation on maize meal agar, and isoenzyme polymorphisms. The wheat type is more pathogenic to wheat and barley than to rye and has faster-growing, even-edged colonies. The rye type is as pathogenic to rye as to wheat and barley and is slower growing and has feathery-edged colonies and more profuse sporulation.

Tapesia yallundae survives for several years as mycelia in infested residue. Sporulation occurs on straw at the soil surface during cool, damp weather or when humidity is near saturation and temperatures are 8° to 12°C in the autumn or spring. Dispersal or dissemination is by splashing rain.

Development of epiphytotics is dependent on production of primary inoculum on residue remaining from the previous rye crop. The coleoptile is most susceptible to infection during the seedling stage, but with the decay of the coleoptile, the susceptibility of leaf sheaths increases.

Secondary inoculum is not important in the development of an epiphytotic since spores occur on developing lesions too late in the growing season to cause infection. However, late infections do add to the amount of inoculum available to infect plants of succeeding crops.

Distribution. Widespread where rye is grown in a cool, moist climate. Rye is less prone to disease than wheat.

Symptoms. The most conspicuous symptom is near the end of the growing season when diseased plants lodge. Lodging caused by eyespot disease causes straw to fall in all directions; lodging caused by wind or rain causes straw to fall primarily in one direction.

Eye-shaped or ovate lesions with white to tan centers and brown margins develop first on the basal leaf sheath. Similar spots that form on stems directly beneath those on the sheath weaken the stems and eventually cause them to lodge.

Crown and basal culm tissue is diseased, but roots are not, although a necrosis occurs around roots in the upper crown nodes. Under moist conditions, lesions enlarge and a black, stroma-like mycelium that develops over the surface of crowns and the bases of culms gives tissues a charred appearance. Stems shrivel and collapse or plants become yellowish to pale green and heads are reduced in size and number. When infected early in their growth, individual culms and weaker plants are killed before maturity.

Management
1. Rotate rye with noncereals.
2. Grow the most resistant cultivars. However, resistant cultivars often become diseased under severe disease conditions.
3. Apply fungicides to residue to decrease primary inoculum. Under experimental conditions, foliar fungicides control eyespot.
4. Reduced tillage limits foot rot incidence.

Fusarium Head Blight

Fusarium head blight is also called scab.

Cause. *Gibberella zeae,* anamorph *Fusarium graminearum,* overwinters as mycelia and spores on seed, and as mycelia and perithecia in infested residue. Ascospores produced in perithecia and conidia produced from mycelia during warm, moist weather are disseminated by wind to plant hosts. Diseased seed causes seedling blight, and soilborne inoculum causes seedling blight and root rot. Fusarium head blight develops in warm, humid weather during formation and ripening of the kernels. Prolonged wet weather during and after anthesis favors infection and subsequent disease development. Conidia produced on infected heads serve as secondary inoculum.

Grain that becomes wet in swaths favors Fusarium head blight development. Fusarium head blight is more severe when rye follows maize.

Distribution. Wherever rye is grown.

Symptoms. Disease begins in flowers and spreads to other parts of the head to give the appearance of premature ripening. Initially, one or more spikelets of emerged immature heads are water-soaked, die, and become light brown, bleached, or white, beginning at the spikelet base. Eventually, hulls change from light brown to dark brown. When infection occurs late in the development of seed, only the base of hulls is brown, but in severe cases, the entire kernel may become shrunken and brown. If the rachis is diseased, the entire head is dwarfed and compressed. All spikelets above that point are closed rather than spreading, bleached, and sterile or contain only a partially filled seed. Diseased kernels are grayish and light in weight. The interior of the kernel is floury and discolored. During humid or wet weather, pink to pink-red mycelium and spores grow on infected spikelets. Later, small, black perithecia grow in the same diseased area. As in Fusarium head blight of wheat, mycotoxins, such as 3-acetyldeoxynivalenol and deoxynivalenol (vomitoxin), are assumed to be produced in diseased heads. Therefore, diseased grain may be harmful when fed to hogs, dogs, and humans.

Management
1. Rotate rye with other crops. Do not grow rye after barley, maize, or wheat. Do not grow rye next to maize.
2. Plow under infested residue to hasten decomposition of residue and prevent spores from being windborne.
3. Do not spread manure that contains infested straw or maize stalks on soil where rye is growing.
4. Sow early to allow rye to escape warm, moist weather.
5. Treat seed with a seed-protectant fungicide.
6. The Food and Drug Administration recommends 2 ppm or less of de-

oxynivalenol in grain entering the milling process, 1 ppm for finished rye products for human consumption, and 4 ppm for animal feed.

Fusarium Root Rot

Cause. *Fusarium culmorum.* Primary inoculum is soilborne chlamydospores and infested residue in the upper 10 cm of soil. Entry into the crown is gained 2 to 3 cm below the soil surface through openings around crown roots or by the infection of newly emerging crown roots. Disease is severe under dry soil conditions.

Distribution. The United States.

Symptoms. Diseased plants rarely show outward symptoms until after heading. The crown and basal stem tissues are decayed and have a brown discoloration and spongy texture. Internodes become chocolate brown; however, leaf sheaths remain symptomless. Pink or burgundy mycelium is seen in hollow stems. Diseased plants die prematurely, resulting in white heads.

Management
1. Do not rotate rye with wheat and oat.
2. Till fields after harvest to improve water infiltration.
3. Establish a dust and stubble mulch in spring.
4. Apply the proper amount of nitrogen.

Halo Spot

Cause. *Pseudoseptoria donacis* (syn. *Selenophoma donacis*) overwinters as pycnidiospores, pycnidia, and mycelia in infested residue, seed, and overwintering rye and wheat. During cool, moist weather, pycnidiospores are exuded from pycnidia and disseminated by wind and splashing rain to hosts.

Distribution. In cool, moist climates of Great Britain, northern Europe, and the United States. Halo spot rarely causes much damage to rye.

Symptoms. Numerous elliptical or diamond-shaped spots less than 4 mm long occur on leaves and sometimes the culms of winter rye in the spring after the snow has melted. Spots have purple-brown margins that eventually fade as the spots age. In time, the centers of spots become gray from the presence of small, black pycnidia. Sometimes spots become so numerous that much of the leaf surface is destroyed.

Management. Not necessary.

Karnal Bunt

Karnal bunt is also called partial bunt.

Cause. *Neovossia indica* (syn. *Tilletia indica*) survives as teliospores in soil for more than 2 years and on seed. Most primary inoculum is derived from seedborne inoculum, but soilborne teliospores also serve as a source of in-oculum. Teliospores germinate in the spring within 2 mm of the soil sur-face and produce a promycelium that grows to the soil surface. Sporidia are produced on promycelia and disseminated by wind to host plants, where florets are infected. Sporidia are disseminated a long distance by air cur-rents and survive up to 12 hours at relative humidities of 95% and above. Maximum germination of teliospores in vitro occurs at temperatures of 15° to 20°C in continuous light. Infection is favored by cool, wet weather.

During the early stages of infection, intercellular hyphae are present among parenchyma and chlorenchyma cells in the distal to midportions, but not the basal portions, of the glume, lemma, and palea. Hyphae are ab-sent from the ovary, subovarian tissue, rachilla, and rachis. Hyphae later grow intercellularly toward the floret base to the subovarian tissue and en-ter the pericarp of the ovary through the funiculus. Hyphae are found in the rachis only during the later stages of infection. The epidermis of the ovary is not penetrated, even after prolonged contact with germinating secondary sporidia. Kernels are wholly or partially converted into chla-mydospores that are disseminated by wind at harvest time to healthy seeds and soil.

Distribution. Afghanistan, India, Iraq, Lebanon, Mexico, Nepal, Pakistan, Syria, Sweden, Turkey, and the United States. Karnal bunt is uncommon on rye.

Symptoms. Normally, only a few random kernels per head are diseased and become wholly or incompletely converted to smut sori. Symptoms first ap-pear at the soft-dough stage in the form of blackened areas surrounding the base of the diseased grain and extend upward on the kernel for various lengths. In the most severe symptoms, glumes may be spread apart and ex-pose the bunted grains; however, this is not a common symptom. Most kernels are broken or partially eroded at their embryo end. On severely dis-eased plants, grain is reduced to a fragile black membranous sack of teliospores with a dead, shriveled embryo. A foul odor of trimethylamine accompanies diseased kernels. Diseased seed has a lower survival rate in storage than healthy seed.

Management
1. Grow resistant cultivars.
2. Treat seed with a fungicide seed treatment.
3. Hot water (60°–80°C) with the addition of NaOCl reduces seedborne teliospores and contamination in storage facilities.

Leaf Rust

Leaf rust is also called brown rust.

Cause. *Puccinia recondita* (syn. *P. rubigo-vera* var. *secalis*), anamorph *Aecidium clematidis,* is a heteroecious long-cycle rust that has species of alkanet (*An-*

chusa spp.) as alternate hosts to rye and other *Secale* spp. However, the pycnial and aecial stages are rarely found in the United States.

Puccinia recondita commonly overwinters as urediniospores and mycelium in leaf tissue. Because of this mode of overwintering, losses are greatest in southern rye-growing areas, where urediniospores overwinter in greater numbers. During warm, wet weather in the spring, urediniospores produced within uredinia are disseminated by wind to host plants. If warm, moist weather conditions continue, new uredinia will be produced in 7 to 10 days. As plants mature, telia are formed in which teliospores are produced. The teliospores either germinate upon maturity in the autumn and produce basidiospores or overwinter and germinate in the spring. In the United States, *Anchusa* spp. have rarely been observed to be infected in nature. Therefore, the teliospore stage is not considered to be important in the life cycle of the fungus, and mycelia and urediniospores perform the overwintering function.

In Europe, teliospores also either germinate in autumn and produce basidiospores or overwinter and germinate the following spring. Aecia are observed on *Anchusa* spp. from spring until autumn. Therefore, pycnia and pycniospores are probably produced in autumn or, more likely, in the spring. The resulting aeciospores are disseminated by wind to rye and uredinia are produced on diseased rye leaves. However, mycelia and urediniospores also overwinter in Europe and the aecial stage also is not thought to be essential. Leaf rust is severe during moist conditions, such as cool nights with abundant dew or warm, wet days.

Distribution. Wherever rye is grown.

Symptoms. Leaf rust of rye is similar in appearance to leaf rust of wheat and barley, and crown rust of oat. Uredinia are small, oval, orange-brown pustules on both leaf surfaces. As uredinia rupture, powdery, red-brown urediniospores are exposed. Similar pustules that are more elongated also develop on leaf sheaths and stems. Severely rusted plants may also have pustules on necks and glumes.

Telia appear toward plant maturity as small, elongated, dark gray pustules that do not immediately break through the epidermis. Several telia may form a circle on a diseased leaf. The pycnial and aecial stages on *Anchusa* spp. resemble those found on the alternate hosts of other leaf rusts.

Management. There is no satisfactory management. Some rye cultivars apparently have more resistance than others. However, since rye is cross-pollinated, cultivars are not uniformly pure and do not display a high amount of resistance.

Leaf Streak

Cause. *Cercosporidium graminis* (syn. *Scolicotrichum graminis*) likely survives as mycelia in infested residue. During wet weather in the spring, conidia are produced and disseminated by wind to hosts. Sporulation ceases during dry weather.

Distribution. Generally distributed in most rye-growing areas, but leaf streak on rye is uncommon.

Symptoms. Oblong to linear, red-brown to brown-purple blotches with definite margins develop on diseased leaves. The necrotic area is dry and sunken. A good diagnostic characteristic is the rows or tufts of conidiophores that emerge through stomata. These structures can be seen with the aid of magnification.

Management. Management measures ordinarily are not practiced, but sanitation will reduce inoculum.

Leptosphaeria Leaf Spot

Cause. *Phaeosphaeria herpotrichoides* (syn. *Leptosphaeria herpotrichoides*) overwinters as mycelia in infested residue and as ascospores in asci of pseudothecia on infested residue. Free water must occur on leaves for more than 48 hours for infection to proceed.

Distribution. Canada, Europe, and the United States. Leptosphaeria leaf spot is considered a minor disease on rye.

Symptoms. Irregular-shaped, diffuse, yellow to tan spots occur on diseased leaves.

Management. Some cultivars are more resistant than others.

Loose Smut

Cause. *Ustilago tritici* is seedborne and survives as dormant mycelium in seed embryos. Smutted heads emerge from the boot 1 to 2 days earlier than healthy heads. Chlamydospores are disseminated by wind to flowers of host plants and germinate during cool temperatures of 16° to 22°C and moisture provided by dews or light rain showers. The resulting germ tubes penetrate flower ovaries and stigmas and the subsequent mycelial growth is in the embryos of developing seed. As grain matures, *U. tritici* becomes dormant in the seed embryos until the following growing season. When infected seed germinates, mycelium grows systemically within plants and replaces healthy kernels with smutted heads filled with chlamydospores.

Distribution. Wherever rye is grown, but loose smut on rye is relatively uncommon.

Symptoms. Diseased seed does not have external symptoms and germination is not affected. Diseased seed gives rise to plants with smutted heads, but only a few kernels, and not the entire head, are usually diseased. Smutted heads emerge 1 to 2 days earlier than healthy heads. The brown to dark brown spore mass is enclosed within a fragile, gray membrane that soon ruptures and releases spores, leaving an erect, naked rachis. At maturity, the barren rachis remains erect and protrudes above the reclining heads of healthy plants.

Management. Loose smut of rye is uncommon and management usually is not necessary. However, treating seed with a systemic fungicide will destroy dormant mycelium in seed ovaries.

Pink Snow Mold

Pink snow mold is also called Fusarium patch.

Cause. *Microdochium nivale* (syn. *Fusarium nivale* and *Gerlachia nivalis*), teleomorph *Monographella nivalis*. *Microdochium nivale* oversummers as perithecia on diseased lower leaf sheaths. In the autumn, leaf sheaths and blades near the soil surface and roots are infected by ascospores from perithecia and by mycelium growing from infested residue. During cool, wet periods, with or without snow cover, mycelia grow from diseased to healthy plants. In the spring, secondary infection may occur from conidia or ascospores. Pink snow mold is most severe when host plants are grown under a heavy snow cover.

Distribution. Central and northern Europe, and North America.

Symptoms. Damage usually corresponds to the pattern of snow cover. At snowmelt, pink mycelia and sporodochia are visible on living and dead plant tissue and on the soil surface. Somewhat rectangular-shaped, brown lesions surrounded by a darker brown band on the first and second leaves cause leaves to be chlorotic, then necrotic. Necrotic leaves remain intact and do not disintegrate. Unless crowns are diseased, plants recover during warm, dry weather despite extensive disease symptoms on leaves.

Management
1. Sow winter rye later in the autumn. Winter rye sown late in the autumn and rye sown in the spring will develop plant growth under conditions unfavorable to infection.
2. Rotate rye with noncereal crops such as legumes.

Platyspora Leaf Spot

Cause. *Platyspora pentamera* (syn. *Clathrospora pentamera*) survives as perithecia in infested residue. During wet spring weather, ascospores are produced and disseminated by wind to hosts. There must be 24 to 72 hours of continual moisture for infection to occur. New perithecia are produced on diseased leaves as rye plants mature.

Distribution. Canada and the north central United States.

Symptoms. Nondescript yellow-brown spots are randomly produced on diseased leaves. As plants mature, dark perithecia are produced on diseased tissue.

Management. No specific management is necessary.

Powdery Mildew

Cause. *Blumeria graminis* f. sp. *secalis* (syn. *Erysiphe graminis* f. sp. *secalis*) overwinters as cleistothecia on infested residue. In areas where leaf tissue survives as green tissue throughout the winter, the fungus may overwinter as mycelium. Ascospores formed within cleistothecia in the spring are disseminated by wind to host plants and serve as primary inoculum in northern rye-growing areas. Conidia are produced as mycelium becomes established on diseased leaf surfaces, especially during humid, cool weather but not in the presence of free water. Conidia account for most of the secondary inoculum and spread of powdery mildew during the growing season. Cleistothecia form on diseased leaf surfaces as plants approach maturity.

Powdery mildew ceases to be a problem when weather becomes dry and warm later in the growing season.

Distribution. Wherever rye is grown.

Symptoms. Disease is most severe on tender, rank-growing plants that have been heavily seeded or have had heavy applications of nitrogen fertilizer. Superficial mycelia and conidia appear as light gray or white spots on the upper surface of diseased leaves, sheaths, and floral bracts. Diseased plant parts appear to be dusted with a gray powder. Spots enlarge, yellow, and eventually darken in color as diseased plants mature. Fungal growth occurs most often on the upper leaf surfaces and infrequently on lower surfaces. The numerous small, round, dark cleistothecia that develop on diseased areas are easily seen under magnification.

Management. Most cultivars of rye are resistant to powdery mildew, but foliar fungicides can be applied in unusual cases where disease becomes severe. Rye is the most resistant cereal to powdery mildew.

Scald

Scald is also called Rhynchosporium leaf scald.

Cause. *Rhynchosporium secalis* overwinters as stroma in lesions on perennial grasses or winter rye infected in the autumn, and in infested residue. During cool, humid spring or autumn weather, conidia produced on stroma are disseminated by wind to plant hosts. Secondary inoculum is conidia that are produced during cool, humid weather.

Distribution. Scald occurs to some extent wherever rye is grown.

Symptoms. Young spots are dark blue-gray, water-soaked blotches that occur mainly on leaves but occasionally on leaf sheaths. Spots become oval or lens-shaped and have a light tan or white center surrounded by a straw-colored border that, in turn, is surrounded by a yellow-green halo whose margin gradually fades into the normal green of the leaf. Leaf spots enlarge in cool weather and may have a zonate appearance due to successive enlargements.

Spots at the base of a diseased leaf may extend across the blade and down into the sheath, causing death of the entire leaf. Elongated spots cause the entire leaf to lose color and wide spots cause death of leaf tissue.

Management
1. Rotate rye with resistant crops.
2. Plow under infested residue.
3. Destroy perennial grasses along edges of fields.
4. Grow resistant or tolerant cultivars.

Septoria Leaf Blotch

Cause. *Septoria secalis* survives as conidia within pycnidia and as mycelia in infested residue, in lesions on live diseased plants, and in seed. There is some question whether or not *S. secalis* is the same as *S. tritici.* Symptoms are similar but some reports state no secondary infection occurs with *S. tritici.*

In the autumn, conidia are disseminated by wind and, under moist conditions, infect winter rye. During moist, cool conditions in the spring, rye is infected by conidia from pycnidia produced the previous season or the current spring.

Distribution. Generally distributed wherever rye is grown.

Symptoms. The first symptoms are light green to yellow spots that occur mostly between diseased leaf veins. Eventually, spots become light brown, irregular-shaped blotches with a speckled appearance due to pycnidia that are sprinkled throughout the spot. Pycnidia are very small but visible to the naked eye as tiny black or dark brown specks. After leaves die, blotches are lighter than the surrounding tissue. Under moist conditions, leaves may die and plant crowns become diseased, resulting in weakened or dead plants.

Management
1. Apply a foliar fungicide before disease becomes severe.
2. Plow under infested residue to aid in decomposition and prevent conidia from being placed on the soil surface, where they are easily wind-borne.
3. Rotate rye with a resistant crop, preferably a legume.

4. Treat seed with a systemic seed-protectant fungicide to kill fungi on seed or any mycelium that has grown into the seed coat.

Septoria tritici Blotch

Septoria tritici blotch is also called speckled leaf blotch.

Cause. *Septoria tritici,* teleomorph *Mycosphaerella graminicola.* The fungi survive as mycelia in live rye plants and as pycnidia on infested residue. Conidia (pycnidiospores) from residue presumably are the primary source of inoculum. *Septoria tritici* is also seedborne; however, this is uncommon.

During temperatures of 15° to 25°C and 100% relative humidity in the autumn or spring, pycnidiospores are exuded from pycnidia in a gelatinous drop called a cirrhi that protects spores from radiation and drying out. Pycnidiospores are disseminated to lower leaves by splashing and blowing rain.

Infection requires 6 or more hours of wetness and subsequent disease development is favored by temperatures of 18° to 25°C. New pycnidia and pycnidiospores are eventually produced on diseased tissue.

As diseased rye matures, ascospores produced in pseudothecia are disseminated by wind to host plants. However, ascospores apparently do not cause much damage because of the maturity of the host plants when infected. *Mycosphaerella graminicola* is readily found on residue in the United Kingdom.

Distribution. Generally distributed wherever rye is produced.

Symptoms. Initially, small, light green to yellow spots occur between veins of the diseased lower leaves of a host plant, especially if the leaves are in contact with soil. Spots rapidly elongate and form tan to red-brown, irregular-shaped lesions that are often partly surrounded by a yellow margin. Lesions age and become light brown to almost white and have small, dark specks (pycnidia) in the lesion center. The presence of pycnidia is a good diagnostic characteristic.

Infection of stem nodes, leaf sheaths, and tips of glumes also occurs. Severely diseased leaves turn yellow and die prematurely. Occasionally an entire plant may be killed. Pycnia are produced in all diseased areas. Autumn and winter infections cause a reduction in root weight and mass.

Management
1. Sow cleaned, certified, disease-free seed treated with a seed-protectant fungicide.
2. Plow under infected residue.
3. Grow the most resistant cultivars.
4. Apply a foliar fungicide if environmental conditions favor disease development.
5. Rotate rye every 3 to 4 years with a resistant crop. Most small grains are susceptible.

Sharp Eyespot and Rhizoctonia Root Rot

Cause. *Rhizoctonia cerealis,* teleomorph *Ceratobasidium cereale,* survives as sclerotia in soil or as mycelium in infested residue. Some researchers also consider *R. solani* as a causal agent. Rye may be infected any time during the growing season. Sclerotia germinate to form mycelium or mycelium grows from a precolonized substrate to infect roots and culms, particularly in cool and dry soils with less than 20% moisture-holding capacity. Only *R. solani* AG-4 has been reported to be pathogenic to rye seedlings.

Distribution. Wherever rye is grown.

Symptoms. Diamond-shaped lesions that resemble those of eyespot occur on lower leaf sheaths. Lesions are more superficial than eyespot lesions and have light tan centers with dark brown margins. Dark mycelium frequently is visible on lesions. Dark sclerotia may develop in lesions and between culms and leaf sheaths.

Seedlings may be killed, but plants produce new roots to compensate for those rotted off. When roots are diseased, plants may lodge and produce white heads. Diseased plants appear stiff, have a grayish cast, and are delayed in maturity.

Management. No management is truly effective. Vigorous plants growing in well-fertilized soil are likely to become severely diseased.

Snow Scald

Snow scald is also called Sclerotinia snow mold.

Cause. *Myriosclerotinia borealis* (syn. *Sclerotinia borealis*) oversummers as sclerotia. In autumn, the sclerotia germinate during damp, cool weather and form cup-shaped apothecia. Ascospores produced in a layer of asci on the upper portion of the apothecium are disseminated by wind to seedlings. Snow scald occurs when cool, damp autumns are followed by a deep snow cover that lasts 5 or more months over unfrozen or slightly frozen soil.

Distribution. Canada, Europe, Japan, Scandinavia, and the former USSR.

Symptoms. Snow scald occurs in scattered patches throughout a field. Sparse, gray mycelia cover bleached, dead plants exposed after the snow has melted. Leaves wrinkle on exposure to light and eventually turn dark because of saprophytic growth of secondary fungi, and crumble. Sclerotia are globular, elongated, or "flake-like" (0.3 to 7.0 mm long) and black at maturity. Sclerotia are found only in and on leaves, leaf sheaths, and crowns of dead plants.

Management
1. Rotate rye with a legume to help reduce inoculum.

2. Plowing under infested residue will bury inoculum and prevent sclerotia from germinating, hastening their decomposition.

Speckled Snow Mold

Speckled snow mold is also called gray snow mold and Typhula blight.

Cause. *Typhula incarnata, T. idahoensis, T. ishikariensis,* and *T. ishikariensis* var. *canadensis.* Some researchers consider *T. ishikariensis* and *T. idahoensis* to be two distinct causal fungi, while others consider these two fungi to be synonymous.

The causal fungi oversummer as sclerotia in infested residue and soil and as mycelia on live plants, but neither fungus survives well saprophytically on infested residue. Optimum infection occurs at temperatures of 1° to 5°C. Sclerotia germinate during wet weather in the autumn and form either basidiocarps or mycelia. Infection occurs either by mycelium growing from sclerotia to nearby host plants or by basidiospores formed on the basidiocarp that are disseminated by wind to hosts. Sclerotia are formed within necrotic tissue or in mycelium. Further infection occurs by mycelial growth under the snow cover. *Typhula* spp. growth is more closely associated with snow cover than with *Microdochium nivale* (syn. *Fusarium nivale* and *Gerlachia nivalis*), the cause of pink snow mold.

Distribution. Canada, central and northern Europe, Japan, and the northwestern United States.

Symptoms. Gray-white mycelia are present on plants and soil that was exposed to the air as the snow melts. Numerous sclerotia occur in living tissues and are scattered in mycelium growing over diseased plant surfaces, giving plants a speckled appearance. *Typhula incarnata* can infect plants aboveground and below the soil surface. Dead leaves are common, but unless the plant crowns are infected, diseased plants recover during warm, dry weather, although they may never be as vigorous as healthy plants. Leaves killed by *T. incarnata* and *M. nivale* crumple easily. Leaves killed by *T. incarnata* are covered with gray to white mycelium, while leaves killed by *M. nivale* are covered by pink mycelium.

Management. Rotation with a legume will help to reduce soil inoculum.

Spot Blotch and Associated Seedling and Crown Rots

Cause. *Bipolaris sorokiniana* (syn. *Helminthosporium sativum* and *H. sorokinianum*), teleomorph *Cochliobolus sativus,* is seedborne and overwinters as mycelia in infested residue and on seedling leaves of winter rye. During moist spring weather, conidia are produced on infested residue and disseminated by wind to seedling leaves. Spot blotch is more severe in the presence of pollen, possibly because causal fungi use pollen as an exogenous source of energy.

Distribution. Wherever rye is grown, but damage is usually not severe.

Symptoms. The first symptoms occur after warm and moist weather. Round to oblong, dark brown to black spots of various sizes with definite margins appear on diseased lower leaf sheaths near the soil line and eventually extend into green leaf blades. Spots coalesce and form lesions or blotches that cover large areas of leaves. Older lesions are olive-colored due to sporulation of fungi. Severely diseased leaves will completely dry up.

Seedborne inoculum results in seedling blight, and root and crown rot in dry, warm soils. Seedlings are dwarfed and have excessive tillering; crowns and roots may be dark brown and rotted. Disease may progress, causing seedlings to become yellow or killing them preemergence or postemergence. The heads of severely diseased live plants either do not emerge completely or have kernels that are poorly filled. Early head blight causes sterility or death of individual kernels soon after pollination. Lesions on floral bracts and kernels vary from small black spots to a dark brown discoloration of the entire surface, which results in blackened ends of kernels.

Management
1. Treat seed with a seed-protectant fungicide.
2. Apply a foliar fungicide to manage the leaf blotch phase before disease becomes severe.
3. Rotate rye with a resistant crop.

Stagnospora Blotch

Stagonospora blotch is also called glume blotch.

Cause. *Stagonospora nodorum* (syn. *Septoria nodorum*), teleomorph *Phaeosphaeria nodorum* (syn. *Leptosphaeria nodorum*), overwinters as mycelia in live plants and seed and survives 2 to 3 years as pycnidia on infested residue. During wet weather at temperatures of 20° to 27°C in the autumn or spring, pycnidiospores are exuded from pycnidia within a gelatinous drop called a cirrhi that protects the spores from radiation and drying up. Pycnidiospores are disseminated by splashing and blowing rain to lower leaves of host plants. Leaves and stems, but not heads, become infected. Infection requires 6 or more hours of wetness. New spores are produced in 10 to 20 days. Ascospores are produced in perithecia as rye matures in late summer or early autumn and are disseminated by wind to plant hosts. However, ascospores normally do not cause much damage due to the advanced maturity of plants when they are infected.

Distribution. Wherever rye is grown.

Symptoms. Nodes turn brown, shrivel, and are speckled with black pycnidia. Straw bends over and lodges just above nodes. Light brown spots, similar in appearance to Septoria leaf blotch, occur on leaves. A brown margin may surround the leaf spots, and pycnidia are present on both surfaces of diseased leaves. If a flag leaf is diseased, the head may be deformed. Infection of the leaf sheath causes a dark brown lesion that may include most of the sheath. Severely diseased plants are stunted.

Management
1. Sow certified, disease-free seed that has been cleaned and treated with a seed-protectant fungicide.
2. Grow resistant rye cultivars.
3. Plow under infected residue where this is feasible.
4. Apply a foliar fungicide.
5. Rotate rye with a resistant crop, such as legumes, every 3 to 4 years.

Stalk Smut

Stalk smut is also called flag smut, leaf smut, stem smut, and stripe smut.

Cause. *Urocystis occulata* overwinters as chlamydospores on seed, which provides most of the primary inoculum. Winter rye seed that is sown into dry soil may be infected infrequently by soilborne chlamydospores. Chlamydospores on seed germinate and produce basidiospores that, in turn, germinate and infect seedlings. The resultant mycelium grows along with the culm. Sori, which are composed of large numbers of chlamydospores, form in parenchyma tissue located between the veins of culms and, infrequently, in leaf blades. At first, sori are covered by an epidermis that ruptures at harvest time, liberating chlamydospores that are disseminated by wind to contaminate seed or soil. If soils are moist, chlamydospores in soil germinate and infect rye seedlings, but if no seedlings are present, the population of chlamydospores in the soil decreases.

Distribution. Wherever rye is grown.

Symptoms. Stalk smut is evident just before heading. Diseased plants are stunted and may be overlooked or hidden among the healthy plants. Diseased plants are darker green than healthy plants and have light green streaks in the upper leaves. Shortly, the light green streaks extend along leaves, sheaths, and stems as long, lead-colored stripes called sori, which contain the spore mass of chlamydospores. The stripes eventually become black and the epidermis splits to expose the mass of chlamydospores, which resembles dark brown or black dust. Leaves eventually split along the stripes.

Rarely does a diseased plant head out. If heads are produced, sori may be found on the chaff. Normally, every stem of a diseased plant is smutted and the affected parts are twisted and distorted.

Management
1. Treat seed with a seed-protectant fungicide to kill chlamydospores that have contaminated the seed surface.
2. Rotate rye with another crop. If stalk smut is a problem, do not grow rye continuously in the same field.

Stem Rust

Cause. *Puccinia graminis* (syn. *P. graminis* f. sp. *secalis*) is a heteroecious long-cycle rust that has barberries (*Berberis canadensis, B. fendleri, B. vulgaris,* and *Mahonia* spp.) as alternate hosts.

Puccinia graminis in North America causes stem rust of rye by one of two life cycles. Urediniospores, or the repeating spores, are produced in uredinia on rye during spring and summer and disseminated by wind to rye plants, where they infect and produce new urediniospores under moist conditions and moderate temperatures of 15° to 25°C. Secondary urediniospores are produced in 7 to 10 days. This cycle will continue indefinitely as long as growing rye plants are available. Urediniospores produced in northern rye-producing areas are blown south in summer and autumn to infect rye in southern rye-growing areas. The urediniospores will then recycle on rye in southern areas until they are disseminated by wind north during late winter and spring to reinfect host plants growing in the northern rye-growing areas.

The second life cycle is as follows: As rye ripens, teliospores form that survive through the winter. In the spring, teliospores germinate and form basidiospores (sporidia) that are disseminated by wind to the young leaves of barberry. Pycnia (spermogonia) formed on the upper leaf surfaces of barberry function in the exchange of genetic material, thereby creating new races of the fungus. Each pycnium produces pycniospores and special mycelia called receptive hyphae. Pycniospores are exuded out of the pycnium in a thick, sticky, sweet liquid that is attractive to insects. Pycniospores are splashed by water or carried by insects from one pycnium to another, where they become attached to receptive hyphae. The pycniospore germinates and the nucleus from the spore enters into the receptive hypha, which results in the formation of an aecium on the lower leaf surface directly under the pycnium. Aeciospores, which differ genetically from both pycniospores and receptive hyphae, are produced in the aecia and disseminated by wind to rye. Each infection gives rise to a uredinium in which urediniospores are formed. The urediniospores then serve as the repeating stage and are windborne to rye, where new generations of urediniospores are formed. Telia are once again produced as rye matures, thus completing the life cycle. Epiphytotics develop during moist weather, but disease is not severe during dry weather.

Distribution. Wherever rye is grown.

Symptoms. Uredinia and telia occur on stems, leaf sheaths, leaf blades, glumes, and beards of rye. Uredinia are red-brown and oblong. The epidermis of leaves and culms becomes ruptured and pushed back around the pustule, giving it a jagged or ragged appearance and exposing the urediniospores.

Just prior to plant maturation, telia appear mostly on leaf sheaths and culms. Telia are oblong to linear and dark brown to black. Teliospores are exposed when the epidermis ruptures.

Pycnia appear on barberry leaves in the spring as bright orange to yellow spots that in the middle sometimes have what appears to be a drop of moisture, which is the liquid containing pycniospores. On the other side of the leaf, opposite the pycnial spots, are aecia, which resemble raised, orange-colored, bell-shaped clusters.

Management
1. Grow resistant rye cultivars.
2. Eliminate the common barberry from rye-producing areas. The common barberry should not be confused with the Japanese barberry, which is immune to stem rust. There are several characteristics to differentiate between the two. The common barberry has a saw-toothed leaf edge, gray outer bark, bright yellow inner bark, berries borne in bunches, and spines with usually three in a group. The Japanese barberry has a smooth leaf edge, red-brown outer bark, bright yellow inner bark, berries borne in ones or twos, and usually a single spine.

Strawbreaker

Strawbreaker is also called Cercosporella foot rot, stem break, eyespot, foot rot, and culm rot.

Cause. *Pseudocercosporella herpotrichoides* (syn. *Cercosporella herpotrichoides*) survives for several years as mycelia in infested residue. Conidia are produced during cool, damp weather in the autumn or spring and infect crown and basal culm tissue but not roots. Winter rye is more likely to be infected than spring rye.

Distribution. Where rye is grown in cool, moist climates.

Symptoms. Eyespot is most conspicuous near the end of the growing season by the lodging of diseased plants and the presence of white heads. Initially, eye- or ovate-shaped lesions with white to tan centers and brown margins develop on diseased basal leaf sheaths. Similar spots form on stems directly beneath those on sheaths and cause lodging. Roots are not infected, but necrosis occurs around upper crown nodes. Under moist conditions, lesions enlarge and the black stroma-like mycelium that develops over crown surfaces and the bases of culms gives tissues a charred appearance. Stems shrivel and collapse or plants are yellowish or pale green and their heads are reduced in size and number. Early infection causes death of individual culms and of plants before maturity.

Management. Rotate rye with noncereals, such as legumes, or grass crops.

Stripe Rust

Stripe rust is also called yellow rust.

Cause. *Puccinia striiformis,* anamorph *Uredo glumarum,* is a rust fungus that is not known to have an alternate host. *Puccinia striiformis* oversummers between harvest and emergence of autumn-sown rye as urediniospores on residual green cereals and grasses. Mycelia and, infrequently, urediniospores overwinter on barley, grasses, rye, and wheat. Urediniospores are formed during cool, wet weather and disseminated by wind to plant hosts. Little infection occurs above 15°C in the summer. Teliospores are formed but are not known to function as overwintering spores.

Distribution. In North America, stripe rust occurs at the higher elevations and cooler climates along the Pacific Coast and intermountain areas from Canada to Mexico. Stripe rust also occurs in the same environment in South America and in the mountainous areas of central Europe and Asia. Stripe rust is not as common on rye as it is on wheat and barley.

Symptoms. The most severe symptoms occur during cool, wet weather in the early growth period of rye. Symptoms occur early in spring, before the symptoms of other rusts appear, especially in areas with mild winters. Yellow uredinia appear on autumn foliage and new spring foliage. Uredinia coalesce and produce long stripes between veins of diseased leaves and sheaths. Small linear lesions occur on floral bracts. Telia develop as narrow, linear, dark brown pustules covered by epidermis.

Management. Grow resistant cultivars. Many rye cultivars have at least partial resistance.

Take-All

Cause. *Gaeumannomyces graminis* var. *tritici* survives as mycelia in live plants or as mycelia and perithecia in infested residue. However, ascospores are not considered important in the life cycle of take-all. In the autumn or spring, seedlings become infected when their roots grow into the vicinity of precolonized residue. Plant-to-plant spread of *G. graminis* var. *tritici* occurs when hyphae grow through soil from diseased to healthy plants or when healthy roots come in contact with diseased roots. Ascospores are produced in perithecia during wet weather but are not disseminated very far by either splashing water or wind. Take-all is usually more severe on rye grown in alkaline soils.

Distribution. Wherever rye is grown. Take-all is less severe on rye than on wheat and barley.

Symptoms. Take-all becomes most obvious as plants approach heading. The first symptoms are light to dark brown necrotic lesions on roots. By the time diseased plants reach the jointing stage of maturity, most roots are brown to blackened, sparse, brittle, and dead. When pulled from the soil, the crowns of diseased plants can easily be broken off. A very dark discoloration of stems just above the soil line and a mat of dark brown fungus mycelium under lower sheaths between stems and inner leaf sheaths are visible. At this point many plants die, or if plants are still alive, they are stunted and leaves are yellow.

Stands are uneven in height and plants appear to be in several stages of maturity. At heading, diseased plants have few tillers, ripen prematurely, and have bleached, sterile heads.

Management. Rotate rye with a nongrass crop since *G. graminis* var. *tritici* is not a good saprophyte and does not survive long in soil in the absence of a host. Maize roots can become infected and carry over inoculum for subsequent crops of rye.

Tan Spot

Tan spot is also called blight, leaf spot, and yellow leaf spot.

Cause. *Pyrenophora tritici-repentis,* anamorph *Drechslera tritici-repentis* (syn. *Helminthosporium tritici-repentis*), overwinters as pseudothecia on infested residue. During wet spring weather, ascospores are released and disseminated by wind to plant hosts and serve as primary inoculum. Conidia and hyphal fragments may also serve as primary inoculum. During wet weather in the growing season, conidia produced in older lesions and disseminated by wind serve as secondary inoculum. In the autumn, pseudothecia are produced on diseased culms and leaf sheaths. Symptom development is favored by frequent rains and cool, cloudy, humid weather early in the growing season.

Distribution. Wherever rye is grown, but rye is usually not severely affected.

Symptoms. Symptoms first appear in the spring on lower leaves and continue throughout the early summer onto the upper leaves of diseased plants. Initially, tan flecks appear on both sides of diseased leaves. The flecks eventually become tan, diamond-shaped lesions up to 12 mm long surrounded by a yellow border with a dark brown spot in the center caused by the sporulation of *P. tritici-repentis*. Lesions may coalesce, starting from the leaf tips, and cause large areas of leaves to die. Pseudothecia will eventually develop on infested residue as dark raised bumps.

Management
1. Apply a foliar fungicide before disease becomes severe and if weather conditions favor development of the disease.
2. Plow under residue.

Diseases Caused by Nematodes

Cereal Cyst

Cause. *Heterodera avenae* survives for a year or more as cysts in soil. Larvae emerge in spring from eggs contained in overwintered cysts and enter plant roots to begin feeding. Female nematodes swell as eggs develop in their bodies and break through the root surface but remain attached by a thin neck. Males revert to a vermiform shape. Eventually, females form cysts that detach from roots. *Heterodera avenae* is also a pathogen of barley, oat, wheat, and numerous annual and perennial grasses. Populations of nematodes increase in sandy soils. Other soil characteristics, such as the death of nematodes, when roots die, and fungal parasites, also cause fluctuations in nematode populations.

Distribution. Africa, Australia, southeastern Canada, Europe, Japan, the United States (Oregon), and the former USSR.

Symptoms. The first symptom is the general poor growth of diseased plants in areas within a field. Leaf tips of diseased plants are red or purple. The discolored leaves die off and plants become yellow. The roots are thickened and more branched than those of healthy plants. Heavy infestations cause wilting (particularly during times of water stress), stunted growth, poor root development, and early plant death. Lemon-shaped cysts are white at first, then gradually become dark brown as they harden. Cysts are visible to the naked eye.

Management
1. Rotate rye with a legume crop.
2. Apply a nematicide to soil before sowing.
3. Sow rye in the autumn.

Leaf and Stem

Cause. *Ditylenchus dipsaci.* Occurrence is associated with heavy soils, high rainfall, cool growing seasons, and winter grains. Free moisture permits the nematode to migrate to and feed on aerial plant parts. The nematode penetrates leaves and stems.

Distribution. Worldwide. Damage is confined to small areas within fields.

Symptoms. Cell hypertrophy and hyperplasia occur. Stunting, distortions, and swollen stems are common symptoms. In severely infested areas within a field, growing points of diseased plants are destroyed and the plants may die. Other plants have reduced spike growth and reduced grain yields.

Management
1. Rotate with noncereal crops.
2. Grow winter cereals to reduce damage.

Root Gall

Cause. *Subanguina radicicola* survives in roots. Larvae penetrate roots, develop in cortical tissue, and form root galls in 2 weeks. The mature females begin egg production within a gall. Eventually, galls weaken and release larvae into the soil that establish secondary infections. Each generation is completed in about 60 days.

Distribution. Canada and northern Europe.

Symptoms. Diseased seedlings frequently have reduced top growth and chlorosis. Galls on roots tend to be inconspicuous and vary in diameter from 0.5 to 6.0 mm. Roots may be bent at the gall site. At the center of larger galls is a cavity filled with nematode larvae.

Management. Rotate rye with noncereal crops.

Root Knot

Cause. *Meloidogyne* spp.

Distribution. Widespread.

Symptoms. Plant growth and dry shoot weight is reduced.

Management. Not reported.

Root Lesion

Cause. *Pratylenchus* spp. overwinter as eggs, larvae, and adults in host tissue or soil. Both larvae and adults penetrate roots, where they move through cortical cells and the females deposit eggs as they migrate. Older roots are abandoned and new roots sought as sites for penetration and feeding.

Distribution. Wherever rye is grown.

Symptoms. Diseased plants in areas of a field will appear yellow and under moisture stress. Roots and crowns will rot when *Rhizoctonia solani* infects through nematode wounds. New roots become dark and stunted, causing yield loss.

Management
1. Sow in autumn when soil temperatures are below 13°C.
2. Soil fumigants could be used where their high costs are warranted.

Seed Gall

Cause. *Anguina tritici* survives as larvae in seed galls for several years. When galls are sown along with rye seed, larvae are released into the moist soil and move upward on plants in a water film to the flower primordia. Nematodes then mature, copulate, and produce eggs. The seed galls develop from undifferentiated flower tissue interacting with the nematodes. If gall development is retarded, larvae may be present in healthy-appearing seed. Galls are mixed with normal seed or fall to the soil, where nematodes become dormant under dry conditions.

Distribution. Eastern Asia, parts of Europe, India, and the southeastern United States. Seed gall nematode is not a common disease of rye.

Symptoms. Prior to heading, diseased plants are swollen near the soil line and leaves will be twisted, wrinkled, or rolled. After heading, distortions are not as obvious but plants are stunted and mature slowly. Heads are small and dark seed-like galls force glumes to spread apart. Galls are dark brown and do not have the brush or embryo markings of normal seed.

Management
1. Sow clean seed.
2. Seed may be soaked in hot (54°C) water for 10 minutes.
3. Rotate rye with a nonhost crop for 2 years. Wheat is also susceptible.

Stubby Root

Cause. *Paratrichodorus* spp. survive in soil or on roots as eggs, larvae, and adults. These nematodes feed only on the outside of roots and move relatively rapidly through fine, sandy soil at a speed of 5 cm/hr. Rye sown early in autumn in sandy soils is most severely diseased.

Distribution. Widely distributed in most agricultural soils.

Symptoms. Diseased roots are thickened, short, and stubby and have brown lesions on the root tips. Tops of diseased plants grow poorly, and entire plants may be easily pulled from the soil due to the lack of a fibrous root system.

Management. Not reported.

Disease Caused by Phytoplasmas

Aster Yellows

Cause. Aster yellows phytoplasma survives in several dicotyledonous plants and leafhoppers. Aster yellows phytoplasma is transmitted primarily by the aster leafhopper, *Macrosteles fascifrons,* and less commonly by *Endria inimica* and *M. laevis.* The leafhoppers acquire the phytoplasma by feeding on diseased plants, then fly to healthy plants, where they transmit the phytoplasma by feeding. Symptoms become most obvious at temperatures of 25° to 30°C.

Distribution. Eastern Europe, Japan, and North America.

Symptoms. Diseased seedlings either die 2 to 3 weeks after infection or if they survive are stunted and have leaves that are completely yellowed or have yellow blotches on them and heads that are sterile and distorted. Infection of older plants causes leaves to become somewhat stiff and discolored shades of yellow, red, or purple, starting from the tip or margin inward. Root systems may not be well developed.

Management. Not reported.

Diseases Caused by Viruses

African Cereal Streak

Cause. African cereal streak virus (ACSV) is transmitted by the planthopper *Toya catilina,* but the virus is not mechanically or seed-transmitted. The natural virus reservoir presumably is native grasses. ACSV is limited to the phloem. Disease development is aided by high temperatures.

Distribution. East Africa.

Symptoms. Initially, faint, broken, chlorotic streaks begin near leaf bases and extend upward on the leaves. Later, definite alternate yellow and green streaks develop along entire leaf blades. Eventually, leaves become completely yellow. New leaves tend to develop a shoestring habit and die. ACSV induces a necrosis in the phloem.

   Young diseased plants become chlorotic, are severely stunted, and die. Seed yield is almost completely suppressed. Plants become soft, flaccid, and velvety to the touch.

Management. Not reported.

Barley Yellow Dwarf

Cause. Barley yellow dwarf virus (BYDV) is in the luteovirus group. BYDV survives in autumn-sown small grains, such as barley, oat, and wheat and in annual and perennial grasses. In general, however, local grasses, winter wheat, and maize are of little importance as primary sources for BYDV.

BYDV is transmitted by several species of aphids but is not transmitted through eggs, newborn aphids, seed, or soil or by mechanical means. Once an aphid acquires BYDV, it is capable of transmitting BYDV for the rest of its life. Some strains of BYDV are transmitted equally well by all species of aphids, but some strains display a high degree of vector specificity. Aphid flights are local or, if assisted by wind, extend for hundreds of kilometers, carrying virus inoculum long distances. Thus, the major source of aphids in spring may be from distant plants; however, in autumn, aphids are both distant and local.

Epiphytotics occur during cool temperatures (10° to 18°C) and moist seasons that favor grass and cereal growth together with aphid multiplication and migration. Infections occur through the growing season but are most numerous in areas that support populations of aphids.

Distribution. Africa, Asia, Australia, Europe, New Zealand, North America, and parts of South America. BYDV on rye is not considered to be a serious disease.

Symptoms. Symptoms are not striking and tend to be confused with nutritional disorders or weather-related problems but become more pronounced at cool temperatures of 16° to 20°C and cloudless days. Single plants or groups of plants within a field are yellow and stunted. Seedling infection slows plant maturity and causes older leaves to turn bright yellow; however, these striking symptoms are considered unusual on rye. Leaves tend to be stiff and discolored various shades of yellow, red, or purple, starting from the blade edge. A later infection causes the flag leaf to become discolored yellow or red. Frequently, the feeding of some vectors produces tiny, brown-black spots on leaves and culms; adjacent tissues first turn yellow, then tan. Roots are not well developed, and phloem tissues are darkened.

Management. Not reported for rye. However, it is suggested that winter cereals be sown later in the autumn.

Soilborne Mosaic

Soilborne mosaic is also called wheat soilborne mosaic.

Cause. Wheat soilborne mosaic virus (WSBMV) survives in soil in *Polymyxa graminis,* a soilborne plasmodiophoraceous fungus that is an obligate parasite of several higher plants. WSBMV is spread inside of *P. graminis* by any means that disseminates soil containing *P. graminis*. *Polymyxa graminis* enters root hairs and epidermal cells of roots as motile zoospores that "swim" through wet soil at temperatures of 10° to 20°C. Once inside plants, *P. graminis* replaces cell contents with plasmodial bodies that either segment into additional zoospores or develop into thick-walled resting spores 2 to 4 weeks after infection. WSBMV is most common in low-lying areas of fields that tend to be wet. Warm weather prevents development of disease symptoms. Wheat soilborne mosaic is most severe on rye sown in the autumn, but symptoms are not as common on rye as on wheat.

Distribution. Argentina, Brazil, Egypt, Italy, Japan, and the eastern and central United States.

Symptoms. Symptoms, which are most prominent in the spring on older leaves, range from a light green to yellow mosaic on diseased leaves. The youngest leaves and sheaths are mottled and develop parallel spots or streaks. Diseased plants may be severely to slightly stunted.

Management
1. Rotate rye with noncereal crops.
2. Sow rye later in autumn.

Wheat Streak Mosaic

Cause. The wheat streak mosaic virus (WSMV) is in the potyvirus group. WSMV survives in live cereals and grasses and is transmitted in late summer or early autumn to early-planted rye by feeding of the wheat curl mite, *Aceria tulipae*. Mites and WSMV cannot survive longer than 1 to 2 days in the absence of a living plant. Only young mites acquire WSMV by feeding 15 or more minutes. Once a mite has picked up WSMV, it is carried internally for several weeks. As winter rye plants mature, mites migrate to nearby cereals, grasses, or maize and infect those plants. However, some grasses are hosts for mites and not for WSMV, and vice versa; some are susceptible to both; and some are resistant to both. WSMV has a wide host range that includes barley, maize, oat, rye, and several annual and perennial grasses.

Distribution. Eastern Europe, western and central North America.

Symptoms. Initially, light green to light yellow blotches, dashes, or streaks occur parallel to leaf veins. Diseased plants become stunted, have a general yellow mottling, and develop an abnormally large number of tillers that vary considerably in height. Stunted plants with sterile heads remain after harvest, standing at the same height or shorter than stubble. As diseased plants mature, streaked leaves turn brown and die.

Heads may be sterile or be partially sterile but have shriveled kernels.

With severe disease, plants die before maturity. Feeding mites often cause leaf edges to curl tightly in toward the upper midvein.

Management
1. Destroy all volunteer cereals and grasses in the subject field 3 to 4 weeks before sowing and in the adjacent fields 2 to 3 weeks before sowing the subject field.
2. Sow rye as late as practical after the Hessian fly–free date. Rye often escapes infection if it emerges in October or later.

Selected References

Boewe, G. H. 1960. Diseases of wheat, oats, barley, and rye. Illinois Nat. Hist. Surv. Circ. 48.

Bragard, C., Singer, E., Alizadeh, A., Vauterin, L., Maraite, H., and Swings, J. 1997. *Xanthomonas translucens* from small grains: Diversity and phytopathological relevance. Phytopathology 87:1111-1117.

Brooks, F. T. 1953. Plant Diseases, Second Edition. Oxford University Press, London.

Cunfer, B. M., and Schaad, N. W. 1976. Halo blight of rye. Plant Dis. Rptr. 60:61-64.

Cunfer, B. M., Schaad, N. W., and Morey, D. D. 1978. Halo blight of rye: Multiplicity of symptoms under field conditions. Phytopathology 68:1545-1548.

Daniels, A., Lucas, J. A., and Peberdy, J. F. 1991. Morphology and ultrastructure of W and R pathotypes of *Pseudocercosporella herpotrichoides* on wheat seedlings. Mycol. Res. 95:385-397.

Degenhardt, K. J., Harper, F. R., and Atkinson, T. G. 1982. Stem smut of fall rye in Alberta: Incidence, losses and control. Can. J. Plant Pathol. 4:375-380.

Duczek, L. J. 1990. Sporulation of *Cochliobolus sativus* on crowns and underground parts of spring cereals in relation to weather and host species, cultivar, and phenology. Can. J. Plant Pathol. 12:273-278.

Farr, D. F., Bills, G. R., Chamuris, G. P., and Rossman, A. Y. 1989. Fungi on Plants and Plant Products in the United States. American Phytopathological Society, St. Paul, MN. 1252 pp.

Harder, D. E., and Bakker, W. 1973. African cereal streak, a new disease of cereals in East Africa. Phytopathology 63:1407-1411.

Hosford, R. M. 1978. Effects of wetting period on resistance to leaf spotting of wheat, barley, and rye by *Leptosphaeria herpotrichoides*. Phytopathology 68:591-594.

Jedlinski, H. 1983. Predisposition of winter rye (*Secale cereale* L.) to ergot (*Claviceps purpurea* (Fr.) Tul.) by barley yellow dwarf virus infection. (Abstr.) Phytopathology 73:843.

Leukel, R. W., and Tapke, V. F. 1954. Cereal smuts and their control. USDA Farmers Bull. No. 2069.

Mains, E. B., and Jackson, H. S. 1924. Aecial stages of the leaf rusts of rye, *Puccinia dispersa* Eriks. and Henn. and of barley, *P. anomala* Rostr., in the United States. J. Agric. Res. 28:1119-1126.

McBeath, J. H. 1985. Pink snow mold on winter cereals and lawn grasses in Alaska. Plant Dis. 69:722-723.

McKay, R. 1957. Cereal Diseases in Ireland. Arthur Guiness Son & Co. Ltd. Dublin.

Ploetz, R. C., Mitchell, D. J., and Gallaher, R. N. 1985. Characterization and pathogenicity of *Rhizoctonia* species from a reduced-tillage experiment multicropped to rye and soybeans in Florida. Phytopathology 75:833-839.

Richardson, M. J., and Noble, M. 1970. *Septoria* species on cereals—A note to aid their identification. Plant Pathol. 19:159-163.

Roane, C. W. 1984. Atypical symptoms in rye caused by *Pseudomonas syringae* pv. *coronafaciens*. (Abstr.) Phytopathology 74:792.

Simmonds, P. M. 1955. Root diseases of cereals. Canada Dept. Agric. Pub. 952.

Walker, J. 1975. Take-all disease of Graminae: A review of recent work. Rev. Plant Pathol. 54:113-144.

Western, J. H. 1971 Diseases of Crop Plants. The Macmillan Press Ltd., London.

16. Diseases of Safflower *(Carthamus tinctorius)*

Diseases Caused by Bacteria

Bacterial Leaf Spot and Stem Blight

Bacterial leaf spot and stem blight is also called bacterial blight.

Cause. *Pseudomonas syringae* survives in soil and infested residue and as a parasite and pathogen on several other plants. Seeds, weeds, residue, and soil are possible sources of primary inoculum. *Pseudomonas syringae* is disseminated by seed, flowing water, and any means that moves soil. Bacteria are splashed or blown onto plant surfaces. Bacterial blight is most severe during warm rain or under sprinkler irrigation.

Distribution. The western United States.

Symptoms. Dark, water-soaked lesions occur on diseased leaves, stems, and leaf petioles. Spots on leaves become red-brown necrotic spots with pale margins. Other descriptions state tissue in the center of leaf spots becomes translucent and is surrounded by a dark brown to black margin. The terminal bud is often necrotic. Rotting of interior tissues of petioles often extends below the soil line into roots.

Severely diseased plants die, but plants that are not severely diseased will recover during dry conditions. Loss of the initial stem and branches can be compensated for by increased lateral branching.

Management. Not reported.

Stem Soft Rot

Cause. *Erwinia carotovora* pv. *carotovora.* Stem soft rot occurs during wet weather.

Distribution. Mexico.

Symptoms. Diseased plants wilt. Stems have a soft internal rot.

Management. Not reported.

Diseases Caused by Fungi

Alternaria Leaf Spot

Cause. *Alternaria carthami. Alternaria alternata* has also been reported as a causal agent. Both fungi are seedborne and survive in residue for at least 2 years. Leaf spots in the early stages of plant development have been reported to yield only *A. carthami*; whereas *A. alternata* was the predominant fungus isolated from leaf spots on diseased plants that were nearing maturity. Under greenhouse conditions, *A. carthami* is pathogenic on safflower at all growth stages. *Alternaria alternata* also infects healthy safflower plants, but infections remain dormant until leaf senescence. *Alternaria solani* is also reported to penetrate healthy leaves, but it does not cause leaf spots and mycelia remain dormant until leaf senescence. Secondary spread is by windborne conidia. Disease development is favored by abundant moisture from dew or rain and temperatures of 25° to 30°C.

Distribution. Wherever safflower is grown.

Symptoms. Aboveground symptoms first appear on seedling cotyledons as brown spots that reach a maximum size of 5 mm in diameter. On diseased leaves of older plants, small brown to dark spots (1 to 2 mm in diameter) with concentric rings enlarge to about 1 cm and coalesce to form large lesions. The centers of these spots are light in color and have alternate light and dark brown rings. Fully mature spots tend to develop holes, which cause a cracking of leaves. Conidia produced in the spots give them a dusty appearance.

When stems of emerging seedlings are diseased, a brown discoloration appears on the stems just above the soil level. Stem infection may cause the complete collapse of a plant.

Flower heads are first infected at the base of the calyx. Diseased buds shrink and dry up instead of opening. Seed coats may have brown, sunken lesions. Diseased seeds that are sown may rot before germination; if seed does germinate, the resulting seedlings may damp off.

Management
1. Sow seed from disease-free plants.
2. Treat seed with a fungicide.
3. Place seeds in water at 50°C for 30 minutes.
4. Grow the most resistant cultivar.

Anthracnose

Cause. *Colletotrichum capsici* and *Gloeosporium carthami.*

Distribution. *Colletotrichum capsici* is reported from India and *G. carthami* from Australia, Japan, Kenya, and the United States.

Symptoms. *Colletotrichum capsici* is reported to cause plant wilting. *Gloeosporium carthami* causes a dieback and premature death. Light brown, elongated lesions may occur on diseased stems.

Management. Not reported.

Botrytis Head Rot

Botrytis head rot is also called Botrytis flower blight, and gray mold.

Cause. *Botrytis cinerea,* teleomorph *Botryotinia fuckeliana,* survives as sclerotia in soil and infested residue. Primary inoculum is sclerotia that either support conidiophores and conidia on their surfaces or germinate and form mycelium. Conidia are disseminated by wind and water and sclerotia are disseminated by any means that moves soil. Secondary inoculum is provided by conidia that are abundantly produced on diseased tissue during moist weather and disseminated by wind for long distances to the flowers of plant hosts. Botrytis head rot is most severe in cool, wet weather during or after flowering.

Distribution. Wherever safflower is grown in moist climates.

Symptoms. Diseased seed heads initially change color from dark to light green, then become completely brown. During cool, wet weather, diseased floral parts are covered with a gray, fuzzy mold consisting of mycelium, conidiophores, and conidia. The fungus progresses through the seed head and into the stem, causing diseased heads to be easily detached from the stem. Infection at the early stage of bloom causes seed to be light in weight; later infection results in seed that is of almost normal weight.

Management. Not reported, although anecdotal evidence suggests it is not advisable to grow safflower in moist areas where environmental conditions are conducive to Botrytis head rot.

Cercospora Leaf Spot

Cause. *Cercospora carthami* survives in infested residue. Dew condensation early in the morning appears to be critical for spore germination and germ tube elongation. Conidia require 3 to 4 days to penetrate leaves but are able to withstand three to four intervening 20-hour dry periods. Spore germination, germ tube penetration, and subsequent disease development are greatest under moist, humid conditions with little wind.

Distribution. Australia, Africa, India, Middle East, southern Russia, and the southern United States. Cercospora leaf spot is considered of minor importance and generally occurs in warmer safflower-growing areas.

Symptoms. Symptoms can occur at almost any stage of growth. Round (3–10 mm in diameter) to irregular-shaped brown spots occur mostly on the lower leaves of diseased plants. Spots are slightly sunken and frequently have concentric rings and a yellowish margin. Diseased leaves become distorted and have interveinal necrosis. During moist conditions, both sides of spots become grayish because of the sporulation of conidia. In Australia, leaf spots remain confined to older leaves and are reported to have a whitish center with a gray-brown margin.

Along with bracts, stems and nodes may become infected. Flower buds turn brown, wither, and die; no seed is produced.

Management. Not necessary.

Charcoal Rot

Cause. *Macrophomina phaseolina* (syn. *Sclerotium bataticola* and *Rhizoctonia bataticola*) is seedborne as pycnidia and sclerotia. Charcoal rot is most severe at high soil temperatures (30°C and above) and low soil moisture.

Distribution. Wherever safflower is grown in warmer climates. Charcoal rot rarely occurs in northern safflower-growing areas.

Symptoms. Symptoms usually occur after flowering. Diseased stalks ripen prematurely; the bases of the stalks become gray and only the vascular bundles or fibers remain, which gives the inside of the stems a shredded appearance. Vascular fibers become covered with small, black sclerotia that resemble pepper or "flecks" of charcoal. Stems infected with *M. phaseolina* and parasitized by stem weevil larvae have a brown to black discoloration that may be with or without typical gray areas on the stem surface. Portions of the lower stems and upper taproots are hollow and contain mixtures of insect frass and fragments of pith tissue.

Flowers ripen prematurely, which results in small, poorly filled, distorted heads with a zone of aborted flowers. Infected seed that is sown either does not germinate or if it does germinate, roots, hypocotyls, and cotyledons are discolored. Pycnidia and sclerotia eventually develop on diseased plant parts.

Management
1. Rotate safflower with a resistant crop, such as small grains. Soybean and sunflower also are susceptible.
2. Plow under residue.
3. Grow cultivars that have resistance.
4. Irrigate at flowering or at flowering and ripening stages of plant growth.

Colletotrichum Stem Rot

Cause. *Colletotrichum orbiculare.*

Distribution. Australia.

Symptoms. Brown, necrotic lesions occur on the bases of diseased stems. Diseased plants may die.

Management. Not necessary.

Damping Off

Cause. Several fungi, including *Alternaria* spp., *Fusarium* spp., *Penicillium* spp., *Phytophthora* spp., and *Pythium* spp.

Distribution. Widespread. Wherever safflower is grown under moist soil conditions.

Symptoms. Diseased plants fail to emerge or die shortly after emergence.

Management. Treat seed with a seed-treatment fungicide.

Downy Mildew

Cause. *Bremia lactucae* f. *carthami.*

Distribution. Cyprus, Iran, Israel, and the former USSR.

Symptoms. Typical mildew symptoms characterized by the downy growth of mycelium, conidiophores, and conidia occur on leaves.

Management. Not reported.

Fusarium Wilt

Cause. *Fusarium oxysporum* f. sp. *carthami,* which survives as chlamydospores in soil and infested residue, is seedborne as internal mycelium and conidia contaminating the outside of seeds. Mycelium and conidia may also survive in infested residue for a short period of time. Abundant conidia are produced on residue and disseminated by wind to other areas. Some conidia are converted to chlamydospores when they fall to the soil. Mycelia from residue, germinating chlamydospores, and conidia penetrate roots and enter into the water-conducting vascular tissue (xylem tissue). Microconidia and conidia move up the xylem tissue and microconidia lodge at nodes, where they germinate. Eventually, water movement within a plant is inhibited by tyloses, gums, and physical blockage.

Fusarium wilt severity is greatest during high temperatures and moist weather in plants grown in acidic, light-textured soils that are high in nitrogen. Different physiologic races of *F. oxysporum* f. sp.*carthami* exist.

Distribution. India and the United States.

Symptoms. A typical symptom is the yellowing of leaves on one side of the diseased plant. Unilateral yellowing also occurs on leaves and some seed heads. Yellowing, which begins on lower leaves, is followed by wilting that progresses up the plant. Plants infected when young are usually killed, and older plants that become infected may also die but usually only branches on one side of the diseased plant are killed. Seed heads are frequently blighted and distorted due to the fungus growing up the vascular system into the seed head. Diseased heads have aborted seed. The vascular system of the roots and stems has a brown discoloration.

Management
1. Do not sow seed from diseased plants.
2. Treat seed with a seed-protectant fungicide.
3. Rotate safflower with other crops.
4. Some safflower selections are resistant to race 4.

Phyllosticta Leaf Spot

Cause. *Phyllosticta carthami.*

Distribution. The Philippines.

Symptoms. Scattered tan spots occur on diseased leaves. Later, black specks, which are pycnidia, can be observed scattered throughout the spots.

Management. Not reported.

Phytophthora Root and Stem Rot

Phytophthora root and stem rot is also called Phytophthora root rot.

Cause. *Phytophthora cactorum, P. cryptogea, P. drechsleri,* and *P. nicotianae* var. *parasitica* (syn. *P. parasitica*) have been reported to cause disease of safflower. All species are thought to survive in infested residue and soil. A general life cycle for all *Phytophthora* spp. is the sporangia are formed that germinate and produce either mycelium or zoospores that "swim" to roots, then encyst and germinate. Plants may be infected at all stages of growth. Phytophthora root rot is most severe in wet soils, particularly in surface-irrigated fields or poorly drained areas, and at soil temperatures of 25° to 30°C. Water stress before infection predisposes plants to root rot caused by *P. cryptogea.* Physiologic races of the different *Phytophthora* spp. are thought to exist.

Distribution. Australia and the western United States.

Symptoms. Lower stems of seedlings collapse, causing plant death. The first aboveground symptom on diseased older plants is that leaves become light green or yellow, then they wilt, die, and turn brown. In the early stages of wilting, roots become discolored reddish. As Phytophthora root and stem rot progresses, the roots and lower stems become dark brown to black. This discoloration extends upward above the soil line on stems and is a diagnostic symptom.

Early inoculation of plants reduces the length of fine roots, which reduces their capacity to take up water. However, this has only a slight overall effect on water uptake because of the increased water uptake by the remaining roots. Lesions that develop on stem bases or taproots greatly reduce water uptake by diseased plants.

Management
1. Grow the most resistant cultivar.
2. Provide good drainage. Where flood irrigation is used, avoid ponding of water.
3. Grow surface-irrigated safflower on beds under furrow irrigation instead of flood irrigation.

Powdery Mildew

Cause. *Erysiphe cichoracearum* (*E. cichoracearum* f. sp. *carthami*).

Distribution. Afghanistan, India, Israel, and Russia.

Symptoms. A gray, powdery mass of mycelium, conidia, and conidiophores occurs on diseased leaf surfaces.

Management. Not reported.

Pythium Damping Off

Cause. Several *Pythium* spp., including *P. acanthicum, P. aphanidermatum, P. debaryanum, P. irregulare, P. myriotylum, P. oligandrum, P. splendens, P. ultimum,* and a divergent form of *P. ultimum* reported from Canada labeled as *Pythium* sp. "group G." Disease is favored by wet soil conditions.

Distribution. Worldwide.

Symptoms. The initial symptom is the hypocotyl becomes flaccid and discolored a light brown. Eventually, seedlings are killed.

Management. Apply fungicide seed treatments.

Pythium Root Rot

Cause. *Pythium* spp., particularly *P. ultimum* and *P. splendens.* The severity of Pythium root rot is greatest on safflower grown under irrigation. Only young plants are susceptible. Susceptibility of host plants is related to the elongation of hypocotyls and the first internode tissue.

Distribution. The western United States.

Symptoms. The hypocotyl and first internode become water-soaked, soft, and discolored a light brown. Plants topple over and tissues continue to rot and collapse.

Management
1. Treat seed with a seed-protectant fungicide.
2. Plants do not become diseased when they are not grown under irrigation.

Ramularia Leaf Spot

Ramularia leaf spot is also called brown leaf spot.

Cause. *Ramularia carthami.* Disease is favored by irrigation but does not occur on rain-fed safflower crops.

Distribution. France, India, Israel, and Russia.

Symptoms. Round spots (2–10 mm in diameter) occur on both sides of diseased leaves. Spots are characterized by a whitish, dense mass of mycelium, conidiophores, and conidia in their centers. Dried up and necrotic spots are discolored brown. Yields are reduced and seed quality is affected.

Management. Rotate safflower with other crops.

Rhizoctonia Blight and Stem Canker

Rhizoctonia blight and stem canker is also called Rhizoctonia blight.

Cause. *Rhizoctonia solani,* teleomorph *Thanatephorus cucumeris.* The optimum soil temperatures for disease development are 20° to 25°C.

Distribution. The United States (New Mexico and Texas).

Symptoms. Dark cortical lesions occur slightly below or at the soil line on diseased seedling stems. In the advanced stages of disease, lesions extend upward on stems beyond the attachment point of the lowest leaves. Lesions frequently girdle stems. Root development is reduced.

Management. Grow resistant cultivars.

Rust

Cause. *Puccinia calcitrapae* var. *centaureae* (syn. *P. carthami*). *Puccinia verruca* is also reported to be a causal fungus. *Puccinia calcitrapae* var. *centaureae* is an autoecious rust fungus whose five spore stages occur only on safflower. Teliospores are seedborne and overwinter on infested residue. In the spring, a teliospore germinates and produces a promycelium on which four basidiospores form. The basidiospores, which are violently discharged upward from diseased or infested tissue, are disseminated by wind to nearby host plants. Basidiospores germinate and pycnia are formed at the infection site. Most of each pycnium is submerged below the leaf surface. However, the opening of each pycnium is above the epidermis surface of cotyledons, hypocotyls, tender stems, and true leaves. Aecia, which form in 4 to 7 days, are in close association or closely grouped with pycnia. In gross morphology, an aecial pustule is similar to a uredinium and has been referred to as a primary uredinium and a uraecium. Uredinia are scattered over host tissue and, unlike aecia, are not closely grouped. As uredinia pustules mature, telia form in the uredinia or in separate telia. Several physiologic races occur.

Distribution. Wherever safflower is grown.

Symptoms. *Puccinia calcitrapae* var. *centaureae,* unlike most rust fungi, has two distinct pathological stages: a seedling phase and a foliage phase. The seedling phase results from infection by basidiospores originating either from soil or from seedborne teliospores. The foliage phase occurs at any growth stage and results from infection by windborne urediniospores.

Infection of seedling hypocotyls causes swelling and often the diseased stem bends and twists to one side. Chestnut-brown pustules, which commonly are aecia and rarely uredinia, form on diseased hypocotyls and less often on cotyledons and young leaves. Aecia are grouped in clusters, but uredinia are scattered. Diseased seedlings may collapse at or slightly below the soil surface and die. On older plants, a pronounced girdling and hypertrophy of stem bases occurs, which causes diseased plants to break at this point during wind- and rainstorms.

Aeciospores or urediniospores from seedling infections initiate foliar infections. Small, powdery, chestnut-brown pustules develop abundantly on the lower leaf surfaces and less abundantly on the upper leaf surfaces, cotyledons, and flower bracts. On highly susceptible varieties, pustules may reach a diameter of 1 to 2 mm and be surrounded by a secondary and frequently a tertiary ring of uredinia. Small chlorotic or necrotic flecks develop on resistant varieties, but no sporulation occurs. Toward the end of the growing season, the chestnut-brown pustules become black from the formation of teliospores in them.

Management
1. Treat seed with a seed-protectant fungicide.
2. Rotate safflower with other crops.
3. Plow under residue.
4. Flood soils for 7 days to reduce inoculum in soil.
5. Grow resistant cultivars; however, cultivars resistant to one race of *P. calcitrapae* var. *centaureae* may be susceptible to other races.

Sclerotinia Stem Rot and Head Blight

Sclerotinia stem rot and head blight is also called Sclerotinia stem rot.

Cause. *Sclerotinia sclerotiorum* survives in soil and infested residue as sclerotia. During wet weather, sclerotia germinate and form either several apothecia or mycelium. Ascospores are formed in a layer of asci on the apothecium and are disseminated by wind to plant hosts. Disease is most severe during humid conditions around the base of the host plant.

Distribution. Wherever safflower is grown.

Symptoms. Diseased leaves turn yellow, wilt, become brown, and shrivel. The entire plant may die, causing the top to curve downward. A white, cottony growth, which is the fungus mycelium, occurs on the base of the diseased stem. Sclerotia appear as black, elongated objects (2–12 mm long) on and in stems, and on adjoining roots. Cortical tissue of lower stems becomes shredded. Roots generally are not infected or discolored.

When head rot occurs, the diseased flower heads sometimes fall from the plant, leaving the outer bracts in place. Sclerotia also form in flower heads. At maturity, diseased heads contain few or no seeds, which often are replaced by a dusty-appearing residue.

Management
1. Do not include safflower in a rotation with other plants susceptible to *S. sclerotiorum.*
2. Use soil fumigation on small areas.
3. Fields may be flooded for 4 weeks, then dried up prior to planting.

Septoria Leaf Spot

Cause. *Septoria carthamicola* and *S. carthami.* The fungi are presumably seed-borne.

Distribution. Australia, Bulgaria, Canada, Morocco, Turkey, and the former USSR.

Symptoms. Tan to dark leaf spots occur on diseased leaves.

Management. Not reported.

Verticillium Wilt

Cause. *Verticillium dahliae* survives as microsclerotia in soil and infested residue. Microsclerotia germinate and produce mycelia that infect nearby host plants. Verticillium wilt is more severe during cool, wet weather. Plants may be infected at any stage of growth.

Distribution. Wherever safflower is grown.

Symptoms. Young plants are stunted. Diseased leaves are darker green than healthy leaves and may be crinkled between the veins. Cotyledons are yellowish, flaccid, and quickly dry up.

Symptoms on plants infected at a later growth stage first occur on the lower leaves. Diseased leaves may display unilateral growth because the healthy half of the leaf grows faster than the diseased half, causing the entire leaf to grow in a "C" or comma shape. The "healthy-appearing" half of a diseased leaf is a normal green, while the diseased half is tan or light brown. A characteristic symptom is chlorotic areas on leaf margins and principal veins that give leaves a mottled appearance. As diseased plants grow older, the chlorotic areas enlarge and become light tan or whitish. A dark discoloration occurs in vascular strands, particularly at the nodes, petioles, and leaf traces. Plants may die prematurely and usually do not produce viable seed.

Management. Verticillium wilt is ordinarily considered a minor disease. Do not rotate safflower with cotton, peanut, or other susceptible crops.

Disease Caused by Nematodes

Root Knot

Cause. *Meloidogyne arenaria, M. incognita,* and *M. javanica* do not survive if the temperature averages below 3°C during the coldest month. Disease is favored by warm soil temperatures.

Distribution. The United States.

Symptoms. Typical root galls form on diseased roots.

Management
1. Sources of resistance are available.
2. Do not rotate with cotton, sorghum, and sugar beet.

Disease Caused by Phytoplasmas

Phyllody

Phyllody is also called safflower phyllody.

Cause. Phyllody phytoplasma is transmitted by the leafhopper *Neoaliturus fenestratus.* Other species in the family Compositae, including tricolor chrysanthemum, *Chrysanthemum carinatum;* strawflower, *Helichrysum bracteatum;* common sunflower, *Helianthus annuus;* and cape marigold, *Dimorphotheca sinuata,* also are susceptible to the phytoplasma. Periwinkle, *Vinca rosea,* is also susceptible.

Distribution. Israel.

Symptoms. Initially, abnormal axillary budding occurs, starting when the diseased plant reaches a height of 20 to 30 cm. The main stem of a diseased plant produces numerous secondary shoots, beginning at the top and spreading rapidly throughout the entire main stem. These secondary shoots are very thin and may produce new branchings. Very small, yellowish leaves appear, and after several weeks, witches' broom symptoms develop. Many short stems appear on the infected flower heads instead of the typical tubular flowers of healthy plants. Flowers on diseased plants either have no seeds or the seeds are sterile. In contrast, healthy plants produce one main stem and few lateral ones on the upper part of the central stem and from one to five colored flower heads.

Management. Keep safflower fields free of the weed *Carthamus tenuis* because the leafhopper breeds in this plant.

Diseases Caused by Viruses

Alfalfa Mosaic

Cause. Alfalfa mosaic virus (AMV).

Distribution. The United States.

Symptoms. Dark streaks occur on diseased roots and stems and eventually encircle the stem. Mosaic symptoms may occur on leaves.

Management. Not reported.

Chilli Mosaic

Cause. Chilli mosaic virus (CMV) is transmitted from diseased to healthy plants by several species of aphids.

Distribution. India.

Symptoms. Light and dark green patches are scattered over diseased leaves.

Management. Not reported.

Cucumber Mosaic

Cause. Cucumber mosaic virus (CMV) is in the cucumovirus group. CMV is transmitted from diseased to healthy plants early in the growing season by the green peach aphid, *Myzus persicae*. CMV has a wide host range.

Distribution. Wherever safflower is grown.

Symptoms. A light and dark green mosaic pattern occurs on diseased upper leaves and bracts. Some leaves may become blistered and distorted. The mosaic pattern is less distinct on lower leaves and becomes a scattering of light and dark green flecks. Diseased plants may be stunted but usually develop to maturity.

Management. Not reported.

Lettuce Mosaic and Necrosis

Cause. Lettuce mosaic virus (LMV). LMV is mechanically transmissible through sap and by the aphid, *Myzus persicae.*

Distribution. The United States (Arizona and California).

Symptoms. Diseased plants have a terminal necrosis of stems, leaves, flower buds, and roots. Dark streaks that occur on roots and stems may eventually encircle the stem. A transverse section through the stem reveals the necrosis of cortical, xylem parenchyma, and cambium tissues.

Inoculated safflower varieties have displayed local lesions due to the development of systemic mosaic symptoms. Inoculated young plants die from necrosis but the systemic spread of the virus is delayed or does not occur in older, inoculated plants.

Management. Resistance has been identified in some safflower cultivars.

Tobacco Mosaic

Cause. Tobacco mosaic virus (TMV) is in the tobamovirus group.

Distribution. Morocco.

Symptoms. After being mechanically inoculated, small, reddish, local lesions occur on cotyledons. During temperatures of 28° to 32°C and high light intensity, necrotic ring and line patterns appear on cotyledons. Later, typical blotchy light and dark green mosaic patterns occur on all subsequent leaves until flowering.

Management. Not reported.

Turnip Mosaic and Necrosis Severe Mosaic

Turnip mosaic and necrosis is also called severe mosaic.

Cause. Turnip mosaic virus (TuMV) is in the potyvirus group. TuMV is mechanically transmitted by the aphid, *Myzus persicae.* Species of *Brassica* serve as sources for TuMV.

Distribution. The United States (the Sacramento Valley in California).

Symptoms. Diseased plants are stunted and their leaf and seed heads are reduced in size. Leaves and bracts of seed heads have small dark green and pale green areas, pale green vein banding, distortion, and bronzing. Seed ovules rot, resulting in a reduction in seed yield. Some cultivars develop necrotic lesions and do not display mosaic symptoms.

Management. Not reported.

Selected References

Ashri, A. 1961. The susceptibility of safflower varieties and species to several foliage diseases in Israel. Plant Dis. Rptr. 45:146-150.

Beech, D. F. 1960. Safflower—An oil crop for the Kimberleys. Western Australia Dept. Agric., Perth, Bull. 2720.

Chakrabarti, D. D., and Basuchaudhary, K. C. 1978. Incidence of wilt of safflower caused by *Fusarium oxysporum* f. sp. *carthami* and its relationship with the age of the host, soil and environmental factors. Plant Dis. Rptr. 62:776-778.

Classen, C. E., Schuster, M. L., and Ray, W. W. 1949. New diseases observed in Nebraska on safflower, Plant Dis. Rptr. 33:73-75.

Duniway, J. M. 1977. Predisposing effect of water stress on the severity of Phytophthora root rot in safflower. Phytopathology 67:884-889.

Erwin, D. C., Starr, M. P., and Desjardins, P. R. 1964. Bacterial leaf spot and stem blight of safflower caused by *Pseudomonas syringae*. Phytopathology 54:1247-1250.

Farr, D. F., Bills, G. R., Chamuris, G. P., and Rossman, A. Y. 1989. Fungi on Plants and Plant Products in the United States. American Phytopathological Society, St. Paul, MN. 1252 pp.

Heritage, A. D., and Harrigan, E. K. S. 1984. Environmental factors influencing safflower screening for resistance to *Phytophthora cryptogea*. Plant Dis. 68:767-769.

Howard, R. J., and Moskaluk, E. R. 1990. Seedling blight of safflower in southern Alberta. (Abstr.). Can. J. Plant Pathology 12:334.

Huang, H. C., Morrison, R. J., Nuendel, H.-H., Barr, D. J. S., Klassen, G. R., and Buchko, J. 1992. *Pythium ultimum* "group G," a form of *Pythium ultimum* causing damping-off of safflower. Can. J. Plant Pathol. 14:229-232.

Irwin, J. A. G., and Jackson, K. J. 1977. Safflower diseases in Queensland. Queensland Agric. J. 103:516-520.

Jacobs, D. L., Bergman, J. W., and Sands, D. C. 1982. Etiology and epidemiology of bacterial leaf spot and stem blight of safflower in Montana. (Abstr.) Phytopathology 72:961.

Klein, M. 1970. Safflower phyllody—A mycoplasma disease of *Carthamus tinctorius* in Israel. Plant Dis. Rptr. 54:735-738.

Klisiewicz, J. M. 1974. Assay of *Verticillium* in safflower seed. Plant Dis. Rptr. 58:926-927.

Klisiewicz, J. M. 1975. Survival and dissemination of *Verticillium* in infected safflower seed. Phytopathology 65:696-698.

Klisiewicz, J. M. 1977. Effect of flooding and temperature on incidence and severity of safflower seedling rust and viability of *Puccinia carthami* teliospores. Phytopathology 67:787-790.

Klisiewicz, J. M. 1977. Identity and relative virulence of some heterothallic *Phytophthora* species associated with root and stem rot of safflower. Phytopathology 67:1174-1177.

Klisiewicz, J. M. 1980. Safflower germplasm resistant to Fusarium wilt. Plant Dis. 64:876-877.

Klisiewicz, J. M. 1981. Isolation of turnip mosaic virus from safflower. (Abstr.) Phytopathology 71:886.

Klisiewicz, J. M. 1983. Etiology of severe mosaic and its effect on safflower. Plant Dis. 67:112-114.

Klisiewicz, J. M., Houston, B. R., and Peterson, L. J. 1963. Bacterial blight of safflower. Plant Dis. Rptr. 47:964-966.

Kolte, S. J. 1985. Diseases of Annual Edible Oilseed Crops. Volume II: Sunflower, Safflower, and Nigerseed Diseases. CRC Press Inc., Boca Raton, FL. 154 pp.

Liddell, C. M., and Duniway, J. M. 1987. Changes in water uptake and root distribution occurring during the development of Phytophthora root rot in safflower. (Abstr.) Phytopathology 77:1745.

Lockhart, B. E. L., and Goethans, M. 1977. Natural infection of safflower by a tobamovirus. Plant Dis. Rptr. 61:1010-1012.

Lopez, C. J. A., and Fucikovsky, Z. L. 1986. Safflower *Carthamus tinctorius* L. stem soft rot at Chapingo, Mexico. (Abstr.) Phytopathology 76:374.

Mortensen, K., and Bergman, J. W. 1983. Cultural variance of *Alternaria carthami* isolates and their virulence on safflower. Plant Dis. 67:1191-1194.

Mortensen, K., Bergman, J. W., and Burns, E. E. 1983. Importance of *Alternaria carthami* and *A. alternata* in causing leaf spot diseases of safflower. Plant Dis. 67:1187-1190.

Raccah, B., and Klein, M. 1982. Transmission of the safflower phyllody mollicute by *Neolaiturus fenestratus*. Phytopathology 72:230-232.

Rathaiah, Y., and Pavgi, M. S. 1971. Ability of germinated conidia of *Cercospora carthami* and *Ramularia carthami* to survive desiccation. Plant Dis. Rptr. 55:846-847.

Sackston, W. E. 1953. Foot and root infection by safflower rust in Manitoba. Plant Dis. Rptr. 37:522-523.

Sackston, W. E. 1960. *Botrytis cinerea* and *Sclerotinia sclerotiorum* in seed of safflower and sunflower. Plant Dis. Rptr. 44:664-668.

Stovold, G. 1973. *Phytophthora drechsleri* Tucker and *Pythium* spp. as pathogens of safflower in New South Wales. Australian J. Exp. Agric. and Animal Husbandry 13:455-459.

Thomas, C. A. 1970. Effect of seedling age on Pythium root rot of safflower. Plant Dis Rptr. 54:1010-1011.

Thomas, C. A., Rubis, D. D., and Black, D. S. 1960. Development of safflower varieties resistant to Phytophthora root rot. Phytopathology 50:129-130.

Thomas, C. A., Klisiewicz, J., and Zimmer, D. 1963. Safflower diseases. USDA Agric. Res. Serv. ARS 34-52.

Weiss, E. A. 1971. Diseases of Safflower in Castor, Sesame and Safflower. Barnes and Noble Inc., New York.

Zazzerini, A., Cappelli, C., and Panattoni, L. 1985. Use of hot-water treatment as a means of controlling *Alternaria* spp. on safflower seeds. Plant Dis. 69:350-351.

Zimmer, D. E. 1963. Spore stages and life cycle of *Puccinia carthami*. Phytopathology 53:316-319.

Zimmer, D. E., and Thomas, C. A. 1967. Rhizoctonia blight of safflower and varietal resistance. Phytopathology 57:946-949.

Zimmer, D. E., Klisiewicz, J. M., and Thomas, C. A. 1963. Alternaria leaf spot and other diseases of safflower in 1962. Plant Dis. Rptr. 47:643.

17. Diseases of Sorghum *(Sorghum bicolor)*

Diseases Caused by Bacteria

Bacterial Leaf Blight

Cause. *Acidovorax avenae* subsp. *avenae* (syn. *Pseudomonas avenae* and *P. albo-precipitans*) does not survive well in infested residue or soil. Vaseygrass, *Paspalum urvillei*, is considered a primary source of inoculum in Florida. Warm, rainy weather favors disease development but leaf blight will occur below 18°C. Bacteria are splashed or blown onto a host plant and enter the stalk through hail wounds or other injuries. Disease symptoms were severe following heavy rains in Illinois.

Distribution. The central and southern United States.

Symptoms. Symptoms occur on diseased leaves from late April to early June and are similar to symptoms on maize except that lesions on sorghum may be more reddish than those found on maize. Water-soaked brown lesions occur on leaves as they emerge from whorls, then the diseased leaves become elliptical-shaped and are gray or white until plant maturity. Lesions may coalesce during wet weather and form large necrotic areas that cause leaves to shred in the wind. Diseased plants may be stunted.

Management. Grow the most resistant cultivar.

Bacterial Leaf Spot

Bacterial leaf spot is also called bacterial spot.

Cause. *Pseudomonas syringae* (syn. *P. syringae* pv. *syringae*) overwinters in soil, on seed, and presumably on infested residue, johnsongrass, and sorghum plants that remain in the field after harvest. Bacteria are disseminated by wind, rain, and insects. Infection occurs through stomates and wounds. The optimum conditons for bacterial leaf spot are wet weather and a temperature of 12°C.

575

Distribution. Wherever sorghum is grown.

Symptoms. Spots first appear on lower leaves and progress upward as diseased plants approach maturity. Spots are circular to elliptical (1–10 mm in diameter). Spots are dark green and water-soaked initially, but soon become tan with a red border. Small lesions may be entirely red and have somewhat sunken centers. Numerous spots coalesce, form large diseased areas, and cause leaves to die. Lesions also occur on leaf sheaths and seeds.

Management
1. Sow healthy seed that has been treated with a seed-protectant fungicide.
2. Rotate sorghum with resistant crops.
3. Plow under infested residue.
4. Destroy johnsongrass and sorghum plants left after harvest.

Bacterial Leaf Streak

Bacterial leaf streak is also called bacterial streak.

Cause. *Xanthomonas campestris* pv. *holcicola* (syn. *Bacterium holcicola, Phytomonas holcicola, Pseudomonas holcicola,* and *X. holcicola*). *Xanthomonas campestris* pv. *holcicola* overwinters in infested residue and standing plants left in a field after harvest. In Kansas, bacteria have been reported to survive up to 24 months in seed in an unheated building. Bacteria are disseminated locally by water and wind and over long distances by residue and seed.

    Xanthomonas campestris pv. *holcicola* enters plants primarily through wounds and stomates. The bacterium is systemic in sorghum and is probably translocated in xylem tissue. Development of bacterial leaf streak is favored by warm, wet weather.

Distribution. Argentina, Australia, India, Mexico, Nigeria, New Zealand, the Philippines, Russia, South Africa, and the United States.

Symptoms. Water-soaked, translucent streaks (3 × 25–150 mm) occur on diseased leaves of all ages. Initially, light yellow drops of bacterial exudate present on translucent streaks dry to thin white or cream-colored scales. Later, red-brown blotches enlarge and become red throughout the streaks, causing the water-soaked and translucent areas to disappear. Portions of streaks may broaden into elongated oval spots with tan centers and narrow red margins. Numerous streaks may coalesce and form long, irregular-shaped areas that cover large portions of diseased leaves. Necrotic tissue between red-brown streaks is bordered by dark, narrow margins.

Management
1. Rotate sorghum with other crops.
2. Plow under infested residue.
3. Sow healthy seed treated with a fungicide.

Bacterial Leaf Stripe

Bacterial leaf stripe is also called bacterial stripe.

Cause. *Pseudomonas andropogonis* is seedborne and survives in infested residue and soil and on sorghum plants remaining in the field after harvest. Local dissemination is by wind, water, and insects; long-range dissemination is by seed or residue. Infection occurs through stomates and injuries caused by wind and insects. Optimum development of bacterial leaf stripe occurs at temperatures of 25° to 29°C and wet weather, such as cloudy, humid days following rain.

Distribution. Argentina, Australia, China, Taiwan, Nigeria, and the United States.

Symptoms. Initially, linear, interveinal lesions that are approximately 1 cm long develop on leaves. Later, the stripes that develop are restricted to interveinal areas of apical portions of the lower leaves and leaf sheaths. Depending on the diseased cultivar, stripes are a continuous color that varies from tan-red to brick red to dark purple–red. Stripes, which vary in length from 25 mm to the length of the leaf blade, sometimes coalesce and cover large areas of diseased leaves. Water-soaking of tissue adjacent to lesions normally does not occur.

Bacterial exudate found on the underside of affected portions of leaves and along leaf margins dries and forms red crusts or thin scales that are washed off by rain. Lesions also occur on kernels, peduncles, and rachis branches and in the interior of stalks.

Management
1. Rotate sorghum with resistant crops.
2. Sow healthy seed treated with a seed-protectant fungicide.
3. Delay sowing to avoid wet climatic conditions.
4. Control weeds, such as johnsongrass.
5. Plow under residue.
6. Grow resistant or tolerant sorghum cultivars.

Erwinia Soft Rot

Cause. *Erwinia chrysanthemi.* Mode of entry is by contaminated water entering whorls of young, growing plants, a situation that occurs during mild flooding. Development of Erwinia soft rot is favored by temperatures of 30°C and higher.

Distribution. The United States (Nebraska).

Symptoms. Symptoms appear in about the fourth-leaf stage of growth. Severe symptoms include death of the top four to five leaves, but the lower leaves appear normal. Dead tissue can be easily pulled from whorls. The inside of whorls is composed of wet, necrotic tissue. An unpleasant odor is associated with this "wet" tissue. Less severe symptoms include necrotic stripes, pigmented stripes, or blotches on the upper leaves without stalk rot or top necrosis.

Management. Not reported. Erwinia soft rot is not thought to pose a significant threat except in poorly drained fields where prolonged periods of high temperatures occur during the early stages of seedling development.

Pseudomonas fuscovaginae Disease

Cause. *Pseudomonas fuscovaginae* is an opportunistic pathogen whose development is favored by low temperatures above an altitude of 1300 m.

Distribution. Burundi.

Symptoms. Symptoms occur at the boot stage of plant development. Initially, brown-black water-soaked spots (2 mm in diameter) occur on the adaxial side of leaf sheaths; later they coalesce and become as long as 20 cm. Inside the leaf sheaths, light brown patches with withered centers and sharp, reddish-brown borders occur on the abaxial side. Occasionally there is an inhibition of the panicle emergence, and large grayish, water-soaked lesions occur on the diseased flag leaf sheath.

Management. Not reported.

Yellow Leaf Blotch

Cause. *Pseudomonas* sp. is seedborne.

Distribution. West Africa.

Symptoms. Scattered cream to yellow to beige lesions (8–10 × 25–35 mm) occur on diseased leaves. Occasionally lesions are 30 mm wide by 200 mm long, slightly water-soaked, and have a tendency to follow leaf veins. Young plants infected at the two- to three-leaf stages sometimes are severely stunted and killed.

Management. Not reported.

Diseases Caused by Fungi

Acremonium Wilt

Cause. *Acremonium strictum* (syn. *Cephalosporium acremonium*). Infection occurs in the leaf blade or sheath, where systemic colonization occurs. Soilborne infection has been reported from Egypt.

Distribution. Argentina, Egypt, Honduras, India, and the United States.

Symptoms. Diseased plants are stunted and chlorotic. The vascular system has a brownish discoloration that extends from the leaf sheaths into the vascular bundles of stalks. The lower leaves become chlorotic initially, then develop red-brown streaks and eventually die. Sometimes large areas of wilted tissue develop along one axis of diseased leaves on either side of a midrib. In severely diseased plants, the upper leaves and shoots die and foliar wilting occurs.

Management. Most cultivars are resistant.

Anthracnose Leaf Blight

Anthracnose is also called red leaf spot.

Cause. *Colletotrichum graminicola,* teleomorph *Glomerella graminicola,* is seed-borne, surviving as conidia and mycelia in seed. Diseased plants are produced from seed with no visible symptoms, such as the presence of acervuli. Conidia and mycelia also survive in infested residue, soil, johnsongrass, and other weeds. In Texas, sclerotia have been reported to survive for 18 months in stalk residue on the soil surface. Germination of sclerotia decreased faster in stalk residues buried at a depth of 10 and 20 cm in soil than in infested residue placed on the soil surface, suggesting the sclerotia on the soil surface may act as a primary source of inoculum for initiating anthracnose in the field.

Spores are produced during warm, wet weather in midsummer. During moist weather later in the growing season, conidia produced in lesions are disseminated by wind to other plant hosts. In Africa, *C. graminicola* has been reported to rapidly destroy sorghum plants as the diseased plants approach maturity. Different physiologic races or pathotypes exist.

Distribution. Wherever sorghum is grown, but anthracnose is most severe in warm, humid climates.

Symptoms. Symptoms occur on leaves at approximately midsummer as well-defined circular to oval-shaped spots that normally are 3 to 6 mm in diameter but may grow as large as 25 mm in diameter. Spots develop small, circular, light tan centers with wide tan, orange, red, or black-purple margins, depending upon the diseased cultivar. During periods of high humidity, spots increase in number and cover much of the diseased leaf area; elliptical to elongate lesions may cover the entire length of the midrib. Blackish growths that may be present in spots are setae, which under magnification appear to be short, stiff hairs. Pink spore masses may develop in these blackish areas during humid weather.

Diseased plants may eventually become defoliated and, under severe disease conditions, die before maturity. Yield losses for susceptible cultivars have been reported to be 50% in a severe anthracnose epiphytotic. On broomcorn, production of heads may be poor and the sugar content of sweet sorghum may be reduced.

The stalk rot phase is discussed in "Red Rot" later in this chapter.

Management
1. Grow resistant cultivars. Dilatory resistance is identified in commercial sorghum hybrids
2. Sow healthy seed that has been treated with a seed-protectant fungicide.
3. Control weeds in a field.
4. Plow under residue.

Banded Leaf and Sheath Blight

Banded leaf and sheath blight is also called Rhizoctonia blight.

Cause. *Rhizoctonia solani* AG-1, teleomorph *Thanatephorus cucumeris*. The primary inoculum originates from mycelia and sclerotia in infested residue or soil. Secondary spread of inoculum is by mycelia contacting host plant leaves or by mycelia and sclerotia disseminated by wind and splashing rain. Basidiospores may also be important in initiating disease. Development of banded leaf and sheath blight occurs during moist conditions with high relative humidities and high temperatures.

Distribution. Tropical areas of the world, but banded leaf and sheath blight is not considered an important disease.

Symptoms. Leaves and sheaths initially develop water-soaked, gray-green lesions that eventually develop into irregular tan to red-brown lesions (2–8 mm wide) with white centers and purple borders. One or several lesions form horizontal bands across diseased leaf blades and leaf sheaths. Lesions may be covered with fluffy, white mycelium or brown mycelial mats of sclerotia. On older lesions, individual sclerotia look like white to dark brown objects that are 1–5 mm in diameter.

Management. Do not rotate sorghum with rice or soybean.

Charcoal Rot

Cause. *Macrophomina phaseolina* (syn. *Botryodiplodia phaseoli, M. phaseoli, Rhizoctonia bataticola,* and *Sclerotium bataticola*) survives primarily as sclerotia in infested residue. Although pycnidia do not commonly occur, they serve as another means of overwintering. *Macrophomina phaseolina* is not a good soil saprophyte and is inhibited in saprophytic growth in residue by other soil microflora. However, growing successive crops of sorghum builds up inoculum, in the form of sclerotia, in the soil. In India, sclerotia counts increased when sorghum was intercropped continuously with pigeon pea in rainy and dry seasons, and when sorghum followed safflower and chickpea in the rainy season. Charcoal rot generally occurs in localized areas. Therefore, it is likely that inoculum is primarily sclerotia disseminated by any means that moves soil or residue rather than propagules that may be disseminated a long distance by wind.

During soil temperatures of 30°C and above and available soil moistures of 80% or less, sclerotia germinate and form infection hyphae that penetrate belowground tissue of host plants. Stress created by drought is necessary for infection by *M. phaseolina*.

Disease severity increases during high nitrogen conditions and high plant populations growing in moisture stress conditions. A high plant population increases competition for available soil moisture, while high nitrogen promotes luxuriant shoot growth relative to root development. Therefore, under moisture stress the lack of a sufficient root system reduces the ability of plants to obtain moisture when the water requirement is increased by luxuriant shoot growth. However, in reports from the Philippines, deep plowing and proper weeding limited the severity of infection regardless of nitrogen, phosphorous, or potassium levels and population density.

Distribution. Wherever sorghum is grown under hot, dry conditions.

Symptoms. Charcoal rot first occurs on host plants growing in the driest soil in a field. Diseased plants may be destroyed in 2 to 3 days from the onset of symptoms. Symptoms move upward on a diseased stalk from the original infections on the crown. After the initial general water-soaking of piths, diseased stalk tissues turn bright red to black, then die and the color fades. The lower stems dry up and shred and numerous small black sclerotia form on the remaining vascular bundles. Diseased plants then lodge.

Management
1. Sweet sorghums and many forage sorghums are resistant, but grain sorghums tend to be susceptible. Resistance is associated with exodermis lignification.
2. Rotate sorghum with other crops.
3. Sow the recommended plant population.
4. Follow cultural practices that conserve soil moisture.

Cochliobolus Leaf Spot

Cause. *Cochliobolus bicolor.*

Distribution. India.

Symptoms. Lesions on diseased leaves are circular and have alternating concentric straw-colored bands. In severe cases, spots coalesce and cover a large area of the leaf surface.

Management. Not reported.

Covered Kernel Smut

Cause. *Sporisorium sorghi* (syn. *Sphacelotheca sorghi*) overwinters primarily as teliospores (chlamydospores) on the outside of infested seed. Teliospores germinate in two ways: (1) When contaminated seed is sown, teliospores germinate along with the seed and form a four-celled promycelium that bears lateral sporidia. Sporidia then germinate, infect developing seedlings, and form mycelia that grow systemically within the infected plants. (2) Teliospores germinate by directly producing germ tubes that infect seedlings.

At the heading stage of plant development, kernels are replaced by teliospores enclosed in a membrane which ruptures at harvest time and releases teliospores that contaminate seed or soil. Teliospores in soil are not important in the infection of seedlings. Smut incidence decreases when seed is planted in wet soils at temperatures of 15.5° to 32.0°C. Several physiologic races of *S. sorghi* exist.

Distribution. Wherever sorghum is grown.

Symptoms. Growth of plants is not affected and diseased plants appear normal until the heading stage of plant development. Although an occasional healthy kernel may occur, usually all kernels are replaced by dark brown, powdery masses of teliospores enclosed in gray or brown membranes. The teliospores and membrane is called a sorus. Sori vary in size from being small enough to be covered by glumes to being over 1 cm long. The membrane ordinarily remains intact until harvest time, when it is ruptured by harvest equipment.

Management
1. Treat seed with a protectant fungicide.
2. Sow seed from smut-free plants.
3. All commercial cultivars of sorgo, kaffir, durra, sudangrass, and broomcorn are susceptible. Hegari, milo, gurno, and feterita types of sorghum are resistant to one or more races.

Crazy Top

Cause. *Sclerophthora macrospora* (syn. *Sclerospora macrospora*) survives as oospores in residue and other grasses. Oospores germinate in saturated soil in 24 to 48 hours and form sporangia in which zoospores are produced. Zoospores are then liberated from sporangia and "swim" to preemergent or postemergent seedlings, where they germinate, infect the host plant, and produce systemic mycelium inside the infected plant. Oospores are produced in and adjacent to the veins of diseased leaves. Crazy top occurs over a wide range of temperatures.

Distribution. Wherever sorghum is grown.

Symptoms. Diseased leaves are mottled, thickened, stiffened, and twisted or curled, and their surfaces are covered with bumps. Although sorghum heads usually are not produced, some may have numerous leafy shoots.

Management. Not necessary, since crazy top is not serious on sorghum.

Curvularia Kernel Rot

Cause. *Curvularia lunata.* Infection is favored by rain, high humidity, and temperatures of 24° to 30°C.

Distribution. Mexico.

Symptoms. Symptoms are observed on diseased spikelets at maturity. Kernels are covered with a dense, "velvet-appearing" fungal growth consisting of mycelium, conidia, and conidiophores.

Management. Not reported. However, sources of resistance to general grain molds have been identified in some accessions.

Ergot

Ergot is also called honeydew and sugary disease.

Cause. *Claviceps sorghi* and *C. africana. Sphacelia sorghi* is considered the anamorph of both *C. sorghi* and *C. africana.*

Sexual reproduction by *C. africana,* or the formation of sclerotia that produce stromatic heads, has only been observed in laboratory experiments and is believed to be unimportant to its survival in nature. The dispersal of secondary conidia over long distances (300 km/day) is believed to be the primary reason for the rapid spread of *C. africana* through the Western Hemisphere. In the absence of sclerotia, survival of *C. africana* in the United States may only occur in southern sorghum-producing areas, thereby reducing the incidence of ergot in the northern sorghum-producing areas.

Male sterility or fertility problems maximize susceptibility through lengthening the time a sorghum flower ovary remains unfertilized. Commercial grain sorghum hybrids have negligible incidence because they are highly self-fertile and rapid pollination prevents infection. Male-sterile (female parent) sorghums are generally the sorghums most heavily damaged by ergot because they require an external pollen source for fertilization. Pollen sterility due to causes such as cool temperatures encourages infection.

Claviceps sorghi sclerotia are present but do not readily germinate in nature. Typically, most ergots germinate and produce stromatic heads at ends of stipes (stalks). Perithecia form in stromatic heads, then ascospores produced in perithecia are disseminated by wind to host florets. However, because the ascospores are not readily formed in nature, their role in ergot epidemiology is unknown. Development of the anamorph, *S. sorghi* also is little understood. It has been suggested that conidia from alternate grass hosts and infested panicle debris in soil may serve as primary inoculum.

Infection occurs through unfertilized florets. Infection of florets is favored by temperatures of 20° to 25°C and high humidity from the time of panicle emergence through the fertilization of ovaries. Conidia germinate and produce germ tubes that penetrate the stigma, grow down the style, and colonize the ovary. The ovary is converted into a fungal mass and stroma; approximately 7 to 10 days after infection, a sweet, sticky substance (honeydew) consisting of liquid and conidia exudes from diseased florets during rainy, cloudy weather.

Honeydew contains three types of conidia: macroconidia, secondary conidia, and microconidia. Macroconidia, which are elliptical-shaped, are the first to be released in the honeydew. Under humid conditions some macroconidia on the surface of the honeydew germinate by germ tubes that enmesh and form a hyphal mat. Others germinate and form erect conidiophores on which pyriform secondary conidia form outside the honeydew surface. Later, small obovate microconidia are found in the honeydew. All three conidial forms germinate on and penetrate the stigma.

Stromata develop at temperatures of 14° to 35°C. Honeydew and conidial production occur at temperatures of 14° to 28°C and relative humidities above 90% for 12 to 16 hours a day. Sclerotia develop at temperatures of 28° to 35°C and relative humidities below 90% for 2 hours a day. Honeydew may be disseminated from flower to flower by insects, rain, and wind. Later, the stroma is transformed into a hard sclerotium.

Honeydew also contaminates the grains, and above 90% relative humidity, stromata and honeydew become colonized by saprophytic fungi. Consequently, sclerotia are not formed, but after honeydew dries, secondary conidia of *S. sorghi* are spread primarily by wind and germinate during rainy, cloudy weather.

Long-range spread possibly is by sclerotia mixed with seed.

Distribution. *Claviceps sorghi* causes ergot in India. *Claviceps africana* causes ergot in Australia, Central America, North America, South Africa, and South America.

Symptoms. Initially, drops of sticky, pinkish to brownish liquid (honeydew), which is sweet to the taste and attracts numerous insects, exudes from ovaries. The saprophytic fungus *Cerebella volkensii* may grow over the honeydew and convert it into a sticky, matted, black mass. If the mycoparasite is not present, fungal tissue in ovaries grows into hard, grayish, elongated and slightly curved, horn-like sclerotia.

Ergot is an ovary replacement disease that creates two major problems. First, infection of male-sterile (A-line) ovaries precludes normal seed-set or development. Thus, infections are directly responsible for seed yield losses. Second, seeds developing adjacent to infected florets often become contaminated with the sugary, sphacelial honeydew and support growth of saprophytic surface molds. Such seeds are of poor and downgraded quality and because of their stickiness may be difficult to remove during harvesting.

Management
1. Sow seed lots free of sclerotia.
2. Locate seed fields in areas not conducive to disease.
3. Apply a foliar fungicide from the flag leaf stage to the end of anthesis.
4. Practice pollen management techniques.
5. Ergot-resistant lines from Ethiopia have been identified.

False Mildew

Cause. *Beniowskia sphaeroidea.* Disease is favored by humid weather.

Distribution. Malawi and Uganda.

Symptoms. Typical white growth consisting of mycelium, conidia, and conidiophores occurs on diseased leaves.

Management. Not reported.

Fusarium Head Blight

Cause. *Fusarium moniliforme,* teleomorph *Gibberella fujikuroi.* According to Nelson et al. (1983), *F. subglutinans* and *F. proliferatum* share many morphological characteristics with *F. moniliforme* and may be involved in the same disease syndrome. *Fusarium subglutinans* is separated taxonomically from *F. moniliforme* because *F. subglutinans* bears microconidia on polyphialides only in false heads and not on microconidial chains. *Fusarium moniliforme* bears microconidia both in false heads and in long chains. *Fusarium proliferatum* is separated from *F. moniliforme* in that *F. proliferatum* also bears microconidia in chains of various lengths and in false heads but it forms microconidia in chains on polyphialides, whereas *F. moniliforme* forms microconidia in chains on monophialides. It is likely these fungi have been misidentified or confused with each other in the past. Some researchers still consider the three fungi synonymous, presumably on the basis of their morphology. It has been reported that *Fusarium napiforme* also has been isolated from sorghum heads.

   Gibberella fujikuroi consists of at least seven distinct mating populations, which are designated A through G. Populations of A, D, and E are the most common on maize and correspond to the particular anamorph species. Both A and F mating populations are considered to be *F. moniliforme*; population A contains many prolific fumonisin-producing strains, while members of the F population produce little or no fumonisin. Most strains of *F. moniliforme* found on maize are population A and most found on sorghum are F. Population D is considered to be *F. proliferatum* and contains many strains that produce copious amounts of fumonisins. *Fusarium proliferatum* was demonstrated to produce beauvericin, fumonisin B 1, and moniliformin in maize grown in Italy. Populations B and E are considered to be *F. subglutinans* and produce little or no fumonisin. In maize grown in temperate regions, *F. moniliforme* and *F. proliferatum* represent the greatest threat for fumonisin production.

Although little is known of the survival or dissemination of *F. moniliforme* as it relates to head blight of sorghum, it is likely similar to Fusarium stalk rot of maize. Therefore, *F. moniliforme* probably overwinters as mycelia in sorghum or maize residue. Conidia are then produced and disseminated by wind to peduncles, where they enter through cracks or wounds in the peduncle, rachis, or panicle branches. Prolonged wet weather preceding anthesis favors Fusarium head blight.

Distribution. Areas where sorghum is grown under warm, humid conditions. *Fusarium napiforme* has been reported from Namibia.

Symptoms. Several to all the florets in a diseased seed head may be killed. Whole seed heads may be covered with a white to pinkish mycelial growth. When panicles are split, a red, brown, or black discoloration occurring in the upper peduncles extends into branches of heads. Sometimes the discoloration extends into stalks and discolors the rind. Peduncles may be necrotic and break over. The rachis and rachis branches may become necrotic or have an external red discoloration.

Management. Grow resistant cultivars. Resistance to general grain molds has been identified in some accessions.

Fusarium Root and Stalk Rot

Cause. *Fusarium moniliforme* is considered the dominant causal organism, but other *Fusarium* spp., specifically *F. graminearum,* also have been reported as causal fungi. The fungi survive as chlamydospores, chlamydospore-like structures, and mycelia in maize or sorghum residue. *Fusarium moniliforme* is also seedborne. Conidia are produced on residue and disseminated by wind to host plants. Mycelium also grows through soil to roots and enters through natural openings or wounds made by insects, machinery, and other means.

Near plant maturity, host plants must be predisposed by insect wounds, disease injury, poor fertility, or hot, dry weather for Fusarium root and stalk rot to occur. Usually disease is most severe during cool, wet weather following hot, dry weather as plants near maturity.

Fusarium root and stalk rot is more severe in conventional tillage than in ecofallow systems. High nitrogen to potassium ratios and high plant populations also increase disease severity. In the Philippines, however, deep plowing and proper weeding limited disease severity regardless of nitrogen, phosphorous, or potassium levels and population density.

Distribution. Wherever sorghum is grown under hot, dry conditions. In the United States, the area from Kansas to Texas is most severely affected.

Symptoms. Although stalk rot is ordinarily accompanied by root rot, symptoms may not be observed when irrigated plants are grown in soils with adequate fertility. Root cortical tissue, then the vascular tissue decompose. Older roots are often totally destroyed, leaving diseased plants with little anchorage. Such plants are easily pulled from soil or toppled over by wind. Newly formed roots have lesions of various sizes and shapes that vary in color from light tan to brick red to pink to black, depending on the diseased cultivar.

Stalk rot symptoms resemble those of maize. The fungi grow from diseased roots and crowns into the plant stalk. Initially, leaves turn brown and the outsides of nodes have dark lesions. Inside the stalk, the pith is water-soaked initially and later disintegrates, leaving only pink, red, or red-purple vascular bundles intact. Eventually, diseased interior stalk tissue becomes dark red. White to pink mycelia may grow on the outsides of nodes during damp weather. Stalks may break over near the soil surface.

Management
1. Provide a full moisture profile at sowing and ensure adequate irrigation during the growing year.
2. Practice good weed and insect control.
3. Maintain a balanced soil fertility.
4. Grow recommended plant populations of hybrids that have good stalk strength.
5. Practice an ecofallow system, especially in a winter wheat–grain sorghum–fallow rotation.

Fusarium thapsinum Infection

Cause. *Fusarium thapsinum,* teleomorph *Gibberella thapsina. Fusarium thapsinum* and *F. moniliforme* are similar morphologically and may have been confused in the literature. Distinguishing *F. thapsinum* from *F. moniliforme* is accomplished by crosses with standard female-fertile tester strains. *Fusarium thapsinum* produces a yellow pigment in culture; *F. moniliforme* does not. *Fusarium thapsinum* does not produce fumonisins, but *F. moniliforme* does. *Fusarium thapsinum* produces moniliformin, whereas *F. moniliforme* does only rarely. *Fusarium thapsinum* is sensitive to hygromycin; *F. moniliforme* is not. Therefore, the numerous reports in the literature that assign the predominant *Fusarium* spp. associated with head blight of sorghum to *F. moniliforme* and consider this organism to be the same one responsible for stalk and ear rot in maize may be erroneous.

The nonproduction of fumonisins or the production of large amounts of moniliformin by isolates of *F. moniliforme* from sorghum probably refers to strains of *F. thapsinum* rather than to strains of *F. moniliforme.* This difference also suggests that sorghum is more likely to be naturally contaminated with moniliformin than with fumonisins and maize is more likely to be naturally contaminated with fumonisins instead of moniliformin.

Distribution. Egypt, South Africa, the Philippines, Thailand, and the United States (Alabama, Arkansas, Georgia, Illinois, Kansas, Mississippi, Missouri, Ohio, and Texas).

Symptoms. *Gibberella thapsina* has been reported to cause a seedling blight. Symptoms consisted primarily of the formation of reddish-brown lesions at the bases of root laterals and, after 3 days, death of the infected lateral. On some plants the entire root system was necrotic.

   Fusarium thapsinum was isolated from asymptomatic seeds, stems, roots, and crowns, which suggests *F. thapsinum* may be endophytic. The pathogenicity has not been tested under field conditions, but under greenhouse conditions, some strains can cause lesions in stalks following toothpick inoculations.

Management. Not reported.

Gray Leaf Spot

Gray leaf spot is also called angular leaf spot and cercosporiosis.

Cause. *Cercospora sorghi* survives as mycelia in infested residue, in sorghum plants that remain standing after a field has been harvested, in grasses as a parasite and saprophyte, and in infested seed. Primary inoculum is conidia produced during warm, wet weather and disseminated by wind and splashed by rain to plant hosts. Later in the season, conidia produced in lesions provide inoculum for secondary infections.

Distribution. Wherever sorghum is grown in warm, humid, or wet climates.

Symptoms. Gray spot normally occurs too late in the growing season to cause injury, but if humid weather persists in midseason, considerable damage may occur. Initially, small, circular to elliptical, dark purple or red spots occur on diseased leaves late in the growing season after plants are mature. Later, spots elongate to 30 mm until they are limited by leaf veins and become rectangular. Spots may be light to dark red, purple, or tan. Thick, gray mycelia, together with numerous conidia, eventually cover the spots; hence, the name gray leaf spot. Sporulation is more prevalent in spots on the diseased lower leaf surfaces. During periods of high humidity, large areas of leaves, including the spots, are covered with a thick mycelial growth. In severe disease conditions, leaf sheaths and upper stems are also affected.

Management
1. Grow tolerant or resistant cultivars.
2. Rotate sorghum with small grains.
3. Plow under infested residue.
4. Eliminate wild sorghum and grasses that act as pathogen hosts.

Green Ear

Cause. *Sclerospora graminicola* overwinters as oospores in soil and on seed. Oospores germinate and form germ tubes that infect new leaves. Temperatures of 24° to 32°C for 2 days after sowing are conducive to heavy infection. Secondary spread is by the dissemination of sporangia that formed on diseased leaves by wind to host plants. Sporangia germinate indirectly and release three or more zoospores, which then germinate and form germ tubes that penetrate host leaves. Sporangia infrequently germinate directly and form a germ tube. Oospores eventually form in diseased tissue and return to the soil or contaminate seed when the diseased tissue decays. Optimum development of green ear occurs with a high humidity and a temperature of 17°C.

Distribution. Africa, Asia, Europe, and North America.

Symptoms. The head is completely or partially transformed into a mass of small, twisted leaves; hence, the name green ear. The malformation is due to upper segments of the floral axis that normally develop into grain being converted into small, leafy shoots. Sporangia that resemble a "fuzz" are produced on diseased leaves during periods of high humidity.

Management. Not necessary.

Head Smut

Head smut is also called ear and tassel smut.

Cause. *Sporisorium holci-sorghi* (syn. *Sphacelotheca reiliana, Sorosporium reilianum,* and *Ustilago reiliana*), according to Farr et al. (1989). *Sporisorium holci-sorghi* survives as chlamydospores in soil. When seed is sown, chlamydospores in the dry soil germinate at temperatures of 27° to 31°C and form four-celled or branched promycelia that bear sporidia terminally and near septa. Sporidia may then sprout and form secondary sporidia or germinate and form germ tubes that penetrate meristematic tissue of seedlings. *Sporisorium holci-sorghi* develops only in actively growing meristematic tissue. Seedborne chlamydospores apparently are not important in infection. Physiologic races of *S. holci-sorghi* are present.

Distribution. Generally wherever sorghum is grown.

Symptoms. Disease symptoms first appear when a young head enclosed in a boot is completely replaced by a large smut gall surrounded by a whitish membrane. The membrane ruptures before the head emerges and exposes a mass of dark brown to black, powdery chlamydospores intermingled with a network of long, thin, dark, broom-like structures. Sometimes chlamydospores are not present, but witches' brooms consisting of many small, rolled leaves protrude from heads of suckers at nodes or joints.

All or part of the panicle becomes incorporated into a single sorus. Parts of a diseased panicle that are not included in a sorus usually are blasted or have a proliferation of individual florets. Sori may develop on the foliage and culms in some sweet sorghums and sudangrass cultivars.

Management
1. Grow resistant cultivars and hybrids.
2. Rotate susceptible sorghum cultivars and hybrids with resistant crops. Grow susceptible sorghum in the same field every 4 years to genetically stabilize the race or races of *S. holci-sorghi*.

Ladder Leaf Spot

Confusion about the common name exists in the literature. It is sometimes referred to as "latter" leaf spot.

Cause. *Phaeoramularia fusimaculans* (syn. *Cercospora fusimaculans*), according to Farr et al. (1989).

Distribution. Brazil, Colombia, Cuba, El Salvador, Honduras, Malawi, Mexico, Rwanda, the United States (Georgia and Texas), Venezuela, and Zambia.

Symptoms. Sclariform, or ladder-like, lesions occur on diseased leaves. Lesions are elliptical to rectangular, pale brown, and segmented by dark-bordered markings that are limited by secondary and tertiary veins and give lesions a "ladder-like" appearance.

Management. Levels of resistance exist in *Sorghum bicolor*.

Leaf Blight

Leaf blight is also called Helminthosporium leaf blight, Helminthosporium blight, and northern leaf blight.

Cause. *Exserohilum turcicum* (syn. *Bipolaris turcica, Drechslera turcica,* and *Helminthosporium turcicum*), teleomorph *Setosphaeria turcica* (syn. *Trichometasphaeria turcica*). *Exserohilum turcicum* is seedborne and overwinters in infested residue and soil as mycelium and as conidia that may be converted into chlamydospores. The optimum production of conidia on infested residue at the soil surface occurs during moist weather at temperatures of 18° to 27°C; conidia are disseminated by wind to plant hosts. Heavy dews particularly favor development of leaf blight. Seed rot and seedling blight can occur in cold, wet soil. During humid weather, secondary inoculum is conidia that are abundantly produced in lesions and disseminated by wind to other hosts.

Distribution. Wherever sorghum is grown, but leaf blight is not common in drier climates.

Symptoms. Infected seedlings either die or develop into stunted plants. The small red-purple or yellow-tan leaf spots that develop on leaves of diseased seedlings enlarge, coalesce, and turn the leaves a purple-gray color.

Spots on older leaves occur as yellowish, grayish, or tan, elliptical lesions several centimeters long. Lower leaves are infected first and the disease symptoms progress up the plant. During humid conditions, lesions enlarge and kill and wither large parts of leaves, giving diseased plants the appearance of being injured by frost. Depending on the cultivar, the centers of lesions are usually grayish to tan and have a red-purple or tan margin. In warm, humid weather, profuse sporulation gives lesions a dark green or black appearance. Grain is not infected but yields are reduced.

When seedlings were inoculated, *Setosphaeria turcica* caused rot of roots, crowns, and stalks. Leaves eventually wilted and plants died.

Management
1. Grow resistant cultivars. Most grain and sweet sorghum cultivars have some resistance. In humid areas of the southern United States, most cultivars will sustain some leaf blight. Two types of resistance are known: polygenic resistance is characterized by a few small lesions, and monogenic resistance is characterized by a hypersensitive fleck with little or no lesion development.
2. Infested residue should be plowed into soil. Leaf blight is more severe under minimum-tillage conditions.

Leaf Spot

Although both leaf spot and red spot are reported to be caused by *Exserohilum rostratum*; the descriptions of the symptoms are sufficiently different to warrant describing them as two different diseases.

Cause. *Exserohilum rostratum* (syn. *Helminthosporium rostratum*) survives as mycelium in infested residue and may be seedborne. The production of conidia on residue is optimum during wet conditions at a temperature of 30°C; conidia are windborne to hosts. Conidia produced in spots during warm, wet conditions provide secondary inoculum.

Distribution. Africa and the United States, particularly in warm areas.

Symptoms. Older leaves are more susceptible to infection than young leaves. Spots (1–2 × 2–5 mm) with tan centers and purple margins develop on diseased leaves. The spots are limited laterally by leaf veins but may coalesce and form large necrotic areas.

Management. Leaf spot is of minor importance on sorghum and no disease management is necessary.

Long Smut

Cause. *Tolyposporium ehrenbergii* survives as teliospores that are disseminated by seed, soil, and wind. However, infection only occurs from airborne spores during flowering, and not from seed or soilborne spores or mycelium. Spores on the soil surface are the most important initial source of infection when they are picked up by wind during dry periods, deposited on flag leaves, and then washed down behind the leaf sheaths. From boot to anthesis stages of plant growth, teliospores germinate directly and produce germ tubes or germinate indirectly and produce sporidia that infect floral parts. Seed or seedling infection does not occur.

Distribution. Asia, Botswana, Malawi, Tanzania, Zambia, and Zimbabwe.

Symptoms. Sori in ovaries are irregularly distributed, somewhat cylindrical-shaped (0.5–1.0 × 4.0 cm) with tapered ends, and surrounded by a firm false membrane of fungal tissue.

Management. Not practiced, but resistant germplasm for breeding programs exists.

Loose Kernel Smut

Cause. *Sporisorium cruentum* (syn. *Sphacelotheca cruenta*) overwinters as teliospores on the outside of seed. Teliospores also survive in soil, but these are apparently not important in seedling infection. When seed is sown, teliospores germinate, at the same time seed does, and form four-celled promycelia that bear lateral sporidia. The sporidia germinate and infect the developing seedling. Optimum infection occurs over a wide range of soil moistures and pHs at temperatures of 20° to 25°C. Mycelium grows systemically in the plant. At heading, kernels are replaced by teliospores enclosed in a membrane that soon disintegrates and releases the spores. Secondary infection may occur when the released teliospores infect late-developing heads, causing them to become smutted. Physiologic races exist.

Distribution. Wherever sorghum is grown, but loose kernel smut is not a common disease.

Symptoms. Diseased plants are stunted and head out early. Abundant side branches and tillers may develop. Normally all kernels in an infected panicle are smutted, but some kernels may be transformed into leafy structures or escape infection completely. Individual kernels are replaced by smut galls that are 2.5 cm or longer, pointed, and covered with a thin membrane that usually breaks when galls are full size. The powdery, dark brown to black teliospores are soon blown away, leaving a long, dark, pointed, curved structure called a columella in the center of what was the gall. Sometimes the primary head is not smutted, but the tillers or the ratoon crop are.

Management
1. Treat seed with a protectant fungicide.
2. Sow seed from smut-free plants.
3. All commercial cultivars of sorgo, kaffir, durra, sudangrass, and broom-corn are susceptible. Hegari, milo, gurno, and feterita types of sorghum are resistant to one or more races.

Milo Disease

Milo disease is also called Periconia root rot and milo root rot.

Cause. *Periconia circinata* survives as conidia in soil and as chlamydospores or thick-walled mycelial cells in roots and soil. Chlamydospores and conidia are disseminated by any process that moves or transports soil. Chlamydospores germinate and produce both conidiophores, on which conidia are borne, and infection hyphae that penetrate roots. Conidia also germinate and produce infection hyphae. *Periconia circinata* produces a toxin that kills plant tissue in advance of mycelial growth. Chlamydospores are eventually produced in diseased tissue and returned to the soil during harvest and upon decomposition of infested plant tissue.

A pathogenic population of *P. circinata* depends on continuous growth of susceptible sorghum cultivars. The populations of toxin-producing isolates are higher in the roots of diseased susceptible genotypes than in the soil. However, pathogenic toxin-producing isolates are not isolated from the roots of resistant genotypes.

Distribution. Primarily the sorghum-growing areas of the southern United States.

Symptoms. The first symptoms occur 3 to 4 weeks after sowing susceptible sorghum seeds into heavily infested soil. Initially, plants are stunted and there is a slight rolling of diseased leaves; the older leaves turn a light yellow at the tips and margins. Leaves continue to yellow and dry up until all leaves are affected and plants die at heading. Sometimes symptoms do not appear until a rapid development of symptoms occurs at heading. A dark red color occurs in the centers of diseased stalks and extends into roots. Roots may become severely rotted and slough from the rest of the plant. Diseased plants usually produce smaller heads.

Management. Grow resistant cultivars and hybrids.

Oval Leaf Spot

Cause. *Ramulispora sorghicola* survives as sclerotia associated with leaf residue on the soil surface. After periods of high humidity or rain, conidia are produced and disseminated by wind or splashed by rain to host leaves. Plants are infected at all stages of growth.

Distribution. Africa, Asia, the Caribbean, and Central America.

Symptoms. Initially, small, water-soaked spots with tan, brick-red, or purple borders occur on diseased leaves. The small spots develop into roughly circular-shaped lesions (1.5–4.0×3.0–8.0 mm) with pink-gray to yellowish centers and dark red or tan borders (1 mm wide). Sometimes a few scattered, small, black sclerotia appear in spots on the underside of diseased leaves.

Oval leaf spot can be confused with anthracnose leaf blight. The diseases may be distinguished by the setae. In oval leaf spot, the setae are clear brown and arise from sclerotia. In anthracnose leaf blight, setae are black and do not arise from sclerotia.

Management. Rotate sorghum with resistant crops.

Panicle and Grain Anthracnose

Cause. *Colletotrichum graminicola,* teleomorph *Glomerella graminicola,* is seedborne and survives in infested residue, soil, johnsongrass, and other weeds. Spores are produced during warm, wet weather in midsummer. During moist weather later in the growing season, conidia are produced in lesions and disseminated by wind to other plant hosts. Different physiologic races or pathotypes exist.

Panicle and grain anthracnose occurs on mature plants during cloudy, warm, and humid weather. Conidia are produced on leaves and washed behind leaf sheaths by rain. Mycelia from germinating conidia enter the panicle.

Distribution. Wherever sorghum is grown, but it is most severe in warm, humid climates.

Symptoms. Normally only mature plants display disease symptoms. Initially, symptoms appear on diseased panicles as water-soaked, discolored lesions that become tan to purple-black. Young lesions appear as elliptic pockets or bars immediately beneath the epidermis. The interior of the panicle has a mottled appearance caused by red-brown tissue interspersed with white tissue. Diseased panicles are small and light in weight and mature early. Small black streaks (acervuli) may appear and extend to the seed.

Diseased seeds have encircling dark brown to black streaks. Severely diseased seeds may be totally discolored and either germinate poorly or cause seedling blight from seedborne inoculum. Diseased plants are produced from seed with visible symptoms, such as the presence of acervuli.

Management
1. Sow disease-free seed.
2. Apply seed-treatment fungicides.
3. Grow resistant cultivars. Resistance to general grain molds has been identified in some accessions.

Phoma Leaf Spot

Cause. *Phoma sorghina* (syn. *P. insidiosa*) overwinters as pycnidia on infested residue and seed. Conidia are produced in pycnidia during warm, wet weather in the spring and disseminated by wind and splashing rain to leaves. Secondary spread is by formation of pycnidia in infected tissue and subsequent production of conidia.

Distribution. Wherever sorghum is grown in warm temperatures. Phoma leaf spot is considered a minor disease.

Symptoms. Irregular or subcircular spots 1 cm or more in diameter develop on leaves. Spots are yellow-brown or tan to gray and may either be indefinite in outline or have a narrow, red-purple margin. Pycnidia develop as small, black dots scattered over dead leaf tissue or in groups or lines between veins. Pycnidia also develop on glumes and seed.

Management. Phoma leaf spot is generally considered to be very minor, but the following aid in managing the disease.

1. Sow clean, healthy seed treated with a seed-protectant fungicide.
2. Plow under residue.

Pokkah Boeng

Pokkah boeng is also called top rot and twisted top.

Cause. *Fusarium subglutinans* (syn. *F. moniliforme* var. *subglutinans*). *Fusarium subglutinans* overwinters as mycelia and perithecia in infested residue and is seedborne as mycelia and conidia. Conidia produced from mycelia growing on residue during wet weather are disseminated by wind and splashing rain to plant hosts, where they lodge in leaf sheaths or whorls and germinate. During wet weather, *F. subglutinans* also grows upward on the outside of stalks. Metabolites produced by *F. subglutinans* cause distortions in plants.

Distribution. Tropical or semitropical areas where humidity is high.

Symptoms. Leaves near the top of plants are deformed and discolored and have wrinkled bases and numerous transverse cuts in the margins. Often leaves are so wrinkled that they do not unfold properly, resulting in a plant with a ladder-like appearance. In severe cases, *F. subglutinans* may grow from whorls and sheaths into stalks, causing the tops of diseased plants to die. Mild symptoms on leaves may be a mosaic that resembles a virus symptom.

Stalks may bend over at narrow, uniform, transverse cuts in the rind as if a knife had cut out portions. The latter symptom may be covered by leaf sheaths and may not be apparent until stalks are broken over at the cut.

Management. Pokkah boeng is not considered a serious disease of sorghum and no management measures are necessary.

Pythium Root Rot

Cause. *Pythium arrhenomanes* and *P. graminicola* are considered the primary causal organisms. Other *Pythium* spp. associated with root rot are *P. periplocum* and *P. myriotylum*. Pythium root rot is favored by high soil temperatures and moisture throughout the growing season. However, infections increase at the boot stage and later stages of plant growth; plant death normally occurs during stress conditions at and after plant maturity. Seedling blight caused by *P. arrhenomanes* has been reported to be optimum at a soil temperature of 15°C.

Distribution. Not known for certain, but Pythium root rot is thought to be widespread.

Symptoms. Symptoms usually occur on larger adventitious roots. Red-brown to blackish, sunken lesions occur on diseased roots. Frequently, at root death, the entire root becomes discolored.

    Pythium arrhenomanes has been reported to cause preemergence and postemergence seedling death.

Management. Highly susceptible genotypes have been identified.

Pythium Seed and Seedling Blight

Cause. Several *Pythium* spp., including *P. debaryanum, P. irregulare, P. paroecandrum, P. rostratum, P. splendens, P. ultimum,* and *P. vexans.* The fungi survive as oospores, sporangia, or mycelia in infested residue.

    Disease is severest in wet soils at temperatures of 10° to 13°C. Under these conditions, seed germination and growth is slow, making plants more susceptible to infection. Sporangia either germinate directly and form germ tubes or germinate indirectly by forming vesicles in which zoospores are formed. Similarly, oospores either germinate directly and form germ tubes or germinate indirectly by forming zoosporangia in which zoospores are formed. Zoospores "swim" by means of flagella through soil water to host seeds and roots. Once in contact with a susceptible plant, a zoospore will encyst, germinate, and produce a germ tube that penetrates the host plant. As mycelium grows through a plant, sporangia and oospores are produced on and in the diseased plant.

Distribution. Wherever sorghum is grown.

Symptoms. Seeds may be rotted before germination, and seedlings may be infected before or after emergence. Seeds are brown, soft, and overgrown with mycelium. It is difficult to find infected seeds in soil because of their rapid decomposition and the adhering soil that is bound to the seed by mycelia, making seeds indistinguishable from the surrounding soil. Roots are water-soaked, discolored brown, and have a sloughed appearance.

Aboveground symptoms are yellowing, wilting, and death of leaves. Lesions on mesocotyls initially are brown and sunken. Eventually, mesocotyls become soft and water-soaked.

Management. Resistance to *P. arrhenomanes* has been identified in a grain sorghum cultivar.

Red Rot

Red rot is also called anthracnose stalk rot, Colletotrichum rot, and red stalk rot. Although red rot is the stalk rot phase of anthracnose leaf blight, it is discussed as a separate disease.

Cause. *Colletotrichum graminicola* is discussed earlier in the chapter in "Anthracnose Leaf Blight." Conidia are washed down behind leaf sheaths, germinate, and infect stalks any time after the jointing stage of plant growth.

Distribution. Wherever sorghum is grown.

Symptoms. Red rot occurs primarily in the stalks of mature plants. The basal portion of a diseased stalk becomes reddened or purple. When a stalk is split lengthwise, the alternate discolored and white areas in the pith give a marbled appearance throughout the diseased area. Depending on the cultivar, discolored areas range from tan to purplish red. A similar symptom occurs when the peduncle or the upper stem below the head becomes diseased. Cankers on stalks are bleached and the surrounding areas are reddish, orange, or purple.

Stalks frequently break near the middle of the stalk or just below the seed head. Diseased plants that do not break produce small heads which often have small seeds.

Management
1. Grow resistant cultivars and hybrids.
2. Rotate sorghum with resistant crops.

Red Spot

Although both leaf spot and red spot are reported to be caused by *Exserohilum rostratum*, the descriptions of the symptoms are sufficiently different to warrant describing them as two different diseases.

Cause. *Exserohilum rostratum* (syn. *Helminthosporium rostratum*).

Distribution. The United States (Mississippi).

Symptoms. Symptoms resemble those of anthracnose leaf blight, but unlike anthracnose leaf blight, the lesions of red spot are not sunken into the diseased leaf. On sweet sorghum, red spot lesions on the leaf midrib are red-brown to black with tan or brown centers, irregular-shaped, and longitudinal (8–12 × 50 mm).

Management. Red spot is of little economic importance, but the symptoms could be mistaken for those of red rot in breeding programs that screen for anthracnose leaf blight.

Rhizoctonia Stalk Rot

Cause. *Rhizoctonia solani* survives in soil and infested residue as sclerotia. Sclerotia are disseminated by any means that moves soil, such as wind and running water. Sclerotia germinate and form infective hyphae that penetrate the plant host. Plants lacking in vigor are more susceptible to infection.

Distribution. Widely distributed but of minor importance.

Symptoms. Initially, the pith becomes reddish but vascular bundles remain as light streaks. Later, brown sclerotia (1–5 mm in diameter) form on the outside of stalks but under the leaf sheaths.

Management. Not necessary.

Rough Leaf Spot

Rough leaf spot is also called Ascochyta spot and rough spot.

Cause. *Ascochyta sorghi* is seedborne and overwinters as pycnidia in soil and as pycnidia and mycelia in the infested residue of sorghum and perennial weeds. During wet weather, conidia are produced in pycnidia and ooze out in a cirrhi, where they are disseminated by splashing rain and by wind to host plants. Later, pycnidia are produced in spots and conidia are formed during wet weather and disseminated to other host plants as secondary inoculum.

Distribution. Africa, Asia, southern Europe, and the United States, particularly the humid sorghum-growing areas of the southeastern states. Rough leaf spot is generally considered to cause minor losses.

Symptoms. Initially, small, circular to oblong, light-colored spots with well-defined margins develop near the ends of diseased leaves. The spots enlarge, coalesce, and form large blotches which are covered by hard, black, raised pycnidia that give the spots a rough surface and appearance. Depending on the diseased cultivar, these lesions may be grayish, yellow-brown, or purple-red and elongated parallel to the veins. As diseased leaves mature, the pycnidia may fall off or be washed off by rain, leaving only large areas of tan necrotic tissues that have either a reddish or tan border, or an indistinct border. Lesions may also occur on leaf sheaths and stalks. Sometimes, pycnidia develop on healthy-appearing, green parts of leaf surfaces and on glumes.

Management
1. Grow resistant grain sorghum and sweet sorghum cultivars.
2. Rotate sorghum with other crops.
3. Plow under infested residue.

Rust

Cause. *Puccinia purpurea* is a rust fungus that overwinters as urediniospores and mycelia in infested sorghum residue in warmer sorghum-growing areas, or on johnsongrass. However, spores are relatively short-lived in the absence of a living host but thrive on ratooned and successively planted sorghum. *Oxalis corniculata* may be an alternate host and has been reported to become infected and produce abundant aecia. Infected perennial and collateral hosts and scattered diseased sorghum plants serve as infection foci. Urediniospores are disseminated by wind to host plants, and during periods of high humidity, urediniospores produced within uredinia become secondary inoculum. Light drizzles and heavy dew especially favor infection.

The function of teliospores is not considered important in the life cycle of *P. purpurea*. Two-celled teliospores, which develop as leaves mature, either germinate immediately or overwinter before germinating to produce one basidiospore per cell.

Distribution. Wherever sorghum is grown, but rust is most severe in areas of high humidity or rainfall.

Symptoms. Rust symptoms normally appear on diseased plants nearing maturity. The first symptoms are small purple, red, or tan flecks or spots on both leaf surfaces. In susceptible cultivars, the spots enlarge and may cover much of the leaf surface. Small, raised, brownish pustules (uredinia) develop in the spots. Uredinia are elliptical (2 mm in length) and form in interveinal tissue parallel to veins. Often a purple, reddish, or tan area surrounds each uredinium. The epidermis of a uredinium ruptures and exposes reddish to dark brown urediniospores.

Telia develop later in the season, frequently in old uredinia and commonly on the undersides of leaves. Telia are dark brown, elliptical to linear-shaped and often observed as streaks or stripes 1 to 3 mm in length. The epidermis may stay intact for a long time.

Management. Forage sorghum types are usually more severely diseased than grain sorghums. Rust usually appears too late in the growing season to cause much injury to some cultivars. Wilson et al. (1991) noted no consistent differences in yield or digestibility between disease-free plots of susceptible and resistant cultivars. They further observed that dry-matter concentration was unaffected by disease but the rate of loss of digestible dry-matter yield was greater at low rust severities than at higher rust severities. Some accessions have resistance.

Sooty Stripe

Cause. *Ramulispora sorghi* survives for several years primarily as sclerotia in soil and residue. Sporodochia, the structures upon which conidia are produced, also are important in survival when infested leaf residue is left on or above the soil surface throughout the winter.

During warm, moist conditions in the growing season, conidia are produced on the surface of sclerotia or on sporodochia and dispersed by wind and splashed by rain to host plant leaves and sheaths. *Ramulispora sorghi* infects plants at all growth stages, but the oldest leaves and leaf sheaths are infected first. Eventually, conidia produced from the stromata in lesions and disseminated by wind function as secondary inoculum. In time, sclerotia are produced on lesion surfaces, where they remain attached to residue or are easily dislodged and fall into the soil.

Distribution. Africa, Asia, South America, and the United States, particularly in warm, humid, sorghum-growing areas.

Symptoms. Initially, small, water-soaked, oblong, red-purple or tan spots with a yellow halo appear on diseased leaf laminae or leaf sheaths. The spots develop into elongated to elliptical lesions with tan necrotic centers and red-purple to tan borders. Diseased leaves often turn bright yellow from the broad yellow margins that surround lesions. Mature lesions average about 5 cm or longer in length. Lesion centers become grayish from the production of conidia during warm, wet weather and, later, appear sooty due to the formation of numerous small, black, superficial sclerotia. Sclerotia are easily wiped off and cling to fingers like soot.

Management. Generally, sooty stripe occurs too late in the growing season to cause severe plant damage.

1. Hybrids vary greatly in tolerance to sooty stripe. Plant resistance is the most practical means to manage this disease.
2. Rotate sorghum with other crops.
3. Treat seed with a seed-protectant fungicide.
4. Plow under residue.

Sorghum Downy Mildew

Cause. *Peronosclerospora sorghi* (syn. *Sclerospora sorghi*) survives for several years as oospores in infested residue and in soil after infested residue has decomposed. *Peronosclerospora sorghi* is also seedborne as oospores on the surface of seeds and mycelium within seeds. Oospores germinate and infect young plant roots under low soil moisture conditions to a minimum temperature of 10°C. Mycelium then grows systemically inside the infected plant and, eventually, oospores are formed within the diseased tissue. Penetration of leaves occurs by germ tubes from oospores penetrating through stomatal openings. Conidia generally cause local lesions but can also cause systemic infections if leaves are infected before they are fully developed.

Conidia are produced only at night on diseased leaf surfaces when dew periods of several hours are associated with cool nights at temperatures of 18° to 21°C. Viable conidia may be carried several thousand meters by the wind, but most remain within a few centimeters or meters from where they were produced and play no role in long-distance dissemination.

Disease severity increases more rapidly from year to year on plants grown on sandy soils rather than on clay soils. Optimum disease incidence apparently occurs at a soil temperature-moisture combination of 25°C and −0.2 bar, respectively, and a soil texture of 80% sand. Some herbicides and herbicide antidotes increase severity of downy mildew damage in susceptible sorghum genotypes.

Three pathotypes of *P. sorghi* known in Texas are referred to as P1, P2, and P3.

Distribution. Africa, southeastern Asia, India, and the United States, particularly the Texas Gulf Coast and midwestern states. P1, the original pathotype, is widely distributed in Texas. P2 is rare and reported from just two locations. P3 is widely distributed along the upper coastal plains of southern Texas.

Symptoms. Seedlings are infected at germination and display a chlorotic mottle or are generally chlorotic and stunted. Conidia and conidiophores, which look like white "fuzz," are produced on the undersides of chlorotic leaves during humid weather.

As diseased plants grow, vivid green and white stripes appear on diseased leaves in the late spring or early summer. Sometimes whole leaves either become chlorotic or have one or two narrow green stripes. Conidia are usually not produced on these leaves.

Local infections on diseased leaves appear as small, rectangular, chlorotic spots that soon become necrotic and give leaves a stippled or speckled appearance. Conidia are also produced in these local lesions.

Systemically diseased plants that develop late in the season have striped or mottled leaves. Diseased plants (1) fail to head, (2) produce sterile heads, or (3) form partially diseased heads. As interveinal tissue is destroyed, leaf shredding occurs and oospores are released.

Management
1. Apply a systemic fungicide seed treatment.
2. Grow resistant sweet and grain sorghum hybrids.
3. Do not grow grain sorghum after sudangrass.
4. Sow only high-quality seed.
5. Do not grow sudangrass and sorghum sudan hybrids in areas where downy mildew is present because very susceptible hosts tend to increase the population of oospores in the soil.

6. Rotate sorghum with other crops not susceptible to *P. sorghi*.
7. Destroy sorghum stubble after harvest.
8. Deep plow to place oospores below seed depth.
9. Sow as early as possible in the growing season.

Southern Leaf Blight

Cause. *Cochliobolus heterostrophus,* anamorph *Bipolaris maydis* (syn. *Drechslera maydis* and *Helminthosporium maydis*), overwinters as conidia and mycelia in infested residue. In the spring, conidia are disseminated by wind and splashed by rain to growing plants, where infection occurs in free moisture and temperatures of 15.5° to 26.5°C. *Bipolaris maydis* does not sporulate readily on sorghum. Disease is most severe when susceptible cultivars are grown in continuous sorghum culture in a minimum-tillage system under overhead irrigation.

Distribution. Wherever sorghum is grown.

Symptoms. Symptoms are primarily lesions on diseased leaves. Occasionally lesions merge and cover large areas of the leaves. Lesions are tan (2–6 × 3–22 mm) and have parallel sides and buff to brown borders. Often lesions have dark reddish-brown borders. Lesions may coalesce and kill leaves.

Management. Disease management measures are not necessary since southern leaf blight is generally not considered an important disease on sorghum.

Southern Sclerotial Rot

Cause. *Sclerotium rolfsii* survives as sclerotia in soil and residue and as mycelia in residue. However, *S. rolfsii* only survives in warmer sorghum-growing climates because low winter temperatures kill both sclerotia and mycelia. Sclerotia germinate and form mycelia that infect seeds and roots. Under continuous wet soil conditions, *S. rolfsii* is reported to grow saprophytically from soil to the lowest leaf sheath.

Distribution. Warmer sorghum-growing areas of the world. Southern sclerotial rot is not an important disease on sorghum.

Symptoms. Typically one to three sheaths are affected on plants nearing the bloom stage or later stages of plant development. Occasionally all leaf sheaths may be diseased and the plant killed. Initially, a water-soaked lesion that becomes red, purple, or brown forms at the base of the lowest leaf sheath that is in contact with soil. Leaf blades eventually wilt, become necrotic, and may shred. Mats of white mycelium grow inside leaf sheaths to approximately 30 cm above the soil surface. Fans of mycelium merge from sheaths and spread over the base of the leaf for approximately 13 mm. Young axillary shoots become blighted and covered with mycelium.
    Numerous sclerotia (1–2 mm in diameter) that are white initially, then

become brown are produced inside diseased leaf sheaths. Numerous sclerotia also are produced on dead shredded leaves at the base of a diseased plant.

Management. Not necessary because southern sclerotial rot is of minor importance on sorghum.

Tar Spot

Cause. *Phyllachora sacchari* (syn. *P. sorghi*) survives on weed hosts such as johnsongrass and wild sorghums. Ascospores are formed in stromata on leaf surfaces and disseminated by wind to other leaf surfaces. Tar spot is most severe during warm, wet weather.

Distribution. India, Madagascar, Malaysia, the Philippines, and Thailand.

Symptoms. Small (1 mm or less in diameter), black, round to elongated, raised spots occur on diseased leaves. The spots, composed of hard stromatic masses of fungus, extend through the diseased leaf to both leaf surfaces. The black stroma may be surrounded by a yellowish, reddish, or necrotic ring. Lesions may coalesce and become large enough to harm the leaf.

Management. Grow resistant cultivars.

Target Leaf Spot

Cause. *Bipolaris cookei* (syn. *Helminthosporium cookei*) survives primarily as mycelia in infested residue, particularly in leaf veins within lesions, and on leaves as chlamydospores formed from conidia. Sporulation occurs under humid conditions and conidia are disseminated by wind to host plants. Development of target leaf spot is favored by moist, humid weather and temperatures of 19° to 33°C.

Distribution. Cyprus, India, Israel, South America, Sudan, and the United States. Target leaf spot is of minor importance.

Symptoms. Symptoms vary according to sorghum genotype. Initially, symptoms are red or gray dots that later develop into definite spots. The well-defined spots on diseased leaves vary in size from small spots (2–3 mm in diameter) to larger elongated lesions (10–15 mm wide) that are delimited by leaf veins. The shapes of spots vary from rectangular to narrow and the colors can be brown, reddish with light-colored centers, tan, purple, or light brown. There is a distinct circular center in each leaf spot lesion. Spots coalesce and form large areas of necrotic tissue. Lesions of target leaf spot somewhat resemble those of gray leaf spot caused by *Cercospora sorghi*; however, the two leaf spots can often be distinguished by the presence of the distinct circular center in each target leaf spot lesion.

Management. Not necessary.

Zonate Leaf Spot

Zonate leaf spot is also called Gloeocercospora leaf spot.

Cause. *Gloeocercospora sorghi* overwinters as sclerotia in infested residue on or above the soil surface. Primary inoculum is sclerotia which germinate and produce conidia that are disseminated by splashing rain and by wind to host plants. Both leaves and sheaths may be infected. *Gloeocercospora sorghi* is also seedborne as mycelium and possibly sclerotia. During warm, wet weather, sporulation occurs in lesions as pink sporodochia that form above stomates. The conidia are borne in a pinkish, gelatinous mass and are splashed to healthy host tissue. Sclerotia are eventually formed in lines parallel to veins in leaf lesions.

Distribution. Africa, Asia, Central and South America, the United States, and the West Indies. Zonate leaf spot is most severe in warm, wet climates.

Symptoms. Initially, small, red-brown, water-soaked spots that sometimes have a narrow chlorotic halo occur on leaves near the soil surface. Spots enlarge, elongate parallel to the veins, and become dark red or tan. Eventually, the spots occur as circular areas on diseased leaves or semicircular patterns along the leaf margins. Typically, circular red-purple bands alternate with tan areas in a concentric or zonate pattern with irregular borders. Spots vary in diameter from a few millimeters to several centimeters and may cover the entire width of the diseased leaf.

During warm, wet weather, conidia are produced in lesions as a pinkish gelatinous spore mass that forms above a stomate. Sclerotia eventually appear as small, black, raised bodies in lines parallel to the veins.

Plants that are severely diseased in the seedling stage may become defoliated or die. Plants infected when older may have their foliage destroyed before they mature, which results in poorly filled seed.

Management
1. Grow resistant or tolerant cultivars or hybrids.
2. Sow only clean seed treated with a seed-protectant fungicide.
3. Rotate sorghum with other crops.
4. Plow under residue.

Diseases Caused by Nematodes

Awl

Cause. *Dolichodorus* spp. are ectoparasitic nematodes that are limited to wet soils. Some dissemination occurs by water. Feeding is at the root tips.

Distribution. The United States.

Symptoms. The aboveground symptom of diseased plants is plant stunting.

Belowground symptoms are stubby and thick roots with some lesions present.

Management. Not reported.

Cereal Root Knot

Cause. *Meloidogyne naasi* overwinters as eggs. Second-stage larvae emerge from eggs and penetrate the root epidermis just above the root cap. Nematode secretions transform pericycle cells into giant cells. Two to three generations of nematodes are produced each growing season.

Distribution. Thailand and the United States (Kansas).

Symptoms. Plants in irregular-shaped areas within a field are stunted and chlorotic. Root galls may be elongated swellings or discrete knots. These galls are normally smaller than those caused by other root-knot nematodes. Roots are often curved in the shape of a hook, horseshoe, or a complete spiral.

Management. Not reported.

Dagger

Cause. *Xiphinema americanum* is an ectoparasite.

Distribution. Not known.

Symptoms. An aboveground symptom on diseased plants is slight to moderate stunting. Belowground symptoms are a reduction in feeder roots accompanied by necrosis of a portion of the root system.

Management. Not reported.

Lesion

Cause. *Pratylenchus hexincisus* and *P. zeae* are endoparasitic.

Distribution. Worldwide.

Symptoms. Diseased plants are stunted and chlorotic. Roots are poorly developed and have necrotic lesions. *Pratylenchus hexincisus* increases the severity of charcoal rot.

Management. Not reported.

Needle

Cause. *Longidorus africanus* and other *Longidorus* spp.

Distribution. Not known for certain but likely the same as for maize: Sandy soils in temperate regions.

Symptoms. Symptoms on sorghum have not been well described but are considered similar to those on maize. Diseased plants are severely stunted and chlorotic. Belowground symptoms include yellow discoloration, slightly swollen root tips, stubby roots, pruning of lateral roots, and a scarcity of small feeder roots.

Management. Not known.

Pin

Cause. *Paratylenchus* spp.

Distribution. Not known.

Symptoms. Brown necrotic areas occur on roots. Large populations can kill roots of other crops.

Management. Not reported.

Reniform

Cause. *Rotylenchulus* spp.

Distribution. Tropics.

Symptoms. Diseased roots are slightly swollen. Damage on sorghum is not thought to be severe.

Management. Not reported.

Ring

Cause. *Criconemella* spp. is an ectoparasite.

Distribution. The southeastern United States.

Symptoms. The aboveground symptom for diseased plants is a mild stunting. Belowground symptoms are lesions on roots and some decay.

Management. Not reported.

Root Knot

Cause. *Meloidogyne* spp.

Distribution. Not known for certain, but root-knot nematodes are presumably widely distributed on sorghum.

Symptoms. Diseased plants in irregular-shaped areas within a field are stunted and chlorotic and have delayed blooming and reduced yields. Roots have galls, elongated swellings, or discrete knots or swellings. Root proliferation is common.

Management. Not practiced. Certain grain sorghum cultivars have been reported to be highly susceptible to *M. incognita*.

Spiral

Cause. *Helicotylenchus* spp. are either ectoparasites, semi-endoparasites, or endoparasites.

Distribution. Likely wherever sorghum is grown.

Symptoms. An aboveground symptom of diseased plants is plant stunting. Belowground symptoms are a reduction in the number of feeder roots and some root rot.

Management. Not reported.

Sting

Cause. *Belonolaimus longicaudatus* is an ectoparasite that produces a phytotoxic enzyme. The nematode is found only in sandy soils.

Distribution. Generally wherever sorghum is grown.

Symptoms. Symptoms on sorghum have not been well described but are thought to be similar to those on maize. Aboveground symptoms are stunting and chlorosis. Belowground symptoms include deep necrotic lesions on roots. Root tips are frequently destroyed, which results in thick, stubby roots.

Management. Not reported.

Stubby Root

Cause. *Paratrichodorus minor* is an ectoparasite found primarily in sandy soils. Most nematode feeding is around root tips.

Distribution. Not known.

Symptoms. Aboveground symptoms of diseased plants include severe stunting and chlorosis. Diseased roots are stubby and sometimes appear thicker than normal.

Management. Not reported.

Stunt

Cause. *Merlinius brevidens, Quinisulcius* spp., and *Tylenchorhynchus* spp. are ectoparasites.

Distribution. Not reported.

Symptoms. Diseased plants may be stunted and chlorotic. The root system is poorly developed and has fewer than normal feeder roots. Numerous root tips may be short and thickened.

Management. Not reported.

Disease Caused by Phytoplasmas

Yellow Sorghum Stunt

Cause. A phytoplasma. It is not known how the causal agent overseasons or is transmitted.

Distribution. The southern and midwestern United States, including Alabama, Georgia, Kentucky, Louisiana, Mississippi, Ohio, and Texas.

Symptoms. Diseased plants are generally more prevalent at or near the edges of fields. Plants are stunted, about half of normal height, and leaves are bunched together at the tops. Diseased leaves are rigid and curled adaxially about the blade axis. Leaves have pronounced puckering, resulting in undulating margins, and a yellow-tinged cream color that makes plants conspicuous from a distance.

   Diseased plants seldom produce panicles; if panicles are produced, they are normally barren. Plants infected early in the growing season when they are less than 0.3 m tall remain at that height but stalks continue to increase in diameter. These plants usually also display other disease symptoms.

Management. Grow resistant cultivars.

Diseases Caused by Viruses

Australian Maize Stripe

Cause. Australian maize stripe agent is thought to be a virus. The agent is vectored by the planthopper *Peregrinus maidis*. The major reservoir host is *Sorghum verticilliflorum*. Acquisition and inoculation threshold periods were approximately 1 hour each.

Distribution. Australia (Queensland). Incidence may be high in sweet maize grown in coastal districts where the vector and reservoir are common, but the disease is rare in the major inland agricultural areas.

Symptoms. Broad yellow stripes occur on diseased leaves.

Management. Not reported.

Brome Mosaic

Cause. Brome mosaic virus (BMV) is in the bromovirus group. BMV persists in cereal crops and perennial grasses and can survive for several months in dried infested tissue. BMV is primarily sap-transmitted. Local spread presumably occurs through plant contact and transfer of sap by machinery and wind. Soil transmission by the dagger nematodes *Xiphinema coxi* and *X. paraelongatum* has been reported from Europe.

Distribution. Europe and the United States.

Symptoms. Diseased leaves display narrow, interveinal chlorotic streaks that appear mottled from a distance. Mechanically inoculated plants may be stunted.

Management. Grow resistant cultivars.

Cucumber Mosaic

Cause. Cucumber mosaic virus (CMV) is transmitted by the aphid *Aphis gossypii*.

Distribution. The United States.

Symptoms. Chlorotic lesions and stripes occur on diseased leaves.

Management. Not reported.

Elephant Grass Mosaic

Cause. Elephant grass mosaic virus (EGMV). EGMV has been tentatively identified as a potyvirus that causes a mosaic of elephant grass but has a very limited host range. In greenhouse tests, the virus was transmitted mechanically but not by aphids.

Distribution. Brazil.

Symptoms. Chlorotic spots and streaks developed on leaves 1 to 2 weeks after inoculation.

Management. Not reported.

Johnsongrass Mosaic

Cause. Johnsongrass mosaic virus (JGMV) is in the potyvirus group. JGMV was formerly the sugarcane mosaic virus, strain JG, from Australia and the maize dwarf mosaic virus, strain O, from the United States. JGMV is transmitted mechanically and nonpersistently by the corn leaf aphid, *Rhopalosiphum maidis*.

Distribution. Presumably wherever sorghum is grown near johnsongrass.

Symptoms. Initially, the youngest leaves of diseased plants have a stippled mottle or mosaic of light and dark green that develops into narrow streaks. Symptoms may occur on all leaves, leaf sheaths, and husks that develop after infection. As diseased plants mature, the mosaic disappears and leaves become yellowish green and frequently show blotches or streaks of red.

Management
1. Grow resistant cultivars.
2. Destroy johnsongrass or other overwintering hosts.

Maize Chlorotic Dwarf

Cause. Maize chlorotic dwarf virus (MCDV) overwinters in johnsongrass and is transmitted in a semi-persistent manner by the leafhoppers *Graminella nigrifrons* and *G. sonora*.

Distribution. The United States (Mississippi, Louisiana, and Texas).

Symptoms. Slight stunting and tertiary veinal chlorosis occurs on diseased plants.

Management. Maize chlorotic dwarf is not of economic importance at present; therefore, disease management is not warranted.

Maize Chlorotic Mottle

Cause. Maize chlorotic mottle virus (MCMV) is in the sobemovirus group. MCMV survives in maize residue and is transmitted by six species of beetles: the cereal leaf beetle, *Oulema melanopa*; the corn flea beetle, *Chaetocnema pulicaria*; the flea beetle, *Systena frontalis*; the southern corn rootworm, *Diabrotica undecimpunctata*; the northern corn rootworm, *D. longicornis;* and the western corn rootworm, *D. virgifera*. MCMV is also transmitted mechanically and possibly by soil.

Distribution. Brazil, Peru, and the United States (Hawaii, Kansas, Nebraska, and Texas).

Symptoms. Diseased leaves are chlorotic and mottled.

Management. Not reported.

Maize Dwarf Mosaic

Cause. Maize dwarf mosaic virus (MDMV) now consists of former MDMV strains A, D, E, and F; former sugarcane mosaic virus, strain J, infective to johnsongrass in the United States; and maize mosaic virus, European-type, from Yugoslavia.

MDMV persists in several annual and perennial grasses. In the spring, aphids, particularly the corn leaf aphid, *Rhopalosiphum maidis,* and the greenbug, *Schizaphis graminum,* feed on overwintering hosts and acquire the virus. Twenty-three species of aphids are known to be vectors of strain A. Aphids can retain infective virus for 20 minutes after feeding and transmit it in a styleborne nonpersistent manner during their movements between diseased and healthy plants. High aphid populations often induce a high disease incidence through secondary infection.

Increases in levels of natural infection by MDMV occur in insecticide-treated (carbofuran) sorghum in the field. In greenhouse tests, carbofuran-treated plants increased infection in *S. graminum* transmission tests. In mechanical inoculation tests, virus infection was reduced in carbofuran-treated sorghum.

Distribution. Potentially wherever sorghum is grown.

Symptoms. Symptoms of MDMV infection can vary according to sorghum genotype, virus strain, and temperature. Plants infected as seedlings are more severely diseased than plants infected later in the growing season. Symptoms are most evident on the upper two or three leaves as an irregular mottling of dark and light green areas interspersed with longitudinal white or light yellow streaks. Some sorghums that have the gene for red pigmentation may have the mottling replaced by a red leaf symptom, especially on abnormally cool nights. This causes a red leaf or a red stripe symptom that appears as elongated stripes with necrotic centers and reddish margins or, infrequently, as round spots.

Severely diseased plants may die. Diseased plants that are still living may be stunted and have delayed flowering and subsequent failure to head or set seed.

Maize dwarf mosaic, strain A, causes pigmented necrotic lesions on panicle branches, followed by excessive shrinkage of seed. These symptoms occur when temperatures of 16°C or less prevail for 4 or more consecutive days while seed is in mild to soft dough stage.

Management
1. No sorghum cultivar is resistant to all strains of MDMV; however some cultivars are resistant to some strains of MDMV.
2. Control grassy weeds, particularly johnsongrass, in a field.

Maize Dwarf Mosaic Head Blight

Cause. Maize dwarf mosaic virus (MDMV) is discussed in "Maize Dwarf Mosaic" in this chapter.

Distribution. Potentially wherever sorghum is grown.

Symptoms. Plants infected by MDMV at or near heading may fail to fill properly. Symptoms may be absent on vegetative parts. Normal and underdeveloped seed occur together in the same spikelet or panicle. The small seed may be restricted to one side of a head or follow patterns of anthesis. Diseased seeds are light in weight, have chalky endosperms, shatter easily, and may be black due to growth of fungi such as *Alternaria* spp. Pigmented, necrotic lesions appear on branches of panicles and are most likely to develop on cool nights on sorghums that carry the gene for red pigmentation. Similar symptoms may occur from root or stalk rots except there are no lesions on branches of panicles.

Management
1. Grow resistant hybrids.
2. Destroy johnsongrass or other overwintering hosts.
3. Sow sorghum early.

Maize Rough Dwarf

Cause. Maize rough dwarf virus (MRDV) is in the fijivirus group. MRDV is disseminated only by the planthopper *Laodelphax striatellus.*

Distribution. Southern and southeastern Europe.

Symptoms. Symptoms on sorghum are not clearly understood. On maize, early infection results in severely stunted plants with numerous enations on the leaf underside and sheath. Root systems are reduced in size and discolored. Ears either are not formed or tend to be small and malformed. Later infection does not result in any noticeable stunting, but enations are still formed.

Management. Not reported.

Maize Streak

Cause. Maize streak virus (MSV) is in the geminivirus group. MSV is transmitted mainly by the leafhopper *Cicadulina mbila,* but other leafhoppers are also vectors. MSV is not seedborne or mechanically transmitted in maize and, presumably, not in sorghum.

Distribution. Kenya, Nigeria, and South Africa.

Symptoms. Chlorotic streaks and stripes occur on diseased leaves.

Management. Some sorghum cultivars have levels of resistance.

Maize Stripe

Cause. Maize stripe virus (MStpV) is in the tenuivirus group. MStpV is transmitted in a persistent manner by the planthopper *Peregrinus maidis.*

Distribution. Australia and possibly India and Africa.

Symptoms. Diseased plants are chlorotic, flower early, and tiller excessively.

Management. Not reported.

Maize Yellow Stripe

Cause. Maize yellow stripe virus (MYSV) is associated with tenuivirus-like filaments. MYSV is transmitted in a persistent manner by both nymphs and adults of the leafhopper *Cicadulina chinai.* Acquisition and inoculation threshold times are 30 minutes each, with a latent period of 4.5 to 8.0 days if temperatures remain between 14°C minimum and 25°C maximum for the time of the latent period. The maximum retention period is 27 days.

Barley, wheat, and graminaceous weeds are winter hosts. Different strains of MYSV exist.

Distribution. Egypt.

Symptoms. Symptoms on sorghum are not well described. Three symptom types exist on maize: fine stripe, coarse stripe, and chlorotic stunt. Each type may represent different MYSV strains. Experimentally, the fine stripes that appear on the first leaves are followed by coarse stripes on younger leaves; however, some leaves have both symptom types.

Management. Not reported.

Peanut Clump

Cause. Peanut clump virus (PCV) is in the furovirus group. PCV is soilborne and the soil fungus *Polymyxa graminis* is the suspected vector. PCV is also mechanically transmitted.

Distribution. Upper Volta.

Symptoms. Great millet, *Sorghum arundinaceum,* is a symptomless natural host.

Management. Not reported.

Red Stripe Disease

Cause. Johnsongrass-infecting strain of sugarcane mosaic virus (SCMV). SCMV is in the potyvirus group.

Distribution. Australia (New South Wales).

Symptoms. Symptoms vary with the diseased cultivar. An initial mosaic pattern that develops on diseased leaves eventually becomes interspersed with necrotic areas. Some cultivars develop a severe necrosis. Systemically infected leaves may also become necrotic. Diseased plants may be stunted and have a severe yellow mosaic.

Management. Control johnsongrass.

Rice Stripe

Cause. Rice stripe virus (RStV) is in the tenuivirus group. Transmission is by the planthoppers *Laodelphax striatellus, Ribautodelphax albifascia,* and *Unkanodes sapporonus* in a persistent and a transovarial manner. RStV multiplies and persists in vectors; disease incidence is correlated to percentage of viruliferous insects. Several other plants are virus hosts.

Distribution. Japan, Korea, and Taiwan.

Symptoms. Diseased leaves are chlorotic or have chlorotic striping.

Management. Not reported.

Sorghum Mosaic

Cause. Sorghum mosaic virus (SrMV), which is in the potyvirus group, consists of former sugarcane mosaic virus strains H, I, and M from the United States. SrMV is soilborne but the actual mechanism of soil transmission is unknown. SrMV is transmitted in a nonpersistent manner by at least seven aphid species. Aphids that normally acquire SrMV from sugarcane or perennial grasses may become widely dispersed by wind. SrMV is also mechanically transmitted.

Distribution. Anywhere that sorghum and sugarcane are grown in the same area.

Symptoms. Virus-caused symptoms are related to the distribution of aphid vectors. Edges of fields are frequently the first and only areas affected. Symptoms vary according to sorghum cultivar, SCMV strain, and environment.

Diseased plants have a light green mottling or mosaic, often with red lesions that occur only at temperatures of 16°C or less. Diseased plants are stunted and the number of heads and seeds and the head length are reduced. Flowering is delayed.

Management. Grow tolerant or resistant cultivars.

Sorghum Stunt Mosaic

Cause. Sorghum stunt mosaic virus (SSMV) is in the rhabdovirus group. SSMV is transmitted by the leafhopper *Graminella sonora*, but is not mechanically transmitted. Minimum acquisition and inoculation times are 6 and 1 hours, respectively. Transmission efficiency is highest at temperatures of 24° to 36°C.

Other hosts of SSMV include maize and wheat.

Distribution. The United States (Arizona and California).

Symptoms. Diseased plants are severely stunted. Leaves have chlorotic and necrotic mottling and streaking.

Management. Not reported.

Sorghum Yellow Banding

Cause. Sorghum yellow banding virus (SYBV) is a sap-transmissible, small isometric virus. A soilborne mechanism is the possible mode of transmission for SYBV.

Distribution. The United States (California and Texas).

Symptoms. The first symptoms are yellow streaks that coalesce into bands. Later, a severe stunting of diseased plants occurs. Naturally infected sudangrass, *Sorghum sudanense,* displayed chlorotic streaks on the leaves.

Symptoms of inoculated plants in the greenhouse were similar and appeared 15 to 20 days after inoculation. In some cases, the plants continued to grow while older leaves lost their symptoms and yellow streaks appeared on new leaves. In other cases, the plants became dwarfed and chlorotic and eventually died. Symptoms on plants in the field occurred 20 to 25 days after inoculation.

Management. Not reported.

Sugarcane Chlorotic Streak

Cause. Sugarcane chlorotic streak virus is transmitted by the insect *Draeculacephala portola.*

Distribution. The United States (Louisiana).

Symptoms. Pale yellow steaks less than 5 cm in length occur on the second and third leaves. Streaks are irregular in width and are not continuous.

Management. Not reported.

Sugarcane Fiji Disease

Cause. A virus that is transmitted by the leafhopper *Perkinsiella saccharicida.*

Distribution. The Philippines and possibly Nigeria.

Symptoms. Diseased leaves have galls and are stiff and malformed. Diseased plants have shortened stalks with poorly formed heads or no heads and become prematurely brown.

Management. Not reported.

Wheat Streak Mosaic

Cause. Wheat streak mosaic virus (WSMV). WSMV is vectored by the wheat curl mite, *Aceria tosichella* (syn. *A. tulipae*).

Distribution. The United States (Kansas).

Symptoms. Symptoms on diseased leaves range from a mosaic with inter-veinal reddening to a mosaic with interveinal reddening that becomes necrotic.

Management. Not reported.

Unknown Virus

Cause. Unknown virus consisting of flexuous rods of approximately 10 nm in diameter and a modal length of 422 nm. The virus is mechanically trans-mitted.

Distribution. The United States (Kansas).

Symptoms. Severe mosaic symptoms occur on diseased leaves.

Management. Not reported.

Disease Caused by Unknown Factors

Weak Neck

Cause. Originally, weak neck was associated with physiologic disorders re-sulting from inherent plant characteristics and environmental stresses in combine types of grain sorghum. The peduncle and rachis do not develop sufficient thick-walled tissue to ripen and support the developing head. Secondary organisms invade such tissue and further weaken it.

Distribution. The United States.

Symptoms. The culm breaks below the head, resulting in poorly developed heads with lightweight seed. The upper culm tissues dry out, become spongy, and bleach to a tan color. Frequently the peduncle may be water-soaked and have an accumulation of a sticky exudate during wet weather.

Management. Grow resistant cultivars in which grain ripens before the culm weakens.

Selected References

Ali, M, E. K., and Warren, H. L. 1987. Physical races of *Colletotrichum graminicola* on sorghum. Plant Dis. 71:402-404.

Ali, M. E. K., Warren, H. L., and Latin, R. X. 1987. Relationship between anthracnose leaf blight and losses in grain yield of sorghum. Plant Dis. 71:803-806.

Anonymous. 1975. Prevent and identify grain sorghum diseases. World Farming 17:11-14.

Bandyopadhyay, R., Mughogho, L. K., and Satyanarayana, M. V. 1987. Systemic infection of sorghum by *Acremonium strictum* and its transmission through seed. Plant Dis. 71:647-650.

Bandyopadhyay, R., Mughogho, L. K., and Prasada Rao, K. E. 1988. Sources of resistance to sorghum grain molds. Plant Dis. 72:504-508.

Bandyopadhyay, R., Mughogho, L. K., Manohar, S. K., and Satyanarayana, M. V. 1990. Stroma development, honeydew formation, and conidial production in *Claviceps sorghi*. Phytopathology 80:812-818.

Bell, K. K., Harris, H., and Wells, H. D. 1973. Rhizoctonia blight of grain sorghum foliage. Plant Dis. Rptr. 57:549-550.

Birchfield, W., and Anzalone, L., Jr. 1982. Grain sorghum root-knot and reniform nematode host reactions. (Abstr.) Phytopathology 72:355.

Borges, O. L. 1983. Pathogenicity of *Drechslera sorghicola* isolates on sorghum in Venezuela. Plant Dis. 67:996-997.

Cardwell, K. F., Hepperly, P. R., and Frederiksen, R. A. 1989. Pathotypes of *Colletotrichum graminicola* and seed transmission of sorghum anthracnose. Plant Dis. 73:255-257.

Casela, C. R., and Frederiksen, R. A. 1993. Survival of *Colletotrichum graminicola* sclerotia in sorghum stalk residues. Plant Dis. 77:825-827.

Castor, L. L., and Frederiksen, R. A. 1980. Fusarium head blight occurrence and effects on sorghum yield and grain characteristics in Texas. Plant Dis. 64:1017-1019.

Cook, G. E., Boosalis, M. G., Dunkle, L. D., and Odvody, G. N. 1973. Survival of *Macrophomina phaseoli* in corn and sorghum stalk residue. Plant Dis. Rptr. 57:873-875.

Craig, J., Frederiksen, R. A., Odvody, G. N., and Szerszen, J. 1987. Effects of herbicide antidotes on sorghum downy mildew. Phytopathology 77:1530-1532.

Creamer, R., He, X., and Styer, W. E. 1997. Transmission of sorghum stunt mosaic rhabdovirus by the leafhopper vector *Graminella sonora* (Homoptera: Cicadellidae). Plant Dis. 81:63-65.

Cuarezma-Teran, J. A., Trevathan, L. E., and Bost, S. C. 1984. Nematodes associated with sorghum in Mississippi. Plant Dis. 68:1083-1085.

Dean, J. L. 1968. Germination and overwintering of sclerotia of *Gloeocercospora sorghi*. Phytopathology 58:113-114.

Doupnik, B., Jr., Boosalis, M. G., Wicks, G., and Smika, D. 1975. Ecofallow reduces stalk rot in grain sorghum. Phytopathology 65:1021-1022.

Duveiller, E., Snacken, F., Maraite, H., and Autrique, A. 1989. First detection of *Pseudomonas fuscovaginae* on maize and sorghum in Burundi. Plant Dis. 73:514-517.

Edmunds, L. K., and Niblett, C. L. 1973. Occurrence of panicle necrosis and small seed as manifestations of maize dwarf mosaic virus infection in otherwise symptomless grain sorghum plants. Phytopathology 63:388-392.

Edmunds, L. K., and Zummo, N. 1975. Sorghum Diseases in the United States and Their Control. USDA Agric. Res. Serv. Agric. Hdbk. No. 468.

Edmunds, L. K., Futrell, M. C., and Frederiksen, R. A. 1969. Sorghum diseases. *In* Sorghum Production and Utilization, J. S. Wall and W. M. Ross (eds.). AVI Publishing Co., Westport, CT.

El-Shafey, H. A., Abd-El-Rahim, M. F., and Refaat, M. M. 1979. A new Cephalosporium-wilt disease of grain sorghum in Egypt. Egypt Phytopathol. Congr. 3rd. Proc., pp. 514-532.

Farr, D. F., Bills, G. R., Chamuris, G. P., and Rossman, A. Y. 1989. Fungi on Plants and Plant Products in the United States. American Phytopathological Society, St. Paul, MN. 1252 pp.

Forbes, G. A., and Collins, D. C. 1985. A seedling epiphytotic of sorghum in south Texas caused by *Pythium arrhenomanes*. Plant Dis. 69:726.

Forbes, G. A., Ziv, O., and Frederiksen, R. A. 1987. Resistance in sorghum to seedling disease caused by *Pythium arrhenomanes*. Plant Dis. 71:145-148.

Frederiksen, D. E., and Leuschner, K. 1997. Potential use of benomyl for control of ergot (*Claviceps africana*) in sorghum A-lines in Zimbabwe. Plant Dis. 81:761-765.

Frederiksen, R. A. 1980. Sorghum downy mildew in the United States: Overview and outlook. Plant Dis. 64:903-908.

Frederiksen, R. A. (ed.). 1986. Compendium of Sorghum Diseases. American Phytopathological Society, St. Paul, MN. 82 pp.

Frederiksen, R. A., and Rosenow, D. T. 1971. Disease resistance in sorghum, pp. 71-82. 26th Ann. Corn and Sorghum Res. Conf. Proc. Chicago.

Janke, G. D., Pratt, R. G., Arnold, J. D., and Odvody, G. N. 1983. Effects of deep tillage and roguing of diseased plants on oospore populations of *Peronosclerospora sorghi* in soil and on incidence of downy mildew in grain sorghum. Phytopathology 73:1674-1678.

Jardine, D. J. 1989. Seedling blight of grain sorghum caused by *Gibberella thapsina*. (Abstr.). Phytopathology 79:1193.

Jensen, S. G., Mayberry, W. R., and Obrigawitch, J. A. 1986. Identification of *Erwinia chrysanthemi* as a soft-rot-inducing pathogen of grain sorghum. Plant Dis. 70:593-596.

Jones, B. L. 1971. The mode of *Sclerospora sorghi* infection of *Sorghum bicolor* leaves. Phytopathology 61:406-408.

Jones, B. L. 1978. The mode of systemic infection of sorghum and sudangrass by conidia of *Sclerospora sorghi*. Phytopathology 68:732-735.

Klaassen, V. A., and Falk, B. W. 1989. Characterization of a California isolate of sorghum yellow banding virus. Phytopathology 79:646-650.

Klittich, C. J. R., Leslie, J. F., Nelson, P. E., and Marasas, W. F. O. 1997. *Fusarium thapsinum (Gibberella thapsina):* A new species in section Liseola from sorghum. Mycologia 89:643-652.

Lakshmanan, P., Mohan, S., and Jeyarajan, R. 1987. A new leaf spot of sorghum caused by *Cochliobolus bicolor* in Tamil Nadu, India. Plant Dis. 71:651.

Langham, M. A., Toler, R. W., Alexander, J. D., and Miller, F. R. 1984. Evaluation of *Sorghum bicolor* (L.) Moench accessions under natural infection with yellow sorghum stunt mycoplasma. (Abstr.) Phytopathology 74:868.

Leon-Gallegos, H. M., and Sanchez Castro, M. A. 1977. The occurrence in Mexico of *Curvularia lunata* on sorghum kernels. Plant Dis. Rptr. 61:1082-1083.

Mabry, J. E., and Lightfield, J. W. 1974. Long smut detected on imported sorghum seed. Plant Dis. Rptr. 58:810-811.

Manzo, S. K. 1976. Studies on the mode of infection of sorghum by *Tolyposporium ehrenbergii,* the causal organism of long smut. Plant Dis. Rptr. 60:948-952.

Marasas, W. F. O., Rabie, C. J., Lubben, A., Nelson, P. E., Toussoun, T. A., and Van Wyk, P. S. 1987. *Fusarium napiforme,* a new species from millet and sorghum in southern Africa. Mycologia 79:910-914.

Martins, C. R. F., and Kitajima, E. W. 1993. A unique virus isolated from elephant grass. Plant Dis. 77:726-729.

Mayhew, D. E., and Flock, R. A. 1981. Sorghum stunt mosaic. Plant Dis. 65:84-86.

McLaren, N. W. 1992. Quantifying resistance of sorghum genotypes to the sugar disease pathogen *(Claviceps africana)*. Plant Dis. 76:986-988.

Munkvold, G. P., and Desjardins. A. E. 1997. Fumonisins in maize. Can we reduce their occurrence? Plant Dis. 81:556-565.

Natural, M. P., Frederiksen, R. A., and Rosenow, D. T. 1982. Acremonium wilt of sorghum. Plant Dis. 66:863-865.

Nelson, P. A., Toussoun, T. A., and Marasas, W. F. O. 1983. *Fusarium* Species: An Illustrated Manual for Identification. The Pennsylvania State University Press, University Park. 193 pp.

Odvody, G. N., and Dunkle, L. D. 1973. Overwintering capacity of *Ramulispora sorghi*. Phytopathology 63:1530-1532.

Odvody, G. N., and Dunkle, L. D. 1975. Occurrence of *Helminthosporium sorghicola* and other minor pathogens of sorghum in Nebraska. Plant Dis. Rptr. 59:120-122.

Odvody, G. N., and Dunkle, L. D. 1979. Charcoal stalk rot of sorghum: Effect of environment on host-parasite relations. Phytopathology 69:250-254.

Odvody, G. N., and Forbes, G. 1984. Pythium root and seedling rots, pp 31-35. *In* Sorghum Root and Stalk Rots: A Critical Review. L. K. Mughogho (ed.). International Crops Research Institute for the Semi-Arid Tropics, Pantancheru, India.

Odvody, G. N., and Madden, D. B. 1984. Leaf sheath blights of *Sorghum bicolor* caused by *Sclerotium rolfsii* and *Gloeocercospora sorghi* in south Texas. Phytopathology 74:264-268.

O'Neill, N. R., and Rush, M. C. 1982. Etiology of sorghum sheath blight and pathogen virulence on rice. Plant Dis. 66:1115-1118.

Pande, S., Mughogho, L. K. Seetharama, N. and Karunakar, R. I. 1989. Effects of nitrogen, plant density, moisture stress and artificial inoculation with *Macrophomina phaseolina* on charcoal rot incidence in grain sorghum. J. Phytopathol. 126:343-352.

Penrose, L. J. 1974. Identification of the cause of red stripe disease sorghum in New South Wales (Australia) and its relationship to mosaic virus in maize and sugarcane. Plant Dis. Rptr. 58:832-836.

Pratt, R. G., and Janke, G. D. 1978. Oospores of *Sclerospora sorghi* in soils of south Texas and their relationships to the incidence of downy mildew in grain sorghum. Phytopathology 68:1600-1605.

Pratt, R. G., and Janke, G. D. 1980. Pathogenicity of three species of *Pythium* to seedlings and mature plants of grain sorghum. Phytopathology 70:766-771.

Raghava Reddy, J., Claflin, L. E., and Ramundo, B. A. 1991. Systemic colonization of grain sorghum plants by *Xanthomonas campestris* pv. *holcicola*. (Abstr.) Phytopathology 81:1193.

Schuh, W., Jeger, M. J., and Frederiksen, R. A. 1987. The influence of soil temperature, soil moisture, soil texture, and inoculum density on the incidence of sorghum downy mildew. Phytopathology 77:125-128.

Seifers, D. L., and Harvey, T. L. 1989. Effect of carbofuran on transmission of maize dwarf mosaic virus in sorghum mechanically and by the aphid *Schizaphis graminum*. Plant Dis. 73:61-63.

Seifers, D. L., Harvey, T. L., Kofoid, K. D., and Stegmeier, W. D. 1996. Natural infection of pearl millet and sorghum by wheat streak mosaic virus in Kansas. Plant Dis. 80:179-185.

Shafie, A. E., and Webster, J. 1979. *Trichometasphaeria turcica* as a root pathogen of *Sorghum bicolor* var. *feterita*. Plant Dis. Rptr. 63:464-466.

Sifuentes, J., and Frederiksen, R. A. 1988. Inheritance of resistance to pathotypes 1, 2, and 3 of *Peronosclerospora sorghi* in sorghum. Plant Dis. 72:332-333.

Singh, D. S., and Pavgi, M. S. 1982. Perpetuation of two foliicolous fungi parasitic on sorghum in India. Phytopathol. Medit. 21:41-42.

Szerszen, J. B., Frederiksen, R. A., Craig, J., and Odvody, G. N. 1988. Interactions between and among grain sorghum, sorghum downy mildew, and the seed herbicide anti-dotes Concep II, Concep, and Screen. Phytopathology 78:1648-1655.

Tangonan, N. G., and Quimio, T. H. 1986. Effects of various cultural management prac-tices and weather factors on the development of sorghum stalk rot in Mindanao. Philipp. Phytopathol. 22:49-57.

Tarr, S. A. 1962. Diseases of Sorghum, Sudan Grass and Broom Corn. Commonwealth My-cological Institute, Kew, England. University Press, Oxford.

Tegegne, G., Bandyopadhyay, R., Mulatu, T., and Kebede, Y. 1994. Screening for ergot re-sistance in sorghum. Plant Dis. 78:873-876.

Toler, R. W. 1985. Maize dwarf mosaic, the most important virus disease of sorghum. Plant Dis. 69:1011-1015.

Tosic, M., Ford, R. E., Shukla, D. D., and Jilka, J. 1990. Differentiation of sugarcane, maize dwarf, johnsongrass, and sorghum mosaic viruses based on reactions of oat and some sorghum cultivars. Plant Dis. 74:549-552.

Trimboli, D. S., and Burgess, L. W. 1983. Reproduction of *Fusarium moniliforme* basal stalk rot and root rot of grain sorghum in the greenhouse. Plant Dis. 67:891-894.

Tuleen, D. M., Frederiksen, R. A., and Vudhivanich, P. 1980. Cultural practices and the in-cidence of sorghum downy mildew in grain sorghum. Phytopathology 70:905-908.

Wall, G. C., Mughogho, L. K., Frederiksen, R. A., and Odvody, G. N. 1987. Foliar disease of sorghum species caused by *Cercospora fusimaculans*. Plant Dis. 71:759-760.

Williams, R. J., Frederiksen, R. A., and Mughogho, L. K. (eds.). 1978. Proceedings of the In-ternational Workshop on Sorghum Diseases. ICRISAT, Patancheru, India. 469 pp.

Wilson, J. M., and Frederiksen, R. A. 1970. Histopathology of the interaction of *Sorghum bicolor* and *Sphacelotheca reiliana*. Phytopathology 60:828-832.

Wilson, J. P., Gates, R. N., and Hanna, W. W. 1991. Effect of rust on yield and digestibility of pearl millet forage. Phytopathology 81:233-236.

Young, G. V., Lefebvre, C. L., and Johnson, A. G. 1947. *Helminthosporium rostratum* on corn, sorghum and pearl millet. Phytopathology 47:180-183.

Zummo, N. 1986. Red spot (*Helminthosporium rostratum*) of sweet sorghum and sugarcane, a new disease resembling anthracnose and red rot. Plant Dis. 70:800.

Zummo, N., and Broadhead, D. M. 1984. Sources of resistance to rough leaf spot disease in sweet sorghum. Plant Dis. 68:1048-1049.

Zummo, N., Bradfute, O. E., Robertson, D. C., and Freeman, K. C. 1975. Yellow sorghum stunt: A disease symptom of sweet sorghum associated with a mycoplasmalike body in the United States. Plant Dis. Rptr. 59:714-716.

Zvoutete, P., Claflin, L. E., and Ramundo, B. A. 1991. Portal of entry of *Xanthomonas campestris* pv. *holcicola* into grain sorghum plants. (Abstr.) Phytopathology 81:1193.

Zvoutete, P., Claflin, L. E., and Ramundo, B. A. 1991. Seedborne role and longevity of *Xan-thomonas campestris* pv. *holcicola* in grain sorghum. (Abstr.) Phytopathology 81:1194.

18. Diseases of Soybean *(Glycine max)*

Diseases Caused by Bacteria

Bacillus Seed Decay

Cause. *Bacillus subtilis* is an aerobic, gram-positive, spore-forming rod (0.7 × 1.0–2.0 μm).

Seed losses are highest during moist and warm conditions at temperatures of 25° to 35°C in storage, in the field, and under experimental conditions.

Distribution. Not known for certain, but likely occurs wherever soybean is grown under warm and wet conditions.

Symptoms. Seeds have a soft, "mushy" decay and are often covered with a slimy, rough to smooth, whitish to grayish bacterial growth.

Management. Not reported.

Bacterial Blight

Cause. *Pseudomonas syringae* pv. *glycinea* has several synonyms. The bacterium has straight or slightly curved gram-negative rods that occur either singly or in chains of a few cells. Polar flagella are either monotrichous or multitrichous, rarely nonmotile. Levan is produced but gelatin and arbutin are not hydrolized.

Pseudomonas syringae pv. *glycinea* is seedborne and overwinters in infested residue and possibly directly in the soil. Bacteria associated with infested leaf tissue overwinter on the soil surface if the weather is cold and dry. The major spread of bacteria occurs during windy, cool, and wet weather when wind and water spread bacteria from soil, cotyledons, and leaves of diseased plants. Diseased cotyledons from seedborne infection may be a major source of secondary inoculum. Bacteria are also spread by leaves rubbing together and during cultivation when foliage is wet. Bacteria in the presence of free water enter plants through wounds and stomates.

Cool, rainy weather favors the development of bacterial blight. Optimum infection occurs at temperatures of 24° to 26°C. Because bacteria become dormant during warm, dry weather, little or no disease development occurs during higher temperatures later in the growing season.

The incidence and severity of bacterial blight are greater in reduced tillage than in conventional tillage systems. However, a report from Missouri said bacterial blight was less in no-till soybeans. The yields from diseased plants were reported to be greater at a row spacing of 45 cm than at 60 or 90 cm. Several races of *P. syringae* pv. *glycinea* exist.

Distribution. Wherever soybean is grown.

Symptoms. Symptoms may appear 5 to 7 days after a storm has occurred and are most common from early to midseason. Lesions first appear on cotyledon margins. As the lesions enlarge, the entire cotyledon collapses and becomes discolored dark brown. Diseased seedlings are stunted, and if the growing point becomes diseased, the seedling may be killed.

Leaf symptoms start as small, angular, water-soaked spots that become yellow. The centers dry out and become red-brown to black as the tissue dies. The spot is usually surrounded by a water-soaked margin bordered by a yellow halo. As spots increase in size, wind and rain may cause large areas of the diseased leaf to fall out, which gives a ragged appearance to diseased leaves. Severely diseased leaves may drop off plants.

Pod lesions, which are small and water-soaked initially, enlarge to cover much of the pod and become brown to black. Diseased seeds may be shriveled, sunken, and discolored, or they may display no symptoms. Severely diseased seeds in the field may be covered with a "slimy" bacterial growth. Seed may also be infected during harvest or storage.

Systemic symptoms have been reported to consist of chlorotic mottling; yellowing; the formation of narrow, elongated leaves; and a general plant stunting.

Management
1. Sow disease-free seed.
2. Rotate soybean with other crops.
3. Plow under residue.
4. Do not cultivate soybeans when the foliage is wet because machinery may disseminate bacteria in a water film.
5. Apply a foliar fungicide during the growing season if conditions, such as growing soybeans for seed, warrant it. However, applying fungicides to control bacterial blight normally is not a practical means of disease management.
6. Treat seed with a seed-protectant fungicide.
7. Avoid growing highly susceptible cultivars.

Bacterial Crinkle Leaf

Bacterial crinkle leaf is also known as bacterial crinkle leaf spot.

Cause. The causal agent of bacterial crinkle leaf spot has been tentatively identified as *Pseudomonas syringae* pv. *syringae.*

Distribution. The United States.

Symptoms. Young diseased leaflets are severely distorted or crinkled and have a necrosis of the veins and interveinal tissues. Lesions are not water-soaked and do not develop chlorosis or a halo around "spots."

Management. Disease management normally is not practiced because bacterial crinkle leaf, to date, is not an economically important disease.

Bacterial Pustule

Cause. *Xanthomonas campestris* pv. *glycines* (syn. *X. phaseoli* var. *sojensis* and several others). The species name sometimes appears in the literature as "*glycinea.*"

Xanthomonas campestris pv. *glycines* is gram-negative, has single, straight rods ($0.4-0.7 \times 0.7-1.6$ μm), is motile by a single polar flagellum, and is obligately aerobic. Colonies on nutrient media are smooth, circular, butyrous or viscid, and usually yellow. Gelatin, starch and casein are hydrolyzed.

Xanthomonas campestris pv. *glycines* overwinters in infested residue on the soil surface and buried in soil, on seeds, and in the rhizosphere of wheat roots. Some weeds, such as redvine, *Brunnichia cirrhosa,* and hyacinth bean, *Dolichos biflorus,* are also hosts.

Bacterial pustule develops during wet, warm weather (optimum temperatures are 30° to 33°C). New infections may occur throughout the growing season because bacteria continue to be active at high temperatures, unlike bacterial blight, where the causal bacterium becomes dormant. Bacteria are disseminated by splashing water, leaf contact, and machinery during cultivation when soybean foliage is wet. The bacteria disperse from a source equally in all directions and at the same rate on resistant or susceptible soybeans. After 35 days, external bacterial populations are 20- to 50-fold greater on susceptible than resistant plants. Bacteria enter plants through leaf wounds or natural openings, such as stomates, and multiply intercellularly.

Distribution. Wherever soybean is grown in a warm, wet climate.

Symptoms. The first symptoms on diseased leaves are small, yellow-green spots that have brown centers and are most conspicuous on the upper leaf surface. Spots lack the water-soaked appearance typical of most bacterial infections. Spots may merge and form larger dead areas in which the diseased tissue falls out, giving leaves a ragged appearance. In cases of severe disease, defoliation may occur. The best diagnostic symptom is the presence of small raised pustules that develop in the center of lesions on the lower leaf surfaces. Pustules eventually rupture and dry. The presence of a pustule and the absence of a water-soaked appearance before the spot becomes yellow distinguishes bacterial pustule from bacterial blight.

Management

1. Grow resistant cultivars. The rxp gene confers resistance by making it necessary to increase the number of bacteria cells necessary to infect a soybean plant.
2. Rotate soybean with other crops.
3. Plow under infested soybean residue.
4. Grow disease-free seed.
5. Do not cultivate when foliage is wet because machinery will spread the bacteria.

Bacterial Tan Spot

Cause. *Curtobacterium flaccumfaciens*. Confusion surrounds the identity of the causal organism, which may be *C. flaccumfaciens* pv. *flaccumfaciens* (syn. *Corynebacterium flaccumfaciens* pv. [or subsp.] *flaccumfaciens*). *Curtobacterium flaccumfaciens* is internally seedborne.

When infected seed is grown, bacteria may move up the vascular system of the diseased plant during temperatures of 25° to 30°C and infect the unifoliate and first trifoliate leaves. *Curtobacterium flaccumfaciens* is spread throughout the growing season by windblown rain and infects plants through wounds resulting from leaf-to-leaf contact caused by wind. Infection is reduced after flower set.

Distribution. The United States (Iowa).

Symptoms. An oval to elongated chlorotic pattern starts at the diseased leaf margins and progresses to the midribs. Symptoms continue on to petioles, leaving large necrotic areas that fall out during high winds. Leaflets with large or multiple lesions defoliate from diseased plants.

Seedlings grown from diseased seed may be stunted and have fused leaflets. Plants that become infected when older will have empty pods and the diseased leaves develop a marginal necrosis or necrotic spots on the interior portions of the laminae.

Plant damage is a function of the number of infected seedlings arising from diseased seed. Yield losses may occur with susceptible cultivars.

Management

1. Grow resistant cultivars.
2. Sow healthy seed.
3. Sow early in the growing season when temperatures are below 25°C to avoid leaf infection via the vascular system.

Bacterial Wilt (*Curtobacterium flaccumfaciens*)

Bacterial wilt may be the same disease as or a phase of bacterial tan spot. Because the symptoms have different descriptions in the literature, they are listed as two separate diseases here.

Cause. *Curtobacterium flaccumfaciens* (syn. *Corynebacterium flaccumfaciens*) is seedborne and may overwinter in infested crop residue.

Distribution. Russia and Ukraine.

Symptoms. Diseased leaves are yellowed and eventually wilt and dry out but remain attached to plants. The vascular system is discolored and diseased plants weaken, become stunted, and eventually die.

Diseased seeds may either appear to be normal or will have a bright yellow discoloration caused by bacteria under the seed coat. Infrequently, a small amount of yellow exudate may form on the hilum of a diseased seed.

Management. No specific control measures are reported in the literature. However, it is suggested those disease management measures used for other bacterial diseases of soybean may be useful in managing bacterial wilt.

1. Sow disease-free seed.
2. Rotate soybean with other crops.
3. Plow under residue because the bacterium does not survive well in soil.
4. Do not cultivate when foliage is wet because bacteria may be spread by machinery.
5. Apply a foliar fungicide during the growing season if conditions, such as growing soybean for seed, warrant it. However, applying fungicides to control bacterial wilt normally is impractical.
6. Treat seed with a seed-protectant fungicide.
7. Avoid growing highly susceptible cultivars.

Bacterial Wilt (*Curtobacterium* sp.)

Cause. *Curtobacterium* sp. (syn. *Corynebacterium* sp.). *Curtobacterium* sp. resembles *Curtobacterium flaccumfaciens;* therefore, this bacterial wilt may be the same as previously described. However, while there appears to be some similarity between the two diseases, the symptoms are dissimilar enough to treat them as different diseases.

Distribution. The United States (Iowa).

Symptoms. Diseased plants are wilted and severely stunted. Pods are empty and abnormally formed. Lower leaves have marginal necrosis. Wilt does not occur after flower set.

Management. Not necessary, but the following measures suggested for bacterial blight may apply.

1. Sow disease-free seed.
2. Rotate soybean with other crops.
3. Plow under residue.
4. Do not cultivate when foliage is wet.
5. Apply a foliar fungicide during the growing season if conditions, such as growing plants for seed, warrant it. However, applying fungicides to control bacterial wilt normally is impractical.
6. Treat seed with a seed-protectant fungicide.
7. Avoid growing highly susceptible cultivars.

Bacterial Wilt (*Pseudomonas solanacearum*)

Cause. *Pseudomonas solanacearum* has several synonyms. It is occasionally seedborne and overwinters in infested crop residue.

Distribution. Ukraine and the United States (North Carolina).

Symptoms. Diseased leaves become chlorotic and eventually develop small, dark brown spots that elongate into lesions with dark borders. Eventually, lesions dry up and disintegrate, giving leaves a ragged appearance. The entire plant then progressively wilts. Young diseased plants may develop a rapid, severe wilt. Other diseased plants wilt only slightly and are stunted.

Management
1. Sow disease-free seed.
2. Rotate soybean with other crops.
3. Plow under residue.
4. Do not cultivate when foliage is wet.
5. Apply a foliar fungicide during the growing season if conditions, such as growing plants for seed, warrant it. However, applying fungicides to control bacterial wilt normally is impractical.
6. Treat seed with a seed-protectant fungicide.
7. Avoid growing highly susceptible cultivars.

Pseudomonas Seed Decay

Cause. *Pseudomonas syringae* pv. *glycinea* has several synonyms. *Pseudomonas syringae* pv. *glycinea* is seedborne and overwinters in infested residue and possibly directly in the soil. Bacteria associated with infested leaf tissue overwinter on the soil surface if the weather is cold and dry. The major spread of bacteria occurs during windy, cool, and wet weather when wind and water spread bacteria from soil, cotyledons, and leaves of diseased plants. Diseased cotyledons from seedborne infection may be a major source of secondary inoculum. Bacteria are also spread by leaves rubbing together and during cultivation when foliage is wet. Bacteria in the presence of free water enter plants through wounds and stomates.

Cool, rainy weather favors the development of Pseudomonas seed decay. Optimum temperatures for infection are 24° to 26°C. Because bacteria become dormant during warm, dry weather, little or no disease development occurs during higher temperatures later in the growing season.

Distribution. Wherever soybean is grown.

Symptoms. Seeds may be symptomless. If symptoms occur, seeds may be shriveled, develop raised or sunken lesions, and become slightly discolored. Under high moisture conditions, seeds may be covered with a "slimy" growth.

Management
1. Sow disease-free seed.
2. Rotate soybean with other crops.
3. Plow under residue.
4. Do not cultivate when foliage is wet.
5. Apply a foliar fungicide during the growing season if conditions, such as growing plants for seed, warrant it. However, applying fungicides to control seed decay normally is impractical.
6. Treat seed with a seed-protectant fungicide.
7. Avoid growing highly susceptible cultivars.

Wildfire

Cause. *Pseudomonas syringae* pv. *tabaci* (syn. *P. tabaci* and *Bacterium tabacum.*) Other synonyms are noted in the literature. *Pseudomonas syringae* pv. *tabaci* overseasons in infested residue and seed and on the root surfaces of many plants, but alternate freezing and thawing kills the bacterium. *Pseudomonas syringae* pv. *tabaci* is spread by wind and splashing water. Water congestion of plant tissues caused by beating rains is often required for invasion and infection by the bacteria.

Bacterial pustule lesions provide the entrance for *P. syringae* pv. *tabaci*; consequently wildfire is closely associated with bacterial pustule and a pustule can usually be found in the center of a wildfire lesion.

Distribution. Brazil and the southern United States.

Symptoms. Light brown, necrotic spots, varying in size and shape, form on diseased leaves and are usually surrounded by a large yellow halo with a definite margin. Sometimes smaller dark brown to black lesions form without the halo. Lesions may enlarge during wet weather and form large dead areas that eventually tear away. During severe disease conditions of wet weather, almost complete defoliation of diseased plants may occur.

Management
1. Grow resistant cultivars.
2. Rotate soybean with other crops.
3. Plow under residue.

4. Sow disease-free seed.
5. To prevent spread of bacteria, do not cultivate soybeans when the foliage is wet.

Diseases Caused by Fungi

Acremonium Wilt

Cause. *Acremonium* sp.

Distribution. Not known for certain but presumed to be the same as for *Phialophora gregata*. (*See* "Brown Stem Rot.")

Symptoms. Similar to those caused by *P. gregata*. A brown discoloration of vascular and pith tissues occurs. Often the browning occurs only at the nodes; internodal tissue may be white. Later in the growing season, the lower part of the diseased stem may show external browning, wilting, and premature leaf drop. Severely diseased plants may lodge, presumably because of extensive stem injury.

Leaf symptoms are not ordinarily a reliable diagnostic tool since they may be confused with other leaf disorders. Leaf symptoms develop about 3 weeks before physiological maturity when diseased plants are subjected to high temperatures or drought stress following a period of cool weather. Leaves may wilt and tissue between the veins turns brown and dries rapidly, but tissue adjacent to veins remains green. Eventually, the whole leaf dies.

Management. Not reported.

Alternaria Leaf Spot

Alternaria leaf spot is also called Alternaria leaf spot and pod necrosis.

Cause. *Alternaria* spp. *Alternaria atrans* is known to be a weak parasite that infects leaves through wounds made by a mechanical injury such as aphid punctures and sunburn injury. *Alternaria tenuissima* and *A. alternata* are frequently associated with seeds and with injury diseases caused by the bean leaf beetle, *Ceratoma trifurcata*. *Alternaria tenuissima* has also been reported to cause a leaf spot of soybeans in India and a wilt of soybeans in Kenya.

Alternaria spp. are seedborne. Conidia produced on saprophytic mycelia are disseminated by wind to soybean plants. Infection occasionally occurs on young plants, but Alternaria leaf spot is generally a disease of plants nearing maturity. High levels of seed infection have been associated with wet years, frost injury, and stink bug and bean leaf beetle damage.

Distribution. Wherever soybean is grown.

Symptoms. Spots on leaves and pods are brown and have concentric rings that are 6 mm or larger in diameter. Leaf spots may merge and form large dead areas that cause diseased leaves to dry up and defoliate prematurely. Diseased seeds are shrunken, green to brown, and often fail to germinate. Seed infected by *A. alternata* is small and shrunken and has light to dark brown lesions.

Management. Alternaria leaf spot usually goes unnoticed, occurs too late in the growing season to cause yield reductions, and is not considered severe enough to warrant control measures.

Anthracnose

Cause. *Colletotrichum truncatum* (syn. *C. dematium* var. *truncatum*, *C. glycines*, *C. caulicola*, *C. viciae*, and *Vermicularia polytricha*). *Colletotrichum gloeosporioides* has been reported from Spain as a cause of anthracnose, but *C. gloeosporioides* is not associated with most phases of the disease. Other fungi mentioned as causal agents are *Glomerella glycines* (anamorph *C. destructivum*), *C. capsici*, *C. caulivorum*, *C. graminicola*, *V. truncata*, and *G. singulata*. *Colletotrichum capsici* is reported to be weakly pathogenic to soybeans, but survival in soybeans may be a source of inoculum for infecting cotton bolls.

Colletotrichum truncatum overwinters as mycelia and microsclerotia in seeds, seed coats, infested residue, and weeds. Conidia produced from mycelia growing on diseased cotyledons or in residue in soil during warm, wet weather are disseminated by wind to host plants. Isolates may be separated into microsclerotia-forming and non-microsclerotia-forming types. *Colletotrichum truncatum* infects soybeans in all stages of development. Preemergence or postemergence damping off occurs during wet springs. Infection occurs up to temperatures of 30°C, then infection declines as the temperature increases. Anthracnose tends to be more severe under narrow row conditions when plants are irrigated. Roots may also be infected.

Distribution. Wherever soybean is grown. Anthracnose tends to be economically important in the humid tropics and subtropics and is widespread and severe in the midwestern and southeastern United States.

Symptoms. Symptoms generally appear during growth stages V1–V3 (early seedling), R7–R8 (late reproductive), and senescence. *Colletotrichum truncatum* grows from diseased cotyledons onto stems and causes either numerous small, shallow, elongated, red-brown lesions or a few large dark lesions, both of which kill the young plant. In humid weather, cotyledons become water-soaked, wither, and fall off.

Stems, pods, and leaves of young plants may be infected without showing external symptoms until environmental conditions become more favorable for disease development. Symptoms on stems, pods, and petioles are irregular-shaped, brown to black discolored areas that somewhat resemble symptoms of pod and stem blight. In advanced stages of disease, diseased tissue is covered with acervuli that, under magnification, resemble tiny pincushions covered with black spines. Diseased plants may be stunted.

Foliar symptoms that develop after prolonged periods of high humidity include leaf rolling, necrosis of laminar veins, petiole cankering, and premature defoliation. Diseased lower branches and leaves that may be killed during wet weather cause the death of mature plants.

When pods or pedicels are infected early, a few small seeds or no seeds develop and mycelium may completely fill the pod cavity. Diseased seeds are dark brown and shriveled, or if less severely diseased, seeds may be symptomless. When sown, diseased seed results in reduced emergence.

Following inoculation in an aeroponic growth chamber, root symptoms initially were light yellow discolorations that developed into irregular-shaped brown lesions. Later, the lesions became water-soaked, dark brown to black, and extended above and below the point of inoculum introduction. Acervuli and sclerotia developed on secondary roots misted for 1 second every 30 minutes, but neither acervuli or sclerotia formed on roots misted for 1 second every 5 minutes.

Management
1. Sow seed from disease-free plants.
2. Treat seed with a seed-protectant fungicide.
3. Apply a foliar fungicide to plants during the growing season.
4. Rotate soybean with other crops.
5. Plow under residue.
6. Resistance has been identified.

Ascochyta Leaf Spot

Cause. *Ascochyta phaseolorum, A. pinodella, A. sojae,* and *A. sojaecola. Ascochyta pinodella* and *A. sojaecola* are reported to be seedborne. Ascochyta leaf spot is favored by higher humidities and frequently begins in the lower canopy of older plants.

Distribution. Europe, Iran, Japan, and Zambia.

Symptoms. Seedling blight may occur, but the most common symptom is leaf spots that first occur on the lower leaves of older diseased plants. Leaf spots are round and discolored brown and have a dark margin. Stems have brown elongated streaks. Pod symptoms vary from small spots to a general browning of poorly developed pods.

Management. Resistant cultivars have been developed in Russia.

Aspergillus Seedling Blight

Cause. *Aspergillus melleus* is seedborne.

Distribution. The United States.

Symptoms. Diseased seedlings are stunted and chlorotic. Cotyledons are curled and have deep necrotic lesions on the lower side. Russetting occurs on hypocotyls of a few plants.

Management. Not reported.

Brown Spot

Brown spot is also called Septoria brown spot.

Cause. *Septoria glycines* overwinters as pycnidia and mycelia in infested residue and as mycelium in diseased seeds. During warm, wet weather, conidia that are produced on infested residue, cotyledons, and diseased unifoliate leaves infected by seedborne inoculum are disseminated by wind and splashing rain to host leaves. Disease development is optimum at a temperature of 28°C and high relative humidity. Dry weather halts the spread of the disease.

Septoria glycines produces the host-specific, macromolecular, uronic acid–rich, polysaccharide pathotoxin, which causes typical brown spot symptoms on soybeans. A report from Missouri suggested brown spot incidence was lower in soybeans grown in reduced tillage systems than in conventional tillage. Several other plants also are susceptible to *S. glycines*.

Distribution. Asia, Brazil, Canada, Europe, and the United States, particularly the Midwest.

Symptoms. Two distinct types of lesions have been described. The most common type is associated with plants grown from yellow seeds. In the spring, angular, red to brown spots that vary from the size of a pinpoint to 5 mm in diameter and are surrounded by a chlorotic halo appear on both surfaces of unifoliate leaves but are more pronounced on the underleaf surface. Leaves become chlorotic and are prematurely defoliated.

The less common type of lesion, which is associated with plants grown from green seeds, has the typical angular, red to brown spots, but the spots do not have the surrounding chlorotic halo.

Spores produced on the primary leaves may spread and infect the trifoliate leaves, stems, and pods. Trifoliate leaves have numerous irregular tan lesions that gradually turn dark brown. Later in the growing season, entire leaves may become rusty brown and defoliate prematurely. Defoliation of diseased plants occurs from the bottom up, causing the lower portions of diseased plants to be bare of leaves before plant maturation. Brown lesions without a characteristic size or shape form on pods, stems, and petioles. Petioles that have fallen to the soil surface may be covered with pycnidia that appear as scattered tiny black spots. Seed weight is reduced by disease, but the number of pods per plant and the number of seeds per pod are not.

Management
1. Apply a foliar fungicide to plants during the growing season.
2. Grow a less susceptible cultivar.
3. Rotate soybean with other crops.
4. Sow seed from disease-free plants.
5. Plow under residue.

Brown Stem Rot

Brown stem rot is also called Phialophora stem rot.

Cause. *Phialophora gregata* (syn. *Cephalosporium gregatum*) survives as mycelia in infested residue, particularly in woody stem tissue. The population density of *P. gregata* is greater in infested residue on the soil surface than in buried residue. *Phialophora gregata* survives up to 30 months in residue on the soil surface, but the fungus cannot be detected in buried residue after 11 to 17 months.

Conidia, the main source of primary inoculum, are produced from mycelia that precolonized residue buried to a depth of 30 cm in the soil. *Phialophora gregata* has been reported to be seedborne as mycelium within the seed coat but has not been detected in the embryo or cotyledon.

Infections occur through roots and the lower stem early in the growing season. Mycelium then grows upward in the xylem vessels, producing conidia that are carried along, and interrupts water and nutrient flow in the diseased plant.

Development of brown stem rot is greatest at temperatures of 15° to 27°C, but temperatures above 27°C suppress disease development. Adequate moisture early in the growing season and moisture stress later in the growing season has been reported to increase disease severity. However, contradictory results suggest dry soil conditions during the preflower period were associated with lower disease incidence, and postflower moisture deficits were associated with reduced internal stem browning and foliar symptoms. The latter results are consistent with the hypothesis that fungal pathogens of the vascular system advance upward in the stem largely by spores carried in the transpiration stream; therefore, symptom severity is greatest if soil water is readily available. There reportedly is a greater development of brown stem rot symptoms in plants grown in low-fertility soils.

Two forms of *P. gregata* have been reported. Type I isolates cause both internal stem browning and foliar interveinal necrosis. Type II isolates cause only stem symptoms. However, evidence suggests that physiologic specialization exists.

Distribution. Canada, Egypt, Mexico, and the United States.

Symptoms. Internal pith and vascular tissue become brown. The browning often occurs only at the nodes and the internodal tissue remains white. Later in the growing season, the lower part of the stem may show external browning, wilting, and premature leaf defoliation. Severely diseased plants may lodge, presumably because of extensive stem injury.

Leaf symptoms are ordinarily not a reliable diagnostic tool because they may be confused with other leaf disorders. Leaf symptoms develop about 3 weeks before physiological maturity when diseased plants are subjected to high temperatures or drought stress following a period of cool weather. Leaves may wilt and the tissue between the veins becomes brown and dries rapidly, but tissue adjacent to veins remains green. Eventually, the entire leaf dies. In contrast to the yellow-green color of a normally maturing field, an entire field of diseased soybeans may turn brown, as if an early frost had occurred.

Pod and seed abortion is a major effect of brown stem rot. Usually the more extensive the internal stem browning is, the greater the yield reduction. However, some results indicate that greater yield losses occur when both foliar and stem symptoms develop than when only internal stem browning develops. Both type I and type II isolates cause browning of internal stem tissues, but they differ in their ability to cause foliar symptoms. Type I isolates cause chlorosis, necrosis, and wilt of foliage; type II isolates do not cause foliar symptoms. Inoculation with type I isolates significantly reduces yields; inoculation with type II isolates does not affect yields.

Management
1. Do not grow soybean, alfalfa, or red clover in infested soil for 3 years. Infested residue decomposes within this time and *P. gregata* cannot survive in soil outside of precolonized residue.
2. Grow resistant cultivars.

Cercospora Leaf Spot and Blight

Cause. *Cercospora kikuchii* (*see* "Purple Stain"). Cercospora leaf spot and blight is favored by extended periods of high humidity and temperatures of 28° to 30°C. However, no germination occurs below 92% relative humidity, and disease does not occur with leaf wetness periods of less than 18 hours or at temperatures above 30°C.

Distribution. Japan, Taiwan, Uganda, and the United States.

Symptoms. Cercospora leaf spot and blight is more severe on early-maturing than late-maturing cultivars. In Arkansas, the initial symptoms were observed at the beginning of seed set through full seed set stages of plant growth.

Diseased upper leaves exposed to the sun have a light purple, leathery appearance. Later, red-purple, angular to irregular-shaped lesions occur on the upper and lower leaf surfaces. These lesions, which vary in size from a pinpoint to irregular patches up to 1 cm in diameter, may coalesce and form large necrotic areas. Veinal necrosis also may occur. Severe disease may cause rapid chlorosis and necrosis that results in defoliation starting with the upper diseased leaves. Green leaves may occur below the defoliated areas. This obvious blighting occurs over large geographic areas, including entire fields.

Red-purple, slightly sunken lesions several millimeters long occur on petioles and stems. On susceptible cultivars, red-purple lesions that later become purplish black occur on pods.

Management
1. Grow the least susceptible cultivars.
2. An increase in the amount of potassium in fertilizer decreased incidence and severity of Cercospora leaf spot and blight.

Cercospora Seed Decay

Cause. *Cercospora sojina* (syn. *C. daizu*) overwinters as mycelium in seed, primarily in the parenchymatous region of the seed coat, and in infested residue. Conidia produced in cotyledonary lesions when infected seed germinates and on residue during warm, moist conditions are disseminated by wind to young leaves, pods, stems, and seeds. Disease severity is greatest under warm, humid conditions. Different races of *C. sojina* exist.

Distribution. In warm, humid areas of the soybean-growing areas of the world.

Symptoms. Seeds have conspicuous light to dark gray and brown areas that vary from specks to large blotches covering the entire seed coat. Some lesions show alternating bands of light and dark brown. Sometimes brown and gray lesions diffuse into each other. Normally the seed coat cracks or flakes.

Management
1. Grow cultivars resistant to frogeye leaf spot.
2. Sow disease-free seed.
3. Rotate soybean with other crops. Leave fields out of soybean for at least 2 years.
4. Apply foliar fungicides at the R3–R5 growth stages.
5. Plow under infested residue.

Charcoal Rot

Charcoal rot is also called dry weather wilt and summer wilt.

Cause. *Macrophomina phaseolina* (syn. *Botryodiplodia phaseoli, M. phaseoli, Rhizoctonia bataticola,* and *Sclerotium bataticola*) survives as sclerotia and mycelia in infested residue and in dry soil, but the fungus does not survive longer than a few weeks in wet soil. The depth of burial in soil apparently has little effect on fungus survival. *Macrophomina phaseolina* is a good colonizer of plant debris in soil but it cannot persist long in the presence of other soil microflora. *Macrophomina phaseolina* is also seedborne as ectophytic and endophytic sclerotia and hyphae; the hyphae are found in the seed coats, endosperms, and embryos. Pycnidia are rarely produced on soybeans.

Seedling disease is greatest at high temperatures (28° to 35°C optimum), and both sclerotia and mycelia are equally effective as infective propagules. Sclerotia germinate on root surfaces and form several germ tubes that penetrate roots and grow as mycelium within xylem tissues. Disease development in older infected plants is also associated with hot, dry weather, and when unfavorable environmental conditions stop plant growth. Charcoal rot occurs on irrigated soybeans when water is withheld to promote plant maturity.

Both infected seedling and older plants wilt due to the physical plugging of the xylem by mycelium or by the production of toxins or enzymes by the fungus, which stops the upward movement of water in the xylem tissue.

Infection by the soybean cyst nematode, *Heterodera glycines,* can increase root colonization by *M. phaseolina.* Charcoal rot becomes more severe when soybeans are grown in the same field in successive years. Long-term tillage does not affect charcoal rot incidence and severity.

Distribution. Wherever soybean is grown.

Symptoms. Red-brown lesions that later become ashy gray to black occur on seedlings at the soil line and extend upward on diseased stems. Leaves of severely diseased plants turn yellow, wilt, and remain attached to the plant. Diseased seedlings are more likely to die in warm and dry soils than in cool and moist soils.

The best diagnostic symptom occurs after midseason. When the epidermis of diseased stems is removed, small black sclerotia which can be observed under the epidermis are so numerous that diseased tissues have a gray-black color resembling a "sprinkling" of powdered charcoal. When the lower portion of diseased plants is split open, there are black streaks in the woody portion. Diseased seeds have indefinite black spots and blemishes on the seed coat.

Management
1. Rotate soybean with other crops. Soybeans should not be grown in the same soil for at least 2 years to significantly reduce the number of propagules.

2. Do not oversow. Crowded seedlings are more subject to infection.
3. Follow a good fertility program.
4. Plow under residue.

Choanephora Leaf Blight

Cause. *Choanephora infundibulifera* may survive in the infested residue of different host plants. Choanephora leaf blight is severe under humid conditions or abundant rainfall.

Distribution. Thailand and the United States (Louisiana and Mississippi).

Symptoms. Disease symptoms first occur on the oldest leaves nearest the soil surface. Distal halves of diseased leaves initially develop a grayish color. In approximately 5 days, the diseased areas become darkened and necrotic. Later, the diseased portion dries up and curls.

Sporangiophores and sporangioles may be observed on blighted leaves and the edges of necrotic areas during periods of high humidity. Defoliation of diseased leaves also may occur during periods of high humidity. If humidity is low, only the diseased area of the leaf drops off, leaving the unaffected leaf portion intact. Severely defoliated plants may be stunted and seeds may be smaller than normal.

Management. Not reported.

Cotyledon Spot

Cotyledon spot is also called Curvularia cotyledon spot.

Cause. *Curvularia lunata* var. *aeria* and *C. lunata* are seedborne.

Distribution. Brazil, India, and the United States (Puerto Rico).

Symptoms. Diseased seeds show a slight discoloration. Upon germination, seed coats have a reddish tint and the underlying cotyledonary tissue becomes light brown. Lesions, which are superficial and irregular-shaped, do not enlarge once cotyledons have expanded. Up to 70% of the surface of diseased cotyledons may be affected.

Management. Not reported. Cotyledon spot is considered of minor importance.

Downy Mildew

Cause. *Peronospora manshurica* (syn. *P. sojae*) overwinters as oospores in infested residue and on seed. When adverse dry conditions occur during seed maturation, *P. manshurica* survives as thick-walled mycelium between the hourglass layer and the parenchymatous tissue of the seed coat and it may also survive on the seed surface as resistant, thick-walled resting or sclerotized mycelium.

Under cool conditions, diseased seeds give rise to systemically infected seedlings. Seedling hypocotyls may be infected by soilborne oospores, which also results in systemically infected seedlings. Conidia produced on the undersides of diseased leaflets and disseminated by wind to other host plants serve as secondary inoculum. Older leaves eventually become resistant to infection.

Development of downy mildew is favored by temperatures of 20° to 22°C and high humidity. Several physiologic races of *P. manshurica* exist.

Distribution. Wherever soybean is grown.

Symptoms. An early symptom is indefinite yellow-green areas on the upper leaf surfaces. In time, diseased areas enlarge, become grayish to dark brown, and are surrounded by yellow-green margins. A gray "fuzz" consisting of mycelium and spores develops in the diseased areas on the undersides of leaves, usually during periods of frequent dews or abundant rain. Severely diseased leaves become yellow, then brown, and defoliate prematurely.

Pod infections occur, but external symptoms may not be evident. The inside of the pod and diseased seeds are encrusted with a white "coating" that consists of thick-walled oospores and mycelium. Diseased seeds often are a dull white, have cracked seed coats, and may be smaller or lighter in weight than healthy seeds.

Systemically infected plants remain small and have mottled gray-green leaves that curl downward at the edges. The undersides of diseased leaves are covered with the typical gray fungal growth consisting of mycelium and spores. Diseased seeds and the seedlings that develop from them usually are not killed, but they may serve as infection foci from which the disease spreads to other host plants.

Management
1. Grow resistant cultivars.
2. Rotate soybean with other crops.
3. Plow under infested refuse.
4. Treat seed with a fungicide.

Drechslera Blight

Cause. *Drechslera glycini.*

Distribution. India.

Symptoms. Circular to angular brown spots form near diseased leaf margins. As the lesions enlarge, the centers turn gray and have a dark brown margin that is sometimes surrounded by a chlorotic halo. Lesions may develop along the veins of diseased leaves and cause leaves to curl and dry.

Management. Not reported.

Essex Disease

Essex disease has some similarities to soybean severe stunt. The two diseases may be part of the same disease syndrome.

Cause. *Fusarium oxysporum, F. solani,* and possibly *Rhizoctonia solani. Fusarium solani* and *F. oxysporum* are isolated from the hypocotyl lesions of 'Essex' seedlings at –0.01 MPa water potential and temperatures of 15°, 20°, or 25°C. *Fusarium oxysporum* is also isolated, at a high incidence, from cotyledon lesions at a temperature of 20°C.

    Rhizoctonia solani is isolated from hypocotyl lesions at a temperature of 25°C; thus it may produce symptoms when soil temperatures are relatively high and soil moisture is low at sowing time.

    A soilborne virus-like agent also was found associated with Essex disease.

Distribution. The United States (Delaware and Virginia). The cultivar 'Essex' is severely diseased, but other cultivars are also affected.

Symptoms. Damping off and reduced seedling emergence occur but are not always observed. Symptoms have been reported to occur when the first true leaves appear bunched, thickened, and darker green than healthy leaves. Diseased plants are stunted and have reduced numbers of flowers and pods.

    Brown or reddish-brown, small, discrete lesions that occur on cotyledons, hypocotyls, and roots of seedlings are uncommon on the roots of older, stunted plants. Brown, elongated lesions occur on the hypocotyl-root transition zone. *Fusarium oxysporum* is isolated from small, light brown lesions on roots, and *F. solani* is isolated from darker brown lesions, particularly on the lateral roots.

    Stunting of diseased plants may be common. In low rainfall years, chlorosis and early plant death may occur.

Management. Not reported.

Frogeye Leaf Spot

Frogeye leaf spot is also called Cercospora leaf spot and frogeye.

Cause. *Cercospora sojina* (syn. *C. daizu*) overwinters as mycelium in seed, primarily in the parenchymatous region of the seed coat, and in infested residue. Conidia are produced in cotyledonary lesions when infected seed germinates, and on infested residue during warm, moist conditions and disseminated by wind to young host leaves. All leaves on a host plant may become infected together with pods, stems, and seeds. Severity of frogeye leaf spot is greatest under warm, humid conditions. Different races of *C. sojina* exist.

Distribution. In warm, humid areas of the soybean-growing areas of the world.

Symptoms. Frogeye leaf spot is primarily a disease of the foliage, but stems, pods, and seeds may also become diseased. Small, circular to angular, red-brown spots form on diseased upper leaf surfaces. On lower leaf surfaces, the centers of spots are a darker brown or gray because of clusters of dark gray conidiophores. Spots are discrete, enlarge (1–5 mm in diameter), and become surrounded by a narrow red-brown border but do not have a chlorotic halo. Older leaf spots become thin, "paper-like," and transparent. Several spots may coalesce and form large irregular-shaped areas on diseased leaves. Severely diseased leaves defoliate prematurely.

Infection of stems and pods occurs later in the growing season. Initially, lesions on stems are elongated and reddish and have a black border. Lesions later become brown, then light gray, and finally black when the causal fungus starts to sporulate. The circular pod lesions are brown to gray and have a narrow dark brown ring. The fungus may grow through the pod wall to infect the seed. (*See* "Cercospora Seed Decay.")

Management
1. Grow resistant cultivars.
2. Sow disease-free seed.
3. Rotate soybean with other crops. Leave field out of soybeans for at least 2 years.
4. Apply foliar fungicides at the R3–R5 growth stages.
5. Plow under residue.

Fusarium graminearum Seed Infection

Cause. *Fusarium graminearum* is reported to be a secondary colonist of oospore-encrusted seeds infected by *Peronospora manshurica*. Prolonged rainy weather and mild temperatures during the autumn after soybeans have matured in the field apparently favors colonization.

Distribution. The midwestern United States.

Symptoms. Pericarps are stained pink to red. Diseased seeds are contaminated with deoxynivalenol (vomitoxin) and zearalenone.

Management. Not reported.

Fusarium oxysporum Seed Infection

Cause. *Fusarium oxysporum*. Fungus hyphae occur in all layers of the seed coat of diseased seeds but not in the endosperm or cotyledons. Terminal and intercalary chlamydospores of *F. oxysporum* form in hyphae growing on the underside of seed coats.

Distribution. Presumably this disease is widespread.

Symptoms. Diseased seeds become shrunken, are slightly irregular-shaped, often have cracks in the seed coat, and have light to dark pink discolored areas over most of the seed surface. Heavily diseased seeds do not germinate. Diseased seeds that do germinate damp off before and after emergence and seedlings may have a root rot.

Management. Not reported.

Fusarium pallidoroseum Seed Infection

Cause. *Fusarium pallidoroseum* (syn. *F. semitectum*). Disease occurs in warm and humid conditions. Frequently, the fungus *Phomopsis longicolla* is also isolated from diseased tissue.

Distribution. Where soybean is grown in tropical and subtropical conditions.

Symptoms. Diseased seed is pink to reddish in color. Germination is somewhat reduced and there are reductions in the root weight and in the dry shoot weight of seedlings.

Management. Apply a fungicide seed treatment.

Fusarium Root Rot

Cause. *Fusarium* spp., including *F. solani* and *F. oxysporum,* are common soil fungi that survive as chlamydospores in soil and infested residue, and as saprophytic mycelia in infested residue. Survival of *Fusarium* spp. is better at low soil moistures. Seeds and roots of seedlings are infected during dry soil conditions.

Fusarium spp. are also seedborne, infecting seeds under conditions such as insect injury, especially stink bug damage, and high amounts of precipitation before harvest. A higher incidence of *F. solani* infection is associated with poor-quality seeds than with healthy seeds, possibly because the growth of *F. solani* is enhanced by the greater amount of exudation from poor-quality seeds than from healthy seeds. Poor stands caused by Fusarium root rot are usually associated with poor seed quality, heavy rains, soil compaction, or wet soil after sowing. Trifluralin predisposes soybean seedlings to infection by *F. oxysporum*.

Distribution. Wherever soybean is grown.

Symptoms. Infected seed may have poor germination, which results in preemergence or postemergence damping off, or late-emerging seedlings and stunted plants.

Fusarium solani has been reported to cause light to dark brown, sunken lesions on primary and secondary roots. Other *Fusarium* spp. may cause complete destruction of the root systems of severely diseased plants. Wilting is most frequently observed on seedlings or young plants when roots are rotted and soil moisture is low. Older plants are seldom killed, but they wilt when soil moisture is low, then recover turgidity at night or when soil moisture again becomes adequate.

Management
1. Seedlings infected with *Fusarium* spp. and showing signs of wilting or death of lower leaves should not be cultivated until adequate soil moisture is available.
2. When plants are cultivated, soil should be ridged around the bases of the plants to promote development of roots from the stem above the diseased area. These roots are not as easily infected by *Fusarium* spp. and help the plants to recover rapidly.

Fusarium Wilt

Fusarium wilt is also called Fusarium blight.

Cause. *Fusarium oxysporum* f. sp. *tracheiphilum* race 1 (syn. *F. tracheiphilum*), *F. oxysporum* f. sp. *vasinfectum* race 2, and *F. oxysporum* f. sp. *glycines*. *Fusarium* spp. survive as chlamydospores in soil and infested residue and as saprophytic mycelia in infested residue and weed hosts. Mycelium, regardless of its origin, penetrates roots and grows up the xylem tissue, thereby inhibiting flow of water to the top of the diseased plant. Seedlings and young plants are predisposed to infection by the soybean cyst, root-knot, and sting nematodes.

Distribution. Wherever soybean is grown.

Symptoms. Symptoms generally appear about midseason when the weather becomes warm (28°C has been reported as the optimum temperature) and is accompanied by moisture stress. The best diagnostic characteristic is a browning or blackening of the vascular system. Wilting of stem tips and the presence of flaccid leaves are the most common symptoms on young plants. Areas of diseased plants within a field are randomly scattered, round to elongated patches. Leaves on diseased plants may either become chlorotic or wither and defoliate.

Management. Some cultivars have a level of resistance.

Leptosphaerulina Leaf Spot

Cause. *Leptosphaerulina trifolii* (syn. *L. briosiana*). *Leptosphaerulina briosiana* has also been reported to be a causal agent.

Distribution. *Leptosphaerulina trifolii* has been reported from India, and *L. briosiana* from the United States (Maryland and Missouri).

Symptoms. Lesions on both young and old diseased leaves initially are circular, purplish spots whose centers eventually become gray-tan and contain black perithecia. Later, lesions become irregular-shaped and dull white to gray. In Maryland, small necrotic spots appear on leaves. Symptoms from Missouri were reported to be similar to those described for soybean sudden death syndrome.

Management. Not reported.

Mycoleptodiscus Root Rot

Cause. *Mycoleptodiscus terrestris* (syn. *Leptodiscus terrestris*) survives as sclerotia in infested residue and soil. Infection occurs in warm, wet soils. Sclerotia presumably germinate directly and form mycelia that penetrate roots. Acervuli and conidia are produced in foliar infections, but it is not known what function they perform in the disease cycle.

Distribution. The United States. *Mycoleptodiscus terrestris* is also a pathogen of alfalfa, red clover, and birdsfoot trefoil.

Symptoms. Postemergence damping off of seedlings is characterized by a brown to black rotting of the roots. Older plants may also develop a root rot that is characterized by a red-brown decay of the taproot cortical tissues and a destruction of the secondary root system. This symptom is similar to that caused by *Rhizoctonia solani.* Because *M. terrestris* occurs in association with other root rot fungi, distinct symptoms are difficult to diagnose and Mycoleptodiscus root rot may be confused with other rots.

Leaf infections have occurred in inoculation experiments. Leaf spots are red to light brown and up to 3 mm in diameter. Acervuli form on diseased leaves.

Management. Not reported.

Myrothecium Disease

Cause. *Myrothecium roridum.* Disease development is favored by high temperature and humidity.

Distribution. Southeast Asia, Brazil, India, and the United States.

Symptoms. Spots on diseased leaves initially are small (1 mm in diameter), round, and brownish. Spots enlarge (8–10 mm in diameter), become dark brown, and are surrounded by translucent concentric zones. Sporodochia appear in the translucent areas as small white erumpent structures that gradually turn dark green and ultimately appear as black dots surrounded by a slight chlorosis. The necrotic tissue in the center of the spots falls out and gives a shot-hole effect to the diseased leaves.

Small white dots, which later turn black due to the formation of sporodochia, appear on flowers. Diseased flowers turn brown and drop from the plant. Sporodochia on pods first appear as white dots, similar to those on leaves, but later turn black and coalesce to form a black mass. Mature pods are less severely infected than young pods.

Management
1. Do not sow infected seed.
2. Treat seed with a seed-protectant fungicide.

Neocosmospora Stem Rot

Neocosmospora stem rot is also called Neocosmospora wilt.

Cause. *Neocosmospora vasinfecta* survives either as ascospores in perithecia or mycelium embedded in infested organic matter in soil with a low moisture content. Severity of Neocosmospora stem rot is greater at higher air temperatures than at lower temperatures. There may be a relationship between Neocosmospora stem rot and soil nematode infections.

Distribution. Japan, Nigeria, and the southeastern United States. Neocosmospora stem rot is not considered to be of economical importance.

Symptoms. Internal symptoms may resemble those of brown stem rot. The primary symptom of diseased stems is a red-brown to dark brown discoloration in the pith and xylem that extends 20 to 25 cm upward from the soil surface. Red-brown lesions of 20 mm or less in length infrequently occur on the outsides of stems. Orange to red perithecia may occur on the outsides of the lower stems of dead plants. Interveinal areas of diseased leaves are brown and have chlorotic margins, which results in the infrequent yellowing and defoliation of lower leaves.

Management. Not reported.

Phomopsis Seed Rot

Phomopsis seed rot is also called Phomopsis seed decay.

Cause. *Phomopsis longicolla* is reported to be the predominant causal organism. However, *Diaporthe phaseolorum* (syn. *D. phaseolorum* var. *caulivora, D. phaseolorum* var. *sojae*, and *D. sojae*), anamorph *P. phaseoli* (syn. *P. batatae, P. glycines, P. sojae*, and *Phoma subcircinata*) also is reported to be a causal fungus and may well be the dominant causal organism. Other reported names for causal fungi are *P. phaseolorum* f. sp. *sojae* and *P. phaseoli* f. sp. *caulivora*; however, these names are not listed in Farr et al. (1989). The nomenclature of the causal fungi remains somewhat unclear, but the causal fungi likely may be the same fungi that are attributed to being the cause of pod and stem rot. (*See* "Pod and Stem Rot.")

Diaporthe phaseolorum is seedborne but most primary inoculum originates from mycelia, perithecia, and pycnidia that have overwintered in infested residue. Plants are infected at all growth stages. Ascospores produced in perithecia and conidia in pycnidia during wet weather are disseminated by wind and splashed by rain to host plants. Two types of conidia—alpha and beta—are produced. Germ tubes from conidia penetrate leaves and cotyledons through stomata but not directly through the leaf cuticle.

Disease development of pods and seeds is more severe at temperatures above 20°C during wet or humid weather in late summer and autumn when the plant growth stage is between physiological maturity (R6–R8) and harvest. Disease development may also be severe if plants remain unharvested during moist weather after maturity. Seed infections are usually concentrated at the lower plant nodes but may occur throughout the plant under wet conditions or when harvest is delayed. Close to 100% relative humidity for prolonged or cumulative periods is essential for seed infection to take place. A cumulative period involves periods of high relative humidity interrupted by periods of low relative humidity. Secondary infection occurs from conidia produced in pycnidia within diseased tissue.

Disease tends to be more severe in early-sown or early-maturing cultivars. Plants infected with bean pod mottle virus (BPMV) are more likely to have higher levels of seed infection than plants not infected with BPMV. This probably is because BPMV delays the rate at which soybean plants lose moisture during the final stages of seed maturation.

Distribution.　Wherever soybean is grown.

Symptoms.　Diseased seed may have white mycelium growing over it but often does not display any external sign of disease. Severely diseased soybean seeds are shriveled and elongated, and their coats are cracked and whitish or "chalky-appearing."

Diseased seeds often fail to germinate in wet and cool soils at temperatures of 15° to 20°C. However, researchers do not agree on the exact temperature parameters. Disease may also reduce emergence and stand establishment when seeds are incubated in dry soil at intermediate temperatures. Soil temperature and moisture probably influence preemergence damping off by affecting the duration of time that damage occurs to the seed or seedling between sowing and germination or seedling emergence. Some researchers think low water potential (−15 bars) inhibits seedling growth more than the growth of the causal fungi does; however, growth of *Phomopsis* spp. still continues at this soil water potential. If diseased seeds germinate, they often give rise to diseased seedlings, which serve as a source of inoculum.

Management.　Resistance to *P. phaseoli* has been reported to be due to physical characteristics, such as the presence or absence of pores and open or closed microphyles, of impermeable seed coats.

1. Harvest seed as promptly as possible after maturity.
2. Apply a foliar fungicide from midflowering (R3–R4) to late pod stage (R6–R7).
3. Treat seed with a seed-protectant fungicide.
4. Sow seed from disease-free plants.
5. Plow under infested residue.
6. Rotate soybean with other crops.

Phyllosticta Leaf Spot

Cause. *Phyllosticta sojicola* (syn. *P. glycinea*). In the literature, the specific names *"sojicola"* and *"glycinea"* sometimes are spelled "sojaecola" and "glycineum," respectively. *Phyllosticta sojicola* may be seedborne and also overwinters as pycnidia and mycelia in infested residue. During wet weather, conidia are produced in pycnidia on residue and disseminated by wind to plant hosts. Phyllosticta leaf spot is favored by cool, wet weather.

Distribution. Borneo, Bulgaria, Canada, China, Europe, Japan, the United States, and the former USSR.

Symptoms. Symptoms of Phyllosticta leaf spot can occur on plants of all ages but are most common on the leaves of younger plants. Spots form on the leaf margins of young plants and grow inward. Spots are dull green initially but later become gray to tan with a dark brown to purple border. The shapes of the spots can be round to oval (2 cm in diameter), irregular, or V-shaped. Numerous pycnidia appear as small, black specks in the older lesions.

Small, narrow, gray lesions form on petioles and stems; these lesions and purple to red-brown bordered, circular lesions form on pods. Symptoms are often confused with those of herbicide injury and drought stress.

Management
1. Sow seed from disease-free plants.
2. Rotate soybean with other crops.
3. Plow under infested residue.

Phyllosticta Pod Rot

Cause. *Phyllosticta sojicola*. The specific name *"sojicola"* sometimes is spelled "sojaecola" in the literature. *Phyllosticta sojicola* may be seedborne and also overwinters as pycnidia and mycelia in infested residue. During wet weather, conidia are produced in pycnidia on residue and disseminated by wind to plant hosts. Phyllosticta pod rot is favored by cool, wet weather.

Distribution. Germany and the United States.

Symptoms. Spots (8 mm in diameter) on pods have dark purple-red borders that surround lighter, brownish centers on which numerous dark pycnidia are found. Infection is most severe on pods developing in the upper half of diseased plants.

Management
1. Sow seed from disease-free plants.
2. Rotate soybean with other crops.
3. Plow under infested residue.

Phymatotrichum Root Rot

Phymatotrichum root rot is also called cotton root rot.

Cause. *Phymatotrichopsis omnivora* (syn. *Phymatotrichum omnivorum*) survives for several years in soil as sclerotia and hyphal strands. Early in the growing season, when soil temperatures have risen sufficiently, sclerotia germinate and form hyphal strands that make contact with the developing root systems of host plants. Root contact is not necessary for plant-to-plant spread of *P. omnivora* because mycelial growth through soil occurs in advance of diseased roots. Plant-to-plant spread within rows is more common than spread across rows. Phymatotrichum root rot is most prevalent in alkaline soils.

Distribution. Northern Mexico and southwestern United States.

Symptoms. Plants wilt rapidly and die. The root cortex is rotted and brown strands of fungal mycelium can be seen growing on the roots.

Management
1. Plowing under a green manure crop, such as sweet clover, lessens the disease severity in succeeding years.
2. Rotate soybean with plants in the grass family.

Phytophthora Root and Stem Rot

Phytophthora root and stem rot is also called Phytophthora root rot.

Cause. *Phytophthora megasperma* f. sp. *glycinea* (syn. *P. megasperma* var. *sojae* and *P. sojae*). *Phytophthora macrochlamydospora* has been ascribed to be the cause of a root and stem rot of soybeans in Australia.

Phytophthora megasperma f. sp. *glycinea* survives primarily as oospores and sometimes as mycelia in soil and infested residue. It is not known for certain how oospores germinate in wet soils and form sporangia in which numerous zoospores form. The zoospores "swim" through soil water to roots, where they encyst and germinate at an optimum soil temperature of 15°C. Leaf infection also occurs when infested soil particles are splashed or blown onto host plant leaves during a storm. Oospores eventually form in diseased roots.

Plants may be infected at any stage of growth, but the disease generally occurs early in the growing season. Phytophthora root rot is most common in "wet areas," such as the low areas of a field; poorly drained soils; compacted soils; and soils with a high clay content. Disease may also occur on well-drained soils during a "wet" growing season.

Other root rot fungi and northern root-knot nematodes may increase severity of root rot. Trifluralin herbicide and nitrogen added to soil have both been reported to increase disease severity. On the basis of differing virulences on eight soybean genotypes, 26 physiologic races have been described. Presumably, more races will be described in the future.

Distribution. Asia, Australia, Canada, and the United States.

Symptoms. Distinctive symptoms do not occur during seed rot and preemergence damping off. The most characteristic symptom is the stem rot phase, which is typified by a brown discoloration of the lower stem and branches. The stem discoloration extends from below the soil line to 20 cm or more above the soil line and may involve lateral branches near the soil. The taproot is dark brown and the entire root system may be rotted. In cases where the roots do not appear to be rotted the symptoms can be confused with those of stem canker.

Leaves of diseased plants yellow and eventually become brown but remain attached to the plant. Sometimes there are no obvious symptoms except stunted plants that are reduced in vigor. This latter symptom is often difficult to recognize without a side-by-side comparison with a resistant cultivar.

The disease pattern varies within a field but is often roughly circular, corresponding to the poorly drained areas, but can be linear from dead or dying individual or groups of plants in a row. Diseased plants may often be found more easily at the end of rows than within a field.

Management

1. Grow resistant cultivars; however, several physiologic races of *P. megasperma* f. sp. *glycinea* exist and cultivars resistant to one race may be susceptible to others. Soybean isolines with the Rpsl-k gene for resistance are also tolerant to the herbicide metribuzin. Several genes that confer resistance to specific races at the low optimum temperature of 24°C are defeated by these races at the high optimum temperature of 32°C.
2. Apply systemic fungicide as a seed treatment and/or to the soil either as a band or an in-furrow treatment. Greater amounts of fungicide are needed at the high optimum temperature of 32°C to reduce disease severity than at lower temperatures (24° to 28°C).
3. Sow in well-drained soils that are warmer than 18°C.
4. Avoid growing susceptible cultivars in soils where disease has occurred.

Pod and Stem Rot

Although the cause and general etiology of pod and stem rot are similar to Phomopsis seed rot, they are usually discussed as separate diseases. Pod and stem rot is also called pod and stem blight.

Cause. *Diaporthe phaseolorum* (syn. *D. phaseolorum* var. *caulivora, D. phaseolorum* var. *sojae,* and *D. sojae*), anamorph *Phomopsis phaseoli* (syn. *P. batatae, P. glycines, P. sojae,* and *Phoma subcircinata*). *Phomopsis longicolla* is also a causal fungus and many researchers consider it to be the primary cause of seed decay. Other reported names for causal fungi are *P. phaseolorum* f. sp. *sojae* and *P. phaseoli* f. sp. *caulivora.* The nomenclature of the causal fungi remains somewhat unclear.

   Diaporthe phaseolorum is seedborne but most primary inoculum originates from mycelia, perithecia, and pycnidia that have overwintered in infested residue. Plants are infected at all growth stages. Ascospores produced in perithecia and conidia in pycnidia during wet weather are disseminated by wind and splashed by rain to plant hosts. Two types of conidia—alpha and beta—are produced. Germ tubes from conidia penetrate leaves and cotyledons through stomata but not directly through the cuticle. The fungus infects all parts of the soybean seed, including the seed coat, cotyledons, and embryo.

   Diseased seeds often fail to germinate in wet soils at temperatures of 15° to 20°C. However, disease may also reduce emergence and establishment when seeds are incubated in dry soil and intermediate temperatures (*see* "Phomopsis Seed Rot"). In the latter case, low water potential probably inhibits seedling growth more than does the growth of the causal fungi, particularly *Phomopsis* spp. If infected seeds germinate, they often give rise to infected seedlings that serve as a source of inoculum.

   Disease development of pods and seeds is more severe if temperatures are above 20°C and wet weather occurs when plants are between physiological maturity (R6–R8) and harvest, or if plants remain unharvested after maturity. *Phomopsis longicolla* infects seeds during or after the yellow pod stage (R7). Seed infections are usually concentrated at lower plant nodes but may occur throughout the plant when conditions are wet or harvest is delayed. Close to 100% relative humidity for prolonged or cumulative periods is essential for seed infection to take place. A cumulative period involves periods of high relative humidity interrupted by periods of low relative humidity. Secondary infection occurs from conidia produced in pycnidia.

   Disease tends to be more severe in early-sown or early-maturing cultivars. Plants infected with bean pod mottle virus (BPMV) are more likely to have high levels of seed infection than plants uninfected with BPMV.

Distribution. Wherever soybean is grown.

Symptoms. Although soybeans are infected earlier in the growing season, symptoms of disease appear on pods and stems of diseased plants nearing maturity, especially if wet weather occurs at this time. The first symptom is the presence of pycnidia on petioles that have fallen from the lower portions of plants to the soil surface. This is a good indication that pod and stem blight may be a potential problem during the current growing season. The best symptom is the presence of numerous small, black pycnidia arranged in linear rows on stems but scattered randomly over pods. During dry weather, pycnidia may be present only on lower stems or nodes.

Diseased seed may have white mycelium growing over the seed coat but often may not display any external signs of disease. Severely diseased seeds are shriveled, elongated, have cracked seed coats, and are whitish and "chalky-appearing." Infected seeds germinate poorly and may damp off either before or after emergence, resulting in a poor plant stand. The fungus may also become systemic but fail to induce symptoms in seedlings grown from infected seeds.

Management. Resistance to *P. phaseoli* has been reported to be due to physical characteristics, such as the presence or absence of pores and open or closed microphyles, of impermeable seed coats.

1. Harvest seed as promptly as possible after maturity.
2. Apply a foliar fungicide from midflowering (R3–R4) to late pod stage (R6–R7).
3. Treat seed with a seed-protectant fungicide.
4. Sow seed from disease-free plants.
5. Plow under infested residue.
6. Rotate soybean with other crops.

Pod Rot and Collar Rot

Cause. *Fusarium pallidoroseum* (syn. *F. semitectum*) is seedborne.

Distribution. India (from seed received from the United States).

Symptoms. Plants having rotted pods do not show any other symptoms than less vigorous growth compared to healthy plants. All pods on diseased plants show drying symptoms characterized by a black discoloration that starts from the distal end. Severely diseased pods produce no seeds, or if seeds are present, they are black, rotted, and shriveled. Diseased pods are blackish on both inner sides, especially where the seed rests. Seed coats and cotyledons of diseased seeds also are black.

Seedlings initially have water-soaked, depressed, cream-colored lesions on cotyledons and hypocotyls. Lesions soon become dark brown to black and sometimes coalesce and form enlarged spots on the diseased cotyledons. Spots enlarge longitudinally on the hypocotyl to the radicle. Seedlings eventually die.

Management. Not reported.

Powdery Mildew

Cause. *Microsphaera diffusa* overwinters as cleistothecia on infested residue. Primary infection occurs from ascospores produced in cleistothecia and disseminated by wind to host plants. Secondary infection occurs from conidia produced on diseased leaves and disseminated by wind to plant hosts. Development of powdery mildew occurs later in the growing season during humid weather but without the presence of free water. Cooler than normal temperatures have been observed to favor disease.

Distribution. Brazil, Canada, China, Peru, South Africa, and the United States.

Symptoms. Powdery mildew becomes evident about the middle of the growing season. Cotyledons, stems, pods, and leaf surfaces are diseased. Small colonies of thin, light gray mycelia are seen on the upper surface of diseased leaves. In time, diseased areas enlarge, cover much of the diseased leaf surface, and have a white, powdery appearance due to large numbers of conidia and conidiophores. Diseased plants may become chlorotic and display green islands and rusty patches on diseased leaves. Luxuriant fungal growth may occur without the other symptoms.

Yield losses of up to 26% have been reported.

Management
1. Grow resistant cultivars. Some cultivars are susceptible as seedlings but not as adult plants. Reaction of soybeans is controlled by three alleles at the Rmd locus: Rmd-c confers resistance throughout the life of the soybean plant, Rmd provides resistance at the adult-plant stages, and rmd conditions susceptibility.
2. Apply a foliar fungicide.

Purple Stain

Purple stain is also called Cercospora blight and leaf spot, lavender spot, purple blotch, purple speck, purple spot, and purple seed stain.

Cause. *Cercospora kikuchii* (syn. *Cercosporina kikuchii*) overwinters as mycelium in seed coats and infested residue but rarely in embryos and cotyledons. Infected seed gives rise to seedlings with infected cotyledons. Often seed coats slip off germinating seed before the cotyledons become infected. During warm, humid weather, conidia are produced on diseased cotyledons and disseminated by wind and splashing rain to leaves, stems, and pods of host plants.

Seeds become infected when pods are inoculated with a conidial suspension. No infection of seeds occurs when flowers in full bloom or postbloom (desiccated petals) are inoculated. In general, disease incidence increases with increasing pod wetness periods up to 30 hours. Some researchers have reported a minimum of 8 hours of dew is necessary for infection to occur, with 24 hours of dew being the optimum. However, other reports have stated no disease occurs when pod wetness is less than 24 hours. Various optimum temperatures for infection have been reported in the 20° to 24°C range and at 25°C. Schuh (1992) reported 25°C was the optimal temperature for infection.

Mycelium grows through pods and into seed coats through the funiculus. The fungus grows intercellularly in plant tissues, where nutrients are acquired through membrane leakage caused by cercosporin.

Isolates from the weeds cocklebur, *Xanthium strumarium*; pitted morning glory, *Ipomoea lacunosa*; prickly sida, *Sida spinosa*; sicklepod, *Cassia obtusifolia*; small flower morning glory, *Jacquemontia tamnifolia*; and spotted spurge, *Euphorbia maculata* caused lesions on the hypocotyls, cotyledons, stems, and leaves of soybeans, and purple stain on seeds of soybeans. Conidia are dispersed by water from dead weeds to developing soybean seedlings. Weeds were colonized and conidia produced on dead weed tissue, but no disease symptoms occurred on weeds.

Differences in pathogenicity apparently exist among isolates. *Cercospora kikuchii* isolates are divided into three groups based on colony coloration caused by cercosporin, a red photoreactive polyketide, on potato-dextrose agar. *Cercospora kikuchii* cp isolates produce a purple coloration, *C. kikuchii* cy isolates produce a yellow background due to lipids, and *C. kikuchii* cc isolates produce a red coloration. Cercosporin may facilitate soybean seed coat tissue colonization by *C. kikuchii*.

Distribution. Wherever soybean is grown.

Symptoms. Reduced seed size and yield losses result from purple stain. The best diagnostic symptom is a pale to dark purple discoloration, caused by cercosporin, that is thought to be restricted only to seed coats. Some researchers have reported that *C. kikuchii* penetrates embryonic seed tissues and causes necrosis of cotyledonary cells and vascular elements. However, this discoloration may be uncommon because others have stated that cotyledons are generally not infected or discolored.

Infected seeds can produce diseased seedlings with cotyledons that are shriveled and dark purple and defoliate prematurely. In the laboratory, germination is reduced and stunted, low vigor seedlings result from soybean seeds when 50% or more of the seed coat is stained purple.

The fungus grows from cotyledons into stems, causing a necrotic area that may kill plants. Late in the season, thickened, and crusty-appearing red-brown spots develop on diseased leaves. There is no effect on maturity, lodging, and plant height. Yields are not reduced, but the value of the seed is lowered from a grading standpoint. The purple discoloration does not affect seed used for processing since heating causes the discoloration to disappear.

Certain isolates will reduce seed density and weight, increase free fatty acid content and protein, and reduce oil content when compared to inoculation with other isolates.

Seedling emergence from purple-stained seeds is lower than from healthy seeds. Prior infection with *C. kikuchii* reduces infection by other seedborne fungi, particularly *Diaporthe phaseolorum* and *Phomopsis* spp. This may be explained, in part, by competition for nutrients or space. Another explanation is *C. kikuchii* and other *Cercospora* spp. produce a nonspecific toxic compound, cercosporin, which is reported to be antagonistic to seedborne fungi of soybean.

There has been reported to be no difference in the incidence of purple stain on seed between plants arising from seeds with purple stain and from "healthy-appearing" seed free of purple stain.

Management
1. Apply a foliar fungicide to plants during the growing season.
2. Treat seed with a seed-protectant fungicide.
3. Grow late-maturing cultivars, which in general are less susceptible than earlier-maturing cultivars.
4. Some cultivars are moderately resistant.

Pyrenochaeta Leaf Spot

Pyrenochaeta leaf spot is also called Dactuliophora leaf spot, Pyrenochaeta leaf blotch, and red leaf blotch.

Cause. *Pyrenochaeta glycines* (syn. *Dactuliochaeta glycines* and *Dactuliophora glycines*) overseasons as pycnidia and sclerotia. Optimum pycnidiospore germination is at temperatures from 20° to 25°C.

Sclerotia are disseminated in contaminated soil during farming operations and may be splashed onto leaf surfaces by rainfall, where they germinate by forming infective mycelium. Incidental dissemination may occur through seed lots contaminated with infested residue and soil peds carrying sclerotia. The disease cycle is not fully understood, but pycnidia may develop from mycelium or directly on sclerotia. Both pycnidia and sclerotia apparently develop in lesions and are returned to soil upon the defoliation and decomposition of diseased leaves. Red leaf blotch occurs during abundant rainfall and high humidity.

Distribution. Cameroon, Ethiopia, Malawi, Nigeria, Rwanda, Uganda, Zaire, Zambia, and Zimbabwe.

Symptoms. Lesions occur on diseased leaves, petioles, pods, and stems throughout the growing season and are often associated with primary leaf veins. Initially, dark red to brown, circular to angular lesions (1–3 mm in diameter) appear on unifoliate leaves. Later, dark red spots develop on the upper surfaces and red-brown spots with dark borders develop on the lower surfaces of diseased trifoliate leaves. A diffuse mycelial growth may surround lesions during high humidity. Lesions enlarge, coalesce, and form irregular-shaped blotches (3–10 mm in diameter) with buff-colored centers and dark margins. Older blotches merge and form large necrotic blotches (up to 2 cm in diameter) that are surrounded by chlorotic halos. Necrotic tissue frequently falls out and gives a shot-hole appearance to diseased leaves. Diseased plants defoliate prematurely.

Lesions occur on stem petioles and pods below the uppermost leaf with symptoms. Petiole and pod lesions are ovoid (1–5 mm long) and mauve to red-purple. Sclerotia develop primarily on the diseased lower leaf surfaces, and pycnidia develop in blotches on the upper leaf surfaces. Yield loss results from reduced seed size.

Management. No measures are recommended.

Pythium Damping Off and Rot

Cause. *Pythium aphanidermatum, P. debaryanum, P. irregulare, P. myriotylum, P. torulosum,* and *P. ultimum. Pythium aphanidermatum, P. debaryanum,* and *P. ultimum* survive as oospores in soil and infested residue and saprophytically as mycelia in infested residue. At low temperatures (10° to 15°C) and wet soil, oospores of *P. debaryanum* and *P. ultimum* germinate and form sporangia. Zoospores form in sporangia and "swim" through soil water to seed or to the roots of seedlings. At higher temperatures, oospores may germinate directly and form germ tubes. Although soybean seed sown in cool and wet soil is most subject to infection by *Pythium* spp., *P. aphanidermatum* is reported to cause disease at temperatures of 25° to 36°C.

Seedlings are most susceptible to infection. Soybeans become progressively more resistant as they grow older.

Distribution. Wherever soybean is grown. Disease is most prevalent under wet soil conditions.

Symptoms. Pythium damping off and rot may occur from the time seed is sown to the end of plant flowering. However, the disease is usually associated with seed rotting in the soil or young seedlings being killed. Diseased plants occur singly, in small circular groups in the low, wet spots within the field, or uniformly over an entire field if there has been a period of rain.

Diseased seeds may not germinate and become soft and overgrown with other fungi and bacteria, which give them a "fuzzy" appearance. Seedling rot or blight results from infection after seed has germinated but before or just after the seedling has emerged through the soil. Roots infected by *P. ultimum* are brown and have a "wet" or water-soaked appearance. Other symptoms attributed to *P. ultimum* include swollen hypocotyls and lesions at the junction of the hypocotyl and primary root. Diseased seedlings may also have a curling growth habit and reddish to brown lesions on hypocotyls and cotyledons. Plants infected after emergence, wilt, and leaves are gray-green initially but become brown after one or two days. Such diseased plants are easily pulled from the soil. As soil dries up, a diseased root resembles a "shoestring" attached to the cotyledon.

Pythium debaryanum causes the growing point to be retarded, which results in a "baldhead" symptom.

Pythium torulosum causes a seed rot or death of germinated seeds. On 2-week-old or younger soybean seedlings, root tip discoloration and yellow lesions occurred on the roots and hypocotyl base.

Management
1. Treat seed with a seed-protectant fungicide.
2. Do not sow carryover seed or seed that has a high percentage of broken seed coats.
3. Sow high-quality, high-germinating seed.
4. Grow soybean cultivars that are more resistant than others.

Red Crown Rot

Red crown rot is also called black root rot, Calonectria root rot, and Cylindrocladium root rot.

Cause. *Calonectria crotalariae,* anamorph *Cylindrocladium crotalariae. Cylindrocladium clavatum, Cylindrocladium floridanum,* and *Cylindrocladium scoparium,* and *Calonectria ilicicola* are also reported to be causal fungi on soybean.

Calonectria crotalariae overwinters as small (35–425 μm in diameter), irregular-shaped microsclerotia that serve as primary inoculum. Microsclerotia germinate and form hyphae that infect roots. The role of ascospores and conidia in the life cycle of red crown rot is not known. Colonization of lateral and taproots is optimum at a temperature of 25°C.

Nematodes have been reported to increase injury by the causal fungus.

Distribution. Possibly Brazil, Cameroon, Japan, Korea, and the southern United States.

Symptoms. In Brazil, *Cylindrocladium clavatum* causes disease symptoms on soybeans identical to those caused by *Cylindrocladium crotalariae.* Individual plants or groups of plants may show symptoms. The first symptom at early pod set (R3–R4) is a yellowing of the top leaves of diseased plants. Later, interveinal tissue becomes brown and defoliation occurs without wilting of the leaves. The insides of stems are gray-brown and entire root systems become rotted. After stem tissue is killed, red-orange perithecia develop from 2.5 to 7.5 cm above the soil line.

Management
1. Grow resistant varieties.
2. Delayed sowing will reduce initial inoculum.

3. Experimentally, the application of glyphosate has been shown to reduce the incidence of disease.

Rhizoctonia Aerial Blight

Rhizoctonia aerial blight is also called aerial blight, foliage blight, Rhizoctonia foliage blight, and web blight.

Cause. *Rhizoctonia solani* AG-1 is a common soil fungus and an excellent saprophyte. *Rhizoctonia solani* survives as sclerotia and as saprophytic mycelium in soil and infested residue for several years. In greenhouse inoculation studies, *Rhizoctonia solani* AG-1 also infected several weed hosts and grew from the diseased weed to a nearby soybean plant, demonstrating a potential overwintering site for the fungus.

Rhizoctonia solani AG-1 IA produces sasakii-type sclerotia on diseased tissue (aerial blight) and *R. solani* AG-1 IB and IC produce abundant microsclerotia on diseased tissue (web blight) that function as airborne propagules and cause secondary infection. During prolonged warm, humid weather, soil or plant parts that contain propagules of *R. solani* are splashed onto the lower plant leaves. Uninterrupted free moisture is most conducive to disease development. High humidity and warm temperatures also favor disease development but encourage mycelial growth and sclerotial formation within lesions on diseased leaves.

Within-row plant populations have no significant effect on disease or yield. Row spacings of 50 cm or more results in decreased disease but yields may be less than in narrower rows because higher plant populations compensate for the increase in disease.

Rhizoctonia solani AG-1 also causes seed and seedling rot. Infected seedlings serve as an inoculum source for the formation of disease foci. Isolates that cause foliage blight on soybeans may not necessarily cause root and stem rot. (For further information, see "Rhizoctonia Damping Off, Root, and Stem Rot.")

Distribution. Brazil, China, India, Japan, Malaysia, Mexico, the Philippines, and the United States (including Puerto Rico).

Symptoms. Aerial blight is characterized by the production of sasakii-type sclerotia on diseased tissue. Aerial blight first appears on diseased lower leaves, pods, and stems and gradually progresses up the plant. Mycelia of the pathogen spread in the canopy by means of mycelial bridges between leaves, pods, and stems.

Lesions on diseased leaves vary from small spots to entire leaves being affected. Symptoms usually start at the bases of leaflet petioles with subsequent spread in a fan-shaped pattern to the rest of the leaflet. Lesions initially are water-soaked and gray-green, and the mycelium advances on the leaf surface ahead of the lesion margin. Lesions become necrotic and dark brown to tan with a red-brown margin. Diseased areas that generally fall out during dry weather give leaves a ragged appearance. Diseased leaves droop down and severely diseased plants may be defoliated.

Web blight is characterized by the production of abundant microsclerotia on diseased tissues during the growing season. The microsclerotia function as the airborne propagules that cause secondary infection. Microsclerotia infections serve as numerous foci for small circular lesions on diseased leaves.

Most stem infections occur at or below the soil line before the stems turn green. Lesions on stems and petioles initially appear as reddish, oval to linear, water-soaked areas that become light brown, green-brown, or red-brown. Later, lesions change color to tan, brown, or black, sometimes have red-brown margins, and dry up. Sasakii-type sclerotia form when hypocotyls or stem lesions cease expanding.

Pods have small brown spots on them or the entire pod may be diseased. Symptoms of seed and seedling rot are circular lesions on the hypocotyls or the growing points. Infection of the growing point stops further growth of the seedlings.

Management
1. Adjust time of sowing to avoid periods of high moisture.
2. Do not sow plants too thickly.
3. Apply foliar fungicide during reproductive phases.
4. Grow resistant cultivars.

Rhizoctonia Damping Off, Root, and Stem Rot

Cause. *Rhizoctonia solani* AG-4, AG-5, and AG-2-2, teleomorph *Thanatephorus cucumeris,* are common soil fungi and excellent saprophytes. *Rhizoctonia solani* survives as sclerotia or saprophytic mycelia in soil and infested residue for several years. During wet soil conditions, sclerotia germinate and form mycelia, or mycelia grow from precolonized residue and infect seeds, roots, and hypocotyls. Following wet soil conditions, root rot is favored by low temperatures that are followed by warm temperatures (25°–29°C optimum). Disease development is most severe on young plants.

Rhizoctonia solani AG-1, the cause of Rhizoctonia foliar blight (aerial blight or web blight), also infects soybean seedlings. Diseased seedlings then become sources of inoculum by producing mycelia that infect neighboring host plants. Seedling infection at an early stage thus has a significant effect on subsequent disease development.

Distribution. Wherever soybean is grown.

Symptoms. Rhizoctonia root rot is first noticed as plants wilt and die during warm weather early in the growing season. The disease pattern within a field will vary from a single plant or groups of dead plants in a row or in circular areas where the soil moisture is higher than in other areas of the field.

Typical symptoms are decay of lateral roots and localized brown to red-brown lesions on diseased hypocotyls and lower stems that do not extend above the soil line. The red-brown color is a good symptom to diagnose the disease, but it is best observed immediately after removing plants from soil because the color fades upon exposure to air. The discoloration is usually limited to the cortical layer of main roots and hypocotyls and does not extend into the roots or stems. Diseased stems remain firm and dry. Although Rhizoctonia root rot is associated with young plants, older plants may die if there is moisture stress and the diseased hypocotyls and roots are sufficiently decomposed to limit uptake of water.

Management
1. Ridge soil around the bases of plants during cultivation to promote root growth above the diseased portions on the plant stem.
2. Apply seed-protectant fungicide.
3. Use good soil drainage.

Rust

Cause. *Phakopsora pachyrhizi* is the causal agent of rust in Asia and Australia, and *P. meibomiae* is the causal agent of rust in the Americas. The pycnidiospore and aeciospore stages are unknown. Teliospores form but have not been observed to germinate. Urediniospores are the overseasoning structures and are found on other leguminous host plants the entire year in warm climates. Urediniospores are disseminated by splashing rain and wind. Upon germination, the resulting germ tube penetrates directly through the cuticle or through stomates.

Infection occurs with a minimum of 6 hours of free moisture (8 hours is optimum) and optimal temperatures of 18° to 26.5°C. Urediniospores on dry foliage lose infectivity in 1 to 2 days during sunny conditions but maintain infectivity under cloudy conditions. *Phakopsora pachyrhizi* has a wide host range that includes several leguminous plants. Races of *P. pachyrhizi* exist.

Distribution. Australia, Brazil, China, Colombia, Costa Rica, Japan, Korea, Taiwan, and the United States (Puerto Rico).

Symptoms. The most commonly observed symptom is the sporulating uredinia on diseased lower leaf surfaces. Two infection types, tan (TAN) and reddish brown (RB), are recognized. The RB lesion type signifies at least partial resistance. At the onset of disease, chlorotic or gray-brown spots appear on diseased leaves and less frequently on petioles and young stem tissues. On leaves, spots enlarge to about 1 mm square and become red-brown or tan lesions that are delineated by secondary or tertiary leaf veins. Type RB has 0-2 uredinia on the abaxial surface, while TAN has 2-5 uredinia on the abaxial surface.

Young lesions of soybean rust are identical to those of bacterial pustule except for the chlorotic ring around the bacterial pustule lesion. As bacterial pustules mature, a crack or split occurs down the middle of the pustule. In contrast, rust uredinia have a distinct pore that has whitish urediniospores, which have been forced through the pore opening, clumped around it. The lesion area and number of uredinia per lesion is greater on younger than on older leaves. Severely diseased plants are defoliated and have reduced pod formation, seed numbers, and seed weights.

Management
1. Apply a foliar fungicide during the growing season.
2. Some accessions of soybean have resistance. Resistance has also been identified in accessions of the perennial *Glycine* spp., *G. argyrea*, *G. canescens*, *G. clandestina*, *G. latifolia*, *G. microphylla*, and *G. tomentella*.

Scab

Cause. *Sphaceloma glycines*.

Distribution. Japan.

Symptoms. Somewhat circular (4 mm in diameter), buff-colored fading to gray, and slightly raised lesions center on the veins of both leaf surfaces. Stem lesions are tiny, elliptical spots that eventually coalesce and form large buff-colored areas (up to 2 cm long) that sometimes have red-brown margins. Pod lesions are initially red to brown but become black with pale centers and red-brown margins at maturity. Seeds do not develop in diseased pods.

Management. Soybean accessions with resistance are known.

Sclerotinia Stem Rot

Sclerotinia stem rot is also called Whetzelinia stem rot and white mold.

Cause. *Sclerotinia sclerotiorum* (syn. *Whetzelinia sclerotiorum*). *Sclerotinia minor* has been associated with a Sclerotinia stem rot in Virginia. *Sclerotinia sclerotiorum* survives as sclerotia for at least 18 months at soil depths of 5 to 30 cm and for about 12 months on the soil surface. *Sclerotinia sclerotiorum* is also seedborne after a wet autumn.

During periods of high moisture, sclerotia germinate and either form several apothecia or mycelium. An apothecium forms at the end of a stipe and asci, each containing eight ascospores, form in a layer on the apothecial surface. Ascospores disseminated by wind to host plants, germinate under moist conditions and moderate air temperatures (20° to 28°C), and infect host plants of all ages. However, infection increases when the plant canopy closes, causing moderate temperatures and higher humidity within the canopy.

Spores initially colonize senescing flowers to obtain energy to parasitize living tissue on the same plant. When plants are crowded or become lodged, the fungal mat may spread to other host plants and infect them.

Eventually, sclerotia form in piths, on stems, and, rarely, in pods. Since sclerotia are roughly the same size as soybean seed, they are "harvested" along with seed and are "sown" with seed the following spring. The incidence of Sclerotinia stem rot in a field is determined primarily by the quantity of sclerotia in the field. Disease development is favored by wet weather.

Several plant species are hosts.

Distribution. Brazil, Canada, Hungary, India, Nepal, Nigeria, South Africa, and the United States.

Symptoms. Seedlings damp off and some cultivars have brown lesions. Infection of older plants originates at stem nodes 10 to 50 cm above the soil line. Water-soaked lesions progress acropetally and basipetally from the infected node along the stem and side branches. A white, cotton-like growth of mycelium occurs over the stem surface. Mycelia and numerous round to oblong, dark brown to black sclerotia are produced in diseased stems and on plant surfaces. Many sclerotia are approximately the same size and shape as seed.

Plants are killed when *S. sclerotiorum* girdles stems. On older plants (R2–R3 growth stage), leaves wilt and remain attached to the plant for a long time. Diseased seed is discolored, flattened, and smaller than healthy seed. Pods are rarely formed on the lower portions of diseased stems.

Management
1. Practice any cultural method that promotes drying out of the canopy, including growing soybean cultivars that do not readily lodge and sowing in rows wider than 50 cm apart. Sclerotia buried less than 2.5 cm in the soil often do not remain viable for more than 1 year. Deeply buried sclerotia may remain viable for 8 to 10 years. Deep plowing may remove sclerotia from providing inoculum the next year, but in future years, they may be returned to the surface, where they could germinate.
2. Do not rotate soybean with other susceptible plants.
3. Reduce irrigation before the R1 growth stage.
4. Grow the least susceptible cultivars.
5. Apply a fungicide seed treatment.

Southern Blight

Southern blight is also called sclerotial blight and southern stem blight.

Cause. *Sclerotium rolfsii* overwinters as sclerotia in soil and as mycelia in infested residue and seeds. Mycelia from sclerotia, residue, or seeds serve as primary inoculum. High soil temperatures, relatively dry soil, and a low soil pH of 3 to 6 favor disease development. High relative humidity stimulates aerial growth of mycelia on stems, branches, and leaves. Cool temperatures in the winter kill sclerotia and mycelia.

Distribution. Wherever soybean is grown in tropical and subtropical areas, but southern blight has also been reported from temperate soybean-growing areas.

Symptoms. Disease is most severe in sandy soils during high soil and air temperatures. Preemergence damping off may occur. The best symptom is a white cotton-like mycelial growth on diseased stems near the soil surface. Abundant tan to dark brown sclerotia the size of mustard seeds form on the surfaces of the mycelia. Disease on leaves is infrequent but occurs when infested soil particles are disseminated to leaves. Leaf spots, which are circular (6–12 mm in diameter) and brown with concentric markings, may have a clump of white mycelium with a sclerotium in the middle.

Management
1. Grow soybean cultivars that have resistance or tolerance.
2. Rotate soybean with other crops.
3. Plow infected soybean residue 12 cm deep or deeper.
4. Apply lime to acid soils to obtain a pH of 6.5 to 7.0.

Stem Canker

Stem canker is also called soybean stem canker and southern stem canker when caused by southern isolates of *Diaporthe phaseolorum*.

Cause. *Diaporthe phaseolorum* (syn. *D. phaseolorum* var. *caulivora*), anamorph *Phomopsis phaseoli*. However, many researchers prefer to distinguish between the northern and southern isolates of *D. phaseolorum*. Some refer to *Diaporthe phaseolorum* var. *meridionalis* as the southern or southeastern pathogen and retain the name *D. phaseolorum* var. *caulivora* for the midwestern pathogen. Others refer to the southern or northern biotypes of *D. phaseolorum* var. *caulivora*.

Since the two pathogens are in genetically distinct vegetative compatibility groups, they exhibit different behavior. Isolates of *D. phaseolorum* from Iowa differ in morphology and pathogenicity compared to isolates from the southern United States. At temperatures from 10° to 35°C, mycelial growth of southern isolates equals or exceeds that of northern isolates. Mycelial growth of northern isolates is inhibited at 30°C, while southern isolates maintain maximum growth rates. Pathogenicity of a southern isolate on soybean seedlings is not affected at temperatures of 21° to 30°C, but northern isolates are almost completely nonpathogenic at 30°C. The two pathogens differ in their sensitivity to benomyl and triadimefon. At least two physiologic races are reported to exist.

The causal fungi are seedborne and overwinter as perithecia in infested residue and seeds. Seed infection is reported to be primarily internal. Infected seed gives rise to seedlings with diseased cotyledons, which serve as sources of further infection. However, because of the low levels of infected seed, poor emergence of infected seed, and no effective secondary dissemination, there is little chance of an epiphytotic resulting directly from the sowing of infected seed. Also, several weeds are hosts and may serve as potential sources of inoculum.

In the spring, ascospores are produced in perithecia during wet weather and disseminated by wind to host plants. Both perithecia and pycnidia of *D. phaseolorum* var. *caulivora* are reported to develop on stems incubated at temperatures of 15°, 20°, 25°, and 30°C. Once spore deposition has taken place, either a prolonged continuous wetting event or several discontinuous wetting periods will facilitate infection and subsequent disease development. It is reported that more than 24 hours (120 to 144 hours maximum) of free moisture preceding a dry period are required for disease development.

Optimum temperatures for infection vary somewhat. Infection by both ascospores and conidia occurs at temperatures from 10° to 34°C (the optimum is 21°C). In greenhouse trials in Arkansas, optimum infection occurred at temperatures from 22° to 30°C accompanied by 40 hours of leaf wetness. Because *D. phaseolorum* is a splash-dispersed pathogen, infection is likely to occur after a rain that is followed by dew periods of at least 24 hours and temperatures from 22° to 30°C.

Younger plant parts are more susceptible to infection by *D. phaseolorum* f. sp. *meridionalis* than older plants. Specifically, plants exposed to inoculum from emergence to V3 stage become more diseased than plants exposed from the V3 through V10 stage. Most infections occur on lower leaf blades and petioles and not directly through leaf scars or stem wounds.

Early-sown soybeans are more susceptible to *D. phaseolorum* var. *caulivora* than later-sown soybeans. Seed infection occurs at 20°C and warmer and high relative humidities up to 100% for prolonged periods at growth stages R6 to R8.

Stem canker is most severe under no-till conditions because infested residue is in close proximity to aboveground plant parts. Disease incidence and severity were reported to be significantly greater in soybean/wheat double-cropping than in soybean monocultures in the southern United States. *Diaporthe* and *Phomopsis* isolates from cotton leaves and seedlings may also incite stem cankers on soybeans. Thus, cotton may serve as an alternate source of inoculum in warmer climates. However, cropping system effects may relate to symptom expression rather than plant infection.

In greenhouse tests, a 50% defoliation by the soybean looper, *Pseudoplusia includens,* reduced canker length and mortality. Canker lengths were also reduced on soybean cyst nematode-infected plants.

When introduced into soybean plants, *Diaporthe phaseolorum* is reported to have produced a phytotoxin that caused symptoms characteristic of stem canker.

Distribution. Canada and the United States.

Symptoms. Stem canker appears about midseason and continues until plant maturity. Plants infected during vegetative growth remain asymptomatic until they enter the reproductive phase of growth 60–70 days after sowing. Patches of scattered dead plants that still have their leaves attached occur in fields.

At first, small brown lesions occur on cotyledons, then extend into stems and kill diseased plants. On the lower part of diseased stems, small red-brown lesions occur on leaf scars (after the petioles have fallen) and on nodes. Lesions progress acropetally, becoming sunken necrotic cankers that eventually girdle stems and cause the upper part of a diseased plant to die. Cankers are usually unilateral and delimited by healthy tissue.

The lesions near the soil line can be confused with symptoms of Phytophthora root rot, but stem canker lesions can occur higher on the stem than those of Phytophthora root rot. After plant death, perithecia may be seen as tiny black dots aligned in rows on stems and scattered randomly over pods.

Yield loss occurs when plants are killed before complete pod fill.

Management
1. Sow seed from disease-free plants.
2. Plow infested residue in the autumn, if possible, to hasten the decay of residue and reduce the level of inoculum.
3. Rotate soybean with other crops.
4. Treat seed with a seed-protectant fungicide.
5. Grow resistant cultivars.

Stemphylium Leaf Blight

Cause. *Stemphylium botryosum.*

Distribution. India.

Symptoms. Initially, small, circular, necrotic spots with dark brown margins and gray centers appear on diseased leaves of all ages. Later, spots enlarge and coalesce, especially along the leaf margins. A dark green fungal growth occurs on the surfaces of spots. Eventually, leaves dry up and defoliation occurs.

Management. Not reported.

Sudden Death Syndrome

Cause. *Fusarium solani* f. sp. *glycines* is a blue-purple to bluish, slow-growing isolate of *F. solani,* which formerly was called strain A or form A (FSA or FS-A). The teleomorph is *Nectria haematococca.*

Fusarium solani f. sp. *phaseoli,* which formerly was called strain B or form B (FSB or FS-B), also is found associated with soybean roots but is now considered to be the cause of seedling disease and root rot only. However, some researchers have considered sudden death syndrome–inciting isolates of *F. solani* to be *F. solani* f. sp. *phaseoli,* a subgroup within *F. solani* f. sp. *phaseoli,* or possibly a separate forma specialis.

Production of macroconidia by *F. solani* f. sp. *glycines* occurs on lower stems and roots of symptomatic plants. Sporulation occurs on taproots 10 cm or more below the soil line. It is presumed that such sporulation contributes to the inoculum density of *F. solani* f. sp. *glycines* in soils. It is likely conidia produced on plants are washed into soil and convert to chlamydospores. The fungus then resides in soil and in root debris mainly as chlamydospores, which constitute the primary inoculum. *Fusarium solani* f. sp. *glycines* likely penetrates roots directly. Colonization appears to be limited primarily to cortical tissue of the root and lower stem. Hyphae grow mainly intracellularly.

Fusarium solani f. sp. *glycines* infects roots in high-yield environments. Cool temperatures during early soybean reproduction favor disease development. Symptom development is greater when inoculated plants are incubated at 25°C than at 35°C. However, some researchers determined foliar symptoms were most severe from 22° to 24°C and root symptoms were most severe at 15°C. Disease is more severe on most soybean cultivars when they are sown early; however, the reverse was true for some cultivars in Kentucky, but other cultivars were unaffected. High soil moisture favors disease development. Sudden death syndrome is prevalent in irrigated fields or in fields with high soil moisture following rain.

Disease development may increase during the change from vegetative growth to reproductive growth because of the change in host physiology. As plants enter reproductive development, net root growth is dramatically reduced and photosynthates are transported to the developing pods and seeds instead of to the vegetative parts of the plants, which may reduce plant defenses.

Sudden death syndrome leaf symptoms can be incited independently of root symptoms, which suggests a phytotoxin(s) is translocated from the roots and is responsible for foliar symptoms. A phytotoxic polypeptide and the phytotoxin monordin from culture filtrates of sudden death syndrome–inciting isolates reproduced foliar symptoms, but non-inciting isolates did not.

High populations of the soybean cyst nematode (SCN), *Heterodera glycines,* have been thought to be associated with sudden death syndrome. However, in some areas, this relationship does not always occur. *Heterodera glycines* possibly hastens the onset and increases the severity of foliar symptoms. The fungus is capable of surviving in nematode cysts in vitro at a temperature of 10°C for a long enough time to overwinter in the field.

This may influence its inoculum potential and dispersal. Typically, the fungus grows within individual cysts, where the probability of its survival until spring may be greater than being free in soil or in infested soybean residue where competition with other microorganisms could be a factor. It is also possible that cysts provide a food base for germination of chlamydospores and growth. *Fusarium solani* f. sp. *glycines* has infrequently been isolated from SCN eggs, but the effect of this interaction has not been determined.

Fusarium solani isolates associated with sudden death syndrome are not host-specific and readily infect other bean hosts. Sudden death syndrome incidence is greater in no-till than in conventional tillage systems.

Distribution. The United States (Arkansas, Illinois, Indiana, Iowa, Kansas, Kentucky, Mississippi, Missouri, and Tennessee).

Symptoms. Sudden death syndrome is a mid- to late-season disease favored by high-yield environments. Symptoms are first noticed after flowering (R3–R4) and continue through maturity. If symptoms develop before the R6 stage, it is likely that complete defoliation and pod abortion will occur in 3 to 6 weeks.

The interveinal chlorosis of foliage is followed in 5 to 10 days by necrosis of the tissue. Necrosis continues to develop until only tissue near the major leaf veins remains green. These symptoms appear throughout the plant but are most severe on the top leaves of a diseased plant. Leaves abscise at the apex of the petiole, leaving leafless stems with attached petioles; later, the petioles may also abscise, causing premature plant defoliation.

Flower and pod abortion occurs. The abscission of pods is most pronounced at the top of the plant. Plants that develop symptoms at a later maturity produce smaller, lighter seed than healthy plants.

Stem and root vascular tissues of diseased plants become reddish brown, but the pith remains a "normal" white color. The discoloration progresses outward from vascular tissue. Roots of diseased plants appear normal at first, but plants with advanced symptoms have a discoloration of roots and lower stems that occurs together with a lateral root rot. After considerable root degradation, chlamydospores may be observed in degraded, sloughed cortical tissue, and blue to blue-green masses of macroconidia may be visible on diseased roots near the soil line. Sporulation on root surfaces is always associated with root rot and is more frequent during or immediately following periods of high soil moisture.

A seed and seedling rot is also reported to occur.

Management. Significant differences in disease development between cultivars occurs at the R3 growth stage and are more distinct at the R6 growth stage. A single dominant gene, Rfs, in the cultivar 'Ripley' controls resistance.

Research has shown that differences appear to be associated with response of some cultivars to race 6 of the SCN. Cultivars susceptible to race 6 were susceptible to SDS, and those that are moderately resistant or resistant to race 6 had low levels of resistance to SDS. With other cultivars, susceptibility to soybean cyst nematode was not related to susceptibility to SDS.

Target Spot

Cause. *Corynespora cassiicola* (syn. *Cercospora melonis, Cercospora vignicola, Helminthosporium vignae,* and *H. vignicola*) survives for 2 or more years as mycelia and chlamydospores in infested residue and soil, and grows saprophytically on different kinds of plant residue in soil. Infection of roots and stems occurs early in the growing season in moist soils at temperatures of 15° to 18°C. Later in the growing season, conidia splashed and blown to leaves cause infection when free moisture is present or the relative humidity is 80% or more. A different race of *C. cassiicola* is thought to infect hypocotyls, roots, and stems than the race that infects leaves, pods, and seeds.

Distribution. Eastern Asia, Canada, Central America, and the United States.

Symptoms. No symptoms develop at temperatures above 20°C. Target spot is most noticeable on diseased leaves later in the growing season. Lesions on leaves are red-brown and vary from small specks to spots 12 mm or larger in diameter. The larger spots are generally zonate, suggesting a target-like appearance. Spots are generally surrounded by a light green to yellow-green halo. Narrow, elongated spots may develop along veins of the upper leaf surface.

Spots on petioles and stems are dark brown and vary in shape from a small speck to an elongated, spindle-shaped lesion. Spots on pods are generally round (1 mm in diameter) with a slightly depressed, purple-black center and brown margins. Sometimes small dark brown spots are formed on seeds.

Taproots, hypocotyls, and larger lateral roots have oval red-brown lesions that later become purple due to *C. cassiicola* sporulating. Lesions will grow larger and coalesce as the plant matures.

Management
1. Grow resistant cultivars.
2. Treat seed with a seed-protectant fungicide.
3. Rotate soybean with other crops.

Thielaviopsis Root Rot

Thielaviopsis root rot is also called black root rot.

Cause. *Thielaviopsis basicola* survives as chlamydospores in infested residue. Chlamydospores germinate during moist soil conditions and temperatures from 16° to 20°C and infect hypocotyls and roots early in the growing season. Chlamydospores are formed in diseased tissues and returned to soil at harvest time as diseased tissues disintegrate. Seedborne infection is uncommon.

Distribution. Canada, Germany, and the midwestern United States, particularly Michigan.

Symptoms. The cortex of the hypocotyl below the soil line and taproots and fibrous roots become dark brown and necrotic. Frequently a severe root rot develops, causing subsequent wilting of foliage or plant death.

Management
1. Grow resistant cultivars.
2. The herbicide chloramben has been reported to increase injury.

Top Dieback

Cause. *Diaporthe phaseolorum* (syn. *D. phaseolorum* var. *caulivora*), anamorph *Phomopsis phaseoli*.
    The causal fungi are seedborne and overwinter as perithecia in infested residue and seeds. Seed infection is reported to be primarily internal. Infected seed give rise to seedlings with diseased cotyledons that serve as sources for the further infection of host plants.

Distribution. Presumed to be similar to stem canker; however, distribution of top dieback may differ from stem canker.

Symptoms. Disease develops late in the growing season and is distinct from the disease stem canker. The five or six uppermost internodes of diseased plants die prematurely.

Management
1. Sow seed from disease-free plants.
2. Plow infested residue in the fall, if possible, to hasten decay of residue and reduce the level of inoculum.
3. Rotate soybean with other crops.
4. Treat seed with a seed-protectant fungicide.

Twin Stem Abnormality Disease

Cause. *Sclerotium* sp. and *Macrophomina phaseolina*.

Distribution. Brazil and the United States.

Symptoms. Symptoms, which appear after cotyledons have opened, have three distinct severity forms. The most severe form is type I. Elongation of the first internode that bears the primary leaves is completely inhibited. The apical meristem resembles a convex knob between the cotyledons. The cotyledons are large and spongy and have a thickened hypocotyl. In 7 to 10 days, two pairs of primary leaves with rudimentary petioles and internodes form. At maturity, the plant is stunted and bushy and may have two main stems originating from the cotyledonary node.

In type II, the moderately severe form, the first internode with underdeveloped primary leaves elongates to a limited extent. The original primary leaf primordials are necrotic. Two secondary pairs of primary leaves develop from the cotyledonary node and continue to grow. Adult plants are stunted and bushy and may have two main stems originating from the cotyledonary node.

In type III, the less severe form, the first internode with underdeveloped primary leaves elongates normally or excessively. Leaves remain wrapped around each other and are bleached at the apex. When leaves open, the bleaching extends to the margin and inward, the apex is curved inward, and the entire leaf is wrinkled. The first trifoliate leaf is formed with little or no elongation of the second internode. One or two pairs of leaves may form at the cotyledonary node but generally remain rudimentary.

Management. Not reported.

Verticillium Pod Loss

Cause. *Verticillium nigrescens.* High soil moisture levels are related to increased infection. Increasing incidence is related to a long history of cotton production in the same field in which soybeans are grown.

Distribution. The United States (Georgia).

Symptoms. *Verticillium nigrescens* normally grows from one end of a pedicel and rarely from other parts of the pod. The fungus can be isolated from flowers and pods at all stages of development. The numbers of pods per plant and seed weight are reduced but there is no vascular discoloration or wilting.

Management. Not reported.

Verticillium Wilt

Cause. *Verticillium dahliae* is possibly seedborne.

Distribution. Not known for certain, but *V. dahliae* has been isolated from soybeans growing in fields in the United States (Iowa).

Symptoms. Vascular tissue in the lower portion of the diseased stem is discolored. Leaves die but remain attached to diseased plants that are scattered randomly throughout a field.

Inoculated plants display sudden wilt symptoms 7 to 10 days after inoculation. Early wilt symptoms occur as apical, marginal, or unilateral wilting of primary leaves. Young tissues remain green and dry without initially
becoming chlorotic. No pith discoloration occurs but a red-brown to black
discoloration is restricted to the vascular tissue.

Management. Some cultivars are evidently resistant. Verticillium wilt is not
considered an important disease, but possibly other diseases have been
mistaken for it.

Yeast Spot

Yeast spot is also called Nematospora spot.

Cause. *Nematospora coryli* is a yeast fungus that survives in and is transmitted
primarily by the adult green stinkbug, *Acrosternum hilare.* Other species of
stinkbugs known to transmit *N. coryli* are the southern green stinkbug,
Nezara viridula, and several *Euschistus* spp., including the brown stinkbug,
E. servus. The stinkbug pierces the pod wall and feeds on seed contents liq-
uefied by salivary enzymes. During feeding, only yeast cells located in the
region of the mouthparts are forced into the developing seed.

Distribution. Africa, Brazil, and the United States.

Symptoms. Yeast spot is primarily a disease of seeds. Seeds infected early may
not mature or remain small and shriveled. Where infection occurs during
pod formation, seeds fail to develop and pods drop prematurely. On developing green seeds, diseased tissue is slightly depressed and varies from a
light cream or yellow to brown. There may be varying amounts of dead tissue extending into diseased embryos. On the surfaces of the seed coats, lesions appear as very small, discolored punctures. Seed shriveling caused by
other diseases may be differentiated from that caused by yeast spot by observing these punctures. The outside of the pods display small, pinpoint,
discolored areas.

Management
1. Sow disease-free seed.
2. Control stinkbugs by insecticides when nymphs average one per 30 cm
   of row.

Diseases Caused by Nematodes

Columbia Lance

Cause. *Hoplolaimus columbus* can survive in dry soil for a long time and has a
life cycle similar to the *Meloidogyne* spp. that cause root knot. *Hoplolaimus
columbus* is an ectoparasite that prefers to feed in the cortex area.

Distribution. The southeastern United States, particularly on sandy soils.

Symptoms. Diseased plants are stunted and chlorotic. Roots are rotted and sparse. In areas of high nematode populations, root systems may be almost completely destroyed as plants near maturity.

Management
1. Rotate soybean with other crops.
2. Plow under infested residue.
3. Clean soil from equipment moved between fields.
4. Apply nematicides.
5. Subsoil or chisel plow to a 36-cm depth where hardpan conditions exist.

Lance

Cause. *Hoplolaimus galeatus.*

Distribution. The southern United States.

Symptoms. Diseased plants are stunted and chlorotic. The taproot becomes necrotic and large numbers of secondary and tertiary roots are produced.

Management
1. Grow tolerant varieties.
2. Sow early.
3. Use nematicides where practical.
4. Subsoil plow where hardpan conditions exist.

Lesion

Cause. *Pratylenchus* spp. are endoparasitic nematodes that enter roots as larvae and attack the root cortex.

Distribution. Wherever soybean is grown.

Symptoms. Diseased plants may be yellowed and stunted. Initially, roots have small dark lesions. Other soilborne microorganisms then may enter the root through wounds made by *Pratylenchus* spp. and cause root rot and a browning of the root system.

Management. Grow resistant cultivars.

Reniform

Cause. *Rotylenchulus reniformis* is a nematode species in which only females invade roots after feeding on the epidermal tissues. *Rotylenchulus reniformis* is semi-endoparasitic because the posterior portion of the female remains outside the root and swells to a kidney shape. Eggs are laid outside the root. Optimum root invasion and nematode development occurs at a temperature of 29°C.

Distribution. Africa and the southeastern United States.

Symptoms. Diseased plants are stunted and chlorotic. Small galls or root swellings sometimes occur on damaged roots.

Management
1. Grow resistant cultivars.
2. Apply nematicides where feasible.

Ring

Cause. *Criconemella ornata* is an ectoparasitic nematode.

Distribution. The United States, probably when soybean is grown in rotation with peanut.

Symptoms. Diseased plants are chlorotic and have necrotic lesions on roots. The number of lateral roots is reduced.

Management. Nematicides are reported to be effective on other crops.

Root Knot

Cause. *Meloidogyne* species, including *M. arenaria, M. hapla, M. incognita,* and *M. javanica. Meloidogyne arenaria, M. incognita,* and *M. javanica* do not survive if the soil temperature averages below 3°C during the coldest month. Infested soil contains eggs that hatch and produce larvae that move through soil and penetrate root tips. Larvae migrate inside the infected root to the vascular tissues, where they become sedentary and commence feeding. Excretions of the larvae stimulate cell proliferation, resulting in the development of syncytia, from which larvae derive their food. Females lay eggs that may be pushed out of the root into the soil.

Distribution. Wherever soybean is grown in sandy, light-textured soils in warm climates.

Symptoms. Diseased plants will be stunted. Leaves turn yellow and wilt during hot, dry weather. Roots have knots or galls of different sizes that disrupt the flow of water and nutrients to the tops of plants.

Management
1. Rotate with an alternate crop for 3 years if the species of nematode is known. Maize can be rotated with soybean to manage *M. hapla.* Cotton can be rotated to manage all species except *M. incognita.*
2. Grow resistant cultivars; however, different races of *M. incognita* are known to occur.
3. Apply nematicides.

Sheath

Cause. *Hemicycliophora* spp. are ectoparasitic nematodes that feed primarily on the root tips.

Distribution. Widespread in sandy soils.

Symptoms. Root tips are galled. Diseased plants are generally unthrifty.

Management. Nematicides are reported to be effective.

Soybean Cyst

Cause. *Heterodera glycines* survives in soil as eggs contained in cysts formed by female bodies. Spread of *H. glycines* in a field or over long distances is by any agent capable of moving infested soil and residue. Larvae hatch from eggs, move through soil, and enter plants near the root tips. Nematodes migrate inside the root to undifferentiated cortical and stelar tissues. Syncytia are formed by salivary excretions. As the female matures, it swells and breaks through the epidermis but its neck remains in the root. Females are then inseminated by males and produce up to 500 eggs that are released into the soil or retained in the female's body until a cyst is formed. Eggs have been reported to survive in cysts for up to 11 years. Eggs are deposited externally and larvae hatch to begin a new life cycle. Nematode development is most rapid at 30°C. Different races exist.

Distribution. Asia, Brazil, and the southeastern and midwestern United States.

Symptoms. Injury to host plants is most serious in dry, sandy soils that are deficient in nutrients and organic matter. Not all plants display aboveground symptoms. Diseased plants are stunted and chlorotic and may lose their leaves prematurely. Affected leaves are distributed over the plant. A yellowing starts at the leaflet margin and gradually includes all of the blade.

Light to moderate infection will stimulate overproduction of lateral roots, but severely diseased plants have a small, necrotic root system with few or no *Rhizobium* nodules. White to yellow females and brown cysts that developed from female bodies are attached to diseased roots and appear somewhat smaller than the head of a pin.

Management
1. Grow resistant cultivars. Resistance to race 3 is conditioned by one dominant and two recessive genes in each parent.
2. Rotate soybean with other crops for 2 to 3 years.
3. Practice sanitation; clean soil from equipment moved between fields.
4. Apply nematicides to soil if feasible.
5. Irrigate to reduce water stress to plants; however, nematode damage is not reduced.

6. Delay in sowing associated with double-cropping wheat in Missouri allowed for a significant decline in numbers of nematodes between early May and late June.
7. Tillage systems. A report from Missouri said populations of *H. glycines* were lower the first year in soybeans grown under no-till tillage than in soybeans grown under conventional tillage practices. No differences in populations were noted the second year.
8. In greenhouse tests, numbers of cysts and juveniles were reduced on plants cankered by *Diaporthe phaseolorum* var. *caulivora*.

Spiral

Cause. *Helicotylenchus* spp. are ectoparasites.

Distribution. Widespread.

Symptoms. Symptoms on soybeans are relatively unknown. Under low- to mid-level populations, a few root cells are destroyed, but roots are thought to be severely affected when infected by high populations of nematodes.

Management. Nematicides are effective.

Stem

Cause. *Ditylenchus dipsaci* is an endoparasite. High humidity and rainy weather favor injury by *D. dipsaci*.

Distribution. The United States.

Symptoms. Diseased stems are stunted, swollen, and distorted. Young leaves are also infected and have a chlorotic spotting.

Management. Stem nematode is not considered a serious problem on soybean.

Sting

Cause. *Belonolaimus gracilis* and *B. longicaudatus* are nematodes that overseason in soil as eggs and larvae. Feeding by nematodes occurs only on root surfaces (not internally). Disease occurs on sandy soils.

Distribution. The southeastern United States.

Symptoms. Damage is confined to plants growing on sandy soils. Seedlings may be killed, which results in a stand loss. Diseased plants other than seedlings usually are not killed but display stunted, chlorotic, or discolored symptoms similar to those of moisture stress.

Diseased roots initially have small, dark, sunken lesions that later enlarge and girdle root tips. Meristematic tissues are destroyed and root proliferation above the area of discoloration results in a stubby and dark root system. Eventually, roots become severely rotted.

Management
1. Use nematicides.
2. Rotate soybean with other crops.
3. Plow under infested residue.
4. Clean soil from equipment moved between fields.

Diseases Caused by Phytoplasmas

Machismo

Machismo is also called amachamiento and proliferacion de yemas.

Cause. A phytoplasma. The causal agent, which is transmitted by the leafhopper *Scaphytopius fuliginosus,* is not known to be mechanically transmitted or seedborne. Several weeds also serve as hosts.

Distribution. Colombia and Mexico.

Symptoms. Symptoms are most severe on plants infected while young. Symptoms first occur at the time of flowering or when pod formation begins. Frequently a branch that is close to the ground will show symptoms but the rest of the plant will appear normal.

Pods may be in an upright position, rigid, curved, flat, and thin with no seed produced, or they may be transformed into corrugated, leaf-like structures resembling phyllody. Once floral parts have been transformed into leaf-like structures, buds proliferate from leaf axils anywhere on the plant, causing a witches' broom.

Sepals may be larger and duller than normal and have the appearance of small leaves. Flowers are small and remain closed. In these cases, the pods or leaf-like structures grow through the tips of the unopened flowers.

Pods formed on diseased plants are slightly corrugated and thicker and softer than those on healthy plants. The roots of immature green seeds that germinate in the pod rupture through the pod wall.

Diseased plants have a darker green color and remain green longer than healthy plants, which causes difficulty in harvesting.

Management
1. Control weed hosts.
2. Rogue infected plants.
3. Apply tetracycline.

Soybean Bud Proliferation

Cause. A phytoplasma transmitted by seed and the leafhopper *Scaphytopius acutus.* It is not mechanically transmitted.

Distribution. The United States (Arkansas and Louisiana).

Symptoms. Symptoms occur at the onset of flowering. Diseased plants have delayed senescence and remain green after healthy plants have matured. There is a proliferation of adventitious buds. Pods are small, underdeveloped, and few in number, and many contain only one large bean.

Management. Not reported.

Witches' Broom

Cause. A phytoplasma-like organism transmitted by the leafhoppers *Orosius orientalis* and *O. argentatus*.

Distribution. Brazil, India, Indonesia, Japan, Mozambique, Nigeria, and Tanzania.

Symptoms. Symptoms resemble those of machismo. Symptoms include a phylloid disorder of floral organs, shoot proliferation, and reduced leaflets.

Management. Not reported.

Diseases Caused by Viruses

Alfalfa Mosaic

Cause. Alfalfa mosaic virus is seedborne and is transmitted by several aphid species.

Distribution. Argentina, Brazil, Japan, Romania, South Africa, and the United States.

Symptoms. Early infections result in the mottling or chlorotic spotting of diseased leaves. Later infections result in a yellow mottle, star-shaped flecks, or slight crinkling.

Management. Not reported.

Bean Pod Mottle

Cause. Bean pod mottle virus (BPMV) is in the comovirus group. BPMV is transmitted by several insects, particularly the bean leaf beetle, *Cerotoma trifurcata*. BPMV is commonly associated with soybean mosaic virus in diseased plants.

Distribution. Ecuador and the United States.

Symptoms. Young diseased leaves have a green to yellow mottling, particularly during rapid growth and cool weather, that disappears as leaves mature. Diseased plants under moisture stress are less turgid than healthy plants and may have a yield loss if stress occurs during the reproductive phase.

Management
1. Grow at least four rows of maize or sorghum between soybean and other legume fields and noncultivated areas.
2. Rogue out diseased plants from seed fields.
3. Control weeds, particularly in noncrop areas next to soybean fields.

Brazilian Bud Blight

Brazilian bud blight is also called tobacco streak.

Cause. Tobacco streak virus (TSV) is in the ilarvirus group. TSV is seedborne and sap-transmissible but is not known to be transmitted by insects. Symptoms appear sooner on inoculated plants grown at 30°C than at lower temperatures, and under continuous light for 24 hours than under 12 hours of dark followed by 12 hours of light.

Distribution. Brazil and the United States (Iowa and Oklahoma).

Symptoms. Symptoms resemble those of bud blight caused by tobacco ringspot virus and are normally not seen on young infected plants. Some plants remain green after maturity and have unexpanded and unfilled pods. Initially, irregular yellow spots develop on diseased leaves. Diseased plants then develop numerous stunted axillary branches that produce dwarfed leaves. Mosaic symptoms and necrotic streaks develop in the stem pith, especially at the stem nodes. Necrotic blotches appear on pods. Infection of the plant at any age delays seed maturation, but early infection prevents seed formation altogether. Diseased plants are stunted and stem tips curve and die.

Management
1. Grow at least four rows of maize or sorghum between soybean and other legume fields and noncultivated areas.
2. Rogue out diseased plants from seed fields.
3. Control weeds, particularly in noncrop areas next to soybean fields.

Bud Blight

Cause. Tobacco ringspot virus (TRSV) is in the nepovirus group. TRSV is seedborne and overwinters or is reservoired in a large number of legumes. TRSV is sap-transmitted by nematodes, particularly the dagger nematode, *Xiphinema americanum;* nymphs of *Thrips tabaci;* grasshoppers, *Melanoplus differentialis;* and possibly other insects. TRSV is also transmitted by root grafts.

Distribution. Australia, Canada, China, and the United States.

Symptoms. Bud blight occurs sporadically, appearing only in certain years, and is frequently more severe in fields adjacent to legume-grass pastures or fencerows, where insects transmit the virus from perennial plants to soybeans. Symptoms of bud blight vary with the stage of growth at which the plants became infected. When plants are infected before flowering, apical buds and shoots turn brown, curve downward and form a crook, and become dry and brittle; hence, the name bud blight. Young diseased leaves develop a rusty flecking and the plants are stunted and produce little seed below 25°C. Several strains of TRSV cause severely dwarfed and barren plants.

Plants infected while less than 5 weeks old are severely stunted because of shortened internodes or fewer nodes. Sometimes the stem below the blighted terminal bud is internally discolored at the nodes. Brown streaks occasionally occur on petioles and large leaf veins. Petioles of the youngest trifoliate leaves are often thickened, shortened, and occasionally curved, distorting shoot tips. Diseased leaf blades are rugose, bronzed, dwarfed, and cupped.

When plants are infected during flowering, small underdeveloped pods are produced. Infection after flowering results in conspicuous dark blotches on poorly filled pods that eventually fall to the ground.

A good diagnostic symptom is the presence of diseased plants that remain green after normal plants have matured. These plants are easily found in a field in the autumn. Losses of 100% have been reported.

Management
1. Do not grow soybean adjacent to legume fields. If this is not feasible, grow a buffer strip of a few rows of maize or forage sorghum between legumes and the soybean field.
2. Control broadleaf weeds in noncrop areas adjacent to soybean fields before soybeans emerge.
3. Rogue infected plants from seed production fields.

Cowpea Chlorotic Mottle

Cause. Cowpea chlorotic mottle virus, soybean-infecting strain (CCMV-S) is in the bromovirus group. CCMV-S is transmitted by sap, the bean leaf beetle *(Cerotoma trifurcata),* and the spotted cucumber beetle *(Diabrotica undecimpunctata).*

Distribution. The United States.

Symptoms. Diseased leaves are mottled, slightly crinkled, and tend to stand upright. Mottling is most intense on the diseased younger leaves. Plants are stunted and produce fewer seeds, which are small in size and reduced in quality. Local necrotic lesions may be produced on resistant cultivars and veinal necrosis may occur on cultivars with intermediate resistance.

Management. Grow resistant cultivars.

Cowpea Mild Mottle

Cause. Cowpea mild mottle virus (CPMMV) is in the carlavirus group. CPMMV is transmitted mechanically and by the whitefly *Bemisia tabaci* but has been demonstrated not to be seedborne.

Distribution. Ivory Coast and Thailand.

Symptoms. Diseased plants are stunted and show a light green mosaic. Inoculated seedlings display vein clearing that evolves into a yellow mosaic and occasional crinkling of leaves. Leaves may also be curled downward or cupped upward. Some cultivars display a vein and top necrosis. Vein enations have also been reported as a symptom.

Management. Not reported.

Cowpea Mosaic

Cause. Cowpea mosaic virus (CPMV) is in the comovirus group. CPMV is transmitted by sap and the bean leaf beetle, *Cerotoma ruficornis*.

Distribution. Central America and the United States (including Puerto Rico).

Symptoms. Inoculated primary leaves have chlorotic and necrotic lesions. A systemic light green mosaic occurs on diseased trifoliate leaves together with very severe leaf malformation and bud blight.

Management. Not reported.

Cowpea Severe Mosaic

Cause. Cowpea severe mosaic virus (CPSMV) is in the comovirus group. CPSMV is transmitted by the beetle *Cerotoma arcuata*.

Distribution. Brazil.

Symptoms. Symptoms are identical to those of bud blight. Significant reductions occur in plant height, yield, number of pods per plant, and seed germination.

Management. Not reported.

Delayed Maturity

Cause. Possibly a virus-like agent that is seedborne and also spread by insects.

Distribution. The United States (Louisiana).

Symptoms. Diseased plants remain green and have pronounced longitudinal ribs after healthy plants have matured. Diseased seeds often germinate inside the pod and rot. Pods often are rough textured and have pronounced wrinkles and veins.

Management. Not reported.

Indonesian Soybean Dwarf

Cause. Indonesian soybean dwarf virus is transmitted in a persistent manner by the aphid *Aphis glycines*.

Distribution. Indonesia.

Symptoms. Diseased plants are dwarfed and dark green with shortened leaf petioles and internodes. Upper leaves are small and curl upward. Lower leaves are rugose and brittle and have shortened veins and interveinal white necrosis. Occasionally diseased leaves have numerous holes caused by their brittleness. Diseased plants produce few pods.

Management. Not reported.

Mung Bean Yellow Mosaic

Cause. Mung bean yellow mosaic virus is transmitted by the whiteflies *Bemisia tabaci* and *B. gossypipderda*.

Distribution. Asia.

Symptoms. A yellowing on the veins of diseased leaflets later develops into a severe yellow mosaic.

Management. Resistance has been identified.

Peanut Mottle

Cause. Peanut mottle virus (PeMoV) is in the potyvirus group. PeMoV is transmitted by sap and in a nonpersistent manner by several species of aphids, but PeMoV is not seedborne in soybeans. PeMoV can be spread from nearby peanut fields to soybean fields.

Distribution. The southeastern United States.

Symptoms. A general mosaic that is similar to symptoms of diseased plants caused by other viruses occurs. Small chlorotic areas that enlarge but do not become continuous leave striking, dark green islands on diseased leaves. Later, chlorotic patches, line patterns, or ring patterns occur on the third and fourth leaves after infection. Leaves that develop even later display a general mosaic without distinct characteristics. Diseased leaves may pucker and curl down at the edges.

Management
1. Grow resistant cultivars.
2. Do not grow soybean following peanut.
3. Separate soybeans from peanut fields by 50 m or more.
4. Rogue volunteer peanut plants from soybean fields.

Peanut Stripe

Cause. Peanut stripe virus (PStV) is in the potyvirus group. PStV is transmitted mechanically and by aphids, but it is not seed-transmitted in soybeans. Infective virus particles can be recovered from seed coats of immature soybean seeds, but as the seeds mature, the virus loses its infective properties. A possible source of inoculum for soybeans may be diseased peanut seed sown in adjacent fields. PStV is then spread from the diseased peanuts to soybeans.

Distribution. China and the southeastern United States.

Symptoms. Symptoms vary according to cultivar and virus isolate. Diseased leaves develop necrotic local lesions that eventually develop into systemic necrosis. Some cultivars develop only a mild mottle.

Management
1. Resistant lines have been identified.
2. Do not grow soybean adjacent to peanut.

Peanut Stunt

Cause. Peanut stunt virus (PSV) is in the cucumovirus group. PSV overwinters in several legumes, such as lespedeza, crown vetch, and peanut, and is disseminated by aphids.

Distribution. The southeastern and central United States.

Symptoms. Symptoms range from diffuse chlorotic lesions to necrotic local lesions on inoculated primary leaves. Vein clearing occurs, followed by a general mosaic of diseased trifoliate leaves. Some diseased plants may be symptomless.

Management. Not reported.

Soybean Chlorotic Mottle

Cause. Soybean chlorotic mottle virus (SbCMV) is in the caulimovirus group. SbCMV is transmitted by sap but not by aphids or seed.

Distribution. Japan.

Symptoms. Diseased plants have mosaic symptoms and are stunted due to shortened internodes. Mechanically inoculated plants initially display vein clearing, chlorosis, and leaf curling of diseased young leaves, which are also smaller than healthy leaves. Eventually, diseased leaves have mottled symptoms.

Management. Not reported.

Soybean Crinkle Leaf

Cause. Possibly a virus that is transmitted by the whitefly *Bemisia tabaci*.

Distribution. Thailand.

Symptoms. Diseased leaves are twisted or curled, and the undersides have veinal enations on them. Plant foliage of diseased plants is darker green than that of healthy plants. In the greenhouse, diseased plants have a yellow netting of veins 10 to 14 days after inoculation.

Management. Not reported.

Soybean Dwarf

Cause. Soybean dwarf virus (SDV) is in the luteovirus group. Another luteovirus, subterranean clover red leaf virus (SCRLV), is considered a strain of SDV. SDV is transmitted by grafting and in a persistent manner by the aphids *Aulacorthum solani* and *Acyrthosiphon pisum*. SDV is not seed- or sap-transmitted. The temperature for optimum transmission is 20°C.

Based on the symptomatology of SDV-infected plants, two strains exist: dwarfing (SDV-D), and yellowing (SDV-Y). The two strains have small differences in host range and physiochemical properties. SDV-D is transmitted more efficiently than SDV-Y by the foxglove aphid, *Aulacorthum solani*.

Distribution. Japan. If SCRLV is a strain, SDV also occurs in Australia and New Zealand.

Symptoms. Diseased plants are severely stunted. Leaves are rugose, curl down, and may have interveinal yellowing.

Management. Not reported.

Soybean Mosaic

Soybean mosaic is also called soybean crinkle.

Cause. Soybean mosaic virus (SMV) is in the potyvirus group. SMV is seed-borne and most primary inoculum consists of diseased seedlings derived from SMV infected seed. SMV also overwinters in perennial weeds. Dissemination within a field, or secondary spread, occurs by the activities of several aphids that transmit SMV in a nonpersistent manner. SMV is also readily sap-transmissible.

Infection before the onset of flowering results in higher levels of seed transmission than in later infections. The percentage of SMV-infected seed will gradually decrease when secondary spread is minimal and if SMV spread occurs after the onset of flowering.

Several strains of SMV have been reported to exist.

Distribution. Wherever soybean is grown.

Symptoms. Symptoms vary from normal-appearing leaves, in which veins are deep in the lamina, to rugose leaves, in which portions of the diseased upper leaf surface has become invaginated into vesicles. The best diagnostic symptom is rugose leaves that appear on the third leaf formed after infection. If diseased plants are grown under cool conditions, the rugose symptoms increase in severity on successive leaves, and the characteristic mosaic pattern of alternate light and dark green patches of leaf tissue may appear. The mosaic symptom is more obvious on some cultivars than others. Leaves become leathery to the touch, coarse, and brittle as diseased plants near maturity. Severely diseased leaves are narrow and have a willow-leaf appearance. If diseased plants are grown under warm conditions at temperatures of 31°C or more, symptoms are masked and leaves develop normally. When very susceptible cultivars are diseased, a yellowish vein clearing develops in the small branching veins of developing leaves 6 to 14 days after infection.

Infected seed may not germinate. Seedlings from infected seed are spindly and have crinkled unifoliate leaves that curl downward and become chlorotic. Subsequent trifoliate leaves are severely stunted and mottled and are more crinkled than the unifoliate leaves.

On pods, SMV has been reported to cause irregular-shaped, depressed black lesions that are similar to those caused by *Cercospora sojina* and *Colletotrichum truncatum.* Diseased seed is discolored; however, mottling of seed is not always indicative of the presence of SMV.

Yields have been reported to be significantly reduced only when 60% or more of the plants are diseased. The largest decreases in yield occurred in plants inoculated with SMV 20 days after emergence. Decreased oil content and nodulation also occur.

Symptoms are enhanced when plants are dually infected with SMV and either cowpea mosaic virus or bean pod mottle virus.

Management
1. Sow seed from disease-free plants.
2. Rogue out diseased plants from seed fields.
3. Practice good weed control in and around soybean fields.

Soybean Severe Stunt

This disease has some similarities, including symptoms, to Essex disease. The two diseases may be portions of the same disease syndrome.

Cause. Soybean severe stunt virus (SSSV). SSSV is transmitted via soil, and the dagger nematode, *Xiphinema americanum,* is strongly associated with the occurrence of the disease within a field. Soybean severe stunt has been associated with continuous cropping of the cultivar 'Essex.'

Distribution. The United States (Delaware and Virginia). The cultivar 'Essex' is severely diseased but other cultivars are also affected.

Symptoms. Initial symptoms in the field occur on the first true leaves at the V1 stage of development. Diseased plants typically have thickened, dark green leaves; shortened internodes, which cause a bunching of leaves and severe stunting; pith discoloration; and superficial stem lesions, as well as fewer flowers and fewer mature pods. A reduction in seedling emergence and stand may also be attributable to infection with SSSV. Yields may be reduced within an affected area by as much as 90%.

Management
1. Grow resistant cultivars.
2. Rotate 2 years with a nonhost crop.

Soybean Yellow Shoot

Cause. Soybean yellow shoot virus (VABS) is in the potyvirus group. VABS is mechanically transmitted.

Distribution. Brazil.

Symptoms. A golden mosaic and veinal and stem necrosis occurs on diseased young leaves and upper stems.

Management. Not reported.

Tobacco Mosaic

Cause. Tobacco mosaic virus soybean strain (TMV-S). The host range of TMV-S differs from that of TMV.

Distribution. Yugoslavia.

Symptoms. Diseased leaves exhibit vein clearing and a mild chlorotic mosaic.

Management. Not Reported.

Tobacco Necrosis

Cause. Tobacco necrosis virus is in the necrovirus group.

Distribution. The United States.

Symptoms. Small necrotic spots surrounded by a chlorotic halo, which occur on diseased secondary leaves, enlarge toward smaller leaf veins. Spots spread from small veins to larger veins, midribs, and petioles and cause a systemic infection of diseased plants. Petioles and stems have a necrosis of phloem tissues and adjacent parenchyma.

Management. Not reported.

Yellow Mosaic

Cause. Soybean yellow mosaic virus (SYMV) is in the potyvirus group. SYMV is disseminated by the aphids *Aphis fabae, Macrosiphum gei, M. pisi, M. solanifolii,* and *Myzus persicae.* SYMV is not thought to be spread by infected seed, but it is easily sap-transmissible. SYMV has a large host range.

Distribution. Canada, Europe, and the United States.

Symptoms. The symptoms caused by SYMV resemble those caused by soybean mosaic virus. Young diseased leaves develop a yellow mottling that is scattered in random spots over leaves or is in indefinite bands along the major leaf veins. Rusty, necrotic spots develop in the yellowed areas as diseased leaves mature. Some strains of SYMV produce severe mottling and crinkling of diseased leaves.

Management
1. Sow seed from disease-free plants.
2. Rogue diseased plants from seed fields.
3. Practice good weed control.
4. Grow cultivars that have some resistance.

Disease Caused by Unknown Factors

Pod Necrosis

Cause. Unknown. *Alternaria alternata* is isolated consistently from necrotic areas on diseased pods, but there was no significant correlation between incidence and reduced seed weight among the 10 soybean cultivars that were tested. Species of *Phomopsis* and *Diaporthe* were isolated from seed in the upper and middle pods of affected plants, but there was no correlation between incidence of these organisms and reduced seed weight.

Distribution. Canada (southeastern Ontario).

Symptoms. Affected areas of diseased plants within fields are circular to oblong and up to 0.5 ha in size. Symptoms appear on soybeans growing in sandy loam soils following dry, warm weather. Diseased leaves are chlorotic and stunted. Internodes are shortened and necrotic spots appear on pods at the tops of diseased plants. Seed weight is reduced.

Management. Not reported.

Selected References

Abney, T. S., and Ploper, L. D. 1994. Effects of bean pod mottle virus on soybean seed maturation and seedborne *Phomopsis* spp. Plant Dis. 78:33-37.

Abney, T. S., Richards, T. L., and Roy, K. W. 1993. *Fusarium solani* from ascospores of *Nectria haematococca* causes sudden death syndrome of soybean. Mycologia 85: 801-806.

Adee, E. A., Grau, C. R., and Oplinger, E. S. 1997. Population dynamics of *Phialophora gretata* in soybean residue. Plant Dis. 81:199-203

Almeidia, A. M. R., and Kihl, R. 1981. Necrose das vagens um novo sintoma causada pelo virus do mosaico da soja. Fitopathologia Brasileria 6:281-283 (in Portuguese).

Almeidia, A. M. R., and Silveira, J. M. 1983. Efeito da idade de inoculacao de plants de soja com o virus do mosaico comum da soja e da percentagem de plantas infecladas sobre o rendimento e algumas caracteristicas economicas. Fitopathologia Brasileira 8:229-236 (in Portuguese).

Anaele, A. O., Bishnoi, U. R., and Pacumbaba, R. P. 1987. Bacterial blight of soybean incidence at three row spacings in no-till and conventional planting systems. (Abstr.) Phytopathology 77:1690.

Anahosur, K. H., and Fazalnoor, K. 1972. Studies on foliicolar fungi of soybean crop. Indian Phytopathol. 25:504-508.

Anderson, T. R. 1987. Pod necrosis: A new disease affecting soybean on droughty soils in southwestern Ontario. (Abstr.) Can. J. Plant Pathol. 9:272.

Anjos, J. R., and Ghabrial, S. A. 1991. Studies on the synergistic interactions between soybean mosaic virus (SMV) and two comoviruses in mixed infections in soybeans. Phytopathology 81:1167.

Anjos, J. R., and Lin, M. T. 1984. Bud blight of soybeans caused by cowpea severe mosaic virus in central Brazil. Plant Dis. 68:405-407.

Backman, P. A., Williams, J. C., and Crawford, M. A. 1982. Yield losses in soybeans from anthracnose caused by *Colletotrichum truncatum*. Plant Dis. 66:1032-1034.

Balducchi, A. J., and McGee, D. C. 1987. Environmental factors influencing infection of soybean seeds by *Phomopsis* and *Diaporthe* species during seed maturation. Plant Dis. 71:209-212.

Black, B. D., Padgett, G. B., Russin, J. S., Griffin, J. L., Snow, J. P., and Berggren, G. T., Jr. 1996. Potential weed hosts for *Diaporthe phaseolorum* var. *caulivora,* causal agent for soybean stem canker. Plant Dis. 80:763-765.

Blackmon, C. W., and Musen, H. L. 1974. Control of the Columbia (lance) nematode *Hoplolaimus columbus* on soybeans. Plant Dis. Rptr. 58:641-645.

Bonde, M. R., Peterson, G. L., and Dowler, W. M. 1988. A comparison of isozymes of *Phakopsora pachyrhizi* from the Eastern Hemisphere and the New World. Phytopathology 78:1491-1494.

Bonde, M. R., Peterson, G. L., Nester, S. E., and Hartwig, E. E. 1996. Virulence of isolates of *Phakopsora pachyrhizi* and *P. meibomiae* on soybean genotypes. (Abstr.). Phytopathology 86:S92.

Boosalis, M. G., and Hamilton, R. I. 1957. Root and stem rot of soybean caused by *Corynespora cassicola* (Berk. & Curt.) Wei. Plant Dis. Rptr. 41:696-698.

Bradbury, J. F. 1986. Guide to plant pathogenic bacteria. C.A.B. International, Farnham Royal, United Kingdom. 332 pp.

Bristow, P. R., and Wyllie, T. D. 1986. *Macrophomina phaseolina,* another cause of the twin-stem abnormality disease of soybean. Plant Dis. 70:1152-1153.

Bromfield, K. R., Melching, J. S., and Kingsolver, C. H. 1980. Virulence and aggressiveness of *Phakopsora pachyrhizi* isolates causing soybean rust. Phytopathology 70:17-21.

Canady, C. H., and Schmitthenner, A. F. 1979. The effect of nitrogen on Phytophthora root rot of soybeans. (Abstr.) Phytopathology 69:539.

Carson, M. L., Arnold, W. E., and Todt, P. E. 1991. Predisposition of soybean seedlings to Fusarium root rot with trifluralin. Plant Dis. 75:342-347.

Chambers, A. Y. 1991. Effects of date of planting on severity of soybean stem canker. (Abstr.) Phytopathology 81:1135.

Cheng, Y. H., and Schenck, N. C. 1978. Effect of soil temperature and moisture on survival of the soybean root rot fungi *Neocosmospora vasinfectum* and *Fusarium solani* in soil. Plant Dis. Rptr. 62:945-949.

Daft, G. C., and Leben, C. 1972. Bacterial blight of soybeans: Epidemiology of blight outbreaks. Phytopathology 62:57-62.

Daft, G. C., and Leben, C. 1973. Bacterial blight of soybeans: Field-overwintered *Pseudomonas glycinea* as possible primary inoculum. Plant Dis. Rptr. 57:156-157.

Dale, J. L., and Walters, H. J. 1985. Soybean bud proliferation of unknown etiology in Arkansas. Plant Dis. 69:811.

Damicone, J. P., Berggren, G. T., and Snow, J. P. 1987. Effect of free moisture on soybean stem canker development. Phytopathology 77:1568-1572.

Damsteegt, V. D. 1985. Vector relationships of two strains of soybean dwarf virus. (Abstr.) Phytopathology 75:1349.

Datnoff, L. E., Naik, D. M., and Sinclair, J. B. 1987. Effect of red leaf blotch on soybean yields in Zambia. Plant Dis. 71:132-135.

Derrick, K. S., and Newsom, L. D. 1984. Occurrence of a leafhopper-transmitted disease of soybeans in Louisiana. Plant Dis. 68:343-344.

Derrick, K. S., Newsom, L. D., and Brlansky, R. H. 1984. Delayed maturity of soybeans. (Abstr.) Phytopathology 74:627.

Dhingra, O. D., and Muchovej, J. J. 1980. Twin-stem abnormality disease of soybean seedlings caused by *Cylindrocladium clavatum* in central Brazil. Plant Dis. 70:977-980.

Donald, P. A., Niblack, T. L., and Wrather, J. A. 1993. First report of Fusarium solani blue isolate, a causal agent of sudden death syndrome of soybeans, recovered from soybean cyst nematode eggs. Plant Dis. 77:647.

Duncan, D. R., and Paxton, J. D. 1981. Trifluralin enhancement of Phytophthora root rot of soybean. Plant Dis. 65:435-436.

Dunleavy, J. M. 1963. A vascular disease of soybeans caused by *Corynebacterium* sp. Plant Dis. Rptr. 47:612-613.

Dunleavy, J. M. 1966. Factors influencing spread of brown stem rot of soybeans. Phytopathology 56:298-300.

Dunleavy, J. M. 1980. Yield losses in soybeans induced by powdery mildew. Plant Dis. 64:291-292.

Dunleavy, J. M. 1985. Transmission of *Corynebacterium flaccumfaciens* by soybean seed. (Abstr.) Phytopathology 75:1295.

Ellis, M. A., Ilyas, M. B., and Sinclair, J. B. 1974. Effect of cultivar and growing region on internally seedborne fungi and *Aspergillus melleus* pathogenicity in soybeans. Plant Dis. Rptr. 58:332-334.

Fagbenle, H. H., and Ford, R. E. 1970. Tobacco streak virus isolated from soybeans, *Glycine max*. Phytopathology 60:814-820.

Farias, G. M., and Griffin, G. J. 1989. Roles of *Fusarium oxysporum* and *F. solani* in Essex disease of soybean in Virginia. Plant Dis. 73:38-42.

Farr, D. F., Bills, G. R., Chamuris, G. P., and Rossman, A. Y. 1989. Fungi on Plants and Plant

Products in the United States. American Phytopathological Society, St. Paul, MN. 1252 pp.

Fetzer, J., Kennedy, B., and Denny, R. 1988. Temperature and light regime effects on incubation time and leaf movements of soybean seedlings inoculated with tobacco streak virus (TSV). (Abstr.) Phytopathology 78:1524.

Fletcher, J., Irwin, M. E., Bradfute, O. E., and Granada, G. A. 1984. A Machismo-like disease of soybeans in Mexico. (Abstr.) Phytopathology 74:857.

Fletcher, J., Irwin, M. E., Bradfute, O. E., and Granada, G. A. 1984. Discovery of a mycoplasmalike organism associated with diseased soybeans in Mexico. Plant Dis. 68:994-996.

Fortnum, B. A., and Lewis, S. A. 1983. Effects of growth regulators and nematodes on Cylindrocladium black root rot of soybean. Plant Dis. 67:282-284.

Gangopadhyay, S., Wyllie, T. D., and Luedders, V. D. 1970. Charcoal rot disease of soybean transmitted by seeds. Plant Dis. Rptr. 54:1088-1091.

Garcia-Jiminez, J., and Alfaro, A. 1985. *Colletotrichum gloeosporioides*: A new anthracnose pathogen of soybean in Spain. Phytopathology 69:1007.

Garzonio, D. M., and McGee, D. C. 1981. Pod and stem blight of soybean: The relative importance of seed-borne and soil-borne inoculum. (Abstr.) Phytopathology 71:218.

Garzonio, D. M., and McGee, D. C. 1983. Comparison of seeds and crop residues as sources of inoculum for pod and stem blight of soybeans. Plant Dis. 67:1374-1376.

Gillaspie, A. G., Jr., and Hopkins, M. S. 1991. Spread of peanut stripe virus from peanut to soybean and yield effects on soybean. Plant Dis. 75:1157-1159.

Gleason, M. L., and Ferriss, R. S. 1985. Influence of soil water potential on performance of soybean seeds infected by *Phomopsis* sp. Phytopathology 75:1236-1241.

Gleason, M. L., and Ferriss, R. S. 1987. Effects of soil moisture and temperature on Phomopsis seed decay of soybean in relation to host and pathogen growth rates. Phytopathology 77:1152-1157.

Granada, G. A. 1979. Machismo disease of soybeans. I. Symptomatology and transmission. Plant Dis. Rptr. 63:47-50.

Granada, G. A. 1979. Machismo disease of soybeans. II. Suppressive effects of tetracycline on symptom development. Plant Dis. Rptr. 63:309-312.

Grau, C. R., Radke, V. L., and Gillespie, F. L. 1982. Resistance of soybean cultivars to *Sclerotinia sclerotiorum*. Plant Dis. 66:506-508.

Gray, F. A., Rodriguez-Kabana, R., and Adams, J. R. 1980. Neocosmospora stem rot of soybeans in Alabama. Plant Dis. 64:321-322.

Gray, L. E. 1971. Variation in pathogenicity of *Cephalosporium gregatum* isolates. Phytopathology 61:1410-1411.

Gray, L. E. 1972. Recovery of *Cephalosporium gregatum* from soybean straw. Phytopathology 62:1362-1364.

Gray, L. E. 1991. Alternate hosts of soybean SDS strains of *Fusarium solani*. (Abstr.) Phytopathology 81:1135.

Groth, D. E., and Braun, E. J. 1989. Survival, seed transmission, and epiphytic development of *Xanthomonas campestris* pv. *glycines* in the north-central United States. Plant Dis. 73:326-330.

Grybauskas, A. P. 1986. First report of soybean naturally infected with *Leptosphaerulina briosiana*. Plant Dis. 70:1159.

Guy, S. O., Grau, C. R., and Oplinger, E. S. 1989. Effect of temperature and soybean cultivar on metalaxyl efficacy against *Phytophthora megasperma* f. sp. *glycinea*. Plant Dis. 73:236-239.

Hartman, G. L., and Sinclair, J. B. 1988. *Dactuliochaeta,* a new genus for the fungus causing red leaf blotch of soybeans. Mycologia 80:696-706.

Hartman, G. L., Manandhar, J. B., and Sinclair, J. B. 1986. Incidence of *Colletotrichum* spp. on soybeans and weeds in Illinois and pathogenicity of *Colletotrichum truncatum.* Plant Dis. 70:780-782.

Hartman, G. L., Datnoff, L. E., Levy, C., Sinclair, J. B., Cole, D. L., and Javaheri, F. 1987. Red leaf blotch of soybeans. Plant Dis. 71:113-118.

Heinrichs, E. A., Lehman, P. S., and Corss, I. C. 1976. *Nematospora coryli* yeast-spot disease of soybeans in Brazil. Plant Dis. Rptr. 60:508-509.

Helbig, J. B., and Carroll, R. B. 1982. Weeds as a source of *Fusarium oxysporum* pathogenic on soybean. (Abstr.) Phytopathology 72:707.

Helbig, J. B., and Carroll, R. B. 1984. Dicotyledonous weeds as a source of *Fusarium oxysporum* pathogenic on soybean. Plant Dis. 68:694-696.

Hershman, D. E., Hendrix, J. W., Stuckey, R. E. Bachi, P. R., and Henson, G. 1990. Influence of planting date and cultivar on soybean sudden death syndrome in Kentucky. Plant Dis. 74:761-766.

Higley, P. M., and Tachibana, H. 1987. Physiologic specialization of *Diaporthe phaseolorum* var. *caulivora* in soybean. Plant Dis. 71:815-817.

Hill, J. H., Bailey, T. B., Benner, H. I., Tachibana, H., and Durand, D. P. 1987. Soybean mosaic virus: Effects of primary disease incidence on yield and seed quality. Plant Dis. 71:237-239.

Hill, J. H., Lucas, B. S., Benner, H. I., Tachibana, H., Hammond, R. B., and Pedigo, L. P. 1980. Factors associated with the epidemiology of soybean mosaic virus in Iowa. Phytopathology 70:536-540.

Hirrel, M. C. 1983. Sudden death syndrome of soybean: A disease of unknown etiology. (Abstr.) Phytopathology 73:501.

Hirrel, M. C. 1984. Influence of overhead irrigation and row width on foliar and stem diseases of soybeans. (Abstr.) Phytopathology 74:628.

Hobbs, T. W., Schmitthenner, A. F., and Ellett, C. W. 1981. Diaporthe dieback of soybean caused by *Diaporthe phaseolorum* var. *caulivora.* (Abstr.) Phytopathology 71:226.

Hobbs, T. W., Schmitthenner, A. F., Ellett, C. W., and Hite, R. E. 1981. Top dieback of soybean caused by *Diaporthe phaseolorum* var. *caulivora.* Plant Dis. 65:618-620.

Irwin, J. A. G. 1991. *Phytophthora macrochlamydospora,* a new species from Australia. Mycologia 83:517-519.

Iwaki, M., Roechan, M., Hibino, H., Tochihara, H., and Tantera, D. M. 1980. A persistent aphidborne virus of soybean, Indonesian soybean dwarf virus. Plant Dis. 64:1027-1030.

Iwaki, M., Thongmeearkom, P., Prommin, M., Honda, Y., and Hibi, T. 1982. Whitefly transmission and some properties of cowpea mild mottle virus on soybean in Thailand. Plant Dis. 66:365-368.

Iwaki, M., Thongmeearkom, P., Honda, Y., and Deena, N. 1983. Soybean crinkle leaf: A new whitefly-borne disease of soybean. Plant Dis. 67:546-548.

Iwaki, M., Isogawa, Y., Tsuzuki, H., and Honda, Y. 1984. Soybean chlorotic mottle, a new caulimovirus on soybean. Plant Dis. 68:1009-1011.

Jain, R. K., McKern, N. M., Tolin, S. A., Hill, J. H., Barnett, O. W., Tosic, M., Ford, R. E., Beachy, R. N., Yu, M. H., Ward, C. W., and Shula, D. D. 1992. Confirmation that fourteen potyvirus isolates from soybean are strains of one virus by comparing coat protein peptide profiles. Phytopathology 82:294-299.

Jeger, M. J., Kenerly, C. M., Gerik, T. J., and Kock, D. O. 1987. Spatial dynamics of Phymatotrichum root rot in row crops in the Blackland region of north central Texas. Phytopathology 77:1647-1656.

Jin, H., Hartman, G. L., Nickell, C. D., and Widholm, J. M. 1996. Phytoxicity of culture filtrate from *Fusarium solani,* the causal agent of sudden death syndrome of soybean. Plant Dis. 80:922-927.

Joye, G. F., Berggren, G. T., and Berner, D. K. 1990. Effects of row spacing and within-row plant population on Rhizoctonia aerial blight of soybean and soybean yield. Plant Dis. 74:158-160.

Keeling, B. L. 1987. Pathogenicity of two *Diaporthe* spp. isolates from soybean plants with "sudden death syndrome" disease symptoms. (Abstr.) Phytopathology 77:1738.

Keeling, B. L. 1988. Influence of temperature on growth and pathogenicity of geographic isolates of *Diaporthe phaseolorum* var. *caulivora.* Plant Dis. 72:220-222.

Kennedy, B. W., and Tachibana, H. 1973. Bacterial diseases, pp. 491-504. *In* B. E. Caldwell (ed.), Soybeans: Improvement, Production and Uses. American Society of Agronomy, Madison, WI.

Khan, M., and Sinclair, J. B. 1992. Pathogenicity of sclerotia- and nonsclerotia-forming isolates of *Colletotrichum truncatum* on soybean plants and roots. Phytopathology 82:314-319.

Killebrew, J. F., Roy, K. W., Lawrence, G. W., McLean, K. S., and Hodges, H. H. 1988. Greenhouse and field evaluation of *Fusarium solani* pathogenicity to soybean seedlings. Plant Dis. 72:1067-1070.

Kuhn, C. W., Demski, J. W., and Harris, H. B. 1972. Peanut mottle virus in soybeans. Plant Dis. Rptr. 56:146-147.

Kulik, M. M. 1988. Observations by scanning electron and bright-field microscopy on the mode of penetration of soybean seedlings by *Phomopsis phaseoli.* Plant Dis. 72:115-118.

Kunwar, I. K., Manandhar, J. B., and Sinclair, J. B. 1986. Histopathology of soybean seeds infected with *Alternaria alternata.* Phytopathology 76:543-546.

Kunwar, I. K., Singh, T., Machado, C. C., and Sinclair, J. B. 1986. Histopathology of soybean seed and seedling infection by *Macrophomina phaseolina.* Phytopathology 76:532-535.

Kuruppu, P. U., and Russin, J. S. 1997. Soil temperature effects on survival and infectivity of soybean red crown rot fungus, *Calonectria ilicicola.* (Abstr.). Phytopathology 87: S55.

Lakshminarayana, C. S., and Joshi, L. K. 1978. Myrothecium disease of soybean in India. Plant Dis. Rptr. 62:231-234.

Lalitha, B., Snow, J. P., and Berggren, G. T. 1989. Phytotoxin production by *Diaporthe phaseolorum* var. *caulivora,* the causal organism of stem canker of soybean. Phytopathology 79:499-504.

Laviolette, F. A., and Athow, K. L. 1971. Relationship of age of soybean seedlings and inoculum to infection by *Pythium ultimum.* Phytopathology 61:439-440.

Leath, S., and Carroll, R. B. 1982. Screening for resistance to *Fusarium oxysporum* in soybean. Plant Dis. 66:1140-1143.

Lin, M. T., and Hill, J. H. 1983. Bean pod mottle virus: Occurrence in Nebraska and seed transmission in soybeans. Plant Dis. 67:230-233.

Lockwood, J. L., Yoder, D. L., and Smith, N. A. 1970. *Thielaviopsis basicola* root rot of soybeans in Michigan. Plant Dis. Rptr. 54:849-850.

Lohnes, D. G., and Nickell, C. D. 1994. Effects of powdery mildew alleles Rmd-c, Rmd, and rmd on yield and other characteristics in soybean. Plant Dis. 78:299-301.

Manandhar, J. B., Hartman, G. L., and Sinclair, J. B. 1986. *Colletotrichum destructivum,* the anamorph of *Glomerella glycines.* Phytopathology 76:282-285.

Manandhar, J. B., Hartman, G. L., and Sinclair, J. B. 1988. Soybean germplasm evaluation for resistance to *Colletotrichum truncatum.* Plant Dis. 72:56-59.

Marchett, M. A., Vecker, F. A., and Bromfield, K. R. 1975. Uredial development of *Phakopsis pachyrhizi* in soybeans. Phytopathology 65:822-823.

Martin, K. F., and Walters, H. J. 1982. Infection of soybean by *Cercospora kikuchii* as affected by dew, temperature and duration of dew periods. (Abstr.) Phytopathology 72:974.

McDaniel, L. L., Maratos, M. L., Goodman, J. E., and Tolin, S. A. 1995. Partial characterization of a soybean strain of tobacco mosaic virus. Plant Dis. 79:206-211.

McGawley, E. C., Winchell, K. L., and Berggren, G. T. 1984. Possible involvement of *Hoplolaimus galeatus* in a disease complex of 'Centennial' soybean. (Abstr.) Phytopathology 74:831.

McGee, D. C. 1992. Soybean Diseases: A Reference Source for Seed Technologists. American Phytopathological Society, St. Paul, MN. 151 pp.

McGee, D. C., and Biddle, J. 1985. A comparison between isolates of *Diaporthe phaseolorum* var. *caulivora* from soybean seeds in Iowa and stem-cankered soybeans in southern states. (Abstr.) Phytopathology 75:1332.

McGee, D. C., and Biddle, J. 1987. Seedborne *Diaporthe phaseolorum* var. *caulivora* in Iowa and its relationship to soybean stem canker in the southern United States. Plant Dis. 71:620-622.

McLaughlin, M. R., Thongmeearkom, P., Goodman, R. M., Milbrath, G. M., Ries, S. M., and Royse, D. J. 1978. Isolation and beetle transmission of cowpea mosaic virus (severe subgroup) from *Desmodium canescens* and soybeans in Illinois. Plant Dis. Rptr. 62:1069-1073.

McLean, K. S., and Roy, K. W. 1988. Purple seed stain of soybeans caused by isolates of *Cercospora kikuchii* from weeds. Can. J. Plant Pathol. 10:166-171.

Melching, J. S., Dowler, W. M., Koogle, D. L., and Royer, M. H. 1988. Effect of plant and leaf age on susceptibility of soybeans to soybean rust. Can. J. Plant Pathol. 10:30-35.

Melching, J. S., Dowler, W. M., Koogle, D. L., and Royer, M. H. 1989. Effects of duration, frequency, and temperature of leaf wetness periods on soybean rust. Plant Dis. 73:117-122.

Mengistu, A., and Grau, C. R. 1986. Variation in morphological, cultural, and pathological characteristics of *Phialophora gregata* and *Acremonium* sp. recovered from soybean in Wisconsin. Plant Dis. 70:1005-1009.

Mengistu, A., and Grau, C. R. 1987. Seasonal progress of brown stem rot and its impact on soybean productivity. Phytopathology 77:1521-1529.

Meyer, W. A., and Sinclair, J. B. 1972. Root reduction and stem lesion development on soybeans by *Phytophthora megasperma* var. *sojae.* Phytopathology 62:1414-1416.

Meyer, W. A., Sinclair, J. B., and Khare, M. N. 1974. Factors affecting charcoal rot of soybean seedlings. Phytopathology 64:845-849.

Milbrath, G. M., and Tolin, S. E. 1977. Identification host range and serology of peanut stunt virus isolated from soybean. Plant Dis. Rptr. 61:637-640.

Mishra, B., and Prakash, O. 1975. Alternaria leaf spot of soybean from India. Indian J. Mycol. Plant Pathol. 5:95.

Moots, C. K., Nickell, C. D., and Gray, L. E. 1988. Effects of soil compaction on the incidence of *Phytophthora megasperma* f. sp. *glycinea* in soybean. Plant Dis. 72:896-900.

Muchovej, J. J., and Kimura, S. 1988. Cotyledon spot of soybean caused by seedborne *Curvularia lunata* var. *aerea* in Brazil. Plant Dis. 72:268.

Narayanasamy, P., and Durairaj, P. 1971. A new blight disease of soybeans. Madras Agric. J. 58:711-712.

Nelson, B., Helms, T., Christianson, T., and Kural, I. 1996. Characterization and pathogenicity of *Rhizoctonia* from soybean. Plant Dis. 80:74-80.

Nicholson, J. F., Dhingra, O. D., and Sinclair, J. B. 1972. Internal seedborne nature of *Sclerotinia sclerotiorum* and *Phomopsis* sp. and their effects on soybean seed quality. Phytopathology 62:1261-1263.

O'Neill, N. R., Rush, M. C., Horn, N. L., and Carver, R. B. 1977. Aerial blight of soybeans caused by *Rhizoctonia solani*. Plant Dis. Rptr. 61:713-717.

Otazu, V., Epstein, A. H., and Tachibana, H. 1981. Water stress effect on the development of brown stem rot of soybeans. (Abstr.) Phytopathology 71:247.

Pacumbaba, R. P. 1995. Seed transmission of soybean mosaic virus in mottled and non-mottled soybean seeds. Plant Dis. 79:193-195.

Padgett, G. B., Snow, J. P., and Berggren, G. T. 1991. Effects of temperature on production of perithecia and pycnidia by *Diaporthe phaseolorum* var. *caulivora*. (Abstr.) Phytopathology 81:1143.

Park, E. W., and Lim, S. M. 1985. Overwintering of *Pseudomonas syringae* pv. *glycinea* in the field. Phytopathology 75:520-524.

Pataky, J. K., and Lim, S. M. 1981. Effects of Septoria brown spot on yield components of soybean. (Abstr.) Phytopathology 71:248.

Pathan, M. A., Sinclair, J. B., and McClary, R. D. 1989. Effects of *Cercospora kikuchii* on soybean seed germination and quality. Plant Dis. 73:720-723.

Pearson, C. A. S., Todd, T. C., and Schwenk, F. W. 1987. Influence of soybean cyst nematode (SCN) on root colonization by *Macrophomina phaseolina*. (Abstr.) Phytopathology 77:1774.

Peterson, D. J., and Edwards, H. H. 1982. Effects of temperature and leaf wetness period on brown spot disease of soybeans. Plant Dis. 66:995-998.

Phillips, D. V. 1971. Influence of air temperature on brown stem of soybean. Phytopathology 61:1205-1208.

Phillips, D. V. 1972. A soybean disease caused by *Neocosmospora vasinfecta*. Phytopathology 62:612-615.

Phillips, D. V., Vesper, S. J., and Turner, J. T., Jr. 1983. *Verticillium nigrescens* associated with soybeans. (Abstr.) Phytopathology 73:504.

Ploetz, R. C., and Shokes, F. M. 1989. Variability among isolates of *Diaporthe phaseolorum* f. sp. *meridionalis* in different vegetative compatibility groups. Can. J. Bot. 67:2751-2755.

Reis, C. H., Alves, A. M. C., and Kitajima, E. 1991. Studies with soybean yellow shoot virus: New potyvirus detected in Brazil. (Abstr.) Phytopathology 81:693.

Roane, C. W., and Roane, M. K. 1976. *Erysiphe* and *Microsphaera* as dual causes of powdery mildew of soybeans. Plant Dis. Rptr. 60:611-612.

Roongruangsree, U-Tai, Olson, L. W., and Lange, L. 1988. The seedborne inoculum of *Peronospora manshurica*, causal agent of soybean downy mildew. J. Phytopathol. 123:233-243.

Rosenbrock, S. M., and Wyllie, T. D. 1987. Leptosphaerulina leaf spot of soybeans in Missouri. (Abstr.) Phytopathology 77:1689.

Rothrock, C. S., Phillips, D. V., and Hobbs, T. W. 1988. Effects of cultivar, tillage, and cropping system on infection of soybean by *Diaporthe phaseolorum* var. *caulivora* and
southern stem canker symptom development. Phytopathology 78:266-270.

Roy, K. W. 1982. Seedling diseases caused in soybean by species of *Colletotrichum* and
Glomerella. Phytopathology 72:1093-1096.

Roy, K. W. 1997. *Fusarium solani* on soybean roots: Nomenclature of the causal agent of
sudden death syndrome and identity and relevance of *F. solani* form B. Plant Dis.
81:259-266.

Roy, K. W. 1997. Sporulation of *Fusarium solani* f. sp. *glycines,* causal agent of sudden death
syndrome, on soybeans in the midwestern and southern United States. Plant Dis.
81:566-569.

Roy, K. W., and Abney, T. S. 1977. Antagonism between *Cercospora kikuchii* and other seedborne fungi of soybeans. Phytopathology 67:1062-1066.

Roy, K. W., and McLean, K. 1984. Epidemiology of soybean stem canker in Mississippi.
(Abstr.) Phytopathology 74:632.

Roy, K. W., and Miller, W. A. 1983. Soybean stem canker incited by isolates of *Diaporthe*
and *Phomopsis* spp. from cotton in Mississippi. Plant Dis. 67:135-137.

Roy, K. W., and Ratnayake, S. 1997. Frequency of occurrence of *Fusarium pallidoroseum,* effects on seeds and seedlings, and associations with other fungi in soybean seeds and
pods. Can. J. Plant Pathol. 19:188-192.

Roy, K. W., Rupe, J. C., Hershman, D. E., and Abney, S. T. 1997. Sudden death syndrome of
soybean. Plant Dis. 81:1100-1111.

Rupe, J. C., and Ferriss, R. S. 1984. The effect of moisture on infection of soybean seed by
Phomopsis sp. (Abstr.) Phytopathology 74:632.

Rupe, J. C., Gbur, E. E., and Marx, D. M. 1991. Cultivar responses to sudden death syndrome of soybean. Plant Dis. 75:47-50.

Rupe, J. C., Sutton, E. A., Becton, C. M., and Gbur, E. E. Jr., 1996. Effect of temperature and
wetness period on recovery of the southern biotype of *Diaporthe phaseolorum* var.
caulivora from soybean. Plant Dis. 80:255-257.

Russin, J. S., Layton, M. B., McGawley, E. C., Boethel, D. J., Berggren, G. T., and Snow, J. P.
1987. Interactions between soybean looper, stem canker fungus, and soybean cyst
nematode on soybean. (Abstr.) Phytopathology 77:1690.

Russin, J. S., Orr, D. B., Layton, M. B., and Boethel, D. J. 1988. Incidence of microorganisms in soybean seeds damaged by stink bug feeding. Phytopathology 78:306-310.

Saharan, G. S., and Gupta, V. K. 1972. Pod rot and collar rot of soybean caused by *Fusarium
semitectum*. Plant Dis. Rptr. 56:693-694.

Schiller, C. T., Ellis, M. A., Tenne, F. D., and Sinclair, J. B. 1977. Effect of *Bacillus subtilis* on
soybean seed decay, germination, and stand inhibition. Plant Dis. Rptr. 61:213-
217.

Schiller, C. T., Hepperly, P. R., and Sinclair, J. B. 1978. Pathogenicity of *Myrothecium roridum* from Illinois soybeans. Plant Dis. Rptr. 62:882-885.

Schlub, R. L., and Lockwood, J. L. 1981. Etiology and epidemiology of seedling rot of soybean by *Pythium ultimum*. Phytopathology 71:134-138.

Schlub, R. L., Lockwood, J. L., and Komada, H. 1981. Colonization of soybean seeds and
plant tissue by *Fusarium* species in soil. Phytopathology 71:693-696.

Schneider, R. W., Dhingra, O. D., Nicholson, J. F., and Sinclair, J. B. 1974. *Colletotrichum
truncatum* borne within the seed coat of soybean. Phytopathology 64:154-155.

Schuh, W. 1991. Influence of temperature and leaf wetness period on conidial germination in vitro and infection of *Cercospora kikuchii* on soybean. Phytopathology 81:1315-1318.

Schuh, W. 1991. Relationship between pod development stage, temperature, pod wetness duration and incidence of purple seed stain of soybeans. (Abstr.) Phytopathology 81:1181.

Schuh, W. 1992. Effect of pod development stage, temperature, and pod wetness duration on the incidence of purple seed stain of soybeans. Phytopathology 82:446-451.

Schwenk, F. W., and Nickell, C. D. 1980. Soybean green stem caused by bean pod mottle virus. Plant Dis. 64:863-865.

Schwenk, F. W., and Paxton, J. D. 1981. Trifluralin enhancement of Phytophthora root rot of soybean. Plant Dis. 65:435-436.

Seaman, W. L., Shoemaker, R. A., and Peterson, E. A. 1965. Pathogenicity of *Corynespora cassiicola* on soybeans. Can. J. Bot. 43:1461-1469.

Sherwood, J. L., and Jackson, K. E. 1985. Tobacco streak virus in soybean in Oklahoma. Plant Dis. 69:727.

Short, G. E., Wyllie, T. D., and Bristow, P. R. 1980. Survival of *Macrophomina phaseolina* in soil and in residue of soybean. Phytopathology 70:13-17.

Shortt, B. J., Sinclair, J. B., and Kogan, M. 1981. Soybean seed quality losses associated with bean leaf beetles and *Alternaria tenuissima*. (Abstr.) Phytopathology 71:1117.

Sinclair, J. B. (ed.) 1982. Compendium of Soybean Diseases, Second Edition. American Phytopathological Society, St. Paul, MN. 104 pp.

Sinclair, J. B. 1993. Phomopsis seed decay of soybean—A prototype for studying seed disease. Plant Dis. 77:329-334.

Sinclair, J. B., and Backman, P. A. 1989. Compendium of Soybean Diseases, Third Edition. American Phytopathological Society, St. Paul, MN. 106 pp.

Smith, E. F., and Backman, P. A. 1989. Epidemiology of soybean stem canker in the southeastern United States: Relationship between time of exposure to inoculum and disease severity. Plant Dis. 73:464-468.

Song, H. S., Lim, S. M., and Clark, J. M., Jr. 1993. Purification and partial characterization of a host-specific pathotoxin from culture filtrates of *Septoria glycines*. Phytopathology 83:659-661.

Sortland, M. E., and MacDonald, D. H. 1986. Development of a population of *Heterodera glycines* race 5 at four soil temperatures in Minnesota. Plant Dis. 70:932-935.

Spilker, D. A., Schmitthenner, A. F., and Ellett, C. W. 1981. Effects of humidity, temperature, fertility and cultivar on the reduction of soybean seed quality by *Phomopsis* sp. Phytopathology 71:1027-1029.

Stall, R. E., and Kucharek, T. A. 1982. A new bacterial disease of soybean in Florida. (Abstr.) Phytopathology 72:990.

Stewart, R. D. 1957. An undescribed species of *Pyrenochaeta* on soybean. Mycologia 49:115-117.

Stuckey, R. E., Ghabrial, S. A., and Reicosky, D. A. 1982. Increased incidence of *Phomopsis* sp. in seeds from soybeans infected with bean pod mottle virus. Plant Dis. 66:826-829.

Subba Rao, K. V., Padgett, G. B., Berner, D. K., Berggren, G. T., and Snow, J. P. 1990. Choanephora leaf blight of soybeans in Louisiana. Plant Dis. 74:614.

Tachibana, H. 1971. Virulence of *Cephalosporium gregatum* and *Verticillium dahliae* in soybeans. Phytopathology 61:565-568.

Tachibana, H., and Shih, S. 1965. A leaf-crinkling bacterium of soybean. Plant Dis. Rptr. 49:396-397.

Tachibana, H., Jowett, D. D., and Fehr, W. R. 1971. Determination of losses in soybeans caused by *Rhizoctonia solani*. Phytopathology 61:1444-1446.

Thompson, T. B., Athow, K. L., and Laviolette, F. A. 1971. The effect of temperature on the pathogenicity of *Pythium aphanidermatum, P. debaryanum,* and *P. ultimum* on soybean. Phytopathology 61:933-935.

Thouvenel, J. C., Monsarrat, A., and Fauquet, C. 1982. Isolation of cowpea mild mottle virus from diseased soybeans in the Ivory Coast. Plant Dis. 66:336-337.

Vakili, N. G., and Bromfield, K. R. 1976. Phakopsora rust on soybean and other legumes in Puerto Rico. Plant Dis. Rptr. 60:995-999.

Velicheti, R. K., and Sinclair, J. B. 1989. Detection of *Fusarium oxysporum* in symptomatic soybean seeds. (Abstr.) Phytopathology 79: 1006.

Velicheti, R. K., and Sinclair, J. B. 1994. Production of cercosporin and colonization of soybean seed coats by *Cercospora kikuchii*. Plant Dis. 78:342-346.

Velicheti, R. K., Kollipara, K. P., Sinclair, J. B., and Hymowitz, T. 1992. Selective degradation of proteins by *Cercospora kikuchii* and *Phomopsis longicolla* in soybean seed coats and cotyledons. Plant Dis. 76:779-782.

Vesper, S. J., Turner, J. T., Jr., and Phillips, D. V. 1983. Incidence of *Verticillium nigrescens* in soybeans. Phytopathology 73:1338-1341.

Walters, H. J. 1980. Soybean leaf blight caused by *Cercospora kikuchii*. Plant Dis. 64:961-962.

Walters, H. J., and Martin, K. F. 1981. *Phyllosticta sojaecola* on pods of soybeans in Arkansas. Plant Dis. 65:161-162.

Warwick, D., and Demski, J. W. 1988. Susceptibility and resistance of soybean to peanut stripe virus. Plant Dis. 72:19-21.

Weldekidan, T., Carroll, R. B., and Evans, T. A. 1988. Association of *Fusarium oxysporum* with a new virus-like disease affecting Delaware soybeans. (Abstr.) Phytopathology 78:1579.

Welderkidan, T., Evans, T. A., Carroll, R. B., and Mulrooney, R. P. 1992. Etiology of soybean severe stunt and some properties of the causal virus. Plant Dis. 76:747-750.

Wicklow, D. T., Bennett, G. A., and Shotwell, O. L. 1987. Secondary invasion of soybeans by *Fusarium graminearum* and resulting mycotoxin contamination. Plant Dis. 71:1146.

Wilcox, J. R., and Abney, T. S. 1973. Effects of *Cercospora kikuchii* on soybeans. Phytopathology 63:796-797.

Wilmot, D. B., Nickell, C. D., and Gray, L. E. 1989. Physiologic specialization of *Phialophora gregata* on soybean. Plant Dis. 73:290-294.

Wrather, J. A., Anderson, S. H., and Wollenhaupt, N. C. 1991. Effects of tillage and row spacing on soybean foliar diseases, and *Heterodera glycines* population dynamics. (Abstr.) Phytopathology 81:1185.

Wrather, J. A., Kendig, S. R., and Tyler, D. D. 1998. Tillage effects on *Macrophomina phaseolina* population density and soybean yield. Plant Dis. 82:247-250.

Wyllie, T. D., and Rosenbroek, S. M. 1985. Crop rotation as a means of controlling populations of *Macrophomina phaseolina* in Missouri soils. (Abstr.) Phytopathology 75:1348.

Yang, X. B., Berggren, G. T., and Snow, J. P. 1988. Effect of free moisture and plant growth stage on disease focus development in Rhizoctonia aerial blight of soybean. (Abstr.) Phytopathology 78:1538.

Yang, X. B., Berggren, G. T., and Snow, J. P. 1988. Seedling infection of soybean by *Rhizoctonia solani* AG-1, causal agent of aerial blight. Plant Dis. 72:644.

Yang, X. B., Berggren, G. T., and Snow, J. P. 1990. Seedling infection of soybean by isolates of *Rhizoctonia solani* AG-1, causal agent of aerial blight and web blight of soybean. Plant Dis. 74:485-488.

Yang, X. B., Berggren, G. T., and Snow, J. P. 1990. Types of Rhizoctonia foliar blight on soybean in Louisiana. Plant Dis. 74:501-504.

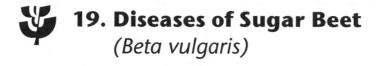

19. Diseases of Sugar Beet
(Beta vulgaris)

Diseases Caused by Bacteria

Bacterial Leaf Spot

Cause. *Pseudomonas syringae* (syn. *P. apatata*) survives on living plants, seed, and infested organic matter in the soil. Dissemination is by seed, but wounds are necessary for infection to occur. Some research suggests that bacteria penetrate roots through wounds and move upward in the infected plant. Bacterial growth is optimum at temperatures of 25° to 30°C.

Distribution. Europe, Japan, and the United States.

Symptoms. A seedling blight may occur. Dark brown to black streaks and spots occur on diseased leaves and infrequently on petioles and seed stalks.

Management
1. Sow disease-free seed.
2. Apply seed-treatment fungicides.

Bacterial Pocket

Bacterial pocket is also called bacterial canker.

Cause. *Xanthomonas beticola* survives in soil for long periods of time and is disseminated by any means that moves soil, such as irrigation water and machinery. *Xanthomonas beticola* enters roots through wounds at or near the crown level. Normally only scattered plants are affected but occasionally an entire field of plants shows symptoms if a hailstorm has occurred. The hail stones injure crowns and provide entrance wounds for bacteria to infect the injured plants.

Distribution. The United States (Colorado, Maryland, Michigan, New Mexico, Utah, Virginia, Wisconsin, and Wyoming).

Symptoms. Galls develop on diseased crowns at the soil surface and also on the petioles and roots. The central portion of galls are water-soaked and discolored yellow due to the presence of bacteria. Galls become rough and fissured as they enlarge, and the top of the diseased root becomes larger. The root surface below the galls also becomes rough, ridged, and fissured. In the later stages of bacterial pocket, an abnormal number of leaves may develop.

Management
1. Grow sugar beet every 4 to 5 years in the same soil.
2. Provide proper soil fertility.
3. Avoid mechanical injury to plants.

Bacterial Vascular Necrosis and Soft Rot

Bacterial vascular necrosis and soft rot is also called Erwinia soft rot.

Cause. *Erwinia carotovora* pv. *betavasculorum* survives in unharvested sugar beet and, for a short time, in soil and on weeds. *Betavasculorum* is sometimes referred to as a subspecies. Bacteria are disseminated with soil and deposited in crowns, where infection begins as bacteria enter the plant, primarily through wounds. The vascular tissue (xylem) of petioles and roots is invaded. Bacteria then multiply in the xylem tissue and progress toward the root tips. Optimum disease development occurs at temperatures of 25° to 30°C. Young plants are the most susceptible to infection.

Disease occurs in low, poorly drained areas of fields and in furrow-irrigated fields. Other factors that aid in the predisposition of sugar beet to disease are surplus or adequate nitrogen applied to soils; a wide plant spacing, which results in rapid and succulent plant growth; and the use of sprinkler irrigation, which keeps wounds moist for a long time and disseminates the causal bacteria.

Differences in aggressiveness exist among strains.

Distribution. The United States (Arizona, California, Idaho, Texas, and Washington).

Symptoms. Diseased beets are not easily recognized in the field until they are severely rotted. Symptoms include a gray to black, watery, internal rot of diseased taproots. The vascular bundles in diseased roots are necrotic or discolored. Black longitudinal lesions run up the petioles. A black or foamy white exudate may occur on split crowns of roots and on petiole lesions. Root tissue surrounding diseased vascular bundles turns pink to red-brown when it is exposed to air. Severely diseased roots become hollow but the plant normally remains alive.

Management
1. Avoid practices that deposit soil in crowns or cause injury to crowns and petioles.

2. Provide adequate drainage.
3. Grow resistant cultivars.
4. Sow early in the growing season and maintain optimum stands.
5. Use judicious amounts of nitrogen fertilizer.

Corynebacterium Infection

Cause. *Corynebacterium sepedonicum* (syn. *Clavibacter michiganense* subsp. *sepedonicum*) is seedborne. *Corynebacterium sepedonicum* may be systemic and move from roots into seeds. Strains of *C. sepedonicum* isolated from sugar beet seed were identical to strains causing potato bacterial ring rot.

Distribution. The United States (Oregon).

Symptoms. Sugar beet roots are symptomless. However, diseased sugar beets may be a source of inoculum for infecting potatoes grown in the field following sugar beet, thus preventing the successful and permanent eradication of the potato bacterial ring rot disease.

Management. Not reported.

Crown Gall

Cause. *Agrobacterium tumefaciens* is a soil inhabitant and pathogen of several plant hosts. Bacteria enter roots through wounds caused by several agents. The production of indoleacetic acid by *A. tumefaciens* is associated with gall formation and hypertrophy of diseased sugar beet tissues.

Distribution. Wherever sugar beet is grown.

Symptoms. Initially, small, wart-like growths form on diseased roots and soon develop into galls of various sizes. Diseased sugar beet plants are rarely killed but growth may be stunted and the sugar content lowered. Galls are harvested along with roots; consequently, there is little loss in total weight.

Management. Not necessary, since crown gall is seldom serious.

Scab

Cause. *Streptomyces scabies* (syn. *Actinomycetes scabies*) is an actinomycete that survives in soil and infested plant residue as mycelia and spores. Infection occurs most frequently in dry, light, sandy soils that have a neutral or slightly alkaline pH by hyphae growing into minute wounds or lenticels.

Distribution. Wherever sugar beet is grown, but Fusarium head blight is a minor problem.

Symptoms. Small round spots that occur on diseased roots enlarge, turn brown, and rupture the epidermis. Eventually, raised corky areas occur that vary in size and shape to resemble "miniature raised terraces." The color of the scabby area varies from gray-white to dark tan. Sometimes the scabby area may be pitted or pocked, but the injury is usually superficial and no decomposition occurs. Occasionally, taproots have brown, corky lesions that can restrict growth and cause roots to be "turnip-shaped."

Management. Not necessary, since the appearance rather than the actual value of diseased beets is affected.

Diseases Caused by Fungi

Alternaria Leaf Spot

Cause. *Alternaria alternata* (syn. *A. tenuis*) overwinters in infested plant residue as conidia and saprophytic mycelial growth. In the spring or during favorable growth conditions for *A. alternata,* conidia, which are produced on residue and diseased plant tissue and disseminated by wind, primarily infect host plants that are deficient in magnesium, manganese, phosphorous, or potassium. *Alternaria brassicae* infects yellowed tissue and is a foliar pathogen of some hybrids. Plants infected with western yellows virus (WYV) are more likely to become infected than sugar beets without WYV. Infection occurs at temperatures of 7° to 10°C and high humidity.

Distribution. Wherever sugar beet is grown. Alternaria leaf spot is most common in Europe.

Symptoms. Symptoms occur late in the growing season. Initially, lesions are dark brown, zonate, circular to irregular-shaped areas (2–10 mm in diameter). The dark brown necrotic spots coalesce and form irregular-shaped areas on diseased leaves.

Management. Vigorous healthy plants do not ordinarily become diseased.

Aspergillus fumigatus Storage Rot

Cause. *Aspergillus fumigatus* invades roots at temperatures of 30°C or more. The slime-producing bacterium *Leuconostoc mesenterioides,* a secondary invader, is often found in association with *A. fumigatus.*

Distribution. Widespread. *Aspergillus fumigatus* was previously known to be a common saprophyte on beets in storage.

Symptoms. A generic rot has been described.

Management
1. Reduce temperature of stored beets to below 30°C.

2. Hybrids and breeding lines with resistance to *Rhizoctonia solani* are also resistant to *A. fumigatus* at temperatures of 30°C or less.

Beet Rust

Cause. *Uromyces betae* is an autoecious rust that produces pycnia, aecia, uredinia, and telia only on sugar beet and other varieties of *Beta vulgaris*. *Uromyces betae* overwinters as uredinia and telia on volunteer host plants and infested residue, and may be seedborne. Uredinia and telia are produced during damp, cool weather in the summer. Aecia may be produced in the autumn or in the spring on sugar beet plants, usually the second-year beets grown for seed production. Optimum disease development occurs during moist weather at temperatures of 15° to 22°C.

Distribution. Asia, Europe, and the United States (Arizona, California, New Mexico, and Oregon). Beet rust is of minor economic importance.

Symptoms. Initially, pustules, which appear as slight raised areas surrounded by a chlorotic halo, are randomly dispersed or aggregated in rings on the green tissue of leaves, seed stalks, and petioles. The epidermis ruptures and exposes urediniospores and teliospores in red-brown pustules that average 2 mm in diameter. Pustules become dark brown toward the end of the growing season.

Management. Grow resistant cultivars.

Beet Tumor

Beet tumor is also called crown wart.

Cause. *Physoderma leproides* (syn. *Urophylyctis leproides*) survives as resting sporangia in infested residue in soil. Sporangia are disseminated primarily by splashing water. Only older leaves become diseased. In Europe, beet tumor is associated with wet soils in irrigated areas.

Distribution. Argentina, Europe, North Africa, and the United States (California). Beet tumor is not an economically important disease.

Symptoms. Galls occur on diseased leaf blades, petioles, and crowns of beets. Galls on leaf blades and petioles are green-brown, rough-textured, and less than 1 cm in diameter. Eventually the small galls coalesce and form larger galls that extend over the entire diseased leaf blade. Diseased leaves are stunted and malformed.

Galls on diseased crowns are attached by a narrow base and are sometimes 8 to 10 cm in diameter on mature sugar beet. These galls vary from red to green-brown. When galls are cut open, small cavities filled with brown spores can be seen. No galls form on taproots.

Management. Not reported.

Black Root

The damping off phase is also called Aphanomyces seedling disease. A *Pythium* sp. and a *Rhizoctonia* sp., two fungi implicated in the black root complex, cause a damping off of seedlings. However, since *Aphanomyces cochlioides* infects plants throughout the growing season, it will be the causal agent discussed here.

Cause. *Aphanomyces cochlioides* survives as oospores in infested soil residue. Some weeds may aid in the survival and the increase of inoculum in soil. Oospores germinate in wet soil at an optimum temperature that varies from 22° to 28°C and form sporangia in which sporangiospores are produced. After 3 to 4 hours, each sporangiospore produces a papilla through which a zoospore is released. The zoospore "swims" to roots and germinates, forming a germ tube that penetrates the host root. Oospores eventually form in diseased tissue.

In Spain, *Aphanomyces cochlioides* was associated with late-sown sugar beet.

Distribution. Canada, Europe, Japan, Spain, and the United States. Black root is most severe in the Great Lakes region of the United States.

Symptoms. In some areas, Aphanomyces seedling disease is considered one of the most common seedling diseases of sugar beet. Seedlings commonly are killed, but disease is also considered to be severe at times on older plants.

Initially, diseased leaves are yellowed and mottled, and plants remain stunted. Most lateral roots are dead and blackened. New roots are killed shortly after they are produced, which causes the main roots to have a "tasseled" appearance due to the groups of dead, black, lateral roots. The ends of main roots may also have dark lesions.

Symptoms from Spain were described as a postemergence damping off. A brown discoloration of the hypocotyl began at the soil surface and frequently extended to the base of the cotyledons.

Management
1. Do not rotate sugar beet with legumes.
2. Control weeds.
3. Grow resistant cultivars.
4. Maintain well-drained soil. Limit irrigation during the seedling stage. In greenhouse studies, seedling disease incidence was reduced when soils were irrigated prior to sowing.

Botrytis Storage Rot

Botrytis storage rot is also called clamp rot.

Cause. *Botrytis cinerea* survives as sclerotia in soil or in infested residue that is incorporated into soil or is lying on the soil surface. Sclerotia germinate and produce mycelia on which conidiophores and conidia are borne.

Conidia disseminated by wind infect sugar beet roots through wounds. Conidia are abundantly produced on diseased tissue. Eventually, sclerotia develop in mycelium on the outside of diseased roots and either dislodge and fall into the soil or remain attached to diseased plant tissue.

Distribution. Wherever sugar beet is grown.

Symptoms. Rotted tissue is normally dark brown or black. In the latter case, *Phoma betae* may also be present. Dark gray fungal growth consisting of mycelium, conidia, and conidiophores are present on rotted tissue. Eventually, dark brown to black round sclerotia (2–5 mm in diameter) form in mycelia on the outsides of diseased roots.

Management. Resistance is present in some germplasms.

Cercospora Leaf Spot

Cause. *Cercospora beticola* overwinters as spores and stromata on infested residue in the soil, and, following severe disease, on seed. *Cercospora beticola* also infects several weed hosts, including red-root pigweed, *Amaranthus retroflexus*; lambsquarters, *Chenopodium album*; prickly lettuce, *Lactua scariola*; common mallow, *Malva neglecta*; sweet clover, *Melilotus* spp.; common plantain, *Plantago major*; curly dock, *Rumex crispus*; and dandelion, *Taraxacum* spp. *Cercospora beticola* also overwinters on infested weed residue.

In the spring or early summer, conidia are produced from mycelium and disseminated by wind, splashing rain, and insects to the leaves of host plants. Conidia germinate optimally in the presence of free water and temperatures of 25° to 35°C and produce germ tubes that penetrate through the stomata on leaf surfaces. Secondary infection occurs from conidia produced through the stomata within leaf spots in 7 to 21 days.

Distribution. Wherever sugar beet is grown; however, Cercospora leaf spot is important only in areas of high relative humidity or abundant moisture.

Symptoms. Spots occur mostly on leaves and occasionally on petioles. Initially, small, light brown spots appear on diseased older outer leaves. The circular spots rapidly increase in size to 3 to 5 mm in diameter and become a brownish or purplish color. Numerous spots coalesce and form large necrotic areas on the diseased leaves. As the spots mature, their centers become ash gray to black with brown to bright purple borders. Similar lesions appear on petioles; however, these lesions are normally longer and more elliptical than the lesions on leaves. Conidia and conidiophores appear as black, dot-like structures on the centers of spots on the upper surfaces of diseased leaves and sometimes in spots on the lower surfaces. During periods of several hours of leaf wetness and high humidity, necrotic spots become gray due to the production of conidiophores and conidia. At this time, the "silver" conidia may be seen in the spots by using magnification. Numerous spots cause leaves to become yellow and die. Severe disease causes the destruction of the older, outer leaves and the production of new leaves, which results in elongated and cone-shaped crowns. Blighted leaves collapse and fall to the ground but remain attached to the crowns.

The upper portion of root, which is not covered by soil, may also have spots. Such spots are slightly sunken, circular to oval (2–5 mm in diameter), and initially brown-purple but later become gray in the center. Tissue beneath the lesions is brown to a depth of about 1.5 mm.

Cercospora leaf spot will reduce tonnage and recoverable sugar.

Management

1. Grow resistant cultivars.
2. Apply foliar fungicides during the growing season before disease becomes severe. Economic losses occur when Cercospora leaf spot exceeds 3% severity by harvest.
3. Rotate to nonhost crops for 2 to 3 years.
4. Plow under residue.
5. Locate new fields at least 100 m from old fields.

Charcoal Rot

Cause. *Macrophomina phaseolina* (syn. *M. phaseoli* and *Sclerotium bataticola*) survives as sclerotia in soil and infested residue. Sclerotia of *M. phaseolina* are most commonly disseminated by any means that moves soil. Sclerotia germinate and form germ tubes that penetrate stressed, weakened, or injured host plants. Sclerotia are eventually formed in diseased tissue. Pycnidia infrequently may be produced on diseased plant tissue and the subsequent conidia disseminated by wind to host plants. Charcoal rot is most severe during high soil and air temperatures (31°C is the optimum air temperature).

Distribution. India, Russia, Ukraine, and the United States (the hot interior valleys of California).

Symptoms. Symptoms initially occur on half-grown and mature plants. Leaves rapidly wilt, then gradually turn brown and die but remain attached to crowns. Brown-black, irregular-shaped, necrotic areas develop only on crowns or the upper portions of roots. Diseased areas have a silvery sheen, and the outer portion of the oldest diseased area is thin and loosely attached to the underlying tissue. Under this outer layer, the numerous black sclerotia resemble a sprinkling of charcoal. An entire diseased root becomes discolored brown-black and has masses of sclerotia in cavities beneath the outer layer. Diseased roots may shrink and become mummified.

Management. Not reported.

Downy Mildew

Cause. *Peronospora farinosa* (syn. *P. farinosa* f. sp. *betae* and *P. schachtii*) overwinters as oospores and mycelia in infested plant residue, wild and volunteer *Beta* spp., and infested seed. Sporangia are borne two ways: (1) under
cool, moist conditions, oospores germinate and produce mycelia on which
sporangia are borne and (2) at temperatures of 4° to 10°C, sporangia are
produced from overwintered mycelia. Sporangia are disseminated by wind
to sugar beet foliage, where they germinate at temperatures of 4° to 10°C.
Infection occurs at temperatures of 7° to 15°C in the presence of relative
humidities of 80% and above. Secondary inoculum is provided by the production of sporangia that are disseminated by wind to leaves and germinate during high relative humidity. Oospores are eventually produced in
diseased leaves.

Distribution. Europe and the United States, primarily California, Oregon,
and Washington.

Symptoms. Seedlings may be killed. The primary bud is usually destroyed in
older diseased plants and plant growth stops. Symptoms occur on cotyledons, primary leaves, and growing points. Primary leaves in the centers of
crowns become infected, which results in small, distorted, thickened, light
green, and puckered leaves with downward curled margins. Both leaf surfaces are covered with a gray downy growth consisting of mycelium, conidia, and conidiophores. Diseased leaves may appear lighter green than
normal or have a distinct purple cast. In either case, the leaves eventually
wilt and die. If conditions become unfavorable for the development of
downy mildew, a secondary heart rot may occur together with a chlorosis
of older diseased leaves that resembles virus yellows infection.

When older leaves become infected, light green, irregular-shaped spots
(12–25 mm in diameter) appear on the diseased upper leaf surfaces. Opposite the spots on the upper leaf surface, the lower leaf surface is covered
with a gray growth consisting of mycelium, conidia, and conidiophores.
Diseased seed beets have stunted and distorted young lateral branches.
Mycelium and oospores may develop within seed clusters.

Management
1. In the United States, grow resistant cultivars.
2. In Europe, separate seed crops from root crops by 1.0 to 1.5 km.

Dry Rot Canker

Dry rot canker and Rhizoctonia crown rot are both caused by *Rhizoctonia
solani.* However, symptoms are distinct and the two diseases are discussed separately.

Cause. *Rhizoctonia solani* survives as bulbils, or thickened hyphae, in infested residue. The fungus resumes growth and forms mycelium that penetrates the host plant. "Dry rot" is caused by a weak strain of *R. solani* that infects under "dry" conditions and temperatures of 30° to 35°C. The strain of *R. solani* causing "dry rot" is apparently different from those causing crown rot and root rot.

Distribution. The United States.

Symptoms. Initially, leaves of diseased plants wilt during the warmest part of the day but regain their turgidity at night. Eventually, leaves permanently wilt and turn brown. Root surfaces have circular brown lesions with alternate dark and light concentric rings that develop into cavities filled with the brown pithy remains of the diseased root tissue. There is a sharp brown line between diseased and healthy tissue. In the latter stages of disease, all the vascular tissues become brown and darker than normal.

Management
1. Follow recommended tillage and fertility practices.
2. Rotate sugar beet with small grains and maize.
3. Avoid hilling soil up around plants.

Fusarium Root Rot

Fusarium root rot is known locally as tip rot in some sugar beet–growing areas.

Cause. The causal organism is temporarily referred to as *F. oxysporum* f. sp. *betae*, Texas strain (syn. *F. oxysporum* f. sp. *spinaciae*). However, isozyme analyses show the Texas isolates to be distinct from other isolates of *F. oxysporum* f. sp. *betae*; therefore, the Texas sugar beet isolates may be a new forma specialis or a new race of *F. oxysporum* f. sp. *betae*. A suggested but not yet proposed name for the causal organism is *F. oxysporum* f. sp. *radicis-betae*. The causal fungus is presumed to survive as chlamydospores in soil and infested residue.

Distribution. The United States (Texas).

Symptoms. Fusarium root rot is similar, symptomatically, to Fusarium yellows; however, it is distinct in that a severe root rot also occurs that is not associated with Fusarium yellows. The disease is first observed as a single rust-colored streak in the central stele of diseased young plants. No external root symptoms are apparent at this time. However, during the hottest period of the day, diseased plants may wilt slightly, but no other foliar symptoms are evident and disease would not be detected by a casual observer.

Obvious foliar symptoms of wilting and leaf chlorosis become apparent about 60 days after emergence and coincide with ambient daytime temperatures above 27°C. At this time, diseased roots show vascular necrosis and discoloration typical of Fusarium yellows, but the external portion of the root frequently is also rotted at the distal end and has black streaks traveling upward. At first, only the tip of the diseased sugar beet is discolored. Eventually, half of the main taproot may become black and rotted. In extreme instances, the diseased portion of the root completely rots, leaving only remnants of vascular bundles. The root rot is distinguished from bacterial vascular necrosis and soft rot in that the bacterial disease causes a watery rot throughout the root and crown.

Management. Not reported.

Fusarium Tip Rot

Cause. *Fusarium javanicum* (syn. *F. radicicola*) survives as chlamydospores in soil or as mycelia in infested plant residue.

Distribution. The United States.

Symptoms. The first symptoms appear about midseason, when leaves start to wilt. Root tips are rotted with a white fungus mycelium growing on the diseased root surfaces. Eventually, the interior of roots has a brown rot but the exterior does not show any symptoms.

Management. Not reported.

Fusarium Yellows

Cause. *Fusarium oxysporum* f. sp. *betae* (syn. *F. oxysporum* f. sp. *spinaciae*) survives in soil and infested residue as chlamydospores, mycelia, and spores. In the spring or summer, chlamydospores germinate in the presence of host roots and form mycelia that penetrate the roots and grow into the vascular (xylem) system.

    Fusarium oxysporum f. sp. *betae* forms tyloses and produces toxic substances that affect xylem tissue, preventing water from reaching the tops of diseased plants. After death or maturity of the diseased plant, the fungus grows out of the vascular tissue and sporulates on the plant surface. Chlamydospores are formed either from conidia or mycelium and survive in soil for several years.

Distribution. Europe, India, and the United States.

Symptoms. About midseason, yellowing occurs between the veins of the largest leaves; occasionally only half of the diseased leaf is affected. The large veins and a narrow border surrounding the veins usually remain green. As the disease progresses, diseased leaves wilt during warm weather, droop down, and touch the soil surface. Eventually, parts of these leaves will turn brown and dry out. As diseased leaves die, leaf blades pull upward along the midribs, margins dry up and become discolored brown, and leaf tips bend to one side. Leaves ultimately become "heaped" around the crown.

The yellowing and progression of disease symptoms will eventually occur on younger heart leaves. Roots may not show any external symptoms, but in cross section, vascular tissue has a gray-brown discoloration. The seed stalks of seed beets will wilt and die prematurely.

Management. Grow sugar beet in the same field once every 4 years or longer.

Fusarium, Miscellaneous Species

Several *Fusarium* spp. reported isolated from sugar beet piles produce the following mycotoxins: acetylneosolaniol, deoxynivalenol, HT-2, T-2, zearalenone, and 15-acetyldeoxynivalenol.

Penicillium Storage Rot

Cause. *Penicillium claviforme.* Conidia are disseminated by wind to plant hosts and enter roots through wounds and tissue rotted by *Phoma betae. Penicillium aurantiogriseum* (syn. *P. cyclopium*), *P. brevicompactum* (syn. *P. stoloniferum*), *P. bordzilowskii, P. duclauxii, P. expansum, P. funiculosum, P. purpurogenum* (syn. *P. rubrum*), and *P. variabile* have also been reported as storage rot pathogens.

Distribution. Wherever sugar beet is grown.

Symptoms. A white mycelial growth that resembles "tufts" and consists of conidiophores, conidia, and sparse mycelia grows on brown, rotted tissue.

Management. Research is being conducted on breeding for resistance and on the use of fungicides in storage piles.

Phoma Leaf Spot

Cause. *Phoma betae* infects sugar beet during all stages of growth. The teleomorph, *Pleospora betae* (syn. *P. bjoerlingii*), is found in Europe but apparently is not common in the United States and Canada.

Phoma betae overwinters as mycelium and conidia on and in infected seed, and for up to 26 months as mycelia and pycnidia in soil and infested residue in the soil. In Europe, *Pleospora betae* overwinters as perithecia in infested residue. Conidia produced in pycnidia during moist weather are exuded in a gelatinous matrix, then are disseminated primarily by splashing rain to plant hosts. Ascospores of *Pleospora betae* are produced in perithecia and disseminated by wind to plant hosts. Only older, outer leaves are infected by both types of spores. Pycnidia eventually develop in spots and produce conidia that function as secondary inoculum. Phoma leaf spot is most severe during periods of high humidity and temperatures of 15° to 32°C.

Distribution. Wherever sugar beet is grown. Damage is usually slight on sugar beet grown for sugar.

Symptoms. Initially, small light-colored spots that soon become light brown develop on diseased leaves. As spots enlarge, alternate light and dark brown concentric rings develop with at least one dark brown ring being prominent. Black pycnidia develop in the dark rings. Spots, which are round to oval (1–2 cm in diameter), may coalesce and form large necrotic areas.

Management
1. Treat seed with a seed-protectant fungicide.
2. Control common lambsquarters, *Chenopodium album,* because this weed is also a host for *P. betae.*
3. Grow sugar beet in infested soil once every 4 years.

Phoma Root Rot and Storage Rot

Phoma root rot and storage rot is also called blackleg.

Cause. *Phoma betae* overwinters as mycelia and conidia on and in seed, and for up to 26 months as mycelia and pycnidia in soil and infested residue in the soil. Conidia are produced in pycnidia during moist weather, exuded in a gelatinous matrix, and disseminated by splashing rain to the upper root near the crown. Infection also occurs by soilborne inoculum, presumably mycelia growing from infested residue.

Weather and soil conditions unfavorable to normal growth of sugar beet favor infection of host plants by *P. betae.* Beets that do not die in the field and are delivered into storage piles act as centers of infection within the pile. Temperatures for optimum root rot and storage rot are 5° to 20°C. Sugar beet is infected during all growth stages.

Distribution. Wherever sugar beet is grown.

Symptoms. Sugar beets affected by Phoma root rot occur randomly over a field. Occasionally, several plants in a limited area will be diseased. The first visible symptom is a wilting of leaves, which was preceded by small brown spots on the upper root near the crown. Spots are soft and watery and spread to cover a large area. Often, the centers of diseased areas are dark brown and "pithy," but margins of diseased areas are soft and watery. The outer surface of diseased roots remains unbroken while rot develops just beneath it. Black vertical growth fissures that may occur on the hypocotyl are, likely, the result of soilborne inoculum.

The centers of crowns are very susceptible to rot and removal of portions of crowns during harvest exposes this tissue to further infection by *P. betae.* About 80 days after roots have been harvested and stored, rot begins. The first indication of rot is in the center of the crown. From there, it spreads in a cone-like pattern into the adjacent crown and down into the main taproot. Rotted tissue is black or dark brown and has occasional pockets lined with the white mycelium of the fungus.

Management
1. Treat seed with a seed-protectant fungicide to prevent seedborne infection.
2. Practice cultural methods that promote vigorous growth of sugar beet.
3. Grow sugar beet in infested soil once every 4 years.
4. Resistance to storage rot occurs in some germplasms.

Phymatotrichum Root Rot

Phymatotrichum root rot is also called Texas root rot.

Cause. *Phymatotrichopsis omnivora* (syn. *Phymatotrichum omnivorum. Phymatotrichum omnivora* is considered a synonym in some literature.). *Phymatotrichopsis omnivora* survives as sclerotia in soil. Infection occurs when soil temperatures exceed 28°C.

Distribution. Northern Mexico and the southwestern United States.

Symptoms. Initially, diseased leaves become yellowed or bronzed, which is followed by a sudden wilting of the entire plant. Eventually, roots develop a yellow to tan rot. A thin, felt-like layer of yellowish mycelium develops on root systems and, occasionally, crust-like spore mats develop on soil surfaces.

Management. Do not grow sugar beet in infested soil.

Phytophthora Root Rot

Cause. *Phytophthora drechsleri* overwinters as oospores in soil or infested residue. Oospores germinate in the presence of moisture to form sporangia in which zoospores are borne. Zoospores "swim" to roots, where they germinate and form a germ tube that penetrates the host's roots. Phytophthora root rot occurs in wet or poorly drained soils; temperatures of 28° to 31°C are optimum.

Phytophthora megasperma has been reported to cause a similar root rot in England, and *P. cryptogea* has been reported as a cause of Phytophthora root rot in Wyoming.

Distribution. England, Iran, and the United States.

Symptoms. The first noticeable symptom is a wilting of diseased leaves during the warmest part of the day. Leaves may recover their turgidity at night. Necrotic spots appear on roots either some distance below the crown or only on the tips of roots. Later, the entire lower portions of diseased roots become rotted and adventitious lateral roots often grow above the diseased areas. Frequently, the rotted portion breaks down completely so that the lower portions of roots are a mass of discolored vascular strands that resemble the ends of a frayed rope. The advancing edge of the necrosis is a narrow band of black-brown discolored tissue that sometimes is separated from healthy tissue by a narrow, light buff–colored band. Small empty cavities frequently occur in necrotic tissue.

Phytophthora cryptogea was reported to cause a firm brown decay on the distal 3 to 5 cm of taproots and a sharp demarcation between healthy and diseased tissues.

Management. Grow sugar beet in well-drained soils. The cultivar R2 may be susceptible to infection by *P. cryptogea*.

Powdery Mildew

Cause. *Erysiphe polygoni* (*E. betae* is a synonym in some literature) overwinters as mycelia or haustoria in the crowns of *Beta* spp. in sugar beet–growing areas of the southwestern United States. In the northwestern states of the United States, *E. polygoni* survives in axillary bud tissue of seed beets. During the growing season, leaves of sugar beet initially become infected by conidia disseminated by wind north and east from the southwestern sugar beet–growing areas. Secondary inoculum is provided by conidia produced on diseased leaves.

Powdery mildew is favored by warm temperatures (25°C optimum) and relatively dry weather. Conidia will germinate at a high relative humidity but production and viability of conidia are optimum at 30% to 40% relative humidities.

Powdery mildew is more severe on older plants, probably due to reduced air circulation, light, and temperature and increased humidity under the denser canopy of older sugar beets. Susceptibility to *E. polygoni* is increased in plants previously infected by one of the beet-yellowing viruses. Different races of *E. polygoni* exist on sugar beet.

Distribution. Europe and the United States.

Symptoms. Symptoms of powdery mildew on sugar beet in the United States appear 2 to 6 months after sowing. The most distinguishing symptom is a white powdery-appearing material consisting of mycelium, conidiophores, and conidia on diseased leaf surfaces. This fungus growth is superficial and does not penetrate far into leaf cells. Prematurely killed leaves become brown but retain the white fungal growth. Cleistothecia, which appear as tiny dark specks, are found in mycelia in Europe, but they have been found only once in the United States.

Management
1. Apply foliar fungicides during the period of rapid root growth.
2. Grow cultivars that are less susceptible to infection. *Beta maritima* provides a source of resistance.

Pythium Damping Off

Cause. *Pythium* spp., including *P. aphanidermatum, P. intermedium, P. purpurogenum* (syn. *P. rubrum*) and *P. ultimum.* The life cycles of most *Pythium* spp. are as follows: Survival is as oospores in soil. Oospores germinate in the presence of water by releasing zoospores that "swim" to roots. Zoospores germinate and form germ tubes that penetrate the host. Eventually, conidia, sporangia, and oospores are formed in rotted tissue. Secondary infection is by further production of zoospores that are spread to adjacent plants by water. Infrequently, sporangia and conidia germinate directly and form germ tubes that penetrate the host.

Pythium damping off is most severe at temperatures of 12° to 20°C and wet soil conditions. The exception is infection by *Pythium aphanidermatum,* which infects host plants in warmer, wet soils. Thickly sown seeds are more susceptible to infection than seeds sown at the recommended plant population.

Distribution. Wherever sugar beet is grown.

Symptoms. Diseased seeds are soft, light brown, and often overgrown with the white mycelia of *Pythium* or secondary fungi. Seedlings killed before emergence have a light brown, soft, water-soaked rot. Seedlings may also be killed after emergence. Plants that survive will recover when secondary thickening starts in the hypocotyls.

Management
1. Treat seed with a seed-protectant fungicide.
2. In experimental tests, osmopriming seed in NaCl or polyethylene glycol reduced damping off.

Pythium Root Rot

Cause. *Pythium aphanidermatum* and *P. deliense. Pythium irregulare* and *P. sylvaticum* also have been reported pathogenic to sugar beet. The life cycles of most *Pythium* spp. are as follows: Survival is as oospores in soil. Oospores germinate in the presence of water by releasing zoospores that "swim" to roots. Zoospores germinate and form germ tubes that penetrate the host. Eventually, conidia, sporangia, and oospores are formed in rotted tissue. Secondary infection is by further production of zoospores that are spread to adjacent plants by water. Infrequently, sporangia and conidia germinate directly and form germ tubes that penetrate the host.

Root rot caused by *P. aphanidermatum* is most severe at soil temperatures greater than 27°C at a soil depth of 10 cm for at least 12 hours a day. Excessive soil moisture and poorly drained soils with a high pH are also favorable for disease development.

Distribution. Iran and the United States (Arizona, California, Colorado, and Texas).

Symptoms. Pythium root rot symptoms become evident later in the growing season. Initially, older leaves become yellow, wilt, and die as the disease progresses to the younger leaves. Leaf petioles have brown, sunken necrotic areas that extend from the petiole base to the leaf blade. Eventually, all diseased leaves die and the crowns and roots become diseased. The diseased portion of roots extends internally from the crown down toward the root tip, and from secondary infections along the roots. Diseased root tissue is somewhat firm.

Management
1. Avoid growing in infested fields where the inoculum threshold is known to be above the desirable level.
2. Increase the frequency and decrease the duration of irrigation.

Ramularia Leaf Spot

Cause. *Ramularia beticola* overwinters in milder climates as mycelia or conidia in infested residue in the soil and, possibly, on seed. Conidia are produced during humid weather conditions at temperatures of 17° to 20°C and disseminated by wind to infect the older leaves of host plants.

Distribution. Canada (British Columbia), Europe, and the United States (primarily northern California, Colorado, Oregon, and Washington).

Symptoms. Spots resemble those caused by *Cercospora beticola* but are larger and more angular. Spots are circular (4–7 mm in diameter) and white to light tan, are surrounded by a brown or purplish border, and have tufts of white mycelium growing in the center. Diseased leaves become yellow, then necrotic, and die.

Management
1. Grow cultivars resistant to Cercospora leaf spot. These cultivars are also resistant to Ramularia leaf spot.
2. Apply copper foliage fungicides before disease becomes severe.

Rhizoctonia Crown and Root Rot

Rhizoctonia crown and root rot is also called brown rot. Dry rot canker and Rhizoctonia crown rot and root rot are both caused by *Rhizoctonia solani*. However, the symptoms are distinct and the two diseases are discussed separately.

Cause. *Rhizoctonia solani*. The teleomorph *Thanatephorus cucumeris*, anamorph *R. solani* AG-3 and AG-5, is reported from Minnesota. In the Red River Valley of the North, AG-1 through AG-5 were isolated along with multinucleate isolates that did not anastomose with any tester isolates. Anastomosis groups AG-4 and AG-5 predominated on seedlings and AG-2-2 on older plants. AG-2-2 is considered to cause crown and root rot, and AG-1, AG-2-2, AG-4, and AG-5 cause damping off.

In Ohio, most multinucleate isolates are in either anastomosis group

AG-2 or AG-4. AG-2 isolates predominate in fine-textured soils, whereas AG-2 and AG-4 isolates are either equal in number or AG-4 predominates in coarser-textured soils. AG-4 isolates are more virulent to sugar beet seedlings, and AG-2 isolates are more virulent to 6- to 8-week-old plants.

In Texas, the predominate isolates from mature sugar beet were AG-2-2, but most isolates obtained from seedlings were AG-4 and AG-5.

Rhizoctonia solani survives as bulbils, or thickened hyphae, and sclerotia in soil and in infested residue in the soil. Sclerotia survive for less than 2 years in soil in the Red River Valley of the North. When *R. solani* resumes growth, mycelium is formed that penetrates host plants through the crowns or the bases of leaf petioles. *Rhizoctonia solani* moves from plant to plant within a row by mycelial growth. Propagules are disseminated by water and tillage equipment.

Disease is most severe in poorly drained soils at temperatures from 25° to 33°C. Depositing soil in and around sugar beet crowns (hilling) increases the incidence and severity of root rot. In Texas, disease is mainly a problem on sugar beet that either is sown late in the growing season or is resown.

Distribution. Wherever sugar beet is grown.

Symptoms. The first symptom is a darkening of the petiole base on half-grown to nearly mature sugar beet roots. The petiole is weakened at the point of infection and leaves fall to the soil and form a rosette of dead leaves around the crown. Foliage wilts as crowns become rotted. Eventually, roots become diseased below the diseased crowns and have a dark brown to black discoloration on the surfaces and a light to dark brown dry rot of internal tissue. Deep cracks often occur at or near diseased crowns. Infrequently, just the root tip may be rotted off 15 to 20 cm below the soil surface.

Symptoms of *T. cucumeris* are a superficial white to gray, dusty hymenial growth on sugar beet petioles.

Management
1. Follow recommended tillage and fertility practices.
2. Rotate with small grains and maize as a general practice. However, research in Texas demonstrated that sugar beet following alfalfa has a high incidence of disease, sugar beet following sorghum and winter wheat has the next highest disease incidence, and sugar beet following cotton, fallow, and sunflower had the least incidence of disease. Therefore, it is recommended that sugar beet not follow wheat in a rotation in Texas because AG-4 is highly virulent to each species. In Minnesota, AG-2-2 was pathogenic to sugar beet and common bean; therefore, avoid rotating sugar beet with common bean.
3. Avoid hilling up soil around plants.

4. Some breeding lines display levels of resistance. Disease on moderately resistant sugar beet cultivars decreases as plant age increases.

Rhizoctonia Foliage Blight

Cause. *Rhizoctonia solani,* teleomorph *Thanatephorus cucumeris.* AG-3 and AG-5 were found on petioles of sugar beet in sugar beet fields in the Red River Valley of the North. Potato debris (discarded potato culls or soilborne root residue of the previous crop) served as a source of inoculum of AG-3 for infection of the subsequent sugar beet crop. AG-5 was reported on other crops grown in the region, including soybean and wheat. AG-3 formed sclerotia on roots, but isolates of AG-5 did not.

Windels et al. (1997) reported that the occurrence of hymenia of *T. cucumeris* on sugar beet was not an indicator of disease on these plants, at least in the Red River Valley of the North. The presence of hymenia on sugar beet in the field was significant in that genetic recombination and inoculum production of *R. solani* occurred on a nonhost crop. In the Windels (1994) study, infection of sugar beet by *T. cucumeris* was related to AGs of *R. solani* associated with the previous crop, i.e., AG-3 for potato and AG-5 for wheat.

Rhizoctonia solani survives as resting mycelia and sclerotia in infested residue and soil. Optimum formation of the teleomorph in the field requires high relative humidity and temperatures of 20° to 30°C. During moist weather in the early part of the growing season, fungal propagules are splashed by rain and infect the leaves of young plants. Half-grown and mature plants are resistant to infection. Disease occurs at temperatures of 21° to 25°C and 100% relative humidity. About 3 weeks after infection, basidia and basidiospores of *T. cucumeris* are produced on leaves. The basidiospores function as secondary inoculum and are disseminated by wind and splashed by rain to surrounding host plants. Dry weather in midseason prevents further disease development.

Distribution. The United States (Colorado, Maryland, Michigan, Nebraska, Ohio, and Virginia).

Symptoms. Black spots (6–12 mm in diameter) form on diseased leaves. Spots become necrotic, dry up, and break away from healthy tissue, giving diseased leaves a ragged appearance. A circular zone of secondary spots surrounds the original spot or the place where the spot had been. About 3 weeks after infection, a filmy gray-white growth consisting of mycelium, basidia, and basidiospores of *T. cucumeris* can be observed on leaf surfaces. The spread of secondary inoculum causes plants to become diseased in a circular pattern (5 to 10 m in diameter), with the original diseased plant serving as the disease focus in the center of the circle.

Management. Grow sugar beet in the same soil once every 4 to 5 years.

Field Crop Diseases

Rhizopus Root Rot

Cause. *Rhizopus arrhizus* (syn. *R. oryzae*) and *R. stolonifer* survive in soil and infested residue as spores and mycelia. The fungi are generally considered to
be weak parasites that infect through crown wounds of host plants that
have been predisposed by excess soil moisture. Larvae of insects, such as
Eumerus sp., wound healthy tissue, thereby providing infection courts for
Rhizopus spp. to enter the plant. *Rhizopus stolonifer* infects at temperatures
of 14° to 16°C, and *R. arrhizus* infects at temperatures of 30° to 40°C. Sugar
beet plants are especially vulnerable to infection after thinning.

Distribution. Canada, Italy, Russia, Ukraine, and the United States.

Symptoms. Leaves wilt during the warmest part of the day but normally recover their turgidity at night. As disease progresses, leaves permanently
wilt and are transformed into a dry, brittle rosette. Superficial gray-brown
necrotic areas appear on the bases of diseased roots and gradually spread
upward toward the crowns. Root tissues become dark and spongy and,
eventually, entire roots become discolored black. White mycelium, which
later turns dark, appears on diseased root surfaces. Eventually, the causal
fungi grow deeper into roots, causing internal cavities that become filled
with a clear liquid that is rich in acetic-like acid.

Symptoms of root rot caused by *R. arrhizus,* have been described as a
foamy white exudate that often exudes from the crowns of dead and dying
plants. Diseased root tissue has a soft to spongy texture and varies from tan
to black.

Management
1. Maintain proper soil drainage.
2. Avoid injuring plants.

Sclerotium Root Rot

Sclerotium root rot is also called southern Sclerotium root rot and southern
root rot.

Cause. *Sclerotium rolfsii* survives as sclerotia in soil and is disseminated by
anything that moves sclerotia-infested soil, such as water and machinery.
Sclerotia germinate by producing germ tubes that penetrate a host. Later in
the growing season, sclerotia are formed in mycelia growing over diseased
plant surfaces. Disease development is favored by temperatures of 25° to
35°C and moist soil.

Distribution. Czechoslovakia, Japan, Korea, Mediterranean countries, Middle East, and the southwestern United States.

Symptoms. Initially, the tops of diseased plants have an unthrifty appearance; eventually, the leaves wilt. Root surfaces are covered with a white,
cottony, mycelial growth that extends into the soil surrounding the dis-

eased roots. Scattered throughout the mycelial growth are small, round sclerotia that resemble white, tan, or brown mustard seeds.

Management
1. Do not grow sugar beet in soils with high numbers of sclerotia.
2. Do not put dump dirt on fields to be cropped to sugar beet.
3. Apply heavy amounts of nitrogen to soil.

Seedling Rust

Cause. *Puccinia subnitens* is a dioecious rust. The pycnial and aecial stages occur on sugar beet; the uredinial and telial stages occur on saltgrass, *Distichlis stricta*.

Distribution. Russia, Ukraine, and the United States (Rocky Mountain states).

Symptoms. Symptoms generally occur only on the diseased lower surfaces of cotyledons, but infrequently can be found on the first true leaves. Bright yellow-orange aecial pustules are aggregated into rings on the lower leaf surfaces. Pycnia may be present on the upper leaf surfaces.

Management. Not necessary, since seedling rust is of no economic importance.

Verticillium Wilt

Cause. *Verticillium albo-atrum* survives as dark resting mycelia in infested residue and soil. *Verticillium albo-atrum* enters through lateral roots and grows into the vascular system. Conidia are formed on dead plant tissue but are not considered to be significant in the spread of *V. albo-atrum*.

Distribution. Europe and the United States (Colorado, Idaho, Nebraska, and Washington).

Symptoms. Diseased leaves initially are yellow. Eventually, the outer leaves wilt and dry up, while inner leaves become twisted and deformed. In cross section, the vascular system may show a slight to dark brown discoloration. Although lateral roots through which the fungus entered into the host plant are usually black and water-soaked, little or no root rot occurs.

Management. Rotate sugar beet with other crops for several years.

Violet Root Rot

Cause. *Rhizoctonia crocorum*, teleomorph *Helicobasidium brebissonii* (syn. *H. purpureum*). *Rhizoctonia crocorum* is referred to as the mycelial stage and overwinters as sclerotia in soil and on the roots of weed hosts. Sclerotia germinate late in the growing season and form mycelia that grow over and through soil to sugar beet roots. *Rhizoctonia crocorum* spreads from plant to plant by mycelial growth through soil. During this time, basidia form on the mycelial mat. *Rhizoctonia crocorum* contacts a root and forms an infection cushion from which mycelium penetrates into the host root and eventually grows over it. Eventually, sclerotia form in the mycelium.

Reports of the effects of soil moisture and pH are conflicting. Some researchers state disease is more prevalent under wet soil conditions and high soil pH. Others maintain these conditions are not a factor in disease development. Fungal growth is optimum at a temperature of 13°C.

Distribution. Europe and the United States.

Symptoms. Violet root rot is a late season disease in the United States. Usually all beets within an area will be diseased. Initially, plants wilt and leaves become light green. Purplish spots occur on root surfaces, and a purple fungal growth develops near the root tip and eventually covers the entire root. The area on the root covered by the fungus is somewhat depressed. In cross section, the root shows a sharp line between healthy tissue and the surrounding brown diseased tissue. Diseased roots carry excessive amounts of soil because fungus mycelium binds the soil to the roots.

Management. Most measures are relatively ineffectual; however, the following may partially aid in managing violet root rot:

1. Rotate sugar beet with other crops.
2. Control weeds in fields.

Diseases Caused by Nematodes

Clover Cyst

Cause. *Heterodera trifolii.* Females reproduce parthenogenetically. The life cycle is similar to that of *H. schachtii.* (*See* "Sugar Beet" in this section.)

Distribution. Europe.

Symptoms. Sugar beet stands may be lost or have uneven growth. Outer leaves of diseased plants become chlorotic, then wilt. Storage roots are not well formed and have branched root systems and excessive growth of fibrous roots.

Management
1. Rotate sugar beet every 4 to 5 years with crops such as small grains. It has been reported that rotation with resistant crops is warranted economically when the infestation is about 1000 eggs or larvae per 100 g of air-dried soil.
2. Equipment should be cleaned between fields.
3. Tare dirt should be returned to nonagricultural soil.
4. Sow as early as feasible.
5. Apply nematicides where feasible. However, nematicide effectiveness may be affected by soil temperature and soil type.

False Root Knot

Cause. *Nacobbus aberrans* and *N. dorsalis.*

Distribution. The United States. *Nacobbus aberrans* is found in Colorado, Kansas, Montana, Nebraska, South Dakota, and Wyoming; *N. dorsalis* is found in California.

Symptoms. Necrosis and gall-like swellings occur on diseased roots.

Management. No specific management has been developed. Presumably, measures used to manage other nematodes will be of some value.

Nebraska Root Gall

Cause. *Nacobbus batatiformis* overwinters as eggs in soil and as adults in galls. Nematodes are disseminated by any mode that transports soil, such as irrigation water and tillage equipment. Eggs are deposited in a gelatinous matrix outside galls and hatch into larvae that "swim" in moisture to penetrate surrounding roots and rootlets. Galls form at the point of penetration and contain from one to several egg-producing females. Growth and reproduction of nematodes is optimum at soil temperatures of 24° to 35°C.

Distribution. The United States (western Nebraska).

Symptoms. Severely diseased sugar beets are stunted and wilt during warm weather. Numerous galls up to 1 cm in diameter are formed, predominantly on lateral roots. Many small rootlets are produced on each gall, creating a hair-like or "whiskery" appearance.

Management
1. Rotate sugar beet at least every 4 years.
2. Sow sugar beet early in growing season because of nematode inactivity in cool soils.
3. Control weeds, especially lambsquarters and kocia, in sugar beet fields since they may harbor *N. batatiformis.*
4. Apply nematicides in the form of fumigants to soil.

Needle

Cause. *Longidorus attenuatus, L. caespiticola, L. elongatis,* and *L. leptocephalus.*

Distribution. England and the United States. However, needle nematodes are not a problem in the United States.

Symptoms. Tips of seedling taproots may be killed. Lateral roots have stubby ends that later become brown and die.

Management. Follow general management practices used to manage nematodes.

Root Knot

Cause. *Meloidogyne arenaria, M. incognita, M. javanica, M. hapla,* and *M. naasi. Meloidogyne incognita* and *M. javanica* cause the greatest damage in the United States. Nematodes overwinter as eggs or second-stage juveniles in soil, galls, and root tissue. *Meloidogyne arenaria, M. incognita,* and *M. javanica* do not survive if the temperature averages below 3°C during the coldest month. *Meloidogyne hapla* is limited by temperatures of 27°C or greater.

Larvae move through soil and penetrate root tips of host plants. The nematodes then migrate through infected roots into the vascular tissues, where they become sedentary and commence feeding. Larval excretions stimulate cell proliferation, resulting in development of syncythia from which larvae derive their food. Females lay eggs that may be pushed out of roots into the soil.

Distribution. Wherever sugar beet is grown.

Symptoms. Initially, small, light tan galls form on fibrous roots and taproots. Later, galls are a darker brown and may be of different shapes: Rounded galls are caused by *M. hapla*; elongated or spiral-shaped galls are caused by *M. naasi*; irregular-shaped, club-like swellings are caused by *M. incognita*. The galls or swellings along the fine roots resemble a string of beads. Root rot may occur.

Plants may be severely diseased without obvious aboveground symptoms. The common aboveground symptom is the presence of small, yellow, stunted plants. Plants under moisture stress may wilt and collapse.

Management
1. Rotate sugar beet for 4 to 5 years with other crops, such as small grains. However, this is not considered effective.
2. Control weeds.
3. Apply nematicides in the form of fumigants to soil.

Stem and Bulb

Cause. *Ditylenchus dipsaci.*

Distribution. Europe.

Symptoms. Diseased seedlings have a swelling of the epicotyl, hypocotyl, midribs, and main veins of leaves and the occasional formation of galls. Severely diseased plants are stunted and have multiple crowns. Crown cankering and girdling may occur at the scars of diseased petioles. Root rot occurs later in the growing season.

Management
1. Rotate sugar beet with nonhost plants.
2. Control weeds.
3. Remove infested residue from fields.

Stubby Root

Cause. *Paratrichodorus anemones, P. pachydermus, P. teres, Trichodorus cylindricus, T. primitivus,* and *T. viruliferus.*

Distribution. Europe and the United States. However, stubby root nematodes are not considered economically important in the United States.

Symptoms. Tips of the taproots of diseased seedlings may be killed. Lateral roots have stubby ends that later turn brown and die.

Management. Use general practices recommended for managing nematodes.

Sugar Beet

Sugar beet nematode is sometimes called the sugar beet cyst nematode.

Cause. *Heterodera schachtii* survives in soil as cysts filled with eggs. Eggs hatch over an extended period of time, sometimes up to several years. The resulting larvae penetrate roots and take up positions in the cortex. The female bodies break through the epidermis, leaving only the head embedded in the root. Eventually, eggs develop in the female and largely displace other organs. Upon reaching maturity, female bodies are transferred into a brown cyst. Optimum growth and multiplication of *Heterodera schachtii* occur at temperatures of 13° to 28°C. *Heterodera schachtii* can mature, increase in number, and continue egg production on respiring, excised roots.

Distribution. Wherever sugar beet is grown.

Symptoms. The initial disease symptoms is stunted or killed plants in small, distinct areas within a field. Young plants wilt and die shortly after thinning. Small white cysts, about the size of a pinhead, are attached to the diseased feeder roots. Later, these cysts turn dark orange-brown and are difficult to see and to differentiate from the host root. The taproot may be small and extremely "hairy" due to the proliferation of fibrous roots.

An aboveground symptom is the presence of small, yellow, stunted plants with sprawling, wilted leaf petioles. In nematode-infested fields, sugar beet is more susceptible to leaf spot and other diseases.

Management
1. Rotate sugar beet for 4 to 5 years with crops such as small grains. It has been reported that rotation with resistant crops is warranted economically when the infestation is about 1000 eggs or larvae per 100 g of air-dried soil.
2. Equipment should be cleaned between fields.
3. Tare dirt should be returned to nonagricultural soil.
4. Sow as early as feasible.
5. Apply nematicides where feasible. However, nematicide effectiveness may be affected by soil temperature and soil type. Nematode populations increase in soils treated with carbamate herbicides.

6. Grow a "catch" crop. Radishes have been demonstrated to be successful.

Disease Caused by Phytoplasmas

Rosette Disease

Rosette disease is also called witches' broom.

Cause. A phytoplasma-like organism.

Distribution. Italy.

Symptoms. A typical rosette-type growth of diseased foliage has been reported. A prolific growth of small leaves and stems occurs together with an epinasty, which gives foliage a "bushy" appearance.

Management. Not reported.

Diseases Caused by Rickettsia

Beet Latent Rosette

Cause. A rickettsia-like organism transmitted by nymphs and adults of the beet lace bug, *Piesma quadratum,* and by dodder, *Cuscuta campestris.* The organism is persistent in vectors and remains infectious for 10 to 30 days.

Distribution. Germany and the United States.

Symptoms. Initially, there is a leaf twisting, downward turning of leaf tips and chlorosis of diseased young leaves. Later in the rosette stage, a cluster of terminal and axillary shoots with strap-like leaves is formed that resembles a witches' broom. The rosette remains after the death of mature, healthy leaves.

Management. No management measures have been developed because it is not considered a serious disease.

Yellow Wilt

Cause. A rickettsia-like organism transmitted by the leafhopper, *Paratamus exitiosus,* and by dodder, *Cuscuta californica* and *C. campestris,* and by grafting.

Distribution. Argentina and Chile.

Symptoms. Leaves yellow or diseased plants wilt and collapse. New leaves may be dwarfed and often turn downward at the tips. Frequently only portions of leaves or the veins of some diseased leaves become yellow. Leaves of plants that have been diseased for a long time may be necrotic or become strap-like. Root tips may become necrotic and form tufts on roots, which cause plants to wilt during periods of high temperatures and low soil moisture, often resulting in plant death.

Management
1. Avoid growing sugar beet in soils where disease has caused losses.
2. Grow resistant varieties when they become available.

Diseases Caused by Viruses

Beet Crinkle

Beet crinkle is known as leaf curl in England and as kopfsalat in parts of Germany.

Cause. A suspected virus. The virus is transmitted by adult lacebugs *(Piesma quadratum)*, but the nymphs also can pick up the virus. The insect hibernates in grass and under trees and hedges around sugar beet fields. The virus multiplies in both larval and adult stages of the insect. Mechanical transmission has been done by grafting from diseased beet to beet.

Three forms of the disease have been recognized: a severe and progressive form that starts early in the year; a similarly severe form interrupted by periods of normal growth; and a slight form that starts much later. It is not known whether these disease forms are due to different virus strains.

Distribution. Central Europe. The disease is now confined to Germany and Poland.

Symptoms. Diseased leaf and petiole veins become crooked, translucent, glassy, and swollen. Veins do not grow as fast as the rest of the leaf, causing both veins and leaves to have a markedly "crinkly" appearance. New leaves formed from crowns remain small and curved inward, forming a compact bunch that gives the diseased plant the appearance of a cabbage. Older diseased leaves die.

Management
1. Control insects with insecticides.
2. Grow an early trap crop of beet around field edges and plow it under before the main sugar beet crop has germinated.

Beet Curly Top

Beet curly top is also called curly top, sugar beet curly top, sugar beet curly leaf, western yellow blight virus, and tomato yellows.

Cause. Beet curly top virus (BCTV) is in the geminivirus group. BCTV overwinters in a large number of perennial plants and is transmitted hundreds of kilometers by the beet leafhopper, *Circulifer tenellus,* in a persistent man-

ner. The beet leafhopper can feed for a minute on a diseased plant and acquire BCTV, then carry BCTV for a month or more. However, BCTV does
not multiply in the leafhopper body, and if an insect has fed only once on
a diseased plant, its ability to transmit the virus diminishes during successive feedings. BCTV is not transmitted through the egg.

Distribution. Canada, the Mediterranean basin, and the United States.

Symptoms. Plants infected prior to or immediately after thinning will not
continue to grow. The most common field symptoms are dwarfing of diseased leaves, with the leaf edges rolling and curling inward, together with
vein clearing, swelling, and spine-like growths. Eventually, swollen veins
give rise to small nipple-like swellings along veins and veinlets. Sometimes
a sticky brown fluid may be exuded and collect in droplets along petioles
and the edges of diseased leaves. In cross section, roots have concentric
black rings together with phloem tissue that becomes necrotic and cracks.
There may be an increase in the number of rootlets.

Infection late in the growing season produces few symptoms except for
a slight vein clearing and swelling of leaves.

Management
1. Grow resistant or tolerant cultivars.
2. Control vectors.

Beet Distortion Mosaic

Cause. Beet distortion mosaic virus (BDMV) is soilborne and mechanically
transmitted. There is some evidence that the vector is the fungus *Polymyxa
betae*. BDMV particles are long flexuous rods (12 × 200–2400 nm) that are
similar in size to those of wheat spindle streak mosaic virus.

Distribution. The United States (Texas).

Symptoms. A distortion and mosaic of diseased leaves occurs.

Management. Not reported.

Beet Leaf Curl

Cause. Beet leaf curl virus (BLCV) is in the geminivirus group. BLCV overwinters in its vector, the beet lace bug *(Piesma quadratum)*, which hibernates in protected areas. In the spring, vectors move into sugar beet fields
and transmit BLCV to young plants. BLCV is transmitted in a persistent
manner by the beet lace bug, which remains infective for the rest of its life.

Distribution. Europe.

Symptoms. Initially, vein clearing occurs in the youngest leaves, which crinkle and curl inward, forming a structure similar to a lettuce head. Diseased
leaves and roots are badly stunted.

Management
1. Do not grow sugar beet in areas where disease occurs.
2. Apply insecticides to sugar beet near overwintering areas of the vector.

Beet Marble Leaf

The causal virus may be a strain of beet mosaic virus.

Cause. Marble leaf virus is transmitted by the aphids *Myzus persicae, M. fabae,* and *M. eurphorbiae.* The virus is also transmitted by rubbing juice from diseased plants onto healthy plants.

Distribution. The United States (Oregon).

Symptoms. Local lesions on diseased plants appear 9 days after inoculation as chlorotic spots (1 mm in diameter). Lesions slowly increase in size to a diameter of 2 to 3 mm. A small necrotic spot that sometimes is surrounded by one or more rings ranging in color from yellow to green may develop in the center of the lesion. If lesions are numerous, inoculated leaves die prematurely.

Systemic symptoms appear 10 to 30 days after inoculation. Initially, symptoms consist of a vein chlorosis or mottling on diseased young leaves and various patterns of chlorosis on half-grown leaves. Chlorotic areas are wider than veins, may be continuous or broken, and in some cases are accompanied by indefinite mottle. As leaves mature, vein chlorosis becomes less conspicuous and the leaves become mottled and the chlorotic areas are not conspicuous or well defined. As leaves age further, mottling becomes indefinite and the tissue between the main veins turns yellow. Leaves appear dry and papery.

Management. Not reported.

Beet Mild Yellowing

Cause. Beet mild yellowing virus (BMYV) is in the luteovirus group. BMYV survives in perennial weeds and overwintered sugar beets. BMYV is transmitted by the peach aphid, *Myzus persicae,* which remains infective throughout its life after once acquiring the virus. Isolates of BMYV are serologically indistinct from beet western yellows virus.

Distribution. The British Isles, Russia, and possibly the western United States.

Symptoms. Diseased leaves have an orange-yellow discoloration.

Management. Insecticides are used to control the vector in Great Britain.

Beet Mosaic

Cause. Beet mosaic virus (BtMV) is in the potyvirus group. BtMV survives in perennial plants and overwintered sugar beets and is transmitted in a nonpersistent manner by several species of aphids. All stages and forms of the aphids can carry BtMV. Aphids feed a few seconds to acquire BtMV but rapidly lose it in a few hours during subsequent feeding. Consequently, an aphid must frequently feed on diseased plants in order to reacquire the virus. BtMV is not transferred to young aphids by the female aphid. Disease is most severe where crops of two growing seasons overlap or where diseased plants have overwintered.

Distribution. Wherever sugar beet is grown. Incidence of the disease depends on the buildup of the aphid population.

Symptoms. Initially, circular chlorotic spots with sharply defined margins occur on young diseased leaves. Spots often occur as chlorotic rings with green centers. Usually the mosaic pattern consists of irregular-shaped patches that are various shades of green.

Management
1. Do not overlap crops of two growing seasons.
2. Destroy wild and escaped beets in the vicinity of newly sown fields.

Beet Pseudo-Yellows

Cause. Beet pseudo-yellows virus (BPYV) is in the closterovirus group. BPYV is transmitted by the common greenhouse whitefly *Trialeurodes vaporariorum*. BPYV is acquired in 1 hour of feeding and is transmitted in a 1 hour feeding period. The latent period in *T. vaporariorum* is less than 6 hours. BPYV is retained in the vector for 6 days. Several other plants are also hosts.

Distribution. The United States (the Salinas Valley in California).

Symptoms. Chlorotic spotting or splotching occurs uniformly on older and intermediate-aged leaves. Eventually, the yellowing becomes more intense and general, with older diseased leaves almost entirely chlorotic except for small scattered islands of green tissue. Irregular-shaped, bright yellow areas (10–15 mm in diameter) may also form on older leaves. Leaves become thickened and brittle.

Management. Not reported.

Beet Ring Mottle

Beet ring mottle virus may be a strain of cucumber mosaic virus.

Cause. Beet ring mottle virus is transmitted mechanically and by a number of aphid species.

Distribution. The United States (the Salinas Valley in California). The disease is not of any economic importance.

Symptoms. Inoculated plants initially display chlorotic rings (2–3 mm in diameter) on diseased lower leaves. The rings enlarge to 4 to 5 mm in diameter and tend to coalesce. A systemic symptom is a distortion of the young diseased leaves, which are stunted and markedly narrowed to a lanceolate form. Leaf margins are wavy, and in some cases, leaves have more blade tissue on one side of the midrib than on the other, which tends to twist and distort the diseased leaves. The distribution of chlorotic pinpoint spots gives leaves a distinct mottle. Diseased plants become stunted.

Management. Not reported.

Beet Soilborne

Cause. Beet soilborne virus (BSBV) is in the furovirus group. BSBV is a rigid, rod-shaped, bipartite virus. Two serogroups of BSBV, Ahlum and Wierthe, have been identified in Europe.

BSBV is transmitted by the soilborne fungus *Polymyxa betae*. *Polymyxa betae* survives in field soil as cystosori, each of which is a mass of resting spores. Cystosori give rise to zoospores (one zoospore per resting spore), which "swim" through free soil water until they contact a host root and encyst. Encysted zoospores produce a structure called a stachel through which zoosporic cytoplasm enters the host cell and becomes a plasmodium. Only primary root tissue of young roots is infected; optimum infection occurs at a temperature of 25°C. The host cell then becomes infected with BSBV if *P. betae* is viruliferous. After a time, the plasmodium develops into a zoosporangium, which releases additional zoospores that repeat the infection cycle. However, some plasmodia develop into cysts, and often nearly every cell in the small feeder roots will contain a cyst. As root cells senesce, cysts are eventually released into the soil, where they can remain viable for years without loss of virulence.

Sugar beet is the only host.

Distribution. Western Europe and the United States.

Symptoms. Sugar beet frequently is a symptomless host. The actual affect of BSBV on sugar beet is unknown, but in greenhouse studies, yield losses have been reported in mechanically inoculated sugar beet.

Management. Not reported.

Beet Soilborne Mosaic

Cause. Beet soilborne mosaic virus (BSBMV) is a furo-like virus. BSBMV is morphologically similar to beet necrotic yellow vein virus (BNYVV) but is seriologically distinct and was originally designated as TX7 and TX8. Particles are 19 nm wide and from 50 to 400 nm long.

BSBMV is a rigid, rod-shaped virus transmitted by the soil fungus

Polymyxa betae. Polymyxa betae survives in field soil as cystosori, each of which is a mass of resting spores. Cystosori give rise to zoospores, one zoospore per resting spore, which "swim" through free soil water until they contact a host root and encyst. Encysted zoospores produce a structure called a stachel through which zoosporic cytoplasm enters the host cell and becomes a plasmodium. Only primary root tissue of young roots is infected; optimum infection occurs at a temperature of 25°C. The host cell then becomes infected with BSBMV if *P. betae* is viruliferous. After a time, the plasmodium develops into a zoosporangium, which releases additional zoospores that repeat the infection cycle. However, some plasmodia develop into cysts, and often nearly every cell in the small feeder roots will contain a cyst. As root cells senesce, cysts are eventually released into the soil, where they can remain viable for years without loss of virulence.

Distribution. The United States (California, Colorado, Idaho, Minnesota, Nebraska, Texas, and Wyoming).

Symptoms. Foliar symptoms include slight leaf distortion, subtle overall mottling, and light green or yellow blotches and bands that follow primary leaf veins. As diseased leaves age, the bands can progress to broad chlorotic areas that usually remain associated with the veins. The bands may become bright yellow but typically are broader than the yellow-vein symptom caused by BNYVV.

Occasionally, systemically infected leaves exhibit a mottled or mosaic pattern or symptoms that are very similar to the vein-banding symptom of BNYVV. Some BSBMV isolates cause bright yellow veinal chlorosis more typically associated with infection by BNYVV. Systemic foliar symptoms caused by BSBMV appear in field-grown sugar beet more frequently than do those caused by BNYVV.

Roots are often asymptomatic, but some beets systemically infected with BSBMV exhibit root symptoms more typically associated with rhizomania, including stunting, proliferation of lateral roots, and constriction of the main taproot.

In greenhouse studies, some isolates of BSBMV have significantly reduced growth of sugar beet seedlings.

Management. Not reported.

Beet Western Yellows

Cause. Beet western yellows virus (BWYV) is in the luteovirus group. BWYV overwinters in several weed hosts or overwintered sugar beet plants and is transmitted mainly by the green peach aphid, *Myzus persicae,* and infrequently by other aphid species. Known hosts are *Brassica napus* var. *napobrassica, Capsella bursa-pastoris, Cardamine oligosperma, Cardaria draba, Erodium cicutarium, Lactuca scariola, Matricaria maritima, Polygonum lap-*

athifolium, Raphanus raphanistrum, Senecio vulgaris, Sisymbrium officinale, and *Stellaria media.*

Myzus persicae can acquire BWYV after a 5-minute feeding on a diseased plant and transmit it after a 10-minute feeding on a healthy plant. An aphid can remain infective for several days. Different strains of BWYV exist.

An ST9-associated RNA is a newly discovered type of plant infectious agent that depends on BWYV for encapsidation but not for replication. The agent is not a satellite virus.

Distribution. Canada, Israel, and the United States (California, Illinois, and Oregon).

Symptoms. It is difficult to differentiate between the symptoms of beet western yellows and beet yellows on sugar beet. At first, the light chlorotic spotting that occurs on diseased older and middle-aged leaves gradually becomes more intense in color. Chlorotic areas may be sharply delimited by veins. Eventually, diseased older leaves become thick, brittle, and yellow except for green areas adjacent to the veins. Such leaves are frequently attacked by *Alternaria* spp.

Management
1. Grow resistant varieties.
2. Eliminate weeds near and adjacent to sugar beet fields.

Beet Yellow Net

Beet yellow net is sometimes identified as yellow net and the causal virus as yellow net virus (YNV).

Cause. Beet yellow net is thought to be caused by the beet yellow net virus (BYNV) and tentatively is placed in the luteovirus group. BYNV overwinters in sugar beet and possibly in weed hosts. BYNV is transmitted mainly by the green peach aphid, *Myzus persicae*. Once an aphid has acquired BYNV, it often remains infective for life. It is possible that BYNV cannot be transmitted without beet mild yellowing virus as a carrier virus.

Distribution. England and the United States (California).

Symptoms. An intense yellow chlorosis of veins and veinlets of diseased leaves looks like a yellow network against a background of green interveinal tissue. The vein bands may coalesce until nearly all of the leaf is chlorotic. Diseased plants are conspicuous in the field and contrast with the surrounding healthy green plants. Leaf texture is smooth and soft, not brittle as with beet yellows virus. Usually, the first symptoms are expressed on the youngest plant leaf. As the diseased leaf expands and matures, chlorosis involves more of the veinlet network. It has been reported, in some circumstances, that early symptoms fade as new leaves appear normal and that infected plants are difficult to locate.

Occasionally a severely diseased plant will have a complete chlorosis of older leaves. No stunting or malformation occurs.

Management. Not necessary.

Beet Yellow Stunt

Cause. Beet yellow stunt virus (BYSV) is in the closterovirus group. BYSV is primarily reservoired in annual sow-thistle, *Sonchus oleraceus,* and is most efficiently transmitted by the sow-thistle aphid, *Nasonovia lactucae.* However, *N. lactucae* apparently does not reproduce on sugar beet. The green peach aphid, *Myzus persicae,* is abundant on sugar beet but is a relatively inefficient vector. The potato aphid, *Macrosiphum euphorbiae,* also is an inefficient vector. For this reason, spread of BYSV in a field is marginal. Disease incidence is high in rows adjacent to areas where sow-thistle is present but becomes progressively less with increased distance from the virus source.

Distribution. The United States (California).

Symptoms. Initially, there is a severe twisting, cupping, and epinasty of one or two diseased intermediate-aged leaves. Petioles are shortened, and leaves become mottled and yellow. Young diseased leaves are dwarfed, malformed, twisted, and slightly mottled. As leaves age, mottling becomes more intense, and at times, leaves become completely chlorotic. Diseased plants are severely stunted and sometimes collapse and die.

Management. Control sow-thistle adjacent to sugar beet fields.

Beet Yellow Vein

Cause. There is no evidence that the causal agent of beet yellow vein is a virus. The causal organism, however, can be transmitted by grafting, juice inoculation, and the leafhopper *Aceratagallia calcaris.*

Distribution. The United States.

Symptoms. Very young diseased leaves are dwarfed and the main veins are yellow. The main veins of all affected leaves are either a continuous or broken yellow, with chlorosis that extends 1 mm or more into surrounding tissue and is visible on both sides of leaves. Commonly, half of a diseased plant may be severely stunted, while the other half grows normally.

Management. Not reported.

Beet Yellows

In some literature the "s" is left off of "yellows" and the causal organism is described as beet yellow virus.

Cause. Beet yellows virus (BYV) is in the closterovirus group. BYV likely survives in overwintered sugar beets and is transmitted by the green peach aphid, *Myzus persicae;* the black bean aphid, *Aphid fabae;* and infrequently by other aphids in a semi-persistent manner. BYV has a retention time in *A. fabae* of 24 to 48 hours. The maximum acquisition time is reached in 6 hours, and inoculation periods occur within 1 hour but maximum transmission efficiency is reached after a 6-hour inoculation access. BYV is not passed on to the progeny of vectors nor is it retained after molting.

Date of sowing apparently does not influence the pattern or distribution of BYV in a field.

Distribution. Wherever sugar beet is grown.

Symptoms. Younger diseased leaves have a vein clearing or vein yellowing that may appear bright yellow or necrotic. Secondary and intermediate veins frequently appear sunken and develop an etch symptom. Older leaves become thickened and brittle, followed by a chlorosis that begins at the leaf margins and tips and spreads downward between veins. Chlorosis may vary from pale green to orange or red, depending on the variety. Small reddish or brown spots develop on many of the older yellowed leaves. Chlorotic areas feel waxy or, in severely diseased plants, dry, causing plants to rustle when shaken or brushed. Such leaves do not wilt easily during dry weather and will splinter when crushed. Early infected leaves eventually become necrotic starting from where they first became chlorotic. The necrotic spots and yellow color often give leaves a bronze cast.

Management
1. Based on a forecasting system, insecticides are used to control vectors in England.
2. Eliminate escaped and overwintered beets.
3. Separate new beet fields from known sources of infection by 2 to 3 km.

Cucumber Mosaic

Cause. Cucumber mosaic virus (CMV) is in the cucumovirus group. CMV overwinters in several perennials, has a large host range, and is disseminated by more than 60 species of aphids. Aphids only have to feed a few seconds to acquire the virus, but they rapidly lose it during subsequent feedings. Consequently, an aphid must frequently feed on diseased plants in order to reacquire the virus. CMV is not transferred to young aphids by the female aphid. CMV is transmitted by seeds of some plants but not by sugar beet seed.

Distribution. Because of the large host range, CMV is generally distributed wherever sugar beet is grown. However, it is not considered a serious disease.

Symptoms. Large, irregular-shaped patches of bright yellow tissue contrast with the dark green of the leaf. Symptoms are of a mosaic type but the pattern is coarser and more contrasting than that found in beet mosaic. Leaves commonly are dwarfed and distorted and have blister-like areas of green tissue. Necrosis often occurs in large areas of the diseased older leaves.

Management. Control weeds around sugar beet fields.

Lettuce Chlorosis

Cause. Lettuce chlorosis virus (LCV) is in the closterovirus group. LCV is transmitted by the sweet potato whitefly, *Bemisia tabaci* "B" biotype.

Distribution. The United States (California).

Symptoms. Interveinal yellowing and thickened leaves occur in plants that are naturally infected. Interveinal reddening occurs in the leaves of inoculated plants.

Management. Not reported.

Lettuce Infectious Yellows

Cause. Lettuce infectious yellows virus (LIYV) is in the closterovirus group. LIYV survives in a wide range of weeds and commercial crops and is transmitted in a semi-persistent manner by the sweet potato whitefly, *Bemisia tabaci*. Cotton is a major source of high populations of this whitefly.

Distribution. The United States (California).

Symptoms. A very mild mottling develops into an interveinal yellowing of diseased leaves. A necrosis eventually develops in the chlorotic areas of the diseased leaves.

Management
1. Alter the sowing of cucurbits to provide a 1- to 3-week period in July or August when cucurbits are not present.
2. Do not grow sugar beet near areas of diseased plants.
3. Destroy infested residue of cucurbits and lettuce after harvesting.
4. Control weeds in the field being sown and in nearby fields.

Rhizomania Disease

Rhizomania disease is also called beet necrotic yellow vein.

Cause. Beet necrotic yellow vein virus (BNYVV) is in the furovirus group. Wild type isolates typically contain four single-stranded RNA species unlike most furoviruses, which possess bipartite, single-stranded RNA genomes.

BNYVV is transmitted by the soilborne fungus *Polymyxa betae* and by inoculation of sap. However, BNYVV is not present in all fungal populations. Movement of *P. betae* and BNYVV is primarily by the physical movement of soil during tillage and harvest operations. A lesser means of spread is through furrow irrigation.

Polymyxa betae is an obligate parasite that survives in field soil as cystosori, each of which is a mass of resting spores that can survive up to 10 years in the soil. Cystosori give rise to zoospores (each resting spore produces a single zoospore), which "swim" through free soil water until they contact a host root and encyst. Encysted zoospores produce a structure called a stachel through which zoosporic cytoplasm enters the host cell and becomes a plasmodium. Only primary root tissue of young roots is infected; optimum infection occurs at a temperature of 25°C. The host cell then becomes infected with BSBMV if *P. betae* is viruliferous. After a time, the plasmodium develops into a zoosporangium, which releases additional zoospores that repeat the infection cycle. However, some plasmodia develop into cysts, and often nearly every cell in the small feeder roots will contain a cyst. As root cells senesce, cysts are eventually released into the soil, where they can remain viable for years without loss of virulence.

Polymyxa betae infects late-sown plants sooner than early-sown plants. Temperatures at early sowing favor plant growth over fungus growth, but the numerous fibrous roots in older plants favor late infection. Disease is most severe at a soil temperature of 20°C for an extended time and wet soils that are neutral to slightly alkaline. *Polymyxa betae* is unable to infect sugar beet roots in soil when the matric potential is −0.4 bar or less.

Greater levels of infection by *P. betae* occur with viruliferous isolates than with nonviruliferous isolates. BNYVV causes roots to proliferate, resulting in more susceptible tissue for *P. betae* to infect. The increase in susceptible tissue does not result in increased infection per unit of root length but may maintain adequate inoculum for subsequent infection by *P. betae*. Most of the damage caused by the fungus-virus complex is due to the virus, but the aggressiveness of the vector is also important. BNYVV infrequently moves systemically in diseased plants and depends on the fungus vector for most cell-to-cell movement.

Different isolates of BNYVV have been found in California. Three other virus entities also have been found, but their relationship to Rhizomania disease is not known. Differences in vectoring ability occur among isolates of *P. betae*.

Distribution. Rhizomania has been found in France, Italy, Japan, and the United States.

Symptoms. Diseased roots are stunted and have a proliferation of lateral rootlets on the main taproot, which gives it a "bearded" look. Roots are frequently rotted and constricted below the soil level. Both viruliferous and nonviruliferous isolates of *P. betae* reduce total dry root mass of sugar beets. Much of the direct damage to roots by BNYVV only is root tip death, which results in a loss of apical dominance and an increased amount of root branching and lower accumulation of root mass.

The vascular system is discolored, and diseased leaves are upright and slightly chlorotic. Leaves proliferate, resulting in excessive crown tissue.

Leaves may also lose turgidity and wilt without becoming discolored. Rarely, distinct veinal yellowing of diseased leaves with necrotic lesions is evident. Nonviruliferous isolates of *P. betae* have been shown to significantly reduce the top mass of sugar beets, but viruliferous isolates did not significantly reduce the top mass.

Management. No economically effective control measure is known. Resistance to the virus has been identified in a wild beet species, *Beta maritima*, and in several sugar beet breeding lines.

Savoy

Savoy is also called beet savoy.

Cause. Beet savoy virus (BSV) overwinters in its vector, the pigweed bug *(Piesma cinerea)*, in grassy or woody areas. Infection occurs mainly at the edges of sugar beet fields. BSV is not mechanically transmitted nor is it seedborne. There is a possibility that the causal agent may be a phytoplasma.

Distribution. The United States, primarily east of the Continental Divide.

Symptoms. The more pronounced effects are found on the innermost diseased leaves, which are dwarfed, curled downward, and savoyed. This gives the lower leaf surface a netted appearance due to veinlets first clearing, then thickening. Later, roots will be generally discolored and phloem will appear necrotic. Both leaves and roots are markedly stunted.

Management
1. Sugar beet fields should be located several hundred meters from woods or uncultivated land that could serve as an insect source.
2. Grow a barrier crop between the sugar beet field and a potential insect source.

Yellows

Cause. Soybean dwarf virus (SDV). The subterranean clover red leaf virus is considered a strain of SDV.

Distribution. Australia (Tasmania).

Symptoms. An interveinal chlorosis initially occurs on diseased older leaves, then develops into a bright yellow color that is occasionally accompanied by a veinal necrosis. Yellow leaves sometimes develop orange tints. Symptoms rarely extend to the younger leaves except those on severely diseased plants that have a mottled appearance. Diseased leaves generally are thickened and brittle.

Management. Not reported.

Selected References

Altman, J. 1981. Increase in cyst nematode populations in soil treated with cycloate and diallate. (Abstr.) Phytopathology 71:199.

Boxch, U., Abbas, H. K., and Mirocha, C. J. 1988. Production of mycotoxins by *Fusarium* cultures on sugar beet substrate. (Abstr.) Phytopathology 78:1552.

Bugbee, W. M. 1975. Dispersal of *Phoma betae* in sugar beet storage yards. Plant Dis. Rptr. 59:396-397.

Bugbee, W. M. 1975. *Penicillium claviforme* and *Penicillium variabile*: Pathogens of stored sugar beets. Phytopathology 65:926-927.

Bugbee, W. M. 1979. Resistance to sugar beet storage rot pathogens. Phytopathology 69:1250-1252.

Bugbee, W. M. 1983. Infection and movement of endophytic bacteria in sugar beet plants. (Abstr.) Phytopathology 73:806.

Bugbee, W. M., and Campbell, L. G. 1990. Combined resistance in sugar beet to *Rhizoctonia solani, Phoma betae,* and *Botrytis cinerea.* Plant Dis. 74:353-355.

Bugbee, W. M., and El-Nashaar, H. M. 1983. A newly recognized symptom of sugar beet root infection caused by *Phoma betae.* Plant Dis. 67:101-102.

Bugbee, W. M., and Gudmestad, N. C. 1988. The recovery of *Corynebacterium sepedonicum* from sugar beet seed. Phytopathology 78:205-208.

Bugbee, W. M., and Nielsen, G. E. 1978. *Penicillium cyclopium* and *Penicillium funiculosum* as sugar beet storage rot pathogens. Plant Dis. Rptr. 62:953-954.

Bugbee, W. M., and Soine, O. C. 1974. Survival of *Phoma betae* in soil. Phytopathology 64:1258-1260.

Bugbee, W. M., Gudemestad, N. C., Secor, G. A., and Nolte, P. 1985. Sugar beet: A natural host for *Corynebacterium sepedonicum.* (Abstr.) Phytopathology 75:1379.

Bugbee, W. M., Gudmestad, N. C., Secor, G. A., and Nolte, P. 1987. Sugar beet as a symptomless host for *Corynebacterium sepedonicum.* Phytopathology 77:765-770.

Duffus, J. E. 1965. Beet pseudo-yellows virus, transmitted by the greenhouse whitefly *(Trialeurodes vaporariorum).* Phytopathology 55:450-453.

Duffus, J. E. 1972. Beet yellow stunt, a potentially destructive virus disease of sugar beet and lettuce. Phytopathology 62:161-165.

Duffus, J. E., and Liu, H. Y. 1987. First report of Rhizomania of sugar beet from Texas. Plant Dis. 71:557.

Duffus, J. E., Larsen, R. C., and Liu, H. Y. 1986. Lettuce infectious yellows virus—A new type of whitefly-transmitted virus. Phytopathology 76:97-100.

Ellis, P. J. 1992. Weed hosts of beet western yellows virus and potato leafroll virus in British Columbia. Plant Dis. 76:1137-1139.

Engelkes, C. A., and Windels, C. E. 1992. Formation and survival of sclerotia of AG-2-2 of *Rhizoctonia solani* on sugar beet roots. (Abstr.) Phytopathology 81:1186.

Engelkes, C. A., and Windels, C. E. 1996. Susceptibility of sugar beet and beans to *Rhizoctonia solani* AG-2-2 IIIB and AG-2-2 IV. Plant Dis. 80:1413-1417.

Gerik, J. S., and Duffus, J. E. 1988. Differences in vectoring ability and aggressiveness of isolates of *Polymyxa betae.* Phytopathology 78:1340-1343.

Gerik, J. S., and Hubbard, J. C. 1989. Effect of soil water matric potential on infection by *Polymyxa betae* and beet necrotic yellow vein virus. (Abstr.). Phytopathology 79:1223.

Giannopolitis, C. N. 1978. Lesions on sugar beet roots caused by *Cercospora beticola*. Plant Dis. Rptr. 62:424-427.

Giunchedi, L., and Langenberg, W. G. 1982. Beet necrotic yellow vein virus transmission by *Polymyxa betae* Keskin zoospores. Phytopathol. Medit. 21:5-7.

Griffin, G. D. 1988. Comparative nematicidal control of *Heterodera schachtii* on sugar beet as affected by soil temperature and soil type. Plant Dis. 72:617-621.

Halloin, J. M., and Roberts, D. L. 1991. A parasitic storage rot of sugar beets caused by *Aspergillus flavus*. Plant Dis. 75:751.

Harveson, R. M., Rush, C. M., and Wheeler, T. A. 1996. The spread of beet necrotic yellow vein virus from point source inoculations as influenced by irrigation and tillage. Phytopathology 86:1242-1247.

Heidel, G. B., Rush, C. M., Kendall, T. L., Lommel, S. A., and French, R. C. 1997. Characteristics of beet soilborne mosaic virus, a furo-like virus infecting sugar beet. Plant Dis. 81:1070-1076.

Herr, L. J., and Roberts, D. L. 1980. Characterization of *Rhizoctonia* populations obtained from sugar beet fields with differing soil textures. Phytopathology 70:476-480.

Hills, F. J., and Worker, G. F., Jr. 1983. Disease thresholds and increases in fall sucrose yield related to powdery mildew of sugar beet in California. Plant Dis. 67:654-656.

Hine, R. B., and Ruppel, E. G. 1969. Relationship of soil temperature and moisture to sugar beet root rot caused by *Pythium aphanidermatum* in Arizona. Plant Dis. Rptr. 53:989-991.

Kontaxis, D. G., Meister, H., and Sharma, R. K. 1974. Powdery mildew epiphytotic on sugar beets. Plant Dis. Rptr. 58:904-905.

Langenberg, W. G., and Kerr, E. D. 1982. *Polymyxa betae* in Nebraska. Plant Dis. 66:862.

Limburg, D. D., Mauk, P. A., and Godfrey, L. D. 1997. Characteristics of beet yellows closterovirus transmission to sugar beets by *Aphis fabae*. Phytopathology 87:766-771.

Lindsten, K., and Rush, C. M. 1994. First report of beet soilborne virus in the United States. Plant Dis. 78:316.

Liu, H. Y., and Duffus, J. E. 1985. The viruses involved in Rhizomania disease of sugar beet in California. (Abstr.) Phytopathology 75:1312.

Liu, H. Y., Duffus, J. E., and Gerik, J. S. 1987. Beet distortion mosaic virus—A new soilborne virus of sugar beet. (Abstr.) Phytopathology 77:1732.

Maas, P. W. T., and Heijbroek, W. 1982. Biology and pathogenicity of the yellow beet cyst nematode, a host race of *Heterodera trifolii* on sugar beet in the Netherlands. Nematologica 28:77-93.

MacDonald, J. D., Leach, L. D., and McFarlane, J. S. 1976. Susceptibility of sugar beet lines to the stalk blight pathogen *Fusarium oxysporum* f. sp. *betae*. Plant Dis. Rptr. 60:192-196.

Marco, S. 1984. Beet western yellows virus in Israel. Plant Dis. 68:162-163.

Martyn, R. D., Rush, C. M., Biles, C. L., and Baker, E. H. 1989. Etiology of a root rot disease of sugar beet in Texas. Plant Dis. 73:879-884.

Mukhopadhyay, A. N. 1987. CRC Handbook on Diseases of Sugar Beet. Volume II. CRC Press, Inc., Boca Raton, FL. 177 pp.

Ramirez, M. C., Raposo, R., and Mateo-Sagasta, E. 1994. Occurrence of *Aphanomyces cochlioides* damping-off of sugar beet in Spain. Plant Dis. 78:102.

Ruppel, E. G., and Tomasovic, B. J. 1977. Epidemiological factors of sugar beet powdery mildew. Phytopathology 67:619-621.

Ruppel, E. G., Harrison, M. D., and Nielson, A. K. 1975. Occurrence and cause of bacterial vascular necrosis and soft rot of sugar beet in Washington. Plant Dis. Rptr. 59:837-840.

Ruppel, E. G., Hills, F. J., and Mumford, D. L. 1975. Epidemiological observations on the sugar beet powdery mildew epiphytotic in western USA in 1974. Plant Dis. Rptr. 59:283-286.

Ruppel, E. G., Jenkins, A. D., and Burtch, L. M. 1980. Persistence of benomyl-tolerant strains of *Cercospora beticola* in the absence of benomyl. Phytopathology 70:25-26.

Rush, C. M. 1987. Root rot of sugar beet caused by *Pythium deliense* in the Texas panhandle. Plant Dis. 71:469.

Rush, C. M., and Heidel, G. B. 1995. Furovirus diseases of sugar beets in the United States. Plant Dis. 79:868-875.

Rush, C. M., and Winter, S. R. 1990. Influence of previous crops on Rhizoctonia root and crown rot of sugar beet. Plant Dis. 74:421-425.

Rush, C. M., Carling, D. E., Harveson, R. M., and Mathieson, J. T. 1994. Prevalence and pathogenicity of anastomosis groups of *Rhizoctonia solani* from wheat and sugar beet in Texas. Plant Dis. 78:349-352.

Schneider, C. L., and Robertson, L. S. 1975. Occurrence of diseases on sugar beet in a crop rotation experiment in Saginaw County, Michigan in 1969-1971. Plant Dis. Rptr. 59:194-197.

Schneider, C. L., Ruppel, E. G., Hecker, R. J., and Hogaboam, G. J. 1982. Effect of soil deposition in crowns on development of Rhizoctonia root rot in sugar beet. Plant Dis. 66:408-410.

Stanghellini, M. E., and Kronland, W. C. 1977. Root rot of mature sugar beets by *Rhizopus arrhizus*. Plant Dis. Rptr. 61:255-256.

Stanghellini, M. E., Von Bretzel, P., and Kronland, W. C. 1981. Epidemiology of *Pythium aphanidermatum* root rot in sugar beets. (Abstr.) Phytopathology 71:905.

Stanghellini, M. E., Von Bretzel, P., Olsen, M. W., and Kronland, W. C. 1982. Root rot of sugar beet caused by *Pythium deliense*. Plant Dis. 66:857-858.

Staples, R., Jansen, W. P., and Anderson, L. W. 1970. Biology and relationship of the leafhopper *Aceratagallia calcaris* to yellow vein disease of sugar beets. J. Econ. Entomol. 63:460-463.

Tamada, T. 1975. Beet Necrotic Yellow Vein Virus. Commonwealth Mycological Institute. Descriptions Plant Viruses. Set. 9, No. 144.

Thompson, S. V., Hills, F. J., Whitney, E. D., and Schroth, M. N. 1981. Sugar and root yield of sugar beets as affected by bacterial vascular necrosis and rot, nitrogen fertilization, and plant spacing. Phytopathology 71:605-608.

Timmerman, E. L., D'Arcy, C. J., and Splittstoesser, W. E. 1984. Beet western yellows virus in Illinois. (Abstr.) Phytopathology 74:1271.

Vaughn, K. M., and Rush, C. M. 1991. Reducing Aphanomyces seedling disease of sugar beets by limited irrigation. (Abstr.) Phytopathology 81:1164.

Vincelli, P. C., Wilcox, W. F., and Beaupr'e, C. M-S. 1990. First report of *Phytophthora cryptogea* causing root rot of sugar beet in Wyoming. Plant Dis. 74:614.

von Bretzel, P., Stanghellini, M. E., and Kronland, W. C. 1988. Epidemiology of Pythium root rot of mature sugar beets. Plant Dis. 72:707-709.

Whitney, E. D. 1971. The first confirmable occurrence of *Urophlyctis seproides* on sugar beet in North America. Plant Dis. Rptr. 55:30-32.

Whitney, E. D. 1987. Identification and aggressiveness of *Erwinia carotovora* subsp. *betavasculorum* on sugar beet from Texas. Plant Dis. 71:602-603.

Whitney, E. D., and Duffus, J. E. (Ed.). 1986. Compendium of Beet Diseases and Insects. American Phytopathological Society, St. Paul, MN. 76 pp.

Windels, C. E., and Nabben, D. J. 1987. *Rhizoctonia solani* and Rhizoctonia-like binucleates associated with sugar beet plants in the Red River Valley. (Abstr.) Phytopathology 77:1715.

Windels, C. E., Kuznia, R. A., and Call, J. 1994. First report of *Thanatephorus cucumeris* (*Rhizoctonia solani* AG-3 and AG-5) on sugar beet. (Abstr.). Phytopathology 84:1161.

Windels, C. E., Kuznia, R. A., and Call, J. 1997. Characterization and pathogenicity of *Thanatephorus cucumeris* from sugar beet in Minnesota. Plant Dis. 81:245-249.

Wisler, G. C., and Duffus, J. C. 1997. First report of lettuce chlorosis virus naturally infecting sugar beets in California. Plant Dis. 81:550.

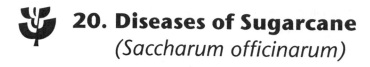

20. Diseases of Sugarcane
(Saccharum officinarum)

Diseases Caused by Bacteria

Bacterial Mottle

Bacterial mottle is also called leaf stripping and chlorotic mottle.

Cause. *Pectobacterium carotovorum* var. *graminarium* persists in diseased standing cane, cane residue, and grasses, specifically elephant grass, *Pennisetum
purpureum*; para grass, *Brachiaria mutica*; and guinea grass, *Panicum maximum*. Bacteria are disseminated by flooding water, primarily during the
hot, wet season. Bacteria are exuded through stomata of diseased cane or
grasses and carried by water to other hosts, where they enter through
wounds in the stem or the buds of growing points. Bacteria are not efficiently disseminated by wind or rain. Seed cane that is infected ordinarily
fails to germinate.

Distribution. Australia. The disease is most common in low-lying areas that
are subject to flooding.

Symptoms. Initially, one to many creamy-white stripes (1–2 mm wide)
extend from or near the bases of leaf blades upward and parallel to leaf
veins. As diseased leaves age, stripes enlarge and the orange to brown-
red discolored areas developing within them often involve most of the
stripe.

   Systemic infection causes diseased leaves to have a chlorotic mottling,
often without any distinct striping. Sometimes shoots become almost
completely chlorotic. As the diseased chlorotic leaves become older, the
abundant small brown-red flecks and short, narrow stripes that appear
cause the chlorotic leaves to look pink.

   Diseased shoots are stunted, and leaf margins wither, causing diseased
leaves to curl inward. Eventually the entire shoot may die. Diseased stalks
may tiller excessively due to the production of several side shoots at their

bases. A witches' broom effect may occur several centimeters above the soil level when older cane has been infected and characteristic leaf symptoms are present. During warm, humid weather, small whitish drops of bacterial exudate may occur on the undersides of diseased leaves.

Management
1. Grow the most resistant cultivars.
2. Plant only healthy seed cane.
3. Destroy diseased plants in a field.
4. Control infested grasses in waterways adjacent to cane fields.

Gumming Disease

Gumming disease is also called gummosis, Cobb's disease of sugarcane, and gum disease.

Cause. *Xanthomonas campestris* pv. *vasculorum.* A second bacterium, *X. officinarum* sp. nov., has been implicated in gumming disease. *Xanthomonas campestris* pv. *vasculorum* survives from one planting to the next in cuttings and may infect healthy sets through contaminated knifes or tools. Dissemination over a large distance is ordinarily through the movement of infected cuttings. Dissemination also occurs by machinery, the brushing of diseased plants against healthy ones, and insects. Bacteria enter sugarcane leaves through small wounds caused by machinery, insects, or leaves brushing against each other. Secondary infections are primarily by bacteria or gum oozing from wounds of diseased plants during wet weather and being disseminated from diseased to healthy plants by windblown rain.

Distribution. Africa, Australia, South America, the United States (Puerto Rico), and the West Indies.

Symptoms. Symptoms start 2 to 3 weeks after wet weather has occurred. Longitudinal streaks (3–6 mm wide) follow the course of the vascular bundles on mature leaves that have not started to discolor. Most streaks start at the diseased leaf apex or margin and vary in length from a few millimeters to the full length of the leaf, but streaks do not extend into the leaf sheath. Young lesions have well-defined margins, but margins of older streaks become diffuse. The streaks are yellow to orange flecked with patches of red. Tissue starts to die at the point of infection and progresses along the length of the streak, which becomes grayish. Symptoms on leaves may seem to disappear during dry weather, particularly if diseased leaves are shed as they age.

On the youngest diseased leaf blades and the undersides of midribs, systemic infection causes short, narrow, well-defined dark red streaks that extend into leaf sheaths. Sometimes creamy-white streaks almost half the width of the leaf start from the leaf base. Red blotches are lacking at first but may develop as systemic diseased leaves mature. Chlorotic areas speckled with red dots may occur on portions of one or many diseased leaves. This symptom may involve the entire top of the diseased plant and usually causes the growing point to die. Although some diseased shoots may recover, a large percentage of diseased stalks wilt and die.

A yellow to orange gummy mass exudes from cut ends of severely diseased stalks. Gum pockets that frequently form in the growing point region can be observed when diseased stalks are cut longitudinally.

Management

1. Grow resistant cultivars.
2. Plow out diseased crop; however, ratoons usually suffer less loss than plant crops.

Leaf Scald

Cause. *Xanthomonas albilineans* is a gram-negative bacterium that colonizes the vascular system of leaves, stalks, and roots. Bacteria enter through the wounded buds of cuttings and are transmitted primarily by planting the infected stalks into noninfested fields but also are transmitted mechanically on contaminated cutting implements, particularly cane knives. Spread may also occur by aerial means, leaf-to-leaf contact, root-to-root contact, and soil infestation. Tolerant cultivars may disseminate bacteria over considerable distances since symptoms of leaf scald do not occur or go unnoticed on tolerant cultivars.

Distribution. Worldwide.

Symptoms. Two types of symptoms have been described: acute and chronic. The chronic, or characteristic, symptom is the presence of one or more straight, well-defined, narrow (3 mm wide), white or chlorotic stripes that extend over and around the vascular bundles the full length of diseased leaf blades and sheaths. The narrow stripes may become necrotic and the resulting necrosis can expand to encompass the entire diseased leaf. Young shoots and stalks may die, and shoots exhibiting symptoms similar to diseased leaves may develop from axillary buds on mature stalks.

Acute symptoms are the sudden wilting and death of diseased stalks. Frequently the progressive withering of diseased leaves gives the plant a scalded appearance. Entire leaves become chlorotic on severely diseased plants, and young shoots may also be chlorotic as they emerge from the soil. Internodes of stalks are shortened, weak, and produce side shoots from the germination of lateral buds.

Diseased vascular bundles have a light red discoloration at the nodes, the juncture of the lateral side shoots, and throughout the stalk. Some vessels are plugged with a reddish gum-like substance, but there is no bacterial ooze.

Some diseased plants may be symptomless. Symptoms may also disappear as diseased plants continue to grow but, conversely, may become more severe as cane approaches maturity if plant growth is retarded for any reason.

Large reductions in yield and sugar content occur, and juice purity is affected.

Management
1. Quarantine seed cane into areas where leaf scald does not occur.
2. Grow resistant cultivars. Two mechanisms may play an important role in resistance: resistance to colonization of the apex, which is characterized by the absence of symptoms, and resistance to colonization of the upper and lower parts of the stalk.
3. Maintain disease-free seed cane nursery programs. These rely on heat therapy, sanitation, and a rigid monitoring system to ensure disease-free seed cane.

Mottled Stripe

Cause. *Pseudomonas rubrisubalbicans* survives in standing cane and infested residue. Dissemination is by wind and rain.

Distribution. Africa, Australia, the Caribbean region, South America, and the United States. Most cane cultivars are susceptible but mottled stripe is not considered a serious disease.

Symptoms. Disease symptoms occur only on leaves. Creamy-white stripes (1–4 mm wide and a maximum of 1 m in length) with distinct edges occur parallel to leaf veins. Stripes frequently develop within reddish discolored areas on a diseased leaf to give a red and white mottled effect to the leaf. If the red areas are large, stripes have a general reddish appearance.

Management. Not reported.

Ratoon Stunting Disease

Ratoon stunting disease is also called Q.28 disease.

Cause. *Clavibacter xyli* subsp. *xyli* is a small, xylem-limiting coryneform bacterium. *Clavibacter xyli* subsp. *xyli* infects a wide variety of graminaceous plants, including johnsongrass, sorghum, maize, sweet sudangrass, and bermudagrass.

Clavibacter xyli subsp. *xyli* is disseminated only mechanically by cutting knives, mechanical harvesters, tools, and cuttings taken from diseased plants. The bacteria are easily spread in susceptible cultivars but not in resistant cultivars. Disease spread within rows of susceptible plants will follow the direction of harvest from a diseased plant source. The incidence of ratoon stunting disease in test plots increased with the number of ratoon crops harvested.

Distribution. Wherever sugarcane is grown. Ratoon stunting disease is widely regarded as the most important disease affecting commercial sugarcane production in the world.

Symptoms. The only external symptoms are stunting and unthriftiness of diseased plants. Ratoons display more pronounced disease symptoms than does older cane. Ratoons are slow to start, particularly in dry weather, and continue to be retarded in growth. Disease is usually not lethal but yield is reduced due to the production of thinner, shorter stalks. Stunting is not uniform from stool to stool and a stand of plants within a field may display uneven growth. Diseased cane wilts sooner than healthy cane.

Internal symptoms may include a salmon-pink discoloration below the growing point of young cane. In mature cane, a yellow, orange, pink-red, or red-brown discoloration of vascular bundles occurs in the nodes below the region of leaf sheath attachment. However, this is not considered a reliable symptom to diagnose the disease.

Management
1. Grow resistant cultivars. No cultivars are immune, but resistance may be conferred by physical features of the vascular structure at the stalk nodes. These features include a more profuse branching of the metaxylem and fewer xylem vessels that pass directly through stalk nodes without terminating.
2. Plant only cuttings from healthy plants.
3. Treat infected cuttings with hot water or hot air. Treated cuttings should have a protectant fungicide applied.
4. Destroy all volunteer cane.
5. Sterilize cutting tools.

Red Stripe and Top Rot

Cause. *Pseudomonas avenae* (syn. *P. rubrilineans*) persists in older, withered leaves and in soil for at least 32 days. Bacterial exudate seeps to the diseased leaf surface during warm, wet weather and forms droplets of bacterial ooze. Bacteria, which are disseminated primarily by wind and rain and rarely by cuttings or any other mechanical means, enter plants through wounds or stomates. All portions of the plant may become diseased, including leaves, stalks, and roots. Bacterial exudate forms galls on the lower portion of the plants or runs down the leaf and causes stem infection.

Distribution. Wherever sugarcane is grown.

Symptoms. Initially, water-soaked stripes surrounded by a chlorotic zone develop on diseased leaves. Stripes are 1–4 mm × 15–40 cm but become wider when they extend down into the leaf sheaths. Stripes later become discolored dark red to maroon and eventually coalesce, forming bands of alternating maroon stripes and chlorotic areas.

The vascular bundles near the growing point of the stalk become slightly red. This discoloration extends down the stalk as a red ring one-fourth to one-half the distance from the rind to the center. The center of the ring is water-soaked and rapidly decomposes, but the outside portion does not change. Eventually, rot extends down to the base of a stalk, leaving a hollow central cylinder. The terminal bud and spindle leaves die, which often results in the growth of lateral buds that also display a red discoloration. Diseased stalks in an advanced stage of decomposition emit an unpleasant odor that can be detected for a considerable distance.

Management
1. Grow resistant cultivars.
2. Rogue diseased stools from seedling nurseries.

Diseases Caused by Fungi

Alternaria Leaf Spot

Cause. *Alternaria alternata* (syn. *A. tenuis*) is a common saprophyte.

Distribution. Cuba, India, and Taiwan.

Symptoms. Spots on diseased leaves that initially appear as minute, water-soaked areas eventually elongate and become elliptical to irregular-shaped and have red-brown to dark brown margins. The leaf area between the spots becomes necrotic and sometimes the entire diseased leaf dies.

Management. Not reported.

Banded Sclerotial Leaf Disease

Cause. *Thanatephorus cucumeris* (syn. *Pellicularia sasakii*), anamorph *Rhizoctonia solani,* survives as sclerotia in soil. Other grasses, particularly bermudagrass, *Cynodon dactylon,* are also diseased. Only older leaves of cane become infected by coming in contact with a diseased grass leaf or infested soil during periods of high humidity.

Distribution. Africa, Australia, Asia, Caribbean, Central America, and the United States. Banded sclerotial leaf disease is considered of minor importance.

Symptoms. Older diseased leaves and, occasionally, diseased leaf sheaths have a series of broad bands across the leaf or sheath. The initial brown-green irregular-shaped areas that occur become discolored brown, then yellow to light tan and have definite red-brown borders. Dark brown to black, irregular to spherical-shaped (2–5 mm in diameter) sclerotia are present in the diseased areas of the plant.

Management. Not necessary.

Black Rot

Cause. *Ceratocystis adiposa* (syn. *C. major, Ceratostomella adiposa, Endoconidiophora adiposa,* and *Ophiostoma adiposa*) persists as perithecia and mycelia in infested residue. Conidia produced from mycelia are disseminated by wind, splashing water, insects, and by any means that moves soil. Black rot is more prevalent in loose, cloddy soils that have large air pockets. In the United States, black rot is most common in seed cane bedded in the autumn, possibly because of air spaces that form around the seed cane at this time. Infection occurs through the cut ends of seed cuttings.

Distribution. Australia, Brazil, China, Dominican Republic, India, Indonesia, Panama, Peru, Taiwan, and the United States.

Symptoms. Seed cuttings become soft and watery at the ends. The interior of diseased cuttings initially is dark purple, then becomes black and emits an odor that resembles fermenting pineapples. Black mycelial growth of *C. adiposa* occurs on the cut ends of seed cane and is the obvious external symptom.

Management
1. Grow resistant cultivars.
2. Plant cuttings in the autumn rather than storing them in beds.

Black Stripe

Cause. *Cercospora atrofiliformis.*

Distribution. Taiwan.

Symptoms. Initially, small, yellow, round or oval spots occur on diseased leaves. Eventually, the spots enlarge to $0.5–1.2 \times 5.0–36.0$ mm and become brown-black.

Management. Not necessary.

Brown Rot

Cause. *Corticium* sp. is a saprophyte and a minor pathogen of trees.

Distribution. Australia. Brown rot has been found only in sugarcane planted in soil recently cleared of trees.

Symptoms. Diseased stools initially are unthrifty. Eventually, the leaves wilt and die and the death of the entire stool follows. A thick layer of brown mycelium extends a few centimeters above the soil line on the diseased stalk and binds the leaf sheaths together. Mycelium also grows over the lower leaf sheaths and belowground plant parts, binding soil to plants and causing dead roots and stems to appear several times thicker than they actually are. Stalks below the soil level are killed and turn brown. Dead tissue dries out quickly and is separated from healthy tissue by a narrow dark brown band.

Management. Not necessary, since brown rot is of minor importance.

Brown Spot

Cause. *Cercospora longipes* persists in infested leaf residue and diseased leaves of standing cane. Conidia are produced on both diseased leaf surfaces but are more abundant on the lower leaf surfaces. Dissemination of conidia is by wind, rain, and cuttings contaminated with conidia.

Distribution. Presumed to be present wherever sugarcane is grown; however, it has not been reported from Australia or Taiwan. Brown spot is not considered an important disease.

Symptoms. Numerous spots first appear on both surfaces of diseased older leaves and progress up the plant. Spots are red-brown and surrounded by a narrow yellow halo, oval to linear, and vary in size from specks to 13 mm in length. The centers of the older spots become tan and surrounded by a red zone and a yellow halo. Spots may coalesce and form large, irregular-shaped red-brown patches on diseased leaves. Severely diseased leaves may die prematurely, which gives diseased plants and entire fields a "fiery" appearance.

Management
1. Grow resistant cultivars.
2. Treat seed cane cuttings with a seed-protectant fungicide.

Brown Stripe

Cause. *Cochliobolus stenospilus*, anamorph *Bipolaris stenospila*. It is unknown how *C. stenospilus* survives in the absence of cane in subtropical areas where cane is not grown year-round. Presumably, *C. stenospilus* may either infect other members of the grass family or mycelia may survive in infested residue.

Conidia develop in lesions on older dead leaves and are disseminated by wind for a considerable distance. Conidia germinate on leaves in the presence of free moisture and the resulting germ tubes infect leaves through the stomates. *Cochliobolus stenospilus* is a weak facultative parasite and is more likely to infect plants predisposed by other causes, such as dry weather.

Distribution. Wherever sugarcane is grown.

Symptoms. Symptoms of brown stripe are most severe in dry weather and when growing conditions for the host are suboptimal. Initially, young leaves have small (0.5 mm in diameter) watery spots that soon become elongated, red-brown to brown, and sometimes are surrounded by a narrow chlorotic halo. Frequently, spots occur in a row across a leaf, which is the result of conidia germinating in the moisture within the spindle. Eventually, the spots elongate and form definite stripes (2–10 mm in length) with a narrow yellow margin. As lesions mature, they become longer, sometimes up to 75 mm in length, but are only 4 to 5 mm wide. When disease is severe, stripes coalesce and prematurely kill the diseased leaves. Infrequently, top rot develops.

Management
1. Grow resistant cultivars.
2. Maintain plants in a vigorous growing condition with proper fertility and moisture.

Collar Rot

Cause. *Hendersonina sacchari.*

Distribution. Argentina, Bangladesh, India, Indochina, Mauritius, the Philippines, and Sri Lanka.

Symptoms. The top leaves of a diseased plant become necrotic and dry from the diseased leaf margin inward, leaving only a green midrib. The interior of the upper nodes is dried up, often with cavities of various sizes that eventually extend throughout the stalk. The tissue of the lower diseased internodes is watery and brown with patches of red, which becomes the dominant color in the basal internodes. Roots growing from the diseased basal internodes are rotted and have a dark brown to blackish discoloration. Diseased stalks are lighter in weight than the stalks of healthy plants.

Management. Not necessary, because collar rot is of minor importance.

Common Rust

Common rust is also called sugarcane rust and sugarcane leaf rust.

Cause. *Puccinia melanocephala.* Urediniospores and teliospores are produced on sugarcane. Other spore stages and alternate hosts are not known. Urediniospores are disseminated by wind and water. Infection occurs under humid conditions at temperatures of 16° to 29°C. In Puerto Rico, the opti-

mum rainfall for infection is 13 to 16 cm/month and plants are most susceptible to infection when they are 2 months old.

Disease severity has been associated with edaphic conditions in Florida.
Disease severity is negatively correlated with soil pH. High levels of soil
phosphorus are associated with high rust severity. At two locations, high
levels of soil magnesium and potassium were associated with lower rust
severity.

Distribution. Wherever sugarcane is grown.

Symptoms. Symptoms occur mostly on diseased leaves. The uredinia are tiny,
brownish, elongated spots that occur on both leaf surfaces but are more
common on the lower leaf surfaces than the upper leaf surfaces. Spots increase in length to 2 to 10 mm, become orange-brown to brown, and are
surrounded by a small yellowish halo. Orange urediniospores initially are
subepidermal but later rupture the leaf epidermis. The surrounding tissue
is killed, which sometimes results in the death of young diseased leaves.
Uredinia eventually darken in color on the lower diseased leaf surfaces due
to the formation of telia either in the uredinium or in separate telia.

Management. Grow resistant cultivars.

Covered Smut

Covered smut is also called kernel smut.

Cause. *Sphacelotheca macrospora* infects ovules in the spikelet, replacing seed
with smut spores.

Distribution. Taiwan.

Symptoms. Diseased plants appear normal until flowering time, when
ovaries are slightly swollen and elliptical or ovoid in shape. The gray-green
membrane that initially covers the ovaries later turns brown. When the
membrane ruptures it exposes a blackish powdery mass of chlamydospores surrounded by a stick-like central column.

Management. Not necessary.

Culmicolous Smut

Culmicolous smut is also called smut, sugarcane smut, and inflorescence
smut.

Cause. *Ustilago scitaminea* survives as chlamydospores in dry soil and infested
residue, as mycelia in latent infections in planting materials, and as spores
on cuttings. Spore viability was lost after 7 to 9 weeks in five different soils
at three moisture levels in Louisiana. Therefore, spores produced during
one season will not persist in soil to the next season under conditions in

Louisiana. Chlamydospores and sporidia in soil may be disseminated by water and infect cuttings.

Each chlamydospore germinates under moist conditions to form a three- or four-celled promycelium. The promycelium produces a hypha that functions as an infection thread, or sporidia bud from each of its cells. The sporidia, in turn, either germinate and form a hypha or bud off and form more sporidia.

Ustilago scitaminea, a parasite of meristematic tissues, enters only through the lower part of the bud below the scales. Wounding increases infection, and when an infected bud begins to grow, the fungus grows just behind the growing tip. After a latent period of 4 to 7 months, the apical meristem is stimulated to produce a long, unbranched, whip-like appendage in which chlamydospores are produced. The enveloping membrane is ruptured and the chlamydospores are exposed and passively released to become windborne. A smut whip may release millions of spores per day during an infectious period that lasts up to 3 months.

Diseased buds on standing cane may either give rise to smutted whips the same season or mycelia may remain dormant within the buds. Severe winters in northern sugarcane-growing areas damage or kill sugarcane buds, causing a decrease in disease incidence. In tropical areas, more than one disease cycle can occur per growing season. True seedlings may also become infected.

Sugarcane smut severity is greater during a dry season, when plants are predisposed to infection.

Distribution. Wherever sugarcane is grown.

Symptoms. The typical symptom, a long, whip-like structure varying in size from a few to several centimeters, is produced from the apex of the diseased stalk and doubles back. Depending on the cultivar, diseased shoots either grow at a faster rate than healthy shoots, resulting in the emergence of typical smut whips above the crop canopy, or produce short, curved whips in the canopy.

The narrow unbranched whip, or sorus, contains dark brown to blackish chlamydospores that are surrounded by a thin silvery membrane. Eventually, the membrane ruptures and exposes a dark mass of chlamydospores that are blown away by the wind, leaving only a core parenchyma and fibro-vascular elements. When infection occurs early, diseased canes have small, narrow leaves and slender stalks with widely spaced nodes, which result either in clumps of spindly canes or small grassy shoots that are followed by smutted whips. The whips are usually straight when short but become irregularly curved when they are 1.0 to 1.5 m long.

After the production of smutted tops, buds lower on the diseased stalks begin to grow and terminate in a black whip. Sometimes smutted side shoots occur on an apparently healthy stalk.

Some unusual symptoms have been described from Hawaii. Whips often remain confined under the leaf sheaths and have a characteristic serpentine form. Buds from which lateral bud sori originated are flattened, elongated, and larger than the typical symptoms displayed by diseased buds.

Callous outgrowths, an atypical symptom, are most abundant at a position on stalks between normal-appearing tissue and areas displaying "typical" smut symptoms. Galls in this region are generally limited to nodal areas and are initiated either in the region of the root band and intercalary meristem, or from a leaf sheath scar. Outgrowths sometimes appear randomly on internodes positioned near or adjacent to the bases of whips. Initially, the callus is smooth and lacks chlorophyll. Eventually, buds, or leaf-like structures, are differentiated. It is not uncommon to have 40 to 60 buds on a gall. Buds continue to grow and terminate by forming a whip.

Symptoms on true seedlings are the presence of early-developed, thin, grassy stems with short, slender internodes and no visible buds or root initials. In the uppermost part of the diseased plant, the leaf sheaths are absent and leaf blades are subtended by long bristle-like hairs. Most seedlings die from severe lodging, but those that live produce a short, whip-like structure containing chlamydospores.

Management

1. Rogue out diseased shoots or stools.
2. Select healthy planting material.
3. Treat planting stock with a protectant fungicide.
4. Avoid ratooning of affected cane fields.
5. Rotate sugarcane with a resistant crop.
6. Grow resistant cultivars.

Other Fungi Reported to Cause Smut

Sorosporium indicum
Sphacelotheca consimilis
S. cruenta (syn. *S. chrysopogonis*, *S. holci*, and *Ustilago cruenta*)
S. erianthi (syn. *U. erianthi*)
S. papuae
S. pulverulenta (syn. *Cintractica pulverulenta*, *U. pulverulenta*, and
        U. pulverulenta)
S. sacchari (syn. *U. sacchari* and *U. sacchari-ciliaris*)
S. schweinfurthiana (syn. *U. schweinfurthiana*)
S. schweinfurthiana var. *minor*
Ustilago consimilis
U. courtoisii

Curvularia Leaf Spot and Seedling Blight

Cause. *Curvularia lunata.*

Distribution. Africa, Argentina, Asia, India, and the United States (Hawaii).

Symptoms. Leaves have scattered circular to oval red-brown spots that enlarge into irregular, dark brown patches. Leaves become chlorotic, necrotic, and dry up. Seedlings also become diseased and die.

Management. Not reported.

Downy Mildew *(Peronosclerospora sacchari)*

Downy mildew is also called leaf stripe and Sclerospora leaf stripe.

Cause. *Peronosclerospora sacchari* (syn. *Sclerospora sacchari*), which survives as oospores in leaf residue, is disseminated primarily by diseased seed pieces but also by airborne and waterborne conidia. Conidia are produced on diseased plants at temperatures of 2° to 25°C and a relative humidity of 90% or more. Sporulation also depends on actual sunlight or radiant energy beyond a certain level. Disseminated by splashing water for a short distance and by wind for up to 400 m from their source, conidia germinate in the presence of free moisture. Infection takes place on very young leaf tissues while they are still in the spindle, or in young buds on stalks. Oospores are produced in the mesophyll tissue of diseased leaves as sugarcane matures.

Peronosclerospora sacchari can infect other plants, particularly maize, *Zea mays,* and teosinte, *Euchlaena mexicana.* These plants may be important in the spread of downy mildew but are of little importance in perpetuating the disease from one season to the next.

Distribution. Downy mildew occurs only in the Eastern Hemisphere. It has been reported from Australia, Fiji, India, Japan, New Guinea, the Philippines, Taiwan, and Thailand.

Symptoms. Well-defined chlorotic stripes on leaves as they emerge from the spindle become more prominent as the leaves elongate. Stripes are light green at first but become yellow as the diseased leaves mature. Stripes are separated by green tissue and are confined between the larger leaf veins. Stripes are 1–25 mm wide and vary in length from 2.5 to several centimeters, often extending the length of the blade. During high humidity and temperatures, a velvety, whitish growth consisting of mycelium, conidia, and conidiophores occurs on stripes. As the diseased leaves become older, stripes become dark red.

Following secondary infection, a symptom called the "hump-up" stage occurs. This symptom is manifest by diseased stalks that increase in height until some are almost twice as tall as healthy plants. These stalks become light, brittle, and watery. The spindle leaves of the stalks are few in number, short, and discolored with yellow stripes. Spindle leaves fail to unfold, cling together at the top of the diseased plant, and eventually become twisted. These leaves wither and shred.

In contrast, plants that grow from diseased seed pieces are severely stunted and have twisted and shredded leaves.

Management
1. Grow resistant cultivars.
2. Obtain planting material from a disease-free source.
3. Harvest diseased plants early and plow the field.
4. Rogue out diseased stools in fields that have ratooned early.

Downy Mildew *(Peronosclerospora spontanea)*

Cause. *Peronosclerospora spontanea* (syn. *Sclerospora spontanea*).

Distribution. The Philippines and Taiwan.

Symptoms. Chlorotic stripes on leaves as they emerge from the spindle become more prominent as the leaves elongate. Stripes are light green at first but become yellow as the diseased leaves mature. The stripes are separated by green tissue and are confined between the larger leaf veins. During high humidity and temperatures, a velvety, whitish growth consisting of mycelium, conidia, and conidiophores occurs on the stripes. As the diseased leaves age, the stripes change color from yellow to dark red.

Management
1. Grow resistant cultivars.
2. Obtain planting material from a disease-free source.
3. Harvest diseased plants early and plow the field.
4. Rogue out diseased stools in fields that have ratooned early.

Downy Mildew Leaf-Splitting Form

Downy mildew leaf-splitting form is also called leaf-splitting disease.

Cause. *Peronosclerospora miscanthi* (syn. *Sclerospora miscanthi*). *Mycosphaerella striatiformans* is also reported to be a causal fungus. *Peronosclerospora miscanthi* persists in soil and in infested residue in soil as oospores and, probably, as mycelia from which conidia are produced. Infection is caused by both oospores and conidia. Conidia are produced during periods of high humidity and are disseminated by wind to host plants. Oospores are abundantly produced within leaf tissue as plants mature and are disseminated by any means that moves soil. *Peronosclerospora miscanthi* is also disseminated by planting diseased seed cane.

Distribution. Fiji, Papua New Guinea, the Philippines, and Taiwan. Downy mildew leaf-splitting form is considered to be of minor importance.

Symptoms. Initially, green-yellow stripes are produced that often extend the entire length of diseased leaves. The color of stripes progressively changes from yellow to red-brown and, finally, to dark red. Eventually, the diseased leaf tissue disintegrates and splits along the lesion lines to change leaves into whip-like bundles of fibers. White velvety masses of mycelia, conidia, and conidiophores are produced on the lower surfaces of leaf lesions.

Management
1. Grow resistant cultivars.
2. Place cuttings in water at 46°C for 20 minutes and then at 52°C for 20 minutes to eliminate the causal fungus.

Drechslera Seedling Blight and Leaf Spot

Cause. *Drechslera spicifera* (syn. *Bipolaris spicifera* and *Helminthosporium spicifera*).

Distribution. India.

Symptoms. In the seedling blight phase, diseased plants are stunted and have poor root development. Leaf spot symptoms on leaves of diseased seedlings are small, elongated to elliptical, reddish specks, with dark reddish margins, that enlarge and coalesce. Leaf tips die and dry up; eventually, the entire diseased plant wilts and dies.

Management. Not necessary. Drechslera leaf spot and seedling blight is considered to be an unimportant disease.

Dry Top Rot

Cause. *Ligniera vascularum* (syn. *Sorosphaera vascularum*) survives in infested residue and soil. *Ligniera vascularum* is an obligate parasite; therefore, a possible life cycle based on closely related genera is as follows: *Ligniera vascularum* persists as resting spores in infested residue. Orange-brown spherical spores are 17 to 25 µm in diameter and have walls that are 1.5 µm thick. *Ligniera vascularum* is disseminated in diseased seed cane by any means that transports residue, and by water that moves spores.

As residue decomposes, resting spores are liberated into the soil, where they germinate under favorable conditions and form zoospores. Zoospores "swim" to root hairs, where they encyst and form infection hyphae that penetrate host plants. A plasmodium is formed within the infected plant cells and spreads from cell to cell in the plant. In older-infected cells, several cell walls may form within the plasmodium, transforming it into several zoosporangia. Zoospores form in zoosporangia that are liberated into soil and rupture to release the zoospores. Zoospores infect roots directly or fuse with a compatible zoospore to form a zygote, which then infects host roots. Dry top rot is most severe on sugarcane grown in wet soils.

Distribution. Barbados, Cuba, and the United States (Florida and Puerto Rico).

Symptoms. Initially, spindle leaves wither and leaf tips dry. Leaves appear to lose their green color, then roll, and wilt. Growing points then die and the entire diseased stalk wilts. Internodes at the tops of diseased plants or near the growing point are shortened or stunted, shrunken, flaccid, and thinner, giving the stalk a tapered appearance. Later stages of the disease include a complete drying and necrosis of spindle and upper leaves, death of the shoot apex and stalk, and, possibly, death of the entire stool. The spindle leaves become detached from stalks that die and expose the withered internodes. The most severely diseased plants die.

Longitudinal and cross sections of the vascular bundles in the lower diseased stalk internodes are orange, yellow, pink, or red due to orange-brown spherical spores clogging the vascular system. Spore masses flow from excised vascular bundles placed in a drop of water. In some xylem vessels, usually above the region where spores are found, a gray granular substance or plasmodium can be found.

Management
1. Plant disease-free cane.
2. Take cuttings from the tops of young cane if only infected cane is available for seed.
3. Rotate sugarcane with legumes.
4. There is genetic variability among cultivars.

Ergot

Cause. *Claviceps purpurea* and *C. pusilla*. *Balansia* sp. is also reported to cause ergot. *Claviceps* spp. survive or overseason as sclerotia that germinate in free moisture in darkness at temperatures of 35° to 40°C. Subsequent ergot development occurs in humid weather.

Distribution. *Claviceps purpurea* is the causal fungus of ergot in the "temperate" regions of Australia, India, and the Philippines. *Claviceps pusilla* is the causal fungus in the warmer regions of the world.

Symptoms. Initial symptoms are a stickiness and drooping of arrows when spikelet blooming is most active. Inflorescences rapidly turn black due to growth of saprophytic fungi. Eventually, typical hard, black, sclerotial bodies replace kernels.

Management. Not usually necessary. Elimination of susceptible material in sugarcane fields and adjacent fields is an aid to disease management.

Eyespot

Cause. *Bipolaris sacchari* (syn. *Drechslera sacchari* and *Helminthosporium sacchari*) persists within infested residue in geographical areas where sugarcane is not grown during the winter. Conidia produced on lesions and infested residue are disseminated a considerable distance by wind. *Bipolaris sacchari* is also seedborne. Eyespot is most severe during periods of high moisture and temperatures of 24° to 31°C. Eyespot is most severe during the winter months in the tropics.

Distribution. Wherever sugarcane is grown.

Symptoms. Seedling disease symptoms are red-brown spots on coleoptiles a few days after planting. Leaves then become chlorotic and seedlings die.

Initially, eyespot lesions on older plants are small, watery spots that are darker green than healthy tissue. In 3 to 4 days, spots become oval (6–10 mm in length) and red-brown with a tan margin. Eventually, spots become gray and have a less noticeable halo. The red-brown areas lengthen and form streaks or "runners" from the original spots to the tips of leaves. Occasionally a short red-brown streak develops in the opposite direction. Spots and runners may coalesce. As diseased leaves die, leaf tips become red-brown, which gives a field of diseased plants a brownish cast. Infection of leaves in the spindle may allow the disease to move into the terminal portions of stalks and cause top rot. Stalks of highly susceptible cultivars have elongated brown lesions.

Management
1. Grow resistant cultivars.
2. Do not apply large amounts of nitrogen fertilizer before eyespot is likely to occur.
3. A 2-year cropping system has been suggested: Sugarcane is planted and harvested every 3 years, immediately following the eyespot season.

Foliage Blight of Seedlings

Cause. *Exserohilum rostratum* (syn. *Bipolaris rostrata, Drechslera rostrata,* and *Helminthosporium rostratum*).

Distribution. Widespread.

Symptoms. Narrow, reddish stripes occur on diseased seedling leaves. The stripes eventually enlarge, coalesce, and become dark brown. Leaf sheaths become dark brown to olivaceous due to the production of conidia and conidiophores. Diseased plants are usually stunted.

Management. Not reported.

Fusarium Sett and Stem Rot

Cause. *Gibberella fujikori,* anamorph *Fusarium moniliforme,* and *G. subgluti-nans* persist in a wide range of infested residue and as perithecia in infected cane. Large numbers of conidia are produced on infested residue. Dissemination is mainly by airborne conidia and ascospores. Infection is through the ends of cane cuttings, young adventitious roots, and nodal leaf scars of the stalk portion planted into infested soil. The common sugarcane borer, *Diatraea saccharalis,* and its parasite, *Bassus stigmaterus,* also disseminate spores and aid infection by wounding sugarcane stalks, which provides entrances for spores disseminated by insects and wind. Roots may be infected without being wounded.

Distribution. Wherever sugarcane is grown.

Symptoms. Starting from the cut ends, vascular bundles become red. Young roots become purple, then decay. Buds may swell but not germinate or may germinate but slight growth occurs. When such sets are split, vascular bundles are purple-red and the surrounding parenchyma tissue is brown.

In standing cane, nodes and internodes may be infected at the point of infestation by the cane borer. Vascular bundles are dark red and the surrounding parenchyma tissue is purplish or brown-red. Diseased leaves turn yellow and dry up. Tops of diseased plants die, although the stalk tissue appears healthy. If harvest is delayed, the entire stalk will decompose.

Management
1. Grow resistant cultivars.
2. Dip or spray cuttings with a fungicide.
3. Rotate sugarcane with nonsusceptible crops.

Iliau

Cause. *Clypeoporthe iliau* (syn. *Gnomonia iliau*), anamorph *Phaeocytostroma il-iau.* In the autumn, perithecia and pycnidia-like structures in which conidia are produced are commonly found submerged in the leaf sheath tissue of diseased mature plants. Ascospores are disseminated by wind and water. Conidia are exuded from pycnidia during moist weather and disseminated by splashing water and, less commonly, by wind. *Clypeoporthe iliau* and *P. iliau* are also disseminated by infected seed cane. Primarily young plants are infected during cool, moist weather. Seedlings produced from infected seed cane are infected early in the growth of the plant and may die before stalk elongation. *Clypeoporthe iliau* spreads rapidly from stalk to stalk when seed cane is stored in covered mats or piles.

Distribution. Australia, Brazil, Cuba, the Philippines, and the United States (Hawaii and Louisiana).

Symptoms. Diseased plants remain small and many break over and die. Overlapping leaf sheaths adhere together due to the growth of fungus mycelium that binds the sheaths. Outer leaves die and white fungal mycelium can be seen after the leaf sheaths are stripped. When the fungus grows into stalks, there is a necrotic area inside the rind that is reddish, while the rest of the stalk interior remains white. Diseased stalks shrivel, become weakened, and often break over. The surface of necrotic tissue becomes rough due to the development of perithecia, whose sharp beaks project above the tissue surface.

Management. Grow resistant cultivars.

Leaf Blast

Leaf blast is also called leaf blast of sugarcane and leaf spot of sugarcane.

Cause. *Didymosphaeria taiwanensis* persists as perithecia in infested residue. Leaves injured by frost are most susceptible to infection.

Distribution. Generally widespread, but leaf blast is not known to occur naturally in the United States. Leaf blast is considered to be of minor importance.

Symptoms. Yellow spots (0.5–1.0×3.0–25.0 mm) occur on both surfaces of diseased leaves. Spots become purple-red and coalesce, forming long, narrow streaks that turn entire diseased leaves purple-red. Leaves may wither and die, starting from the leaf tip back. Perithecia appear as small black objects in the dead areas of diseased leaves.

Management. Not necessary.

Leaf Blight

Cause. *Leptosphaeria taiwanensis,* anamorph *Stagonospora taiwanensis,* causes leaf blight throughout the year in high rainfall areas.

Distribution. Japan, Okinawa, the Philippines, and Taiwan.

Symptoms. Small, narrow, elliptical or elongate, spindle-shaped, yellowish spots are produced on both sides of diseased immature leaves that are in the spindle. Spots become red-brown and elongate into streaks that often coalesce. Severely diseased leaves die and dry up, giving distant diseased plants a red-brown appearance. Occasionally purple-red lesions occur on leaf sheaths, especially in the latter stages of disease development. Perithecia appear as small black specks in the margins of the oldest lesions.

Management. Grow resistant cultivars.

Leaf Scorch *(Leptosphaeria bicolor)*

Cause. *Leptosphaeria bicolor.*

Distribution. Kenya.

Symptoms. Spindle-shaped lesions enlarge and coalesce, giving leaves a scorched appearance.

Management. Not reported.

Leaf Scorch *(Stagonospora sacchari)*

Cause. *Stagonospora sacchari* persists in infested leaves as mycelia and pycnidia but does not survive well in soil. During moist weather, conidia are exuded out of pycnidia in a gelatinous mass that sticks firmly to leaf surfaces when the gelatinous mass dries up. Subsequent moisture or rain loosens conidia from the dried gelatinous mass and disseminates them to adjacent leaves or plants. Conidia are rarely airborne. Diseased leaves that adhere to the seed piece also provide a source of inoculum. Disease severity is greatest during moist weather.

 Stagonospora sacchari produces two toxins—toxin I(6-ethylsalicylic acid) and toxin II (epoxydon)— that are capable of inducing leaf necrosis on susceptible cultivars at 30 μg/ml and 25 μg/ml, respectively.

Distribution. Asia, Africa, Central America, and South America.

Symptoms. Very small red or red-brown spots that occur on diseased young leaves ultimately enlarge, become spindle-shaped, and have a yellow margin. Eventually, spots coalesce and extend along vascular bundles for 5–17 cm or more, becoming spindle-like streaks. At first, the streaks are red-brown but later become tan with a dark red border. Numerous pycnidia appear as small, black, pimple-like objects in the older lesions. Occasionally lesions extend to the upper parts of diseased leaf sheaths, but no pycnidia are ordinarily produced in them. During dry weather, severely diseased leaves will appear scorched. Spots on older diseased leaves remain small and do not develop into streaks.

Management. Grow resistant cultivars.

Marasmius Sheath and Shoot Blight

Cause. *Marasmiellus stenophyllus* (syn. *Marasmius stenophyllus*) is saprophytic in infested residue and infects host plants predisposed by other means or through wounds. Dissemination is by any means that moves mycelial strands and spores.

Distribution. Wherever sugarcane is grown. Marasmius sheath and shoot blight is of minor importance.

Symptoms. White mycelium grows over the lower portions of diseased plants, causing leaf sheaths to adhere tightly to stalks. White mushrooms develop at the base of diseased stalks. Mature leaves of diseased plants are covered with red spots that develop during dry weather periods. The underground portion of stems and young shoots may also become infected.

Management. Most cultivars are resistant.

Myriogenospora Leaf Binding

Myriogenospora leaf binding is also called leaf binding and tangle top.

Cause. *Myriogenospora aciculispora* persists as perithecia in infested residue.

Distribution. Argentina, Brazil, and the United States (Louisiana). Myriogenospora leaf binding is not common.

Symptoms. Diseased plants are stunted, and the tips of unfolded leaves adhere to adjacent older leaves. As new leaves elongate, their tips are held firmly along the midrib of the previously unfolded leaf by a black stroma of the causal fungus, giving diseased plants a "whip-like" appearance. Although growing points and shoots may be killed, new shoots commonly develop from the basal buds or buds formed later. A few healthy shoots may develop in diseased stools. Perithecia may occur in the tips of diseased leaves in the spindle.

Management. Not necessary.

Northern Poor Root Syndrome

Northern poor root syndrome is also known as poor root syndrome. Pachymetra root rot is considered a significant component but is described as a separate disease.

Cause. A complex of organisms, including *Pachymetra chaunorhiza, Pythium graminicola, Pythium myriotylum,* and nematodes. *Pachymetra chaunorhiza* causes the primary root rot symptom for the root rot component of the syndrome. Oospores of *P. chaunorhiza* survive in fallow soil for at least 5 years. Inoculum density decreases when resistant cultivars are grown.

Distribution. Australia (Queensland) in the widely grown cultivar 'Q90'. *Pythium myriotylum* has also been identified from sugarcane in Taiwan.

Symptoms. Disease symptoms are premature wilting, unthrifty growth, and a tendency for stools to tip at the ground level, resulting in a reduction of subsequent ratoon stool numbers.

A soft, flaccid rot develops in the root cortex. Both primary and secondary roots develop various lesions and a general discoloration, and development is restricted and the fine root system is poorly developed. Diseased roots contain high numbers of verrucose oogonia of *Pachymetra chaunorhiza*. The underground portions of stalks and the root system can be easily uprooted if diseased sugarcane plants lodge.

Pythium myriotylum is reported to cause red to black elliptical lesions on roots and a rot of root tips. Root dry weight of some cultivars is reduced.

Management. Cultivar resistance has been identified for Pachymetra root rot.

Orange Rust

Orange rust is also called sugarcane leaf rust.

Cause. *Puccinia kuehnii*. Urediniospores and teliospores are produced on sugarcane. Other spore stages and alternate hosts are not known for certain; however, spontaneums, wild sugarcane, and related genera have been suggested as collateral hosts. Urediniospores are disseminated by wind and water. Infection occurs during humid conditions at temperatures of 16° to 29°C.

Distribution. Asia, Australia, and Oceania.

Symptoms. Symptoms are similar to those caused by *P. melanocephala* (common rust) except uredinia are cinnamon to yellow-brown.

Management. Grow resistant cultivars.

Pachymetra Root Rot

Pachymetra root rot is considered to be an important component of northern poor root syndrome and the probable cause of the root rot component of the syndrome.

Cause. *Pachymetra chaunorhiza* does not have a saprophytic growth stage. Oospores, the only known propagules, are produced in rotted roots. The disease occurs in the high rainfall districts. It is suspected that *P. chaunorhiza* colonized sugarcane from a native Australian grass, but the alternative host is yet to be identified. *Pachymetra chaunorhiza* can infect sorghum, but disease is generally restricted and oospore production is sparse. *Pachymetra chaunorhiza* can also infect clones of *Saccharum spontaneum* and *Erianthus* spp.

Distribution. Australia (high rainfall districts of Queensland).

Symptoms. Aboveground symptoms include the loss of plant vigor, poor stooling, uneven stalk height, thin stalks, and a loss of stool anchorage. Lodged stools are often uprooted, and many stools are pulled out of the ground during normal harvesting operations because of the loss of stool anchorage.

Initial root symptoms are a water-soaking of the diseased root section that develops into a soft, flaccid rot of primary and larger secondary roots; lesions and general discoloration of primary and secondary roots; and poor fine root development. The rot occurs in the root tip region and, in susceptible cultivars, may extend the length of the entire root. In field soils that have a high population of *P. chaunorhiza* propagules, root systems may be greatly restricted, with many short, flaccid roots extending from the stool. A red discoloration up to 2 mm long that usually encircles the root accompanies the flaccid rot, indicating the limit of disease. In advanced stages of the disease, diseased roots are held intact only by the root epidermis because all other root structures have disintegrated. Where high inoculum levels occur, 80% to 90% of primary shoot roots in susceptible cultivars may be completely decomposed.

Diseased roots contain high numbers of oospores that have a verrucose oogonial wall.

Management. Grow a resistant or tolerant cultivar.

Philippine Downy Mildew

Philippine downy mildew is also called sleepy disease.

Cause. *Peronosclerospora philippinesis* (syn. *Sclerospora indica* and *S. philippinesis*).

Distribution. India and the Philippines.

Symptoms. Diseased seedling leaves have chlorotic streaks. Diseased mature leaves have necrotic streaks and are frequently deformed.

Management. Not reported.

Phyllosticta Leaf Spot

Cause. *Phyllosticta hawaiiensis.*

Distribution. Widespread.

Symptoms. Symptoms on sugarcane are not well known. General leaf spots are reported to occur on diseased foliage.

Management. Not reported.

Phytophthora Rot of Cuttings

Cause. *Phytophthora megasperma* and *P. erythroseptica* persist in soil and infested residue as oospores. Sporangia normally germinate indirectly by producing zoospores but occasionally germinate directly and produce germ tubes. Zoospores "swim" through soil water to plant host roots, where they encyst and germinate and form germ tubes that penetrate the host plant. Infection occurs more frequently through the nodal areas than through cut ends. Oospores are eventually produced in diseased tissue. Phytophthora rot is most severe when a long, cool spring follows a cold, wet winter.

Distribution. The United States (Louisiana).

Symptoms. Seed pieces planted in infested soil do not produce roots and few buds germinate. Initially, interiors of diseased seed pieces are water-soaked, followed by the occurrence of pink to orange-red streaks. With yellow-stalked cultivars, the discolored streaks appear through the stalk rind of the diseased plant. Streaks may also be seen in the rind of dark-stalked cultivars, but these streaks are not as distinct and are darker colored. Eventually, the entire seed piece becomes water-soaked and the color darkens to red-brown and is accompanied by an ether-like odor.

Management
1. Grow resistant cultivars.
2. Improve soil drainage.

Pineapple Disease

Cause. *Ceratocystis paradoxa,* anamorph *Chalara paradoxa* (syn. *Thielaviopsis paradoxa*), which persists as saprophytic mycelia in infested soil residue, infects seed pieces by conidia growing into the cut ends. Sometimes seed pieces are infected prior to being planted. Standing cane may also become infected by airborne conidia that enter stalks through injuries caused by rodents, insects, or machinery. Pineapple disease is often most severe in poorly drained areas within a field where soils remain wet and relatively cool.

Distribution. Wherever sugarcane is grown.

Symptoms. Primarily the central core sugarcane cuttings are affected. Initially, the central core becomes reddened but stays firm. Later, the interiors of diseased stalks break down and become hollow and discolored black, but vascular bundles remain intact. Cuttings may decay before buds germinate or diseased shoots die back after growing only a few centimeters. In the early stages of decomposition, a pineapple-like odor may occur. Leaves on diseased stalks of standing cane wilt, then the stalks die.

Management
1. Make a cutting of at least three nodes to protect the center node.
2. Do not plant under conditions of low soil temperatures or excessive wetness or dryness.
3. Protect seed piece ends with a fungicide.
4. Do not plant too deep in the soil.

Pokkah Boeng

Pokkah boeng is also called Fusarium sett and stem rot, knife cut, and pokkah boeng chlorosis.

Cause. *Gibberella fujikori,* anamorph *Fusarium moniliforme,* and *G. subglutinans* persist as saprophytic mycelia in infested residue and as perithecia in diseased cane. Large numbers of conidia are ultimately produced on infested residue. Conidia and ascospores are disseminated primarily by wind to host plants. Infection occurs through the spindle, along the margins of partially unfolded leaves, where a small opening is formed during periods of dry weather. Conidia enter the spindle and are carried down to the susceptible region, where they germinate during wet weather. Mycelium grows through the soft cuticle into the inner plant tissues.

Distribution. Wherever sugarcane is grown.

Symptoms. Pokkah boeng occurs when susceptible plants are in a rapid stage of growth, usually when they are 3 to 7 months old. A chlorotic area develops on the basal portion of diseased lower leaves as they emerge from the spindle. Red spots or stripes often develop on the chlorotic areas. Affected areas usually are confined to only a few leaves on one side of the spindle, but sometimes the young stem is penetrated by fungus mycelium. Diseased leaves are frequently deformed and do not unfold normally but remain short. Often diseased areas are torn. Sometimes cavities develop in internodes and present a ladder-like appearance. Occasionally, during wet weather, a soft rot develops in diseased areas. Usually diseased plants recover and new leaves develop normally.

Management. Grow resistant cultivars.

Purple Spot of Sugarcane

Cause. *Pseudocercospora rubropurpurea* (syn. *Cercospora rubropurpurea*).

Distribution. Taiwan.

Symptoms. Red-purple to dark purple and irregular-shaped to circular (2–12 mm in diameter) leaf spots are most distinct on the upper surfaces of diseased leaves.

Management. Not reported.

Pythium Root Rot

Cause. *Pythium aphanidermatum, P. arrhenomanes, P. catenulatum, P. dissotocum, P. graminicola, P. mamillatum,* and *P. splendens.*
  Root rot caused by *P. arrhenomanes* occurs during high temperatures in the early winter followed by alternate wet and dry periods in autumn but

not in summer plantings. Shoots growing in loam soils are more susceptible than those growing in sand.

Weeds, particularly johnsongrass, can serve as hosts for *P. arrhenomanes*
and may be involved in the etiology of Pythium root rot.

Distribution. Widely distributed.

Symptoms. Necrotic dark reddish lesions occur on diseased roots. Root tips
are flaccid, water-soaked, and necrotic. The growth of fine feeder roots is
reduced.

Diseased shoots exhibit stunting and wilting and eventually may be
killed. Poor ratooning may also occur.

Management. Plant sugarcane in well-drained soils.

Red Leaf Spot

Red leaf spot is also called purple spot.

Cause. *Dimeriella sacchari* persists as perithecia in infested residue.

Distribution. Australia, Bangladesh, Cuba, Fiji, Japan, Java, Nepal, New
Guinea, Panama, the Philippines, Taiwan, Tanzania, Tobago, and
Trinidad. Red leaf spot is of minor importance.

Symptoms. Leaf tips are more severely affected than the other parts of diseased leaves. Spots begin as red dots that later become round or elliptical
(0.5–2.0 mm in diameter) and have a yellow margin. Later, the color
changes to purple-red. Perithecia appear as small black specks in the older
spots. When disease is severe, premature death of leaves may occur.

Management. Not necessary.

Red Rot

Cause. *Glomerella tucumanensis* (syn. *Physalospora tucumanensis*), anamorph
Colletotrichum falcatum, persists in diseased plants. Cuttings are a major
source of inoculum because of the conidia and mycelia present on, or in,
seed pieces. Incipient infections may be present in bud scales, leaf scars,
other nodal tissues, and stalk borer tunnels. Conidia and ascospores are
produced in diseased plant tissue and are primarily disseminated by water;
therefore, it takes a relatively long time for leaves higher on plants to become infected. Conidia are the principal source of stalk infections. Conidia are produced directly on mycelia or in acervuli in the lesions on
midribs, then are washed down behind the leaf sheaths, where they germinate and become established in the nodes and internodes. Seed pieces
taken from this cane are internally diseased or have dormant mycelia in
bud scales, leaf scars, and other nodal tissues. Conidia are also present on
the cutting surface.

Disease development decreases with a decrease in temperature from 31.0° to 21.1°C. Disease development is optimum at temperatures from 29.4° to 31.0°C.

Injuries caused by insects and machinery predispose sugarcane to infection. Disease severity can be increased by the occurrence of drought conditions during the initial growth processes of vegetatively propagated sugarcane stalks.

Distribution. Wherever sugarcane is grown.

Symptoms. All plant parts may be diseased, but red rot is most important on standing stalks, planted cuttings, and leaf midribs. In the early stages of disease, stalks display no external symptoms. Tissues inside diseased stalks become a dull red and give a mottled appearance to stalk interiors. Red vascular bundles pass through reddened areas and extend into healthy tissues. The reddish discoloration and cavities that contain either mycelium or a clear liquid may occur through the length of diseased stalks. Eventually, diseased internal tissues become dark, shrink, and dry out. The stalk exteriors of severely diseased plants may become discolored red-brown and have acervuli that appear as dark tufts or black hair-like objects under magnification.

Seed pieces may rot during periods adverse to sugarcane growth, such as too wet or too dry conditions. Diseased seed pieces are rotted and have various shades of red, brown or gray discoloration. Gray mycelia develop in pith cavities and acervuli develop on the exteriors of seed pieces. Nodal tissue of susceptible cultivars becomes red to almost black.

Initial symptoms on leaves are small red spots on the upper surfaces of midribs. Spots enlarge and produce elongated lesions which fuse together and form lesions that extend the entire length of diseased leaves. Spots become tan, have dark red margins, and are covered with acervuli that look like numerous black tufts. Diseased leaves may wilt, then dry up.

Management
1. Grow resistant cultivars.
2. Use only healthy cane for seed pieces.
3. Plant when conditions are proper for rapid germination. Maintain proper soil moisture.
4. Harvest susceptible cultivars before they have passed the peak of maturity.

Red Rot of Leaf Sheath and Sprout Rot

Cause. *Athelia rolfsii* (syn. *Pellicularia rolfsii*), anamorph *Sclerotium rolfsii*, persists as sclerotia in soil and on infested host plant residue. Red rot of the leaf sheath occurs during warm, wet weather.

Distribution. Wherever sugarcane is grown.

Symptoms. Irregular-shaped, orange-red discolored areas with distinct margins develop on lower diseased leaf sheaths. White mycelium grows into the inner sheaths and binds them loosely together. The underlying rind becomes discolored light brown and has a distinct margin. As diseased stalks mature, the affected area dies and dries up, causing an extensive shallow canker. Small, round, brownish sclerotia develop between sheaths, along the edges or over the surface of diseased areas. In severe cases, leaves gradually dry up and die. Most shoots of a diseased stool are thin and stunted and have many short internodes.

Management. Not necessary.

Red Spot

Cause. *Exserohilum rostratum* (syn. *Helminthosporium rostratum*).

Distribution. The United States (Mississippi). Red spot is thought to be of little economic importance.

Symptoms. Bright red lesions (15–25 × 20–35 mm) occur on the midribs of diseased leaves.

Management. Not reported.

Red Spot of Leaf Sheath

Red spot of leaf sheath is also called red leaf sheath rot of sugarcane.

Cause. *Mycovellosiella vaginae* (syn. *Cercospora vaginae*). Red spot of leaf sheath is favored by warm, moist weather.

Distribution. Wherever sugarcane is grown. Red spot of leaf sheath is of minor importance.

Symptoms. Small, round, bright red spots with sharp margins occur on the diseased upper leaf sheaths. Spots coalesce and form bright red, irregular-shaped patches that extend through the diseased upper leaf sheaths into the inner sheaths. Later, dark brown to black conidia and conidiophores appear as a "sooty" coating on all diseased areas but are the most abundant on the inside of diseased leaf sheaths.

Management. Not necessary.

Rhizoctonia Sheath and Shoot Rot

Cause. *Rhizoctonia solani*.

Distribution. Wherever sugarcane is grown.

Symptoms. Similar to those of banded sclerotial leaf disease (see earlier discussion of this disease).

Management. Not necessary.

Rind Disease

Rind disease is also called sour smell and sour rot.

Cause. *Phaeocytostroma sacchari* (syn. *Melanconium sacchari* and *Pleocyta sacchari*) produces stromata inside diseased tissues. The stroma enlarges and develops enough pressure to eventually split the plant epidermis and form a pustule. During moist conditions, conidia are produced inside stromata and exuded in a long, gelatinous, thread-like structure through the top of pustules. Conidia are disseminated by wind and rain. Infection normally occurs through wounds in the plant host and in plants not growing vigorously because of drought or overmaturity.

Distribution. Australia, the United States (Hawaii), and the West Indies.

Symptoms. Symptoms are most common on stalks but also occur on leaf blades and sheaths. Numerous black, coiled, hair-like fruiting structures exude from pustules that have broken through the surfaces of diseased tissues. Internal diseased stalk tissues have a dark discoloration and a sour smell. Diseased leaves may yellow, dry up, and be covered with the black, coiled structures. A symptom from Hawaii is that the nodes become red-brown and, later, the internodes become light brown.

Management. Harvest mature sugarcane as soon as possible.

Ring Spot

Ring spot is also called ring spot of sugarcane leaf.

Cause. *Leptosphaeria sacchari,* anamorph *Phyllosticta* sp., persists as saprophytic mycelia on dead leaves.

Distribution. Wherever sugarcane is grown. Ring spot is usually of minor importance.

Symptoms. Symptoms occur on diseased leaf blades and leaf sheaths. Spots initially are dark green to brownish, then become red-brown and oval or diamond-shaped and are surrounded by narrow chlorotic borders. Later, the centers of spots become tan and are surrounded by narrow red margins. Often an older spot may still be surrounded by a yellow margin. Older spots have fruiting structures of various fungi that appear as blackish specks.

Management. Grow resistant cultivars.

Ring Spot of Sugarcane

Cause. *Pseudocercospora saccharicola* (syn. *Cercospora saccharicola*).

Distribution. Taiwan.

Symptoms. Initially, leaf spots are circular to elliptical, dark green to brown, and generally surrounded by a narrow red-brown to maroon border. Eventually, spots become gray and are surrounded by a purple-brown margin up to 7.5 mm wide.

Management. Not reported.

Sclerophthora Disease

Sclerophthora disease is also called Sclerospora disease and downy mildew.

Cause. *Sclerophthora macrospora* (syn. *Sclerospora macrospora*) persists in a large number of grass hosts and seed pieces. Transmission is primarily by planting infected seed pieces, which give rise to a high percentage of diseased plants. Young sugarcane plants have been successfully infected by placing sporangia in the spindle. Sporangia or conidia are abundantly produced on the undersides of diseased leaves during periods of heavy dew or fog. Sporangiophores grow through stomates and produce one to five sporangia on their tips. The windborne sporangia germinate indirectly in the presence of free water and produce zoospores that "swim" to host tissue, where they encyst and produce germ tubes that penetrate plant host tissue.

Sclerophthora disease is confined to low areas within fields that are subject to flooding. Oospores are abundantly produced in diseased leaf tissue, but their role in survival and dissemination has not been determined, since they have not been observed to germinate.

Distribution. Australia, Mauritius, Peru, South Africa, and the United States. Sclerophthora disease is usually of minor importance.

Symptoms. Diseased stools are normally severely stunted, but tall plants or a cluster of small, dwarfed shoots frequently arise from a diseased stool. The cluster of small, dwarfed shoots is a profuse tillering that gives diseased plants the appearance of clumps of coarse grass. When these stools produce cane, it is common to find stem galls of various sizes and shapes on them.

Diseased leaves are small, coarse, and brittle and are yellow-green because of the irregular-shaped, yellow-white streaks (one-half to several centimeters long) located between veins. Chlorotic blotches of varying sizes and shapes that also occur on diseased leaves give leaves a mosaic appearance. During rapid plant growth, leaves droop and their edges become wavy. However, during poor growing conditions, diseased leaves become stiff and erect, and margins die and dry out, which cause shredding and give leaf edges a ragged appearance. Leaf tips may curl into one or more loops.

Using magnification, oospores can be observed as numerous white specks along veins on diseased plants. No downy or external mycelial growth can be observed, although sporangia are produced on diseased leaf surfaces. Infection of older cane causes plant tops to develop symptoms, while plant parts below the infection(s) remain healthy. Buds on the diseased upper cane may proliferate and produce clusters of shoots at each node.

Management
1. Plant only seed pieces from healthy cane.
2. Do not plant in poorly drained areas in a field.
3. Rogue infected stools.
4. Destroy wild grasses adjacent to a cane field.

Sheath Rot

Cause. *Cytospora sacchari.*

Distribution. Widespread, but Cytospora sheath rot is considered of minor economic importance.

Symptoms. Diseased leaves die from the leaf tips. Severely diseased shoots are killed.

Management. Not reported.

Sooty Mold

Cause. *Capnodium* sp. and *Fumago* sp. are fungi that grow in insect secretions.

Distribution. Widespread.

Symptoms. A dark brown to black film or crust consisting of fungal growth occurs on leaves and sheaths.

Management. Not necessary.

Sugarcane Stubble Decline

Cause. A complex of factors, including low winter temperatures and freezes, poor soil aeration and drainage, the physiological maturity of plants at time of harvest, the condition of the stubble root system as affected by cultural practices and Pythium root rot, weed competition, a stalk rot caused by *Glomerella tucumanensis,* single and combined effects of ratoon stunting disease caused by *Clavibacter xyli* subsp. *xyli,* and sugarcane mosaic virus.

Distribution. In the cooler sugarcane growing areas of the United States.

Symptoms. Stubble gradually deteriorates, thereby preventing ratoon crops from developing. Yields of the ratoon or stubble crops progressively decline, and the crop cycle is generally limited to the plant cane crop and two ratoon crops.

Management. Pythium root rot is reduced by tolerant interspecific hybrid cultivars.

Tar Spot

Tar spot is also called black spot of leaves.

Cause. *Phyllachora sacchari.* Ascospores in residue are a source of primary infection. Spores are disseminated by wind and rain.

Distribution. Argentina and Asia.

Symptoms. Black fungal stromata on diseased leaves resemble tar.

Management. Not reported.

Target Blotch

Cause. *Helminthosporium* sp.

Distribution. Cuba, South Africa, and the United States (Florida). Target blotch is not economically important.

Symptoms. Symptoms occur only in the winter on diseased leaf blades and midribs of mature cane. Spots are small and red-brown initially, then become tan to brownish necrotic areas with irregular concentric rings that resemble a target. The spots develop first on the rolled leaves in leaf spindles, but the target symptom becomes more pronounced as diseased leaves unroll. Spots cease to develop when leaves are unrolled from the spindle.

Management. Not necessary.

Veneer Blotch

Cause. *Deightoniella papuana.*

Distribution. The British Solomon Islands, New Britain, and Papua New Guinea.

Symptoms. Small, oval, light green to tan spots with a thin red-brown margin occur on diseased leaves. Next, two more similar spots occur, one on each side of the original spot. Then two more spots occur, each flanking the second pair, and so on until up to 12 spots (1.2×61.0 cm), each larger than the preceding spot, have formed on each side of the original spot. The center of each spot finally becomes brown, which gives each spot a beautiful pattern similar to figured veneer of wood. Conidiophores cause a thick, mat-like appearance in the older spots on the undersides of leaves and occasionally on upper leaf surfaces.

Management. Not reported.

White Rash

White rash is also called anthracnose maculata, Elsinoe disease, spotted anthracnose, and white speck.

Cause. *Elsinoe sacchari,* anamorph *Sphaceloma sacchari.*

Distribution. Brazil, Japan, Malaysia, the Philippines, Taiwan, and the United States (Florida and Puerto Rico). White rash is considered to be of minor importance.

Symptoms. Symptoms occur on diseased leaf blades, midribs of leaves, and occasionally leaf sheaths. Lesions are tiny; yellowish (but sometimes described as purplish); round, elliptical, or fusiform spots ($0.1–0.5 \times 0.2–1.0$ mm) that become yellow to tan and finally whitish, sometimes with red-brown margins. These slightly elevated spots with distinct centers coalesce and form narrow, elongated streaks. A diseased area has the appearance of lacework.

Management. Not necessary.

Wilt

Cause. *Fusarium sacchari* (syn. *Cephalosporium sacchari*). *Fusarium sacchari* persists as saprophytic mycelia in infested residue. Abundant microconidia are produced on mycelia and disseminated by wind to plant hosts. *Fusarium sacchari* is also disseminated by infected seed cane and by any means that transports residue. Infection commonly occurs through insect wounds. Wilt development is more severe at a soil pH of 7.0–8.0 and an increase in the carbon:nitrogen ratio.

Distribution. Wherever sugarcane is grown.

Symptoms. Symptoms occur from when cane is half-grown until harvest time. The first noticeable symptom is the wilting of a single cane or an entire clump. Initially, leaves become yellow, then dry up. The diseased stem pith has light purple or reddish streaks and becomes light and hollow, making it worthless for milling. A foul odor occurs after splitting a diseased stalk, and a cottony mycelial growth with abundant microconidia is present in pith cavities. A diseased seed piece usually does not produce eyes and roots. Occasionally, an eye may swell or grow slightly before it dies. The interior of such a seed piece has a brown discoloration.

Management
1. Grow the most resistant cultivars.
2. Plant only healthy seed pieces.

Yellow Spot

Cause. *Mycovellosiella koepkei* (syn. *Cercospora koepkei*) persists in infested residue. Conidia, which are produced on the ends of conidiophores that emerge through stomata in disease spots during wet, humid weather, are disseminated by splashing rain and by wind to new infection sites. Conidia are more numerous on the lower surfaces than the upper surfaces of diseased leaves.

Distribution. Africa, Asia, Australia, Brazil, Colombia, and Cuba.

Symptoms. Yellow-green spots (up to 12 mm in diameter) occur on the youngest diseased leaves. Small, reddened areas first appear on diseased lower leaf surfaces, then increase on both leaf surfaces to give diseased plants and entire fields a general rusty yellow appearance at leaf maturity. A dirty, gray fungal growth develops somewhat profusely on the lower surfaces but sparsely on the upper surfaces of diseased leaves. Spots coalesce and cover large areas of leaves until, at maturity, most of the leaf surfaces may be affected. Consequently, leaves shrivel and die prematurely.

Management
1. Grow resistant cultivars.
2. Apply foliar fungicides during the growing season.

Zonate Leaf Spot

Cause. *Gloeocercospora sorghi*. Presumably *G. sorghi* persists as sclerotia on infested residue.

Distribution. Areas where sorghum and sugarcane are grown in the same culture.

Symptoms. Initially small, red-brown, water-soaked spots that sometimes have a narrow chlorotic halo occur on diseased leaves near the soil surface. Spots enlarge, elongate in a direction parallel with the veins, and become dark red or tan. Eventually, the spots occur as circular areas on diseased leaves or as semicircular patterns along the leaf margins. Spots typically are circular red-purple bands that alternate with tan areas and give a concentric or zonate pattern with irregular borders. Spots vary in diameter from a few millimeters to several centimeters and may cover the entire width of a diseased leaf.

During warm, wet weather, conidia are produced in lesions as a pinkish gelatinous spore mass that forms above a stomate. Sclerotia eventually appear as small, black, raised bodies in lines parallel to the veins.

Plants that were severely diseased in the seedling stage may become defoliated or die. Plants infected when older may have their foliage destroyed before maturity.

Management
1. Rotate sugarcane with a nonhost.
2. Plow under infected residue.

Diseases Caused by Nematodes

Lance

Cause. *Hoplolaimus columbus.*

Distribution. Unknown.

Symptoms. Diseased root systems are coarse and depleted. Top and root weights are not significantly reduced.

Management. Not reported.

Lesion

Cause. *Pratylenchus* spp. feed only in root cortex.

Distribution. Australia, Hawaii, Mauritius, and the United States (Louisiana).

Symptoms. Diseased roots are thickened, have discolored lesions, and have fewer fine roots than healthy roots.

Management. Fumigate soil with a volatile nematicide.

Root Knot

Cause. *Meloidogyne* spp. Root knot nematodes multiply best in moderately dry, well-drained soils. Weeds in cane fields may also serve as hosts.

Distribution. Wherever sugarcane is grown.

Symptoms. Knots or galls on diseased roots form at or near the root tips and appear as nodules or elongated thickenings. Thickened roots are often curled. The apical meristem is affected and growth is retarded or stopped; there is little or no proliferation of the lateral roots.

Management. Fumigate soil with a volatile nematicide.

Spiral

Cause. *Helicotylenchus* spp., *Rotylenchulus* spp., and *Scutellanema* spp.

Distribution. Wherever sugarcane is grown.

Symptoms. In pot experiments, diseased roots were blunted and malformed and the number of branch rootlets was reduced.

Management. Fumigate soil with a volatile nematicide.

Diseases Caused by Phytoplasmas

Grassy Shoot

Cause. A phytoplasma-like organism is transmitted by infected seed pieces, mechanically by cutting knives, and by the aphids *Aphis idiosacchari, A. sacchari,* and *A. maidis.* Seed pieces can be infected by juice from chlorotic leaves. Jowar is another host.

Distribution. India, Taiwan, and Thailand.

Symptoms. Typically, a mass of stunted, crowded shoots grows from diseased seed pieces or ratoon stools. Some shoots are void of chlorophyll, and if enough shoots are affected, the stools may die. Stalks produced in a diseased stool have long, delicate, scaly buds that develop into slender chlorotic shoots.

Symptoms of secondary infection occur in the leaves of numerous tillers at the base of stools. White to yellow well-defined stripes become diffuse or remain distinct on thin and chlorotic leaves.

Management. Plant only disease-free cuttings.

White Leaf

Cause. A phytoplasma-like organism that is transmitted by the leafhopper *Matsumuratettix hiroglyphicus.*

Distribution. Asia.

Symptoms. A diagnostic symptom of white leaf is leaves that are totally chlorotic in the spindle portion of tillers. Early foliar symptoms consist of a single white or cream line parallel to the midrib of one leaf in the spindle. Later, symptoms may involve stripes, mottling, or total chlorosis of leaves. Stripes may extend the entire leaf length but do not involve the leaf sheath. Stripes may vary in color and intensity and have well-defined or slightly diffuse margins. Mottling occurs as irregular-shaped islands of green that may appear as dots, streaks, or patches on a white background.

Symptoms become masked during times of relatively low temperatures. If low temperatures occur when diseased plants are mature, the newly developed leaves become green again. In young plants or those inoculated by insect vectors, symptoms disappear shortly after their initial appearance. However, typical symptoms may again develop on these diseased plants when the temperatures rise again.

Management
1. Sugarcane planted during December through April is less diseased than sugarcane planted during July through October.
2. Rogue diseased plants.
3. Plant disease-free cuttings.

Diseases Caused by Viruses

Banana Streak

Cause. Banana streak virus (BSV) is transmitted from banana to sugarcane by the pink sugarcane mealybug, *Saccharicoccus sacchari*. BSV has viral-like bacilliform particles measuring approximately 30×150 nm. It is serologically related to sugarcane bacilliform virus.

Distribution. Colombia.

Symptoms. Diseased leaves have chlorotic streaks.

Management. Not reported.

Chlorotic Streak

Cause. Sugarcane chlorotic streak virus (SCCSV) is disseminated when healthy and diseased plants are grown together in quartz sand and when healthy plants are grown in soil obtained from diseased plants in fields. This mode of dissemination suggests a soilborne vector and infection through the roots. SCCSV is also disseminated by running water, in stalk cuttings used for seed, and from plant to plant in the field in a manner that suggests an insect vector. The leafhopper *Draeculacephala portola* has been implicated as a possible vector.

The disease is most prevalent in low-lying, poorly drained areas within a field and in plants growing in potassium-deficient soil. Several grasses, including *Arundo donax, Brachiaria mutica, Panicum maximum,* and *Pennisetum purpureum,* also are hosts.

Distribution. Australia, Guyana, Java, and the United States (Puerto Rico).

Symptoms. Symptoms are more common in ratoons than in plant cane and are most prevalent in early summer. One to many yellowish to whitish streaks with wavy, irregular margins occur on both sides of diseased leaves, leaf midribs, and leaf sheaths. Streaks are short, narrow, and faint but later, particularly on older foliage, the streaks are well marked (3–15 mm wide) and extend the entire lengths of diseased leaf blades. Centers of older streaks become necrotic in sections or along the entire streak length. Leaf symptoms may disappear when streaked leaves senesce.

Young diseased plants have stiff, erect leaves and wilt, even in the presence of abundant moisture. Young shoots die or become stunted, eventually causing an entire field to have an uneven appearance. Vascular bundles are entirely discolored through the node.

Management
1. Grow resistant cultivars.
2. Treat cane in a hot water treatment at 52°C.
3. Obtain disease-free planting material.
4. Rogue out infected plants in a seed field if disease incidence is less than 2%.
5. Improve soil drainage.

Dwarf

Cause. Sugarcane dwarf virus (SCDV) is not known to be mechanically transmitted and the vectors are not known. Infection may occur from an alternate host plant or from soil, but no evidence presently exists.

Distribution. Australia.

Symptoms. Diseased plants are stunted and have short, stiff, erect, distorted, fan-like tops. Numerous small (less than 1 mm wide and of variable length), white, longitudinal stripes occur most abundantly on the younger diseased leaves. Wide, diffuse, chlorotic stripes that later become necrotic and have a papery texture occur primarily near leaf margins. Transverse splitting of midribs occurs.

Stools of susceptible cultivars in the advanced stages of disease appear as groups of grassy shoots. Infected stools that ratoon usually produce diseased plants.

Management
1. Rogue out diseased plants.
2. Grow resistant cultivars.
3. Plant only disease-free seed cane.

Fiji Disease

Cause. Sugarcane Fiji disease virus (SFDV) is in the fijivirus group. SFDV is transmitted by at least two leafhoppers, *Perkinsiella saccharicida* and *P. vastatrix*. Both species acquire SFDV as nymphs. Other leafhopper species, such as *P. vitiensis,* may also be vectors. SFDV is not mechanically transmitted. Sugarcane is the only known host for SFDV.

Distribution. Australia, Fiji, Java, the Philippines, and New Guinea.

Symptoms. Elongated swellings or galls occur along the larger veins, the vascular bundles on the undersides of diseased leaves, and the vascular bundles of diseased stalks. The last leaves to unfold from spindles are short-ened or crumpled. Diseased shoots grow to a normal height and produce healthy-looking leaves of normal length and color until the leaves unfold from spindles, then short, deformed, twisted leaves are produced and plant growth ceases completely. The upper half or two-thirds of these leaves appear burned or scalded before expanding, leaving short, crum-pled stumps. Galls are produced on both healthy-appearing and deformed leaves during the advanced stages of Fiji disease prior to plant death. Shoots either remain alive for several months or die in a short time.

Management
1. Grow resistant cultivars.
2. Plant only disease-free cuttings.
3. Harvest diseased plants earlier than normal.
4. Plow under diseased plants.
5. Rogue out diseased stools if disease incidence is low.

Mosaic

Mosaic is also called sugarcane mosaic.

Cause. Sugarcane mosaic virus (SCMV) is in the potyvirus group. SCMV now includes the former SCMV strains A, B, D, E, and MB in one group; the for-mer SCMV strains SC, BC, and Sabi from Australia and the former maize dwarf mosaic virus (MDMV) strain B in the United States are in another group. SCMV has a wide host range and infects a large number of wild and cultivated grasses.

SCMV is transmitted by numerous aphid species, including *Aphis gossypii, Carolinaia cyperi, Hysteroneura setariae, Rhopalosiphum maidis, R. padi,* and *Schizaphis graminum.* Sugarcane infected with SCMV in the first year are not foci for infections in succeeding years. SCMV is also mechani-cally transmitted with difficulty, and it is rarely seedborne in maize.

Distribution. Wherever sugarcane is grown.

Symptoms. Symptoms are most obvious in diseased leaves that have just un-rolled from spindles. Leaves have irregular, oval or oblong, pale green blotches of various sizes that are not limited by veins. Blotches are various widths throughout their length and do not resemble stripes. Diseased leaves ordinarily are not distorted, but some cultivars have stunted shoots that culminate in twisted and distorted leaves. Some cultivars have a stem mottling that causes tissue death and results in a cankered area. Other cul-tivars have small and deformed sticks and some areas of discoloration in internal tissue. Mosaic symptoms sometimes disappear from leaves of growing plants and healthy-appearing plants grow from infected seed cane.

Management
1. Grow resistant cultivars.
2. Rogue out diseased plants when disease incidence is low.
3. Do not plant sorghum next to sugarcane or rotate the two crops.

Sereh

Cause. Sugarcane sereh disease virus (SCSDV) is transmitted through cuttings. SCSDV is not mechanically transmitted nor are vectors known.

Distribution. Australia, India, Indonesia, Sri Lanka, Taiwan, the United States (Hawaii), and the West Indies.

Symptoms. Normally, diseased shoots are of different heights and secondary shoots from basal buds are similarly affected. In some stools, taller stalks that develop have shortened internodes near the top and leaves in a fan-like arrangement. Hairy adventitious roots may be produced on some nodes and abundant side shooting may occur.

A severe symptom is that shoots from infected cuttings cease to grow after reaching a certain height, which causes diseased stools to be converted into a "bunch-like" tuft. A less severe symptom is reddish-colored vascular bundles in the nodes and sometimes in the internodes. Diseased vascular bundles sometimes are narrow, dull red stripes on leaf blades.

Management
1. Grow resistant cultivars.
2. Plant certified disease-free cane.
3. Do not grow ratoon crops.

Streak Disease

Cause. Maize streak virus (MSV) sugarcane strain is transmitted by the leafhoppers *Cicadulina bipunctella zeae, C. mibla,* and other *Cicadulina* spp. Work done with maize strains shows the leafhoppers can acquire the virus as nymphs or adults in a few minutes of feeding and remain infective throughout their lives. MSV is also transmitted in cuttings from diseased plants but not mechanically. Several strains of MSV exist; each is virulent to one or a few hosts, mainly grasses, and avirulent or weakly virulent to other hosts.

Distribution. Africa.

Symptoms. Symptoms are most obvious on the youngest diseased leaves. Diseased leaves have a pattern of straight, narrow, translucent stripes (0.5 × 0.5–1.0 mm long and of uniform width) that follow leaf veins. Streaks may coalesce and be relatively wide on the leaves of young shoots developing from infected cuttings. Diseased leaves may be crinkled and narrower than normal.

Management. Resistant cultivars are being developed.

Striate Mosaic

Cause. Sugarcane striate mosaic virus (SCSMV). It is possible that more than one virus causes this disease, with one virus causing striations and another virus causing stunting. The virus(es) are sett-transmitted. SCSMV is associated with poor growing conditions in a field, such as areas with excessive sand or water-logged soils.

Distribution. Australia (northern Queensland).

Symptoms. A few to numerous short (0.5–2.0 mm long), fine, light green striations develop initially on the youngest diseased leaves but become difficult to see as the leaves mature. Striations are less numerous around larger vascular bundles, giving younger leaves a striping effect. Sometimes the entire tops of leaves are yellowed. Striations occur on lower midribs but not on the top midribs. Stunting and poor stooling usually occur.

Management
1. Plant disease-free cuttings.
2. Maintain plants in vigorous growing conditions.

Sugarcane Bacilliform

Cause. Sugarcane bacilliform virus (SCBV) is closely related serologically to banana streak virus (BSV) and is in the badnavirus group. SCBV and BSV can be considered as strains of the same virus. Virus particles are bacilliform in shape and contain a double-stranded DNA genome.

SCBV is transmitted mechanically, but the major vector is considered to be the pink sugarcane mealybug, *Saccharicoccus sacchari.*

Distribution. Widely distributed where sugarcane is grown.

Symptoms. Symptoms are variable but may include flecks or freckles on diseased leaves or diagnostic symptoms may be absent.

Management. Not reported.

Sugarcane Mild Mosaic

Cause. Sugarcane mild mosaic virus (SCMMV) is a clostero-like virus that occurs in mixed infections with sugarcane bacilliform virus (SCBV). SCMMV is transmitted mechanically and by the pink sugarcane mealybug, *Saccharicoccus sacchari,* in the absence of SCBV.

Distribution. Malawi, Mauritius, and the United States (Florida).

Symptoms. A very mild mosaic occurs on diseased leaves or no apparent foliar symptoms occur.

Management. Not reported.

Sugarcane Red Leaf Mottle

Cause. Peanut clump virus (PCV) is a furovirus transmitted by the soil fungus, *Polymyxa graminis*. The fungus develops within the roots of several species of Gramineae but not within roots of peanuts. The fungus can survive in soil for several years.

Distribution. Burkina Faso, Senegal, and Sudan.

Symptoms. Symptoms on diseased leaves are variable and include a reddish mottle surrounded by a mild mosaic together with diamond-shaped chlorotic spots, discolored stripes, and streaks of varying thickness.

Management. Not reported.

Sugarcane Yellow Leaf Disease

Cause. Sugarcane yellow leaf disease–associated virus (SCYLaV) may have some serological relationship to barley yellow dwarf virus. SCYLaV appears to be an isometric virus with several characteristics of a luteovirus. SCYLaV particles are found only in phloem companion cells and in the cytoplasm, not in the nucleus.
Disease is likely related to a phloem dysfunction.

Distribution. Brazil. Symptoms similar to sugarcane yellow leaf disease have been described for some cultivars in Australia, the United States (Hawaii), and other sugarcane-producing countries.

Symptoms. Symptoms in diseased plants are not visible in the youngest two or three leaves and become evident only in mature leaves. The first symptom is the intense yellowing on the abaxial surface of the midrib. Older leaves, usually the sixth or seventh from the growing tip, show a red discoloration on the adaxial surface of the midrib. Discoloration subsequently spreads to the leaf blade, proceeding from the tip toward the base of the leaf, and is followed eventually by tissue necrosis. Roots and stalks show impaired growth and production is significantly reduced.

Management
1. Plant disease-free cuttings.
2. Maintain plants in vigorous growing conditions.

Disease Caused by Unknown Factors

Yield Decline

Cause. Unidentified soil microorganisms that do not infect roots and may be associated with the organic matter fraction of soil.

Distribution. Australia (Queensland).

Symptoms. Yield is reduced 20% to 40%. Root browning occurs and there is reduced growth of fine roots.

Management. Not reported.

Selected References

Abbott, E. V. 1964. Black rot. *In* Sugar Cane Diseases of the World, Vol. II. C. G. Hughes, E. V. Abbott, and C. A. Wismer (eds.). Elsevier Publishing Company, New York.

Abbott, E. V. 1964. Brown spot. *In* Sugar Cane Diseases of the World, Vol. II. C. G. Hughes, E. V. Abbott, and C. A. Wismer (eds.). Elsevier Publishing Company, New York.

Abbott, E. V. 1964. Collar rot, *In* Sugar Cane Diseases of the World, Vol. II. C. G. Hughes, E. V. Abbott, and C. A. Wismer (eds.). Elsevier Publishing Company, New York.

Abbott, E. V. 1964. Dry top rot. *In* Sugar Cane Diseases of the World, Vol. II. C. G. Hughes, E. V. Abbott, and C. A. Wismer (eds.). Elsevier Publishing Company, New York.

Abbott, E. V. 1964. Myriogenospora leaf binding, *In* Sugar Cane Diseases of the World, Vol. II. C. G. Hughes, E. V. Abbott, and C. A. Wismer (eds.). Elsevier Publishing Company, New York.

Abbott, E. V. 1964. Red leaf spot (purple spot). *In* Sugar Cane Diseases of the World, Vol. II. C. G. Hughes, E. V. Abbott, and C. A. Wismer (eds.). Elsevier Publishing Company, New York.

Abbott, E. V. 1964. Red spot of the leaf sheath. *In* Sugar Cane Diseases of the World, Vol. II. C. G. Hughes, E. V. Abbott, and C. A. Wismer (eds.). Elsevier Publishing Company, New York.

Abbott, E. V., and Hughes, C. G. 1961. Red rot. *In* Sugar Cane Diseases of the World, Vol. I. J. P. Martin, E. V. Abbott, and C. G. Hughes (eds.). Elsevier Publishing Company, New York.

Abbott, E. V., and Matsumoto, T. 1964. Banded sclerotial disease. *In* Sugar Cane Diseases of the World, Vol. II. C. G. Hughes, E. V. Abbott, and C. A. Wismer (eds.). Elsevier Publishing Company, New York.

Abbott, E. V., and Tippett, R. L. 1941. *Myriogenospora* on sugarcane in Louisiana. Phytopathology 31:564-566.

Abbott, E. V., Wismer, C. A., and Martin, J. P. 1964. Rind disease. *In* Sugar Cane Diseases of the World, Vol. II. C. G. Hughes, E. V. Abbott, and C. A. Wismer (eds.). Elsevier Publishing Company, New York.

Ammar, E. D., Kira, M. T., and Abul-Ata, A. E. 1980. Natural occurrence of streak and mosaic diseases on sugarcane cultivars at upper Egypt and transmission of sugarcane streak by *Cicadulina bipunctella zeae* China. Egypt. J. Phytopathol. 12:21-26.

Anderson, D. L., Raid, R. N., Irey, M. S., and Henderson, L. J. 1990. Association of sugarcane rust severity with soil factors in Florida. Plant Dis. 74:683-686.

Astudillo, G. E., and Birchfield, W. 1980. Pathology of *Hoplolaimus columbus* on sugarcane. (Abstr.) Phytopathology 70:565.

Beniwal, M. A., and Satyavir. 1991. Effect of atmospheric temperature on the development of red rot of sugarcane. Indian Phytopathol. 44:333-338.

Birch, R. G., and Patil, S. S. 1983. The relation of blocked chloroplast differentiation to sugarcane leaf scald disease. Phytopathology 73:1368-1374.

Bourne, B. A. 1961. Fusarium sett or stem rot. *In* Sugar Cane Diseases of the World, Vol. I. J. P. Martin, E. V. Abbott, and C. G. Hughes (eds.). Elsevier Publishing Company, New York.

Braithwaite, K. S., Egeskov, N. M., and Smith, G. R. 1995. Detection of sugarcane bacilliform virus using the polymerase chain reaction. Plant Dis. 79:792-796.

Byther, R. W., and Steiner, G. W. 1974. Unusual smut symptoms on sugarcane in Hawaii. Plant Dis. Rptr. 58:401-405.

Cassalett, C., Victoria, J. I., Carrillo, P., and Ranjel, H. 1983. Resistance to sugarcane smut (*Ustilago scitaminea* Sydow). (Abstr.) Phytopathology 73:126.

Chatenet, M., and Saeed, I. 1995. First report of sugarcane red leaf mottle virus in Sudan. Plant Dis. 79:321.

Chu, H. T. 1964. Leaf-splitting disease. *In* Sugar Cane Diseases of the World, Vol. II. C. G. Hughes, E. V. Abbott, and C. A. Wisner (eds.). Elsevier Publishing Company, New York.

Comstock, J. C., Ferreira, S. A., and Tew, T. L. 1983. Hawaii's approach to control of sugarcane smut. Plant Dis. 67:452-457.

Comstock, J. C., Miller, J. D., and Farr, D. F. 1994. First report of dry top rot of sugarcane in Florida: Symptomatology, cultivar reactions, and effect on stalk water flow rate. Plant Dis. 78:428-431.

Comstock, J. C., Shine, J. M., Jr., Davis, M. J., and Dean, J. L. 1996. Relationship between resistance to *Clavibacter xyli* subsp. *xyli* colonization in sugarcane and spread of ratoon stunting disease in the field. Plant Dis. 80:704-708.

Croft, B. J. 1989. A technique for screening sugarcane cultivars for resistance to Pachymetra root rot. Plant Dis. 73:651-654.

Damann, K. E. 1992. Effect of sugarcane cultivar susceptibility on spread of ratoon stunting disease by the mechanical harvester. Plant Dis. 76:1148-1149.

Damann, K. E. Jr., and Benda, G. T. A. 1983. Evaluation of commercial heat-treatment methods for control of ratoon stunting disease of sugarcane. Plant Dis. 67:966-967.

Davis, M. J., Dean, J. L., and Harrison, N. A. 1988. Distribution of *Clavibacter xyli* subsp. *xyli* in stalks of sugarcane cultivars differing in resistance to ratoon stunting disease. Plant Dis. 72:443-448.

Dean, J. L. 1981. Inoculation of wounded and unwounded sugarcane with *Ustilago scitaminea*. (Abstr.) Phytopathology 71:870.

Dissanayake, N., Hoy, J. W., and Griffin, J. L. 1997. Weed hosts of the sugarcane root rot pathogen, *Pythium arrhenomanes*. Plant Dis. 81:587-591.

Eagan, B. T. 1964. Rust. *In* Sugar Cane Diseases of the World, Vol. II. C. G. Hughes, E. V. Abbott, and C. A. Wismer (eds.). Elsevier Publishing Company, New York.

Edgerton, C. W. 1958. Sugarcane and Its Diseases. Louisiana State University Press, Baton Rouge.

Farr, D. F., Bills, G. R., Chamuris, G. P., and Rossman, A. Y. 1989. Fungi on Plants and Plant Products in the United States. American Phytopathological Society, St. Paul, MN. 1252 pp.

Grisham, M. P. 1994. Strains of sorghum mosaic virus causing sugarcane mosaic in Louisiana. Plant Dis. 78:729-732.

Harrison, N. A., and Davis, M. J. 1988. Colonization of vascular tissues by *Clavibacter xyli* subsp. *xyli* in stalks of sugarcane cultivars differing in susceptibility to ratoon stunting disease. Phytopathology 78:722-727.

Hoy, J. W., and Grisham, M. P. 1988. Spread and increase of sugarcane smut in Louisiana.

Phytopathology 78:1371-1376.

Hoy, J. W., and Jiaxie, Z. 1991. Longevity of teliospores of *Ustilago scitaminea* in soil. (Abstr.) Phytopathology 81:1204.

Hoy, J. W., and Schneider, R. W. 1988. Role of *Pythium* in sugarcane stubble decline: Effects on plant growth in field soil. Phytopathology 78:1692-1696.

Hoy, J. W., Zheng, J., Grelen, L. B., and Geaghan, J. P. 1993. Longevity of teliospores of *Ustilago scitaminea* in soil. Plant Dis. 77:393-397

Hughes, C. G. 1961. Gumming disease. *In* Sugar Cane Diseases of the World, Vol. I. J. P. Martin, E. V. Abbott, and C. G. Hughes (eds.). Elsevier Publishing Company, New York.

Hughes, C. G., and Ocfemia, G. O. 1961. Yellow spot disease. *In* Sugar Cane Diseases of the World, Vol. I. J. P. Martin, E. V. Abbott, and C. G. Hughes (eds.). Elsevier Publishing Company, New York.

Humbert, R. P. 1968. Control of pests and diseases. *In* The Growing of Sugar Cane. Elsevier Publishing Company, New York.

Isakeit, T., Magarey, R. C., and Croft, B. J. 1992. Transmission of soilborne yield decline of sugarcane in northern Queensland, Australia. (Abstr.). Phytopathology 82:1089.

Kao, J., and Damann, D. E., Jr. 1978. Microcolonies of the bacterium associated with ratoon stunting disease found in sugarcane xylem matrix. Phytopathology 68:545-551.

Lee, Y. S., and Hoy, J. W. 1992. Interactions among *Pythium* species affecting root rot of sugarcane. Plant Dis. 76:735-739.

Liao, C. H., and Chen, T. A. 1981. Isolation, culture and pathogenicity to Sudangrass of a corynebacterium associated with ratoon stunting of sugarcane and with Bermudagrass. Phytopathology 71:1303-1306.

Liu, L. J. 1983. Sugarcane rust: The components of the epidemic and their usefulness in disease forecasting in Puerto Rico. (Abstr.) Phytopathology 73:796.

Lo, T. T. 1961. Leaf scorch. *In* Sugar Cane Diseases of the World, Vol. I. J. P. Martin, E. V. Abbott, and C. G. Hughes (eds.). Elsevier Publishing Company, New York.

Lockhart, B. E. L., and Autrey, L. J. C. 1988. Occurrence in sugarcane of a bacilliform virus related serologically to banana streak virus. Plant Dis. 72:230-233.

Lockhart, B. E. L., Autrey, L. J. C., and Comstock, J. C. 1992. Partial purification and serology of sugarcane mild mosaic virus, a mealybug-transmitted closterolike virus. Phytopathology 82:691-695.

Magarey, R. C. 1994. Effect of Pachymetra root rot on sugarcane yield. Plant Dis. 475-477.

Magarey, R. C., and Mewing, C. M. 1994. Effect of sugarcane cultivars and location on inoculum density of *Pachymetra chaunorhiza* in Queensland. Plant Dis. 78:1193-1196.

Martin, J. P. 1961. Brown stripe. *In* Sugar Cane Diseases of the World, Vol. I. J. P. Martin, E. V. Abbott, and C. G. Hughes (eds.). Elsevier Publishing Company, New York.

Martin, J. P. 1964. Iliau. *In* Sugar Cane Diseases of the World, Vol. II. C. G. Hughes, E. V. Abbott, and E. A. Wismer (eds.). Elsevier Publishing Company, New York.

Martin, J. P., and Robinson, P. E. 1961. Leaf scald. *In* Sugar Cane Diseases of the World, Vol. I. J. P. Martin, E. V. Abbott, and C. G. Hughes (eds.). Elsevier Publishing Company, New York.

Martin, J. P., and Wismer, C. A. 1961. Red stripe. *In* Sugar Cane Diseases of the World, Vol. I. J. P. Martin, E. V. Abbott, and C. G. Hughes (eds.). Elsevier Publishing Company, New York.

Matsumoto, T., and Abbott, E. V. 1964. Red rot of the leaf sheath. *In* Sugar Cane Diseases of the World, Vol. II. C. G. Hughes, E. V. Abbott, and C. A. Wismer (eds.). Elsevier

Publishing Company, New York.

Purdy, L. H., Liu, L. J., and Dean, J. L. 1983. Sugarcane rust, a newly important disease. Plant Dis. 67:1292-1296.

Qhobela, M., and Claflin, L. E. 1989. *Xanthomonas officinarum* sp. nov., a second organism involved in sugarcane gumming disease. (Abstr.). Phytopathology 79:1191.

Reichel, H., Belalcazar, S., Munera, G., Arevalo, E., and Narvaez, J. 1997. First report of banana streak virus infecting sugarcane and arrowroot in Colombia. Plant Dis. 81:552.

Ricaud, C., Egan, B. T., Gillaspie Jr., A. G., and Hughes, C. G. 1989. Diseases of Sugarcane: Major Diseases. Elsevier Science Publishers. New York, NY. 399 pp.

Shaw, D. E. 1964. Veneer blotch. *In* Sugar Cane Diseases of the World, Vol. II. C. G. Hughes, E. V. Abbott, and C. A. Wismer (eds.). Elsevier Publishing Company, New York.

Sreeramulu, T., and Vittal, B. P. R. 1970. Incidence and spread of red rot lesions on midribs of sugarcane leaves. Plant Dis. Rptr. 54:226-231.

Steib, R. J. 1961. Phytophthora seed piece rot. *In* Sugar Cane Diseases of the World, Vol. I. J. P. Martin, E. V. Abbott, and C. G. Hughes (eds.). Elsevier Publishing Company, New York.

Steindl, D. R. L. 1964. Bacterial mottle. *In* Sugar Cane Diseases of the World, Vol. II. C. G. Hughes, E. V. Abbott, and C. A. Wismer (eds.). Elsevier Publishing Company, New York.

Subramanium, L. S., and Chona, B. L. 1938. Notes on *Cephalosporium sacchari* Butler (causal organism of sugarcane wilt). Indian J. Agric. Sci. 8:189-190.

Taballa, H. A. 1969. Smut on true seedlings of sugarcane. Plant Dis. Rptr. 53:992-993.

Todd, E. H. 1960. Elsinoe disease of sugarcane in Florida. Plant Dis. Rptr. 44:153.

Todd, E. H. 1962. Target blotch of sugarcane in Florida. Plant Dis. Rptr. 46:486.

Tosic, M., Ford, R. E., Shukla, D. D., and Jilka, J. 1990. Differentiation of sugarcane, maize dwarf, johnsongrass, and sorghum mosaic viruses based on reactions of oat and some sorghum cultivars. Plant Dis. 74:549-552.

Van der Zwet, T., and Forbes, I. L. 1961. *Phytophthora megasperma,* the principal cause of seed-piece rot of sugarcane in Louisiana. Phytopathology 59:634-640.

Vega, J., Scagliusi, S. M. M., and Ulian, E. C. 1997. Sugarcane yellow leaf disease in Brazil: Evidence of association with a luteovirus. Plant Dis. 81:21-26.

Venkatasubbaiah, P., Kohmoto, K., Otani, H., Hamasaki, T., Nakajima, H., and Hokama, K. 1987. Two phytotoxins from *Stagonospora sacchari* causing leaf scorch of sugarcane. Ann. Phytopathol. Soc. Japan 53:335-344.

Victoria, J. I., Carrillo, P., and Cassalett, C. 1983. Chemical control of sugarcane smut (*Ustilago scitaminea* Sydow) by seedcane treatment. (Abstr.) Phytopathology 73:126.

Williams, J. R. 1969. Nematodes as pests of sugarcane. *In* Pests of Sugar Cane, Elsevier Publishing Company, New York.

Yen, W. Y. 1964. Leaf blight. *In* Sugar Cane Diseases of the World, Vol. II. C. G. Hughes, E. V. Abbott, and C. A. Wismer (eds.). Elsevier Publishing Company, New York.

Yen, W. Y., and Chi, C. C. 1964. Leaf blast. *In* Sugar Cane Diseases of the World, Vol. II. C. G. Hughes, E. V. Abbott, and C. A. Wismer (eds.). Elsevier Publishing Company, New York.

Yin, Z., and Hoy, J. W. 1997. Effect of stalk desiccation on sugarcane red rot. Plant Dis. 81:1247-1250.

Zummo, N. 1986. Red spot (*Helminthosporium rostratum*) of sweet sorghum and sugarcane, a new disease resembling anthracnose and red rot. Plant Dis. 70:800.

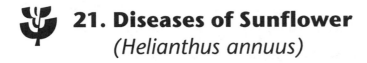

21. Diseases of Sunflower
(Helianthus annuus)

Diseases Caused by Bacteria

Apical Chlorosis

Cause. *Pseudomonas syringae* pv. *tagetis* is seedborne. Other Compositae also are hosts.

Distribution. The United States (Kansas, Minnesota, North Dakota, and Wisconsin).

Symptoms. Diseased plants are scattered throughout a field, occurring singly or in small groups within a row. A common symptom is a leaf chlorosis, but without lesions, that is most frequent and severe on diseased seedlings but may occur on sunflower at all growth stages. Diseased leaves, including veins, are pale yellow to white. Often only a portion of the initially affected leaf is chlorotic, but subsequently formed leaves, including veins, are uniformly chlorotic.

Diseased seedlings are frequently stunted and die; however, infection of older plants rarely results in stunting. Systemic chlorosis of seedlings lasts up to 8 weeks and involves eight to 10 leaves, but chlorosis on older prebloom plants is frequently limited to a few leaves. Chlorotic leaves do not recover. Symptoms apparently do not occur on subapical, fully expanded leaves on plants past the bud stage. Thus, subsequent leaves on diseased plants are normal in color.

Management. Grow the most resistant cultivar. Differences in resistance exist among sunflower cultivars.

Bacterial Leaf Spot *(Pseudomonas spp.)*

Cause. *Pseudomonas cichorii*, *P. syringae* pv. *aptata*, and *P. syringae* pv. *mellea*.

Distribution. Brazil, Canada, Japan, Spain, and the former Yugoslavia.

Symptoms. Generally, diseased leaves have large necrotic lesions surrounded by narrow yellow halos. Light and dark brown lesions occur on diseased stems. The vascular system is discolored and white exudate may be present on leaf axils.

Management. Not reported.

Bacterial Leaf Spot *(Pseudomonas syringae* pv. *helianthi)*

Bacterial leaf spot is also called angular bacterial leaf spot.

Cause. *Pseudomonas syringae* pv. *helianthi.* It is likely that disease is accentuated by frequent rainfall and humid conditions during the growing season.

Distribution. South Africa and likely other sunflower-growing areas.

Symptoms. Initial lesions on diseased leaves, stems and petioles are water-soaked spots. Spots become dark brown, necrotic (1–2 mm in diameter), and angular-shaped. Lesions sometimes coalesce and make diseased leaves dry and brittle.

Management. Not reported.

Bacterial Wilt

Cause. *Pseudomonas solanacearum.*

Distribution. Bulgaria, Mauritius, Tanzania, the United States (Puerto Rico), the former USSR, and Zimbabwe.

Symptoms. The initial symptom is an overall wilting of the diseased plant. A black discoloration at the base of the stem spreads upward (90 to 120 cm) until the diseased plant eventually falls over.

Management. Some cultivars have a higher percentage of tolerant or resistant plants than other cultivars.

Crown Gall

Cause. *Agrobacterium tumefaciens.* Bacteria enter the root system or stem through wounds near the soil line. Once inside the host plant, bacteria invade other plant tissue intercellularly and stimulate the surrounding cells to divide.

Distribution. Widely distributed. However, crown gall is rarely observed under natural conditions.

Symptoms. Initially, small overgrowths occur on diseased stems or roots, particularly near the soil line. This overgrowth later turns into galls that are spherical and white. As the galls enlarge and their surfaces become convoluted, the surfaces become dark brown to black. There is no definite size and shape of galls as they develop in an irregular and disorganized manner. Diseased plants with galls at the crown or main root remain weak and grow poorly. In addition to the formation of galls, diseased plants may remain stunted and have small chlorotic leaves and, in severe cases, a lengthwise splitting of the stem. The diseased plants may show epinasty, suppression of lateral buds, and production of adventitious roots.

Management. No practical disease management strategy is available. Most sunflower cultivars appear to have a high degree of resistance or tolerance.

Erwinia Stalk Rot and Head Rot

Cause. *Erwinia carotovora* pv. *carotovora* and *E. carotovora* pv. *atroseptica*. Bacterial stalk rot occurs after extended wet periods late in the growing season, which suggests that only stressed or senescing plants become infected. Infection occurs at the petiole axils, a site on the host plant that collects water and consequently serves as a favorable environment for bacterial "residence" and insect activity. Infection is aided by wounds caused by insects, hail, or mechanical damage. The sunflower budworm, *Suleima helianthana*, preferentially oviposits in this axil and larvae feed and exit from the same location. In South Africa, the bacteria enter sunflower tissue through lesions caused by *Alternaria helianthi* at the bases of leaf petioles. In the former Yugoslavia, disease has been reported to occur following damage by hailstones.

Distribution. Europe, South Africa, and the United States.

Symptoms. Diseased stems are black on the outside, with the blackening often centered around a petiole axil. An ink-black, watery breakdown of the pith results in a hollowing of stems. This breakdown is odorless unless stems are in an advanced state of decomposition. Plants often lodge under the weight of maturing heads. In some instances, heads will also develop soft rot symptoms.

Symptoms from the former Yugoslavia have been described as follows: Green-black spots occur on stalks and eventually coalesce to cover most of the stalk surface, causing stalks to soften and dry up. Flowers and seeds are also discolored.

Management. Do not rotate sunflower with potatoes.

Diseases Caused by Fungi

Achene Blemish Syndrome

Cause. *Alternaria alternata*, *Cladosporium* sp., and *Ulocladium atrum*. Western flower thrips, *Frankliniella occidentalis*, are associated with infection of the achenes. *Alternaria alternata* induces a significantly higher disease incidence than the other two fungi, but mixtures of all three fungi result in a higher disease incidence than each fungus alone.

Achenes were susceptible to infection only at the time of their development.

Distribution. Israel.

Symptoms. Symptoms become visible just before physiological maturity. Small, scattered lesions that vary in color, size, and shape occur on the surfaces of diseased achene shells. The brown, black, or gray lesions sometimes are surrounded by a dark halo. Lesions range in size from 0.5 to 2.0 mm and are round, oval, elongated, or irregular-shaped. Diseased achenes do not differ in size or weight from healthy achenes, nor is seed taste altered. However, diseased crops are judged by the sunflower industry to be of lower quality than healthy crops.

Management. Not reported.

Alternaria Leaf Blight, Stem Spot, and Head Rot

Alternaria leaf blight, stem spot, and head rot is also called Alternaria leaf and stem spot, and black spot.

Cause. *Alternaria alternata* (syn. *A. tenuis*), *A. chrysanthemi*, *A. helianthi* (syn. *Helminthosporium helianthi*), *A. helianthicola*, *A. helianthinficiens*, *A. leuconthemi*, *A. longissima*, *A. protenta*, *A. tenuissima*, and *A. zinniae*. Most *Alternaria* spp. are seedborne as conidia on seed coats and mycelia within seed coats. *Alternaria* spp. also overwinter as mycelia in infested stem residue on the soil surface and, to a lesser extent, within residue buried in soil. The inoculum from infested residue decreases later in the growing season. Alternaria leaf blight, stem spot, and head rot occurs under warm (25°–28°C), humid conditions that cause extended periods of leaf wetness. Plants are most susceptible to infection at the anthesis or seed-filling stages of plant growth.

Distribution. Generally, wherever sunflower is grown. *Alternaria helianthi* is recognized as the most prevalent and damaging species worldwide.

Symptoms. Early-sown seedlings are more severely diseased than later-sown seedlings. Dark brown, oval, necrotic spots occur on diseased heads, leaves, petals, petioles, and stems. Often leaves of younger plants (in the vegetative or budding stage of plant growth) have dark brown necrotic spots surrounded by a chlorotic halo. Other leaf symptoms are described as gray centers (several millimeters to 1.5 cm in diameter) with dark brown margins. Severely diseased plants are defoliated and frequently lodge. Symptoms caused by *A. alternata* are described as brown zonate spots on basal leaves that progress to the upper leaves. Spots eventually coalesce and diseased leaves dry up and defoliate. Disease severity varies from premature declining of lower leaves to extensive defoliation of the entire diseased plant.

Stem lesions begin as black flecks or streaks that later enlarge to cover large areas of the diseased stems. Stems are girdled and diseased plants are eventually killed. Head diameter, number of seeds per head, and oil content are reduced.

Symptoms specifically ascribed to *A. zinniae* are not noticeable until after flowering; however, seedling blight occurred when seeds were inoculated. Such seedlings either failed to emerge or had dark lesions on the hypocotyls and cotyledons that caused seedlings to damp off after emergence. Leaf spots, which are circular and uniformly dark colored, have a target-like appearance. Severe leaf disease results in defoliation. Brown superficial flecks or streaks occur on stems, petioles, and backs of heads and, when numerous, form large necrotic areas. Severe stem infection causes lodging.

Management
1. Apply a foliar fungicide.
2. Plow under infested residue.
3. Delay sowing.
4. Rotate for 1 year with a nonsusceptible crop.
5. Apply a seed-treatment fungicide.

Ascochyta Leaf Spot

Cause. *Ascochyta* sp. and *A. compositarum*.

Distribution. Japan and Kenya.

Symptoms. Dark spots occur on diseased leaves.

Management. Not reported.

Botrytis Head Rot

Botrytis head rot is also called gray mold and gray mold head rot.

Cause. *Botrytis cinerea* is seedborne and infects host plants during periods of abundant moisture. Botrytis head rot is common during wet weather in the autumn when the harvest is delayed.

Distribution. Europe.

Symptoms. Brown lesions on the backs of diseased sunflower heads later become a soft rot that is covered with gray fungal growth typical of *B. cinerea*. Eventually, only the vascular bundles remain from the decomposed heads. Seedling blight may occur, thereby reducing stands.

Management. Not reported.

Cephalosporium Wilt

Cause. *Cephalosporium acremonium* (syn. *Acremonium strictum*) is weakly pathogenic on young, healthy, growing sunflowers. Infection occurs through insect wounds and mechanical injuries from wind, hail, and cultivation. Plants become more susceptible as they mature and heads form.

Distribution. The United States (Indiana).

Symptoms. The described symptoms were caused by a combination of *C. acremonium* and *Erwinia carotovora* pv. *carotovora*. Brown lesions are present on the outside of diseased stalks. The inside of the stalks is hollow, causing them to bend and collapse under the weight of a maturing head. The head frequently cannot be harvested or decomposes, which reduces the yield. Plants injected with spores at the onset of senescence developed black necrotic areas on stems that were covered with mycelium.

Management. Not reported.

Cercospora Leaf Spot

Cause. *Cercospora bidentis*, *C. helianthi*, and *C. pachypus*.

Distribution. *Cercospora bidentis* is reported from India, *C. helianthi* from Venezuela, and *C. pachypus* from the Philippines and Uganda.

Symptoms. Tan spots develop on diseased leaves.

Management. Not reported.

Charcoal Rot

Cause. *Macrophomina phaseolina* (syn. *Sclerotium bataticola* and *Rhizoctonia bataticola*) is seedborne as pycnidia and sclerotia. Sclerotia develop on the inner surfaces of pericarps and seed coats. Charcoal rot is most severe at high soil temperatures (30°C and above) and low soil moisture. The sunflower stem weevil, *Cylindrocopturus adspersus*, carries *M. phaseolina* as weevils emerge from overwintered roots and stalks and transmit the fungus while feeding and ovipositioning in stalks. Other fungi also contribute to the discoloration of stalks infected by *M. phaseolina* and parasitized by larvae of the stem weevil.

Distribution. Wherever sunflower is grown in warmer climates. Charcoal rot rarely occurs in northern sunflower-growing areas.

Symptoms. Symptoms usually occur after flowering. Premature ripening of diseased stems occurs together with prematurely ripened, poorly filled, small, and distorted heads that have a zone of aborted flowers.

Diseased stems have a brown to black discoloration with or without typical gray areas on their surfaces. Stalk bases are gray and only vascular bundles or fibers remain, giving the insides of stems a shredded appearance. Portions of diseased lower stems are hollow and frequently there is a mixture of frass and fragments of pith tissue both in the lower stems and the upper taproots. Vascular fibers become covered with small, black sclerotia that resemble pepper or flecks of charcoal.

When diseased seed is sown, the seeds either do not germinate, or if germination occurs, there is a discoloration of roots, hypocotyls, and cotyledons. Pycnidia and sclerotia eventually develop on diseased parts.

Management
1. Rotate sunflower with a resistant crop, such as small grains.
2. Plow under residue.
3. Grow cultivars that have resistance.
4. Irrigate at flowering stage or at flowering and ripening stages.

Cladosporium Leaf Spot

Cause. *Cladosporium cladosporoides*.

Distribution. India.

Symptoms. The initial chlorotic spots on diseased leaves later become gray to olive green and are surrounded by a chlorotic halo. The underside of the diseased leaf has a moldy appearance due to the sporulation and growth of *C. cladosporoides*. Under high humidity conditions, the spots elongate along the veins, resulting in extensive chlorotic patches. The upper half of the leaf is usually more severely diseased than the lower half.

Management. Not reported.

Curvularia Leaf Spot

Cause. *Curvularia lunata*.

Distribution. Egypt.

Symptoms. Diseased leaves have tan lesions (up to 2 mm in diameter) with reddish or dark brown margins. Lesions may coalesce and form necrotic areas up to 1 cm long.

Management. Not reported.

Downy Mildew

Cause. *Plasmopara halstedii* and *P. helianthi* f. *helianthi* survive as oospores in the soil and as mycelium and zoosporangium in seed. Sowing infected seed rarely results in systemically infected seedlings, but infected seed is how *P. halstedii* is introduced into soil.

During wet soil conditions in the spring, oospores germinate and produce zoosporangia in which zoospores are formed and released, then "swim" to roots, where they encyst and germinate. Infection occurs primarily in the zone of radicle elongation or adjacent to it. The fungi then become systemic and grow upward inside the plant. Sporulation on belowground and aboveground plant parts occurs in the form of zoosporangia that are disseminated by wind and rain to other host plants, where secondary infection occurs through apical buds or leaves. Oospores are produced in diseased tissues and returned to the soil during harvest. Plants can be infected from germination until the flowering stage of plant growth, but systemic infection usually occurs after plant emergence since host plants become increasingly resistant with age.

At least eight different races of *P. halstedii* exist.

Distribution. Wherever sunflower is grown, except Australia.

Symptoms. Seedlings have light green or yellowish areas that spread out from the leaf midrib in the diseased leaf. During wet weather, a downy, whitish, fungal growth consisting of zoosporangia and mycelium occurs on the undersides of diseased leaves opposite the yellowed areas. These plants are systemically infected and most of them die.

As diseased plants continue to grow, leaves become wrinkled and distorted, and the entire plant is stunted. Diseased plants usually produce normal-sized heads that remain upright and contain empty seeds.

Management
1. Sow only clean, healthy seed.
2. Grow sunflower in the same soil once every 5 years. Rotation should include small grains.
3. Control weeds in sunflower fields that may serve as alternate hosts.
4. Treat seed with a systemic fungicide.

Drechslera Leaf Spot

Cause. *Drechslera hawaiiensis*. Older leaves are more susceptible to infection than are younger ones.

Distribution. India.

Symptoms. The small, brown spots that initially occur on diseased leaves gradually increase in size (0.1–10.0 mm in diameter), are round to irregular-shaped, have a pink-brown center that becomes gray, and are encircled by a chlorotic zone. Leaves become blighted during severe disease. Plant growth and yield are affected.

Management. Not reported.

Fusarium moniliforme Wilt

Cause. *Fusarium moniliforme* likely survives as mycelia in infested residue and by growth on collateral hosts.

Distribution. India.

Symptoms. Diseased leaves are chlorotic and drooping, followed by wilting and death of the plant. Pronounced stunting occurs when diseased plants are infected in the preflowering stage. Vascular elements of roots are brown.

Management. Not reported.

Fusarium oxysporum Wilt

Cause. *Fusarium oxysporum* survives as chlamydospores in infested residue and soil. Young plants up to the age of 6 weeks are highly susceptible to infection and die in a short time. Plants 10 weeks and older are not susceptible.

Distribution. India and the United States.

Symptoms. The lower leaves of diseased plants change from green to yellow, often with a distortion that affects only one side of the leaf. Diseased plants wilt, tend to fall over, and become desiccated. Older diseased plants have a tendency to recover.

Management. Not reported.

Fusarium Pith Canker

Fusarium pith canker is also called pith canker.

Cause. *Plectosphaerella cucumerina*, anamorph *Fusarium tabacinum* (syn. *Microdochium tabacinum*). *Plectosporium* has been described as the new genus for *F. tabacinum*. Optimum growth in culture is at temperatures of 25°–26°C. Disease is only observed when high relative humidity and low temperatures occur in the summer.

Distribution. Italy. However, *M. tabacinum* is a common fungus both in arable soil and on decaying plant material. It has been reported in Australia, other parts of Europe, New Zealand, and the United States.

Symptoms. Symptoms have been reported only on stems, where external necrotic streaks and a pale pinkish-red discoloration of the pith occur.

Management. Not reported.

Helminthosporium Leaf Blight

Cause. *Helminthosporium helianthi.*

Distribution. Argentina, India, Tanzania, Turkey, and Uganda.

Symptoms. Diseased leaves have blackish-brown zonate spots (10 to 30 mm in diameter) that have ashy centers. Spots are irregular-shaped.

Management. Not reported.

Myrothecium Leaf and Stem Spot

Cause. *Myrothecium roridum* and *M. verrucaria*. The fungi likely survive in infested residue and soil. Disease is favored by warm, moist weather.

Distribution. India.

Symptoms. Lesions on diseased leaves tend to be circular and light brown or tan. Lesions are sometimes surrounded by a dark brown halo.

Management. Not reported.

Phialophora Yellows

Phialophora yellows is also called wilt.

Cause. *Phialophora asteris* f. sp. *helianthi* is soilborne and overwinters as saprophytic mycelia in infested residue. The severity of Phialophora yellows increases during periods of moist weather or excess soil moisture.

Distribution. Canada (Manitoba) and Italy.

Symptoms. Phialophora yellows is an inconspicuous disease. Symptoms may be confused with those caused by nitrogen deficiency, other mineral deficiencies, or injury from excess water.

    The first symptoms are obvious as diseased sunflowers approach flowering. First, the lower leaves turn a dull light green, then large areas of leaves become dull yellow, starting at the apex and margins and extending inward. Leaf margins and the angular-shaped patches (5–10 mm long) of yellowed leaf interveinal leaf tissues become necrotic. Symptoms move progressively up diseased plants. The bottom leaves dry and wither, whereas upper leaves remain green. The vascular tissue inside the stalk becomes brown. Diseased plants are stunted, and heads remain small and sterile.

    Microscopic observations reveal hyaline hyphae in the vascular tissues of stems.

Management. Grow cultivars that have some resistance.

Phoma Black Stem

This disease is also called black stem, Phoma black stem, and Phoma girdling. In North Dakota, Phoma girdling is one aspect of premature death.

Cause. *Phoma macdonaldii*, teleomorph *Leptosphaeria lindquistii*, overwinters as perithecia, pycnidia, and mycelia in infested stalk residue. Primary inoculum is ascospores and conidia produced during moist weather. Conidia are disseminated by splashing rain and wind, and ascospores by wind to host plants. Disease occurs under wet conditions during and after plant flowering.

Distribution. Argentina, Canada, and the United States.

Symptoms. Symptoms are most noticeable after flowering. Lesions start at the bases of leaf petioles and spread along stems. Numerous lesions coalesce and form large, black patches on diseased stems, and dark, irregular-shaped lesions occur on leaves and flowers. Leaves infected early are killed, and the stems weaken and lodge. Pycnidia, which resemble tiny, black bumps, are produced in mature lesions. Diseased plants produce small heads with little seed.

Symptoms of premature death are difficult to distinguish from normal senescence, especially in early-maturing cultivars. Premature death most often occurs in circular areas within a field, although scattered individual plants also may die.

Management
1. Rotate sunflower with nonhost plants.
2. Differences in susceptibility exist among sunflower lines.

Phoma Leaf Blight

Cause. *Phoma exigua*.

Distribution. India.

Symptoms. Initially, irregular-shaped yellowish patches or spots develop on diseased leaves. Later, the spots become reddish brown to dark black and coalesce, forming irregular-shaped patches. Symptoms, such as a black crust that is rough to the touch due to the numerous submerged pycnidial bodies, are more pronounced on upper leaf surfaces. Diseased leaves become leathery and brittle and have blighted margins.

Management. Not reported.

Phomopsis Brown Stem Canker

Cause. *Phomopsis helianthi*, teleomorph *Diaporthe helianthi*. *Diaporthe helianthi* overwinters as perithecia on infested stem residue on the soil surface. Ascospores produced during wet, cool weather are the principal source of infection throughout the growing season. Hyphae from germinated ascospores infect leaves, then become systemic and grow into the xylem tissue and, finally, through the stem cortex. There may be more than one species or biotype that attack sunflower.

Distribution. The United States and the former Yugoslavia.

Symptoms. Leaf symptoms are necrotic areas surrounded by chlorotic tissue that starts at the apical ends or edges of diseased leaves. Leaf veins and petioles darken into brown to black cankers that later become ashy gray around the diseased petiole bases. Leaf symptoms reported from Ohio were characterized by interveinal bronzing.

Spots on stalks are initially light brown and about 10 mm wide. Eventually, these spots become darker in color, have a wet appearance, and increase in length (15–20 cm) and in width, girdling the diseased stalks. At the death of epidermal tissue, the spots again become light brown, and they have dark brown borders of different widths. Several internodes may become discolored as disease progresses into the interiors of stalks and destroys pith tissue and causes hollow areas. Pycnidia, which resemble tiny, black bumps, are abundant in diseased tissue. Diseased plants ripen prematurely and the percentage of oil in seeds decreases.

Management. Not reported.

Phymatotrichum Root Rot

Phymatotrichum root rot is also called cotton root rot.

Cause. *Phymatotrichopsis omnivora* (syn. *Phymatotrichum omnivorum*) survives for several years in soil as sclerotia and hyphal strands. Early in the growing season, when soil temperatures have risen sufficiently, sclerotia germinate and form new hyphal strands that make contact with the developing root systems of host plants. Root contact is not necessary for plant-to-plant spread of *P. omnivora* because mycelial growth through soil occurs in advance of diseased roots. Plant-to-plant spread within rows is more common than spread across rows. Phymatotrichum root rot is most prevalent in alkaline soils.

Distribution. Northern Mexico and the southwestern United States.

Symptoms. Diseased plants wilt rapidly and die. The root cortex is rotted and brown strands of fungus mycelium can be seen growing on the roots.

Management
1. Plowing under a green manure crop, such as sweet clover, lessens disease severity in succeeding years.
2. Rotate sunflower with plants in the grass family.

Phytophthora Stem Rot

Cause. *Phytophthora drechsleri*. *Phytophthora cryptogea* has also been identified as a causal fungus; however, some researchers have suggested *P. drechsleri* may be a synonym of *P. cryptogea*. *Phytophthora drechsleri* has been identified as the cause of Phytophthora black stem rot of sunflower in Iran. However, Phytophthora black stem rot is probably the same disease as Phytophthora stem rot.

The fungus overwinters as oospores in infested residue and probably in soil. Oospores germinate and form sporangia in which zoospores are produced. The zoospores are liberated from the sporangia and "swim" to plant hosts, where they encyst and germinate. Secondary inoculum is by sporangia produced on diseased tissue and disseminated by wind and splashing rain to other host plants. Eventually, oospores are formed in diseased tissue. Phytophthora stem rot is most severe under wet soil conditions, when stems contact water or wet soil.

Distribution. Iran and the United States (California).

Symptoms. A dark brown to black discoloration occurs that partially or totally girdles diseased stems, primarily on the area of the stem that is relatively near the soil surface. Decay spreads to internal stem tissues, which become weakened and cause plants to fall over.

Management. No disease management strategy is effective at present. However, some sunflower hybrids in Iran are reported to be more resistant than others.

Powdery Mildew

Cause. *Erysiphe cichoracearum,* anamorph *Oidium asteris-punicei; E. cichoracearum* var. *latispora,* anamorph *Oidium latisporum; Leveillula compositarum* f. *helianthi; L. taurica; Oidiopsis sicula;* and *Sphaerotheca fuliginea.* The fungi survive as cleistothecia on infested residue. During the summer, ascospores are formed in cleistothecia and are disseminated by wind to leaves of host plants, where the lower leaves become more heavily infected than the upper leaves. Conidia are produced on the surfaces of diseased leaves and serve as secondary inoculum by being disseminated by wind to healthy tissue. Cleistothecia are formed in mycelium growing on the surfaces of diseased leaves.

Distribution. Wherever sunflower is grown.

Symptoms. Symptoms occur after full bloom and are more obvious on the lower leaves. Initially, white powdery areas consisting of mycelium and conidia appear primarily on leaves, but all aboveground plant parts may be diseased. Later, the whitish fungal growth becomes gray, and cleistothecia appear as small black objects within the surface fungal growth. Severely diseased leaves become yellow and dry up.

Management. Not necessary, since powdery mildew occurs too late in the growing season to affect yield.

Rhizopus Head Rot

Cause. *Rhizopus oryzae* (syn. *R. arrhizus* and *R. nodosus*), *R. microsporus*, and *R. stolonifer* (syn. *R. nigricans*) are cosmopolitan in the soil as spores and occur as saprophytic mycelia in decaying plant residue.

Sporangia, which require free water to germinate, are readily disseminated by wind and rain and may be transferred by insects and birds. Birds may disseminate the fungi directly from a rotted to a healthy head by inoculum that adheres to their claws or beaks. Heads damaged by mechanical implements, such as irrigation equipment, and by birds, hail, and insects (particularly larvae of the sunflower moth, *Homoeosoma electellum*) are most subject to infection because infective mycelium can only penetrate plant hosts through wounds. In Texas, larvae of the sunflower moth are directly correlated to the incidence of Rhizopus head rot.

The rate of head rot is dependent primarily on the temperature. Diseased heads may become completely rotted within 3 to 7 days at temperatures of 20° to 30°C.

Susceptibility of sunflower to infection by *R. oryzae* is dependent upon the age of the susceptible host plant at infection. At the bud stage, sunflower heads are basically resistant. However, sunflower becomes susceptible at initiation of anthesis, after which the heads become highly susceptible to infection as they age. Wet weather following flowering favors disease development.

Distribution. Wherever sunflower is grown. However, Rhizopus head rot is predominant in countries where flowering and seed maturation occur at high temperatures.

Symptoms. Initial symptoms of Rhizopus head rot in sunflower are small, water-soaked spots on the back of the diseased head. The spots gradually enlarge, and as mycelium invades more of the parenchyma tissue, the receptacle becomes brown, soft, and pulpy. During wet weather, coarse, thread-like strands or woolly masses of whitish fungus mycelium may be seen in the hollow part of the flower receptacle. Later, the woolly masses of whitish mycelium are interspersed with numerous black spore-producing sporangia the size of pinheads. The mycelium and sporangia are evident externally and are especially conspicuous in the vicinity of diseased achenes. At death, sunflower heads shred.

Management. Cultivars with upright heads are more susceptible to infection than cultivars with nodding heads.

Rust

Cause. *Puccinia helianthi*, *P. xanthi*, and *Uromyces junci*. *Puccinia helianthi* is an autoecious, long-cycled rust that produces pycnia, aecia, uredinia, and telia on sunflower. Survival in northern areas is by teliospores that overwinter on infested residue. In the spring and early summer, teliospores germinate and produce basidiospores at temperatures of 6° to 28°C. Basidiospores are disseminated by wind to host plants, where they germinate and infect leaves. In 10 to 12 days, pycnia develop at the point of basidiospore infection. Each pycnium produces pycniospores and receptive hyphae. Pycniospores are carried to compatible receptive hyphae of other pycnia by insects or splashing water. The nuclei of pycniospores fuse with nuclei of receptive hyphae, migrate down the hyphae, and form aeciospores in aecia on the opposite side of the leaf from the pycnia. Aeciospores are disseminated by wind to leaves, where they germinate. Uredinia, in which urediniospores are produced, are formed in the aeciospore infections. Urediniospores are windborne and serve as secondary inoculum, or the repeating stage, and produce new uredinia at each point of infection. Toward host plant maturity or during conditions unfavorable for disease, telia develop in uredinia lesions.

Urediniospore germination occurs at temperatures of 4° to 20°C and requires 6 to 10 hours of leaf wetness. The temperatures for minimum, optimum, and maximum infection are 4°, 10° to 24°, and 30°C, respectively. Temperatures for optimum secondary sporulation are 20° to 35°C.

Generally, warm, humid weather favors disease development. Different physiologic races of *P. helianthi* exist in nature.

Distribution. Wherever sunflower is grown.

Symptoms. Pycnia occur on diseased upper surfaces of leaves as small clusters of pale yellow or orange spots. Aecia appear on the lower surfaces of leaves, opposite the pycnia, as yellow or orange spots. Uredinia first occur in midsummer as dark brown, powdery spots on both surfaces of the lower leaves and sometimes on stems. Severely diseased leaves die and dry up. Telia occur as diseased plants mature and appear as dark brown to black spots.

Management

1. Grow resistant cultivars.
2. Rotate sunflower, and locate fields as far as possible from fields where sunflower was grown the previous year.
3. Sow sunflower early in the season.
4. Destroy volunteer and wild sunflowers in the vicinity of commercial fields.
5. Apply labeled foliar fungicides. The suggested time to apply fungicides is when rust reaches 5% severity on the lower leaves during the R1 stage, when the flower buds are just visible, to the R6 stage, the end of flowering when the ray flowers are wilting.

Sclerotinia Basal Stalk Rot and Wilt

Cause. *Sclerotinia minor* survives in soil and seed lots as sclerotia. During moist soil conditions, sclerotia germinate to produce hyphae that infect the lower stems of plant hosts. Sclerotia are eventually produced on and in the diseased areas of the stem.

Distribution. Australia.

Symptoms. A brownish canker forms on diseased stems at the soil line. Eventually, white mycelium covers the diseased areas and black to dark brown sclerotia form within the mycelium. Diseased plants suddenly wilt.

Management
1. Sow clean, healthy seed that is free of sclerotia.
2. Control weeds in sunflower fields.
3. Plow infested residue deep into the soil.

Sclerotinia Basal Stalk Rot and Wilt, and Midstalk Rot

This disease is also called Sclerotinia white mold.

Cause. *Sclerotinia sclerotiorum* (syn. *S. libertiana* and *Whetzelinia sclerotiorum*) overwinters primarily as sclerotia in soil and intermixed with seeds in storage. The sclerotia are then sown, along with the seeds, the following spring. Survival is also as mycelia in infested stalk residue and rotted or healthy-appearing seed. However, survival of mycelia in infested stems is too low to be a significant source of pathogenic inoculum. Sclerotia are disseminated by any means that moves soil.

Sclerotia on or near the soil surface germinate two ways: (1) Myceliogenically, or directly, by producing mycelia that infect stalks and roots. The taproot-hypocotyl axis is the primary site of infection; therefore, disease incidence is increased when seed is planted adjacent to sclerotia. (2) Carpogenically, or indirectly, by producing apothecia on which ascospores are borne in a layer of asci. The ascospores are disseminated by wind to host plants and are the major cause of midstalk infections.

The initial infected plants serve as primary infection foci from which *S. sclerotiorum* spreads through root contact between adjacent plants when mycelia grow from diseased to healthy plants.

Although infection may happen at any stage of plant growth, there are two times during plant growth when most infections occur. The first time is during seed germination and seedling establishment; it is attributed to myceliogenic germination of incompletely melanized or injured sclerotia. The second cycle, from plant bud formation through seed development, is possibly attributable to carpogenic germination of black sclerotia induced by exogenous nutrients. Disease is most severe when frequent rains and high humidity occur after flowering.

The hyperparasites *Coniothyrium minitans* and *Talaromyces flavus*, anamorph *Penicillium vermiculatum*, parasitize mycelia and sclerotia on root surfaces, thus reducing inoculum in soil.

Distribution. Wherever sunflower is grown.

Symptoms. Wilt symptoms on single diseased plants or groups of diseased plants appear at flowering time after a period of warm weather. Upper leaves droop and, in 2 to 3 days, all diseased leaves dry out and plants die. Young dead plants turn black, but older plants remain tan.

Wilted plants have soft, water-soaked, gray to brown cankers that encircle stems 7 to 25 cm above the soil line. Large, black, irregular-shaped sclerotia are produced in dense white mycelium on canker surfaces or in diseased stalks. Stems of diseased plants become shredded and break over at the soil line.

A similar symptom called middle stalk occurs when cankers extend a maximum of 125 cm above the soil line. This happens when the surface soil is wet for 10 to 14 days, causing sclerotia to germinate.

Management
1. Rotate sunflower with cereals, but avoid rotations with canola, common bean, mustard, and safflower since these plants are also hosts for *S. sclerotiorum*. Susceptible crops should not be sown for 4 to 6 years after a severe outbreak of disease.
2. Sow healthy seed that is free of sclerotia.
3. Control weeds in sunflower fields that serve as alternate hosts.
4. Deep-plow infested residue. Sclerotia buried deeply in soil cannot germinate. However, sclerotia in surrounding fields must also be buried for this to be effective since airborne inoculum may initiate infections.
5. Plant spacings of 36 cm or greater at 26,000–49,000 plants/ha will maximize yield under disease conditions; however, some researchers reported that plant spacing did not make a difference.
6. Leaving stalks undisturbed through the winter in northern areas reduces the survival of sclerotia in stalks, possibly because of mycoparasites. It is suggested that tillage be done in spring.
7. Apply a fungicide seed treatment. Treatment of sunflower seed with benomyl, iprodione, procymidone, or vinclozolin virtually eliminated seedborne *S. sclerotiorum*.
8. Levels of resistance occur in genotypes.

Sclerotinia Head Rot

This disease is also called Sclerotinia white mold.

Cause. *Sclerotinia sclerotiorum* (syn. *S. libertiana* and *Whetzelinia sclerotiorum*) overwinters primarily as sclerotia in soil or among seeds. Survival is also as mycelia in infested stalk residue and rotted or healthy-appearing seed.

However, survival of mycelia in infested stems is too low to be a significant source of pathogenic inoculum. Sclerotia are disseminated by any means that moves soil.

Sclerotia on or near the soil surface germinate two ways: (1) Myceliogenically, or directly, by producing mycelia that infect stalks and roots. The taproot-hypocotyl axis is the primary site of infection; therefore, disease incidence is increased when seed is sown adjacent to sclerotia. (2) Carpogenically, or indirectly, by producing apothecia on which ascospores are borne in a layer of asci. The ascospores are disseminated by wind to host plants and are the major cause of head infections. Circumstantial evidence suggests that senescent florets provide a food base for ascospore infections.

Distribution. Generally distributed wherever sunflower is grown.

Symptoms. The soft, water-soaked spots that initially occur on diseased heads later dry out and give diseased plant tissues a pinkish color. A white, cotton-like mold grows over diseased areas during humid conditions. Diseased heads eventually become partially or entirely rotted, leaving only vascular bundles and fibers, which cause heads to have a shredded or brush-like appearance. Seed hulls are discolored and scurfy. Large black sclerotia develop below the seed layer in the head or around seeds. The presence of sclerotia in seed lots gives high acidity values and is the main reason for poor oil quality.

Management
1. Rotate sunflower with cereals, but avoid rotations with canola, common bean, mustard, and safflower since these plants are also hosts for *S. sclerotiorum*. Susceptible crops should not be sown for 4 to 6 years after a severe outbreak of disease.
2. Sow healthy seed that is free of sclerotia.
3. Control weeds in sunflower fields that serve as alternate hosts.
4. Deep-plow infested residue. Sclerotia buried deeply in soil cannot germinate. However, sclerotia in surrounding fields must also be buried for this to be effective since airborne inoculum may initiate infections.
5. Plant spacings of 36 cm or greater at 26,000–49,000 plants/ha maximize yield under disease conditions; however, some researchers found plant spacing did not make a difference.
6. Leaving stalks undisturbed through winter in northern areas reduces the survival of sclerotia in stalks, possibly because of mycoparasites. It is suggested that tillage be done in spring.
7. Apply a fungicide seed treatment. Treatment of sunflower seed with benomyl, iprodione, procymidone, or vinclozolin virtually eliminated seedborne *S. sclerotiorum*.
8. Levels of resistance occur in genotypes.

Sclerotium Basal Stalk and Root Rot

Sclerotium basal stalk and root rot is also called southern blight.

Cause. *Sclerotium rolfsii*, teleomorph *Athelia rolfsii*, overwinters as sclerotia in soil and grows as saprophytic mycelia in soil residue. Sclerotia germinate in dry soil during periods of high soil temperatures and form mycelia that infect seed and host plants of all ages. Cool soil temperatures kill sclerotia and mycelia. *Sclerotium rolfsii* occurs most commonly in neutral to acidic sandy soils. Delayed sowing due to wet weather may favor disease development.

Distribution. In the warm sunflower-growing areas throughout the world.

Symptoms. Discolored areas on diseased crowns and stems are sometimes accompanied by whitish, fan-shaped mycelia. Diseased plants normally wilt, die, and eventually dry out. Sclerotia are frequently produced at the bases of diseased plants.

Management. Do not rotate sunflower with other susceptible crops.

Septoria Leaf Spot

Cause. *Septoria helianthi*. Little is known of the survival and dissemination of *S. helianthi*, but it is presumed that *S. helianthi* overwinters as pycnidia on infested residue. Conidia are produced in pycnidia and are disseminated by wind and splashing rain to host tissue, which is similar to the survival and dissemination of other *Septoria* species. Humid weather is conducive for disease development, which normally does not occur during dry weather.

Distribution. Canada, China, the United States, and possibly other sunflower-growing areas in the world.

Symptoms. Septoria leaf spot occurs on plants of any age, but symptoms usually begin on the lower leaves after flowering. The water-soaked, circular (6–12 mm in diameter) spots become gray with a dark margin. Other descriptions depict spots as brown with a narrow yellow border that gradually fuses with the surrounding green tissues. Spots coalesce and produce irregular-shaped dead areas on leaves. Tiny, indistinct, dark brown pycnidia form in lesions.

When temperatures are moderately high and moisture is abundant, leaves are dropped progressively from the bottom of the plant upward until only a few leaves are left on a diseased plant. Diseased plants are normally not seriously damaged.

Management. Not reported.

Verticillium Wilt

Verticillium wilt is also called leaf mottle.

Cause. *Verticillium albo-atrum* and *V. dahliae*. Although the causal fungus was originally identified as *V. albo-atrum,* some researchers think *Verticillium dahliae* is the only causal organism. *Verticillium dahliae* survives as microsclerotia in soil, on external and internal pericarp tissue, on testas, and as mycelia in seed. Most infection presumably occurs from microsclerotia that germinate and infect host plant roots. Mycelium grows into the xylem tissue and produces microconidia that spread upward in the water-conducting cells. Eventually, the fungus grows through xylem tissue to all parts of the plant and into the seed. At plant death, microsclerotia are produced from mycelia and returned to the soil at harvest.

Distribution. Canada, China, Russia, South America, Ukraine, and the United States.

Symptoms. Symptoms on single plants or groups of diseased plants first occur on the lower leaves and gradually progress up the plant at flowering time. Tissue between leaf veins becomes chlorotic, then brown, which gives diseased leaves a mottled appearance. Black areas occur on diseased stems, particularly near the soil line. In cross section, the vascular system of diseased stems is brown to black. Severely diseased plants are stunted and either mature prematurely or die before flowering.

Management
1. Grow resistant cultivars.
2. Grow sunflower in the same field once every 5 years.
3. Sow disease-free seed treated with a fungicide seed protectant.

White Rust

White rust is also called gray stem spot and white blister.

Cause. *Albugo tragopogonis* (syn. *Cystopus tragopogonis*) overwinters as oospores in infested soil residue and, in milder climates, as mycelia or sporangia on weeds and volunteer sunflowers. Little is known of the source and dissemination of primary inoculum, but presumably oospores are disseminated to leaves, germinate, and form zoospores. Sporangia may also be produced on overwintered diseased plants and disseminated by wind to host plants, where they germinate and yield zoospores. Zoospores penetrate stomata, encyst, and eventually germinate and form mycelia.

Secondary inoculum is by sporangia that are formed on diseased lower leaf surfaces and are windborne to host plants, where they form zoospores that germinate and infect the plant. Oospores eventually form in diseased tissue. Disease development is favored by temperatures of 10° to 20°C and high rainfall.

Distribution. Argentina, Australia, Russia, Ukraine, and Uruguay. White rust likely occurs in other countries also, but it is not considered a serious disease.

Symptoms. Creamy-white blister-like pustules appear most commonly on the undersides of leaves located near the bottom of diseased plants. Tissue on the upper leaf surfaces, opposite pustules, is raised and yellow-green. No scorching or shriveling of diseased leaves occurs.

Gray localized lesions on diseased stems are caused by oospores that have formed intracellularly in the stem epidermis. Stem lodging occurs following the collapse of tissues of the cambial region and vascular rays.

Management. Not necessary.

Yellow Rust

Yellow rust is also called Coleosporium rust.

Cause. *Coleosporium helianthi* and *C. pacificum* (syn. *C. madiae*). *Coleosporium pacificum* is a heteroecious, long-cycled rust fungus that has Monterey pine, *Pinus radiata*, as an alternate host. The uredinial and telial stages occur on sunflower; the pycnial and aecial stages occur on Monterey pine.

Distribution. The West Coast of the United States.

Symptoms. Uredinia occur as golden yellow pustules on diseased sunflower leaves.

Management. Not reported.

Diseases Caused by Nematodes

Dagger

Cause. *Xiphinema americanum* is an ectoparasite.

Distribution. Not known for certain, but probably widespread.

Symptoms. An aboveground symptom is a slight to moderate stunting. Belowground symptoms are a reduction in the feeder roots accompanied by necrosis of a portion of the root system.

Management. Not reported.

Lesion

Cause. *Pratylenchus hexincisus* is endoparasitic; its larvae enter roots and attack the root cortex.

Distribution. Wherever sunflower is grown. However, sunflower normally is not severely affected.

Symptoms. Diseased plants may be yellowed and stunted. Initially, roots have small dark lesions. Other soilborne microorganisms may enter the root through wounds caused by *Pratylenchus* spp., which results in root rot and a browning of the root system.

Management. Not necessary.

Pin

Cause. *Paratylenchus projectus* has been reported to be a mild parasite on sunflower; therefore, large populations appear necessary for substantial plant growth reduction. Populations of pin nematodes in greenhouse tests developed best at temperatures of 20° to 25°C.

Distribution. Korea and the United States.

Symptoms. Seed yields were reduced in greenhouse tests.

Management. Early sowing may reduce damage.

Reniform

Cause. *Rotylenchulus reniformis* is a nematode in which only females invade roots after feeding on epidermal tissues. *Rotylenchulus reniformis* is semi-endoparasitic because the posterior portion of the female remains outside the root and swells to a kidney shape. Eggs are laid outside the root. Optimum temperature for root invasion and nematode development is 29°C.

Distribution. Africa and the southeastern United States; however, sunflower normally is not severely affected.

Symptoms. Diseased plants are stunted and chlorotic. Small galls or root swellings may sometimes occur on damaged roots.

Management. Not necessary.

Root Knot

Cause. *Meloidogyne arenaria*, *M. incognita*, and *M. javanica*. These nematodes do not survive if the temperature averages below 3°C during the coldest month.

Distribution. Not known but probably widespread.

Symptoms. Diseased plants are reduced in stand and vigor, and wilt during the day but often recover their turgidity at night. Diseased plants are easily pulled from the soil due to the destruction of the root system. Numerous galls are present on diseased roots.

Management. Not reported.

Spiral

Cause. *Helicotylenchus* sp. is an ectoparasite.

Distribution. Widespread.

Symptoms. Symptoms on sunflower are relatively unknown. Under low to midlevel populations, a few root cells are destroyed, but roots are severely affected when infected by high populations of nematodes.

Management. Not reported.

Stunt

Cause. *Tylenchorhynchus nudus* and *Quinisulcius acutus* are ectoparasites.

Distribution. Not known, but presumably these nematodes are widespread on sunflower.

Symptoms. Aboveground symptoms are a moderate plant stunting and chlorosis. The belowground symptom is a reduction in root growth.

Management. Not reported.

Disease Caused by Phytoplasmas

Aster Yellows

Cause. Aster yellows phytoplasma survives in several dicotyledonous plants and in leafhoppers. Transmission is primarily by the aster leafhopper, *Macrosteles fascifrons*.

Distribution. Wherever sunflower is grown.

Symptoms. Either the entire head or a pie-shaped sector of the diseased head displays symptoms. Flowers remain green instead of normally turning yellow. Structures resembling small leaves, but larger than normal flowers, replace the floral parts in diseased flower heads. Eventually, diseased portions turn brown and die; the brown discoloration extends downward as a narrow stripe along the stem. Sometimes head symptoms are not evident, but a black slimy rot occurs on stalks below the diseased heads.

Diseased plants may be stunted or break over. Other diseased plants may set seed only on the normal portion of their heads.

Management. Not reported.

Diseases Caused by Viruses

Sunflower Mosaic

Cause. Sunflower mosaic is caused by three different viruses. Cucumber mosaic virus (CMV) is in the cucumovirus group. CMV is mechanically transmitted and has a wide host range. Sunflower virus and tobacco mosaic virus are the other two causal viruses.

Distribution. The United States (Maryland).

Symptoms. Symptoms of mosaic and chlorotic rings are most severe on diseased leaves that are younger than 2 months. Diseased plants are stunted and sometimes have discrete, narrow, light brown streaks on petioles and stems. Malformed heads that have shriveled seed are produced.

Management. Not reported.

Yellow Leaf Spot Mosaic

Cause. A virus that resembles tomato spotted wilt virus.

Distribution. Ukraine.

Symptoms. A yellow spotted mosaic is produced on diseased leaves.

Management. Not reported.

Disease Caused by Nonparasitic Factors

Bract Necrosis

Cause. Nonparasitic. Disease is caused by temperatures of 40°C or higher.

Distribution. The United States (Texas).

Symptoms. Disk flowers and bracts have a brown discoloration that becomes black after a rain. When bract necrosis occurs during the bud stage, buds remain unopened. Some injured buds may open but produce few or no disk flowers and little pollen.

Management. Not reported.

Selected References

Allen, S. J., Brown, J. F., and Kochman, J. K. 1983. Effects of temperatures, dew period and light on the growth and development of *Alternaria helianthi*. Phytopathology 73:893-896.

Allen, S. J., Brown, J. F., and Kochman, J. K. 1983. The effects of leaf age, host growth stage, leaf injury and pollen on infection of sunflowers by *Alternaria helianthi*. Phytopathology 73:896-898.

Arsenijevic, J. 1970. A bacterial soft rot of sunflower (*Helianthus annuus* L.). Acta Phytopathol. Acad. Sci. Hungary 5:317-326.

Banihashemi, Z. 1975. Phytophthora black stem rot of sunflower. Plant Dis. Rptr. 59:721-724.

Bernard, E. C., and Keyserling, M. L. 1985. Reproduction of root-knot, lesion, spiral, and soybean cyst nematodes on sunflowers. Plant Dis. 69:103-105.

Blanco-Lopez, M. A., and Jimenez-Diaz, R. M. 1983. Effect of irrigation on susceptibility of sunflower to *Macrophomina phaseoli*. Plant Dis. 67:1214-1217.

Campos, V. P., Huang, S. P., Tanaka, M. A. S., and Rezende, A. M. 1982. Ocorrencia de *Meloidogyne incognita* em cultural de girassol no Est. Minas Gerais, Brasil. Fitopathologia Brasileira, pp. 309-310 (in Portuguese).

Carson, M. L. 1985. Epidemiology and yield losses associated with Alternaria blight of sunflowers. Phytopathology 75:1151-1156.

Carson, M. L. 1987. Effects of two foliar pathogens on seed yield of sunflower. Plant Dis. 71:549-551.

De Waele, D., and Bolton, C. 1988. *Trophurus pakendorfi* n. sp. from sunflower in South Africa (Nemata: Telotylenchinae). Phytophylactica 20:153-156.

Donald, P. A., Bugbee, W. M., and Venette, J. R. 1986. First report of *Leptosphaeria lindquistii* (sexual stage of *Phoma macdonaldii*) on sunflower in North Dakota and Minnesota. Plant Dis. 70:352.

Donald, P. A., Miller, J. F., and Venette, J. R. 1986. Reaction of sunflower lines to *Phoma macdonaldii*. (Abstr.). Phytopathology 76:956.

Donald, P. A., Venette, J. R., and Gulya, T. J. 1987. Relationship between *Phoma macdonaldii* and premature death of sunflower in North Dakota. Plant Dis. 71:466-468.

Dwan, J. M., and Sobrino, E. 1987. First report of *Pseudomonas syringae* pv. *helianthi* on sunflower in Spain. Plant Dis. 71:101.

Fakir, G. A., Rao, M. H., and Thirumalachar, J. J. 1976. Seed transmission of *Macrophomina phaseolina* in sunflower. Plant Dis. Rptr. 60:736-737.

Fourie, D., and Viljoen, A. 1996. Bacterial leaf spot of sunflower in South Africa. Plant Dis. 80:1430.

Gudmestad, N. C., Secor, G. A., Nolte, P., and Straley, M. L. 1984. *Erwinia carotovora* as a stalk rot pathogen of sunflower in North Dakota. Plant Dis. 68:189-192.

Gulya, T. J., Urs, R. R., and Banttari, E. E. 1981. Apical chlorosis of sunflower incited by *Pseudomonas tagetis*. (Abstr.) Phytopathology 71:221.

Gulya, T. J., Urs, R. R., and Banttari, E. E. 1982. Apical chlorosis of sunflower caused by *Pseudomonas syringae* pv. *tagetis*. Plant Dis. 66:598-600.

Gulya, T. J., Ooka, J. J., and Mancl, M. K. 1985. Diseases of cultivated sunflower in Hawaii. Plant Dis. 69:542.

Herd, G. W., and Phillips, A. J. L. 1988. Control of seedborne *Sclerotinia sclerotiorum* by fungicidal treatment of sunflower seed. Plant Pathol. 37:202-205.

Hoes, J. A. 1972. Sunflower yellows, a new disease caused by *Phialophora* sp. Phytopathology 62:1088-1092.

Hoes, J. A. 1975. Sunflower diseases in western Canada. *In* Oilseed and Pulse Crops in Western Canada. Modern Press, Saskatoon, Saskatchewan.

Hoes, J. A., and Huang, H. C. 1985. Effect of between-row and within-row spacings on development of Sclerotinia wilt and yield of sunflower. Can. J. Plant Pathol. 7:98-102.

Holley, R. C., and Nelson, B. D. 1986. Effect of plant population and inoculum density on incidence of Sclerotinia wilt of sunflower. Phytopathology 76:71-74.

Huang, H. C. 1976. Importance of *Coniothyrium minitans* in survival of *Sclerotinia sclerotiorum* in wilted sunflower. Can. J. Bot. 55:289-295.

Huang, H. C., and Dueck, J. 1980. Wilt of sunflower from infection by mycelial-germinating sclerotia of *Sclerotinia sclerotiorum*. Can. J. Plant Pathol. 2:47-52.

Huang, H. C., and Hoes, J. A. 1980. Importance of plant spacing and sclerotial position to development of Sclerotinia wilt of sunflower. Plant Dis. 64:81-84.

Huang, H. C., and Kozub, G. C. 1990. Cyclic occurrence of Sclerotinia wilt of sunflower in western Canada. Plant Dis. 74:766-770.

Infantino, A., and Di Giambattista, G. 1997. First report of *Sclerotium rolfsii* on sunflower in Italy. Plant Dis. 81:960.

Jeffrey, K. K., Lipps, P. E., and Herr, L. J. 1984. Effects of isolate virulence, plant age, and crop residues on seedling blight of sunflower caused by *Alternaria helianthi*. Phytopathology 74:1107-1110.

Jeffrey, K. K., Lipps, P. E., and Herr, L. J. 1985. Seed-treatment fungicides for control of seedborne *Alternaria helianthi* on sunflower. Plant Dis. 69:124-126.

Jimenez-Diaz, R. M., Blanco-Lopez, M. A., and Sackston, W. E. 1983. Incidence and distribution of charcoal rot of sunflower caused by *Macrophomina phaseolina* in Spain. Plant Dis. 67:1033-1036.

Jones, B. L., and Van Der Walt, L. 1987. Bacterial stalk rot of sunflower in South Africa. (Abstr.). Phytophylactica 19:121.

Jones, B. L., and Van Der Walt, L. 1988. Occurrence of sunflower stalk rot incited by *Erwinia carotovora* in South Africa. Plant Dis. 72:994.

Kajornchaiyakul, P., and Brown, J. F. 1976. The infection process and factors affecting infection of sunflower by *Albugo tragopogi*. Brit. Mycol. Soc. Trans. 66:91-95.

Klisiewicz, J. M., and Beard, B. H. 1976. Diseases of sunflower in California. Plant Dis. Rptr. 60:298-301.

Kolte, S. J. 1985. Diseases of Annual Edible Oil Seed Crops, Volume III: Sunflower, Safflower, and Nigerseed Diseases. CRC Press, Inc., Boca Raton, FL. 154 pp.

Lagopodi, A. L., and Thanassoulopoulos, C. C. 1998. Effect of a leaf spot disease caused by *Alternaria alternata* on yield of sunflower in Greece. Plant Dis. 82:41-44.

Ljubich, A., Gulya, T. J., and J. F. Miller. 1988. A new race of sunflower downy mildew in North America. (Abstr.). Phytopathology 78:1580.

McDonald, W. C. 1964. Phoma black stem of sunflowers. Phytopathology 54:492-493.

McDonald, W. C., and Martens, J. W. 1963. Leaf and stem spot of sunflowers caused by *Alternaria zinniae*. Phytopathology 53:93-96.

McLaren, D. L., Huang, H. C., Kozub, G. C., and Rimmer, S. R. 1994. Biological control of Sclerotinia wilt of sunflower with *Talaromyces flavus* and *Coniothyrium minitans*. Plant Dis. 78:231-235.

Middleton, K. J. 1971. Sunflower diseases in South Queensland. Queensland Agric. J. 97:597-600.

Mihaljcevic, M., Muntanola-Cvetkovic, M., Vukojevic, J., and Petrov, M. 1985. Source of infection of sunflower plants by *Diaporthe helianthi* in Yugoslavia. Phytopathol. Zeitschr. 113:334-342.

Morris, J. B., Yang, S. M., and Wilson, L. 1983. Reaction of *Helianthus* species to *Alternaria helianthi*. Plant Dis. 67:539-540.

Muntanola-Cvetkovic, M., Mihaljcevic, M., and Petrov, M. 1981. On the identity of the causative agent of a serious *Phomopsis-Diaporthe* disease in sunflower plants. Nova Hedwigia 34:417-435.

Muntanola-Cvetkovic, M., Mihaljcevic, M., Vukojevic, J., and Petrov, M. 1985. Comparison of *Phomopsis* isolates obtained from sunflower plants and debris in Yugoslavia. Brit. Mycol. Soc. Trans. 85:477-483.

Orellana, R. G. 1971. Fusarium wilt of sunflower, *Helianthus annuus*: First report. Plant Dis. Rptr. 55:1124-1125.

Orellana, R. G., and Quacquarelli, A. 1968. Sunflower mosaic caused by a strain of cucumber mosaic virus. Phytopathology 58:1439-1440.

Palm, M. E., Gams, W., and Nirenberg, H. I. 1995. *Plectosporium,* a new genus for *Fusarium tabacinum*, the anamorph of *Plectosphaerella cucumerina*. Mycologia 87:397-406.

Parmelee, J. A. 1972. Additions to the autoecious species of *Puccinia* on Heliantheae in North America. Can. J. Bot. 50:1457-1459.

Piening, L. J. 1976. A new bacterial leaf spot of sunflowers in Canada. Can. J. Plant Sci. 56:419-422.

Putt, E. D. 1964. Breeding behavior of resistance to leaf mottle or Verticillium wilt in sunflowers. Crop Sci. 4:177-179.

Rashid, K. Y. 1993. Incidence and virulence of *Plasmopara halstedii* on sunflower in western Canada during 1988-1991. Can. J. Plant Pathol. 15:206-210.

Reddy, P. C., and Siradhana, B. S. 1978. A new leaf spot disease of sunflower in India. Plant Dis. Rptr. 62:508.

Rich, J. R., and Dunn, R. A. 1982. Pathogenicity and control of nematodes affecting sunflower in north central Florida. Plant Dis. 66:297-298.

Richeson, M. L. 1981. Etiology of a late season wilt in *Helianthus annuus*. Plant Dis. 65:1019-1021.

Rogers, E. C., Thompson, T. E., and Zimmer, D. E. 1978. Rhizopus head rot of sunflower: Etiology and severity in the southern plains. Plant Dis. Rptr. 62:769-771.

Sackston, W. E. 1960. *Botrytis cinerea* and *Sclerotinia sclerotiorum* in seed of safflower and sunflower. Plant Dis. Rptr. 44:664-668.

Sackston, W. E. 1981. The sunflower crop and disease: Progress, problems, and prospects. Plant Dis. 65:643-648.

Sackston, W. E., McDonald, W. C., and Martens, J. W. 1957. Leaf mottle or Verticillium wilt of sunflower. Plant Dis. Rptr. 41:337-343.

Shtienberg, D. 1994. Achene blemish syndrome: A new disease of sunflower in Israel. Plant Dis. 78:1112-1116.

Shtienberg, D. 1997. Rhizopus head rot of confectionery sunflower: Effects on yield quantity and quality and implications for disease management. Phytopathology 87:1226-1232.

Shtienberg, D., and Vinta, H. 1995. Environmental influences on the development of *Puccinia helianthi* on sunflower. Phytopathology 85:1388-1393.

Smolik, J. D. 1987. Effects of *Paratylenchus projectus* on growth of sunflower. Plant Disease 71:975-976.

Stovold, G. E., and Moore, K. J. 1972. Diseases of sunflower. Agric. Gazette of New South Wales 83:262-264.

Styer, D. J., and Durbin, R. D. 1982. Isolation of *Pseudomonas syringae* pv. *tagetis* from sunflower in Wisconsin. Plant Dis. 66:601.

Tosi, L., and Zazzerini, A. 1995. *Phialophora asteris* f. sp. *helianthi*, a new pathogen of sunflower in Italy. Plant Dis. 79:534-537.

Viljoen, A., Kruger, H., and van Wyk, P. S. 1997. Histopathology of sunflower stems infected by *Albugo tragopogonis*. (Abstr.). Phytopathology 87:S100.

Watters, B. L., Herr, L. J., and Lipps, P. E. 1983. A new stem canker disease of sunflower. (Abstr.). Phytopathology 73:798.

Wehtje, G., and Zimmer, D. E. 1978. Downy mildew of sunflower: Biology of systemic infection and the nature of resistance. Phytopathology 68:1568-1571.

Yang, S. M. 1983. Bract necrosis, a nonparasitic disease of sunflower. (Abstr.). Phytopathology 73:844.

Yang, S. M. 1984. Etiology of atypical symptoms of charcoal rot in sunflower plants parasitized by larvae of *Cylendrocopturus adsperus*. Phytopathology 74:479-481.

Yang, S. M. 1988. Diseases of cultivated sunflower in Liaoning Province, People's Republic of China. Plant Dis. 72:546.

Yang, S. M., and Gulya, T. 1986. Prevalent races of *Puccinia helianthi* in cultivated sunflower on the Texas high plains. Plant Dis. 70:603.

Yang, S. M., and Owen, D. F. 1982. Symptomatology and detection of *Macrophomina phaseolina* in sunflower plants parasitized by *Cylindrocopturus adspersus* larvae. Phytopathology 72:819-821.

Yang, S. M., Morris, J. B., Unger, P. W., and Thompson, T. E. 1979. Rhizopus head rot of cultivated sunflower in Texas. Plant Dis. Rptr. 63:833-835.

Yang, S. M., Rogers, C. E., and Luciani, N. D. 1983. Transmission of *Macrophomina phaseolina* in sunflower by *Cylindrocopturus adspersus*. Phytopathology 73:1467-1469.

Yang, S. M., Berry, R. W., Lutterell, E. S., and Vongkaysone, T. 1984. A new sunflower disease in Texas caused by *Diaporthe helianthi*. Plant Dis. 68:254-255.

Yarwood, C. E. 1969. Sunflower, a new host of *Coleosporium madia*. Plant Dis. Rptr. 53:648.

Zazzerini, A., and Tosi, L. 1987. New sunflower disease caused by *Fusarium tabacinum*. Plant Dis. 71:1043-1044.

Zimmer, D. E. 1975. Some biotic and climatic factors influencing sporadic occurrence of sunflower downy mildew. Phytopathology 65:751-754.

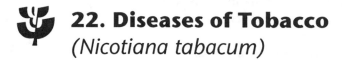

22. Diseases of Tobacco
(Nicotiana tabacum)

Diseases Caused by Bacteria

Angular Leaf Spot

Angular leaf spot is also called blackfire.

Cause. *Pseudomonas syringae* pv. *angulata* (syn. *Pseudomonas angulata*) survives in soil and infested residue, and persists as a parasite on the roots of weeds and winter grains. Dissemination of *P. syringae* pv. *tabaci* over a long distance is by contaminated transplants. Water blown by wind is the most important means of dissemination from plant to plant in the field and over short distances. Bacteria are splashed onto leaves, where they are forced into leaf stomates or wounds, and move into flooded intercellular spaces. Leaf tissues must be water-soaked before infection can occur. Optimum conditions for infection occur during rainstorms accompanied by high winds.

   The major difference between *P. syringae* pv. *tabaci* and *P. syringae* pv. *angulata* is that *P. syringae* pv. *tabaci* produces toxin and *Pseudomonas syringae* pv. *angulata* does not.

Distribution. Wherever tobacco is grown.

Symptoms. Symptoms suddenly appear a few days after rainy periods as angular to irregular-shaped (1–13 mm in diameter) black to brownish water-soaked spots that rapidly kill tissue and form dark brown angular-shaped spots. As the spots age, their centers become light tan with dark borders and fall out during rainy weather. A thin yellow band often occurs around spots. Spots eventually coalesce and cover large areas of a diseased leaf. On rapidly growing plants, diseased leaves become "puckered" and torn. Symptoms in plant beds are similar to those in fields.

Management

1. Plow planting beds in the autumn.
2. Steam or fumigate planting beds plowed in the spring.
3. Sow only seed from healthy plants.
4. Treat seed with bactericides.
5. Apply bactericide sprays to plants in the seedbed and field.
6. Use only disease-free transplants.
7. Grow resistant cultivars.
8. Provide proper drainage for plant beds.
9. Do not grow tobacco in a field for 2 or more consecutive years.
10. Avoid working in a field when leaves are wet.

Barn Rot

Cause. *Erwinia carotovora* subsp. *carotovora* (syn. *Bacillus polymyxa, E. aroideae, E. carotovora,* and *E. chrysanthemi*). The fungi *Alternaria alternata, Botrytis cinerea, Pythium* spp., *Rhizopus arrhizus,* and *Sclerotinia sclerotiorum* also are associated with barn rot.

Overwintering occurs on infested or infected seed. Infected seed is an important source of inoculum and, presumably, is the primary source of infection. Dissemination is by conveyors that move tobacco plants.

Barn rot occurs when wet, mechanically primed leaves are packed tightly into bulk-curing barns that have poor air circulation. Barn rot also occurs in stick barns with poor circulation when they are overfilled with wet tobacco leaves. Mechanical leaf harvesters injure leaves, creating wounds for bacteria to enter.

Distribution. Wherever tobacco is grown.

Symptoms. A watery, soft rot occurs during the yellowing stage of tobacco curing. A water-soaked brownish discoloration begins in the petioles or laminae, darkens to a dark brown or black discoloration, and spreads throughout the leaves, causing them to decompose. Often leaves break away from strings and fall to the barn floor. There is an odor of decaying vegetation within a barn. The damage may vary from a few leaves to the entire contents of a barn.

Management

1. Avoid placing wet tobacco in the barn.
2. Do not crowd leaves on a stick.
3. Do not crowd sticks together in the barn.
4. Use heat and ventilation to dry out excess moisture rapidly if wet tobacco is placed in a barn.
5. Do not injure tobacco.

Black Rust

Cause. *Bacterium pseudozoogloae.* This may be the same bacterium as *Pseudomonas syringae* pv. *tabaci. Bacterium pseudozoogloae* survives within infested residue and is disseminated by splashing rain.

Distribution. Indonesia and possibly Japan and Italy.

Symptoms. Symptoms occur on diseased lower leaves as plants approach maturity. The dark green spots that appear initially become necrotic and dark brown in the center and have zonate concentric rings. A dark green margin that surrounds each spot eventually reaches 1 to 2 cm in diameter. The margins remain dark green after diseased leaves have dried.

Management. No management other than to remove and burn the lower diseased leaves.

Blackleg

Blackleg is the seedling form of hollow stalk.

Cause. *Erwinia carotovora* subsp. *carotovora* (syn. *E. aroidea* and *E. carotovora* subsp. *atroseptica*) survives in soil and infested residue. Cool and "heavy" clay soils with a neutral pH favor survival of causal bacteria. *Erwinia carotovora* subsp. *carotovora* is disseminated to fields on infected transplants and by any means that moves soil. Bacteria may also be spread within a planting bed by maggot flies of *Hylemyia* spp. Host plants become infected by leaves touching infested soil. Bacteria then spread through the petioles and into the stems. Blackleg is most severe during damp, cloudy weather over a wide range of temperatures.

Distribution. Wherever tobacco is grown. However, blackleg is rarely a severe problem.

Symptoms. Blackleg occurs in irregular circular areas approximately 1 m in diameter within plant beds. Rotted petioles and stems become black and eventually split open or rot off.

   Some plants may be slightly affected and display no disease symptoms. When these plants are set in the field, they may grow normally. However, if symptomless plants are removed from beds and kept overnight for planting the next day, all plants may become rotted and slimy.

Management. As soon as blackleg is noticed, the covering should be removed from the plant bed to reduce humidity.

Crown Gall

Cause. *Agrobacterium tumefaciens* (syn. *Rhizobium radiobacter* var. *tumefaciens*) is a soil inhabitant that is disseminated by anything that moves soil and

enters plants through wounds caused by any means. Bacteria then stimulate cells to divide and eventually form a tumor. The following steps are
thought to occur in tumor development: (1) Bacteria enter a wound cell
and bind to a specific intracellular wound site. (2) Bacterial DNA is transferred to the plant cell and causes it to divide. (3) Bacterial DNA increases
and is transferred to plant cells that are free of bacterial cells. These plant
cells also divide and contribute to the tumor development.

Tumor development is optimum at a temperature of 25°C but is inhibited at higher temperatures.

Distribution. Crown gall is distributed worldwide, but it rarely occurs on tobacco.

Symptoms. Round tumors or overgrowths that occur on diseased roots of
shoots are several times larger than the stem or root. The round tumors
have large indentations that make them appear convoluted.

Management. Not necessary.

Frenching

Cause. *Bacillus cereus* is a nonpathogenic bacterium common in soil and
dust, and on plant surfaces. *Bacillus cereus* does not enter or infect
plants. A diffusate is produced in soil by *B. cereus* that acts as a toxin
to tobacco. The toxin disturbs the nitrogen metabolism of plants,
which increases different amino acids, particularly isoleucine. The excess
of free amino acids then causes the tobacco abnormalities. Frenching is
more prevalent on wet soils that sometimes lack available nitrogen or
other materials. Frenching does not occur at soil temperature of 21°C or
less.

Distribution. Wherever tobacco is grown. Frenching rarely causes much damage.

Symptoms. Frenching begins to occur on 2- to 3-week-old plants and continues until flowering. Chlorosis occurs initially along margins of young
leaves and eventually covers all interveinal areas but the leaf veins remain
green. "Diseased" plants are stunted and have a large number of small, narrow, distorted, brittle leaves. Axillary buds of severely diseased plants are
stimulated, which causes a rosette of small, narrow, light green leaves that
gives a witches' broom effect.

Management
1. Ensure that soils are properly drained and fertilized.
2. Do not plant into alkaline soils and avoid heavy applications of
   lime.
3. Apply fall applications of elemental sulfur.

Granville Wilt

Granville wilt is also called bacterial wilt and several other names in countries where the disease occurs.

Cause. *Pseudomonas solanacearum* is a bacterium that survives in soil and infects plants through root wounds caused by several factors. Bacteria are disseminated by infected seedlings. During periods of high soil moisture, bacteria that are released from diseased roots spread to the roots of adjacent plant hosts, particularly where the roots intermingle. Disease severity is greatest during high soil temperatures (30° to 35°C) and excess moisture.

Distribution. Wherever tobacco is grown in warm temperate and semitropical areas of the world. *Pseudomonas solanacearum* is rarely found in areas where the mean temperature in the winter is below 10°C.

Symptoms. Initially, one or two leaves on young, succulent plants droop during the day but recover their turgidity at night. Often only half a leaf wilts. If the disease progresses slowly, diseased leaves become light green and gradually turn yellow. In severe disease conditions, the whole plant wilts and dies within a few days.

Light tan to yellow-brown streaks occur in the vascular tissue. Eventually, the streaks darken, and in advanced stages, blackened areas occur on the surface of diseased stalks. A dirty white to brownish slimy ooze in the form of glistening beads occurs on a cut stem. When the two cut edges of a stem are pulled apart slightly, a bacterial strand stretches between the two pieces.

Eventually, roots become dark brown to black and rotted. In moist soils, roots become soft and slimy.

Management
1. Grow resistant cultivars.
2. Do not transplant plants showing wilting or other disease symptoms.
3. Do not grow tobacco in soil containing *P. solanacearum* for at least 5 years. Crops to include in a rotation are grasses, legumes, small grains, cotton, maize, soybean, and sorghum.
4. Practice good nematode control.

Hollow Stalk

Blackleg is the seedling form of hollow stalk.

Cause. *Erwinia carotovora* subsp. *carotovora* (syn. *E. aroidea* and *E. carotovora* subsp. *atroseptica*) survives in soil and in infested soil residue that is composed primarily of decayed root crowns. Soils high in clay content and of neutral pH favor survival of the bacterium. Disease development is most rapid during wet weather over a wide range of temperatures. Bacteria enter plants through wounds made during topping, suckering, and harvesting. Bacteria are disseminated in contaminated soil on workers' hands during topping and suckering. Mineral oil, which is used to control suckering, apparently favors the reproduction of *E. carotovora* subsp. *carotovora* in and on plants.

Distribution. Wherever tobacco is grown. However, hollow stalk is rarely an important disease.

Symptoms. The first disease symptom appears at topping and suckering time and commonly is a discoloration in the pith at the break made at topping and at any stem wound. Disease may occasionally begin at the bottom of plants. The pith rapidly becomes brown, followed by a "soft" rotting and disintegration. The pith eventually becomes a foul-smelling slimy pulp that dries up to form a hollow stem. The top leaves wilt first, and as disease progresses down the stem, successive leaves wilt, hang down next to the stalk, or fall off and cause the entire diseased stalk to be bare. Black or brown sunken lesions form in leaf axils.

Diseased leaves may rot in the curing barn and smell like decaying vegetation. The leaf petiole rots at the point where the string is attached and the leaf falls to the floor.

Management
1. Do not top and sucker during damp or cloudy weather.
2. Reduce humidity in curing barns when wet, succulent tobacco leaves are hung.
3. Avoid manual topping.
4. Use sucker control agents other than mineral oils.

Leaf Gall

Leaf gall is also called fascination.

Cause. *Rhodococcus fascians* (syn. *Corynebacterium fascians*) is commonly found in soil and does not need a wound to infect a plant. Disease occurs due to an imbalance of growth hormones in diseased plants caused by the bacteria. Optimum growth occurs at temperatures of 25° to 28°C.

Distribution. Not known for certain, but leafy gall rarely occurs on tobacco.

Symptoms. Numerous short, fleshy, thick stems or multiple buds with misshapen leaves develop at or below the soil line; however, the main stem of the diseased plant appears normal but may be somewhat stunted. The proliferated growth may reach a diameter of 2 to 8 cm and resemble a leafy gall.

Management. Not necessary.

Philippine Bacterial Leaf Spot

Cause. *Pseudomonas aeruginosa* is a common soil saprophyte and is seed-borne. Bacteria enter the plant through wounds and stomates of water-soaked leaves. Dissemination is by splashing rain.

Distribution. The Philippines.

Symptoms. Bleached white spots that vary in size from a pinpoint to several millimeters in diameter occur on seedling leaves. Diseased areas develop a wet rot that becomes necrotic and disintegrates. Petioles and stems also become diseased during wet weather, causing death of diseased plants.

Diseased plants in the field have whitish or opaque lesions that eventually become brown and zonate and are bordered by a chlorotic zone that disappears as leaves mature. Lesions, which are most numerous on the bottom leaves, coalesce and form large, irregular-shaped, dead areas. After curing, brown spots disappear but the white spots are still evident.

Management
1. Plow planting beds in the autumn.
2. Steam or fumigate planting beds plowed in the spring.
3. Sow only seed from healthy plants.
4. Treat seed with a bactericide.
5. Apply bactericide to foliage of plants in seedbeds and in fields.
6. Use only transplants that are free of disease.
7. Grow resistant cultivars.
8. Provide proper drainage for plant beds.
9. Do not grow tobacco in infested soil for 2 consecutive years.
10. Avoid working in a field when the tobacco leaves are wet.

Wildfire

Cause. *Pseudomonas syringae* pv. *tabaci* (syn. *P. tabaci*) survives in soil, infested residue, infested dry leaves, and some types of manufactured tobacco, and as a parasite on the roots of weeds, grasses, and winter grains. Bacteria are commonly disseminated by windblown water from plant to plant in the field and over long distances. Dissemination is less common on contaminated transplants. Bacteria are splashed onto leaves, where they infect through stomates or wounds. Water-soaked areas of leaf tissue, which develop as a result of flooding intercellular spaces, must be present before infection occurs. Bacteria are forced into leaf openings and then move into flooded intercellular spaces.

Optimal infection conditions occur during rainstorms accompanied by high winds. Optimum growth in culture is at temperatures of 24° to 28°C. The major difference between *P. syringae* pv. *tabaci* and *P. syringae* pv. *angu-*

lata is that *P. syringae* pv. *tabaci* produces toxin and *Pseudomonas syringae* pv. *angulata* does not.

Distribution. Wherever tobacco is grown.

Symptoms. Wildfire may suddenly appear in areas approximately 1 to 2 m in diameter in the wettest part of plant beds on tender, rapidly growing plants after a cool, rainy period. During high humidity, small, circular, yellow-green, water-soaked spots develop on diseased leaves. Spots rapidly turn brown and are surrounded by yellow-green halos, which give plants a scorched appearance. Severely diseased seedlings, which eventually die, are normally in the middle of an area of diseased plants.

Older diseased plants have spots that are circular, and have a brown and dead center (12–25 mm in diameter) that is surrounded by a yellow-green, water-soaked border or halo. Spots coalesce and form large dead areas on diseased leaves. Leaves that are diseased only on one side become twisted and distorted.

Management
1. Plow planting beds in the autumn.
2. Steam or fumigate planting beds plowed in the spring.
3. Sow only seed from healthy plants.
4. Treat seed with a bactericide.
5. Apply bactericide to foliage of plants in seedbeds and in fields.
6. Use only transplants free of disease.
7. Grow resistant cultivars.
8. Provide proper drainage for plant beds.
9. Do not grow tobacco in infested soil for 2 consecutive years.
10. Avoid working in a field when leaves are wet.

Wisconsin Bacterial Leaf Spot

Cause. *Bacterium melleum*. However, this is probably a strain of *Pseudomonas syringae* pv. *tabaci*. Some researchers suggest *B. melleum* is an intermediate between *P. syringae* pv. *tabaci* and *P. syringae* pv. *angulata*. Therefore, Wisconsin bacterial leaf spot may have been confused with wildfire.

Distribution. Europe and the United States (Wisconsin). There has been no report of the disease from Wisconsin since 1937.

Symptoms. Initially, circular pinpoint-sized specks surrounded by distinct chlorotic zones appear on diseased leaves and enlarge rapidly during periods of high temperatures and humidity to form spots that are approximately 1 cm in diameter. Necrotic tissue in spots becomes brown and lesions on veins are brown and sunken. Irregular-shaped spots form when the smaller spots coalesce.

Diseased seedlings have inconspicuous angular spots that are surrounded by chlorotic margins. The necrotic tissue becomes brown as the spots age. When spots are numerous, foliage appears blighted.

Management
1. Plow planting beds in the autumn.
2. Steam or fumigate planting beds plowed in the spring.
3. Sow only seed from healthy plants.
4. Treat seed with a bactericide.
5. Apply bactericide to the foliage of plants in seedbeds and in fields.
6. Plant only disease-free transplants.
7. Grow resistant cultivars.
8. Provide proper drainage for plant beds.
9. Do not grow tobacco in infested soil for 2 consecutive years.
10. Avoid working in a field when tobacco leaves are wet.

Diseases Caused by Fungi

Anthracnose

Cause. *Colletotrichum destructivum* overwinters as mycelia and acervuli in infested residue and soil, and is seedborne. Conidia are produced in acervuli and disseminated by wind and splashing rain to host plant leaves. Anthracnose develops over a wide temperature range (18° to 32°C) and a high relative humidity.

Distribution. Asia, Africa, Australia, Brazil, and the United States.

Symptoms. Young leaves have small, depressed, light green, water-soaked spots that enlarge to 3 mm in diameter and give the undersides of diseased leaves a "greasy" appearance. As spots age, they dry out, become a gray-white color that resembles paper, and are surrounded by raised water-soaked borders that become brown. Large spots are zonate and have dark brown centers. Diseased leaves become wrinkled and distorted, and die. Small diseased plants are stunted or killed. Large diseased plants have water-soaked areas that later form red-brown cankers up to 12 mm long on leaf midribs and petioles.

Management
1. Apply foliar fungicides to plants in plant beds and in fields.
2. Fumigate infested soils.
3. Treat seed with a seed-protectant fungicide.
4. Destroy all weeds in the vicinity of plant beds.
5. Do not grow peppers and tomatoes in seedbeds.

Ascochyta Leaf Spot

Ascochyta leaf spot is also called ragged leaf spot.

Cause. *Ascochyta phaseolorum.* Survival is presumably as pycnidia in infested residue in the soil. Ascochyta leaf spot is more severe at cool temperatures (around 22°C) during wet weather.

Distribution. Asia, Europe, and the United States.

Symptoms. Symptoms occur on diseased plants of all ages. Spots are circular (2.5 cm in diameter) with gray to brown centers that fall out and give a ragged appearance to the diseased leaf. Spots coalesce and form large dead areas. Scattered black pycnidia are found in the centers of older lesions and on the stems of young diseased seedlings.

Management. Apply foliar fungicides.

Barn Spot

Barn spot is the curing phase of frogeye.

Cause. *Cercospora nicotianae.*

Distribution. Africa, Central America, Southeast Asia, and the United States.

Symptoms. Black spots (1–6 mm in diameter) occur on flu-cured tobacco leaves in the barn.

Management
1. Sow only disease-free seed.
2. Remove residue from plant beds.
3. Apply foliar fungicides before transplanting.
4. Do not overfertilize with nitrogen.
5. Prime tobacco before it becomes overripe.
6. Plow under infested residue in the field.
7. Rotate tobacco with other crops.
8. Cure tobacco at a temperature of 38°C with 100% relative humidity.

Black Mildew

Cause. The causal fungus is similar in morphology to *Diporotheca rhizophila* and has been found only on plants infested with the cyst nematode, *Globodera* spp. However, not every plant growing in nematode-infested soil is infected.

Distribution. The United States (Connecticut). Black mildew is of no economic importance.

Symptoms. Diseased tobacco roots are colonized with dark fungal mycelia.

Management. Not necessary.

Black Root Rot

Black root rot is also called maricume radicale, root rot, and Thielavia root rot.

Cause. *Thielaviopsis basicola* (syn. *Chalara elegans*) survives as chlamydospores and endoconidia in soil and as saprophytes on infested residue of different plants, which allows *T. basicola* to persist indefinitely in the absence of tobacco. However, Hood and Shew (1997) reported that *T. basicola* did not display significant saprophytic ability and should be classified ecologically as an obligate parasite.

Thielaviopsis basicola is disseminated on roots of transplants and by any means that moves soil. Chlamydospores, considered to be the primary infective propagule, and endoconidia germinate and form hyphae that penetrate roots. Chlamydospores and endoconidia produced in mycelium on the outside of diseased roots serve as secondary inoculum.

Black root rot is most severe at relatively cool temperatures (17° to 23°C) and in wet soils. Plants growing in soils with a pH of 6.4 and above are more prone to infection. Black root rot does not develop in soils with an average pH of less than 5.2.

Populations of *T. basicola* in field soils are directly related to the frequency of soybean and tobacco production, with the larger fungal populations occurring in soils on which continuous tobacco is grown. Disease severity is normally positively correlated with inoculum density of *T. basicola*. However, some studies have reported that inoculum density is not related to black root rot severity, and that environmental factors and cultivar resistance have a greater influence on disease development.

Distribution. Wherever tobacco is grown.

Symptoms. Seedlings in plant beds damp off. Usually diseased plants are stunted or uneven in growth only in limited areas of affected fields. Host plants infected later in the growing season have black lesions on roots and the leaves become yellow.

Diseased roots are partially or entirely black; small roots are completely rotted off, but larger main and lateral roots only have roughened black lesions on them. Severely diseased plants have a few blackened, stubby roots attached to a stem. If the weather becomes warm, many diseased plants produce new roots and assume normal growth. Leaves become yellow and wilt during the day, and plants flower prematurely.

Management
1. Grow resistant cultivars. There are two types of resistance. One type is partial resistance, which is characterized by low to moderate levels of resistance derived from *Nicotiana tabacum*. The second type is a high level of monogenic resistance, which is derived from *N. debneyi*. Resistance is related to the presence of the phenolic compounds scopolin and scopoletin. These compounds prevent the melanization of fungi, which is necessary for fungal growth in host tissue.

2. Use a new bed site each year.
3. Treat soil with steam or chemicals if an old bed site is used.
4. Rotate tobacco with resistant crops, such as grass, for 3 years or longer.
5. Plow under cover crops and manure.
6. Do not apply more than 10 tons of manure per acre.
7. Maintain a soil pH of 6.0 to 6.4.
8. Fertilize according to soil tests. Avoid excessive fertilizer.
9. Avoid transplanting when soils are cold or when the air temperature is low.

Black Shank

Cause. *Phytophthora parasitica* var. *nicotianae* (syn. *P. nicotianae*) or *P. nicotianae* var. *nicotianae*. Both names are considered correct but not synonymous.

Phytophthora parasitica var. *nicotianae* survives as chlamydospores in soil and in infested residue, and as mycelia in infested residue. Oospores are formed and overwinter in infested tissue and soil, but their function is not known. Dissemination is on transplants, by water, or by any means that moves soil. Spread of *P. parasitica* var. *nicotianae* is greatest between susceptible plants and least between resistant plants.

Chlamydospores germinate and form germ tubes in which either another chlamydospore or sporangia are produced. The sporangia then germinate and form either hyphae or zoospores. Zoospores and sporangia germinate and produce hyphae that infect plants. Zoospores encyst, germinate, and penetrate roots within 1 hour. Sporulation occurs on stems near the soil surface or on infested residue. Sporangia disseminated some distance by wind and water initiate further infections on host plants. Plants of all ages may be infected, but seedlings are most susceptible.

Black shank is most severe in wet soils at temperatures above 21°C. Thus, disease incidence is correlated with soil texture and drainage class of the parent soil. Disease infection and severity are increased by the nematodes *Meloidogyne incognita* and *Globodera solanacearum*. Different physiologic strains of *P. nicotianae* exist.

Distribution. Africa, Asia, Australia, Caribbean, eastern Europe, North America, and South America.

Symptoms. Symptoms initially occur in wetter areas of fields. Roots become dark brown or black, with the discoloration on the stalk extending several centimeters above the soil line. Plants in the early stages of disease have one or more dead and blackened lateral roots but no discoloration of stems. Tops of diseased plants eventually wilt during the day and do not recover turgidity at night. Usually plants wilt and collapse before leaves mature and can be harvested.

The collapse of diseased plants occurs very suddenly during drought stress or hot weather. The first symptom on vigorously growing plants is a rapid wilting of leaves, which quickly become chlorotic, hang down, die, and become brown. As disease progresses, the entire root system and base of the stalk dies and becomes blackened,.

Older plants that become infected initially have a black discoloration on the stem that extends some distance above the soil line. Soon the leaves become chlorotic, then brown, shrivel, and sometimes have a few small green leaves remaining at the top of the diseased plant. Such diseased plants may bloom prematurely.

One of the most characteristic symptoms of black shank is seen when the stem is split lengthwise. The pith is dry, brown to black, and separated into plate-like disks.

During rainy weather, large lesions may develop on the lower leaves. At first a lesion is pale green; later it becomes brown and grows up to 8 cm in diameter.

Management. Where black shank is present:

1. Grow resistant cultivars.
2. Move plant beds to an area that has not been planted in row crops.
3. Convert an infested field to permanent pasture for at least 5 years. Other alternatives include 2 to 3 years of clean fallow followed by rye grown for grain. Two or 3 years of peanuts, soybeans, and cotton are not as satisfactory but still aid in reducing the population of the pathogen. In South Africa, blue buffalo grass, *Cenchrus ciliaris*, grown for 2 to 3 years in succession significantly reduced the incidence of black shank on tobacco.
4. Use good nematode control practices.
5. Leave areas or a field infested with *P. nicotianae* until last during tillage operations.
6. Disinfect machinery parts with fungicides after use.
7. Clean shoes after working in infested soil. Use a stomp box (a wood box with a sack in the bottom on which fungicide is poured).
8. Apply fungicides.

Where black shank is not present:

1. Grow your own plants. This prevents movement of the fungus on transplants.
2. Do not irrigate from water sources originating in infested fields.
3. Do not borrow equipment. Clean off all equipment with a fungicide solution.

Blue Mold

Blue mold is also called downy mildew.

Cause. *Peronospora tabacina* (syn. *P. hyoscyami* f. sp. *tabacina*) survives as mycelia and sporangia within infested residue in geographical areas with mild winter temperatures. Systemic mycelia overwinter in live roots of host plants whose aerial portion has been previously killed. Such plants produce suckers during favorable weather and become systemically infected. Wild tobacco, *Nicotiana repanda,* also provides a means of overwintering and is a significant source of inoculum.

In colder tobacco-growing areas, oospores overwinter in diseased seedlings. However, it is not known how they germinate to initiate primary infection. Therefore, primary inoculum in cold areas may be conidia that are disseminated by wind from warmer tobacco-growing regions.

Peronospora tabacina is primarily disseminated by wind. However, other dissemination is by any means that residue containing conidia, mycelia, and oospores can be moved from place to place. Conidia (sporangiospores) are produced from mycelia and disseminated by wind for long distances. Conidia germinate in the presence of moisture and produce germ tubes that penetrate a host's leaves. Conidia produced in lesions provide secondary inoculum. Oospores are also formed in diseased tissue.

Temperatures from 15°C to 25°C and 95% or more relative humidity with intermittent rainy weather are optimum conditions for blue mold development. Availability of moisture on leaves is an important factor for infection; consequently, disease spreads rapidly in cool, damp weather. Conversely, *P. tabacina* has been reported to adapt to higher temperatures in some geographical areas. Epiphytotics have occurred during day and night temperatures up to 36° and 25°C, respectively, for 48 hours. Significantly fewer conidia are produced at these higher temperatures, but these conidia could be important as secondary inoculum for subsequent disease development.

Young plants are more susceptible to infection than are older plants.

Distribution. Australia, Canada, Cuba, Europe, the Mediterranean area, Mexico, the Near East, and the United States.

Symptoms. Small diseased plants in seedbeds have erect leaves. Circular groups of larger diseased plants become yellow and the plants in the center of a group have cupped leaves. Some cupped leaves have a bluish fungal growth consisting of mycelia, conidia, and conidiophores on the lower leaf surfaces. Diseased leaves become twisted and turn their lower surface upward, accentuating the bluish color, especially when moisture is available. Diseased plants begin to die and turn light brown. Eventually, the entire plant or all the leaf tissue except the growing tip is killed.

Older diseased plants in seedbeds become deformed and have partially killed leaves or leaves that are twisted and puckered. Irregular-shaped necrotic lesions may develop on leaves. Plants become stunted and roots become dark brown.

Diseased leaves on plants in the field have circular blotches that coalesce and form brown necrotic areas. Diseased leaves also become distorted and puckered and have large areas that disintegrate. Lesions also occur on buds, flowers, and capsules. Diseased plants are stunted and have wilt symptoms and a vascular discoloration that is observed as brown streaks. Stems lodge and roots become dark brown.

Management

1. Apply foliar fungicides to plants in a seedbed. Treatment must begin before disease appears.
2. Destroy beds after setting is complete.
3. Destroy any live plants remaining after harvest.
4. Locate beds in an area where they will not receive morning shade.
5. Grow resistant cultivars. Resistance may be related to the presence of the phenolic compounds scopolin or scopoletin.

Botryosporium Barn Mold

Cause. *Botryosporium longibrachiatum* and *B. pulchrum* survive on infested residue. Host plants are apparently infected or have a latent infection at harvest time. Secondary infection originates from bringing into the barn infested plants on which conidia are produced and spread around the building. Humid, wet conditions favor disease development.

Distribution. Widespread.

Symptoms. White mycelia and conidiophores are confined to lesions (1–20 cm long) on the midribs and on the secondary veins of lower leaves. Lesions initially are water-soaked. Sporulation then occurs on laminae adjacent to the infected areas on the midribs and veins. Lesion development and sporulation occurs after tobacco reaches the brown stage of curing. Eventually, sporulation spreads farther and is evident on all leaf surfaces. Lower leaves drop to the barn floor; those remaining on stalks are severely water-soaked and decayed.

Management. Supply supplemental heating to assist drying.

Brown Spot

Cause. *Alternaria alternata* (syn. *A. tenuis*) survives as mycelia in woody plant residue, such as tobacco stems. During periods of wet weather, conidia are produced and disseminated by water and wind to the lower leaves of host plants. *Alternaria alternata* is also disseminated by infected seed and transplants. Secondary inoculum is conidia that are produced in lesions during moist weather and disseminated by wind to other plant hosts.

Infection and disease development occur at temperatures higher than 21.0°C but are most rapid when temperatures are higher than 26.5°C. Tobacco leaves are infected at a late stage of plant maturity. A metabolite, AT toxin, produced by *A. alternata* induces the "typical" symptoms.

Distribution. Wherever tobacco is grown.

Symptoms. Round brown spots occur on diseased leaves. On inoculated plants, spots are larger on old leaves than on young leaves and frequently are marked by concentric rings. Small spots, which initially are less than 6 mm in diameter, eventually enlarge up to 25 mm in diameter during moist weather, coalesce, and kill large areas of diseased leaves.

Spots occur on stalks late in the growing season and are smaller than on leaves but become more numerous and give diseased stalks a "speckled" appearance. There is a significant decrease in sugar and nicotine alkaloid content in diseased tissue.

Management
1. Sow seed that is free of disease.
2. Fumigate plant beds and remove residue.
3. Rotate tobacco in fields where brown spot was present.
4. Grow tolerant cultivars. In general, burley and dark tobacco are less susceptible than flu-cured tobacco.

Charcoal Rot

Cause. *Macrophomina phaseolina* (syn. *M. phaseoli*) survives as sclerotia in soil and infested residue. Sclerotia germinate and form germ tubes that penetrate the host. Charcoal rot is more severe in dry weather at temperatures of 38°C and above and on plants to which mineral oil was applied during temperatures of 32°C and above.

Distribution. Wherever tobacco is grown in tropical or subtropical areas.

Symptoms. Symptoms are associated with injury caused by mineral oil applied for sucker control. A black lesion develops from which *M. phaseoli* grows inward, causing pith and wood decay, followed by plant death and lodging. Sclerotia are embedded throughout the bark and wood of diseased plants.

Management. Fumigate the soil to kill sclerotia.

Collar Rot

Cause. *Botrytis cinerea* and *Sclerotinia sclerotiorum. Botrytis cinerea* is discussed under "Gray Mold." *Sclerotinia sclerotiorum* survives as sclerotia in infested stalk residue and soil. In the spring, sclerotia germinate and form either mycelia or apothecia. Most primary infections originate from ascospores produced on apothecia and disseminated by wind to plants. However, infrequently infection may originate from the mycelium growing from a sclerotium.

Ascospores produced throughout the summer infect flowers that fall onto leaves and cause dead blossom leaf spot. Injury spots caused by heat and chemicals provide additional infection courts for ascospores to germinate on. Infection also occurs in wounds where the plant tops or suckers were removed. Sclerotia are then produced in the diseased stalks later in the growing season.

Collar rot is more severe during wet weather. Seedlings 35 to 53 days old are the most susceptible. Disease development is enhanced when an external source of nutrient is present on the host leaf surface, such as clipping debris or leaf pieces. Seedlings may be infected up to transplant size.

Distribution. Wherever tobacco is grown.

Symptoms. Tobacco transplants in greenhouses are affected. Initially, a small dark brown lesion with a well-defined margin occurs at the base of a diseased stem. After transplanting, the lesion may circle the stem and kill the plant or become gray, have a less well-defined margin, and develop higher on the stem. A white, cottony, mycelial growth in which small black sclerotia are present grows over a diseased area.

Management
1. Steam or fumigate plant beds.
2. Do not place plant beds in low, wet areas.
3. Provide adequate ventilation.
4. Remove debris from around transplants in the greenhouse.

Corynespora Leaf Spot

Cause. *Corynespora cassiicola* survives as mycelia and chlamydospores in infested tobacco residue and soil for 2 years or more, and as a saprophyte in the residue of other plant hosts. Conidia splashed by water and blown by wind to host leaves cause infection when free moisture is present or the relative humidity is 80% and above. Older leaves are infected first; younger leaves are infected later in the growing season. Disease incidence increases as host plants approach harvest. Several other plant species are hosts.

Distribution. Nigeria.

Symptoms. Dark brown circular spots (2–3 mm in diameter initially) that occur on leaves enlarge to 20 to 30 mm in diameter. Spots tend to be zonated: the dark brown center is surrounded by a light brown zone, a thin dark brown ring, and finally a light brown margin. Spots may coalesce. Petioles and midribs have dark brown spots; later veins become brown, starting from the midrib.

Disease causes a reduction in sugars, phenol, and nicotine alkaloid content of leaves and an increase in nitrogen content. The reduction of phenol content causes an undesirable aroma.

Management
1. Harvest on time since mature leaves are likely to become infected.
2. Do not grow any known alternate host, such as cotton, soybean, cow-

pea, cucumbers, lupines, and watermelons, in the vicinity of tobacco fields.

Curvularia Leaf Spot

Cause. *Curvularia verruculosa.*

Distribution. India.

Symptoms. Spots that appear on diseased leaves as round to oval areas with concentric zones enlarge rapidly and affect the entire lamina. Conidia and conidiophores appear on both leaf surfaces. Diseased leaves initially are yellow but later turn dark brown.

Management. Do not rotate tobacco with rice.

Damping Off

Cause. *Pythium* spp., including *P. aphanidermatum, P. debaryanum,* and *P. ulti-mum.* Other *Pythium* spp. may also be involved. The fungi survive in soil as oospores and chlamydospores, which are sometimes referred to as sporangia in the literature. Although dissemination is by any means that moves soil, water is the most important. Both structures germinate by germ tubes or form zoospores, which in turn, produce germ tubes that infect host stems and roots at or just below the soil line. Oospores are eventually formed in diseased tissue. Damping off is more severe in water-saturated soils at temperatures below 24°C.

Distribution. Wherever tobacco is grown.

Symptoms. Damping off is primarily a problem in plant beds. Commonly a brown, watery, soft rot that develops on hypocotyls soon girdles the hypocotyls and causes seedlings to fall over. Roots may also be diseased and decayed.

Older seedlings become chlorotic. The roots have a soft, watery rot and the cortex eventually peels away from the wooden central cylinder. If plants are infected at the soil line, roots are not rotted and remain white.

Transplants have brown, watery lesions that cause stems to become limp, shrivel, and disintegrate. Plants suddenly wilt and die, especially when weather is cool and damp following setting.

Management
1. Disinfect seedbeds.
2. Avoid wet soils as plant bed sites.
3. Avoid dense stands and provide adequate ventilation during wet, humid periods.
4. Apply fungicide.
5. Disk and reset the field if loss of transplanted seedlings is high. New rows should be placed in the middle between the old diseased ones.

Dead Blossom Leaf Spot

Cause. *Botrytis cinerea* and *Sclerotinia sclerotiorum. (See* "Collar Rot." *Botrytis cinerea* is also discussed under "Gray Mold.")

Distribution. Wherever tobacco is grown.

Symptoms. Diseased flowers fall and adhere to leaves. Small dark spots that develop on leaves where the flowers have fallen eventually become larger and brown to gray. During wet weather, a spot may enlarge enough to cover half of a diseased leaf. If the midrib is rotted, the leaf becomes yellow and hangs down. Brown lesions several centimeters long may occur on stems and cause all leaves above the lesion to die.

Management. No practical management exists for dead blossom leaf spot except late-flowering cultivars are usually less affected than early-flowering cultivars.

Frogeye

The curing phase of frogeye is called barn spot or green spot.

Cause. *Cercospora nicotianae* (syn. *C. apii,* f. sp. *nicotianae*) overwinters as mycelia in infested tobacco residue and in the residue of other plant hosts. In the spring, conidia are produced and disseminated by wind and splashing rain to host leaves. Mature leaves are most susceptible to infection; subsequent lesions usually occur on the lower leaves. Disease development is more severe during warm, wet weather.

Distribution. Africa, Central America, Southeast Asia, and the United States.

Symptoms. Frogeye symptoms may occur on both seedlings and plants in the field. Lesions are small (2–15 mm in diameter) and brown, gray, or tan and have a "paper-like" center. Dark dots, which are groups of conidia and conidiophores, may be found in lesions. Near harvest time, the upper leaves develop large necrotic spots, particularly during wet weather, and are quickly destroyed. During dry weather, frogeye lesions may be only the size of a pinpoint.

Disease results in significant decreases in sugar and nicotine alkaloid content.

Management
1. Sow only seed that is free of disease.
2. Remove residue from plant beds.
3. Apply foliar fungicides before transplanting.
4. Do not overfertilize with nitrogen.
5. Prime tobacco before it becomes overripe.
6. Plow under infested residue in the field.
7. Rotate tobacco with other crops.
8. Cure at a temperature of 38°C and 100% relative humidity.

Fusarium Head Blight

Fusarium head blight is also called blotch and scab.

Cause. *Hymenula affinis* (syn. *Fusarium affine*) was originally identified as *Septomyxa affinis*. Fusarium head blight is most severe in plant beds during wet years or in damp, shaded areas of the plant bed. Light-colored nitrogen-starved plants and those diseased with black root rot are more susceptible to infection by *H. affinis*.

Distribution. The United States and Zimbabwe.

Symptoms. Diseased upper leaf surfaces, petioles, and stems have olive-brown, irregular-shaped blotches and streaks. During wet weather, diseased plant parts develop a soft rot, which results in the leaves having a ragged appearance. The petioles and stems eventually decompose. Slightly diseased plants recover when they are transplanted to the field.

Management. Not reported.

Fusarium Wilt

Cause. *Fusarium oxysporum* (syn. *F. oxysporum* f. sp. *nicotianae*) survives as chlamydospores in soil and infested residue, as mycelia in infested residue, and on roots of symptomless hosts. *Fusarium oxysporum* is disseminated in tobacco transplants and by any means that moves soil. Chlamydospores germinate and form mycelia that grow from the residue to host roots, where entry is gained into the roots through wounds. Mycelium then grows into the xylem, or water-conducting, tissue.

Wilt severity is increased in the presence of the tobacco cyst nematode, *Globodera tabacum*. This phenomenon is most pronounced with high nematode numbers, low fungal population densities, and *Fusarium*-resistant tobacco lines.

Macroconidia and microconidia are produced on the outside of moribund plant tissue. Microconidia are also formed in the vessels of xylem tissue. Mycelial cells, macroconidia, and microconidia are converted into chlamydospores and returned to soil when the diseased tissue decomposes. Fusarium wilt is most severe at soil temperatures from 28° to 31°C.

Distribution. Wherever tobacco is grown.

Symptoms. Initially, leaves on one side of diseased plants are dwarfed and yellowed. Eventually, the leaves wilt and become brown. The plant top usually becomes yellow and bends toward the affected side of the diseased plant. This condition may continue for some time before the diseased plant wilts and dies. When the soft outer bark of the stem is removed, the surface of the exposed wood is dark chocolate brown.

Management
1. Grow resistant cultivars.
2. Fumigate soil with a nematicide where nematodes are a problem.
3. Rotate tobacco with a resistant crop. Flu-cured, but not burley and dark, tobacco can be grown in rotation with cotton. Do not grow sweet potatoes in the same rotation with cotton.

Gray Mold

Cause. *Botrytis cinerea*, teleomorph *Botryotinia fuckeliana*, survives as sclerotia in soil and infested residue, and as saprophytic mycelia on the residue of several plant hosts. Sclerotia germinate during wet weather and produce mycelia on which conidia are borne and disseminated by wind and splashing rain to tobacco. Stalk-cut tobacco is infected through wounds and injuries. Sclerotia are formed on the surface of diseased tissue.

Distribution. Wherever tobacco is grown.

Symptoms. The first symptoms occur as spots on the lower diseased leaves when seedlings are ready to be transplanted. If the weather is dry after infection has been initiated and subsequent disease has started, spots are limited in growth and become brown with a yellow margin. Conversely, during wet weather, a gray mycelial growth covers the spots and diseased leaves collapse but remain attached to stems. The fungus grows from leaf petioles into the stem and causes lesions several centimeters long that are covered with gray mycelial growth. Diseased plants usually die; however, slightly diseased plants recover during dry weather.

Management
1. Place seedbeds on a well-drained site with adequate ventilation.
2. Apply foliar fungicides.

Olpidium Seedling Blight

Cause. *Olpidium brassicae* survives in soil and infested residue as resting sporangia. In the presence of moisture, sporangia discharge zoospores that "swim" to roots, where they encyst. The cyst protoplast moves through the host wall and establishes itself in the host cytoplasm. Resting sporangia eventually form in roots. Olpidium seedling blight is most severe in moist and cool soils at temperatures from 10° to 16°C.

Distribution. Wherever tobacco is grown.

Symptoms. The aboveground portions of a diseased plant will yellow and wither. Roots have a brown decay.

Management. Fumigate seedbeds.

Phyllosticta Leaf Spot

Cause. *Phyllosticta nicotiana. Nicotiana* is sometimes spelled "nicotianae" in the literature. Other *Phyllosticta* species reported on tobacco are *P. tabaci, P. capsulicola,* and *P. nicotianicola.*

    Phyllosticta nicotiana overwinters as pycnidia in infested residue and is seedborne. *Phyllosticta nicotiana* is disseminated into fields on transplants. During wet weather, pycnidiospores are produced in pycnidia and disseminated by wind and splashing rain to "stressed," dead, or dying tobacco leaves, where infection occurs.

Distribution. Africa, the Caribbean, Europe, South America, and the United States.

Symptoms. Leaf lesions are brown irregular-shaped (1–10 mm in diameter) zonate spots. Lesions are dark brown in the center and lighter brown toward the lesion margin and have a chlorotic halo that surrounds the brown or necrotic areas. Pycnidia, which appear as small black dots, are embedded in the necrotic tissue.

    Other symptoms are small white spots whose centers fall out and give the diseased leaf a shot-hole effect. Eventually, large white blotches occur that grow up to 15 mm in diameter. Sometimes all interveinal tissue falls out, which gives a diseased leaf a skeletal appearance. Pycnidia develop in the white tissue surrounding the shot holes.

Management. Phyllosticta leaf spot is rarely severe enough to warrant disease management measures.

Powdery Mildew

Cause. *Erysiphe cichoracearum* overwinters as perithecia on infested residue, and as mycelia in tobacco plants and weed hosts left standing in a field after harvest. Primary inoculum is either ascospores that are released from perithecia or conidia produced on mycelia. Both spore types are disseminated by wind to host leaves. Secondary inoculum is provided by conidia produced on leaves that are airborne to other host plants. As tobacco matures, perithecia form in fungal growth on diseased leaves and stems.

    Erysiphe cichoracearum is tolerant of a wide range of temperatures and humidities. Disease development is more severe under reduced light intensity.

Distribution. Africa, Asia, Central America, Europe, and South America.

Symptoms. Plants in the field do not display symptoms until approximately 6 weeks after planting. The initial felt-like patches that appear on diseased lower leaf surfaces eventually cover the entire leaf surface. Soon, a powdery gray layer consisting of conidia, conidiophores, and mycelium covers both leaf surfaces. Eventually, brown spots appear on the diseased upper leaf surfaces.

Diseased leaves become thin and "papery" in texture. The black, round perithecia that form on diseased leaf surfaces can be viewed with a hand lens.

Management
1. Grow resistant cultivars.
2. Properly fertilize soil.
3. Plant tobacco early.
4. Apply foliar fungicide.

Rhizoctonia Leaf Spot

Cause. *Rhizoctonia solani* AG-2-2 and AG-3, teleomorph *Thanatephorous cucumeris.* Rhizoctonia leaf spot is favored by cool, wet weather.

Distribution. Brazil, Costa Rica, and the United States (North Carolina).

Symptoms. Rhizoctonia leaf spot occurs on flu-cured tobacco. The initial symptom is small, circular, water-soaked spots that rapidly expand into light green to tan lesions (2–6 cm in diameter) with an irregular margin. The tissue within lesions is almost transparent and often displays a pattern of concentric rings. This transparent tissue frequently drops out and gives a shot-hole appearance to diseased leaves. Fungal mycelia may be present at the margins of lesions on the undersides of diseased leaves. Occasionally a hymenial layer and the basidiospores of *T. cucumeris* are observed in these same locations.

Lesions are most common on the lower leaves of a diseased plant but may occur as high on a stalk as 85 cm above the soil.

Management. Not reported.

Rhizopus Stem Rot

Cause. *Rhizopus arrhizus.* Rhizopus stem rot is apparently favored by high temperatures that occur during the summer months.

Distribution. Iraq, on *Nicotiana glauca.*

Symptoms. Initially, pale green, water-soaked lesions appear on the stems immediately below the inflorescence. Disease symptoms extend upward and downward from the original lesion and result in a slimy, wet rot of cortical tissues. Lesions may be confined to one side of the diseased stem or may completely encompass it. Leaf petioles attached to diseased portions of the stem become flaccid and die. Diseased stem tissues appear pale to yellow-brown when dry. Frequently a flower head will topple downward, bending over at the diseased portion of the stem but remaining attached to the plant, or be broken off by strong winds. Severe disease results in plant death. Fluffy mycelial growth of *R. arrhizus* is visible in the piths of diseased stems.

Management. Not reported.

Rust *(Puccinia substriata)*

Cause. *Puccinia substriata.* Aecia are produced on tobacco; uredinia and telia stages occur on various grasses, including *Paspalum* spp., *Digitaria* spp., and *Setaria* spp. *Puccinia substriata* is a weak pathogen that infects tobacco plants during the rainy season.

Distribution. Honduras, Nicaragua, and Zimbabwe.

Symptoms. Diseased tobacco leaves are severely spotted. Round, raised lesions (1 cm in diameter) composed of thick, hard tissue occur on diseased upper leaf surfaces and remain a light green after curing. Aecia, which form on the undersides of lesions, have dense concentric rings that vary in color from cream to orange. Advanced symptoms of what is thought to be the same disease cause tissues to turn black and fall out, leaving a hole in the diseased leaf.

Management. Not necessary.

Rust *(Uredo nicotianae)*

Cause. *Uredo nicotianae.* Uredinia are produced on tobacco.

Distribution. Brazil, Italy, and the United States (southern California).

Symptoms. Brown sori are produced on diseased leaves.

Management. Not necessary.

Sooty Mold

Cause. *Fumago vagans* grows on honeydew secreted on leaves by aphids, but the fungus does not infect the tobacco plant itself.

Distribution. Formosa, Italy, Japan, the United States, and Zimbabwe. Sooty mold rarely occurs on tobacco.

Symptoms. The lower mature leaves are most often affected. Fungus mycelium, chlamydospores, and conidia occur as a superficial black film or sooty layer that can be easily scraped from leaf surfaces with a fingernail.

Management. Control aphids.

Soreshin

Soreshin is also called canker, collar rot, blackleg, rotten stalk, stem rot, and soreshank.

Cause. *Rhizoctonia solani*, teleomorph *Thanatephorus cucumeris*, persists as mycelia and sclerotia in soil, as a saprophyte on the residues of a large number of plant species, and as a parasite and pathogen on the roots of weeds and other crops grown in rotation with tobacco. *Thanatephorus cucumeris* survives for a few weeks as basidiospores in soil.

Roots are infected by mycelia from germinating sclerotia and basidiospores, or directly from mycelia growing in soil and from infested residue. *Rhizoctonia solani* is disseminated as sclerotia and mycelia by any means that moves soil or residue. Disease development can occur over a wide temperature range in either relatively dry or moist soils. Soreshin can occur during temperatures of 20°C and lower in a plant bed and during relatively high temperatures in the field.

Distribution. Wherever tobacco is grown. Soreshin is generally not a serious disease.

Symptoms. A dark brown decayed area occurs at or near the soil line on the stems of seedlings and transplants and extends upward and around the stems until the plant topples over. If slightly diseased plants are set in a field during cool, wet weather, a poor stand will result.

A dark brown lesion occurs at or near the soil line on the stems of older diseased plants in the field and eventually girdles the stem. The canker may extend up the stem to the lower leaves and cause them to drop off. The woody part of the diseased stalk becomes hard and brittle, and the pith is decayed, dried, and brown. Light gray patches of mycelium may be present in the diseased stalk. The entire plant may appear stunted, yellow, and wilted.

Diseased plants are usually not noticed until the wind, or another force, topples the plant over. Roots generally remain healthy until the plant dies.

Management. Management measures can be used in the plant bed but not in the field. Experimentally, *Trichoderma harzianum* added to seedbeds fumigated with methyl bromide before seed was sown controlled *R. solani*. Triadimenol fungicide integrated with *T. harzianum* enhanced disease management.

1. Grow plants on a well-drained soil.
2. Disinfect seedbeds with steam or a fumigant.

Southern Stem and Root Rot

Southern stem and root rot is also called southern blight.

Cause. *Sclerotium rolfsii,* teleomorph *Athelia rolfsii,* survives as sclerotia in soil and on infested residue, and as mycelia in infested residue. *Sclerotium rolfsii* is disseminated as sclerotia in seed lots and by any means that moves soil. Sclerotia germinate and form infection hyphae that penetrate the host plant stalk. Infection also occurs by mycelium growing from precolonized host plant residue. Sclerotia are formed on the outside of diseased tissue.

Southern stem and root rot is more severe at high temperatures (30° to 35°C) and high soil moisture. *Sclerotium rolfsii* is killed by low temperatures.

Distribution. Wherever tobacco is grown in warm climates.

Symptoms. Diseased plants are usually scattered throughout a field but may occur in groups within a field. Initially, a yellowing and wilting of diseased lower leaves occurs that gradually progresses up the plant. Eventually, leaves die and turn brown. The stem at the soil line has a dry, brown, sunken canker that completely girdles the stem. The roots usually do not decay until the entire diseased plant dies. Under moist conditions, a white, cotton-like mycelium occurs on the canker Later, small, white to dark brown sclerotia the size of mustard seeds appear on and in the mycelium.

Management
1. Bury infested residue at least 8 cm deep.
2. Do not place soil containing organic matter against stems when cultivating.

Target Spot

Target spot is also called Rhizoctonia leaf spot.

Cause. *Thanatephorus cucumeris,* anamorph *Rhizoctonia solani* AG-3. Disease is associated with periods of frequent rainfall and below normal temperatures of 20° to 30°C in June and July in North Carolina. Basidiospores are produced either on diseased plants or on infested plant residue on the soil surface and disseminated by wind to other plant hosts.

Distribution. Brazil, Canada (Ontario), Costa Rica, South Africa, the United States (Kentucky, North Carolina, South Carolina, Tennessee, and Virginia), and Zimbabwe. Target spot probably occurs worldwide.

Symptoms. Plants of all ages become diseased. Target spot occurs on plants in the seedbed or greenhouse only after leaves are large enough to close the canopy. The disease occurs in the field after plant leaves are large enough to shade the soil surface and the lower leaves of the plant.

Initially, symptoms are small (2–3 mm in diameter), circular, water-soaked spots. When diseased leaves on seedlings are held up to a light

source, lesions have a "net-like" appearance. The initial lesions usually remain distinct even as they enlarge. During high relative humidity and
moderate temperature, lesions enlarge rapidly, become light green and almost transparent, and have irregular margins and chlorotic halos. In less
humid conditions, lesions expand more slowly and develop a pattern of
concentric rings similar to lesions of Alternaria brown spot. A chlorotic
halo is present around many expanding lesions. The necrotic tissue becomes brittle and often drops out, giving a shot-hole appearance to the
diseased leaf.

   Mycelium is often present at lesion margins on the top and bottom leaf
surfaces. Hymenial layers with numerous basidiospores are produced on
both the diseased upper and lower leaf surfaces.

Management
1. Practice sanitation to prevent introduction of inoculum into seedling
    production beds or greenhouses.
2. Limit wetting of foliage in the greenhouse by proper spacing and venti-
    lation, and by flood irrigating.
3. Eliminate primary inoculum from previously used trays with methyl
    bromide and steam treatments of 80°C for 0.5 to 2.0 hours.

Tobacco Stunt

Cause. *Glomus macrocarpum* is an endogonaceous mycorrhizal fungus that
sporulates profusely on the roots of severely stunted plants.

Distribution. The United States (Kentucky).

Symptoms. Diseased plants are stunted, flowering is delayed, and yield and
quality are reduced. Roots do not become necrotic, and diseased plants are
seldom killed. Since roots apparently are not colonized by arbuscules or external hyphae, interference with the hormonal control of root initiation
may be the mechanism of stunting rather than competition for photosyn-
thate.

   There is a significant reduction in the number of propagules of
Thielaviopsis basicola in soil.

Management. Experimentally, tobacco stunt is managed by soil fumigation.

Verticillium Wilt

Cause. *Verticillium dahliae* and *V. albo-atrum*. *Verticillium* spp. survive as microsclerotia in soil and infested residue. Microsclerotia germinate and produce chlamydospores or conidia, which in turn germinate and produce
hyphae that infect host roots and grow into the xylem tissue. Microsclerotia are abundantly produced in diseased tissue and are released into soil
when residue decays. Microsclerotia and conidia are disseminated by

wind, water, and any means that moves soil. Verticillium wilt is most severe when abundant moisture follows dry weather at temperatures of 22° to 28°C.

Distribution. Wherever tobacco is grown. Verticillium wilt is a severe disease in New Zealand.

Symptoms. Symptoms are not obvious until flowering time. One or more lower leaves on a diseased plant wilt, particularly during hot weather. Eventually, the interveinal areas of lower wilted leaves turn a bright orange or orange-yellow, then become chlorotic. Eventually, the tissue dies, becomes brown, and has an orange border between the living and dead tissue. Wilting sometimes occurs on one side of a diseased leaf or on one side of an entire plant. However, all leaves on a diseased plant eventually die. The vascular system of leaves, petioles, and stems becomes light brown.

Management
1. Fumigate plant beds.
2. Manage nematodes.
3. Grow resistant cultivars.

Diseases Caused by Nematodes

Brown Root Rot

Cause. *Pratylenchus* spp., particularly the lesion nematodes *P. pratensis, P. brachyurus,* and *P. zeae.* The nematodes overwinter as adults, eggs, or larvae in roots and soil. Eggs are deposited in root tissue and hatch in 6 to 17 days. The resulting larvae will either continue to feed in the same root or emerge and migrate to another host root. Eggs may also be deposited into soil by decaying roots. The larvae penetrate the root and migrate within the infected plant but mainly in the cortex.

The different lesion nematodes are favored by different soil temperatures; therefore, disease development can be expected to occur over a wide temperature range. Although lesion nematodes are sensitive to drying, soil moisture ordinarily is not a factor in disease development.

Distribution. Canada and the United States.

Symptoms. Symptoms usually occur in a defined area in a field, but occasionally an entire field may be affected. Diseased plants are stunted and wilt during the day, recover their turgidity at night, but repeatedly wilt again. Eventually, the repeated wilting causes leaves in the middle and lower portions of stalks to have brown and necrotic margins.

Roots, at first, will have colorless and water-soaked lesions. Eventually, cortical lesions vary from pale yellow to dark brown or black and girdle feeder roots. Lesions break open and cortical tissue sloughs off, leaving only the vascular cylinder. Numerous adventitious roots often develop above a lesion and give a diseased root a bushy appearance. Diseased roots shrivel and die, resulting in extensive root pruning. If pruned roots are numerous, the root system will also have a stubby appearance. Severely diseased plants can easily be pulled from the soil because of the almost complete destruction of the root system. The surviving roots are usually grouped near the soil surface.

Management
1. There are no resistant cultivars, but some cultivars are injured less than others.
2. Crop rotation. Do not grow tobacco after timothy, rye, maize, cotton, bluegrass, and legumes since these plants support large populations of *Pratylenchus* spp. Some of these plants also release toxic chemicals that may harm tobacco roots; therefore, crops should be turned over in time to allow for decomposition of residue before planting tobacco.

Bulb and Stem

Bulb and stem is also called stem break.

Cause. *Ditylenchus dipsaci* survives as the fourth larval stage in dry plant material and as larvae and eggs in the soil for several years. Eggs hatch and the resulting larvae migrate to host plant roots and up the plant in a film of water. Larvae, which enter the stem through stomates, lenticels, and wounds, are confined to cortical tissue.

Stem break is favored by cool temperatures (15° to 20°C) and wet weather. Disease incidence is usually more severe on plants grown in moist loam or clay soils. *Ditylenchus dipsaci* has a wide host range.

Distribution. France, Germany, Holland, and Switzerland.

Symptoms. Young plants become more severely diseased than do older ones. Initially, small, yellow swellings or galls on diseased stems may occur up to 40 cm or more above the soil level. Older galls die, causing stems to turn black. Diseased plants stop growing and the upper leaves become yellow, while the lower ones fall off. Eventually, stems break and plants fall over but the vascular system remains intact. Therefore, wilting usually does not occur until just before plants break over.

Management
1. Remove diseased plants from fields.
2. Rotate tobacco with resistant crops for long intervals.
3. Fumigate soil with nematicides.
4. Apply contact nematicides to plants.

Cyst

Cause. *Globodera solanacearum* (syn. *G. virginiae*) and *G. tabacum* survive as eggs in soil and as larvae in cysts. Dissemination is by any means that moves soil. Larvae emerge from cysts and migrate to roots, where they begin to feed with their heads in the stele of the rootlet. Females become spherical and break through the root epidermis with only their heads and necks remaining in the root. The female is fertilized by males and eggs are produced in the female body. Many eggs hatch and reinvade rootlets. When females die, their bodies become resistant cysts containing hundreds of eggs. Injury is most severe under continuous tobacco culture.

Distribution. The eastern United States.

Symptoms. Diseased plants wilt; they are stunted and have a small root system. Dark brown oval cysts (0.5 mm in diameter) are attached to roots.

Management
1. Rotate to a nonhost crop. Fescue has proven successful in Virginia.
2. Apply nematicides.
3. Grow resistant cultivars; however, these cultivars are not tolerant.
4. Practice sanitation with equipment and irrigation water.

Dagger

Cause. *Xiphinema americanum.*

Distribution. The United States (North Carolina).

Symptoms. Diseased plants are stunted and have a reduced root system.

Management. Not reported.

Foliar

Cause. *Aphelenchoides ritzemabosi.*

Distribution. Unknown.

Symptoms. *Aphelenchoides ritzemabosi* attacks buds and foliage, causing retarded growth and small, distorted foliage. On some diseased plants, leaf spots and defoliation occur. Symptoms on tobacco are uncertain.

Management. Not known.

Lesion

Cause. *Pratylenchus brachyurus* and *P. penetrans* are endoparasites. All stages of the nematode overwinter in soil and in living and dead host roots. Dissemination is by any means that transports soil. Eggs are deposited in the cortex of host roots and in soil. Juveniles feed in the same root as the egg was deposited in or migrate to other roots. Eggs released into soil also hatch and the juveniles are attracted to host roots by root exudates. The juveniles penetrate through the root cortex and feed on cells, thereby destroying them.

Distribution. Worldwide.

Symptoms. Diseased plants occur singly or in small groups. Plants are stunted, chlorotic, and wilt during the day. Leaves have chlorotic margins that eventually become necrotic. Water-soaked spots that occur behind the root growing point become tan to black and may girdle the root. Cortex tissues slough off, leaving only the vascular cylinder. Extensive root pruning may give root systems a stubby appearance. Diseased plants may be easily pulled from the soil.

Management
1. Rotate tobacco with nonhost crops.
2. Grow the most tolerant cultivars.

Reniform

Cause. *Rotylenchulus reniformis* is an ectoparasite.

Distribution. Unknown.

Symptoms. Diseased plants are stunted and have a retarded root system.

Management. Unknown, but it is likely that measures for managing other nematodes on tobacco will aid in managing reniform nematodes.

Root Knot

Cause. The root knot nematodes *Meloidogyne arenaria, M. hapla, M. incognita,* and *M. javanica* survive in soil as eggs and larvae, and in galls and roots as adults and eggs. Eggs in soil hatch and produce larvae that move through soil water and enter roots, where they remain the rest of their lives. Nematodes are disseminated by any means that moves soil. When females are mature, eggs are produced in a sac that emerges from female genital openings prior to egg production. Eggs are released into the soil as the diseased roots, together with the galls, decay.

Nematodes survive with difficulty at subzero temperatures. *Meloidogyne arenaria, M. incognita,* and *M. javanica* do not survive if the temperature averages below 3°C during the coldest month. *Meloidogyne hapla* is limited by temperatures greater than 23°C. Disease is ordinarily more severe in lighter sandy soils but may occur in any soil type. Soil moisture is not an important factor in disease development. Different races or strains of *Meloidogyne* spp. exist.

Distribution. Wherever tobacco is grown.

Symptoms. Diseased plants occur at random throughout a field and are noticed by their stunted and yellowed appearance. Plants may occasionally be killed, especially during dry weather. Plants may wilt during the day but recover their turgidity at night; subsequent wilting reoccurs until leaves in the middle and lower part of diseased plants develop necrotic margins and tips.

Galls form on roots and vary in size from a pinhead to several times the thickness of the root. Galls vary in shape from irregular to spherical and are most often found on rootlets, where they resemble beads on a string. Galls may form so close together that they resemble one large gall.

Management
1. Grow resistant cultivars. Resistance to *M. incognita* races 2 and 4 and moderate resistance to *M. javanica* have been identified.
2. Treat soil with a nematicide.
3. Rotate tobacco with resistant crops, such as small grains. Plant tobacco in the same soil only once every 3 to 4 years.
4. Plow or disk as soon after harvest as possible. Exposing roots to the drying action of the sun and air kills many nematodes. As long as tobacco stalks and root systems are left in the field undisturbed, plants will be alive and nematodes will continue to live and reproduce in them.

Spiral

Cause. *Helicotylenchus* spp.

Distribution. Unknown.

Symptoms. Diseased plants are stunted and have a retarded root system.

Management. Unknown, but it is likely measures that aid in managing other nematodes on tobacco will reduce populations of spiral nematodes.

Stubby Root

Cause. *Paratrichodorus* spp. and *Trichodorus* spp. are nematodes with a wide host range. Little is known of their life cycle, but survival likely occurs as eggs, larvae, and adults in soil. An increase in nematode population is favored by soil temperatures of 22° to 30°C and well-aerated, light sandy soils. *Trichodorus* spp. are found at depths down to 100 cm in the soil profile.

Distribution. Throughout the world.

Symptoms. Plant growth is retarded. Diseased plants wilt easily, even in the presence of moisture. Root tips stop growing, which results in a few short, stunted, stubby roots.

Management. Same as for root knot, but stubby root nematodes are more difficult to manage.

Stunt

Cause. *Merlinius* spp. and *Tylenchorhynchus* spp. *Tylenchorhynchus claytoni*, the principal stunt nematode infecting tobacco, survives for several months in soil as eggs, larvae, and adults if the soil does not become dried out. The nematode has a wide host range.

Distribution. Canada and southeastern United States.

Symptoms. Diseased plants are stunted and have a small root system. Diseased roots are shriveled and flaccid, do not elongate normally, and, in general, are poorly developed.

Management
1. Grow resistant cultivars.
2. Treat soil with a nematicide.
3. Rotate tobacco with resistant crops, such as small grains. Grow tobacco in the same soil only once every 3 to 4 years.
4. Plow or disk as soon after harvest as possible. Exposing roots to the drying action of the sun and air kills many nematodes. As long as tobacco stalks and root systems are left in the field undisturbed, plants will be alive and nematodes will continue to live and reproduce in them.

Diseases Caused by Phytoplasmas

The diseases of tobacco caused by phytoplasmas tend to be similar in symptoms and management practices.

Aster Yellows

Cause. Aster yellows phytoplasma infects and persists in more than 175 species of plants. It is disseminated by several species of leafhoppers, including *Aphrodes bicintus, Hyalesthes obsoletus, Euscelis plebejus, Macrosteles fascifrons, M. laevis,* and others. The phytoplasma multiplies in leafhoppers and overwinters in leafhopper eggs and adults. The optimum temperature for multiplication in plant and insect tissue is 25°C. Leafhoppers may be blown by the wind or fly several hundred kilometers.

Distribution. Wherever tobacco is grown.

Symptoms. Diseased plants are stunted. Apical leaves are small, whitish, and curled. Older leaves have interveinal necrosis and hang close to the stem.

Numerous small leaves are present on sucker growth. Flowers change to green leaf-like structures that are intermingled with numerous short, stiff branches. The proliferation of floral parts results in a compact, tufted growth habit.

Management
1. Eliminate weed hosts in tobacco fields and in fields adjacent to tobacco fields.
2. Plant only transplants that are free of disease.
3. Avoid growing tobacco near other host plants, such as potatoes, tomatoes, and peppers.
4. Plant when leafhopper populations are low.
5. Harvest and cure separately if only a few plants are diseased.

Bigbud

Cause. Bigbud phytoplasma infects and persists in several species of plants. The bigbud phytoplasma overwinters in the eggs and adults of the leafhopper *Orosius argentatus*. Later, the phytoplasma multiplies in and is disseminated by the same leafhopper. The leafhopper does not breed on tobacco, so transmission occurs during periods of migration in late spring and early summer. Conditions favoring leafhopper flight are a 9:00 p.m. temperature above 27°C, high humidity, and no wind.

Distribution. Australia.

Symptoms. Symptoms are similar to those of aster yellows. Diseased plants are stunted. Apical leaves are small, whitish, and curled. Older leaves have interveinal necrosis and hang close to the stem. Numerous small leaves are present on sucker growth. Flowers change to green leaf-like structures that are intermingled with numerous short, stiff branches. The proliferation of floral parts results in a compact, tufted growth habit.

Management
1. Eliminate weed hosts in tobacco fields and in fields adjacent to tobacco fields.
2. Plant only transplants that are free of disease.
3. Avoid growing tobacco near other host plants, such as potatoes, tomatoes, and peppers.
4. Plant when leafhopper populations are low.
5. Harvest and cure separately if only a few plants are diseased.

Stolbur

Cause. Stolbur phytoplasma overwinters primarily in the common perennial bindweed, *Convolvulus arvensis,* and less often in several other species of plants and in the eggs and adults of leafhoppers. *Nicotiana glauca* is a symptomless host. The phytoplasma is disseminated and multiplies in several species of leafhoppers that are blown or fly several hundred kilometers. Stolbur phytoplasma strains are vector-specific; some strains are transmitted principally by one species of leafhopper. Leafhoppers inoculate the phytoplasma directly into the phloem.

Distribution. Europe.

Symptoms. Younger plants become more severely diseased than plants infected later in the growing season. Symptoms are less severe during cool weather. Diseased apical leaves are small, whitish, and curled. Other diseased leaves have interveinal necrosis and hang close to the stem. Numerous small leaves are present on sucker growth. Cured diseased leaves are more hygroscopic than are healthy leaves. When bulked with healthy leaves, diseased leaves interfere with fermentation and decrease quality.

Flowers are changed to green, leaf-like structures that result in a compact, tufted growth habit. Diseased plants do not set seed even though the pollen is healthy. Diseased plants in general are stunted and form numerous short, stiff branches. Internal symptoms are a degeneration of the phloem and irregularities in meiotic processes.

Management
1. Eliminate weed hosts in tobacco fields and in fields adjacent to tobacco fields.
2. Plant only transplants that are free of disease.
3. Avoid growing tobacco near other host plants, such as potatoes, tomatoes, and peppers.
4. Plant when leafhopper populations are low.
5. Harvest and cure separately if only a few plants are affected.

Yellow Dwarf

Cause. Yellow dwarf phytoplasma infects and persists in several species of plants and overwinters in the eggs and adults of the leafhopper *Orosius argentatus*. The phytoplasma multiplies and is disseminated only by *O. argentatus*. *Orosius argentatus* does not breed on tobacco, so transmission occurs during periods of migration in late spring and early summer. Infection may occur at any stage of plant growth but small, rapidly growing plants are highly susceptible. Conditions favoring leafhopper flight are a 9:00 p.m. temperature above 26°C, high humidity, and no wind.

Distribution. Australia.

Symptoms. Plants are yellowish and dwarfed. They grow slowly and produce small leaves that are unsuitable for commercial use. Older diseased leaves become thick and wrinkled. Root systems remain undeveloped.

Management
1. Eliminate weed hosts in tobacco fields and in fields adjacent to tobacco fields.
2. Plant only transplants that are free of disease.
3. Avoid growing tobacco near other host plants, such as potatoes, tomatoes, and peppers.
4. Plant when leafhopper populations are low.
5. Harvest and cure separately if only a few plants are affected.

Disease Caused by Viroids

Nicotiana glutinosa Stunt Viroid

Cause. *Nicotiana glutinosa* stunt viroid (NgSVd) is a low molecular weight RNA. NgSVd is mechanically transferred to *N. clevelandii, N. debneyi, N. plumbaginifolia, Chenopodium amaranticolor, Cucumis sativus, Luffa acutagula, Cajanus cajan,* and *Vigna unguiculata.*

Distribution. India.

Symptoms. Diseased plants are characterized by an epinasty of leaves, stunting, bunchiness, and occasional necrosis.

Management. Not reported.

Diseases Caused by Viruses

Alfalfa Mosaic

Cause. Alfalfa mosaic virus (AMV) overwinters in a wide number of hosts and is transmitted by aphids. Several strains of AMV exist and all the strains may be found in one plant.

Distribution. Europe, Japan, New Zealand, and the United States. AMV has not been a serious disease of tobacco.

Symptoms. Initially, chlorotic spots and blotches occur on diseased leaves. The first leaves infected systemically have a vein clearing followed by white rings, arcs, and coalescing line patterns of necrotic tissue. Sometimes bud leaves are distorted and have a bright yellow mosaic.

Management
1. Isolate tobacco fields from sources of inoculum.
2. Control weeds and aphids.
3. Experimentally, salicylate watered into the soil prevented systemic infection by AMV.

4. Some cultivars of *Nicotiana tabacum* display more severe symptoms than others, and some *Nicotiana* spp. display no symptoms when infected. *Nicotiana debneyi* had no symptoms and a negative ELISA reading when inoculated with two AMV isolates.

Beet Curly Top

Cause. Beet curly top virus (BCTV) is in the geminivirus group. BCTV overwinters in more than 244 species of host plants and is transmitted by the leafhoppers *Circulifer tenellus* in North America and by *Agallia albidula* and *Agalliana ensigera* in South America. BCTV can persist up to 85 days in leafhopper vectors. The acquisition period takes up to 2 days and the latent period in leafhoppers is from 4 to 24 hours. BCTV does not multiply in vectors, which causes leafhoppers to lose infectivity in 8 to 10 weeks.

Young plants are most susceptible to infection. Different strains of BCTV exist.

Distribution. Brazil and the United States.

Symptoms. Initially, vein clearing occurs and leaf tips and margins bend downward. Larger veins are restricted in growth, which results in a rolling and crinkling of diseased leaves and veins that are swollen in places. Flowers may also be distorted. Diseased plants in the field are stunted and have small, rugose leaves that curl downward.

Other strains of BCTV in the United States cause similar symptoms. Plants are stunted and have rugose, warty leaves that curl downward. Leaves fully grown at the time of infection yellow and die. Sometimes diseased plants appear to recover.

Management
1. Fumigate plant beds.
2. Control weeds and other hosts in plant beds and fields.
3. Exclude insects from plant beds by using insect-proof covers.
4. Apply insecticides periodically to plants in plant bed.
5. Plow under residue and any plants left after transplanting.

Bushy Top

Cause. A combination of two viruses: the tobacco vein-distorting virus (TVDV) and the bushy top virus (BTV). TVDV is in the luteovirus group. BTV may be a strain of tobacco mottle virus. BTV can be transmitted by aphids only if TVDV is also present in the same plant.

Distribution. Nyasaland and Zimbabwe.

Symptoms. Plants infected early in their growth have severe symptoms. Excessive growth of axillary buds produces numerous brittle shoots and crisp leaves. Flowers are small but set seed.

Management
1. Sow early before aphid populations are too large.
2. Apply systemic insecticides.
3. Control all weeds in tobacco beds and in fields adjacent to tobacco fields.
4. Destroy all plants in beds after transplanting.

Clubroot

Cause. Clubroot virus is transmitted by grafting.

Distribution. The United States (Kentucky, Maryland, and Tennessee). Club root is a minor disease.

Symptoms. Disease symptoms may be confused with those caused by the root-knot nematodes. Root-knot galls are found on small and large roots. Clubroot galls of various sizes are usually only found on older roots. Diseased plants are stunted due to a shortening of the internodes and have curled and otherwise distorted leaves. Infrequently, veins enlarge and enations form on the undersides of diseased leaves.

Management. Not necessary.

Cucumber Mosaic

Cause. Cucumber mosaic virus (CMV) is in the cucumovirus group. CMV overwinters in several perennial and annual host plants. Pollen of at least four *Stellaria* species can be infected with virus; consequently, CMV overwinters in the seed of these plants. Several species of aphids are vectors, with *Myzus persicae* and *Aphis gossypii* being the two principal vectors. CMV is nonpersistent and aphids retain the virus for less than 4 hours. Different strains of CMV exist. Some strains are efficiently transmitted by aphids and other strains are not efficiently transmitted.

Distribution. Wherever tobacco is grown. Disease is especially severe in Japan and wherever vegetables are grown.

Symptoms. Symptoms are similar to tobacco mosaic virus. Pale green circular spots on diseased leaves give the leaves a mottled and mosaic pattern. Systemic infection initially appears as a slight vein clearing that is followed by a mild, general mottling of the leaf. Sometimes diseased leaves are stunted and narrow.

Mild strains cause only a faint mottling of leaves. Severe strains may cause interveinal discoloration and an "oak-leaf" pattern of necrosis on diseased lower leaves. A burn or sun scald sometimes appears on diseased upper leaves.

Management
1. Control weed hosts in tobacco fields and in fields adjacent to tobacco fields.
2. Plant only disease-free or certified transplants.
3. Grow a nonhost crop to act as a barrier between the source of inoculum and tobacco.
4. Do not grow tobacco near potatoes, tomatoes, and peppers.
5. Sow tobacco when aphid populations are low.

Eggplant Mosaic

Cause. Eggplant mosaic virus, tobacco strain (EMV-T), is in the tymovirus group. EMV-T is readily transmitted mechanically and by the chrysomelid beetle *Diabrotica speciosa* from eggplant to other solanaceous hosts, *Chenopodium quinoa* and *C. amaranticolor*.

Distribution. Brazil.

Symptoms. Chlorotic and/or necrotic spots and necrotic rings that develop on diseased plant leaves 5 to 7 days after inoculation are followed by vein clearing, vein banding, and mosaic.

Management. At present, EMV-T is not considered of economic importance and does not interfere with plant productivity.

Leaf Curl

Cause. Tobacco leaf curl virus (TLCV) is in the geminivirus group. TLCV survives in several different species of plants and is transmitted by the whitefly *Bemisia tabaci* and possibly by other insects. TLCV is also graft-transmitted but is not seed- or sap-transmitted. The acquisition period is from 15 to 120 minutes. TLCV persists in the vector for at least 6 days.

Leaf curl is more severe during dry weather and warm temperatures around 30°C because of the increased whitefly activity. Several strains of TLCV exist.

Distribution. Leaf curl is most prevalent in tropical areas of Africa, Asia, Australia, Central America, and South America. Leaf curl has been found in Russia and Switzerland.

Symptoms. The entire diseased plant is stunted and has small, twisted and curled leaves. The most characteristic symptom is the production of leaf outgrowths up to 1 cm wide on the veins of diseased lower leaf surfaces. Outgrowths are a dark green thickening of sections of the veins. The smaller veins have either a vein clearing or a yellow discoloration. In some cases, veins may be greener than normal. Sometimes leaf margins are rolled downward. Flower parts are curled and deformed. The loss of apical dominance gives a broom-like appearance to diseased plants.

Mild strains of TLCV cause little stunting. The uppermost leaves of nearly mature plants may be curled and twisted but other leaves appear normal.

Management
1. Control weeds in and adjacent to tobacco beds and fields.
2. Do not locate tobacco beds and fields near alternate hosts.
3. Apply insecticides or mulches to control whiteflies in seedbeds.
4. Destroy tobacco residue after the completion of harvest.
5. Rogue diseased plants.

Nicotiana velutina Mosaic

Cause. *Nicotiana velutina* mosaic virus (NVMV) is in the furovirus group. NVMV is transmitted by the soilborne fungus *Polymyxa betae,* which survives in field soil as cystosori. Cysts give rise to zoospores that "swim" through free soil water until they contact a host root and encyst. Encysted zoospores produce a structure called a stachel through which zoosporic cytoplasm enters the host cell and becomes a plasmodium. (Only primary root tissue of young roots is infected; the optimum temperature for infection is 25°C.) The host cell then becomes infected with NVMV if *P. betae* is viruliferous. Later, the plasmodium develops into a zoosporangium, which releases additional zoospores that repeat the infection cycle. However, some plasmodia develop into cysts, and often nearly every cell in the small feeder roots will contain a cyst. As root cells senesce, cysts are eventually released into the soil, where they can remain viable for years without loss of virulence.

Distribution. Australia.

Symptoms. A bright yellow mosaic occurs on *N. velutina*. Etched ringspots, followed by a mild systemic mottling, occurs on *N. tabacum,* a systemic chlorosis occurs on *N. glutinosa,* and a mosaic and leaf malformation followed by a necrosis occurs on *N. ebneyi*.

Management. Not reported.

Peanut Chlorotic Streak

Cause. Peanut chlorotic streak virus (PClSV) is in the caulimovirus group. PClSV is mechanically transmissible to several plants in the Leguminosae and Solanaceae. Disease is favored by the very high temperatures of the semiarid tropics. PClSV is confined primarily to inoculated leaves unless plants are grown at temperatures above 28°C.

Distribution. India.

Symptoms. Young leaflets display oval chlorotic streaks along the veins 3 to 4 weeks after inoculation. Streaks are not distinct in older leaflets and can be seen only when viewed against light. In field infections, diseased plants are stunted and have chlorotic streaks on younger leaflets.

Management. Not reported.

Peanut Stunt

Cause. Peanut stunt virus (PSV) is in the cucumovirus group. PSV overwinters in a large number of host plants, primarily clover species and other herbaceous perennials. Although PSV can be transmitted mechanically, the most prevalent means of spread is by aphids. *Aphis craccivora, A. spiraecola,* and *Myzus persicae* transmit PSV nonpersistently.

Distribution. Japan, the eastern United States, and the state of Washington in the United States.

Symptoms. Diseased burley tobacco is stunted. Diseased young leaves have large areas of chlorotic tissue. Older diseased leaves have chlorotic tissue that borders small veins and large veins have spots and rings of chlorotic tissue.

Flu-cured tobacco is stunted. Mosaic symptoms that resemble tobacco mosaic virus and ringspot lesions are present.

Management
1. Control weed hosts in tobacco fields and in fields adjacent to tobacco fields.
2. Plant only disease-free or certified transplants.
3. Grow a nonhost crop to act as a barrier between the source of inoculum and tobacco.
4. Do not grow tobacco near potatoes, tomatoes, and peppers.
5. Sow tobacco when aphid populations are low.

Pepper Veinal Mottle

Cause. Pepper veinal mottle virus (PVMV) is in the potyvirus group. PVMV is transmitted by sap, grafting, and the aphid *Myzus persicae.* PVMV is closely related to tobacco vein mottle virus.

Distribution. Nigeria.

Symptoms. Irregular vein banding of diseased leaves is the primary symptom. A systemic mosaic or chlorotic mottle fades with increasing age of leaves.

Management. Not reported.

Ringspot

Cause. Tobacco ringspot virus (TRSV) is in the nepovirus group. TRSV survives in a wide range of host plants, including cucumbers and other vegetables, soybeans, horse nettle, ground cherry, pokeweed, sweet clover, lespedeza, and alfalfa. TRSV is mechanically transmitted easily and is seed-transmitted in several plants, notably soybeans. Several vectors transmit TRSV, notably all stages of the nematode *Xiphinema americanum.* Other vectors are thrips, *Thrips tabaci*; mites, *Tetranychus* sp.; grasshoppers, *Melanoplus differentialis*; and the tobacco flea beetle, *Epitrix hirtipennis*. Different strains of TRSV exist.

Distribution. Australia, Canada, Europe, Formosa, Japan, Nyasaland, New Zealand, Russia, South Africa, Sumatra, and the United States.

Symptoms. Symptoms may appear on diseased plants in the plant bed or shortly after seedlings are transplanted into the field. Initially, diseased leaves have either a few or numerous necrotic rings, which become blanched or brown after a few days. Frequently, there are line patterns that parallel veins or occur as irregular wavy lines between larger leaf veins. Leaves on only one side of the plant may be diseased. Severely diseased plants may be dwarfed and have small leaves of poor quality. Pollen becomes sterile.

Management
1. Control nematodes in plant beds and fields.
2. Control weeds in and adjacent to plant beds and fields.
3. Do not grow tobacco after or next to other plant hosts for TRSV.
4. Control insects in plant beds and fields.

Tobacco Etch

Cause. Tobacco etch virus (TEV) is in the potyvirus group. TEV overwinters in tobacco roots. Plants that grow from these roots are diseased. TEV also overwinters in several different solanaceous plants, including horse nettle, *Solanum carolinense,* and ground cherry, *Physalis virginiana.* TEV is not seedborne and does not survive in infested residue.

TEV is a nonpersistent virus transmitted by aphids. Because the probability of transmission does not depend on the number of virus particles acquired by the aphid vector, aphid transmission requires relatively few virus particles. Consequently, aphids that make brief acquisition probes into plants diseased by TEV are as likely to transmit virus as aphids that make longer probes or those that feed during the acquisition access. It is hypothesized that only those virus particles retained in the presence of a helper component in the maxillary stylets or in the foregut of the vectors are involved in transmission.

Distribution. Canada, Germany, Japan, and the United States.

Symptoms. Tobacco etch is most important on burley tobacco. Symptoms in the field are first noticed when plants approach the flowering stage of plant growth. Vein clearing is the first symptom to occur, followed by line patterns of necrotic tissue, or etching, and mottling. Young leaves have mild mosaic symptoms but tip leaves are not mottled. On older leaves, chlorotic spots up to 6 mm in diameter occur in the interveinal area and eventually become white and necrotic. Severely diseased plants may be stunted and have fired, chlorotic, and tattered leaves at harvest time.

Management
1. Control perennial weeds.
2. Grow only certified transplants.
3. Destroy live tobacco stalks and roots.
4. Control aphids.
5. Grow resistant cultivars.

Tobacco Leaf Curl

Cause. Tobacco leaf curl virus (TLCV) is in the geminivirus group, subgroup B. TLCV is transmitted by the whitefly *Bemisia tabaci.*

Distribution. Worldwide, but severe disease occurs only in tropical areas. Tobacco leaf curl has caused significant losses in Venezuela.

Symptoms. Diseased leaves are severely crinkled, curled, and dwarfed.

Management
1. Use insecticides to control the whitefly vector.
2. Plant tobacco at the time of year when the vector population is lowest.

Tobacco Mosaic

Cause. Tobacco mosaic virus (TMV) is in the tobamovirus group. TMV over-winters in several perennial weeds, including horse nettle, *Solanum caroli-nense,* and ground cherry, *Physalis angulata.* TMV also persists in air-dried tobacco and in infested residue in the field in the absence of freezing, des-iccation, or complete rotting.

TMV is disseminated mechanically. Workers who chew or smoke to-bacco, or handle infested residue or weed hosts and then touch live to-bacco plants, are primary sources of dissemination. Other means of spread occur during tillage, suckering, or other field operations. A species of large grasshopper may mechanically transmit the virus as well as a leaf miner fly; however, these are not efficient vectors. TMV is transported systemi-cally in the phloem of *Nicotiana tabacum* cv. Xanthi nc. plants when TMV is dually inoculated with potato leafroll virus.

Distribution. Wherever tobacco is grown.

Symptoms. Local lesions do not form on inoculated leaves, but under high temperature and light conditions, small, circular, faintly chlorotic spots may occur. Systemic infection first causes a vein clearing of the youngest diseased leaf, which eventually becomes mottled and distorted. The outer leaf edge may turn slightly upward and form a rim around the leaf. Eventually, large blisters of green tissue and raised, or sunken, yellow areas develop along with a marked mottling of dark and light green. Considerable malformation and distortion also occur. Frequently, the leaf lamina may be so reduced that a "shoestring" effect is produced.

Management
1. Fumigate plant bed soils with a chemical that kills weeds and inactivates the virus.
2. Control weeds in and around plant beds.
3. Do not use tobacco while working in the seedbed.
4. Spray the plant bed with milk 24 hours before pulling. Laborers should dip hands in milk every 20 minutes while working with seedlings.
5. Rotate tobacco in a field every 2 years.
6. Rogue out any diseased plants prior to the first cultivation.
7. Grow resistant cultivars. Resistance has been reported to be related to the presence of the phenolic compounds scopolin or scopoletin.

Tobacco Necrosis

Cause. Tobacco necrosis virus A (TNV-A) is in the necrovirus group. TNV-A infects and persists in a large number of host plant roots but not in the resting sporangia of the soil fungus vector *Olpidium brassicae*. A second virus, tobacco necrosis satellite virus (TNSV), is found only in association with TNV-A. TNSV has not been demonstrated to multiply by itself and appears to be incapable of multiplication unless associated with TNV-A.

The viruses are transmitted by the soil fungus *O. brassicae,* which, in turn, is disseminated by any means that moves soil. Particles of both viruses leak out of diseased roots into soil water, where they are adsorbed to the flagella and plasmalemmae of *O. brassicae* zoospores. When the zoospores encyst, the flagella bearing the virus particles are pulled into the cytoplasm of the zoospores. Upon germination of the encysted zoospore, the virus particles are carried along with zoospore cytoplasm into the host root. Root infection can also occur when wounded roots are in contact with soil water that has a high content of virus particles. Both viruses are also mechanically transmitted. TNV-A is primarily restricted to the roots of various plants that appear to be symptomless. The viruses occasionally move out of the root into the aerial parts of the plant, but they are arrested at a point just above the soil level.

Disease symptoms are more severe during cool temperatures and low light intensities. Different strains of the virus exist.

Distribution. Europe, Japan, New Zealand, and the United States.

Symptoms. Tobacco necrosis only occurs in young seedlings. The first symptoms are necrotic spots that often coalesce along the midribs and veins of diseased lower leaves. Only the lower and oldest leaves become necrotic and dry on older plants, but the rest of the plant appears normal. Very young plants may be killed.

Management
1. Sterilize soil to kill *O. brassicae.*
2. Remove tobacco residue from plant bed areas.
3. Set only healthy transplants in the field.

Tobacco Rattle

Cause. Tobacco rattle virus (TRV) is in the tobravirus group. TRV overwinters in at least 100 different plant species and is transmitted by both adult and juvenile nematodes of the genus *Trichodorus,* including *T. pachydermus, T. primitivus, T. christii,* and *T. alii.* TRV contaminates stylets and enters roots through wounds made by nematode feeding. TRV is rarely transmitted by the seed of some weed species. Transmission is favored by temperatures of 25°C and below.

Acquisition and transmission times are 1 hour each. TRV can be retained for 20 weeks in nematodes that have not fed but does not persist through eggs or molting. Tobacco rattle is most prevalent on sandy soils.

Distribution. Brazil, Denmark, Germany, Holland, Japan, Scotland, and the United States.

Symptoms. Inverted spoon-shaped leaves have spots and ring or line patterns of necrotic tissue. Diseased leaves become curled, have sinuate margins, and break and rattle when touched. Discontinuous brown to gray stripes of sunken necrotic tissue develop on stems, petioles, and leaf veins.

Leaves infected with mild strains of TRV have numerous tan lesions that form rings. Systemically infected leaves are slightly distorted and mottled but there is little stem necrosis. Little or no plant stunting occurs. Diseased plants usually recover and produce normal-appearing leaves that are somewhat elongated and pointed at the tip.

Management
1. Fumigate or use nematicides to control nematodes in the plant beds.
2. Use nematicides in the field. Experimentally, salicylate watered into soil prevented systemic infection of white burley plants. Long-distance spread was likely prohibited.
3. Rotate tobacco with resistant crops.

Tobacco Ringspot

Cause. Tobacco ringspot virus (TRSV) is in the nepovirus group. TRSV is seed-borne and overwinters or is reservoired in a large number of legumes. TRSV is sap-transmitted by nematodes, particularly the dagger nematode, *Xiphinema americanum*; nymphs of thrips, *Thrips tabaci*; grasshoppers,

Melanoplus differentialis; and possibly other insects. TRSV is also transmitted by root grafts.

Distribution. Worldwide.

Symptoms. Normally only isolated plants are diseased, but severe losses may occur in some fields. Diseased plants have chlorotic rings on leaves or chlorotic V-shaped patterns that emanate from secondary leaf veins and give the superficial appearance of an oak-leaf pattern.

Management. Not reported.

Tobacco Rosette Disease

Cause. A combination of two viruses: tobacco vein-distorting virus (TVDV) and tobacco mottle virus (TMoV). TVDV is in the luteovirus group. TMoV overwinters in several plant species in the Solanaceae. TMoV is mechanically transmitted and is transmitted by the aphid *Myzus persicae,* but only if TVDV is also present in the plant.

TVDV also overwinters in several plant species in the Solanaceae. TVDV is transmitted by *M. persicae* but is not transmitted mechanically. TVDV persists in aphid bodies for long periods.

Distribution. Nyasaland and Zimbabwe.

Symptoms. Tobacco rosette may occur in plant beds or fields. Midribs of young diseased leaves are twisted and distorted and form a knot in the center of the plant that looks like a rosette. Diseased leaves curl sharply at the tip. Diseased plants are stunted.

Diseased older plants do not become distorted, but leaves droop. If plants are diseased on only one side, the flower bends over. Splitting of stems and petioles may also occur.

Management
  1. Sow early, before aphid populations are too large.
  2. Apply systemic insecticides.
  3. Control all weeds in tobacco beds and adjacent fields.
  4. Destroy all plants remaining in beds after transplanting.

Tobacco Streak

Cause. Tobacco streak virus (TSV) is in the ilarvirus group. TSV survives in a large number of plants, including tomato, sweet clover, cotton, peanut, and several others. TSV is transmitted mechanically and by dodder. Insects reported to transmit TSV are *Frankliniella* sp., *F. occidentalis,* and *Thrips tabaci.* In Australia, transmission was attributed to pollen from the weed *Ageratum houstonianum* and the insect *Microcephalothrips abdominalis.*

Distribution. Asia, Australia, Canada, Europe, and the United States.

Symptoms. Diseased leaves are smaller than normal, narrow, and slightly crinkled. Leaves may be affected on one side only, with the midrib curling toward the affected side. Necrotic lesions are surrounded by water-soaked lines or rings that later become brown and necrotic. Lesions spread along veins, with parallel necrotic lines appearing in the tissue surrounding the veins. Sometimes midribs and petioles collapse.

Systemic symptoms appear as a net pattern, rings, or partial rings in leaves. These symptoms are brown at first but later become gray-white and are closely associated with the veins and bases of young leaves. Necrotic tissue often falls out of leaves.

Management. No management is routinely practiced. However, other hosts, particularly sweet clover, should be eliminated from the vicinity of tobacco fields before transplanting occurs.

Tobacco Stunt

Cause. Tobacco stunt virus (TStV) infects and persists in a large number of host plants and in the soil fungus *Olpidium brassicae* for long periods of time, perhaps several years. TStV is disseminated on infected transplants and by any means that moves soil. TStV is transmitted primarily by *O. brassicae* and, infrequently, mechanically and by grafts. TStV is carried into the root by the zoospore cytoplasm and then multiplies both in the cytoplasm of the tobacco cells and in the developing zoospores. Disease severity is increased by cool temperatures and low light intensities.

Distribution. Japan.

Symptoms. Symptoms first appear when seedlings have five to eight leaves. The first symptom is a vein clearing; later, the veins become necrotic. Eventually, small brown to white necrotic spots or necrotic ring-like patterns develop on the bases of the lower diseased leaves. Bud leaves become generally yellow and have surface crinkles and tips that curl downward. Internodes fail to develop, causing diseased plants to remain in the rosette stage. A band of necrotic tissue occurs on stems at the soil level. Transplants moved to fields remain yellow and stunted. Frequently severely diseased plants die. Plants with mild symptoms have leaves and stems that are narrow, and flower before normal plants do.

Management
1. Sterilize soil to kill *O. brassicae*.
2. Remove tobacco residue from plant bed areas.
3. Set only healthy transplants in the field.

Tobacco Veinal Necrosis

Tobacco veinal necrosis caused by PVY is sometimes called vein banding.

Cause. Potato virus Y strain N (PVY) is in the potyvirus group. PVY overwinters in several species of plants, including potatoes, tomatoes, and peppers. PVY is nonpersistently transmitted from overwintering hosts to tobacco on the stylet tips of *Myzus persicae* and several other species of aphids.

Two viral proteins are known to be involved in the transmission process: the viral CP; and a nonstructural protein, the helper component (HC). Aphids need to acquire the HC prior to or in combination with the virions for transmission to take place. The biologically active HC is thought to be an oligomer-soluble protein. The HC differs between viral strains.

Distribution. Africa, Canada, Europe, South America, and the United States. Disease tends to be more severe where potatoes are also grown.

Symptoms. Symptoms vary with the tobacco cultivar that is infected. Most cultivars develop a faint mottling and vein clearing of the young expanding leaves. Larger leaves have a slight interveinal chlorosis that leaves a band of dark green tissue along each side of the vein. Leaf epinasty and stunting may occur.

Systemic necrosis occurs in burley tobacco and may cause a complete loss of the crop. Initially, vein banding occurs when the smaller veinlets in the leaves become necrotic, a symptom that is accompanied by a mild mosaic of the leaves. Severely diseased plants have veins that turn dark brown to black. Leaves yellow prematurely and plants may die. Later, there may be a severe leaf necrosis that spreads to the petiole and stems. Eventually, the lower leaves collapse.

Yields and tobacco quality are reduced.

Management
1. Control weed hosts in tobacco fields and in areas adjacent to the fields.
2. Plant only disease-free or certified transplants.
3. Grow a nonhost crop to act as a barrier between the source of inoculum and tobacco.
4. Do not grow tobacco near potatoes, tomatoes, and peppers.
5. Sow tobacco when aphid populations are low.
6. Resistance has been identified.

Tobacco Vein-Banding Mosaic

Cause. Tobacco vein-banding mosaic virus (TVBMV) is in the potyvirus group. TVBMV is transmitted by aphids.

Distribution. Taiwan and the United States (Tennessee).

Symptoms. Vein-banding symptoms are similar to those caused by mild strains of potato virus Y. (*See* "Tobacco Veinal Necrosis.")

Management. Resistance has been identified.

Tobacco Vein Mottling

Cause. Tobacco vein-mottling (or mottle) virus (TVMV) is in the potyvirus group. TVMV overwinters in the weeds dock, horse nettle, and ground cherry and also in living tobacco roots. TVMV is transmitted nonpersistently by the green peach aphid, *Myzus persicae,* and other aphid species, but not mechanically under natural conditions. TVMV is not known to be seed-transmitted.

Distribution. The United States (Kentucky and North Carolina).

Symptoms. Symptoms vary with time of infection and cultivar. Initially, burley types have a slight vein clearing. All diseased plants have an irregular dark green area along the veins of some leaves. Certain cultivars develop a severe systemic necrotic spotting on leaves that becomes increasingly more severe during the season.

   Plants infected early are stunted and are not as green as healthy plants. Tip leaves of these diseased plants may have light green spots.

Management
1. Control overwintering hosts in tobacco fields and in areas adjacent to tobacco fields.
2. Control insects with an insecticide application(s). However, because an aphid can transmit the virus to a healthy plant a few seconds after it starts feeding, insecticides are often a questionable control. Repeated transmissions to several healthy plants may be reduced.
3. Grow barrier rows of maize around a tobacco field to intercept aphids reaching the field.
4. Grow resistant cultivars. No cultivar is immune, but resistance varies greatly between cultivars.

Tomato Spotted Wilt

Cause. Tomato spotted wilt virus (TSWV) is in the tospovirus group. TSWV overwinters in several species of perennial plants in areas that have cold winter temperatures, and in annual plants where winters are not cold enough to kill plants.

   In Louisiana, virus inoculum originates from noncultivated areas. The presence of TSWV has been reported in spiny amaranthus, *Amaranthus spinosus;* blackseed plaintain, *Plantago rugelii;* buttercup, *Ranunculus* spp., with *R. sardous* the species most often associated with natural TSWV infection; coneflower, *Rudbeckia amplexicaulis;* horse nettle, *Solanum carolinense;* dandelion, *Taraxacum officinale;* spiny sow-thistle, *Sonchus asper;* blue vervain, *Verben brasiliensis;* and *Lactucua* spp., with *L. floridana* an important overwintering host.

   TSWV is transmitted by several species of thrips, including potato or onion thrips, *Thrips tabaci*; blossom or cotton thrips, *Frankliniella schultzei*; western flower thrips, *F. occidentalis*; and tobacco thrips, *F. fusca*. The chili thrips, *T. setosus,* has been shown to transmit the virus in Japan. TSWV must first be acquired by the larval form, then the subsequent adults are able to transmit TSWV. TSWV persists in insects as long as they live, but it is not transmitted through the eggs. Tomato spotted wilt is more severe during high temperatures and abundant moisture. Plants of all ages can be infected.

Distribution. Wherever tobacco is grown.

Symptoms. The yellow-green spots that initially occur on diseased young leaves later become red-brown and have concentric necrotic rings or zonate necrotic spots. Spots frequently coalesce and form large irregular-shaped areas. Young leaves infected only on one side are distorted and puckered. Necrotic streaks develop along stems, and dark necrotic areas, or cavities, appear in the cortex and pith. Diseased plants are stunted, and apical buds droop or bend over. Severely diseased plants do not grow for several weeks, and eventually the leaves droop and the entire plant finally dies.

Management
1. Use insecticides to control thrips in the seedbed and for a month after transplanting.
2. Eliminate overwintering hosts and deep-plow all tobacco debris in a field.

Wound Tumor

Cause. Wound tumor virus (WTV) is in the phytovirus group. WTV is transmitted by leafhoppers.

Distribution. Widespread.

Symptoms. Tumors or cell enlargements are produced at sites of wounding on diseased stems and roots.

Management. Not known.

Disease Caused by Higher Plants

Broomrape

Cause. *Orobanche ramosa* is a parasitic flowering plant that survives as seed in the soil for several years. *Orobanche ramosa* grows on the roots of tobacco.

Distribution. Africa, Asia, Europe, and North America.

Symptoms. Diseased plants do not grow well, become yellow, and mature earlier than healthy plants. Large masses of blue-flowered broomrape plants break through the soil around the base of the tobacco plant.

Management
1. Do not grow tobacco in an infested field for several years.
2. Destroy broomrape plants before seed is ripe.

Selected References

Ahmed, W., and Thomas, P. E. 1991. Phloem transport of tobacco mosaic virus in Xanthi nc. tobacco induced by potato leafroll virus. (Abstr.). Phytopathology 81:1167.

Anderson, T. R., and Welacky, T. W. 1983. Barn mold of burley tobacco caused by *Botryosporium longibrachiatum*. Plant Dis. 67:1158-1159.

Anderson, T. R., and Welacky, T. W. 1988. Populations of *Thielaviopsis basicola* in burley tobacco field soils and the relationship between soil inoculum concentration and severity of disease on tobacco and soybean seedlings. Can. J. Plant Pathol. 10:246-251.

Barker, K. R., and Lucas, G. B. 1987. Nematode parasites of tobacco, pp. 213-241. *In* Plant and Insect Nematodes. W. R. Nicle, ed. Marcel Dekker, New York.

Bhattiprolu, S. L. 1991. Studies on a newly recognized disease of *Nicotiana glutinosa* of viroid etiology. Plant Dis. 75:1068-1071.

Bower, L. A., Fox, J. A., and Miller, L.I. 1980. Influence of *Meloidogyne incognita* and *Globodera solanacearum* on development of black shank of tobacco. (Abstr.). Phytopathology 70:688.

Canto, T., Lopez-Moya, J. J., Serra-Yoldi, M. T., Diaz-Ruiz, J. R., and Lopez-Abella, D. 1995. Different helper component mutations associated with lack of aphid transmissibility in two isolates of potato virus Y. Phytopathology 85:1519-1524.

Cole, J. S., and Zvenyika, Z. 1988. Integrated control of *Rhizoctonia solani* and *Fusarium solani* in tobacco transplants with *Trichoderma harzianum* and triadimenol. Plant Pathol 37:271-277.

Cornelissen, A. P. F., and van Wyk, R. J. 1987. *Nicotiana tabacum* breeding lines resistant to various root-knot nematodes (*Meloidogyne* spp.). Nematologica 33:316-321.

Davis, J. M., and Main, C. E. 1984. Meteorological aspects of the spread and development of blue mold on tobacco in North Carolina. (Abstr.). Phytopathology 74:840.

Davis, J. M., Main, C. E., and Bruck, R. I. 1981. Analysis of weather and the 1980 blue mold epidemic in the United States and Canada. Plant Dis. 65:508-512.

Fajola, A. O., and Alasodura, S. O. 1973. Corynespora leaf spot, a new disease of tobacco (*Nicotiana tabacum*). Plant Dis. Rptr. 57:375-378.

Farr, D. F., Bills, G. R., Chamuris, G. P., and Rossman, A. Y. 1989. Fungi on Plants and Plant Products in the United States. American Phytopathological Society, St Paul, MN. 1252 pp.

Ferrin, D. M., and Mitchell, D. J. 1986. Influence of soil water status on the epidemiology of tobacco black shank. Phytopathology 76:1213-1217.

Fortnum, B. A., Csinos, A. S., and Dill, T. R. 1982. Metalaxyl controls blue mold in flu-cured tobacco seedbeds. Plant Dis. 66:1014-1016.

Fortnum, B. A., Krausz, J. P., and Conrad, N. D. 1984. Increasing incidence of *Meloidogyne arenaria* on flu-cured tobacco in South Carolina. Plant Dis. 68:244-245.

Francki, R. I. B., Milne, R. G., and Hatta, T. 1985. Atlas of Plant Viruses. CRC Press, Boca Raton, FL. 222 pp.

Fulton, R. W. 1980. Tobacco blackfire disease in Wisconsin. Plant Dis. 64:100.

Gayed, S. K. 1978. Tobacco Diseases. Canada Dept. Agric. Pub. 1641.

Greber, R. S., Klose, M. J., Teakle, D. S., and Milne, J. R. 1991. High incidence of tobacco streak virus in tobacco and its transmission by *Microcephalothrips abdominalis* and pollen from *Ageratum houstonianum*. Plant Dis. 75:450-452.

Gutierrez, W., and Shew, H. D. 1997. Factors that affect the development of collar rot disease of tobacco. (Abstr.). Phytopathology 87:S37.

Gwynn, G. R., Barker, K. R,. Reilly, J. J., Komm, D. A., Burk, L. G., and Reed, S. M. 1986. Genetic resistance to tobacco mosaic virus, cyst nematodes, root-knot nematodes, and wildfire from *Nicotiana repanda* incorporated into *N. tabacum*. Plant Dis. 70:958-962.

Hartill, W. F. T. 1967. A rust of tobacco in Rhodesia. Rhodesia Zambia Malawi J. Agric. Res. 5:189.

Hendrix, J. W., and Csinos, A. S. 1985. Tobacco stunt, a disease of burley tobacco controlled by soil fumigants. Plant Dis. 69:445-447.

Herr, L. J,. and Sutton, P. 1984. Tobacco black shank control with metalaxyl and cultivars. (Abstr.). Phytopathology 74:854.

Hood, M. E., and Shew, H. D. 1997. Reassessment of the role of saprophytic activity in the ecology of *Thielaviopsis basicola*. Phytopathology 87:1214-1219.

Hopkins, J. C. F. 1956. Tobacco Diseases. Commonwealth Mycological Institute, Surrey, United Kingdom.

Ishida, Y., and Kumashiro, T. 1988. Expression of tolerance to the host-specific toxin of *Alternaria alternata* (AT toxin) in cultured cells and isolated protoplasts of tobacco. Plant Dis. 72:892-895.

Johnson, M. C., Pirone, T. P., and Litton, C. C. 1982. Selection of tobacco lines with a high degree of resistance to tobacco etch virus. Plant Dis. 66:295-297.

Johnson, R. R., Black, L. L., Hobbs, H. A., Valverde, R. A., Story, R. N., and Bond, W. P. 1995. Association of *Frankliniella fusca* and three winter weeds with tomato spotted wilt virus in Louisiana. Plant Dis. 79: 572-576.

Kelman, A., and Sequeira, L. 1965. Root-to-root spread of *Pseudomonas solanacearum*. Phytopathology 55:304-309.

Komm, D. A., Reilly, J. J., and Elliott, A. P. 1983. Epidemiology of a tobacco cyst nematode (*Globodera solanacearum*) in Virginia. Plant Dis. 67:1249-1251.

Ladipo, J. L., and Roberts, I. M. 1979. Occurrence of pepper veinal mottle virus in tobacco in Nigeria. Plant Dis. Rptr. 63:161-165.

LaMondia, J. A., and Taylor, G. S. 1987. Influence of the tobacco cyst nematode (*Globodera tabacum*) on Fusarium wilt of Connecticut broadleaf tobacco. Plant Dis. 71:1129-1132.

Latorre, B. A., and Flores, V. 1987. Wilt of tobacco in Chile caused by *Verticillium dahliae*. Plant Dis. 71:101.

Lucas, G. B. 1974. Diseases of Tobacco, Third Edition. Harold E. Parker & Sons Printers, Fuquay-Varina, NC.

McIntyre, J. L., Sands, D. C., and Taylor, G. S. 1978. Overwintering, seed disinfestation, and pathogenicity studies of the tobacco hollow stalk pathogen, *Erwinia carotovora* var. *carotovora*. Phytopathology 68:435-440.

Meyer, J. R., and Shew, H. D. 1991. Development of black root rot on burley tobacco as influenced by inoculum density of *Thielaviopsis basicola,* host resistance, and soil chemistry. Plant Dis. 75:601-605.

Meyer, J., Shew, H. D., and Shoemaker, P. B. 1989. Populations of *Thielaviopsis* and the occurrence of black root rot on burley tobacco in western North Carolina. Plant Dis. 73:239-242.

Modjo, H. S., and Hendrix, J. W. 1986. The mycorrhizal fungus *Glomus macrocarpum* as a cause of tobacco stunt disease. Phytopathology 76:688-691.

Moss, M. A., and Main, C. E. 1985. Temperature tolerance of *Peronospora tabacina* in the U.S. (Abstr.). Phytopathology 75:1341.

Moss, M. A., and Main, C. E. 1988.The effect of temperature on sporulation and viability of isolates of *Peronospora tabacina* collected in the United States. Phytopathology 78:110-114.

Muller, A. S., and Acedueda, H. A. 1964. Rust on tobacco discovered in Honduras. (Abstr.). Phytopathology 54:499.

Oke, O. A., 1988. Changes in chemical constituents of tobacco leaves infected with *Corynespora cassiicola* and *Colletotrichum nicotianae.* J. Phytopathol 122:181-185.

Pirone, T. P., and Nesmith, W. C. 1994. Survey of Kentucky for potato virus Y strain N and other potyviruses in tobacco. Plant Dis. 78:754.

Pirone, T. P., and Thornbury, D. W. 1988. Quantity of virus required for aphid transmission of a potyvirus. Phytopathology 78:104-107.

Powell, N. T., Melendez, P. L., and Batten, C. K. 1971. Disease complexes in tobacco involving *Meloidogyne incognita* and certain soilborne fungi. Phytopathology 61:1332-1337.

Prasad, S. S., and Acharya, B. 1965. A new leaf-spot disease of tobacco. Current Sci. 34:542.

Ramachandraiah, M., Venkatarathnam, P., and Sulochana, C. B. 1979. Tobacco necrosis virus: Occurrence in India. Plant Dis. Rptr. 63:949-951.

Ramachar, P., and Cummins, G. G. 1965 .The species of *Puccinia* on the Paniceae. Mycopathol. Mycol. Appl. 25:7-60.

Reddick, B. B., Christie, R. G., Gooding, G. V., Jr., and Collins-Shepard, M. H. 1991. Tobacco vein-banding mosaic virus: A new virus in North America. (Abstr.). Phytopathology 81:1243.

Reeleder, R. D., Monette, S., VanHooren, D., and Sheidow, N. 1996. First report of target spot of tobacco caused by *Rhizoctonia solani* (AG-3) in Canada. Plant Dis. 80:712.

Reilly, J. J. 1980. Chemical control of black shank of tobacco. Plant Dis. 64:274-277.

Reuveni, M., Nesmith, W. C., and Siegel, M. R. 1986. Symptom development and disease severity in *Nicotiana tabacum* and *N. repanda* caused by *Peronospora tabacina.* Plant Dis. 70:727-729.

Ribeiro, S. G., Kitajima, E. W., Oliveira, C. R. B., and Koenig, R. 1996. A strain of eggplant mosaic virus isolated from naturally infected tobacco plants in Brazil. Plant Dis. 80:446-449.

Richardson, M. J., and Zillinsky, F. J. 1972. A leaf blight caused by *Fusarium nivale.* Plant Dis. Rptr. 56:803-804.

Rotem, J., and Aylor, D .E. 1984. Development and inoculum potential of *Peronospora tabacina* in the fall season. Phytopathology 74:309-313.

Shew, H. D. 1983. Effect of host resistance level on spread of *Phytophthora parasitica* var. *nicotianae* under field conditions. (Abstr.). Phytopathology 73:505.

Shew, H. D., and Lucas, G. B. 1991. Compendium of Tobacco Diseases. American Phytopathological Society, St. Paul, MN. 68 pp.

Shew, H. D., and Main, C. E. 1985. Rhizoctonia leaf spot of flu-cured tobacco in North Carolina. Phytopathology 69:901-903.

Shew, H. D., and Main, C. E. 1990. Infection and development of target spot of flu-cured tobacco caused by *Thanatephorus cucumeris*. Plant Dis. 74:1009-1013.

Shew, H. D., and Melton, T. A. 1995. Target spot of tobacco. Plant Dis. 79:6-11

Shew, H. D., and Shoemaker, P. B. 1993. Effects of host resistance and soil fumigation on *Thielaviopsis basicola* and development of black root rot on burley tobacco. Plant Dis. 77:1035-1039.

Sidebottom, J. R., and Shew, H. D. 1985. Effects of soil texture and matric potential on sporangium production by *Phytophthora parasitica* var. *nicotianae*. Phytopathology 75:1435-1438.

Sidebottom, J. R., and Shew, H. D. 1985. Effects of soil type and soil matric potential on infection of tobacco by *Phytophthora parasitica* var. *nicotianae*. Phytopathology 75:1439-1443.

Spurr, H. W., Jr., Echandi, E., Haning, B. C., and Todd, F. A. 1980. Bacterial barn rot of flu-cured tobacco in North Carolina. Plant Dis. 64:1020-1022.

Stavely, J. R., and Kincaid, R. R. 1969. Occurrence of Phyllosticta leaf spot on Florida cigar-wrapper tobacco. Plant Dis. Rptr. 53:837-839.

Sun, M. K.C., Gooding, G. V., Jr., Pirone, T. P., and Tolin, S. A. 1974. Properties of tobacco vein-mottling virus, a new pathogen of tobacco. Phytopathology 64:1133-1136.

Taylor, G. S. 1969. A black mildew of the genus *Diporotheca* found on roots of *Nicotiana tabacum* in Connecticut. Plant Dis. Rptr. 53:85-86.

Tisdale, W. B. 1929. A disease of tobacco seedlings caused by *Septomyxa affinis*. (Abstr.). Phytopathology 19:90.

Wilson, K. I., Al-Beldawi, A, S., and Dwazah, K. 1983. Rhizopus stem rot of *Nicotiana glauca*. Plant Dis. 67:526-527.

Wolf, F. A. 1957. Tobacco Diseases and Decays. Duke University Press, Durham, NC.

23. Diseases of Wheat
(Triticum aestivum)

Diseases Caused by Bacteria

Bacterial Leaf Blight

Bacterial leaf blight is also called bacterial leaf necrosis and bacterial blight.

Cause. *Pseudomonas syringae* pv. *syringae* is seedborne and overwinters in infested residue, water, and soil and on the surfaces of different plant species. The bacteria can survive as epiphytes on leaves, where they may become pathogenic under the right conditions. During wet weather and temperatures of 15° to 25°C, bacteria are splashed and blown onto host leaves, which they enter through stomates and wounds. Bacteria may also migrate from infected or infested seed by swimming up the seedling or by being carried along with the growing point.

   Pseudomonas syringae pv. *syringae* is a weak pathogen that requires moist conditions during the incubation period for significant infection to occur. The presence of *P. syringae* pv. *syringae* can increase frost damage to wheat plants grown in high-altitude areas.

Distribution. The north central United States.

Symptoms. Symptoms on winter wheat occur during cool, wet weather in the spring and may be confused with spots caused by other pathogens. At first, water-soaked spots (less than 1 mm in diameter) occur on the top leaves of plants in the boot to early heading stage. Eventually, the spots grow larger and become necrotic, changing from a dull gray-green to light brown. During wet weather, droplets of bacterial ooze develop in a spot. During continued wet weather, spots that grow together into ragged streaks or blotches in 2 to 3 days sometimes kill the entire leaf.

   Lesions on spring wheat during moist, humid weather coalesce into elongated necrotic areas that gradually expand laterally until the entire leaf is destroyed. A chlorotic halo may be present around some lesions.

Typical halo-like symptoms caused by *P. syringae* have been reported from south central Colorado.

Management. Grow resistant cultivars of spring and winter wheat.

Bacterial Leaf Streak

Bacterial leaf streak and black chaff are generally considered to be two phases of the same disease. Bacterial leaf streak refers to water-soaked lesions on leaves; black chaff refers to black lesions on glumes. Because symptom expression tends to be regarded separately in much of the literature, bacterial leaf streak and black chaff are discussed separately in this book.

Bacterial leaf streak, bacterial stripe, black chaff, black chaff and bacterial streak, and Xanthomonas streak are names that refer to basically the same disease syndrome.

Cause. *Xanthomonas campestris* pv. *translucens* (syn. *X. campestris* and *X. translucens* pv. *undulosa*). The taxonomy of the causal organism(s) continues to be a subject in flux. Bacterial leaf streak (and black chaff), also called bacterial blight or bacterial streak on other small grains, is considered to be caused by five of the former *X. campestris* pathovars: pv. *cerealis*, pv. *hordei*, pv. *secalis*, pv. *translucens*, and pv. *undulosa*. These pathovars are often grouped together under the name "translucens group" and are considered to be a single species (*X. translucens)* or pathovar (*X. campestris* pv. *translucens*). Bragard et al. (1997) agreed that *X. translucens* pv. *translucens* and *X. translucens* pv. *hordei* are true synonyms. They also reported that using restriction fragment-length polymorphism and fatty acid methyl esters analyses, the pathovars *X. translucens* pv. *cerealis*, *X. translucens* pv. *translucens,* and *X. translucens* pv. *undulosa* clustered in different groups and corresponded to true biological entities.

The causal bacteria are seedborne and overwinter in infested residue, on overwintered hosts, and in soil. Primary infection is by bacteria that have been disseminated by splashing water and insects, particularly aphids. Injuries from windblown sand and insects provide entrances for infection. Primary infection may also occur by seedborne bacteria, but seed stored 6 months or more is not considered an important source of inoculum. However, some researchers consider seed to be the most important source of primary inoculum. In Arkansas, infested seed is suspected to be the principal source of inoculum. Small spaces and grooves that contain water act as a reservoir for bacteria. Secondary spread occurs by plant-to-plant contact, rain, and insects.

The bacteria can increase frost damage to wheat plants grown at high altitudes. Although frost is not essential to infection, susceptibility of wheat plants increases following frost and infection spreads more rapidly in frosted tissue.

Distribution. Wherever wheat is grown. It is an important disease of soft red winter wheat in the southeastern United States.

Symptoms. Symptoms usually occur after several days of damp or rainy weather. Small water-soaked spots occur on tender green leaves, sheaths of older plants, and sometimes seedlings. The spots enlarge, coalesce, and become glossy, olive-green, translucent stripes or streaks of various lengths that later turn yellow-brown. Stripes may extend the length of a sheaf and are usually narrow from being limited by leaf veins. Occasionally a spot may become large and blotch-like, causing the leaf to die, shrivel, and turn light brown. Severely diseased leaves die back from the tips. Lesions develop most rapidly in plants subjected to frost.

Droplets of milky bacterial exudate are seen on the surface of diseased spots during early morning humid conditions. The droplets dry into hard, yellowish granules that are easily removed from the leaf surface. As plants reach maturity, brown to black stripes are produced on stems below the heads and upper joints. If a flag leaf is diseased, the head may not emerge from the boot but may break through the side of the sheath and be distorted and blighted.

Management
1. Do not sow seed from diseased plants. Seed should be cleaned to remove lightweight infected kernels.
2. Treat seed with a seed-protectant fungicide.
3. Do not rotate wheat with barley.
4. Treat seed with acidified cupric acetate. However, since germination and the resultant stand are affected, this treatment should not be used routinely except for special cases, such as foundation seed health programs.
5. Partial resistance has been identified in diallel crosses of three genotypes.

Bacterial Mosaic

Cause. *Clavibacter michiganensis* subsp. *tessellarius*. The species name *michiganense* is sometimes used in the literature. The bacterium may be seedborne.

Distribution. Canada and the United States (Alaska, Illinois, Iowa, and Nebraska).

Symptoms. The initial small yellow spots, which are most numerous near the midribs of diseased leaves, eventually grow together and form yellow lesions with indefinite margins. Diseased leaves become light to dark brown; severely diseased leaves turn brown and dry up. The typical water-soaking symptoms associated with bacterial infections are lacking.

Management. Do not sow seed from diseased plants.

Bacterial Sheath Rot

Cause. *Pseudomonas fuscovaginae.* Disease becomes severe during high humidity and low temperatures at booting to heading stages.

Distribution. Mexico, in the central highlands at 2249 to 2640 m above sea level.

Symptoms. Angular, blackish-brown lesions (10–20 cm in length) with gray centers and bordered by a purple-black angular area (1–2 mm wide) occur on diseased leaf sheaths of bread and durum wheat.

Management. Not necessary, because the disease is not considered to be either widespread or severe.

Basal Glume Rot

Cause. *Pseudomonas syringae* pv. *atrofaciens* (syn. *P. atrofaciens* and *Phytomonas atrofaciens*) is seedborne and overwinters in infested residue. Bacteria are disseminated on dust particles, insects, and splashing water. Bacteria multiply near glume joints when water is present, but remain dormant when moisture is lacking.

Distribution. Wherever wheat is grown.

Symptoms. The main symptom at heading time is a browning of the bottom one-third of glumes that is more evident on the inside than the outside of the diseased glume. Sometimes the entire glume is affected and displays symptoms. Severely diseased spikelets are slightly dwarfed and lighter than healthy ones. Sometimes the only symptom is a dark line where the glume is attached to the spike. The base of diseased kernels has a light brown to black discoloration and seed filling is sometimes limited.

Diseased leaves have small, dark, water-soaked spots that elongate and become yellow, then brown as diseased tissue dies.

Management
1. Treat seed with a seed-protectant fungicide.
2. Sow seed that has been thoroughly cleaned.
3. Rotate wheat with resistant crops, such as legumes.

Black Chaff

Black chaff and bacterial leaf streak are generally considered to be two phases of the same disease. Bacterial leaf streak refers to water-soaked lesions on leaves; black chaff refers to black lesions on glumes. Because symptom expression tends to be regarded separately in much of the literature, bacterial leaf streak and black chaff are discussed separately in this book.

Black chaff is also called bacterial leaf streak, bacterial streak, bacterial stripe, black chaff and bacterial streak, and Xanthomonas streak.

Cause. *Xanthomonas campestris* pv. *translucens* (syn. *X. campestris* and *X. translucens* pv. *undulosa*). The taxonomy of the causal organism(s) continues to be a subject in flux. Bacterial leaf streak (and black chaff) is considered to be caused by five of the former *X. campestris* pathovars: pv. *cerealis*, pv. *hordei*, pv. *secalis*, pv. *translucens,* and pv. *undulosa*. The pathovars are often grouped together under the name "translucens group" or are considered to be a single species (*X. translucens)* or pathovar (*X. campestris* pv. *translucens*). Bragard et al. (1997) agreed that *X. translucens* pv. *translucens* and *X. translucens* pv. *hordei* are true synonyms. They also reported that using restriction fragment-length polymorphism and fatty acid methyl esters analyses, the pathovars *X. translucens* pv. *cerealis*, *X. translucens* pv. *translucens,* and *X. translucens* pv. *undulosa* clustered in different groups and corresponded to true biological entities.

The causal bacteria are seedborne and overwinter in infested residue, on overwintered hosts, and in soil. Primary infection is by bacteria that have been disseminated by splashing water and insects, particularly aphids. Injuries from windblown sand and insects provide entrances for infection. Primary infection may also occur by seedborne bacteria, but seed stored 6 months or more is not considered an important source of inoculum. However, some researchers consider seed to be the most important source of primary inoculum. In Arkansas, infested seed is suspected to be the principal source of inoculum. Small spaces and grooves that contain water act as a reservoir for bacteria. Secondary spread occurs by plant-to-plant contact, rain, and insects.

The bacteria can increase frost damage to wheat plants grown at high altitudes. Although frost is not essential to infection, susceptibility of wheat plants increases following frost; infection also spreads more rapidly in frosted tissue than in healthy plant tissue.

Distribution. Wherever wheat is grown. Black chaff is an important disease of soft red winter wheat in the southeastern United States. Black chaff is also serious in wheat-growing areas at higher elevations. In eastern Idaho, black chaff is a serious disease in irrigated spring wheat and has caused disease losses of 30% to 40%.

Symptoms. Symptoms usually occur after several days of damp or rainy weather but usually black chaff is not noticed until diseased plants are about two-thirds grown. Diseased plants grow slowly and may be stunted. The chaff or glumes have longitudinal, dark, somewhat sunken stripes or spots that are most abundant on the upper glumes, where they coalesce and form larger spots or blotches. The insides of diseased glumes have brown or black spots. Beards of bearded cultivars are brown, especially at the base. In moist weather, tiny yellow beads of bacteria ooze to the surface of dark lesions and dry as small yellow scales.

Grain is not destroyed, but it may be brown, shrunken, and carry bacteria to infect next year's crop. If a flag leaf is diseased, the head may not emerge from the boot but may break through the side of the sheath and be distorted and blighted.

Management
1. Do not sow seed from diseased plants. Seed should be cleaned to remove lightweight infected kernels.
2. Treat seed with a seed-protectant fungicide.
3. Do not rotate wheat with barley.
4. Treat seed with acidified cupric acetate. However, because germination and the resultant stand are affected, this treatment should not be routinely used except for special cases, such as foundation seed health programs.
5. Partial resistance has been identified in diallel crosses of three genotypes.

Pink Seed

Cause. *Erwinia rhapontici* (syn. *E. carotovora* pv. *rhapontici*) infects only damaged kernels, particularly those injured by the gall midge (Cecidomyidae) or when grain is harvested too early. It is not known how *E. rhapontici* survives and is disseminated to wheat, but the bacterium was isolated from plants grown under sprinkler irrigation in Idaho. *Erwinia rhapontici* causes a crown rot of rhubarb, *Rheum* spp., its only other known host.

Distribution. Canada, Europe, and the United States (Idaho and North Dakota). Pink seed is not serious.

Symptoms. Seeds are pink and appear as if they had been treated with a dye. The endosperm is soft and pink.

Management. Not necessary.

Spike Blight

Spike blight is also called gummosis.

Cause. *Clavibacter tritici* (syn. *Corynebacterium michiganense* pv. *tritici* and *Corynebacterium tritici*) and *C. iranicum*. The species name *iranicum* is sometimes spelled "iranicus" in the literature. *Clavibacter* spp. are seedborne and survive in infested organic matter in wet soils, usually in the lower areas within a field. Dissemination within a field is primarily by the seed gall nematode, *Anguina tritici*. Long-distance dissemination is by seed. Nematode larvae become contaminated with bacteria in the soil and carry them up the stem to the plant apex. Wet weather or wet, low areas in a field favor disease development.

Distribution. Australia, Canada, China, Egypt, Ethiopia, and India. *Clavibacter iranicum* is reported from Iran. Spike blight is not important except in wet areas of a field.

Symptoms. Initially, diseased leaves may be wrinkled or misshapen as they emerge from a whorl. Heads are misshapen and covered with a sticky yellow bacterial exudate. Dried exudate appears as white flecks that distort heads, necks, and upper leaves by inhibiting their elongation.

Management
1. Sow only cleaned seed.
2. Do not grow wheat in nematode-infested soil for at least 2 years.
3. Drain any wet areas in a field.

Stem Melanosis

Cause. *Pseudomonas cichorii* overwinters on several weed species found in wheat fields. Stem melanosis occurs when wheat is grown in copper-deficient soils. Wounds facilitate the entry of bacteria. Disease occurs only with repeated infections over a 2- to 3-week period that coincides with high temperatures and nightly dew formation.

Distribution. Canada.

Symptoms. Symptoms initially occur at the milky ripe stage of growth. The disease pattern in a field is sharply defined, irregular-shaped, dark patches that range in size from several square meters to several hectares.

Small light brown lesions first develop on stems beneath the lower two nodes. During the following 2 weeks, coinciding with the soft-dough stage, lesions darken and coalesce on stems, rachis, and peduncles and an occasional mottling of glumes occurs. The rachis and upper portions of the peduncles beneath the heads and immediately beneath nodes turn dark brown. Diseased heads become bleached and thin and have shriveled kernels.

Copper-deficiency symptoms are a lack of plant vigor, pale stunted vegetative growth, leaf tip injury, delayed maturity, aborted or shriveled grains, head bending, and melanism. These same symptoms occur to a limited extent on oats and barley. *Pseudomonas cichorii* is isolated from all melanotic tissues of wheat, but never from barley or oats.

Management. Stem melanosis is reduced when copper is added to copper-deficient soils.

Wheat Leaf Blight

Wheat leaf blight may be similar to bacterial leaf blight.

Cause. *Pseudomonas syringae.* Many researchers consider the causal organism to be synonymous with *P. syringae* pv. *syringae.* Infection occurs during periods of heavy rainfall.

Distribution. Argentina.

Symptoms. Seedlings infiltrated with the bacterium display limited water-soaking symptoms in 7 days and complete leaf necrosis in 10 days. Water-soaking occurs at diseased leaf margins which eventually become a confluent area of white necrosis.

Management. Not reported.

White Blotch

Cause. *Bacillus megaterium* pv. *cerealis* survives in soil, plant parts, seeds of susceptible cultivars, and insects. Bacteria are spread by water, insects and/or mites, and seed. Infection is favored by high temperatures and a high light intensity.

Distribution. United States (North Dakota) on hard red spring, hard red winter, and durum wheats.

Symptoms. The initial small yellow or white lesions on diseased leaf blades, sheaths, and culms enlarge into white or very light tan, irregular-shaped blotches and streaks. The white blotches are often broader than the streaks symptomatic of bacterial leaf blight (bacterial leaf necrosis) and lack the initial water-soaking symptoms associated with bacterial leaf blight.

Management. Not reported.

Diseases Caused by Fungi

Alternaria Leaf Blight

Cause. *Alternaria triticina* is seedborne, surviving as conidia on seed and as mycelium in seed. *Alternaria triticina* presumably overseasons in soil and within infested soil residue, which probably is the source of primary inoculum. Plants are infected at about 4 weeks old when leaves in contact with soil become infected. Secondary inoculum is provided by conidia produced in lesions and disseminated by wind to other plant hosts. Infection is favored by 10 hours or more of continued leaf wetness and temperatures of 20° to 25°C.

Distribution. India.

Symptoms. Initially, small, oval, chlorotic spots occur on diseased lower leaves. Spots enlarge, become sunken, assume irregular shapes, and turn brown-gray. Symptoms progress up a diseased plant and extend to the heads and leaf sheaths. During humid conditions, spots have a dark, powdery appearance due to sporulation by the fungus.

Management
1. Grow resistant cultivars.
2. Apply foliar fungicides where feasible.

3. Apply seed-protectant fungicides; however, these are frequently not effective.

Anthracnose

Cause. *Colletotrichum graminicola,* teleomorph *Glomerella graminicola,* is a soil-inhabiting fungus that is a successful saprophytic colonizer of a wide range of plant residues. The fungus also survives as mycelia and conidia on cereals and grasses and their residues, and becomes seedborne during harvest. The optimum production of conidia from mycelia is during wet weather and a temperature of 25°C. Conidia are disseminated by wind and water to plant hosts. Infection of all plant parts is most severe during wet weather at 25°C. Anthracnose is most likely to occur when wheat is grown in a rotation with wheat or other cereals on coarse soils that are low in fertility.

Distribution. Wherever wheat is grown.

Symptoms. Symptoms become apparent toward plant maturity as elliptical lesions (1 to 2 cm long) above and below the soil level on diseased stems. Lesions, at first, are water-soaked, then become bleached and necrotic. Diseased crowns and stem bases are bleached and later become brown. Later, when moisture is plentiful, acervuli develop as small, black, raised spots on the surfaces of diseased lower leaf sheaths and culms, and on leaf blades of dead plants. Round to oblong lesions with acervuli also may occur on green leaves. Plants display gross symptoms of a premature ripening or whitening, a general reduction in vigor, death, lodging, and shriveled grain.

Management
1. Rotate wheat with a noncereal, such as a legume or grass.
2. Improve soil fertility.
3. Grow resistant cultivars.
4. Treat seed with a seed-protectant fungicide.

Ascochyta Leaf Spot

Cause. *Ascochyta tritici. Ascochyta graminicola* and *A. sorghi* have also been reported as pathogens. *Ascochyta tritici* overwinters as pycnidia and mycelia in infested residue. During hot weather, pycnidiospores are disseminated a short distance. Factors that enhance infection are high humidity caused by weather or dense foliage, and the contact of leaves with soil. Secondary infection occurs as pycnidiospores from pycnidia produced in older leaf spots.

Distribution. Europe, Japan, and North America.

Symptoms. Ascochyta leaf spot is an inconspicuous disease and can be easily overlooked or confused with other diseases, particularly glume blotch.

Spots tend to be mostly on diseased lower leaves. Initially, spots are yellow, round to elliptical (1–5 mm in width). Later, spots coalesce and cause large areas of diseased leaves to be light brown and necrotic. Pycnidia are often formed in necrotic tissue and appear as minuscule black dots because most of the fungal structure is submerged in the diseased tissue so just the top portion is visible.

Management. Application of foliar fungicides is somewhat effective.

Aureobasidium Decay

Cause. *Microdochium bolleyi* (syn. *Aureobasidium bolleyi* and *Gloeosporium bolleyi*).

Distribution. Widespread, but Aureobasidium decay is of little importance. *Microdochium bolleyi* has been reported as part of the common root rot complex in different areas.

Symptoms. A mild necrosis of the seedling roots occurs.

Management. Not reported.

Black Head Mold

Black head mold is also called sooty head mold.

Cause. Several fungi, including *Alternaria* spp., *Cladosporium* spp., *Epicoccum* spp., *Sporobolomyces* spp., and *Stemphylium* spp. Black head mold occurs when wet weather accompanies maturation, especially if harvest is delayed. The disease becomes more severe on heads of plants damaged from causes such as other diseases, nutrient deficiency, or lodging.

Distribution. Wherever wheat is grown.

Symptoms. Green-black molds develop on diseased heads.

Management. Grain should be sufficiently dried before it goes into storage.

Black Point

Black point is also called kernel smudge.

Cause. Several fungi, including *Aspergillus* spp.(especially *A. alternata)*, *Bipolaris* spp. (especially *B. sorokiniana)*, *Chaetomium* spp., *Cladosporium* spp., *Curvularia* spp., *Fusarium proliferatum*, *Fusarium* spp., *Gloeosporium* spp., *Helminthosporium* spp., *Myrothecium* spp., *Nigrospora* spp., *Penicillium* spp., *Plenodomus* spp., *Rhizopus* spp., and *Stemphylium* spp.

Black point is most severe at a relative humidity of 90% and higher or continuous rainfall during seed maturation. Damage will increase if grain is stored at 95% relative humidity and 25% moisture (wet weight basis). *Alternaria alternata* may use pollen as a nutrient source during the infection process. Other exogenous sources of nutrients may account for higher disease incidence.

Irrigation of soft white spring wheat during the mid- or soft-dough stage significantly increases the incidence of black point. Black point of the spring wheat cultivar 'Yecora Rojo' occurs more frequently in fields irrigated by center-pivot than by rill or wheel line irrigation.

Distribution. Wherever wheat is grown.

Symptoms. Black point describes the darkened and sometimes shriveled embryo end of diseased seed. Diseased kernels are discolored, weathered, black-pointed, or "smudged". Germination decreases when the embryo is invaded. Discolored grain is discounted at the market.

Management
1. Treat seed with seed-treatment fungicide.
2. Store grain at 15% moisture (wet weight basis) or less.
3. Cultivars with some resistance to black point caused by *B. sorokiniana* may not be resistant to other diseases caused by this fungus, such as common root rot and spot blotch.

Brown Root Rot

Cause. *Plenodomus meliloti* (syn. *Phoma sclerotioides*).

Distribution. Canada.

Symptoms. Necrosis occurs where pycnidial initials are attached to roots. Pycnidial initials are green-black, globular or flattened, and up to 0.5 mm in diameter. In the spring, they are found attached to roots and shoot bases by a short attachment process. These stalks develop singly from adaxial surfaces of the protopycnidia, which resemble sclerotia.

Management. Not reported.

Cephalosporium Stripe

Cause. *Hymenula cerealis* (syn. *Cephalosporium gramineum*) is seedborne and survives as conidia and sporodochia in infested residue in the top 8 cm of soil for up to 5 years, and as mycelia in infested host residue in soil for an indeterminate amount of time.

Maximum infection occurs when roots are wounded or broken in the presence of high concentrations of inoculum. Mechanical injuries are not absolutely required for infection, but a higher incidence occurs when there are root wounds usually caused by soil heaving but also, in some cases, by insects.

Infection is related to root growth, root injury, and environmental factors that increase infection sites and concentrations of inoculum. High soil temperatures following sowing make plants more susceptible to infection because of the increase in the size of the root system. The addition of phosphate fertilizer in drill rows further increases root growth and leads to greater disease incidence at high temperatures but not at lower temperatures. Infection incidence and severity increase with decreasing soil pHs from 7.5 to 4.5, especially with highly susceptible cultivars. The effect of a lower pH may be accounted for by the presence of aluminum and hydrogen ions that injure roots by causing root membranes to leak, thereby allowing adsorbed cations to be lost. Decreases in soil pH have also been attributed to the heavy use of ammonia-based nitrogen fertilizers.

Roots are more prone to injury and breakage by heaving soil in the spring. These conditions are likely to happen when a warm autumn is followed by a cool spring and a large number of soil freeze-thaw cycles are accentuated by high soil moisture during the autumn and winter. Excess soil and surface water during early winter also favor development of sporodochia, which in turn increases both sporulation and the dispersal of conidia that serve as primary inoculum. Exudates from cold-stressed but unbroken roots also increase spore germination, conidiogenesis, and hyphal branching.

Thus the greatest amount of inoculum is produced during the winter and early spring, which coincides with a greater incidence of wounds caused by soil heaving and freezing, and by insects. Under these circumstances, crown roots are the primary infection court for the pathogen to colonize wounded and senescent root cortical tissue, then penetrate into the vascular system. The subcrown internode can also be infected just as seed is germinating. Severed or wounded roots have been reported to be susceptible to infection for 16 days after wounding.

Soil freezing is not absolutely necessary for infection in all locales. When temperatures decrease and rainfall begins in the Pacific Northwest of the United States, single-celled conidia produced on colonized residue on or near the soil surface are liberated into the soil, where they serve as the primary inoculum and infect directly through unbroken roots.

After infection, conidia enter xylem vessels and are systemically carried upward, where they lodge and multiply at nodes and leaves but do no further damage to roots. The fungus prevents water movement up the plant and produces harmful metabolites.

At harvest time, *C. gramineum* is returned to soil in residue, where it is a successful saprophytic competitor with other soilborne microorganisms. Cephalosporium stripe is most severe in no-tillage plots and decreases in minimal- and conventional-tillage plots because of a threshold size for plant fragments to be infested with *C. gramineum*. The optimum size of plant fragments varies between soil types because infection is influenced by total nutrients available from residue and soil.

Distribution. Great Britain, Japan, and most winter wheat–growing areas of North America.

Symptoms. Winter wheat is most severely diseased. Spring wheat is susceptible but apparently escapes the disease and usually does not show symptoms. Diseased plants are scattered throughout a field but are usually more numerous in the lower and wetter areas.

Diseased plants are dwarfed. During jointing and heading, normally one or two, but up to four, distinct yellow stripes develop on leaf blades, sheaths, and stems and extend the length of the plant. Thin brown lines consisting of infected veins occur in the middle of each stripe. The stripes eventually turn brown and become highly visible on green leaves and remain noticeable even on yellow straw. Toward harvest, culms at or below diseased nodes become dark due to fungus sporulation. Heads of diseased plants are white and contain shriveled seed or no seed.

Management

1. Grow resistant or moderately resistant cultivars. Disease declines over several seasons with monoculture of moderately resistant cultivars. Thus, higher levels of resistance are not needed to manage Cephalosporium stripe (Shefelbine and Bockus 1989). Inoculum production is not related to the resistance response of the cultivar.
2. Rotate wheat with a nonhost for at least 2 years. However, this recommendation conflicts with information that disease declines over several seasons with monoculture of moderately resistant cultivars.
3. Sow later in the autumn or when the soil temperature at 10 cm below the soil surface is below 13°C. Plants under these conditions have limited root growth, thus reducing the number of infection sites.
4. Infected residue should be plowed deeper than 8 cm.

Ceratobasidium gramineum Snow Mold

Cause. *Ceratobasidium gramineum* (syn. *Corticium gramineum*). Snow mold occurs in well-drained, upland paddy fields but is more common on barley.

Distribution. Japan.

Symptoms. Diseased leaves appear "rotted" and have a dark green appearance just after snowmelt. After leaves have dried, they become pale gray to pale yellow and have ellipsoidal lesions with a narrow brown margin. A pale gray mycelial mat grows in the center of lesions. Leaf sheaths frequently have a brownish appearance.

Symptom development ceases after snowmelt, but later in the growing season, ellipsoidal lesions are found on leaf sheaths near the soil level.

Management. Not reported.

Common Bunt

Common bunt is also called covered smut, European bunt, hill bunt, and stinking smut.

Cause. *Tilletia caries* (syn. *T. tritici*) and *T. laevis* (syn. *T. foetens* and *T. foetida*). The specific name *laevis* is sometimes spelled *"levis"* in the literature. The two fungi are closely related and have similar life cycles that may occur together in the same diseased plant.

Both fungi overwinter as teliospores (chlamydospores) on seed and in soil. When wheat is sown, teliospores located on or near the seed in the soil germinate in the presence of moisture and cool temperatures (5° to 15°C). A promycelium (basidium) is formed on which eight to 16 basidiospores (sporidia) are produced. The basidiospore fuses in the middle with a compatible basidiospore and forms an "H"-shaped structure that germinates and forms yet another structure called a secondary sporidia.

The secondary sporidia germinate and produce mycelium that infects seedlings. Mycelium grows behind the growing point or meristematic tissue of the infected plant, invades the developing head and displaces the grain. Eventually, teliospores form inside the diseased seed. At harvest time, the diseased seed is broken and teliospores disseminated by wind contaminate healthy host kernels and the soil.

Tilletia caries has been reported to be pathogenic for up to 10 years after being incorporated in soil. The spores have been reported to be viable after passage through the digestive tracts of cattle and sheep, but not pigs or poultry.

Distribution. *Tilletia caries* is generally distributed wherever wheat is grown but is most prevalent in the northwestern United States. *Tilletia laevis* is limited to areas of Europe and North America, where it was reported from the midwestern and northwestern United States.

Symptoms. Common bunt symptoms are more pronounced following cool soil temperatures after sowing. Wheat sown in the spring may be less severely diseased. Diseased plants are somewhat stunted but cannot be readily distinguished from healthy plants until heading. Bunted heads are more slender than healthy heads and the glumes of spikelets may be spread apart. A bunted head will often stay green, with a bluish cast, for a longer period of time than will a healthy head. Bunted kernels are about the same size as healthy seed but are light brown and have a more rounded shape. The pericarp ruptures at harvest and teliospores contaminate healthy seeds, giving them a "fishy" odor; hence, the name stinking smut. The smut balls or diseased kernels have an oily feeling when crushed.

Management
1. Treat seed with a systemic seed-treatment fungicide.
2. Grow resistant cultivars.

3. Sow early in the autumn. Seedlings may be far enough advanced and less susceptible to infection by the time secondary sporidia develop.

Common Root Rot

Common root rot is also called crown rot, dry land root rot, and foot rot.

Cause. *Bipolaris sorokiniana,* teleomorph *Cochliobolus sativus,* is considered to be the primary cause of common root rot; however, other fungi reported to cause common root rot symptoms are *Fusarium acuminatum, F. avenaceum, F. crookwellense, F. culmorum, F. graminearum* group 1, and *F. poae.* All are widespread fungi that survive as saprophytic mycelia in infested soil residue. Conidia of *B. sorokiniana* and chlamydospores of *Fusarium* spp. also survive for several months in the soil and are seedborne.

Microdochium bolleyi is also a secondary invader of roots and is sometimes isolated together with *B. sorokiniana* and *Fusarium* spp. Although frequently referred to as a minor pathogen because it does not penetrate the stele of the host plant, *M. bolleyi* is reported to seriously alter root function or affect shoot growth. In Finland, *M. bolleyi* is reported to be commonly isolated together with *B. sorokiniana* and *Fusarium* spp. from diseased roots.

Inoculum density of *B. sorokiniana* in soil is related to the amount of sporulation occurring on infested residues. Some researchers have seen no significant differences on root rot or plant yield due to tillage or straw management. Others have reported that significantly more spores were found with conventional tillage than in a reduced tillage system in Texas. However, the distribution of spores in the soil profile was not significantly affected by the tillage system; the majority of the spores were found in the top 10 cm of soil, regardless of tillage method. Secondary sporulation is greatest on the diseased crowns of mature wheat plants, on crown roots, and on subcrown internodes, seed pieces, and seminal roots; the highest numbers of conidia are present on necrotic or senescent tissue.

Primary infections occur on coleoptiles, primary roots, and subcrown internodes. Plants put under stress by drought, freezing, warm temperatures, lack of nutrition, and Hessian fly injury are most subject to infection. Under these conditions, disease severity increases with sowing depth because of an increased number of infection sites on belowground plant tissues. *Fusarium* spp. infect secondary roots as they emerge from the crown. The diseased plant does not die as long as it can produce new roots. However, early infection reduces yield potential of plants by restricting the production of tillers. Conidia are produced when disease progresses aboveground.

Moisture is required for infection, but once initiated, disease development requires warm temperatures and moisture stress. Disease may be most severe in dry soil when inoculum is seedborne. Infection of seedlings in natural soil by *F. graminearum* group 1 is restricted to soil-water potentials from −0.1 to −1.5MPa. A low soil-water potential predisposes wheat seedlings to colonization and further damage by the fungus.

Crop sequence has an effect on infection: the incidence of common root rot increases under continuous cultivation. Oat stubble favors the buildup of *F. culmorum* in the soil; therefore, the poor grain yields associated with an oat-wheat rotation may be attributed to *F. culmorum* crown and root rot.

Isolates of *C. sativus* from fields in continuous wheat are more pathogenic to wheat than isolates from fields in first-year wheat. Isolates of *C. sativus* from diseased wheat and barley also were highly virulent on their original host species but weakly virulent on alternate host species. In Canada, spring wheat tended to be colonized by higher levels of *B. sorokiniana* than winter wheat did.

Distribution. Wherever dryland wheat is grown.

Symptoms. Disease incidence and severity were usually significantly higher in conventional tillage plots than in reduced tillage plots in Texas. Diseased plants occur in random patches throughout a field, are stunted, and have a lighter green color than healthy plants. Diseased plants mature early, produce few tillers, and have bronzed, bleached, or white heads, which results in shriveled seed. Seedlings may be killed before or after emergence. The surviving seedlings have brown lesions on the coleoptiles, roots, and culms.

Plants with diseased crowns are usually killed. Darkening of subcrown internodes is usually caused by *B. sorokiniana*. *Fusarium* spp. cause roots, culm bases, and lower nodes to dry and darken and cause diseased plants to become brittle and break off easily near the soil line. The browning of the root system can sometimes be seen only when adhering soil is washed from diseased roots. Winter survival of wheat is reduced, particularly when diseased by *B. sorokiniana*.

Management
1. Treat seed with a seed-protectant fungicide.
2. Properly fertilize soil to ensure growth of new plant roots.
3. Sow winter wheat in the late autumn.
4. Grow cultivars with a degree of resistance. At least three genes have been identified that control resistance. Cultivars with some resistance to common root rot may not be resistant to other diseases caused by *B. sorokiniana,* such as black point and spot blotch.
5. Leave field fallow for 3 to 4 years.
6. Winter wheat rotations incorporating canola and flax result in greater wheat survival, superior plant vigor, greater plant height, larger numbers of heads formed per meter of row, higher grain yield, and lower levels of root disease.

7. Tillage method. Disease incidence and severity were usually significantly higher in conventional tillage plots than in reduced-tillage plots
in Texas.

Cottony Snow Mold

Cottony snow mold is also called Coprinus snow mold, LTB (low-temperature
basidiomycete), SLTB (sclerotial low-temperature basidiomycete), snow mold,
and winter crown rot.

Cause. *Coprinus psychromorbidus* is the low-temperature basidiomycete (LTB)
form of cottony snow mold. The nonsclerotial form of cottony snow mold
needs a prolonged and persistent snow cover. Such a snow cover produces
relatively constant soil temperatures of $-1°C$ to $-5°C$, which appear to be
the most conducive for development of LTB. However, a thin snow cover
for a short duration appears sufficient to permit limited penetration of
wheat by *C. psychromorbidus* and development of the fungus within the
plant, even in the absence of a persistent snow cover. Under certain conditions, microsclerotia form in diseased plant tissue and may also be a survival mechanism.

The LTB form of cottony snow mold is the most prevalent snow mold
pathogen of winter cereals in central and northern Alberta and northeastern British Columbia and is responsible for severe mortality in most years.

Distribution. Canada.

Symptoms. After snowmelt, patches of dead plants normally occur where the
snow was deepest in the field. Sometimes there is a sparse mycelial growth
that gives diseased plants a gray sheen; however, often little mycelium can
be found. Irregular-shaped gray to black sclerotia (1 mm in length) are
found inside diseased leaf sheaths and infrequently on roots and subcrown
internodes. The sclerotia are loosely attached to diseased host tissues and
can be easily scraped off.

Management. Not reported.

Crater Disease

Crater disease and Rhizoctonia bare patch are similar, but their symptoms differ sufficiently to consider them as two separate diseases. A disease identified as
patchy stunting in Tanzania is possibly similar to crater disease, except crater
disease occurs exclusively in black montmorillonite clay soils and patchy stunting has been observed in clay loam, silty clay loam, and silty loam soils. Crater
disease is similar to Rhizoctonia bare patch disease, but it is caused by a *R. solani*
that does not anastomose with AG-8, the cause of Rhizoctonia bare patch.

Cause. *Rhizoctonia solani* AG-6. This is the same strain of *R. solani* that also
causes patchy stunting. The umbrella thorn tree, *Acacia tortilis,* is a natural

host for the fungus. Crater disease occurs only on very heavy, black mont-morillonite clay soil that is 40% clay or greater and in which root growth is impeded by soil structure. Disease is also enhanced by dry weather and early infection by *R. solani*. Isolates of *R. solani* that cause crater disease anastomose with each other but not with any other anastomosis group tester strains and not with AG-8.

Distribution. South Africa and Tanzania.

Symptoms. Severely stunted patches of wheat occur within a field. Under continuous cultivation of wheat, patches enlarge each year and assume an irregular-shaped pattern that tends to stretch in the direction of cultivation. Some fields display a rippling effect caused by elongated patches of less obviously stunted plants.

Initially, there is a sudden appearance of yellow areas of plants. About 3 weeks after emergence, plants wilt and the lower leaves soon become dry and white. Most plants die as seedlings, but some survive until maturity and have reduced tillering and incomplete grain filling. Roots of stunted plants have bead-like swellings composed of hyphae that superficially resemble sclerotia. Xylem vessels in roots are brown and occluded with a gel-like material; roots break easily at these points. Frequently the entire root is brown and rotted where the roots are broken.

Management. Stunting is alleviated by thorough and repeated cultivation of soil.

Several experimental methods of disease management have been worked on in South Africa. *Trichoderma harzianum* treatments significantly increased plant growth and reduced root infection. With the addition of a fungicide, *T. harzianum* and sheep manure incorporated into tilled and "ripped" soils increased grain yield.

In other research, soil drenching with phosphorous acid at 50 g per square meter prior to sowing and a foliar application to 2-week-old plants at the rate of .125 dm (cubic) of a 10% solution per square meter significantly reduced infection of wheat by *R. solani* in naturally infested crater disease soil in the field.

Disease was managed by soil solarization and fumigation with methyl bromide by using deep cultivation. However, the researchers were careful to point out this was not practical for commercial application.

Dilophospora Leaf Spot

Dilophospora leaf spot is also called twist.

Cause. *Dilophospora alopecuri* (syn. *D. graminis*) survives as mycelia and pyc-nidia in infested residue, and as conidia on seed. Primary infection is caused by conidia that are produced in pycnidia, released during wet weather, and disseminated by wind and splashing rain to host seedlings.

Secondary infection is caused by conidia produced in pycnidia formed within the disease spots on seedling leaves during moist weather. The twist phase of the disease is caused by secondary conidia disseminated into the whorls by larvae of the seed gall nematode, *Anguina tritici,* as it moves up the plant in a water film.

Distribution. Canada, Europe, India, and the United States. Twist has not been observed for several years and is considered to be a rare disease on wheat.

Symptoms. Initially, small, elongated, yellow spots appear only on diseased leaves. Spots become light brown with black centers that consist of stromata and pycnidia of *D. alopecuri.* Diseased leaves may be killed if spots become numerous. When the whorl is colonized by the fungus, leaves do not emerge or emerge twisted, distorted, and covered with gray mycelium. This phase of the disease rarely occurs in the absence of *A. tritici.* *Dilophospora alopecuri* dries as stromata that appear as dark streaks on leaves and in which pycnidia are produced.

Management
1. Rotate wheat with other crops. Most cereals are not susceptible.
2. Plow under infested residue.
3. Treat seed with a seed-protectant fungicide.
4. Sow certified wheat seed.

Disease Caused by *Fusarium sporotrichioides*

Cause. *Fusarium sporotrichioides.*

Distribution. The United States (Minnesota) on spring wheat.

Symptoms. Foliar and glume necrosis is severe only on very young plants growing under optimum disease conditions. Leaf symptoms are variable and include the following: small tan flecks with dark red borders; oval to elongated tan, red, or red-tan lesions of different sizes that occasionally appear similar to a mechanical abrasion; tan streaking, especially in young leaves; and tip burn. Diseased leaf sheaths occasionally kill leaves.

Head symptoms are identical to those of scab (Fusarium head blight) except kernels are not visibly affected. Diseased culms sometimes kill heads.

Management. Not reported.

Downy Mildew

Cause. *Sclerophthora macrospora* (syn. *Sclerospora macrospora* and *Phytophthora macrospora*) is seedborne as oospores and survives for years as oospores that are embedded in infested leaf and stem tissue or are loose in soil when diseased tissues decay. *Sclerophthora macrospora* is an obligate parasite and cannot grow saprophytically on dead plant tissue. Oospores are disseminated in infested soil residue by wind and water. Oospores germinate in saturated soil and produce sporangia from which zoospores are released 1 to 2 hours after sporangia formation. In some instances, zoospore formation ceases and secondary sporangia are formed. Zoospores "swim" through soil water, settle on the developing seedling, and produce germ tubes that penetrate the host plants. In another mode of infection, oospores survive for months in dry soil and germinate directly and form germ tubes that penetrate hosts, or germinate indirectly by producing sporangia from which zoospores are released. Infection occurs over a wide temperature range (7° to 31°C). Following infection, *S. macrospora* develops systemically in plants, particularly in meristematic tissue.

Distribution. Wherever wheat is grown. Downy mildew is not considered a serious disease except in localized wet areas.

Symptoms. Downy mildew occurs only in localized areas of fields where seedlings have been in flooded or water-logged soil for 24 hours or longer. Wheat plants are dwarfed and deformed, tiller excessively, and have several yellow leaves. Diseased plants have leathery, stiff, thickened leaves, stems, and heads. Diseased leaves, stems, and heads are twisted and distorted, and no seeds are formed in severely diseased plants.

In less severely diseased plants, dwarfing may be slight and one or more of the upper leaves may be stiff, upright, or variously curled and twisted. Heads and stems are not deformed. Numerous round yellow-brown oospores may be found in diseased tissue examined under magnification.

Management. Downy mildew is ordinarily not serious enough to warrant special management measures. However, the following are some general measures that will aid in managing disease should a serious problem occur.

1. Provide proper soil drainage.
2. Control grassy weeds that serve as hosts.
3. Sow cleaned seed from disease-free plants.

Dry Seed Decay

Cause. *Penicillium* spp. infect seed when soil is too dry for seed to germinate and seedlings to emerge.

Distribution. Semiarid regions of the world.

Symptoms. Diseased seed is shriveled and covered with a blue to green *Penicillium* growth.

Management. Treat seed with a seed-treatment fungicide.

Dwarf Bunt

Dwarf bunt is also called dwarf smut, short smut, stunt smut, stubble smut, and TCK smut.

Cause. *Tilletia controversa* survives as teliospores in soil and on the surfaces of infested seed. However, disease occurs only when heavily infested seed, which has an inoculum infestation equal or greater than 1 g of teliospores per kg of seed, is sown in disease-conducive locations. When wheat is sown in moist soil, teliospores on or close to seed germinate under warm, dry conditions and form a promycelium on which eight to 16 basidiospores are borne. A basidiospore fuses in the middle with a compatible basidiospore and forms an "H"-shaped structure that, in turn, germinates and forms a structure called a secondary sporidium. The secondary sporidia germinate and produce infective mycelium.

Tilletia controversa is better adapted to survive in soil and has a lower temperature range (1°–5°C) for optimum germination than the common bunt fungi *T. caries* and *T. laevis*. Protracted moist conditions and temperatures of less than 5°C in the spring induce dormancy in teliospores. The infection process occurs 35 to 105 days after germination, depending on temperature. Infection generally requires a heavy snow cover over unfrozen ground thus providing suitable conditions of moisture and temperature at the soil surface for teliospore germination and infection of host plants. Therefore, disease is restricted to areas where wheat is grown under a persistent snow cover.

After infection, mycelium grows behind the meristematic tissue of growing points and invades developing heads to displace grain. Eventually, teliospores form within the seed. At harvest time, diseased seed is broken and teliospores are disseminated by wind and contaminate healthy wheat kernels or soil.

Distribution. Canada, Europe, and the United States where a heavy snow cover is likely to occur over unfrozen ground.

Symptoms. Dwarf bunt is limited to winter wheat and, infrequently, to winter barley. Diseased plants are generally one-fourth to one-half the size of healthy plants. Viable pollen does not develop, the ovaries of infected florets are larger than those in healthy florets, and there are more bunt-infected kernels per spikelet than seeds per spikelet on a healthy plant.

Smut balls are smaller and rounder than common bunt, and the spore mass feels dry. Spores have a foul odor that is similar to rotten fish.

Management
1. Grow resistant cultivars in adapted areas.
2. Treat seed with a fungicide seed treatment.

Epicoccum Glume Blotch

Cause. *Epicoccum purpurascens* (syn. *E. nigrum*).

Distribution. India.

Symptoms. Symptoms initially appear in the dough stage of plant development. Purple-brown, elliptical spots, which later become gray, occur on glumes and sometimes on awns. Severely diseased glumes are entirely covered with spots that occasionally are dotted with black spore masses. In highly susceptible cultivars, the rachis and peduncle are covered with dark brown spots and the diseased leaves are covered with irregular-shaped spots. Grain is severely shriveled.

Management. Varying degrees of resistance exist among cultivars.

Ergot

Cause. *Claviceps purpurea,* anamorph *Sphacelia segetum. Claviceps purpurea* survives as sclerotia in soil for about 1 year and for longer periods when mixed with grain in storage facilities. In response to moisture and after a cold-temperature treatment, sclerotia from grasses germinate during the spring or early summer and form one or more stalked stromata. Perithecia, which contain numerous asci in which ascospores are borne, are produced in the apex of the stromata. Ascospores are the primary inoculum for floret infection and are disseminated by wind and splashing rain to plant hosts. Ascospores that contact a stigma germinate and penetrate the ovary within 24 hours. In 5 days, conidia form on the ovary surface within a sweet liquid (honeydew) and serve as secondary inoculum. Conidia are disseminated to other florets by plant-to-plant contact, splashing rain, and insects that have been attracted to the honeydew. Conidia production declines over time and the swollen, convoluted ovaries enlarge and become converted from base to tip into sclerotia.

Ergot is favored by wet, cool weather that prolongs the flowering period of wheat plants. These conditions also favor honeydew formation. Florets are most susceptible to infection just before anthesis.

Distribution. Worldwide, but the incidence of ergot on wheat is normally low.

Symptoms. The most conspicuous symptom is the purple-black horn-like sclerotia or ergot bodies that replace one or more seeds in the diseased wheat head. Sclerotia protrude from glumes and are up to 10 times larger than healthy seed. Before sclerotia are formed, the honeydew stage occurs in the form of sticky yellowish droplets that attract insects to congregate on diseased heads. Just before sclerotia develop, infected ovaries swell and their surface becomes convoluted with a superficial layer of conidiophores.

Wheat is considered "ergoty" if it contains more than 0.3% ergot bodies or sclerotia by weight. Sclerotia contain toxic alkaloids that affect warm-blooded animals. Consumption of infested grain or hay may cause blood vessel constriction and smooth muscle contractions. In pregnant animals, continuous doses of ergot can cause uterine contractions that lead to spontaneous abortion. Ingesting small amounts of ergot may cause a reduction or complete cessation in lactation or the loss of extremities, especially ears, tails, and hooves, because of blood vessel constriction followed by the onset of gangrene.

Management
1. Use seed that is free of sclerotia.
2. Rotate wheat with a nongrass crop.
3. Plow residue deep into the soil.

Eyespot

Eyespot is also called Cercosporella foot rot, culm rot, foot rot, stem break, strawbreaker, and strawbreaker foot rot.

Cause. *Pseudocercosporella herpotrichoides* (syn. *Cercosporella herpotrichoides*), teleomorph *Tapesia yallundae.* Two other species, *P. anguioides* and *P. aestiva,* are associated with eyespot, but their significance in the etiology of eyespot has not been established.

Pseudocercosporella herpotrichoides survives for several years as mycelia within infested residue. Apothecia are reported on wheat straw in Europe. Sporulation occurs on infested residue and straw at the soil surface during cool, damp weather or when the humidity is near saturation and temperatures are 8° to 12°C in the autumn or spring. Dispersal is primarily by splashing rain.

Development of epiphytotics is dependent upon the production of primary inoculum on infested residue remaining from the previous wheat crop. Winter wheat is more likely to be infected than spring wheat. The coleoptile is most susceptible to infection during the seedling stage; however, with decay of the coleoptile, the susceptibility of leaf sheaths increases.

Severity of both leaf sheath and stem lesions increases linearly with the amount of time host plants are subjected to favorable disease conditions. Disease severity has been suggested to be a function of accumulated temperatures since the development of epiphytotics is often interrupted by temperatures unfavorable to disease development. Leaf sheath penetration by *P. herpotrichoides* may be delayed by cold weather, and if the rate of leaf sheath death exceeds the rate of leaf sheath penetration, lesions do not occur. The rapid death of basal leaf sheaths prevents the fungus from becoming established in stems. Secondary inoculum is not important in the development of an epiphytotic because spores do not occur on developing lesions until later in the growing season. However, late infections add to the amount of inoculum to infect succeeding wheat crops. On winter wheat in Ontario, some infections occurred in the autumn but the peak of infection occurred in April. Soil or volcanic ash placed around the base of winter wheat in the spring increased disease severity in the state of Washington.

At least two distinct morphological types, corresponding to two patho-types of *P. herpotrichoides,* are present. Daniels et al. (1991) reported the two pathotypes are designated as wheat (W) or rye (R) and are distinguished on the basis of pathogenicity to wheat seedlings, spore morphology and colony morphology, pigmentation on maize meal agar, and isoenzyme polymorphisms. The wheat type is more pathogenic to wheat and barley than to rye and has faster-growing, even-edged colonies in culture. The rye type is as pathogenic to rye as to wheat and barley, and is slower-growing in culture and has feathery-edged colonies and more profuse sporulation.

Distribution. Where wheat is grown in a cool, moist climate.

Symptoms. Symptoms are most conspicuous near the end of the growing sea-son, when diseased plants lodge. In the lodging that is symptomatic of eye-spot, the straw falls in all directions. In contrast, lodging caused by wind or rain causes straw to fall primarily in one direction.

Eye- or ovate-shaped lesions with white to tan centers and brown mar-gins develop first on the basal leaf sheath of diseased plants. Similar spots form on stems directly beneath those of the leaf sheath and weaken stems sufficiently to cause lodging.

Crown and basal culm tissue, but not roots, become diseased. However, a necrosis occurs around roots in the upper crown nodes. Under moist con-ditions, lesions enlarge and the black, stroma-like mycelium that develops over the surface of crowns and the base of culms gives diseased plant tis-sues a "charred" appearance. Stems either shrivel and collapse or plants be-come yellowish to pale green and have heads that are reduced in size and number. When host plants are infected early in their growth, individual culms and weaker plants are killed before maturity. White heads, which are similar to a symptom of take-all, are often present at maturity.

Management
1. Rotate wheat with noncereals.
2. Spring wheat and late-sown winter wheat are less subject to infection.
3. Grow the most resistant cultivars. However, resistant cultivars often be-come diseased under severe disease conditions. Resistance has been de-termined in accessions of *Tricticum tauschii* (syn. *Aegilops squarrosa*).
4. Apply fungicides to residue to decrease primary inoculum. Foliar fungi-cides reduce eyespot under experimental conditions.
5. Reduced tillage limits foot rot incidence.

Flag Smut

Cause. *Urocystis agropyri.* Some researchers consider *U. tritici* to be a synonym. *Urocystis agropyri* survives as teliospores in soil for 3 years and on seed for 4 years. Teliospores germinate and produce sporidia that germinate and in-fect wheat coleoptiles prior to their emergence. Infection is optimum at soil moistures of 10% to 15% and temperatures of 10° to 20°C. *Urocystis* *agropyri* then overwinters as mycelium in seedlings and in the spring grows systemically in the upper portion of the plant, producing sori in which hyphal tips differentiate into teliospores. In a few days or when wheat is harvested, the sori erupt and liberate teliospores that contaminate soil and seed.

Distribution. Localized in most wheat-growing areas. Commonly found in Australia and in the western and certain areas of the midwestern United States.

Symptoms. Symptoms appear in spring and summer on a few tillers or on an entire plant. Smut lesions on leaves, sheaths, and upper parts of the diseased stem are long, light green stripes that soon become gray-black streaks. These streaks are subepidermal smut sori that develop between the leaf veins and cause leaves to become rolled and twisted. Splits along the sori in the diseased leaf liberate the gray-black spore masses. Diseased leaves become weakened, frayed, and split longitudinally. Diseased plants are dwarfed, tiller excessively, and rarely head out. However, if they do head out, glumes and necks also are striped.

Management
1. Grow resistant cultivars.
2. Treat seed with a systemic seed-protectant fungicide.
3. Rotate wheat with a resistant host for at least 2 years and preferably for 3 years.
4. Do not sow seed deep since shallow sowing partially prevents some infection.

Fusarium Glume Spot

Cause. *Fusarium poae* in close association with the mycophageous mite *Siteroptes avenae*. *Fusarium poae* occurs widely in the soils of the temperate regions and is characterized as a saprophyte or weak parasite of grasses and herbaceous plants. Two kidney-shaped, sac-like structures (sporothecae) in female mites contain microconidia of *F. poae* when the mites feed on fungi. Conidia can be released from sporothecae when mites feed on plants, thereby inoculating them. *Fusarium poae* is also seedborne.

Distribution. South Africa, on winter wheat and autumn-sown irrigated, spring wheat.

Symptoms. Diseased glumes have necrotic lesions that are surrounded by dark brown borders. The borders are often more pronounced toward the base of the glume. Generally, a few florets per spike are symptomatic.

When mature wheat plants were inoculated with a suspension of *F. poae* microconidia, water-soaked lesions occurred on leaves, and black chaff-like symptoms and necrosis occurred on awns.

Management. Not reported.

Fusarium Head Blight

Fusarium head blight is also known as blight, Fusarium blight, head blight, pink mold, scab, tombstone scab, wheat head blight, wheat scab, and white head.

Cause. *Fusarium graminearum* group 2, teleomorph *Gibberella zeae*; *F. acuminatum*; *F. avenaceum*, teleomorph *G. avenacea*; *F. crookwellense*; *F. culmorum*; *Microdochium nivale* (syn. *F. nivale*), teleomorph *Monographella nivalis* (syn. *Calonectria nivalis*); *F. equiseti*; *F. moniliforme*; *F. oxysporum*; *F. poae*; *F. proliferatum*; *F. pallidoroseum* (syn. *F. semitectum*); *F. sambucinum*; *F. sporotrichioides*; *F. subglutinans*; and *F. tricinctum. Fusarium graminearum* group 2 is considered to be the principal pathogen; however, other fungi are reported to be the principal causes in some geographical areas.

The fungi overwinter as mycelia and spores on seed, and as mycelia and perithecia in infested residue. Rainfall may be needed for perithecia and ascospores to form and mature on crop residue, but not to trigger the actual release of ascospores. Perithecial drying during the day, followed by sharp increases in relative humidity may provide the stimulus for release of ascospores.

In eastern Canada, ascospores were released during the first 3 weeks of July. Ascospore release during this time was a diurnal pattern beginning around 1600 to 1800 hours and reaching a peak before midnight, then declining to low levels by 0900 hours the following morning. The beginning of ascospore release was correlated with a rise in relative humidity during the early evening hours. Ascospores produced in perithecia and conidia produced from mycelia on infested residue during warm, moist weather are disseminated by wind to plant hosts.

Temperatures for optimum disease development differ among *Fusarium* species. Optimum disease development caused by *F. graminearum* occurs at 26.4°C, but occurs at 13.8°C for *F. crookwellense.*

Infected seed causes seedling blight, and soilborne inoculum causes seedling blight and root rot. Fusarium head blight develops in warm, humid weather during the formation and ripening of the kernels. Prolonged wet weather during and after anthesis favors infection and subsequent disease development. Conidia produced on diseased heads serve as secondary inoculum.

Grain that becomes wet in swaths favors Fusarium head blight development. Disease is more severe when wheat follows maize, when nitrogen and phosphorous fertilization is inadequate, and where weed density is high. Although Fusarium head blight occurs when conditions are warm and humid, disease caused by *M. nivalis* is reported to occur under cool, moist conditions in the Pacific Northwest.

Fusarium graminearum group 2 readily produces perithecia in culture, but *F. graminearum* group 1, which is associated with crown and root rot of wheat, does not.

Distribution. Wherever wheat is grown.

Symptoms. Disease begins in flowers and spreads to other parts of the head, giving the appearance of premature ripening. One or more spikelets of emerged immature heads are initially water-soaked, then die and become light brown, bleached, or white, beginning at the spikelet base. Eventually, hulls change from a light to dark brown discoloration. When infection occurs late in the development of seed, only the base of the hull is brown, but in severe cases, the entire kernel may become shrunken and brown. If the rachis is infected, the entire head is dwarfed and compressed. All spikelets above that point are closed rather than spreading, bleached, and sterile or contain only a partially filled seed. Diseased kernels are gray-brown and light in weight, and the interior of the kernel is floury and discolored. During humid or wet weather, pink to pink-red mycelia and spores grow on diseased spikelets. Later, small black perithecia grow in the same diseased area.

Diseased grain may be harmful when fed to hogs, dogs, and humans. Mycotoxins, such as 3-acetyldeoxynivalenol and deoxynivalenol (vomitoxin), may be produced in diseased heads. Deoxynivalenol and 3-acetyldeoxynivalenol affect protein synthesis at the ribosomes of susceptible wheat cultivars, which suggests these trichothecenes are phytotoxins as well as mycotyoxins.

The fungus might survive poorly in diseased seed because the germination of seed improves after several months in storage. Researchers reported that emergence of diseased seeds was better after storage temperatures of 10° and 20°C than at −10° and 2.5°C.

Management
1. Rotate wheat with other crops. Do not grow wheat after maize, barley, or rye. Do not grow wheat next to maize.
2. Plow under infected residue to hasten decomposition of infested residue and prevent spores from being windborne.
3. Do not spread manure that contains infected straw or maize stalks on soil where wheat is growing.
4. Sow early in the growing season to allow wheat to escape warm, moist weather.
5. Treat seed with a seed-protectant fungicide.
6. Grow resistant cultivars. Resistance is characterized as being of two types: type I is resistance to the initial infection, and type II is resistance to hyphal invasion of the plant tissue. In South Africa, differences in susceptibility to Fusarium head blight caused by *F. graminearum* and *F. crookwellense* have been reported.

   A type III resistance has been suggested. Some wheat cultivars identified as resistant apparently possess the ability to degrade trichothecenes, whereas susceptible cultivars do not have this ability. Compared to susceptible cultivars, resistant cultivars can tolerate

10–1000 times the concentration of trichothecenes with no apparent effect on growth.

7. The Food and Drug Administration recommends 2 ppm and below of deoxynivalenol in wheat entering the milling process, 1 ppm for finished wheat products destined for human consumption, and 4 ppm for wheat products to be used as animal feed.

Fusarium Leaf Blight

Cause. *Fusarium nivale* (syn. *Microdochium nivale*).

Distribution. Scotland.

Symptoms. Lesions occur on diseased upper leaves at the milky ripe growth stage of plant growth. Lesions are extensive, dull, diffuse, and water-soaked and have pustules of *F. nivale* sporulating through the stomata.

Management. Not reported.

Fusarium Rot

Fusarium rot is also called Fusarium crown and foot rot.

Cause. *Fusarium culmorum* and *F. graminearum* group 1 are most commonly cited. *Fusarium graminearum* group 1 does not produce perithecia in culture and is associated with crown and root rot of wheat; *F. graminearum* group 2 readily produces perithecia in culture and is one of the causal fungi of Fusarium head blight. *Fusarium avenaceum, F. acuminatum,* and *F. tricinctum* have been reported as minor components of the disease complex. *Fusarium equiseti* has been reported to cause a crown rot of wheat in South Africa. A new species of *Fusarium* that has characteristics in common with *F. longipes* and *F. scirpi* var. *filiferum* has been reported from the Orange Free State in South Africa. The symptoms caused by the new *Fusarium* are indistinguishable from those caused by *F. graminearum* group 1. However, this is possibly the same disease caused by *F. equiseti*.

Primary inoculum is soilborne chlamydospores and mycelia, conidia, and chlamydospores originating from infested residue in the upper 10 cm of the soil. Entry into the crown is gained 2 to 3 cm below the soil surface through openings around crown roots or is gained by the infection of newly emerging crown roots.

Disease is severe under dry soil conditions; a −40 bar midday leaf water potential is cited as the optimum disease condition. However, infections caused by the new *Fusarium* species are from an area in South Africa that has summer temperatures of 13.2° to 25.8°C, winter temperatures of 0.3° to 15.2°C, and a relatively high rainfall that averages 638.1 mm in the summer and 168.2 mm in the winter.

Distribution. The United States. *Fusarium culmorum* is widely distributed in central and eastern Washington, north central Oregon, and northern

Idaho. *Fusarium graminearum* is limited to south central Washington and *F. equiseti* is reported from South Africa.

Symptoms. Diseased plants rarely show outward symptoms until after heading. The crown and basal stem tissues are decayed and have a brown discoloration and a spongy texture. Internodes become chocolate brown, but leaf sheaths remain symptomless. Pink or burgundy mycelium is seen in hollow stems. Diseased plants die prematurely, resulting in white heads.

Symptoms caused by *F. equiseti* are stunted patches of wheat, a red-brown discoloration of roots and crowns, and, usually, severely rotted crowns. Inoculated plants in the greenhouse have a lower plant mass and grain yield than do control plants.

Management. In greenhouse trials, differences in tolerance to *F. graminearum* group 1 among wheat cultivars have been reported.

1. Do not rotate wheat with oat.
2. Till fields after harvest to improve water infiltration.
3. Establish a dust and stubble mulch in the spring.
4. Apply the proper amount of nitrogen.
5. Sow winter wheat in early September in North America as a compromise between early sowing, which produces larger plants and greater stress, and later sowing, which results in lower yields.

Halo Spot

Cause. *Pseudoseptoria donacis* (syn. *Selenophoma donacis* and *Septoria donacis*) overwinters as pycnidiospores, pycnidia, and mycelia in infested residue, seed, and overwintering wheat. During cool, moist weather, pycnidiospores are exuded from pycnidia and disseminated by wind and splashing rain to plant hosts.

Distribution. In the cool, moist climates of Great Britain, northern Europe, and the United States. Halo spot rarely causes much damage.

Symptoms. Numerous elliptical or diamond-shaped spots (less than 4 mm long) occur on leaves and sometimes the culms of winter wheat in the spring. At first, spots have purple-brown margins, but this discoloration fades as the spots age. In time, the centers of spots become gray from the presence of small black pycnidia. Sometimes spots become so numerous that much of the diseased leaf surface is destroyed.

Management. Not necessary. The cultivar 'Gaines' is very susceptible.

Karnal Bunt

Karnal bunt is also called partial bunt.

Cause. *Neovossia indica* (syn. *Tilletia indica*) survives as teliospores in soil for more than 2 years and on seed. Most primary inoculum is derived from seedborne inoculum, but soilborne teliospores also serve as a source of inoculum. In the spring, teliospores germinate within 2 mm of the soil surface and produce promycelia that grow to the soil surface. Sporidia are produced on promycelia and disseminated by wind to host plants, where the florets become infected. Sporidia are disseminated a long distance by air currents and may survive up to 12 hours if the relative humidity is 95% or higher. The maximum germination of teliospores in vitro occurs at temperatures of 15° to 20°C and in continuous light. Infection is favored by cool, wet weather.

During the early stages of infection, intercellular hyphal growth occurs among parenchyma and chlorenchyma cells in the distal to mid-portion, but not the basal portion, of the glume, lemma, and palea. Hyphae are absent from the ovary, subovarian tissue, rachilla, and rachis. Later, hyphae grow intercellularly toward the floret base to the subovarian tissue and enter the pericarp of the ovary through the funiculus. Hyphae are found in the rachis only during the later stages of infection. The epidermis of the ovary is not penetrated, even after prolonged contact with germinating secondary sporidia. Kernels are wholly or partially converted into teliospores that are disseminated by wind at harvest time and contaminate healthy seeds or soil.

Distribution. Afghanistan, India, Iraq, Lebanon, Mexico, Nepal, Pakistan, Sweden, Syria, Turkey, and the United States.

Symptoms. Normally only a few random diseased kernels per head are wholly or incompletely converted to smut sori. Symptoms first appear at the soft-dough stage of plant development in the form of blackened areas that surround the bases of the grains and extend upward on the grain for various lengths. During the most severe disease, glumes may be spread apart and expose the bunted grains, but this is not a common symptom. Most kernels are broken or partially eroded at their embryo end. On severely diseased plants, grains are reduced to fragile black membranous sacks of teliospores and have dead, shriveled embryos. A foul odor of trimethylamine, characteristic of the bunt fungi of wheat, accompanies diseased kernels. Infected seeds have a lower survival rate in storage than healthy seeds do.

Management
1. Grow resistant cultivars. Most durum wheats are immune.
2. Treat seed with a fungicide seed treatment; however, no chemical seed treatments can guarantee wheat seed is not carrying viable *T. indica* teliospores.
3. Hot water (60°–80°C) with the addition of NaOCl reduces seedborne teliospores and the contamination in grain storage facilities.

Leaf Rust

Leaf rust is also called brown rust, dwarf rust, and orange rust.

Cause. *Puccinia triticina* (syn. *P. recondita* f. sp. *tritici* and *P. recondita*). *Puccinia triticina* is a heteroecious, long-cycled rust fungus that has species of meadow rue, *Thalictrum* spp., and, rarely, *Anchusa* spp. and *Isopyrum* spp. as alternate hosts to wheat. Because meadow rue is found primarily in Europe and is not common in most sections of the United States, the pycnial and aecial stages of *P. triticina* are usually found only in Europe. *Anchusa italica* was reported to be an important alternate host in Morocco.

Puccinia triticina survives the winter in the Great Plains of the United States as sporulating or dormant mycelia in live winter wheat plants and as urediniospores in uredinia on infested dead leaves. Mycelium is apparently capable of surviving all environmental conditions that infected live host tissue can survive. Circumstantial evidence also implies local oversummering and overwintering of certain phenotypes in Pennsylvania. Rust overwinters as urediniospores and mycelia on wheat in the southern United States and Mexico.

Primary inoculum originates from sporulating uredinia, dormant mycelia, or viable urediniospores. These are the most important sources of inoculum for infection because they are likely to be virulent to the cultivar. Urediniospores disseminated by wind to the northern United States and Canada in the spring infect host plants if moisture is present. However, this source of inoculum may not be as important in initiating new infections because these urediniospores may not be virulent to the cultivar.

A new generation of urediniospores is produced every 7 to 14 days if favorable disease conditions of heavy dews, light rains, or high humidity and temperatures of 15° to 22°C occur. The minimum continuous dew period necessary for infection increases from 4 to 6 hours at the optimal temperatures to at least 16 hours for suboptimal temperatures. The optimum temperature for completion of the infection process, as measured by development of appressoria, substomatal vesicles, and infection hyphae with at least one haustorium, is 16°C. Infection occurs over a wide temperature range (5° to 25°C).

Urediniospores are wind-disseminated and cause secondary infections until host plants mature. As plants near maturity, teliospores are formed. If plants are infected near maturity, teliospores may not be produced. Teliospores in North America serve no function, but they may serve as a means of overwintering in Europe. During late summer and autumn in North America, urediniospores in northern wheat-growing areas that are disseminated southward by wind infect winter wheat in the southern United States and Mexico.

Distribution. Wherever wheat is grown. Leaf rust, which may be the most widely distributed wheat disease, is most prevalent where wheat matures late. However, infection often occurs too late to do severe damage.

Symptoms. Pustules first develop on diseased lower leaves and leaf sheaths, and during moist conditions, progress up to the flag leaf. Pustules also occur on stems, and the awns and glumes of the head. Uredinia pustules are small, round to oval, raised, orange-yellow, and contain numerous urediniospores that appear as an orange powder.

As wheat matures, telia develop as dark gray to black flattened pustules that are about the same size as uredinia but do not rupture the epidermis. Leaf rust can be distinguished from stem rust on the basis of these pustules. Leaf rust pustules are smaller than those of stem rust, are orange-yellow rather than the brick red color typical of stem rust pustules, and lack the jagged fragments of wheat epidermis that typically adhere to the sides of stem rust pustules.

Management

1. Grow resistant wheat cultivars. Adult plant resistance is conferred by two complementary genes: Lr 13 and Lr 34. The magnitude of gene interaction is influenced by environment, particularly temperature. Lr 10 and an unknown gene for seedling resistance are present in some cultivars.
2. Soils should have balanced fertility. An excess of nitrogen increases the susceptibility of host plants to infection.
3. Apply a foliar fungicide if conditions warrant it.
4. Experimentally, fungicide seed treatments have proved effective.
5. Resistance is found in some species of *Aegilops* and *Agropyron*.

Leaf Spot (*Ascochyta* spp.)

Cause. *Ascochyta sorghi* and *A. graminicola* have been found in association with *Stagonospora nodorum* (syn. *Septoria nodorum*).

Distribution. The United States (Virginia).

Symptoms. Leaf lesions that initially are elliptical to lanceolate later coalesce into irregular patterns. On diseased upper leaf surfaces, lesions are zonate, with a narrow outer chlorotic band that encloses a light brown necrotic band and dark brown center. The dark brown center is absent on the diseased lower leaf surfaces.

Pycnidia of both fungi are present. Those of *A. sorghi* are golden-brown and submerged, with only the ostiole or neck above the leaf surface. Those of *S. nodorum* are dark brown and superficial.

Management. Not necessary.

Leaf Spot (*Phaeoseptoria urvilleana*)

Cause. *Phaeoseptoria urvilleana* overwinters as pycnidiospores in pycnidia on infested residue. Free water must be present on leaves for 48 hours or longer for infection to occur.

Distribution. Great Britain and the United States.

Symptoms. In the field, irregular-shaped leaf spots form on leaves of diseased plants in the late milk stage of plant development. Plants inoculated with *P. nodorum* have irregular-shaped, diffuse leaf spots that initially are yellow, then become tan.

Management. Some cultivars are resistant or partially resistant.

Leptosphaeria Leaf Spot

Cause. *Phaeosphaeria herpotrichoides* (syn. *Leptosphaeria herpotrichoides*) over-winters as mycelia in, and as ascospores in pseudothecia on, infested residue. Free water must be present on leaves 48 hours or longer for infection to occur.

Distribution. Canada, Europe, and the United States. Leptosphaeria leaf spot is considered a minor disease.

Symptoms. Irregular-shaped, diffuse, yellow to tan spots occur on diseased leaves.

Management. Some cultivars are more resistant than others.

Loose Smut

Cause. *Ustilago tritici* (syn. *U. nuda*) survives as dormant mycelia in seed embryos. When an infected seed germinates, mycelium grows systemically in the plant and produces smutted heads filled with chlamydospores (teliospores) in place of healthy kernels. Smutted heads emerge from the boot 1 to 2 days earlier than healthy heads and the chlamydospores are disseminated by wind to the flowers of host plants. During weather with temperatures of 16° to 22°C and moisture provided by dews or light showers—conditions that are favorable for disease—chlamydospores germinate and form germ tubes that penetrate flower ovaries and possibly the stigmas. Accessibility of germ tubes to embryos appears to be a limiting factor in infection. Subsequent mycelial growth occurs in the germs or embryos of developing seeds. As seeds mature, mycelia in embryos become dormant until the following growing season. Different races of *U. tritici* exist.

Distribution. Wherever wheat is grown.

Symptoms. Diseased seed does not show outward symptoms and germination is not affected. Plants grown from diseased seed have dark green, erect leaves with chlorotic streaks; smutted heads then emerge from boots 1 to 2 days earlier than do heads of healthy plants. The brown to dark brown spore mass is enclosed in a fragile, gray membrane that soon ruptures, releasing spores that are disseminated by wind to healthy flowers. An erect, naked rachis that protrudes above reclining heads of healthy plants is all that remains of a smutted head.

Management
1. Treat seed with a systemic seed-protectant fungicide.
2. Grow resistant cultivars.
3. Sow certified seed known to be smut-free.
4. Hot water treatments have been used to destroy seedborne inoculum.

Microscopica Leaf Spot

Microscopica leaf spot is similar to Leptosphaeria leaf spot.

Cause. *Phaeosphaeria microscopica* (syn. *Leptosphaeria microscopica*) survives as pseudothecia on grass and wheat stubble. Disease occurs under continuous wet periods of 48 or more hours.

Distribution. The United States (Minnesota and North Dakota).

Symptoms. The symptom on diseased leaves is yellow-brown blotches.

Management. Not necessary, since *P. microscopica* is considered a weak parasite.

Patchy Stunting

Cause. *Rhizoctonia solani.* The *R. solani* that causes patchy stunting anastomoses with the *R. solani* that causes crater disease. The umbrella-thorn tree, *Acacia tortilis,* is a natural host for the fungus.

Distribution. Tanzania.

Symptoms. Patchy stunting is characterized by patches of stunted and chlorotic plants typical of Rhizoctonia bare patch disease. However, seminal roots of diseased plants contain nodulose swellings and sclerotial sheaths rather than displaying the girdling and rotting symptoms that are typical of bare patch. The only difference between crater disease and patchy stunting is that crater disease occurs exclusively in black montmorillonite clay soils. Patchy stunting has also been observed in clay loam, silty clay loam, and silty loam soils.

Management. Not reported.

Phialophora Take-All

Cause. *Phialophora graminicola* (syn. *P. radicicola* var. *graminicola*). Phialophora take-all occurs at soil temperatures of 24° to 29°C. Only plants in the seedling stage are susceptible.

Distribution. Unknown.

Symptoms. Seminal and coronal roots of diseased plants are rotted, and culms are discolored black. Both root and shoot weights are reduced.

Management. Not reported.

Phoma Spot

Phoma spot is also called Phoma glume blotch.

Cause. *Phoma glomerata, P. sorghina* (syn. *P. insidiosa*), and other *Phoma* spp. The *Phoma* spp. likely overseason as pycnidia in infested residue. Disease development is aided by long periods of wetness.

Distribution. India, Mexico, and South America.

Symptoms. Symptoms caused by *P. sorghina* occur primarily on glumes and awns, but leaves are infected occasionally. Spots on glumes are brown and oval (2.0–2.5 × 5.0–7.0 mm). Pycnidia are present within spots.

Under humid conditions, purple-brown lesions develop on the tips of glumes of inoculated plants, then enlarge and form a gray center. No yellow halo is present. Grain is discolored and shriveled. Inoculated leaves develop faded green streaks on the upper surfaces. Pycnidia develop in rows below the epidermis.

Phoma glomerata causes dark brown lesions on leaf sheaths.

Management. Not reported.

Pink Snow Mold

Pink snow mold is also called Fusarium patch.

Cause. *Microdochium nivale* (syn. *Fusarium nivale* and *Gerlachia nivalis*), teleomorph *Monographella nivalis,* oversummers as perithecia that developed on diseased lower leaf sheaths during cool, humid weather in the spring. During cool, wet weather in the autumn, with or without snow cover, leaf sheaths and blades that are close to and below the soil surface are infected by ascospores and mycelia growing from residue. Saprophytic mycelial growth from residue may also provide an inoculum source for other infections in the spring. Secondary infection occurs from conidia and ascospores that are disseminated by wind to host plants.

Distribution. Canada, central and northern Europe, and the United States.

Symptoms. Damage corresponds to the pattern of snow cover within a field. Pink mycelium and sporodochia are visible on living and dead plant tissue and on the soil surface after the snow melts. When disease severity is light, lesions with straw-colored centers and dark margins occur on diseased leaves. When disease is severe, leaf sheaths, shoots, and leaves are killed. Severely diseased leaves first become chlorotic, then necrotic, but remain intact and do not disintegrate. White or faintly pink mycelium is sometimes present on the lower parts of diseased plants. Diseased leaves become faintly pink with mycelium when the snow cover melts and the leaves are exposed to light and air. Unless crowns are diseased, diseased plants often recover during warm, dry weather despite extensive leaf infection.

Management
1. Sow winter wheat later in the autumn or in the spring. Most plant growth will be during conditions that are unfavorable to fungus infection.
2. Rotate wheat with a noncereal crop, such as a legume.
3. Do not mow plants late in the autumn. Removal of clippings reduces deleterious effects of mowing.

Platyspora Leaf Spot

Cause. *Clathrospora pentamera* (syn. *Platyspora pentamera*) survives as perithecia in infested residue. Ascospores are produced during wet, spring weather and disseminated by wind to host plants. There must be 24 to 72 hours of continuous moisture for infection to occur. Perithecia are produced in residue as wheat matures.

Distribution. Canada and the north central United States. Incidence of the disease is rare on wheat.

Symptoms. While Platyspora leaf spot is rare on wheat, ascospores of *C. pentamera* are abundant in the air in some wheat-growing areas. Nondescript yellow-brown spots are randomly produced on wheat residue as plants reach maturity.

Management. Grow resistant cultivars. Management measures listed under "Septoria Leaf Blotch" would also be beneficial.

Powdery Mildew

Cause. *Blumeria graminis* f. sp. *tritici* (syn. *Erysiphe graminis* f. sp. *tritici*) overwinters as cleistothecia on infested residue, and as mycelium and conidia in diseased leaf tissue of live plants growing in geographical areas that have mild winter temperatures. Ascospores, which form in cleistothecia in the spring and are disseminated by wind to hosts, serve as primary inoculum in northern wheat-growing areas. Conidia are produced when mycelium becomes established on diseased leaf surfaces during humid, cool weather but without the presence of free water. Conidia account for most of the secondary inoculum and spread of disease during the growing season. Cleistothecia are formed on diseased leaf surfaces as plants mature.

Powdery mildew is more severe on tender, rank plants that are growing at high population densities or have had heavy applications of nitrogen fertilizer. Cool, humid, and cloudy weather conditions are conducive to disease severity. Powdery mildew usually ceases to be a problem when weather becomes dry and warm later in the growing season.

Winter wheat is infected in the autumn; therefore, inoculum becomes available for the initiation of spring epiphytotics. The incidence of autumn infection decreases with later sowing.

Distribution. Wherever wheat is grown.

Symptoms. Superficial mycelium, conidiophores, and conidia appear as powdery, light gray or white spots on the upper surfaces and, infrequently, the lower surfaces of diseased leaf blades, leaf sheaths, and floral bracts. Chlorotic patches appear on leaf surfaces that are opposite the gray fungal colonies. The fungus is superficial on leaf surfaces except for haustoria, which penetrate into epidermal cells. Later, spots enlarge, yellow, then die and turn brown as diseased plants mature. In some cultivars, leaf tissue adjacent to mycelium becomes necrotic and turns brown. Numerous small, round, dark cleistothecia that are easily observed without magnification develop on diseased areas. Severely diseased plants are killed.

Winter wheat infected in the autumn generally does not produce symptoms except in those autumns that have extended warm temperatures. The lowest yields occur when plants become diseased early in autumn.

Management
1. Grow resistant cultivars. Peroxidases function in plant resistance against *B. graminis* f. sp. *tritici*.
2. Apply a foliar fungicide where economically feasible. Early applications of fungicides at plant growth stages 6 to 8 reduce powdery mildew and result in higher yield than fungicides applied later in the season.
3. Rotate wheat with noncereal crops.
4. Plow under volunteer wheat.
5. Under experimental conditions, significant reductions in disease can be achieved in some seasons by applying potassium chloride in the spring.
6. Sow winter wheat later than recommended.
7. Apply the seed-treatment fungicide triadimenol on seed of wheat sown in the autumn.

Prematurity Blight

Prematurity blight in the Canadian prairies is a severe manifestation of common root rot and is similar to Fusarium crown rot in other wheat-growing areas.

Cause. *Cochliobolus sativus* and *Fusarium culmorum*. Disease is favored by moisture stress for 6 to 12 days at the flag leaf or boot stage of plant development.

Distribution. Canada (Prairie Provinces).

Symptoms. Bleached individual plants are scattered among healthy green plants. Usually the entire plant is bleached, but occasionally only one or two tillers are affected and the rest of the plant remains green. Often diseased plants are slightly stunted, but since heads remain erect, the stunting is not obvious.

Brown lesions typical of common root rot usually are present on diseased subcrown internodes and crowns. The brown discoloration may extend up the stem for several centimeters. Occasionally the basal discoloration is pink rather than brown. Heads may contain severely shriveled grain or may lack seeds entirely. Infrequently diseased plants may be found with all symptoms except the discoloration of the basal parts.

Management. Cultivars differ in resistance.

Pythium Root Rot

Pythium root rot is also called browning.

Cause. Several *Pythium* species, including *P. aphanidermatum, P. aristosporum, P. arrhenomanes, P. graminicola, P. heterothallicum, P. irregulare, P. myriotylum, P. sylvaticum, P. torulosum, P. ultimum, P. ultimum* var. *sporangiiferum,* and *P. volutum,* have been associated with root rot of wheat. *Pythium* spp. fungi, which are present in all agricultural soils, survive for 5 years or more in soil and infested residue as oospores. Oospores germinate and form germ tubes that infect the host directly or form a sporangium in which 10 to 40 zoospores are produced. The zoospores "swim" through soil water to a host plant, where they rest and eventually produce germ tubes that infect roots.

Pythium spp. infect wheat embryos within 48 hours after sowing in soil at −0.3 bar or wetter. Optimum fungal growth and infection occur during low temperatures in wet soils in the autumn and spring. At soil temperatures of 10° to 25°C, *P. ultimum* is the most pathogenic of the tested *Pythium* spp., whereas *P. irregulare* is the most pathogenic of all *Pythium* spp. at 5°C. Oospores are produced in diseased tissue and are returned directly to soil upon the decomposition of diseased plant tissue or remain within the diseased, but intact, tissue.

Infection is most likely to occur in the wetter areas of fields that are deficient in phosphorous and organic matter. Wheat seedlings produced from old seed are more susceptible to infection than seedlings produced from new seed.

Distribution. Wherever wheat is grown. Pythium root rot occurs in patches rather than being generally distributed throughout a field.

Symptoms. Young diseased seedlings are stunted and have chlorotic leaves that become tan. Roots lack root hairs, and fine rootlets and tips of the newest roots develop soft, wet, light brown areas. Seedlings grow out of this condition unless low soil temperatures continue, then diseased seeds and seedlings may damp off. Diseased plants have fewer tillers, which results in fewer heads per plant and those heads are slow to mature. Yields are reported to be only 75% to 80% of an average yield in years of normal or above normal precipitation.

Management
1. Treat seed with a seed-protectant fungicide.
2. Maintain adequate phosphate levels in soils. This promotes root growth and allows plants to "grow away" from the disease.
3. Some cultivars are able to tolerate disease because they have a better root system than other cultivars.

Red Smudge or Pink Smudge

The names red smudge and pink smudge are used interchangeably in the literature. Pink smudge is similar to pink seed caused only by *Erwinia rhapontici*, where only damaged kernels are infected, particularly those injured by the gall midge or when grain is harvested too early. The two diseases are considered separately in this book because pink or red smudge has been reported to be caused by more than one microorganism.

Cause. *Erwinia rhapontici, Fusarium* spp., and *Pyrenophora tritici-repentis*. Discoloration by *P. tritici-repentis* only is commonly known as red smudge. Disease occurs during seed development and is favored by low temperatures, high rainfall, and frosts that delay maturity and ripening of the crop. Grain discoloration is favored by high moisture during kernel development and may be more of a problem under irrigation.

Distribution. Widespread.

Symptoms. Symptoms are a pinkish or reddish discoloration of part or all of the kernel.

Management. Not reported.

Rhizoctonia Bare Patch

Rhizoctonia bare patch is considered to be a form of Rhizoctonia root rot. Rhizoctonia bare patch is also called bare patch, no-growth patches, purple patch, Rhizoctonia patch, Rhizoctonia root rot, root rot, and stunting disease. Crater disease is similar but is caused by a *R. solani* that does not anastomose with AG-8. Symptoms differ sufficiently to consider Rhizoctonia bare patch (and root rot) and crater disease separate diseases.

Cause. *Rhizoctonia solani* AG-8, teleomorph *Thanatephorus cucumeris,* is generally considered the causal pathogen. However, some researchers have suggested that a complex of fungi, rather than any one pathogen alone, is the cause of Rhizoctonia bare patch of cereals in western Australia and elsewhere. In Texas, most isolates were found to be AG-4 or AG-5, due to the Texas disease possibly being different from Rhizoctonia bare patch disease in Australia. Other fungi isolated from Rhizoctonia bare patches include a binucleate *Rhizoctonia* sp. AG-C, *Fusarium graminearum, Bipolaris sorokini-*

ana, R. oryzae (teleomorph *Waitea circinata*), *Mortierella* sp., and *Pythium irregulare*. Other Rhizoctonias with the teleomorphs *Ceratobasidium cornigerum* and *Ceratobasidium* spp. have been found to be mildly pathogenic to wheat.

The pathogen is possibly spread by the movement of infested residue or by mycelia growing from plant to plant along the drill row. Disease severity is greater under conservation-tillage conditions than under conventional cultivation. This may be because cultivation fragments pieces of infested residue, which in turn reduces the propagule size and affects the ability of hyphae to ramify through the soil. Fragments of hyphae also may be more prone to attack by other soil microorganisms. Cultivation may also prevent a buildup of organic matter that provides a substrate for the fungus to grow on. A toxin or abiotic factor may also be involved with the stunting of plants within patches that have apparently healthy root systems.

In southern Australia, the disease occurs in highly calcareous, sandy soils; however, it has since occurred on other soil types because of the widespread use of reduced-tillage systems. In New South Wales, the disease is found on calcium-deficient and acid soils, and in western Australia, disease is most severe on plants growing in sandy to loamy sand soils. Disease, in general, is more severe on host plants that are growing in dry soils.

In Texas, disease was observed on wheat sown in August or early September when soil temperatures were high. However, in Australia, the pathogenicity of *R. solani* AG-8 in artificially infested soil was greater at 12°C than at 27° and 32°C. In Oregon, root rot was more severe at low temperatures of 19° and 6°C, day and night, respectively, than at high temperatures of 27° and 16°C, day and night, respectively.

The herbicide chlorsulfuron at 2.5 g/ha significantly increased disease.

Distribution. Australia, Canada, England, Scotland, South Africa, and the United States (Idaho, Oregon, Texas, and Washington). The disease may not be the same in all geographical areas. For example, the disease in Texas may be different from that in Australia.

Symptoms. The descriptions of symptoms in the literature vary between geographical areas. The most characteristic field symptom is the occurrence within an apparently healthy crop of stunted plants in distinct patches that vary in size from less than 30 cm in diameter to large irregular-shaped areas up to 0.4 ha. Patches are sometimes circular but are usually elongated in the direction of sowing. Patches often occur in clusters rather than randomly. Within clusters, the patches often coalesce and form larger, irregular-shaped patches. A related symptom is the killing of all wheat plants when the plants are still small so that by harvest time, the "hole" in the crop is filled with weeds.

Patches normally have a distinct margin that forms between the stunted plants in patches and normal-sized plants in the surrounding crop. Patches are most clearly observed at the tillering stage but can be detected 2 to 3 weeks after sowing.

Plants within the patch are stunted and display a range of symptoms associated with poor nutrition and moisture stress. Plants immediately outside the patch usually do not display the symptoms of nutrient deficiency or moisture stress. Plants within the patch display the following symptoms: yellowing of the lower leaves, a stiff upright leaf habit, rolled leaf blades, spindly growth, failure to produce tillers, and under cold conditions, a dark green coloration and purplish tinge of leaf blades when they are viewed from a distance. In Texas, diseased plants lose the lower leaves and the lower leaf sheaths become dark brown and necrotic.

In some locations, stunted plants will have little or no growth during the growing season, but at other locations, plants display some growth but remain stunted or spindly compared to the rest of the crop. Plants within patches often remain green longer than the surrounding crop, possibly due to a delay in maturity or to greater availability of soil moisture in the patch. If plants within patches are not severely stunted, they may recover but have fewer tillers than healthy plants and produce a poor yield.

The seminal roots of young seedlings have discrete dark brown, water-soaked lesions that are confined to one side of the diseased root or girdle the entire root and cause the cortex to slough off, leaving the stele exposed. The stele eventually rots through and leaves a pointed stub or spear tip. The root tips are yellowish brown but often become dark brown in older plants. Severe disease on some plants causes the entire root system to be truncated. Disease severity may vary among the different seminal roots, and an occasional plant may have apparently healthy seminal roots but severely damaged coronal roots. Some plants within a patch have an apparently intact root system but remain stunted like the surrounding diseased plants. Plants outside a patch may have considerable root rot but no stunting, possibly due to late infection by *R. solani* AG-8 or other *Rhizoctonia* spp.

Management

1. Application of N as ammonium sulfate, urea, or sodium nitrate reduces disease. Calcium applied to calcium-deficient soils reduces disease.
2. Practice the appropriate tillage system. Wheat sown in reduced-tillage situations becomes more severely diseased than wheat sown in conventional tillage conditions. Mixing soil deeply (5 cm or more) reduces root rot, possibly because loosened soil may allow faster root growth. Two cultivations reduce disease more than a single cultivation. Cultivation close to sowing also reduces disease.
3. Practice proper rotation. Wheat following a medic-dominant pasture has more severe root rot than wheat following a grass pasture. In Texas, do not rotate sugar beet with wheat because *R. solani* AG-4 is also highly virulent on sugar beet.

Sclerotinia Snow Mold

Sclerotinia snow mold is also called snow scald.

Cause. *Myriosclerotinia borealis* (syn. *Sclerotinia borealis*) oversummers as sclerotia in infested residue in soil near the soil surface and on the soil surface. In autumn, the sclerotia germinate during damp, cool weather and form cup-shaped apothecia. Ascospores are produced in a layer of asci on the upper portion of the apothecium and disseminated by wind to seedlings. Snow scald occurs when cool, damp autumns are followed by a deep snow cover over unfrozen or slightly frozen soil that lasts for 5 or more months. *Myriosclerotinia borealis* has been identified on cocksfoot, *Dactylis glomerata*.

Distribution. Canada, Europe, Japan, Scandinavia, and the former USSR.

Symptoms. Symptoms of snow scald occur in scattered patches throughout a field. Sparse, gray mycelia cover bleached, dead plants that are exposed to the air as the snow melts. Diseased leaves wrinkle on exposure to light and eventually turn dark, because of the saprophytic growth of other fungi, and crumble away. Sclerotia are globular, elongated, or flake-like (0.3–7.0 mm long) and are black at maturity. Sclerotia are only found in and on leaves, leaf sheaths, and the crowns of dead plants.

Management. While there is no entirely satisfactory management of snow scald, the following measures will be of some benefit:

1. Rotate winter wheat with spring wheat, another spring cereal, or a legume to help reduce inoculum.
2. Plowing under residue will bury sclerotia and prevent them from germinating, thereby hastening their decomposition.

Septoria Leaf Blotch

Septoria leaf blotch is also called leaf blotch, Septoria disease, Septoria leaf spot, *Septoria tritici* blotch, and speckled leaf blotch.

Cause. *Septoria tritici,* teleomorph *Mycosphaerella graminicola.* The fungi overwinter as mycelia in diseased live wheat plants, and survive as pycnidia on infested residue for 2 to 3 years. Residue is the primary source of inoculum. *Septoria tritici* is seedborne; however, seedborne inoculum is uncommon.

Disease development is dependent on temperature rather than on the growth stage of plants at infection. During temperatures of 15° to 25°C and 100% (the optimum) relative humidity in the autumn or spring, pycnidiospores are exuded from pycnidia in a gelatinous drop, or cirrhi, that protects spores from radiation and drying out. Pycnidiospores are disseminated to the lower leaves of host plants by splashing and blowing rain. In Indiana, primary infection takes place in the autumn before the onset of cold weather; consequently, *S. tritici* probably overwinters as asymptomatic infections. In the absence of functional pycnidiospores in California, ascospores were ascribed to be the primary inoculum.

Infection requires 6 or more hours of wetness; subsequent disease development is favored by temperatures of 18° to 25°C. However, response to temperature varies with the amount of cultivar resistance. New pycnidiospores are produced in 10 to 20 days; more pycnidia are produced on susceptible cultivars than on resistant cultivars.

Ascospores produced in pseudothecia as wheat matures in late summer or early autumn and disseminated by wind to host plants contribute to the inoculum that initiates primary disease of wheat sown in the autumn. *Mycosphaerella graminicola* is readily found on residue in the United Kingdom.

Distribution. Generally distributed wherever wheat is produced. Septoria leaf blotch can become a serious disease during environmental conditions that are conducive for disease development.

Symptoms. Septoria leaf blotch has two conspicuous phases: during the winter on basal leaves of wheat sown in the autumn, and during the summer on upper leaves of host plants. Initially, small, light green to yellow spots occur between the leaf veins of lower leaves, especially if the leaves are in contact with soil. Spots rapidly elongate and form tan to red-brown, irregular-shaped lesions that are often partly surrounded by a yellow margin. Lesions age and become light brown to almost white and have small, dark specks (pycnidia) in the center. The presence of pycnidia is a good diagnostic characteristic.

Infection of stem nodes, leaf sheaths, and the tips of glumes also occurs. Severely diseased leaves become yellow and die prematurely. Occasionally an entire plant may be killed. Pycnidia are produced in all diseased areas of a plant. Autumn and winter infections cause a reduction in the root weight and the mass of diseased plants.

Management
1. Sow cleaned, certified seed that is free of disease and has been treated with a seed-protectant fungicide.
2. Plow under infested residue.
3. Grow resistant wheat cultivars. Durum wheat is often reported as relatively resistant; however, durum wheats tested in Tunisia were more susceptible to the local isolates of *S. tritici* than the bread wheat entries were.
4. Apply a foliar fungicide if environmental conditions favor disease development.
5. Rotate wheat with a resistant crop once every 3 to 4 years.
6. Sow winter wheat after the Hessian fly–free date.

Sharp Eyespot

Cause. *Rhizoctonia cerealis* CAG-1, teleomorph *Ceratobasidium cereale*. *Rhizoctonia solani* AG-4 and a binucleate species also have been reported as the causal fungi; the strains that cause sharp eyespot differ from those that cause Rhizoctonia root rot. The strains of AG-4 did not infect roots but killed seedlings in the greenhouse.

The fungi survive as sclerotia in soil and as mycelia in the infested residue of a large number of plant genera. Sclerotia germinate and form mycelia or mycelia grow from precolonized substrates to infect roots and culms any time during the growing season, particularly in dry soils with less than 20% moisture-holding capacity, and during cool soil temperatures.

Distribution. Wherever wheat is grown.

Symptoms. Diseased plants are stiff, have a grayish cast, and delayed maturity. Seedlings may be killed, but older diseased plants may produce new roots to compensate for those that rotted off. When roots are diseased, plants lodge and produce white heads.

Diamond-shaped lesions that resemble those of eyespot occur on diseased lower leaf sheaths. However, sharp eyespot lesions are more superficial than eyespot lesions. Lesions have light tan centers with dark brown margins. Frequently dark mycelium is visible on the lesions. Dark sclerotia eventually develop in lesions and between the culms and leaf sheaths of diseased plants.

Management. Vigorous plants growing in well-fertilized soil do not become as severely diseased as unthrifty plants.

Snow Rot

Cause. *Pythium aristosporum, P. iwayami, P. okanoganense,* and *P. paddicum*. *Pythium ultimum* has also been reported as a causal fungus. Survival over the summer is presumed to be by oospores in soil and infested residue. Oospores germinate and form sporangia in which zoospores are produced. The zoospores are released into the soil and disseminated by snowmelt water to host plant roots. Wheat plants must be predisposed by an extended period of time under snow cover before they become susceptible to infection. Snow rot development depends on extended periods of snow cover and near-freezing temperatures. In Washington, disease is confined to areas where water collects during snow and ice melt. In North Dakota, the absence of ground frost contributes to the disease syndrome.

Distribution. Japan and the United States. In Japan, *P. paddicum* was the dominant pathogen isolated from poorly drained paddy fields and *P. iwayami* was the dominant pathogen isolated from upland fields. *Pythium iwaya-*

mai, P. okanoganense, and *P. paddicum* were all isolated from well-drained paddy fields.

Symptoms. As plants emerge from under a snow cover, older leaves have large, dark green, water-soaked areas. Younger leaves are distorted, water-soaked, dark green, and flaccid. The soft tissue becomes brown or tan and the basal leaf sheaths are dark brown and filled with oospores. Leaves touching the soil are more decayed than those in an upright position. Diseased plants are killed when fungi invade the growing point.

Root tissues are invaded when runoff water washes soil from around crowns. Stele and cortical root tissues rot and the diseased cortical tissues slough from roots. Diseased plants develop few tillers and root systems are reduced. Inoculated plants are severely stunted.

Management. Not reported.

Southern Blight

Southern blight is also called Sclerotium base rot and white rot.

Cause. *Sclerotium rolfsii,* teleomorph *Athelia rolfsii.*

Distribution. Australia and Brazil.

Symptoms. White mycelia and sclerotia occur on the necrotic areas of diseased roots, crowns, and lower portions of stems. Diseased plants turn white and die.

Management. Not reported.

Speckled Snow Mold

Speckled snow mold is also called gray snow mold and Typhula blight.

Cause. *Typhula idahoensis, T. incarnata* (syn. *T. itoana*), *T. ishikariensis,* and *T. ishikariensis* var. *T. ishikariensis* overseason as sclerotia in infested residue and soil or as parasitic growth on live plants. Saprophytic growth on dead tissue provides an inoculum source for increased disease in the spring; however, the fungi are poor saprophytes and depend primarily on parasitism for existence. Sclerotia germinate during wet weather in the autumn and form either basidiocarps or mycelia. Optimum infection occurs at temperatures of 1° to 5°C when basidiospores form on basidiocarps and are disseminated by wind, or mycelia directly infect host plants by growing over the surface of the soil. Further infection occurs by mycelial growth under snow cover. Sclerotia form within necrotic tissue or in mycelium.

Bacteria described as gram-negative white colonies of fluorescent pseudomonads have been reported to be the most antagonistic to *Typhula incarnata* and *T. ishikariensis* at temperatures of 5° to 15°C. This may be one reason why these fungi prevail almost exclusively under snow cover, where low temperatures inactivate most antagonists.

The growth of *Typhula* spp. is more closely related to snow cover than is the growth of *Microdochium nivale,* the cause of pink snow mold. *Typhula ishikariensis* has been identified on cocksfoot, *Dactylis glomerata.*

Distribution. Canada, central and northern Europe, Japan, and the northwestern United States.

Symptoms. At snowmelt, gray-white mycelia are present on the leaves and crowns of diseased plants, and on soil. The numerous sclerotia in diseased plant tissues and scattered in mycelia growing over plant surfaces give plants a "speckled" appearance. Dead leaves are common, but unless the crowns are diseased, plants recover during warm, dry weather but are not as vigorous as healthy plants. Dead leaves crumple easily and are covered with gray to white mycelium.

Management
1. Rotate wheat with a legume to reduce the inoculum in soil since the fungi are not good saprophytes.
2. Disease incidence was decreased in experiments by fungicide seed treatments.

Spot Blotch

Spot blotch is also called Bipolaris leaf spot and Helminthosporium leaf spot.

Cause. *Bipolaris sorokiniana* (syn. *Helminthosporium sativum* and *H. sorokinianum*), teleomorph *Cochliobolus sativus,* survives primarily as conidia and mycelia in infested residue. Conidia are produced from mycelium in residue and disseminated by wind to host plants. Infection occurs primarily during wet weather as host plants approach maturity.

Distribution. Wherever wheat is grown.

Symptoms. Symptoms are evident after senescence. They are most frequent on the lower diseased leaves as distinct, elongated, brown-black lesions that are less than 1 cm in length.

Following mixed inoculation with *Pyrenophora tritici-repentis,* wheat leaves developed less necrosis than the average necrosis produced by the two pathogens alone.

Management
1. Ensure proper soil fertility.
2. Rotate wheat with resistant crops.
3. Grow resistant cultivars. Cultivars with some resistance to spot blotch may not be resistant to other diseases, such as black point and common root rot, caused by *B. sorokiniana.*

Stagonospora Blotch

Stagonospora blotch is also called glume blotch, Septoria blotch, *Septoria nodorum* blotch, *Septoria nodorum* blotch of wheat, and *Septoria nodorum* leaf and glume blotch of wheat. In some reports, *Stagonospora* is spelled *"Stagnospora."*

Cause. *Phaeosphaeria avenaria* f. sp. *triticea,* anamorph *Stagonospora avenae* f. sp. *triticea* (syn. *Septoria avenae* f. sp. *triticea*), and *Stagonospora nodorum* (syn. *Septoria nodorum*), teleomorph *Phaeosphaeria nodorum* (syn. *Leptosphaeria nodorum*). The causal fungi are seedborne, surviving in stored seed for more than 2 years. The fungi also overwinter as mycelia in live diseased plants, and survive for 2 to 3 years as pycnidia on infested residue. Residue is the most important source of primary inoculum.

In the autumn or spring, pycnidiospores are exuded from pycnidia in a cirrhi, or gelatinous drop, during wet weather and temperatures of 20° to 27°C. The cirrhi protects the spores from radiation and drying out until they are disseminated by splashing and blowing rain to the lower leaves of host plants. Infection requires 6 or more hours of continuous wetness. Infection of the coleoptile from seedborne inoculum occurs over a wide range of soil moistures, temperatures, and planting depths. In Indiana, primary infection takes place in the autumn before the onset of cold weather. Consequently, under these conditions *S. nodorum* overwinters as asymptomatic infections in infected plants.

New pycnidiospores, which are produced in 10 to 20 days and disseminated, serve as secondary inoculum. Ascospores are produced in perithecia as wheat matures in the late summer and early autumn. In Ireland, perithecia have been recovered from green leaves of symptomless winter wheat plants collected in January and February.

The severity of disease on seedlings increases until 4 weeks after sowing. Thereafter, disease is usually not severe until plants near maturity. The greatest yield losses occur when rainfall is excessive between flowering and grain harvest. Increasing nitrogen fertilizer has caused an increase in disease severity.

Several grasses are alternative hosts. Some grass isolates are associated with large lesions on wheat, but most isolates produce small lesions. Considering the potential number of isolates available from grasses, their effect on wheat is probably less than would be expected.

Distribution. All major wheat-growing areas.

Symptoms. Glumes, culms, leaf sheaths, and leaves become diseased, but usually little damage occurs until diseased plants near maturity. Two to three weeks after the heads emerge, small, grayish or brown spots occur near the top third of glumes, then enlarge and become a chocolate-brown discoloration. Later, the centers of spots become gray and have small black pycnidia scattered throughout them.

Diseased nodes become brown, shrivel, and are speckled with black pycnidia. These infections often cause straw to bend over and lodge just above nodes.

Diseased leaves have light brown spots that are similar to those of Septoria leaf blotch. Brown margins surround leaf spots and pycnidia are present on both leaf surfaces. If flag leaves are diseased, the heads are deformed. Diseased leaf sheaths have a dark brown lesion that includes most of the leaf sheath. Severely diseased plants are stunted.

A symptom caused by an atypical form of *S. nodorum* was a brown discoloration at the base of coleoptiles on seedlings grown from infected seed. Inoculated leaves developed brown lesions that were surrounded by yellowish halos.

Symptoms may be confused with those of black chaff or basal glume rot. Stagonospora blotch spots do not form streaks and are not as sharply defined or as dark brown as those of black chaff. Stagonospora blotch does not have the water-soaked appearance of basal glume rot.

Yield losses up to 50% have been reported. Yield reductions are associated primarily with reduced kernel weight. Yields are most consistently correlated with disease severity of the leaf below the flag leaf at the Feekes growth stage 11.1.

Management

1. Sow certified, cleaned seed that is free of disease and has been treated with a seed-protectant fungicide. Seed treatment is effective when the primary source of inoculum is from seed, but it is ineffective when the soil is heavily infested with inoculum.
2. Grow resistant cultivars.
3. Plow under infested residue.
4. Apply a foliar fungicide.
5. Rotate wheat with a resistant crop once every 3 to 4 years. However, disease development is not reduced when infected seed is used.
6. Sow winter wheat after the Hessian fly–free date.
7. Grow taller cultivars instead of shorter ones. A less favorable microclimate for disease development occurs with taller cultivars.

Stem Rust

Stem rust is also called black rust and black stem rust.

Cause. *Puccinia graminis* (syn. *P. graminis* f. sp. *tritici*) is a heteroecious, long-cycle rust that has the barberries *Berberis canadensis, B. fendleri, B. vulgaris,* and *Mahonia* spp. as alternate hosts. *Puccinia graminis* f. sp. *tritici* causes stem rust of wheat by one of two life cycles: (1) During the spring and summer, urediniospores (also called repeating spores in the literature) produced in uredinia on wheat are disseminated by wind to wheat, where they infect and produce new (secondary) urediniospores under moist conditions and moderate temperatures (15° to 25°C). Secondary uredin-iospores are produced in cycles every 7 to 10 days throughout the growing season until wheat plants are mature. Urediniospores produced in northern wheat-producing areas are blown south in the summer and autumn and infect wheat in southern wheat-growing areas. The urediniospores then recycle on wheat in the southern wheat-growing areas. Eventually, spores disseminated northward during the late winter and spring infect wheat in the northern wheat-growing areas.

(2) As wheat ripens, teliospores are formed that survive through the winter. In the spring, teliospores germinate and form basidiospores (sporidia) that are disseminated by wind and infect the young leaves of barberry. Pycnia (spermogonia) form on the upper leaf surfaces of barberry and function in the exchange of genetic material, thereby creating new races of the fungus. Each pycnium produces pycniospores and special mycelia called receptive hyphae. Pycniospores are exuded out of the pycnium in a thick, sticky, sweet liquid (honeydew) that is attractive to insects. Pycniospores are splashed by water or carried by insects from one pycnium to another, where they become attached to receptive hyphae. The pycniospore germinates and the nucleus from the spore enters the receptive hypha and forms an aecium on the lower leaf surface, directly under the pycnium. Aeciospores that differ genetically from both pycniospores and receptive hyphae are produced in the aecia, then are windborne to wheat hosts, where infection occurs. Each infection gives rise to a uredinium that produces urediniospores that serve as the repeating stage and are repeatedly disseminated by wind to wheat, where new generations of urediniospores are formed. Telia are once again produced as diseased wheat plants mature, thus completing the life cycle. Epiphytotics develop during moist weather but disease is not severe during dry weather.

Distribution. Wherever wheat is grown.

Symptoms. Uredinia and telia occur on diseased stems, leaf sheaths, leaf blades, glumes, and beards of wheat. Uredinia are red-brown oblong pustules. The plant epidermis ruptures and is pushed back, exposing the urediniospores and giving the pustule a jagged or ragged appearance.

Just prior to plant maturation, telia appear primarily on leaf sheaths and culms. Telia are oblong to linear and dark brown to black. Teliospores are exposed by a rupturing of the host epidermis.

Pycnia appear on barberry leaves in the spring as bright orange to yellow spots with what sometimes appears to be a drop of liquid in the middle that contains the pycniospores and receptive hyphae. On the side of the barberry leaf opposite the pycnial spots are the aecia, which resemble raised, orange, bell-shaped clusters.

Management

1. Grow resistant wheat cultivars. The hypersensitive reaction to stem rust

is closely associated with resistance controlled by the Sr5 gene.

2. Eliminate the common barberry from wheat-producing areas. The common barberry should not be confused with the Japanese barberry, which is immune to stem rust. There are several characteristics to differentiate between the two. The common barberry has a saw-toothed leaf edge, gray outer bark, bright yellow inner bark, berries borne in bunches, and spines with usually three in a group. The Japanese barberry has a smooth leaf edge, red-brown outer bark, bright yellow inner bark, berries borne in ones or twos, and usually a single spine.

3. Late sowing reduces the effectiveness of the slow rusting character that bestows tolerance to rust.

Stripe Rust

Stripe rust is also called glume rust and yellow rust.

Cause. *Puccinia striiformis* f. sp. *tritici* (syn. *P. striiformis*) is a rust fungus that is not known to have an alternate host. The fungus oversummers as urediniospores on residual green cereals and grasses during the period between harvest and the emergence of wheat sown in the autumn. Mycelia and infrequently urediniospores remain alive over the winter in or on different hosts, such as barley, grasses, and rye. Teliospores are formed but are not known to function as overwintering spores.

Urediniospores are formed during cool, wet weather and disseminated by wind to host plants, where infection occurs. Infection is restricted to a relatively narrow range of temperatures (5° to 12°C). Little infection occurs above 15°C. The minimum continuous dew period necessary for infection increases from 4 to 6 hours at the optimal temperatures to at least 16 hours for suboptimal temperatures. Warmer than normal winters and cooler April temperatures favor development of stripe rust epiphytotics.

Distribution. Australia. In North America, stripe rust occurs at the higher elevations and cooler climates along the Pacific Coast and intermountain areas from Canada to Mexico. It also occurs under the same circumstances in South America and in the mountainous areas of central Europe and Asia.

Symptoms. The most severe symptoms occur during cool, wet weather in the early growth stages of wheat. Symptoms occur early in the spring before other rusts are evident, especially in wheat-growing areas that have mild winters. Yellow uredinia appear early in the spring on autumn foliage and new spring foliage. Uredinia coalesce and form long stripes between the veins of diseased leaves and sheaths. Small, linear lesions occur on floral bracts. Telia develop as narrow, linear, dark brown pustules that are covered by the leaf epidermis.

Management
1. Grow resistant wheat cultivars.

2. Ensure soils are well fertilized. Potassium levels must be adequate.
3. Cool spring temperatures delay temperature-sensitive adult plant resistance in some cultivars.
4. Apply foliar fungicide. In Australia it is recommended to spray when 1% of leaf area is affected and predicted yield loss is sufficient to make spraying economical.
5. Experimentally, fungicide seed treatments have proven effective.

Take-All

Take-all is also called Ophiobolus patch.

Cause. *Gaeumannomyces graminis* var. *tritici, G. graminis,* and *G. graminis* var. *avenae* (syn. *Ophiobolus graminis* var. *avenae*) survive as mycelia in diseased plants or as mycelia and perithecia in infested residue. *Gaeumannomyces graminis* var. *tritici* is not a good saprophyte and does not survive long in soil in the absence of a live host. Because the competitive saprophytic ability of *G. graminis* var. *tritici* is low, roots and crowns colonized by the fungus during parasitism act as primary inoculum. The viability of inoculum declines as soil microorganisms subsequently colonize and decompose the parasitized residue. Thus, the pathogen population is characterized by increasing inoculum potential during the parasitic phase followed by decreasing inoculum potential during the saprophytic phase.

Seedlings become infected in the autumn or spring when roots grow in the vicinity of residue previously parasitized by *G. graminis* var. *tritici* and come in contact with mycelia. Plant-to-plant spread of take-all occurs by hyphae growing through soil from a diseased plant to a healthy plant or by a diseased root coming into contact with a healthy root. Ascospores are produced in perithecia during wet weather but are not considered important in disease spread because they are not disseminated a great distance either by splashing water or by wind.

Disease incidence and severity depend on a number of interrelated factors. Disease occurs during low to moderate temperatures and is more severe in reduced-tillage systems than in conventional tillage systems. Tillage practices that fragment the inoculum and expose it to increased microbial degradation also can lower the inoculum potential of the pathogen and limit disease development. Take-all is more severe with direct drilling in the Pacific Northwest due to the large size of inoculum pieces associated with direct-drill conditions.

Inoculum in the form of "propagules" builds up more under pasture than wheat because of a larger mass of fine, dense roots in the top 10 cm of soil under grasses than under wheat. A decline in propagule numbers and infectivity in the absence of a susceptible host has been associated with the wetting of warm soils, which stimulates the activity of competitive microorganisms. High soil moisture favors microbial activity, accelerating the breakdown of pathogen-infested residue and, hence, the destruction of the food base and habitat of *G. graminis*. var. *tritici*. Conversely, some researchers have reported a high incidence and severity of disease was associated with regular rainfall patterns.

However, disease may be more dependent on soil factors than on the number of propagules and infectivity. Disease becomes more severe at higher soil pHs. Increased soil acidity may directly reduce the activity of the fungus or may stimulate activity of microorganisms antagonistic to *G. graminis* var. *tritici*. Host plant predisposition may also result from inadequate supplies of certain essential plant nutrients, such as manganese, at higher pHs. Soil phosphorous and mycorrhizae affect disease development by increasing the phosphorous status of the host. Subsequently, the leaking of exudates from roots is reduced, which in turn reduces pathogen activity. There may also be a relationship between the form of nitrogen fertilizer and disease incidence.

Under humid, somewhat acidic soil conditions, fertilizers containing ammonium nitrogen rather than nitrate nitrogen suppress take-all or increase yield of wheat infested with *G. graminis* var. *tritici*. Fertilizers containing chloride are effective in suppressing take-all and increasing yield of soft winter wheat grown on acidic soils in the Pacific Northwest. Chloride may suppress take-all by decreasing the water potential in the plant or by inhibiting nitrification in unlimed acidic soils. Under alkaline soil conditions in the northern Great Plains, the application of chloride has little effect on disease severity indices, and the nitrogen fertilizer source did not affect root disease scores but did affect the percentage of white heads.

Distribution. Wherever wheat is grown. Take-all is generally considered to be the most important root disease of wheat worldwide.

Symptoms. Take-all symptoms are most severe on wheat grown in alkaline soils. The first symptoms are light brown to dark brown necrotic lesions on roots. By the time a diseased plant reaches the jointing stage, most of its roots are brown and dead. At this point, many plants die and plants that live remain chlorotic and stunted.

Symptoms become most obvious as plants approach the heading stage of plant growth. The stand is uneven in height and plants appear to be in several stages of maturity. Diseased plants have few tillers, ripen prematurely, and have heads that are bleached and sterile. Roots are sparse, blackened, and brittle, and when pulled from the soil, plants easily break free of crowns. A very dark discoloration of the stem is visible just above the soil line. A mat of dark brown mycelium normally can be seen under the lower leaf sheath between the stem and the inner leaf sheaths.

Management
1. Rotate wheat with a nonhost crop. In western Australia, a rotation with lupines, oat, or field peas reduced inoculum and suppressed growth of the pathogen in soil.

2. Treat seed with a systemic seed-treatment fungicide.
3. Sow later in the growing season. This extends the period the fungus must survive as a saprophyte and reduces the inoculum level at planting time. The lower temperatures associated with late sowing also favor a more vigorously growing plant.
4. Apply fertilizers containing ammonium nitrogen, phosphorous, and chloride if soils are acid.
5. Certain macronutrients, such as phosphorous, and micronutrients, such as copper, iron, manganese, and zinc, have the potential for limiting take-all by lessening susceptibility of host tissues, promoting formation of new roots, or both.
6. Severity of take-all declines and wheat yields increase following one or two outbreaks of the disease and continuous monoculture of wheat.
7. Fluorescent pseudomonads in the rhizosphere of wheat produce phenazine antibiotics that are reported to suppress take-all.
8. Management practices that increase soil shading, such as volunteer wheat, double cropping, and no tillage, tend to prevent high soil temperatures and may promote inoculum survival and disease.

Tan Spot

Tan spot is also called leaf spot, blight, and yellow leaf spot.

Cause. *Pyrenophora tritici-repentis,* anamorph *Drechslera tritici-repentis.* Some researchers consider *P. trichostoma* to be synonymous with *P. tritici-repentis* but Farr et al. (1989) do not.

Pyrenophora tritici-repentis produced on straw in the autumn and winter overwinters as mycelia and pseudothecial initials in infested residue that is primarily on the soil surface. *Pyrenophora tritici-repentis* is also seedborne, but airborne inoculum is the primary source of infection. Ascospores are the most important primary inoculum, but conidia and hyphae also serve as primary inoculum. *Pyrenophora tritici-repentis* is also isolated from several grass species. Isolates from these grass hosts differ in their ability to cause disease symptoms on wheat, and some are as aggressive as wheat isolates. Infection and disease development are favored by frequent rains and cool, cloudy, humid weather early in the growing season.

Conidia produced in older lesions on leaves killed by the pathogen and disseminated by wind to plant hosts serve as secondary inoculum. Infection requires a 6- to 48-hour wet period, and the most infections occur in the vicinity of infested host residues. Increasing the post-inoculation wet period of foliage increases lesion size. Pseudothecial initials are produced on infected culms and leaf sheaths in the autumn.

Infection of wheat seed in the field occurs after the early dough stage and is positively correlated with tan spot severity on the flag leaf after anthesis. Disease development is related to the initial amount of primary inoculum, despite multiple infection cycles caused by windborne secondary inoculum. Seed infection predominantly occurs via the lemma and palea, rather than the glume. The presence of anthers adjacent to the seed apparently enhances seed infection.

Epiphytotics are commonly associated with conservation-tillage practices because straw remaining on the soil surface supports development of abundant primary inoculum. *Pyrenophora tritici-repentis* survives poorly in straw that is buried or placed directly on the soil surface. In contrast, it survives well in straw that is slightly above the soil surface, such as standing stubble or straw within a mulch layer. Disease is more severe in tall stubble than in short stubble. Residue in close contact with soil is moist more frequently and remains moist longer than residue above the soil surface. This high moisture environment could result in more intense microbial activity and competitive interactions detrimental to the survival and inoculum potential of *P. tritici-repentis*.

Isolates of *P. tritici-repentis* differ in virulence. Four pathotypes have been identified based on their ability to induce one of four symptoms: tan necrosis followed by extensive chlorosis, tan necrosis only, extensive chlorosis only, and neither necrosis nor chlorosis on appropriate differential cultivars.

In culture, *Pyrenophora tritici-repentis* releases a host-selective toxin designated the Ptr toxin, a protein of low molecular weight. Ptr toxin is produced by tan necrosis isolates only and is found associated with the induction of tan necrosis in the host.

The reaction of wheat cultivars to tan spot differs from one part of the world to another. Resistance occurs first as papilla formation, then as the restriction of lesion size and mycelial growth in the lesion. Wheat genotypes differ in their ability to restrict growth of *P. tritici-repentis* as the period of post-inoculation foliar wetness lengthens and/or the temperature rises.

The herbicide glyphosate inhibits ascocarp formation on straw.

Distribution. Wherever wheat is grown.

Symptoms. Symptoms first appear in the spring on lower diseased leaves and leaf sheaths and progress to the upper leaves in the early summer. Regardless of plant growth stage at infection, the oldest leaf is the most severely spotted and the youngest leaf is the least severely spotted. Diseased plants exhibit one of the following four symptoms:

1. Tan necrosis and extensive chlorosis begins as small brown or tan flecks or spots on both sides of seedling leaves. The flecks eventually become diamond-shaped, necrotic lesions up to 12 mm long, with a yellow border (halo) and a dark brown spot in the center that is due to sporulation by *P. tritici-repentis*. Necrosis and/or chlorosis often extend to cover the entire leaf. Extensive chlorosis may mask the tan necrosis.

2. Necrotic lesions are well defined and consist of tan collapsed tissue that may coalesce and cause large areas of diseased leaves to die from the tip inward.
3. Chlorotic lesions exhibit a gradual yellow discoloration without collapsed tissue on large areas of diseased leaves. Lesions coalesce and cause large areas of diseased leaves to die from the tip inward. Pseudothecia will eventually develop on straw as dark raised bumps.
4. Resistant symptoms are characterized by the absence or presence of slight amounts of necrosis and/or chlorosis.

A red smudge symptom on seed results in a downgrading of quality.

It has been reported that half of the total yield loss is due to disease that occurs before the boot stage. Following mixed inoculation with *Bipolaris sorokiniana*, wheat leaves developed less necrosis than the average produced by each of the two pathogens alone.

Management
1. Grow resistant cultivars. Differences in resistance exist among wheat genotypes. Resistance has been attributed to polygenic or monogenic control. Resistance decreases with leaf aging.
2. If weather conditions are wet, apply a foliar fungicide before the disease becomes severe. Fungicide sprays for tan spot management may have to be applied earlier than for the rusts.
3. Plow under infested residue.
4. Rotate with nonhost crops. Rotation to sorghum was reported to be as effective as plowing under infested residue.
5. Nitrogen fertilizations NH_4SO_4 and $CaNO_3$ reduce tan spot severity by delaying leaf senescence but do not have a direct effect on tan spot.

Tar Spot

Cause. *Phyllachora graminis*, anamorph *Linochora graminis*, overseasons as stromata on residue. Disease is favored by moist, shaded areas.

Distribution. Widespread. Tar spot occurs infrequently on wheat.

Symptoms. Glossy black and somewhat sunken spots occur on diseased leaf blades and sheaths. Spots (0.1–0.2 × 5.0 mm) are the stromata of *P. graminis*.

Management. Not necessary.

Typhula-like Snow Mold

Cause. A basidiomyceteous fungus similar to *Typhula* spp. Disease occurs under a prolonged snow cover. The fungus is considered a weak pathogen and has been identified as OKLA-1 in the literature.

Distribution. The United States (Oklahoma).

Symptoms. Dead plants occur under a prolonged snow cover, especially in fields that slope northward. Small black subepidermal sclerotia, which are flatter than sclerotia of both *T. ishikariensis* and *T. idahoensis,* are present in the basal leaf sheaths of living and dead plants.

Management. Not reported.

Wheat Blast

Cause. *Pyricularia grisea.* Confusion exists in the literature as to the nomenclature of the causal fungus. *Pyricularia oryzae* has been ascribed to be the causal fungus by some researchers. However, Farr et al. (1989) state that *P. oryzae* is the name for the causal fungus of rice blast, but *P. grisea* is an older name for *P. oryzae* and is the proper name for blast or gray spot of grasses. The teleomorph stage is unknown. There may be two distinct entities causing wheat blast in Brazil, and one may be the rice blast strain.

Long periods of leaf wetness and high temperatures are needed for disease to occur. Barley, oat, and rye are also susceptible, which suggests these diseased plants may be potential sources of inoculum. There are five known races of the wheat blast fungi.

Distribution. Brazil.

Symptoms. The greatest damage occurs when symptoms appear during or just after the anthesis stage of plant growth. Diseased spikes are totally or partially whitened and have blue-violet lesions on the rachis. Yields are reduced.

Management
1. There are differences in resistance between cultivars.
2. Apply foliar fungicides. However, the application of foliar fungicides has resulted in incomplete control.
3. Apply fungicide seed treatments.

Zoosporic Root Rot

Cause. *Lagena radicicola, Ligniera pilorum, Olpidium brassicae,* and *Rhizophydium graminis.* The fungi thrive in wet soils.

Lagena radicicola overseasons as resting spores in diseased root hairs. Resting spores give rise to zoospores that "swim" to root epidermal cells and gain entrance by means of a penetration peg. Once a zoospore is inside a cell, a thallus forms and attaches to the cell wall. Thalli develop into either sporangia or thick-walled resting cells. The other fungi have similar life cycles.

Distribution. Widespread.

Symptoms. The fungi are not serious pathogens but are more important as possible vectors of soilborne plant viruses.

Management. Not necessary.

Diseases Caused by Nematodes

Bulb and Stem

Cause. *Ditylenchus dipsaci.* Occurrence of the nematode is associated with heavy soils, high rainfall, cool growing seasons, and winter grains. Free moisture permits the nematode to migrate to and feed on aerial plant parts. The nematode penetrates the leaves and stems.

Distribution. Worldwide. Damage is confined to small areas within fields.

Symptoms. Cell hypertrophy and hyperplasia occur. Stunting, distortions, and swollen stems are common symptoms. In severely infested areas, growing points are destroyed and diseased plants may die. Other plants have reduced spike growth and grain yields.

Management
  1. Rotate wheat with noncereal crops.
  2. Grow winter cereals to reduce damage.

Cereal Cyst

Heterodera avenae is called the cereal cyst nematode and oat cyst nematode.

Cause. *Heterodera avenae* (syn. *H. major*). Other Heterodera species on wheat are *H. bifenestra, H. hordecalis,* and *H. latipons. Heterodera avenae* survives for a year or more as cysts in soil. In the spring, larvae emerge from eggs within overwintered cysts, enter plant roots, and begin feeding. As eggs develop in their bodies, female nematodes swell and break through the root surface but remain attached to the root by a thin neck. Males revert to a veriform size. Eventually, females form cysts that detach from roots. One nematode generation from egg to adult is completed in 9 to 14 weeks. *Heterodera avenae* is also a pathogen of barley, oat, rye, and numerous annual and perennial grasses.

Populations of nematodes increase in sandy soils. Other conditions, such as the death of nematodes when roots die, and nemaphagous fungi that parasitize nematodes also cause a fluctuation in nematode populations.

Distribution. Africa, Australia, southeastern Canada, Europe, Japan, the United States (Oregon), and the former USSR.

Symptoms. The first symptom is poor growth of diseased plants in one or more areas within a field. Leaf tips of heavily infested plants are red or purple. The discolored leaves die and diseased plants become yellow. The roots are thickened and more branched than those of healthy plants. Heavy nematode infestations cause wilting, particularly during times of water stress; stunted growth; poor root development; and early plant death. Lemon-shaped cysts are white at first, then gradually become dark brown as they harden. Cysts are visible without magnification.

Management
1. Rotate wheat with a legume crop.
2. Apply a nematicide to infested soil before sowing wheat.
3. Grow a resistant or tolerant cultivar. Tolerance has been identified in wheat cultivars grown in experimental plots. Significant differences between cultivars have been identified for total grain weight, number of grains, number of fertile spikelets, and number of heads per plot.
4. Sow wheat in the autumn. Because plants from seed sown in the autumn develop larger root systems, they tolerate nematode infection better than wheat sown in the spring.

Columbia Root Knot

Cause. *Meloidogyne chitwoodi.*

Distribution. Widespread.

Symptoms. General growth of diseased plants is reduced. There are less tillers per plant of winter wheat, and the dry shoot weight may be reduced in some cultivars.

Management. Not reported.

Dagger

Cause. *Xiphinema americanum* is an external parasite that prefers feeding on young succulent roots.

Distribution. Widespread. Damage on wheat is likely not very severe.

Symptoms. Moderate swelling occurs on young diseased roots. Severely diseased plants have clusters of short, stubby root branches, and small roots shrivel at their points of attachment. The most obvious symptom is one or several necrotic flecks, which vary from light brown to dark brown or black, on diseased roots.

Management. Not necessary.

Grass Cyst

Cause. *Punctodera punctata* (syn. *Heterodera punctata*). The life cycle is similar to that of *H. avenae*. (*See* "Oat Cyst Nematode" in Chapter 11.)

Distribution. Canada and the United States.

Symptoms. The first symptom is poor growth of diseased plants in areas within a field. Leaf tips of severely diseased plants are red or purple. The discolored leaves die and plants generally become yellow. Diseased roots are thickened and more branched than those of healthy plants. Heavy infestations cause wilting, particularly during times of water stress; stunted growth; poor root development; and early plant death. Lemon-shaped cysts are white at first, then gradually become dark brown as they harden. Cysts are visible without magnification.

Management. Not reported, but presumably disease management is similar to that for the cereal cyst nematode.

Lesion

Cause. *Pratylenchus* species, including *P. minyus, P. neglectus,* and *P. thornei.* Nematodes live free in soil as migratory endoparasites and overwinter as eggs, larvae, or adults in diseased host tissue and in soil. Both larvae and adults penetrate roots and move through cortical cells, where the females deposit eggs as they migrate. Older diseased roots are abandoned and new roots are sought as sites for penetration and feeding.

Distribution. Wherever wheat is grown.

Symptoms. Diseased plants in areas of a field will appear yellow and moisture-stressed. Roots and crowns will be rotted when *Rhizoctonia solani* infects host plants through nematode wounds. Diseased roots become dark and stunted, and there is a loss in grain yield.

Management
1. Sow wheat in autumn when soil temperatures are below 13°C.
2. Use soil fumigants where the high costs warrant them.
3. Application of nitrogen and phosphorous reduced populations of *P. thornei* on roots; however, this occurs to a greater extent in wheat in a rotation than in monoculture.

Ring

Cause. *Criconemella* spp. are ectoparasites.

Distribution. Not reported.

Symptoms. Disease symptoms on wheat are not well known. The feeding of *Criconemella* spp. on the roots of other plant species does not result in necrosis.

Management. Not reported.

Root Gall

The root gall nematode is also called the grass root gall nematode.

Cause. *Subanquina radicicola* (syn. *Ditylenchus radicicola* and *Anguillulina radicicola*) is an endoparasite that survives in host roots. Larvae penetrate roots and develop in the cortical tissue and form a root gall within 2 weeks. Mature females begin egg production in galls that eventually weaken and release larvae to establish secondary infections. Each generation is completed in about 60 days.

Distribution. Canada and northern Europe.

Symptoms. Diseased seedlings frequently have reduced top growth and a general chlorosis. Galls on roots tend to be inconspicuous and vary in diameter from 0.5 to 6.0 mm. At the center of the larger galls is a cavity filled with nematode larvae. Roots may be bent at the gall site.

Management. Rotate wheat with noncereal crops.

Root Knot

Cause. *Meloidogyne* spp. are endoparasitic nematodes that overwinter as eggs in the soil. The three species most commonly found on wheat are *M. arenaria, M. incognita,* and *M. naasi. Meloidogyne arenaria* and *M. incognita* do not survive if the soil temperature averages below 3°C during the coldest month of the year. In the spring, larvae hatch from eggs but enter host roots at any time. By the middle of summer, females inside the root-knot tissue release eggs into the soil.

Distribution. Wherever wheat is grown.

Symptoms. Root knots are found on diseased roots in spring and summer. They are swellings, or thickenings, comprised of swollen cortical cells and the bodies of nematodes containing egg masses. When root knots are cut open, egg masses turn dark.

Management
1. Sow wheat in the autumn.
2. Rotate wheat with root crops.

Seed Gall

Seed gall is also called wheat gall.

Cause. *Anguina tritici* survives for several years as larvae in seed galls. Long-range dissemination is by seed galls that are mixed in with grain. When galls are sown together with seed, larvae released into the moist soil move up host plants in a water film, eventually reaching the flower primordia.

Mature nematodes copulate and produce eggs. Seed galls develop from un-differentiated flower tissue that has been infected by the nematodes. If gall development is retarded, larvae may be present in healthy-appearing seed. Galls are mixed with normal seed or fall to the ground, where the nematodes become dormant under dry conditions.

Distribution. Eastern Asia, parts of Europe, India, and southeastern United States.

Symptoms. Prior to the heading stage of plant growth, wheat plants are swollen near the soil level and diseased leaves are twisted, wrinkled, or rolled. After heading, the distortions are not as obvious, but diseased plants are stunted and mature slowly. Heads are small and the dark, seed-like gall forces the glumes to spread apart. Galls are dark brown and do not have the brush or embryo markings of healthy seed.

Management
1. Sow clean seed.
2. Seed may be soaked for 10 minutes at 54°C.
3. Grow nonhost crops for 2 years in a rotation.

Spiral

Cause. *Helicotylenchus* spp. are ectoparasites.

Distribution. Not reported.

Symptoms. Symptoms on wheat are not well known. Small necrotic flecks on roots and a general decline in plant vigor have been suggested.

Management. Not reported.

Stubby Root

Cause. *Paratrichodorus* species, principally *Paratrichodorus minor* (syn. *P. christiei* and *Trichodorus christiei*) survives in soil or on diseased roots as eggs, larvae, and adults. These nematodes are migratory ectoparasites that move relatively rapidly through soil at 5 cm per hour, especially in fine sandy soils, and feed only on the outside of wheat roots. Feeding is confined to epidermal and external cortical cells of the host roots. Wheat sown early in the autumn in sandy soils is most severely diseased.

Distribution. Widely distributed in most agricultural soils.

Symptoms. Diseased roots are thickened, short, and stubby and have brown lesions on the root tips. The tops of plants grow poorly and the entire plant is easily pulled from soil due to a lack of a fibrous root system.

Management. No practical management is available.

Stunt

Cause. *Merlineus brevidens* survives in soil and in association with host tissue as different morphological stages of the nematode. *Merlineus brevidens* is an ectoparasite that feeds on the outside of wheat roots, often in association with the fungus *Olpidium brassicae.*

Distribution. Presumed to be indigenous to most soils, but the nematode is rarely found at high populations.

Symptoms. The most severe damage occurs on winter wheat that is growing in wet soils. The lower leaves die, and a few short tillers with small seed form on diseased plants. Roots of diseased seedlings are short, dark, and shriveled.

Management
1. Apply nematicides or soil fumigants to infested soils.
2. Grow cultivars that tolerate stunting.

Disease Caused by Phytoplasmas

Aster Yellows

Cause. The aster yellows phytoplasma-like organism (PLO) survives in several biennial or perennial dicotyledonous plants and in leafhoppers. The aster yellows phytoplasma is transmitted primarily by the aster leafhopper, *Macrosteles fascifrons,* and less commonly by the leafhoppers *Athysanus argentarius, Elymana sulphurella, Endria inimica,* and *M. laevis.* Leafhoppers first acquire the phytoplasma by feeding on diseased plants and then transmit the phytoplasma by flying to healthy plants and feeding on them.

Canadian wheat cultivars were not infected by a strain of aster yellows from Oklahoma, which suggests the possibility of different PLO strains or leafhopper biotypes in the two regions.

Distribution. Generally distributed throughout eastern Europe, Japan, and North America. However, aster yellows is rarely severe.

Symptoms. Symptoms become most obvious at temperatures of 25° to 30°C. Diseased seedlings either die 2 to 3 weeks after infection or are stunted. Diseased leaves are yellow or have yellow blotches, and the heads are sterile and have distorted awns.

Infection of older plants causes leaves to become somewhat stiff and discolored shades of yellow, red, or purple from the tip or margin inward. Root systems may not be well developed.

Management. Not reported.

Disease Caused by Viroids

Seedborne Wheat Yellows

Cause. Seedborne wheat yellows viroid (SWTV). SWTV is seedborne and is thought to be mechanically transmitted.

Distribution. China.

Symptoms. Chlorotic spots initially occur on the upper and middle portions of diseased seedling leaves and, subsequently, on new leaves as they emerge. Spots coalesce and form large chlorotic areas that, in turn, become necrotic.

Management. Sow pathogen-free seed.

Diseases Caused by Viruses

African Cereal Streak

Cause. African cereal streak virus (ACSV). The natural virus reservoir is native grasses. ACVS is transmitted by the planthopper *Toya catilina,* but it is not mechanically or seed-transmitted. ACSV is limited to the phloem of the plant host, where it induces a necrosis. Disease development is aided by high temperatures of 20°C and above.

Distribution. East Africa, specifically Kenya and possibly Ethiopia.

Symptoms. Initially, faint, broken, chlorotic streaks begin near the base of the diseased leaf and extend upward. Later, definite alternate yellow and green streaks develop along the entire leaf blade, and eventually diseased leaves become almost completely yellow. New diseased leaves tend to develop a shoestring habit and die.
Plants infected when young become chlorotic and severely stunted, and die. Plants infected when older have distorted yellow heads. Diseased plants become soft, flaccid, and almost velvety to the touch. Yield is almost completely suppressed in diseased plants.

Management. Not reported.

Agropyron Mosaic

Agropyron mosaic is also called Agropyron green mosaic, streak mosaic, and yellow mosaic.

Cause. Agryopyron mosaic virus (AgMV) is in the potyvirus group. AgMV is reservoired in quackgrass and other grass hosts, including *Elymus smithii, Bothriochloa laguroides, Aristida* sp., and *Sorghastrum nutans*; therefore, disease tends to occur near these virus sources. AgMV is transmitted by the mite *Abacarus hystix* and is sap-transmissible between gramineous hosts.

Distribution. Canada, northern Europe, and the northern United States and Oklahoma. Agropyron mosaic is generally not considered to be an important disease.

Symptoms. Symptoms occur as patches of diseased plants within a field or at the edges of fields along grassy borders. Initially, pale green or yellow mosaics, streaks, or dashes occur on leaf blades. Moderate stunting may occur. With the exception of stunting, symptoms become less conspicuous as diseased plants mature.

Management
1. Grow tolerant cultivars.
2. Eliminate volunteer wheat and grasses.
3. Sow winter wheat later in the autumn.

American Wheat Striate Mosaic

Cause. American wheat striate mosaic virus (AWSMV) is in the rhabdovirus group. AWSMV overwinters in wheat and grasses in association with the painted leafhopper, *Endria inimica,* which overwinters as eggs. AWSMV is transmitted mainly by *E. inimica* and occasionally by the leafhopper *Elymana virenscens*. Leafhopper adults are migratory, and all leafhopper stages are dispersed by wind. Symptom development is favored by warm temperatures (25° to 33°C). Disease is associated with heavy infestations of leafhoppers.

Distribution. South central Canada and north central United States.

Symptoms. Diseased leaves have obvious striations consisting of thin, yellow to white, parallel streaks that are more severe on the abaxial leaf surface. Streaks may spread over the entire leaf and become necrotic. Older leaves are stunted and chlorotic, then become necrotic. On some wheat cultivars, AWSMV causes a brown necrotic streaking of culms and glumes.

Management. Not reported.

Australian Wheat Streak Mosaic

Australian wheat streak mosaic is also called chloris striate mosaic.

Cause. Australian wheat streak mosaic virus (AWSMV) is in the geminivirus group. AWSMV is transmitted by the leafhoppers *Nesoclutha obscura* and *N. pallida,* but it is not sap-transmissible. AWSMV is introduced into the mesophyll by the vectors. Barley, maize, and oat are also susceptible.

Distribution. Australia. The disease is not considered important.

Symptoms. Diseased plants are dwarfed. Leaves have broken yellow streaks and fine, grayish striping.

Management. Not necessary.

Barley Stripe Mosaic

Cause. Barley stripe mosaic virus (BSMV) is in the hordeivirus group. BSMV is transmitted through seed, sap, and pollen. BSMV can remain viable in seed for up to 8 years, but seed transmission is relatively uncommon in wheat. When an infected seed germinates, the resulting seedling is diseased. BSMV is disseminated through plant sap when a diseased leaf rubs against a healthy leaf. BSMV can be transmitted by infected pollen but not by insects or other means.

Distribution. Southern Asia, Australia, Europe, Japan, western North America, and the former USSR.

Symptoms. Diseased leaves have yellow mottling or spots that are either narrow or wide, numerous or few, and continuous or broken. The spots may be light green, tan, yellowish, or bleached white, while the rest of the leaf is green. Normally, chlorotic stripes develop on leaves and become increasingly yellow or brown. Virulent strains of the virus cause brown stripes that are either continuous or broken and have irregular margins. Diseased plants are dwarfed, rosetted, and excessively tillered.

   Plants infected by BSMV through sap develop acute and localized symptoms on subsequent leaves. Within a few days, symptoms disperse throughout the plant as a mild systemic mosaic. Symptoms are best expressed in diseased plants grown at temperatures of 22° to 30°C.

Management. Do not rotate wheat with barley.

Barley Yellow Dwarf

Cause. Barley yellow dwarf virus (BYDV) is in the luteovirus group. BYDV survives in autumn-sown small grains, such as barley, oat, and wheat, and in annual and perennial grasses. However, in general, local grasses, winter wheat, and maize are of little importance as primary sources of BYDV.

   BYDV is transmitted by several species of aphids but is not transmitted through eggs, newborn aphids, seed, and soil, or by mechanical means. Once an aphid acquires BYDV, it is capable of transmitting the virus for the rest of its life. Some strains of BYDV are transmitted equally well by all species of aphids; others display a high degree of vector specificity. Aphid flights are local or are assisted by wind and extend for hundreds of kilometers, causing virus inoculum to be carried by aphids for a long distance. Thus, the major source of aphids in the spring may be from distant plants, but in the autumn, viruliferous aphids are both distant and local. In susceptible wheat cultivars, BYDV induces physiological changes that improve their acceptability to the English grain aphid, *Sitobion avenae,* and enhance the vector's effectiveness.

Epiphytotics occur during cool temperatures (10° to 18°C) and moist seasons that favor grass and cereal growth as well as aphid multiplication and migration. Infections occur through the growing season but are most numerous in areas that support populations of aphids.

Distribution. Africa, Asia, Australia, Europe, New Zealand, North America, and parts of South America.

Symptoms. Symptoms are not striking and tend to be confused with nutritional disorders or weather-related problems but become more pronounced at cool temperatures (16° to 20°C) and on cloudless days. Single plants or groups of plants are discolored yellow and stunted. Seedling infection slows plant maturity and causes older diseased leaves to become bright yellow. Leaves tend to be stiff and discolored in various shades of yellow, red, or purple, starting from the leaf blade edge. A later infection causes the flag leaf to turn yellow or red. Frequently, the feeding of some vectors produces tiny, brown-black spots on leaves and culms, with adjacent tissues first turning yellow, then tan. Diseased roots are not well developed, and phloem tissues are darkened. The cold hardiness of diseased winter wheat plants is reduced.

Management
1. Grow tolerant wheat cultivars.
2. Sow winter wheat later in the autumn.

Barley Yellow Streak Mosaic

Cause. Barley yellow streak mosaic virus (BaYSMV) is a novel virus transmitted by the brown wheat mite, *Petrobia latens. Petrobia latens* is a non-web-spinning spider mite that is generally believed to reproduce parthenogenetically. The mite deposits eggs on debris located on the soil surface near a host plant. BaYSMV particles are extremely large for a plant virus (64 × 1000 nm) and appear to be enveloped in and contain single-stranded RNA. BaYSMV has been transmitted mechanically to *Nicotiana benthamiana* and *Chenopodium quinoa.*

Distribution. Canada (Alberta) and the United States (Idaho and north central Montana).

Symptoms. Diseased leaf symptoms consist of a yellow-green mosaic, streaking, color banding, and severe necrosis.

Management. Not reported.

Barley Yellow Striate Mosaic

Cause. Barley yellow striate mosaic virus (BYSMV) is in the rhabdovirus group. BYSMV is reservoired in volunteer diseased wheat plants and is transmitted by the planthopper *Laodelphax striatellus.*

Distribution. Italy, but the incidence and severity are very low.

Symptoms. A mild yellow mottling occurs on diseased basal leaves and striations occur on upper leaves.

Management. Not reported.

Barley Yellow Stripe

Cause. The causal agent is not known for certain, but it is transmitted by the leafhopper *Euscelis plebejus.* Disease tends to occur along field borders and near grassy reservoirs of *E. plebjus.*

Distribution. Italy.

Symptoms. Fine, continuous stripes that occur on diseased leaves sometimes are followed by yellowing and death.

Management. Not reported.

Brome Mosaic

Cause. Brome mosaic virus (BMV) is in the bromovirus group. BMV survives in cereals, perennial grasses, and dry leaf tissues and is seedborne. BMV is primarily sap-transmitted, but the virus is inefficiently transmitted by the nematodes *Xiphenema coxi* and *X. paraelongatum* in Europe and by the Russian wheat aphid, *Diuraphis noxia,* in North America. Transmission also occurs through plant contact and any means that transfers sap.

BMV infects plant embryos and traces of the virus are occasionally found in seed coats. Plants infected early in the growing season have the highest incidence of seedborne infection.

Distribution. North America, Northern Europe, South Africa, the former USSR, and Yugoslavia.

Symptoms. Seedborne virus affects germination. Initially, a streak-like, yellow-green mosaic occurs on diseased leaves, but it becomes less prominent as diseased plants age. Mild stunting and head deformation may also occur. Some cultivars are symptomless carriers.

Management. Control grassy weeds.

Cereal Chlorotic Mottle

Cause. Cereal chlorotic mottle virus is a rhabdovirus transmitted by cicadellids.

Distribution. Australia and northern Africa.

Symptoms. Severe necrotic and chlorotic streaks occur on diseased leaves.

Management. Not reported.

Cereal Tillering

Cause. Cereal tillering virus (CTV) is in the reovirus group. CTV is transmitted by the planthoppers *Laodelphax striatellus* and *Dicranotropis hamata*. CTV is limited to the phloem and is similar to oat sterile dwarf virus, rice black-streaked dwarf virus, and maize rough dwarf virus.

Distribution. Italy and Sweden.

Symptoms. Diseased plants are dwarfed and excessively tillered, and have a general dark green discoloration and poor grain yields. Infrequently, leaves become malformed and have serrated edges.

Management. Not reported.

Cocksfoot Mottle

Cause. Cocksfoot mottle virus (CFMV) is in the sobemovirus group. CFMV is sap-transmitted and also is spread by the cereal leaf beetles *Oulema melanopa* and *O. lenchenis*.

Distribution. Great Britain.

Symptoms. The initial mottling of diseased leaves is followed by a general yellowing and, finally, death.

Management. Not reported.

Cucumber Mosaic

Cause. Cucumber mosaic virus (CMV) is in the cucumovirus group. CMV is reservoired in diseased vegetables and is transmitted to wheat by windborne aphids. CMV has been detected in seed.

Distribution. South Africa.

Symptoms. Symptoms are first noticed at the heading stage of plant growth. An uneven elongation of shoots occurs, with a subsequent appearance of sterile, yellow-white heads. Leaves become chlorotic, and severely diseased plants die.

Management. Not reported.

Eastern Wheat Striate

Cause. Eastern wheat striate virus (EWSV) is transmitted by the leafhopper *Cicadulina mbila,* but is not transmitted by sap, seed, soil, or aphids. EWSV overseasons in perennial naraenga grass.

Distribution. India.

Symptoms. Fine chlorotic stripes on diseased leaves and leaf sheaths become necrotic and more pronounced as diseased plants age. Diseased plants are variously stunted. Stunting and striate symptoms are more pronounced in barley than in wheat, and in plants infected when young. Plants infected while young usually die. Diseased plants produce partially filled heads with shriveled seed of poor quality.

Management. Not reported.

Enanismo

Enanismo is also called cereal dwarf.

Cause. One or more unidentified viruses and a toxin from the leafhopper *Cicadulina pastusae.* Leafhopper adults and nymphs of both sexes transmit the virus or viruses in a circulative mode. Females pass the virus(es) transovarially and are more efficient vectors than males. Barley and oat are also hosts.

Distribution. Colombia and Ecuador.

Symptoms. Diseased seedlings are killed or stunted. Later infections result in less stunting, but diseased leaves have blotches and ear-like enations or galls. Galls appear on the newest leaves formed on an infected plant 1 to 3 weeks after leafhopper feeding, rather than appearing on the leaves that were fed upon. Plants infected at heading time produce distorted heads and incompletely filled seeds.

Management
1. Sow wheat later in the spring after vector activity declines.
2. Grow tolerant or resistant cultivars.

European Wheat Striate Mosaic

European wheat striate mosaic is also called oat striate, red disease, wheat striate, and wheat striate mosaic.

Cause. European wheat striate mosaic virus (EWSMV) is in the tenuivirus group. EWSMV is passed through eggs and is transmitted persistently by the planthopper *Javesella pellucida* and sometimes by *J. dubia.* Nymphs are more efficient vectors than adults. Symptoms occur 2 to 4 weeks after feeding.

Distribution. Central Europe, Great Britain, and Scandinavia.

Symptoms. Diseased seedlings die. Older plants are severely stunted and do not head out. Plants infected near the heading stage of plant growth mature early, develop yellow-white leaf stripes, and have white heads with shrunken seed.

Management. Sow wheat later in the autumn or in the spring to avoid exposure to vectors.

Flame Chlorosis

Cause. The likely disease agent or its replicative intermediate is double-stranded RNAs that range in size from 900 to 2800 base pairs and are present in vesicles rather than virions. The causal agent is soil-transmitted and has been transmitted by growing host plants in soil where diseased plants had grown.

The grassy weeds green foxtail, *Setaria viridis,* and barnyard grass, *Echinochloa crusgalli* are also hosts.

Distribution. Canada (Manitoba) on spring wheat.

Symptoms. A striking flame-like pattern of leaf chlorosis and severe stunting occur in spring wheat. Chloroplasts and mitochondria of diseased cells are hypertrophied and contain an extensive proliferation of fibril-containing vesicles that form within the organellar envelope. Diseased plants continued to produce leaves with symptoms after they were transferred to sterile potting medium.

Management. Not reported.

High Plains Disease

Cause. High Plains virus (HPV) is transmitted by the wheat leaf curl mite, *Aceria tosichella,* which also transmits wheat streak mosaic virus (WSMV) and results in a mixed infection by these two viruses. HPV is associated with a 32-kDa protein that resembles tenuiviruses. However, some researchers have suggested that there is also some resemblance to tospoviruses.

HPV is also transmitted to barley and maize.

Distribution. The midwestern United States (Colorado, Idaho, Kansas, Nebraska, Texas, and Utah).

Symptoms. The first symptom is small, chlorotic spots on diseased leaves. The spots rapidly expand into a mosaic and then into a general yellowing of the entire diseased plant. Severely diseased plants become necrotic. Field infections very often are mixed infections of WSMV and HPV.

Management. Not reported.

Indian Peanut Clump

Cause. Indian peanut clump virus.

Distribution. India.

Symptoms. Diseased plants are severely stunted, have dark green leaves, and show mosaic symptoms on the youngest leaves.

Management. Not reported.

Maize Dwarf Mosaic

Cause. Maize dwarf mosaic virus (MDMV) is in the potyvirus group. MDMV is transmitted from maize to wheat by several aphid species. Infections occur on plants grown near diseased maize.

Distribution. The United States (New York). However, infection of wheat is rare.

Symptoms. Diseased leaves have a mild mottling.

Management. Not necessary.

Maize Streak

Cause. Maize streak virus (MSV) is in the geminivirus I group. MSV is transmitted circulatively by five species of *Cidaulina* leafhoppers, especially *Cidaulina mbila*. All stages of leafhoppers transmit MSV. Wheat is most likely to become infected when grown in association with maize.

Distribution. Africa and Southeast Asia.

Symptoms. Diseased leaves are shortened and curled and have fine, linear, chlorotic streaks. Diseased plants are excessively tillered, are sterile, and have shortened culms.

Management
1. Grow resistant wheat cultivars.
2. Sow winter wheat later in the autumn than the normal time to sow.
3. Do not grow wheat in the vicinity of maize.

Maize White Line Mosaic

Maize white line mosaic is also called white line mosaic and stunt.

Cause. Maize white line mosaic virus (MWLMV) is not mechanically transmitted but has been transferred to healthy roots from diseased roots placed in sterile soil together with zoospores of an *Olpidium*-like fungus that is considered to be the likely vector. Infection by MWLMV is associated with the time of the growing season rather than plant age. Maize white line mosaic often occurs locally in poorly drained areas of a field and along edge rows. A satellite virus serologically related to a satellite-like particle associated with maize dwarf ringspot virus is associated with MWLMV.

Distribution. The United States (Michigan, New York, Ohio, Vermont, and Wisconsin).

Symptoms. Inoculated wheat plants displayed a mild mosaic, but no necrotic lesions, on the leaves.

Management. Not reported.

Maize Yellow Stripe

Cause. Maize yellow stripe virus (MYSV) is associated with tenuivirus-like filaments. MYSV is transmitted in a persistent manner by both nymphs and adults of the leafhopper *Cicadulina chinai*. Acquisition and inoculation threshold times are 30 minutes each with a latent period of 4.5 to 8.0 days, depending on temperatures from 14°C (minimum) to 25°C (maximum). The maximum retention period is 27 days.

Barley, sorghum, and graminaceous weeds are winter hosts. Different strains of MYSV exist.

Distribution. Egypt.

Symptoms. Symptoms on wheat are unclear. Three symptom types exist on maize: fine stripe, coarse stripe, and chlorotic stunt. Each type may represent different MYSV strains. Experimentally, fine stripes appeared on the first leaves and coarse stripes on younger leaves; however, some leaves have both symptom types.

Management. Not Reported.

Mal de Rio Cuarto

Cause. Mal de Rio Cuarto virus (MRCV) is in the fijivirus group. MRCV is transmitted in a persistent, propagative way by the planthopper *Delphacodes kuscheli*. The insects develop mainly on oat and wheat, then migrate to maize during their first stages of growth. Disease is most severe when infection takes place during the early stages of plant development and during periods of high rainfall. MRCV has been detected in 12 species of weeds in the families Poaceae and Cyperaceae.

Distribution. Argentina. Mal de Rio Cuarto is the most important disease of maize in Argentina.

Symptoms. Diseased plants have deformed leaves, spikes, and spikelets; shortened internodes; and leaves with serrated borders. Spikelets may be sterile.

Management. Not reported.

Northern Cereal Mosaic

Cause. Northern cereal mosaic virus (NCMV) is in the rhabdovirus group. NCMV is transmitted by the planthopper *Laodelphax striatellus*. The insects *Unkanodes sapporonus* and *Delphacodes* spp. have also been reported as vectors. The wheat rosette stunt virus (WRSV) reported from northern China may be the same virus. Barley, oat, rice, and rye are also hosts.

Distribution. China (if WRSV and NCMV are the same), Japan, and Korea.

Symptoms. Symptoms on diseased leaves are a yellow mosaic and chlorotic leaf streaks. Diseased plants are stunted.

Management. Grow resistant cultivars.

Oat Sterile Dwarf

Oat sterile dwarf is also called oat base tillering disease and oat dwarf tillering disease.

Cause. Oat sterile dwarf virus (OSDV) is in the reovirus group. OSDV is transmitted persistently by planthoppers, especially *Javesella pellucida* and *Dicranotropis hamata*. Barley, grasses, maize, millet, and oat are also susceptible.

Distribution. Eastern and northern Europe.

Symptoms. Diseased plants are stunted, excessively tillered, and sterile and have helical leaf twisting.

Management
1. Control grassy weeds.
2. Avoid growing wheat adjacent to oat.

Orchardgrass Mosaic

Cause. Orchardgrass mosaic virus is mechanically transmitted. Orchardgrass is a source of inoculum.

Distribution. Canada (Quebec).

Symptoms. Typical mosaic symptoms occur on diseased leaves. Diseased plants are severely stunted and have reduced or delayed heading.

Management. Not reported.

Rice Black-Streaked Dwarf

Cause. Rice black-streaked dwarf virus (RBSDV) is in the reovirus group. RBSDV is transmitted by the planthoppers *Laodelphax striatellus, Unkanodes* *sapporonus,* and *U. albifascia.* RBSDV is retained from season to season in all stages of the planthoppers, but it is not transmitted through planthopper eggs.

Distribution. Japan.

Symptoms. Diseased plants are severely stunted. Leaves are twisted and have waxy, veinal swellings on their undersides and on the culms.

Management. Do not grow wheat adjacent to rice.

Rice Hoja Blanca

Rice hoja blanca is also called white leaf, white spike, and white tip.

Cause. Rice hoja blanca virus (RHBV) is in the tenuivirus group. RHBV is transmitted persistently by planthoppers, especially *Sogata cabana* and *S. orizicola.* RHBV is circulative and passes through eggs for up to 10 generations.

Distribution. The Caribbean, Central and South America, and the southern United States.

Symptoms. Upper diseased leaves and the spike have a gray-white discoloration, while other leaves are mottled, striped, and chlorotic. Diseased plants become sterile and die.

Management. Do not grow wheat in association with susceptible rice cultivars.

Russian Winter Wheat Mosaic

See "American Wheat Striate Mosaic."

Cause. Winter wheat mosaic virus (WWMV) is disseminated by the leafhoppers *Psammotettix striatus* and *Macrosteles laevis. Psammotettix striatus* carries WWMV through all life cycle stages, transovarially passing the virus to successive generations. Barley, rye, and some grasses are also hosts.

Distribution. Eastern Europe, Russia, and Ukraine.

Symptoms. Diseased leaves have mosaic and streak mosaic patterns. Plants are mildly stunted and have distinct yellow dashes and streaks that are oriented parallel to leaf veins.

Severe disease results from seedling infections and includes severe stunting, tillering, and necrosis. Surviving seedlings exist as rosettes and have typical leaf mosaic symptoms.

Management. Grow resistant cultivars.

Tobacco Mosaic

Cause. Tobacco mosaic virus (TMV) is in the tobamovirus group. TMV is associated with wheat soilborne mosaic virus.

Distribution. The United States (Kansas).

Symptoms. Inoculated plants of cultivar 'Pawnee' incubated at 30°C and in bright sunlight developed faint, chlorotic, local lesions that rapidly disappeared, leaving symptomless plants. A faint, transient mosaic occurred on inoculated leaves of 'Pawnee,' 'Arthur,' and 'Michigan Amber' wheats and 'Reno' barley.

Management. Not needed.

Wheat (Cardamom) Mosaic Streak

Cause. A virus transmitted through sap and the aphids *Brachycaudus helichrysi* and *Rhopalosiphum maidis*. Infection occurs near reservoirs of virus in cardamom plants growing in close proximity to wheat.

Distribution. India.

Symptoms. Diseased plants are mildly stunted and predisposed to infection by *Bipolaris sorokiniana*. Leaves have chronic yellow-green mosaic symptoms.

Management
1. Grow resistant wheat cultivars.
2. Grow wheat apart from cardamom.

Wheat Chlorotic Streak

Cause. Wheat chlorotic streak virus (WCSV) is in the rhabdovirus group. WCSV is presumed to be reservoired in *Agropyron repens,* which also maintains the planthopper vector *Laodelphax striatellus*. WCSV is disseminated in a persistent and transovarial manner.

Distribution. France. Wheat chlorotic streak occurs infrequently on wheat.

Symptoms. The growth and yield of diseased individual plants are reduced.

Management. Not reported.

Wheat Dwarf

Cause. Wheat dwarf virus (WDV) is in the geminivirus I group. WDV is transmitted by the leafhoppers *Psammotettix alienus* and *Macrosteles laevis*. Nymphs, which are more efficient vectors than adults, must acquire WDV from diseased plants since the virus is not passed through eggs.

Wheat dwarf occurs in association with heavy infestations of leafhoppers, and disease results from the combined effects of leafhopper feeding and virus infection.

Distribution. The former Czechoslovakia, Russia, Sweden, and Ukraine.

Symptoms. Diseased plants infected as seedlings are severely dwarfed and do not head out. If host plants are infected later, dwarfing is less severe but emerged heads contain little seed or the seed is shriveled. Diseased leaves develop scattered, fine, light green to yellow-brown spots and blotches that coalesce and cause prominent yellowing and necrosis.

Management. Grow resistant cultivars. Soft-wheat cultivars are less susceptible than hard-wheat cultivars.

Wheat Rosette Stunt

Cause. Wheat rosette stunt virus (WRSV) is the rhabdovirus group. WRSV is a strain of northern cereal mosaic virus.

Distribution. China.

Symptoms. A yellow mosaic and/or chlorotic leaf streaks occur on diseased leaves. Diseased plants are stunted.

Management
1. Sow wheat later in the growing season.
2. Eliminate alternate grassy weed hosts.

Wheat Soilborne Mosaic

Wheat soilborne mosaic is also called eastern wheat mosaic, green mosaic, mosaic rosette, soilborne mosaic, soilborne wheat mosaic, and yellow mosaic.

Cause. Wheat soilborne mosaic virus (WSBMV) is in the furovirus group. WSBMV survives in soil in *Polymyxa graminis,* a soilborne plasmodiophoraceous fungus that is an obligate parasite in roots of many higher plants. When soil is water-saturated and at low soil temperatures (10° to 20°C), *P. graminis* enters root hairs and epidermal cells of roots as motile zoospores. Once inside the plant, *P. graminis* replaces plant cell contents with plasmodial bodies that either segment into additional zoospores or develop into thick-walled resting spores 2 to 4 weeks after infection.

WSBMV is spread by any method that disseminates infested soil. Disease is most common in low-lying areas of fields that tend to be wet. Autumn-sown wheat is most severely diseased because abundant plant growth occurs during the low temperatures that favor dissemination and infection by *P. graminis.*

WSBMV combined with wheat spindle streak mosaic virus (WSSMV) causes disease in plants that are resistant to WSBMV. The ratio of WSBMV and WSSMV in infected plants is about 20:1, respectively.

Different strains, labeled the yellow and green strains, have been reported.

Distribution. Argentina, Brazil, Egypt, Italy, Japan, and the United States. Soilborne wheat mosaic is considered a major virus disease of hard red winter wheat in the plains states of the United States.

Symptoms. Two phases, mosaic and mosaic-rosette, are caused by yellow and green strains, respectively. Symptoms are light green to yellow leaf mosaics that are most prominent on diseased lower leaves in the spring. The youngest diseased leaves and leaf sheaths are mottled and develop parallel spots or streaks. Diseased plants are slightly to severely stunted. Warm weather prevents development of disease symptoms.

Symptoms of the combined WSBMV and WSSMV infection are a yellowing and overall bronze appearance to the field and plants that are stunted, have a mosaic appearance, and have reduced tillering. Symptoms are most obvious on plants growing in the lower portions of fields. During warm weather in the late spring, diseased plants lose disease symptoms except for the stunting.

Large, brown, vacuolate inclusions in epidermal cells stained with calcomine-orange and Luxol brilliant green can be seen with a light microscope.

Management

1. Rotate wheat with noncereal crops.
2. Sow wheat later in the autumn.
3. Grow resistant cultivars. Expression of resistance may involve an inhibition of virus movement from roots to foliage. The mechanisms of resistance to WSBMV may involve a reduction in rates of virus particle assembly, movement, and/or replication.

Wheat Spindle Streak Mosaic

Wheat spindle streak mosaic is also called Ontario soilborne wheat mosaic, wheat variegation, and wheat yellow mosaic. (*See* "Wheat Yellow Mosaic.")

Cause. Wheat spindle streak mosaic virus (WSSMV) is in the bymovirus group. WSSMV is transmitted by the soil fungus *Polymyxa graminis* and can survive for several years in soil in association with the fungus. Resting spores (cystosori) of *P. graminis* form in the root cortex and can remain dormant for many years. Virus is thought to remain infectious in cystosori for over 10 years in the absence of the virus's plant host. Some researchers regard WSSMV as a distinct strain of wheat yellow mosaic virus.

WSSMV is a cool-weather disease; the temperature for optimum virus transmission in the soil is 15°C. During favorable temperatures for virus transmission, high soil moisture is also needed for the release of infective zoospores of *P. graminis*. Infection does not occur above 20°C, and disease development does not occur above 18°C. Although symptom expression occurs in the spring, infection probably occurs in the autumn. Optimum disease development is at air or soil temperatures from 5° to 13°C for 60 days.

WSSMV combined with wheat soilborne mosaic virus (WSBMV) causes a disease of plants resistant to WSBMV. The ratio of WSBMV and WSSMV in infected plants is about 20:1, respectively.

Distribution. Canada (southern Ontario), France, India, and the United States. WSSMV is a prevalent and damaging pathogen of winter wheat in central and eastern North America. It is likely that wheat spindle streak mosaic occurs throughout the world.

Symptoms. Symptoms occur on lower, older leaves during cool spring weather and tend to occur on diseased plants uniformly distributed throughout a field. Warmer weather (18°C or more) prevents symptom development on younger leaves. The first leaves produced in the spring have yellow-green mottling, dashes, or streaks. Streaks are oriented parallel with leaf veins and have tapered ends that resemble spindles. As the leaves mature, brown areas replace the yellow-green areas and leaf tips or entire leaves may die. As warm weather continues, new symptomless leaves hide diseased lower leaves.

Diseased plants remain stunted, and their tillering capacity is reduced or lost, which is the major reason yields are reduced. Because both shoot and root growth are reduced and the stress of low temperatures is greater, plants have less cold-hardiness. Milling and baking qualities may also be reduced. Maturity may be delayed, which results in an increase in other foliar diseases, such as leaf rust.

Symptoms of the combined WSBMV and WSSMV infection are a yellowing and overall bronze appearance to a field and plants that are stunted, have a mosaic appearance, and have reduced tillering. Symptoms are most obvious on diseased plants growing in the lower portions of fields. During warm weather in late spring, diseased plants lose their disease symptoms except for the stunting.

Management
1. Grow resistant cultivars.
2. Sow wheat later in the autumn. Spring wheat tends to be unaffected by WSSMV.
3. Crop rotation is of some help, but WSSMV remains infective in soil for several years.

Wheat Spot Mosaic

Wheat spot mosaic is also called wheat spot, and wheat spot chlorosis.

Cause. The causal agent, wheat spot mosaic virus (WSpMA), is thought to be a virus transmitted by the wheat curl mite *Eriophyes tulipae*. WSpMA is retained through mottling but not through the egg stage. WSpMA is not sap-transmissible.

Distribution. Canada (Alberta) and the United States (Ohio). A similar disease has been reported from North Dakota.

Symptoms. About 3 to 8 days after mites have fed, light green spots (0.5–1.0 mm in diameter) develop on the youngest diseased leaves. The spots become necrotic, enlarge, coalesce, and form yellowish areas that may cover entire leaves. Later, diseased plants may develop severe mottling, chlorosis, leaf tip necrosis, and stunting, then die, especially if the wheat streak mosaic virus is present.

Following transmission of WSpMA and incubation periods of 12 to 15 days at the optimum temperature of 20°C, double-membrane-bound bodies (0.1–0.2 μm in diameter) are scattered or found as large aggregates in the cytoplasm of epidermal cells, subepidermal parenchyma cells, and phloem elements.

Management
1. Destroy all volunteer wheat and grasses in adjoining fields 2 weeks before sowing and 2 to 4 weeks before using the sowing field.
2. Sow wheat as late as practical after the Hessian fly–free date. The fly-free date is a guide that assumes mite populations have also been reduced. Wheat often escapes infection if it emerges in October or later.
3. Resistant cultivars are being developed. Some cultivars are tolerant.

Wheat Streak Mosaic

Cause. Wheat streak mosaic virus (WSMV) is in the rymovirus group of the Potyviridae. WSMV is transmitted by the wheat curl mites *Aceria tulipae* and *A. tosichella* and survives in several annual and perennial grasses. No active virus has been detected in dead plants or seed. Mites are also able to retain the virus for several weeks after they have acquired it. Neither the mite nor the virus can survive longer than 1 to 2 days in the absence of a living plant. In many wheat-growing areas, volunteer wheat is the major source of both WSMV and the vector. Maize is an over-summering host for both WSMV and *A. tulipae*.

As winter wheat plants mature, mites migrate to nearby volunteer wheat, grasses, or maize plants and infect them with WSMV. Later, mites move from volunteer wheat to early-sown wheat. Mites are disseminated by wind for a distance of at least 2.4 km and possibly farther. Some grasses are hosts for mites but not for WSMV, and vice versa. Some grasses are susceptible to both mites and WSMV, and some grasses are resistant to both.

Only young mites acquire WSMV by feeding 15 or more minutes on diseased plants. Infection depends on three factors: the population of the mites; the nearness of infected plants, especially volunteer wheat; and enough moisture to keep wheat vigorously growing where mites attain maximum reproduction.

Serologically, different strains of WSMV occur.

Distribution. Eastern Europe, western and central North America (particularly the Great Plains), and the former USSR.

Symptoms. Winter wheat is commonly infected in the autumn, but symptoms ordinarily do not appear until the following spring. The greatest losses occur in plants infected early in the autumn. WSMV is most severe in early-sown wheat; conversely, late sowing decreases symptoms of WSMV. In Oklahoma, red winter wheat cultivars, except 'Rall,' were more severely diseased when they had been inoculated in the autumn. 'Rall' displayed some resistance when it was sown in the autumn during the time recommended for north central Oklahoma. Spring inoculation of wheat sown early in the autumn (September or October) did not consistently result in symptoms. However, spring inoculation of wheat sown late in the autumn (November) resulted in symptoms. Thus, the maturity of plants at the time of infection may affect the severity of symptoms. Wheat sown in November was less mature (Feekes growth stage 5) than wheat sown in September or October (Feekes growth stage 6).

The first symptoms consist of light green to light yellow blotches, dashes, or streaks parallel to leaf veins. As diseased plants mature, yellow-striped leaves turn brown and die. Feeding mites often cause leaf edges to curl tightly in toward the upper midvein. Diseased plants become stunted, have a general yellow mottling, and develop an abnormally large number of tillers that vary considerably in height. Diseased stunted plants may remain standing after harvest at the same height or shorter than stubble. Heads may be sterile or partially sterile and have shriveled seed. In severe cases, diseased plants die before maturity.

Systemic infection of wheat by WSMV interferes with chloroplast development. Synergistic effects that are suspected between wheat streak mosaic virus and other viruses, such as barley yellow dwarf virus, make field identification difficult.

Management

1. Destroy all volunteer wheat and grasses in adjoining fields 2 weeks before sowing and 2 to 4 weeks before using the field.
2. Sow wheat as late as practical after the Hessian fly–free date. The fly-free date is a guide for assuming mite populations have been reduced. Wheat often escapes infection if it emerges in October or later.
3. Resistant cultivars are being developed. Some cultivars are tolerant.

Wheat Yellow Leaf

Cause. Wheat yellow leaf virus (WYLV) is in the closterovirus group. WYLV is transmitted in a semipersistent manner by the aphid *Rhopalosiphum maidis*. Barley is also a host.

Distribution. Japan.

Symptoms. Diseased leaves become discolored yellow and blighted. Diseased plants often die or ripen prematurely.

Management. Not reported.

Wheat Yellow Mosaic

Cause. Wheat yellow mosaic virus (WYMV) is in the potyvirus group. WYMV is soilborne and sometimes is associated with wheat soilborne mosaic virus. WYMV is also sap-transmitted.

Distribution. Japan.

Symptoms. Diseased leaves have a yellow mosaic.

Management. Grow resistant cultivars.

Unknown Virus

Cause. An unknown virus that is possibly in the tenuivirus group.

Distribution. The United States (Kansas).

Symptoms. Diseased plants are stunted, and leaves are mottled and streaked. Symptoms are similar to those of wheat streak mosaic.

Management. Not reported.

Diseases Caused by Unknown Factors

Physiologic Leaf Spot

Physiologic leaf spot is also called no-name disease.

Cause. Unknown. Leaf spot is observed to be more severe in conservation tillage than in low-residue tillage systems, especially when wheat is sown early and grown in a monoculture. Disease is less severe as the date of sowing is delayed and as the rate of nitrogen fertilization increases. Disease is more severe in annual wheat than in rotations of wheat with fallow or peas.

Foliar application of urea plus calcium chloride reduces leaf spot severity and increases grain yield. However, the application of urea plus micronutrients does not reduce disease.

Distribution. On winter and spring wheat in the northwestern United States, particularly the semiarid regions of the inland Pacific Northwest.

Symptoms. Symptoms vary among cultivars. Initially, brown spots appear on some cultivars, necrotic lesions appear on other cultivars, and enlarged necrotic lesions with chlorotic halos appear on yet other cultivars. Diseased lower leaves often become necrotic and senesce early. Up to 60% of the flag leaf area becomes necrotic soon after the flag leaves become fully extended. Grain production is reduced by 10%.

Leaf spots of similar description include Alternaria leaf blight, Septoria leaf blotch, tan spot, and bacterial leaf blight.

Management. Disease severity can be reduced by the management of wheat cultivar selection, crop rotation, sowing date, and plant nutrition.

Unknown

This disease may be the same as wheat spot mosaic reported from Canada (Alberta) and the United States (Ohio).

Cause. Unknown, but it is associated with the wheat curl mite *Aceria tulipae*. The disease was found in association with wheat streak mosaic. Membrane-bound ovoid bodies (130 to 220 nm in diameter) were observed in the mesophyll cells of diseased plants.

Distribution. The United States (eastern North Dakota).

Symptoms. Spotting, mosaic, and yellowing of diseased leaves.

Management. Not reported.

Selected References

Anderegg, J. C., and Murray, T. D. 1988. Influence of soil matric potential and soil pH on Cephalosporium stripe of winter wheat in the greenhouse. Plant Dis. 72:1011-1016.

Arneson, E., and Stiers, D. L. 1977. *Cephalosporium gramineum*: A seedborne pathogen. Plant Dis. Rptr. 61:619-621.

Azad, H., and Schaad, N. W. 1988. The relationship of *Xanthomonas campestris* pv. *translucens* to frost and the effect of frost on black chaff development in wheat. Phytopathology 78:95-100.

Babadoost, M., and Hebert, T. T. 1984. Factors affecting infection of wheat seedlings by *Septoria nodorum*. Phytopathology 74:592-595.

Babadoost, M., and Hebert, T. T. 1984. Incidence of *Septoria nodorum* in wheat seed and its effects on plant growth and grain yield. Plant Dis. 68:125-129.

Bailey, J. E., Lockwood, J. L., and Wiese, M. V. 1982. Infection of wheat by *Cephalosporium gramineum* as influenced by freezing of roots. Phytopathology 72:1324-1328.

Bailey, K. L., Knott, D. R., and Harding, H. 1988. Heritability and inheritance of resistance to common root rot (*Cochliobolus sativus*) in wheat (*Triticum aestivum*). Can. J. Plant Pathol. 10:207-214.

Bateman, G. L. 1988. *Pseudocercosporella anguioides,* a weakly pathogenic fungus associated with eyespot in winter wheat at a site in England. Plant Pathology 37:291-296.

Beddis, A. L., and Burgess, L. W. 1992. The influence of plant water stress on infection and colonization of wheat seedlings by *Fusarium graminearum* Group 1. Phytopathology 82:78-83.

Bockus, W. W. 1983. Effects of fall infection by *Gaeumannomyces graminis* var. *tritici* and triadimenol seed treatment on severity of take-all in winter wheat. Phytopathology 73:540-543.

Bockus, W. W., O'Connor, J. P., and Raymond, P. J. 1983. Effect of residue management method on incidence of Cephalosporium stripe under continuous winter wheat production. Plant Dis. 67:1323-1324.

Bockus, W. W., Davis, M. A., and Norman, B. L. 1994. Effect of soil shading by surface residues during summer fallow on take-all of winter wheat. Plant Dis. 78:50-54.

Bonde, M. R. 1987. Possible dissemination of teliospores of *Tilletia indica* by the practice of burning wheat stubble. (Abstr.) Phytopathology 77:639.

Bragard, C., Singer, E., Alizadeh, A., Vauterin, L., Maraite, H., and Swings, J. 1997. *Xanthomonas translucens* from small grains: Diversity and phytopathological relevance. Phytopathology 87:1111-1117.

Brakke, M. K., Estes, A. P., and Schuster, M. L. 1965. Transmission of soilborne wheat mosaic virus. Phytopathology 55:79-86.

Bretag, T. W. 1985. Control of ergot by a selective herbicide and stubble burning. Trans. Brit. Mycol. Soc. 85:341-343.

Broscious, S. C., and Frank, J. A. 1986. Effects of crop management practices on common root rot of winter wheat. Plant Dis. 70:857-859.

Brown, J. S., and Holmes, R. J. 1983. Guidelines for use of foliar sprays to control stripe rust of wheat in Australia. Plant Dis. 67:485-487.

Brown, W. M., Perotti, L. E., and Hill, J. P. 1985. Wheat take-all in Colorado high country irrigated spring wheat. (Abstr.) Phytopathology 75:1296.

Bruehl, G. W., and Cunfer, B. 1971. Physiologic and environmental factors that affect the severity of snow mold of wheat. Phytopathology 61:792-799.

Bruehl, G. W., and Machtmes, R. 1984. Effects of "dirting" on strawbreaker foot rot of winter wheat. Plant Dis. 68:868-870.

Bruehl, G. W., Peterson, C. J., Jr., and Machtmes, R. 1974. Influence of seeding date, resistance and benomyl on Cercosporella foot rot of winter wheat. Plant Dis. Rptr. 58:554-558.

Bruehl, G. W., Machtmes, R., and Murray, T. 1982. Importance of secondary inoculum in strawbreaker foot rot of winter wheat. Plant Dis. 66:845-847.

Campbell, W. P. 1958. A cause of pink seeds in wheat. Plant Dis. Rptr. 42:1272.

Carling, D. E., Meyer, L., and Brainard, K. A. 1996. Crater disease of wheat caused by *Rhizoctonia solani* AG-6. Plant Dis. 80:1429.

Carlson, R. R., and Vidaver, A. K. 1981. Bacterial mosaic of wheat: Distribution and hosts. (Abstr.) Phytopathology 71:207.

Carlson, R. R., and Vidaver, A. K. 1982. Bacterial mosaic, a new corynebacterial disease of wheat. Plant Dis. 66:76-79.

Cashion, N. L., and Luttrell, E. S. 1988. Host-parasite relationships in Karnal bunt of wheat. Phytopathology 78:75-84.

Chamswarng, C., and Cook, R. J. 1985. Identification and comparative pathogenicity of *Pythium* species from wheat roots and wheat field soils in the Pacific Northwest. Phytopathology 75:821-827.

Clement, D. K., Lister, R. M., and Foster, J. E. 1986. ELISA-based studies on the ecology and epidemiology of barley yellow dwarf virus in Indiana. Phytopathology 76:86-92.

Conner, R. L. 1990. Interrelationship of cultivar reactions to common root rot, black point, and spot blotch in spring wheat. Plant Dis. 74:224-227.

Conner, R. L., and Atkinson, T. G. 1989. Influence of continuous cropping on severity of common root rot in wheat and barley. Can. J. Plant Pathol. 11:127-132.

Conway, K. E., and Williams, E., Jr. 1986. Typhula-like snow mold on wheat in Oklahoma. Plant Dis. 70:169-170.

Cook, R. J. 1980. Fusarium foot rot of wheat and its control in the Pacific Northwest. Plant Dis. 64:1061-1066.

Cook, R. J., and Hering, T. F. 1986. Infection of wheat embryos by *Pythium* and seedling response as influenced by age of seed. (Abstr.) Phytopathology 76:1061.

Cook, R. J., and Waldher, J. T. 1977. Influence of stubble-mulch residue management on Cercosporella foot rot and yields of winter wheat. Plant Dis. Rptr. 61:96-100.

Cook, R. J., Sitton, J. W., and Haglund, W. A. 1987. Influence of soil treatments on growth and yield of wheat and implications for control of Pythium root rot. Phytopathology 77:1192-1198.

Cook, R. J., and Zhang, B. X. 1985. Degrees of sensitivity to metalaxyl within the *Pythium* spp. pathogenic to wheat in the Pacific Northwest. Plant Dis. 69:686-688.

Cook, R. J., Slitton, J. W., and Waldher, J. T. 1980. Evidence for *Pythium* as a pathogen of direct-drilled wheat in the Pacific Northwest. Plant Dis. 64:102-103.

Cook, R. J., Sitton, J. W., and Haglund, W. A. 1987. Influence of soil treatments on growth and yield of wheat and implications for control of Pythium root rot. Phytopathology 77:1192-1198.

Cotterill, P. J., and Sivasithamparam, K. 1988. Reduction of take-all inoculum by rotation with lupine, oats or field peas. J. Phytopathol. 121:125-134.

Cotterill, P. J., and Sivasithamparam, K. 1988. Effect of sowing date on take-all of wheat in Western Australia. Phytophylactica 20:11-14.

Cotterill, P. J., and Sivasithamparam, K. 1988. Inoculum of the take-all fungus in rotations of wheat and pasture: Relationships to disease and yield of wheat. Trans. Brit. Mycol. Soc. 91:63-72.

Cox, D. J., and Hosford, R. M., Jr. 1987. Resistant winter wheats compared at differing growth stages and leaf positions for tan spot severity. Plant Dis. 71:883-886.

Cronje, C. P. R., and Whitlock, V. H. 1987. The effect of germination and seed transmission of brome mosaic virus (BMV) in wheat seeds (CV. Betta). (Abstr.) Phytophylactica 19:128.

Cunfer, B. M., Demski, J. W., and Bays, D. C. 1988. Reduction in plant development, yield, and grain quality associated with wheat spindle streak mosaic virus. Phytopathology 78:198-204.

da Luz, W. C., and Bergstrom, G. C. 1987. Interactions between *Cochliobolus sativus* and *Pyrenophora tritici-repentis* on wheat leaves. Phytopathology 77:1355-1360.

Damsteegt, V. D., and Hewings, A. D. 1988. The Russian wheat aphid, *Diuraphis noxia*, confirmed as vector of brome mosaic virus in North America. Plant Dis. 72:79.

Daniels, A., Lucas, J. A., and Peberdy, J. F. 1991. Morphology and ultrastructure of W and R pathotypes of *Pseudocercosporella herpotrichoides* on wheat seedlings. Mycol. Res. 95:385-397.

Deacon, J. W., and Scott, D. B. 1985. *Rhizoctonia solani* associated with crater disease (stunting) of wheat in South Africa. Trans. Brit. Mycol. Soc. 85:319-327.

Dehne, H. W., and Oerke, E. C. 1985. Investigations on the occurrence of *Cochliobolus sativus* on barley and wheat. l. Influence of pathogen, host plant and environment on infection and damage. Zeitschrift fur Pflazenkrankheiten und Pflanzenschutz 92:270-280.

Delfosse, P., Reddy, A. S., Devi, P. S., Murthy, A. K., Wesley, S. V., Naidu, R. A., and Reddy, D. V. R. 1995. A disease of wheat caused by Indian peanut clump virus. Plant Dis. 79:1074.

de Vallavieille-Pope, C., Huber, L., Leconte, M., and Goyeau, H. 1995. Comparative effects of temperature and interrupted wet periods on germination, penetration, and infection of *Puccinia recondita* f. sp. *tritici* and *P. striiformis* on wheat seedlings. Phytopathology 85:409-415.

Dewan, M. M., and Sivasithamparam, K. 1987. First report of *Sclerotium rolfsii* on wheat in Western Australia. Plant Dis. 71:1146.

Diehl, J. A., Tinline, R. D., Kochhan, R. A., Shipton, P. J., and Rovira, A. D. 1982. The effect of fallow periods on common root rot of wheat in Rio Grande do Sul, Brazil. Phytopathology 72:1297-1301.

Duczek, L. J. 1989. Relationship between common root rot (*Cochliobolus sativus*) and tillering in spring wheat. Can. J. Plant Pathol. 11:39-44.

Duczek, L. J. 1990. Sporulation of *Cochliobolus sativus* on crowns and underground parts of spring cereals in relation to weather and host species, cultivar, and phenology. Can. J. Plant Pathol. 12:273-278.

Duveiller, E., and Maraite, H 1990. Bacterial sheath rot of wheat caused by *Pseudomonas fuscovaginae* in the highlands of Mexico. Plant Dis. 74:932-935.

Edwards, M. C., and McMullen, M. P. 1988. A newly discovered wheat disease of unknown etiology in eastern North Dakota. Plant Dis. 72:362.

Engel, R. E., and Mathre, D. E. 1988. Effect of fertilizer nitrogen source and chloride on take-all of irrigated hard red spring wheat. Plant Dis. 72:393-396.

Eversmeyer, M. G., Kramer, C. L., and Browder, L. E. 1988. Winter and early spring survival of *Puccinia recondita* on Kansas wheat during 1980-1986. Plant Dis. 72:1074-1076.

Eversmeyer, M. G., Kramer, C. L., and Hassan, Z. M. 1988. Environmental influences on the establishment of *Puccinia recondita* infection structures. Plant Dis. 72:409-412.

Eyal, Z., Scharen, A. L., Huffman, M. D., and Prescott, J. M. 1985. Global insights into virulence frequencies of *Mycosphaerella graminicola*. Phytopathology 75:1456-1462.

Ezzahiri, B., and Roelfs, A. P. 1989. Inheritance and expression of adult plant resistance to leaf rust in Era wheat. Plant Dis. 73:549-551.

Ezzahiri, B., Diouri, S., and Roelfs, A. P. 1992. *Anchusa italica* as an alternate host for wheat leaf rust in Morocco. Plant Dis. 76:1185.

Farr, D. F., Bills, G. R., Chamuris, G. P., and Rossman, A. Y. 1989. Fungi on Plants and Plant Products in the United States. American Phytopathological Society, St. Paul, MN. 1252 pp.

Fernandes, J. M., and Picinini, E. C. 1989. *Pyricularia oryzae,* a new pathogen of wheat in Brazil. (Abstr.) Can. J. Plant Pathol. 11:189.

Fernandez, J. A., Wofford, D. S., and Horton, J. L. 1985. Interactive effects of freezing and common root rot fungi on winter wheat. Phytopathology 75:845-847.

Fernandez, M. R., Clark, J. M., DePauw, R. M., Krvine, R. B., and Knox, R. E. 1994. Black point and red smudge in irrigated durum wheat in southern Saskatchewan in 1990-1992. Can. J. Plant Pathol. 16:221-227.

Fitt, B. D. L., and White, R. P. 1988. Stages in the progress of eyespot epidemics in winter wheat crops. Z. Pfl. Krankh. Pfl. Schutz. 95:35-45.

Forster, R. L. 1990. Pink seed of wheat caused by *Erwinia rhapontici* in Idaho. Plant Dis. 74:81.

Forster, R. L., and Schaad, N. W. 1988. Control of black chaff of wheat with seed treatment and a foundation seed health program. Plant Dis. 72:935-938.

Frank, J. A. 1985. Influence of root rot on winter survival and yield of winter barley and winter wheat. Phytopathology 75:1039-1041.

Frank, J. A., Cole, H., Jr., and Harley, O. E. 1988. The effect of planting date on fall infections and epidemics of powdery mildew on winter wheat. Plant Dis. 72:661-664.

Fried, P. M., and Meister, E. 1987. Inheritance of leaf and head resistance of winter wheat to *Septoria nodorum* in a diallel cross. Phytopathology 77:1371-1375.

Fryda, S. J., and Otta, J. D. 1978. Epiphytotic movement and survival of *Pseudomonas syringae* on spring wheat. Phytopathology 64:1064-1067.

Gardner, W. S. 1981. Relationship of corn to the spread of wheat streak mosaic virus in winter wheat. (Abstr.) Phytopathology 71:217.

Gaudet, D. A., Bhalla, M. K., Clayton, G. W., and Chen, T. H. H. 1989. Effect of cottony snow mold and low temperatures on winter wheat survival in central and northern Alberta. Can. J. Plant Pathol. 11:291-296.

Goates, B. J. 1988. Histology of infection of wheat by *Tilletia indica,* the Karnal bunt pathogen. Phytopathology 78:1434-1441.

Goel, R. K., and Gupta, A. K. 1979. A new glume blotch disease of wheat in India. Plant Dis. Rptr. 63:620.

Gough, F. J., and Lee, T. S. 1985. Moisture effects on the discharge and survival of conidia of *Septoria tritici.* Phytopathology 75:180-182.

Gough, F. J., and Merkle, O. G. 1977. The effect of speckled leaf blotch on root and shoot development of wheat. Plant Dis. Rptr. 61:597-599.

Graham, J. H., and Menge, J. A. 1982. Influence of vesicular-arbuscular mycorrhizae and soil phosphorous on take-all disease of wheat. Phytopathology 72:95-98.

Grey, W. E., Mathre, D. E., Hoffman, J. A., Powelson, R. L., and Fernandez, J. A. 1986. Importance of seedborne *Tilletia controversa* for infection of winter wheat and its relationship to international commerce. Plant Dis. 70:122-125.

Griffin, G. D. 1990. Effect of *Meloidogyne chitwoodi* on the growth of wheat. (Abstr.) Phytopathology 80:1006.

Haber, S., and Chong, J. 1993. Flame chlorosis induces vesiculations in chloroplasts and mitochondria: What does it mean? (Abstr.) Can. J. Plant Pathol. 15:57.

Haber, S., and Hardener, D. E. 1992. Green foxtail (*Setaria viridis*) and barnyard grass (*Echinochloa crusgalli*), new hosts of the virus-like agent causing flame chlorosis in cereals. Can. J. Plant Pathol. 14:278-280.

Halfon-Meiri, A., and Kulik, M. M. 1977. *Septoria nodorum* infection of wheat seeds produced in Pennsylvania. Plant Dis. Rptr. 61:867-869.

Hannukkala, A., and Koponen, H. 1988. *Microdochium bolleyi,* a common inhabitant of barley and wheat roots in Finland. Karstenia 27:31-36.

Harder, D. E., and Bakker, W. 1973. African cereal streak, a new disease of cereals in East Africa. Phytopathology 63:1407-1411.

Herrman, T., and Weise, M. V. 1985. Influence of cultural practices on incidence of foot rot in winter wheat. Plant Dis. 69:948-950.

Hess, D. E., and Shaner, G. 1987. Effect of moisture and temperature on development of *Septoria tritici* blotch in wheat. Phytopathology 77:215-219.

Higgins, S., and Fitt, B. D. L. 1984. Production and pathogenicity to wheat of *Pseudocer-cosporella herpotrichoides* conidia. Phytopathol. Zeitschr. 111:222-231.

Hiruki, C. 1989. Characterization of the disease agent of wheat spot mosaic disease in western Canada. (Abstr.) Can. J. Plant Pathol. 11:190.

Hoffman, J. A., and Goates, B. J. 1981. Spring dormancy of *Tilletia controversa* teliospores. (Abstr.) Phytopathology 71:881.

Hoffman, J. A., and Sisson, D. V. 1987. Evaluation of bitertanol and thiabendazole seed treatment and PCNB soil treatment for control of dwarf bunt of wheat. Plant Dis. 71:839-841.

Holmes, R. J., and Dennis, J. I. 1985. Accessory hosts of wheat stripe rust in Victoria, Australia. Trans. Brit. Mycol. Soc. 85:159-160.

Hosford, R. M., Jr. 1971. A form of *Pyrenophora trichostoma* pathogenic to wheat and other grasses. Phytopathology 61:28-32.

Hosford, R. M., Jr. 1975. *Platyspora pentamera,* a pathogen of wheat. Phytopathology 65:499-500.

Hosford, R. M., Jr. 1978. Effects of wetting period on resistance to leaf spotting of wheat, barley and rye by *Leptosphaeria herpotrichoides.* Phytopathology 68:591-594.

Hosford, R. M., Jr. 1978. Effects of wetting period on resistance to leaf spotting of wheat by *Leptosphaeria microscopica* with conidial stage *Phaeoseptoria urvilleana.* Phytopathology 68:908-912.

Hosford, R. M., Jr. 1982. White blotch incited in wheat by *Bacillus megaterium* pv. *cerealis.* Phytopathology 72:1453-1459.

Hosford, R. M., Jr., Hogenson, R. O., Huguelet, J. E., and Kiesling, R. L. 1969. Studies of *Leptosphaeria avenaria* f. sp. *triticea* on wheat in North Dakota. Plant Dis. Rptr. 53:378-381.

Hosford, R. M., Jr., Jordahl, J. G., and Hammond, J. J. 1990. Effect of wheat genotype, leaf position, growth stage, fungal isolate, and wet period on tan spot lesions. Plant Dis. 74:385-390.

Huber, D. M. 1987. Immobilization of Mn predisposes wheat to take-all. (Abstr.) Phytopathology 77:1715.

Huber, D. M., and Hankins, B. J. 1974. Effect of fall mowing on snow mold of winter wheat. Plant Dis. Rptr. 58:432-434.

Inglis, D. A., and Cook, R. J. 1981. *Calonectria nivalis* causes scab in the Pacific Northwest. Plant Dis. 65:923-924.

Ingram, D. M., and Cook, R. J. 1987. Influence of temperature and plant residues on pathogenicity of *Pythium* spp. on wheat barley, peas and lentils. (Abstr.) Phytopathology 77:1239.

Jacobs, D. L., and Bruehl, G. W. 1986. Saprophytic ability of *Typhula incarnata, T. idahoensis,* and *T. ishikariensis.* Phytopathology 76:695-698.

Jacobsen, B. J. 1977. Effect of fungicides on Septoria leaf and glume blotch, Fusarium scab, grain yield, and test weight of winter wheat. Phytopathology 67:1412-1414.

Jardine, D. J., Bowden, R. L., Jensen, S. G., and Seifers, D. L. 1994. A new virus of corn and wheat in western Kansas. (Abstr.). Phytopathology 84:1117.

Jenkyn, J. F., and King, J. E. 1988. Effects of treatments to perennial ryegrass on the development of *Septoria* spp. in a subsequent crop of winter wheat. Plant Pathology 37:112-119.

Jensen, S. G., Lane, L. C., and Seifers, D. L. 1996. A new disease of maize and wheat in the High Plains. Plant Dis. 80:1387-1390.

Kane, R. T., Smiley, R. W., and Sorrells, M. E. 1987. Relative pathogenicity of selected *Fusar-*

ium species and *Microdochium bolleyi* to winter wheat in New York. Plant Dis. 71:177-181.

Kemp, G. H. J., Pretorius, Z. A., and Wingfield, M. J. 1996. Fusarium glume spot of wheat: A newly recorded mite-associated disease in South Africa. Plant Dis. 80:48-51.

Khokhar, L. K., and Pacumbaba, R. P. 1985. Additional alternative grass hosts of *Leptosphaeria nodorum*. (Abstr.) Phytopathology 75:1295.

Krupinsky, J. M. 1985. Leaf Spot diseases of wheat related to stubble height. (Abstr.) Phytopathology 75:1295.

Krupinsky, J. M. 1992. Grass hosts of *Pyrenophora tritici-repentis*. Plant Dis. 76:92-95.

Krupinsky, J. M. 1997. Aggressiveness of *Stagonospora nodorum* isolates from perennial grasses on wheat. Plant Dis. 81:1032-1036.

Lai, P., and Bruehl, G. W. 1966. Survival of *Cephalosporium gramineum* in naturally infected wheat straws in soil in the field and in the laboratory. Phytopathology 56:213-218.

Lamari, L., and Bernier, C. C. 1991. Genetics of tan necrosis and extensive chlorosis in tan spot of wheat caused by *Pyrenophora tritici-repentis*. Phytopathology 81:1092-1095.

Lamari, L., Bernier, C. C., and Smith, R. B. 1991. Wheat genotypes that develop both tan necrosis and extensive chlorosis in response to isolates of *Pyrenophora tritici-repentis*. Plant Dis. 75:121-122.

Latin, R. X., Harder, R. W., and Wiese, M. V. 1982. Incidence of Cephalosporium stripe as influenced by winter wheat management practices. Plant Dis. 66:229-230.

Lawn, D. A., and Sayre, K. D. 1989. Effects of crop rotation, tillage and straw management on common root rot of wheat. (Abstr.) Phytopathology 79:1142.

Lawton, M. B., Burpee, L. L., and Goulty, L. G. 1987. Seed treatments control gray snow mold of winter wheat. (Abstr.) Can. J. Plant Pathol. 9:281.

Lipps, P. E., and Bruehl, G. W. 1978. Snow rot of winter wheat in Washington. Phytopathology 68:1120-1127.

Lipps, P. E., and Herr, L. J. 1982. Etiology of *Rhizoctonia cerealis* in sharp eyespot of wheat. Phytopathology 72:1574-1577.

Lipps, P. E., and Madden, L. V. 1989. Effect of fungicide application timing on control of powdery mildew and grain yield of winter wheat. Plant Dis. 73:991-994.

Lockhart, B. E. L. 1986. Occurrence of cereal chlorotic mottle virus in northern Africa. Plant Dis. 70:912-915.

Lommell, S. A., Willis, W. G., and Kendall, T. L. 1986. Identification of wheat spindle streak mosaic virus and its role in a new disease of winter wheat in Kansas. Plant Dis. 70:964-968.

Loria, R., Weise, M. V., and Jones, A. L. 1982. Effect of free moisture, head development and embryo accessibility on infection of wheat by *Ustilago tritici*. Phytopathology 72:1270-1272.

Los, O., van Wyk, P. S., and Marasas, W. F. O. 1987. A new species of *Fusarium* causing crown rot of wheat in South Africa. (Abstr.) Phytophylactica 19:29.

Love, C. S. 1985. Effect of soil pH on infection of wheat by *Cephalosporium gramineum*. Phytopathology 75:1296.

Love, C. S., and Bruehl, G. W. 1987. Effect of soil pH on Cephalosporium stripe in wheat. Plant Dis. 71:727-731.

Luke, H. H., Pfahler, P. L., and Barnett, R. D. 1983. Control of *Septoria nodorum* on wheat with crop rotation and seed treatment. Plant Dis. 67:949-951.

Luke, H. H., Barnett, R. D., and Pfahler, P. L. 1985. Influence of soil infestation, seed infection, and seed treatment on *Septoria nodorum* blotch of wheat. Plant Dis. 69:74-76.

Luz, W. C. D., and Hosford, R. M., Jr. 1980. Twelve *Pyrenophora trichostoma* races for virulence to wheat in the central plains of North America. Phytopathology 70:1193-1196.

Luzzardi, G. C., Luz, W. C., and Perobom, C. R. 1983. Podridas branca dos cereais causada por *Sclerotium rolfsii* no Brazil. Fitopatologia Brasileira 8:371-375 (in Portugese).

Maas, E. M. C., and Kotze, J. M. 1985. *Fusarium equiseti* crown rot of wheat in South Africa. Phytophylactica 17:169-170.

MacNish, G. C. 1985. Methods of reducing Rhizoctonia patch of cereals in Western Australia. Plant Pathol. 34:175-181.

MacNish, G. C., and Neate, S. M. 1996. Rhizoctonia bare patch of cereals: An Australian perspective. Plant Dis. 80:965-971.

Madariaga, R. B., and Gilchrist, D. G. 1990. Phenology of ascospore release by *Mycosphaerella graminicola* from wheat stubble in relation to crop cycle, and other associated fungi. (Abstr.) Phytopathology 80:1006.

Maloy, O. C., and Specht, K. L. 1988. Black point of irrigated wheat in central Washington. Plant Dis. 72:1031-1033.

Manisterski, J., Segal, A., Levy, A. A., and Feldman, M. 1988. Evaluation of Israeli *Aegilops* and *Agropyron* species for resistance to wheat leaf rust. Plant Dis. 72:941-944.

Martin, J. M., Johnston, R. H., and Mathre, D. E. 1989. Factors affecting the severity of Cephalosporium stripe of winter wheat. Can. J. Plant Pathol. 11:361-367.

Mathieson, J. T., Rush, C. M., Bordovsky, D., Clark, L. E., and Jones, O. R. 1990. Effects of tillage on common root rot of wheat in Texas. Plant Dis. 74:1006-1008.

Matsumoto, N., and Tajimi, A. 1987. Bacterial flora associated with the snow mold fungi, *Typhula incarnata* and *T. ishikariensis*. Ann. Phytopathol. Soc. Japan 53:250-253.

McBeath, J. H. 1981. Bacterial mosaic disease on spring wheat and triticale in Alaska. (Abstr.) Phytopathology 71:893.

McBeath, J. H. 1985. Pink snow mold on winter cereals and lawn grasses in Alaska. Plant Dis. 69:722-723.

Meyer, L., Wehner, F. C., Kuwite, C. A., and Piening, L. 1996. Crater disease and patchy stunting of wheat caused by the same strain of *Rhizoctonia solani*. Plant Dis. 80:1079.

Mihuta-Grimm, L., and Forster, R. L. 1989. Scab of wheat and barley in southern Idaho and evaluation of seed treatments for eradication of *Fusarium* spp. Plant Dis. 73:769-771.

Modawi, R. S., Heyne, E. G., Brunetta, D., and Willis, W. G. 1982. Genetic studies of field reaction to wheat soilborne mosaic virus. Plant Dis. 66:1183-1184.

Moore, K. J., and Cook, R. J. 1984. Increased take-all of wheat with direct drilling in the Pacific Northwest. Phytopathology 74:1044-1049.

Nagaich, B. B., and Sinha, R. C. 1974. Eastern wheat striate: A new viral disease. Plant Dis. Rptr. 58:968-970.

Nelson, K. E., and Sutton, J. C. 1988. Epidemiology of eyespot on winter wheat in Ontario. Phytoprotection 69:9-21.

Nema, K. G., Dave, G. S., and Khosla, H. K. 1971. A new glume blotch of wheat. Plant Dis. Rptr. 55:95.

Nguyen, H. T., and Pfeifer, R. P. 1980. Effects of wheat spindle streak mosaic virus on winter wheat. Plant Dis. 64:181-184.

Nyvall, R. F., and Kommedahl, T. 1973. Competitive saprophytic ability of *Fusarium roseum* f. sp. *cerealis* 'Culmorum' in soil. Phytopathology 63:590-597.

O'Reilly, P., Bannon, E., and Downes, M. J. 1988. *Leptosphaeria nodorum* on wheat in Ireland. Plant Pathol. 37:153-154.

Otta, J. A. 1974. *Pseudomonas syringae* incites a leaf necrosis on spring and winter wheats in South Dakota. Plant Dis. Rptr. 58:1061-1064.

Paliwal, Y. C. 1982. Role of perennial grasses, winter wheat and aphid vectors in the disease cycle and epidemiology of barley yellow dwarf virus. Can. J. Plant Pathol. 4:367-374.

Paliwal, Y. C., and Andrews, C. J. 1979. Effects of barley yellow dwarf and wheat spindle streak mosaic viruses on cold hardiness of cereals. Can. J. Plant Pathol. 1:71-75.

Palmer, L. T., and Brakke, M. K. 1975. Yield reduction in winter wheat infected with soilborne wheat mosaic virus. Plant Dis. Rptr. 59:469-471.

Patykowski, J., Urbanek, H., and Kaczorowska, T. 1988. Peroxidase activity in leaves of wheat cultivars differing in resistance to *Erysiphe graminis* DC. J. Phytopathol. 122:126-134.

Paulitz, T. C. 1996. Diurnal release of ascospores by *Gibberella zea* in inoculated wheat plots. Plant Dis. 80:674-678.

Paulsen, A., Niblett, C. L., and Willis, W. G. 1975. Natural occurrence of tobacco mosaic virus in wheat. Plant Dis. Rptr. 59:747-750.

Peterson, J. F. 1989. A cereal-infecting virus from orchardgrass. Can. Plant Dis. Surv. 69:13-16.

Pfender, W., and Wootke, S. 1985. Overwintering microbial populations in wheat straw infested with *Pyrenophora tritici-repentis*. (Abstr.) Phytopathology 75:1350.

Pfender, W., and Wootke, S. 1987. Production of pseudothecia and ascospores by *Pyrenophora tritici-repentis* in response to macronutrient concentrations. Phytopathology 77:1213-1216.

Piening, L. J., MacPherson, D. J., and Malhi, S. S. 1989. Stem melanosis of some wheat, barley, and oat cultivars on a copper-deficient soil. Can. J. Plant Pathol. 11:65-67.

Piening, L. J., Orr, D. D., and Bhalla, M. 1990. Survival of *Coprinus psychromorbidus* under continuous cropping. Can. J. Plant Pathol. 12:217-218.

Pool, R. A. F., and Sharp, E. L. 1969. Some environmental and cultural factors affecting Cephalosporium stripe of winter wheat. Plant Dis. Rptr. 53:898-902.

Prabhu, A. S., and Prasada, R. 1966. Pathological and epidemiological studies of leaf blight of wheat caused by *Alternaria triticina*. Indian Phytopathol. 19:95-112.

Pumphrey, F. V., Wilkins, D. E., Hane, D. C., and Smiley, R. W. 1987. Influence of tillage and nitrogen fertilizer on Rhizoctonia root rot (bare patch) of winter wheat. Plant Dis. 71:125-127.

Puning, L. J., and MacPherson, D. J. 1985. Stem melanosis, a disease of spring wheat caused by *Pseudomonas cichorii*. Can. J. Plant Pathol. 7:168-172.

Raemaekers, R. H., and Tinline, R. D. 1981. Epidemic of diseases caused by *Cochliobolus sativus* on rainfed wheat in Zambia. Can. J. Plant Pathol. 3:211-214.

Rakotondradona, R., and Line, R. F. 1984. Control of stripe rust and leaf rust of wheat with seed treatments and effects of treatments on the host. Plant Dis. 68:112-117.

Rao, A. S. 1968. Biology of *Polymyxa graminis* in relation to soilborne wheat mosaic virus. Phytopathology 58:1516-1521.

Reis, E. M., and Wunsche, W. A. 1984. Sporulation of *Cochliobolus sativus* on residues of winter crops and its relationship to the increase of inoculum density in soil. Plant Dis. 68:411-412.

Reis, E. M., Cook, R. J., and McNeal, B. L. 1982. Effect of mineral nutrition on take-all of wheat. Phytopathology 72:224-229.

Reis, E. M., Cook, R. J., and McNeal, B. L. 1983. Elevated pH and associated reduced trace-nutrient availability as factors contributing to increased take-all of wheat upon soil liming. Phytopathology 73:411-413.

Rewal, H. S., and Jhooty, J. S. 1986. Physiological specialization of loose smut of wheat in the Punjab state of India. Plant Dis. 70:228-230.

Richardson, M. J., and Noble, M. 1970. *Septoria* species on cereals-A note to aid their identification. Plant Pathol. 19:159-163.

Rivoal, R., and Sarr, E. 1987. Field experiments on *Heterodera avenae* in France and implications for winter wheat performance. Nematologica 33:460-479.

Roane, C. W., Roane, M. K., and Starling, T. M. 1974. *Ascochyta* species on barley and wheat in Virginia. Plant Dis. Rptr. 58:455-456.

Roberts, F. A., and Sivasithamparam, K. 1987. Effect of interaction of *Rhizoctonia* spp. with other fungi from cereal bare patches on root rot of wheat. Trans. Brit. Mycol. Soc. 89:256-259.

Rodriguez Pardina, P. E., Gimenez Pecci, M. P., Laguna, I. G., Dagoberto, E., and Truol, G. 1998. Wheat: A new natural host for the Mal de Rio Cuarto virus in the endemic disease area, Rio Cuarto, Cordoba Province, Argentina. Plant Dis. 82:149-152.

Rovira, A. D. 1986. Influence of crop rotation and tillage on Rhizoctonia bare patch of wheat. Phytopathology 76:669-673.

Rovira, A. D., and McDonald, H. J. 1986. Effects of the herbicide chlorsulfuron on Rhizoctonia bare patch and take-all of barley and wheat. Plant Dis. 70:879-882.

Rowe, R. C., and Powelson, R. L. 1973. Epidemiology of Cercosporella foot rot of wheat: Disease spread. Phytopathology 63:984-988.

Rowe, R. C., and Powelson, R. L. 1973. Epidemiology of Cercosporella foot rot of wheat: Spore production. Phytopathology 63:981-984.

Rush, C. M., Carling, D. E., Harveson, R. M., and Mahieson, J. T. 1994. Prevalence and pathogenicity of anastomosis groups of *Rhizoctonia* solani from wheat and sugar beet in Texas. Plant Dis. 78:349-352

Russell, C. C., and Perry, V. G. 1966. Parasitic habit of *Trichodorus christiei* on wheat. Phytopathology 56:357-358.

Scardaci, S. C., and Webster, R. K. 1982. Common root rot of cereals in California. Plant Dis. 66:31-34.

Schafer, J. F., and Long, D. L. 1986. Evidence for local source of *Puccinia recondita* on wheat in Pennsylvania. Plant Dis. 70:892.

Scharen, A. L., and Krupinsky, J. M. 1971. *Ascochyta tritici* on wheat. Phytopathology 61:675-680.

Schilder, A. M. C., and Bergstrom, G. C. 1992. The process of wheat seed infection by *Pyrenophora tritici-repentis*. (Abstr.). Phytopathology 82:1072.

Scott, D. B., Visser, C. P. N., and Rufenacht, E. M. C. 1979. Crater disease of summer wheat in African drylands. Plant Dis. Rptr. 63:836-840.

Scott, P. R., Benedikz, P. W., Jones, H. G., and Ford, M. A. 1985. Some effects of canopy structure and microclimate on infection of tall and short wheats by *Septoria nodorum*. Plant Pathol. 34:578-593.

Scott, P. R., Sanderson, F. K., and Benedikz, P. W. 1988. Occurrence of *Mycosphaerella graminicola*, teleomorph of *Septoria tritici*, on wheat debris in the U. K. Plant Pathol. 37:285-290.

Seifers, D. L., Harvey, T. L., and Bowden, R. L. 1995. Occurrence and symptom expression of American wheat striate mosaic virus in wheat in Kansas. Plant Dis. 79:853-858.

Seifers, D. L., Harvey, T. L., Martin, T. J., and Jensen, S. G. 1997. Identification of the wheat curl mite as the vector of the High Plains virus of corn and wheat. Plant Dis. 81:1161-1166.

Sellam, M. A., and Wilcoxson, R. D. 1976. Bacterial leaf blight of wheat in Minnesota. Plant Dis. Rptr. 60:242-245.

Shane, W. W., and Baumer, J. S. 1987. Population dynamics of *Pseudomonas syringae* pv. *syringae* on spring wheat. Phytopathology 77:1399-1405.

Shane, W. W., Baumer, J. S., and Teng, P. S. 1987. Crop losses caused by Xanthomonas streak on spring wheat and barley. Plant Dis. 71:927-930.

Shaner, G., and Buechley, G. 1995. Epidemiology of leaf blotch of soft red winter wheat caused by *Septoria tritici* and *Stagonospora nodorum*. Plant Dis. 79:928-938.

Shaner, G., and Powelson, R. L. 1973. The oversummering and dispersal of inoculum of *Puccinia striiformis* in Oregon. Phytopathology 63:13-17.

Shefelbine, P. A., and Bockus, W. W. 1989. Decline of Cephalosporium stripe by monoculture of moderately resistant winter wheat cultivars. Phytopathology 79:1127-1131.

Shefelbine, P. A., and Bockus, W. W. 1990. Host genotype effects on inoculum production by *Cephalosporium gramineum* from infested residue. Plant Dis. 74: 238-240.

Sherwood, J. L., Myers, L. D., and Hunger, R. M. 1991. Replication and movement of wheat soilborne mosaic virus (WSBMV) in hard red winter wheat. (Abstr.) Phytopathology 81: 1216.

Shipton, P. J. 1975. Take-all decline during cereal monoculture, pp. 137-144. *In* Biology and Control of Soil-borne Plant Pathogens, G. W. Bruehl, ed. American Phytopathological Society, St. Paul, MN. 216 pp.

Sinha, R. C., and Benki, R. M. 1972. American Wheat Striate Mosaic Virus. Descriptions of Plant Viruses. Set 6, No. 99. Commonwealth Mycological Institute.

Slykhuis, J. T. 1970. Factors determining the development of wheat spindle streak mosaic caused by a soilborne virus in Ontario. Phytopathology 60:319-331.

Slykhuis, J. T., and Barr, D. J. S. 1978. Confirmation of *Polymyxa graminis* as a vector of wheat spindle streak mosaic virus. Phytopathology 68:639-643.

Smilanick, J. L., Hoffmann, J. A., and Royer, M. H. 1985. Effect of temperature, pH, light and desiccation on teliospore germination of *Tilletia indica*. Phytopathology 75:1428-1431.

Smiley, R. W., Fowler, M. C., and Reynolds, K. L. 1986. Temperature effects on take-all of cereals caused by *Phialophora graminicola* and *Gaeumannomyces graminis*. Phytopathology 76:923-931.

Smiley, R. W., Gillespie-Sasse, L.-M., Uddin, W., Collins, H. P., and Stoltz, M. A. 1993. Physiologic leaf spot of winter wheat. Plant Dis. 77:521-527.

Smiley, R. W., Uddin, W., Zwer, P. K., Wysocki, D. J., Ball, D. A., Chastain, T. G., and Rasmussen, P. E. 1993. Influence of crop management practices on physiologic leaf spot of winter wheat. Plant Dis. 77:803-810.

Smith, E. M., and Wehner, F. C. 1987. Biological and chemical measures integrated with deep soil cultivation against crater disease of wheat. Phytophylactica 19:87-90.

Smith, J. D. 1981. Snow molds of winter cereals: Guide for diagnosis, culture and pathogenicity. Can. J. Plant Pathol. 3:15-25.

Spadafora, V. J., Cole, H., Jr., and Frank, J. A. 1987. Effects of leaf and glume blotch caused by *Leptosphaeria nodorum* on yield and yield components of soft red winter wheat in Pennsylvania. Phytopathology 77:1326-1329.

Stack, R. W. 1989. A comparison of the inoculum potential of ascospores and conidia of *Gibberella zeae*. Can. J. Plant Pathol. 11:137-142.

Stack, R. W., and McMullen, M. P. 1985. Head blighting potential of *Fusarium* species associated with spring wheat heads. Can. J. Plant Pathol. 7:79-82.

Stack, R. W., Jons, V. L., and Lamey, H. A. 1979. Snow rot of winter wheat in North Dakota. (Abstr.) Phytopathology 69:543.

Stanton, J. M., and Fisher, J. M. 1987. Field assessment of factors associated with tolerance of wheat to *Heterodera avenae*. Nematologica 33:357-360.

Stiles, C. M., and Murray, T. D. 1996. Infection of field-grown winter wheat by *Cephalosporium gramineum* and the effect of soil pH. Phytopathology 86:177-183.

Sturz, A. V., and Bernier, C. C. 1987. Incidence of pathogenic fungal complexes in the crowns and roots of winter and spring wheat relative to cropping practice. Can. J. Plant Pathol. 9:265-271.

Sturz, A. V., and Bernier, C. C. 1989. Influence of crop rotations on winter wheat growth and yield in relation to the dynamics of pathogenic crown and root rot fungal complexes. Can. J. Plant Pathol. 11:114-121.

Takamatsu, S. 1989. A new snow mold of wheat and barley caused by foot rot fungus, *Ceratobasidium gramineum*. Ann. Phytopathol. Soc. Japan 55:233-237.

Takamatsu, S., and Ichitani, T. 1987. Detection of Pythium snow rot fungi in diseased leaves of wheat and barley grown in paddy fields and upland fields. Ann. Phytopathol. Soc. Japan 53:56-59.

Taylor, R. G., Jackson, T. L., Powelson, R. L., and Christensen, N. W. 1983. Chloride, nitrogen form, lime, and planting date effects on take-all root rot of winter wheat. Plant Dis. 67:1116-1120.

Teich, A. H., and Nelson, K. 1984. Survey of Fusarium head blight and possible effects of cultural practices in wheat fields in Lambton County in 1983. Can. Plant Dis. Surv. 64:11-13.

Teyssandier, E., and Sands, D. C. 1987. Wheat leaf blight caused by *Pseudomonas syringae* in Argentina. (Abstr.) Phytopathology 77:1766.

Tinline, R. D. 1981. Effect of depth and rate of seeding on common root rot of wheat in Saskatchewan. (Abstr.) Phytopathology 71:909.

Tinline, R. D. 1994. Etiology of prematurity blight of hard red spring wheat and durum wheat in Saskatchewan. Can. J. Plant Pathol. 16:87-92.

Traquair, J. A., and Smith, J. D. 1982. Sclerotial strains of *Coprinus psychromorbidus*, a snow mold basidiomycete. Can. J. Plant Pathol. 4:26-36.

Trione, E. J., Stockwell, V. O., and Latham, C. J. 1989. Floret development and teliospore production in bunt-infected wheat, in plants and in cultured spikelets. Phytopathology 79:999-1002.

Urashima, A. S., Igarashi, S., and Kato, H. 1993. Host range, mating type, and fertility of *Pyricularia grisea* from wheat in Brazil. Plant Dis. 77:1211-1216.

Usugi, T., and Saito, Y. 1979. Relationship between wheat yellow mosaic virus and wheat spindle streak mosaic virus. Ann. Phytopathol. Soc. Japan 45:397-400.

van Wyk, P. S., Los, O., and Kloppers, F. J. 1986. Crown rot of wheat caused by *Fusarium crookwellense*. Phytophylactica 18:91-92.

van Wyk, P. S., Los, O., and Marasas, W. F. O. 1988. Pathogenicity of a new *Fusarium* sp. from crown rot of wheat in South Africa. Phytophylactica 20:73-75.

Vargo, R. H., and Baumer, J. S. 1986. *Fusarium sporotrichioides* as a pathogen of spring wheat. Plant Dis. 70:629-631.

Vargo, R. H., Baumer, J. S., and Wilcoxson, R. D. 1981. *Fusarium tricinctum* as a pathogen of spring wheat. (Abstr.) Phytopathology 71:910.

Velasco, V. R., Ishimaru, C. A., and Brown W. M., Jr. 1991. Halo blight of spring wheat caused by *Pseudomonas syringae*. (Abstr.) Phytopathology 81:1159.

von Wechmar, M. B. 1987. Cucumber mosaic virus causes wheat disease. (Abstr.) Phytophylactica 19:126.

Walker, J. 1975. Take-all disease of Graminae: A review of recent work. Rev. Plant Pathol. 54:113-144.

Wallwork, H., and Spooner, B. 1988. *Tapesia yallundae*—The teleomorph of *Pseudocercosporella herpotrichoides*. Trans. Brit. Mycol. Soc. 91:703-705.

Wang, Y. Z., and Miller, J. D. 1988. Effects of *Fusarium graminearum* metabolites on wheat tissue in relation to Fusarium head blight resistance. J. Phytopathology 122:118-125.

Warham, E. J. 1990. Effect of *Tilletia indica* infection on viability, germination, and vigor of wheat seed. Plant Dis. 74:130-132.

Weller, D. M., Cook, R. J., MacNish, G., Bassett, E. N., Powelson, R. L., and Petersen, R. R. 1986. Rhizoctonia root rot of small grains favored by reduced tillage in the Pacific Northwest. Plant Dis. 70:70-73.

Weste, G. 1972. The process of root infection by *Ophiobolus graminis*. Trans. Brit. Mycol. Soc. 59:133-147.

Wiese, M. V. 1987. Compendium of Wheat Diseases, Second Edition. American Phytopathological Society, St. Paul, MN.

Wilcoxson, R. D., Kommedahl, T., Ozmon, E. A., and Windels, C. E. 1988. Occurrence of *Fusarium* species in scab by wheat from Minnesota and their pathogenicity to wheat. Phytopathology 78:586-589.

Wilkie, J. P. 1973. Basal glume rot of wheat in New Zealand. New Zealand J. Agric. Res. 16:155-160.

Wilkinson, H. T., Cook, R. J., and Alldredge, J. R. 1985. Relation of inoculum size and concentration to infection of wheat roots by *Gaeumannomyces graminis* var. *tritici*. Phytopathology 75:98-103.

Zhang, W., and Pfender, W. F. 1992. Effect of residue management on wetness duration and ascocarp production by *Pyrenophora tritici-repentis* in wheat residue. Phytopathology 82:1434-1439.

Zhang, L., Zitter, T. A., and Lulkin, E. J. 1991. Artificial inoculation of maize white line mosaic virus into corn and wheat. Phytopathology 81:397-400.

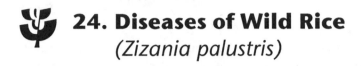

24. Diseases of Wild Rice
(Zizania palustris)

Diseases Caused by Bacteria

Bacterial Brown Spot

Cause. *Pseudomonas syringae* pv. *syringae.*

Distribution. The United States (Idaho and Minnesota).

Symptoms. Diseased leaves initially have small water-soaked spots or short streaks that later become chestnut or tan and are surrounded by a diffuse green-brown halo with a dark brown margin. Water-soaked areas frequently become translucent spots or slits in the centers of lesions and fall out of older lesions, giving leaves a "shot-hole" appearance. Small or narrow lesions are uniformly dark brown. Although typical lesions are roughly elliptical or spindle-shaped (1–10 × 2–100 mm or more), some lesions may be irregular-shaped or diffuse. White exudate rarely is present in the centers of lesions.

Management. Not reported.

Bacterial Leaf Streak

Bacterial leaf streak is also called bacterial streak.

Cause. *Xanthomonas campestris* and *Pseudomonas syringae* pv. *zizaniae.*

Distribution. The United States (Idaho and Minnesota).

Symptoms. Initially, the water-soaked, dark green lesions on leaves are narrow (1–2 cm long), linear, and parallel to veins. Later, lesions become necrotic and dark brown to black, then dry up and are covered with a glistening crust of bacterial exudate.

Management. Not reported.

Diseases Caused by Fungi

Anthracnose

Cause. *Colletotrichum graminicola* (syn. *C. sublineola*). The species name *"sublineola"* is sometimes spelled *"sublineolum"* in the literature.

Distribution. The United States (Minnesota).

Symptoms. Initially, small (1–2 mm in diameter), dark green–brown, water-soaked lesions occur on diseased leaves of aerial plants and on floating leaves of young plants. Later, lesions become elliptical to fusiform in shape (0.1–1.5 × 0.1–0.6 cm) and have light tan centers and dark brown margins. Lesions contain numerous, black, setose, vein-limited acervuli covered with orange-pink masses of conidia.

Management. Not reported.

Ergot

Cause. *Claviceps zizaniae* overwinters as sclerotia. When wild rice flowers are forming in early summer, sclerotia germinate and produce stalks on which stroma are produced. Perithecia containing ascospores form in stromatic heads. Ascospores are disseminated by wind to wild rice flowers and infect them. After infection, insects are attracted to a sweet, sticky liquid containing conidia that is exuded from flowers. Insects then disseminate conidia as secondary inoculum to healthy flowers. Eventually, hard sclerotia form in place of grain, drop into water, and wash to shore, where they survive unfavorable periods.

Distribution. Wherever wild rice is grown, but ergot is rare in commercial wild rice paddies.

Symptoms. Large ergot bodies or sclerotia replace kernels. Initially, sclerotia are pink to purple, when growing on plants, but become black and hard when they mature and dry up. Sclerotia vary in size (3–6 × 3– 20 mm). Prior to formation of sclerotia, drops of the sweet liquid and insects may be seen on plant heads.

Yield losses are minimal. Ergot bodies may be harmful if consumed, but because of their larger size, they are easily separated from grain.

Management. Remove sclerotia from grain during the cleaning process. Chips and pieces of ergot bodies that remain in grain can be floated out with water.

Fungal Brown Spot

Fungal brown spot is also called brown spot, Helminthosporium brown spot, and Helminthosporium blight.

Cause. *Bipolaris oryzae* infrequently overwinters as mycelium in infested whole stems that are not flooded, and in reed canarygrass. It is also seed-borne, but this is likely not a factor in disease development. Conidia are produced in summer and are disseminated by wind to wild rice plants, where infection occurs beginning at the flowering stage of plant growth. Secondary infection is by conidia produced in new lesions and disseminated by wind to host plant tissue.

Infection and disease development are optimum at temperatures of 25° to 30°C and continuous wet periods of 16 to 28 hours. Disease development is greatest between the flowering stage of plant growth and harvest.

Distribution. The United States (Idaho and Minnesota).

Symptoms. Symptoms do not become evident until flowering or later in the growing season. Where environmental conditions are conducive to disease development, symptoms will increase in incidence and severity until harvest time. All plant parts are diseased.

The first symptoms, which are found on leaves, are tiny, brown or purple lesions that develop into different shapes and may be limited by veins. Some lesions remain 1 mm or less in diameter and retain a very dark brown to purple-black color. Frequently, lesions will expand and develop brown centers with chlorotic margins. Some lesions retain a discrete shape and a size of 6 to 10 mm, whereas others lengthen to 3 to 4 cm. As lesions become larger and more numerous, leaves turn brown to yellow-brown and die.

Stem and sheath lesions develop similarly to those on leaves. Large necrotic areas develop on stems, frequently causing them to weaken and break. Infected spikes are bleached and florets and caryopses fail to develop, which results in a reduced number of seeds or no seeds being produced. Yields suffer slight to total (100%) losses.

Management
1. Apply foliar fungicides.
2. Paddies should be isolated from each other by 1.6–3.2 km.
3. Sow only clean, healthy seed.
4. Remove infested crop residue from paddy.
5. Plow paddy soil.
6. Grow wild rice for 2 successive years in the same soil followed by fallow soil or an alternative crop for at least 1 year.

Fusarium Head Blight

Fusarium head blight is also called scab.

Cause. Several *Fusarium* spp., but *F. graminearum* is the dominant species. Fungi overwinter as mycelia and spores on crop residue and other plant hosts. The fungi spread to plants by windborne spores. Infection of wild rice is thought to occur under wet environmental conditions.

Distribution. Unknown. It is presumed to be generally distributed wherever wild rice is grown.

Symptoms. Diseased spikelets are light brown. Diseased seed is light in weight and poorly developed. Diseased seed generally "shatters" from the head and falls to the soil surface. The fungi are then destroyed upon flooding of the paddy.

Management. Flooding paddies controls local sources of inoculum. However, conidia are likely disseminated by wind from other host plants and infested residue.

Phytophthora Crown and Root Rot

Cause. *Phytophthora erythroseptica.* Onset of symptoms coincides with hot, windy weather, which possibly predisposes host plants to disease. This is the only known example of a species of *Phytophthora* causing a serious disease of a mature grass host.

Distribution. The United States (California).

Symptoms. Disease distribution is relatively uniform in some fields and patchy in others, appearing to be confined to a portion of a field. Not all plants in an affected area are diseased. Healthy plants can grow adjacent to diseased ones and can compensate for diseased ones so that severely affected areas appear to be normal at the end of the growing season.

Plants ranging in age from the early-tillering to the grain-filling stages are killed. Initially, the disease progresses rapidly in the field but after 3 to 4 weeks, disease incidence and severity slows dramatically and few additional plants show symptoms.

Typically, diseased plants within fields show drought symptoms even though they are growing in flooded conditions. Leaves of diseased plants are gray to gray-green, then become tan or straw-colored, dry, and brittle. The crown, first internode, portions of leaf sheaths surrounding the internode, and many adventitious roots become necrotic; crowns typically show the most severe necrosis. Crowns frequently become so rotted that tillers separate from them and float to the water surface, leaving roots embedded in the soil.

Management. Not known.

Smut

The resultant galls are known as gau-sun and kah-peh-sung and are sold as vegetables. This host-pathogen combination is intentionally cultivated throughout the Orient for this purpose.

Cause. *Ustilago esculenta.*

Distribution. Asia and the United States (California on Manchurian wild rice, *Zizania latifolia*).

Symptoms. Culms enlarge and form galls up to 3 cm in diameter. Dark sori (1 × 5–10 mm) filled with teliospores are present inside galls.

Management. Practice sanitation.

Spot Blotch

Cause. *Bipolaris sorokiniana* overwinters primarily as mycelia on grasses growing adjacent to cultivated wild rice fields and within infested residue of several plant host species. Conidia are produced on residue or grasses in the spring and early summer and disseminated to host plants, where infection occurs on the first aerial leaves and infrequently on the floating leaves.

Distribution. The United States (Idaho and Minnesota).

Symptoms. All plant parts are infected. The first symptoms are tiny brown or purple lesions on the first aerial leaves and infrequently on the floating leaves. Spots develop into different shapes that are sometimes limited by veins. Some spots remain 1 mm or less in diameter and retain a very dark brown to purple-black color. Some retain a discrete shape and size of 6 to 10 mm, whereas other spots lengthen to 3 to 4 cm. As spots become larger and more numerous, leaves turn brown to yellow-brown and die.

Stem and sheath spots develop similarly to those on leaves. Large necrotic areas develop on stems, frequently causing them to weaken and break. Infected spikes are bleached, and florets and caryopses fail to develop, which results in a reduced number of seeds or no seeds being produced.

Management
1. Apply foliar fungicides.
2. Paddies should be isolated from each other by 1.6–3.2 km.
3. Sow only clean, healthy seed.
4. Remove infested crop residue from paddy.
5. Plow paddy soil.
6. Grow wild rice for 2 successive years followed by fallow soil or an alternative crop for at least 1 year.

Stem Rot

Cause. *Sclerotium* sp. and *Nakataea sigmoidea* (syn. *Helminthosporium sigmoideum*). These two fungi, or similar fungi, also cause a stem rot on white rice, *Oryzae sativa*. On white rice, the sclerotial stage is *S. oryzae* and the conidial stage is *N. sigmoidea*. On wild rice, both *N. sigmoidea* and *Scle-*

rotium sp. produce sclerotia that differ morphologically but are likely different stages of the same causal fungus, as in white rice.

The fungi overwinter as sclerotia in infested residue, paddy water, and soil. Optimum survival is in the residue on top of the soil, but numerous sclerotia can be observed floating on top of the water after paddies have been flooded. Sclerotia of *Sclerotium* sp. are also produced in lesions of white water lily, *Nymphaea odorata,* which serves as a site for overwintering and an increase of inoculum.

The fungi are disseminated from wild rice paddy to paddy primarily by infested residue and paddy water and infrequently by seed lots contaminated with sclerotia. Sclerotia germinate and produce conidia that are disseminated by wind or sclerotia that float to host plants, where they germinate and produce infective mycelium. Plants are infected at the water level or at the soil level and crowns when water is removed from paddies. Disease development is slow until water is removed, then increases rapidly if the weather stays dry and temperatures are above 24°C. Sclerotia form in culms, leaf sheaths, and floating leaves that were cut from plants during thinning operations in the spring or early summer.

Distribution. The United States (Minnesota).

Symptoms. Initially, purplish lesions develop on stems at the water level and on floating leaves and the first, and infrequently the second, aerial leaves. Black, round sclerotia (less than 2 mm in diameter) occur in lesions and on senescent plants. Later, as paddies are drained, brown lesions develop in crowns and the lower 15 cm of stems. Diseased stems become necrotic, dry, and brittle and break easily at the soil line. Diseased culms and leaf sheaths are partially or wholly destroyed and contain masses of black sclerotia embedded in white mycelium. Heads are poorly filled with seed that is light in color and underdeveloped.

Management
1. Remove plant residue from paddies.
2. Paddy soil should be plowed.
3. Paddies should be left fallow for at least 1 year.
4. Sow only cleaned seed.

Stem Smut

Cause. *Entyloma lineatum.*

Distribution. Wherever wild rice is grown.

Symptoms. Symptoms do not occur until plants are almost mature. Glossy black lesions occur on stems, culms, and heads due to numerous chlamydospores produced under the surface of diseased stems and leaves. As plants mature and dry, lesions elongate, coalesce, and girdle stems near the head and change color from black to lead gray. Stem and leaf infections do not cause much damage, but head infection reduces seed production.

Management. Differences in resistance exist among individual breeding lines and selections.

Zonate Eyespot

Cause. *Drechslera gigantea* survives in volunteer wild rice plants when paddies are fallow. Smooth brome, *Bromus inermis;* quackgrass, *Agropyron repens;* and reed canarygrass, *Phalaris arundinacea,* serve as alternate hosts. Zonate eyespot is associated with dense stands and infested residue from the previous crop. Disease occurs following 72 hours of 90% to 100% relative humidity and temperatures of at least 30°C.

Distribution. The United States (Minnesota).

Symptoms. Initial symptoms on leaves are small (1 mm in diameter), water-soaked, gray-green lesions with well-defined brown margins. As lesions enlarge, they become tan to brown in the center and secondary lesions that form around the primary lesions produce typical zonate eyespot lesions. Lesions typically are 0.8 to 1.5 cm long but may enlarge to 1 to 2 cm in length. Lesions coalesce and cover the entire leaf area of susceptible cultivars. Under humid conditions, white, prostrate mycelial strands grow outward from the lesion margins.

Management. Not reported.

Miscellaneous Fungi

Diplodia oryzae was isolated from dead culms.
Doassansia zizaniae causes a stem smut.
Mycosphaerella zizaniae causes a leaf spot.
Ophiobolus oryzinus causes a culm rot and is reported to be synonymous with
    Gaeumannomyces graminis.
Phaeoseptoria sp. causes a leaf spot.
Sclerotium hydrophilum was isolated from stem and foliar lesions.

Diseases Caused by Nematodes

Miscellaneous Nematodes

Radopholus gracilis was isolated from the cortex of roots.
Hirschmanniella gracilis was isolated from paddy soil.

Disease Caused by Viruses

Wheat Streak Mosaic

Cause. Wheat streak mosaic virus wild rice isolate (WSMV-WR) is in the potyvirus group. The life cycle of WSMV-WR is unclear. Dissemination is likely by the mite *Aceria tulipae,* which is disseminated by wind for long distances and is commonly found on wild rice. *Aceria tulipae* retains WSMV-WR for several days.

Distribution. The United States (Minnesota).

Symptoms. Diseased plants have typical streak symptoms. As disease symptoms progress for 10 days, chlorotic areas on the lower diseased leaves become necrotic. The necrotic areas eventually coalesce and cause the death of diseased leaves.

Management. Not reported.

Selected References

Bean, G. A., and Schwartz, R. 1961. A severe epidemic of Helminthosporium brown spot disease on cultivated wild rice in northern Minnesota. Plant Dis. Rptr. 45:901.

Berger, P. H., Percich, J. A., and Ransom, J. K. 1981. Wheat streak mosaic virus in wild rice. Plant Dis. 65:695-696.

Bowden, R. L., and Eschen, D. J. 1986. First report of wild rice diseases in Idaho. Plant Dis. 70:800.

Bowden, R. L., and Percich, J. A. 1981. Bacterial leaf streak of wild rice. (Abstr.) Phytopathology 71:204.

Bowden, R. L., and Percich, J. A. 1981. Bacterial leaf streak of wild rice caused by *Xanthomonas campestris* and *Pseudomonas syringae.* (Abstr.) Phytopathology 71:862.

Bowden, R. L., and Percich, J. A. 1983. Etiology of bacterial leaf streak of wild rice. Phytopathology 73:640-645.

Bowden, R. L., and Percich, J. A. 1983. Bacterial brown spot of wild rice. Plant Dis. 67:941-943.

Bowden, R. L., Kardin, M. K., Percich, J. A., and Nickelson, L. J. 1984. Anthracnose of wild rice. Plant Dis. 68:68-69.

Brantner, J., Malvick, D., Percich, J., and Nyvall, R. F. 1994. Variation in sensitivity to propiconazole in *Bipolaris oryzae* and *B. sorokiniana,* causal organisms of fungal brown spot of wild rice. (Abstr.) Phytopathology 84:1139.

Brantner, J. A., Nyvall, R. F., and Percich, J. A. 1995. Over-wintering sites of *Bipolaris oryzae* and *B. sorokiniana* causing fungal brown spot of wild rice. (Abstr.) Phytopathology 85:1123.

Farr, D. F., Bills, G. R., Chamuris, G. P., and Rossman, A. Y. 1989. Fungi on Plants and Plant Products in the United States. American Phytopathological Society, St. Paul, MN. 1252 pp.

Gunnell, P. S., and Webster, R. K. 1988. Crown and root rot of cultivated wild rice in California caused by *Phytophthora erythroseptica* sensu lato. Plant Dis. 72:909-910.

Johnson, D. A., Stewart, E. L., and King, T. H. 1976. A *Sclerotium* species associated with water lilies in Minnesota. Plant Dis. Rptr. 60:807-808.

Johnson, D. R., and Percich, J. A. 1989. Detection of ophiobolin in culture filtrates of *Bipolaris oryzae* and use in a wild rice root elongation assay. (Abstr.). Phytopathology 79:1209.

Kardin, M. K., Bowden, R. L., Percich, J. A., and Nickelson, L. J. 1981. Zonate eyespot of wild rice in Minnesota. (Abstr.) Phytopathology 71:885.

Kardin, M. K., Bowden, R. L., Percich, J. A., and Nickelson, L. J. 1982. Zonate eyespot on wild rice caused by *Drechslera gigantea.* Plant Dis. 66:737-739.

Kernkamp, M. F., Kroll, R., and Woodruff, W. C. 1976. Diseases of cultivated wild rice in Minnesota. Plant Dis. Rptr. 60:771-775.

Kernkamp, M. F., Kroll, R., and Woodruff, W. C. 1977. Wild rice infected by *Sclerotium* sp. isolated from white water lily. Plant Dis. Rptr. 61:187-188.

Kohls, C. L., Percich, J. A., and Huot, C. M. 1987. Wild rice yield losses associated with growth-stage-specific fungal brown spot epidemics. Plant Dis. 71:419-422.

Moffat, A., Nyvall, R., and Percich, J. 1997. Host range of *Bipolaris oryzae,* cause of fungal brown spot on cultivated wild rice. (Abstr.). Phytopathology 87:S67.

Morrison, R. H., and King, T. H. 1971. Stem rot of wild rice in Minnesota. Plant Dis. Rptr. 55:498-500.

Nyvall, R. F., and Percich, J. 1992. Control of fungal brown spot of cultivated wild rice with propiconazole. (Abstr.) Phytopathology 82:1068.

Nyvall, R. F., Percich, J. A., and Brantner, J.R. 1994. Fungal brown spot of cultivated wild rice is two different diseases. (Abstr.) Phytopathology 84: 1102.

Nyvall, R. F., Percich, J. A., Porter, R. A., and Brantner, J. R. 1994. Comparison of fungal brown spot severity to incidence of seedborne *Bipolaris oryzae* and *B. sorokiniana* and infected floral sites on cultivated wild rice. Plant Dis. 78:249-250.

Nyvall, R.F., Porter, R. A., and. Percich, J. A. 1995. First report of scab on cultivated wild rice in Minnesota. Plant Dis. 79:82.

Pantidou, M. E. 1959. *Claviceps* from *Zizania.* Can. J. Bot. 37:1233-1236.

Percich, J. A., and Huot, C. M. 1989. Comparison of propiconazole and mancozeb applied individually or sequentially for management of fungal brown spot of wild rice. Plant Dis. 73:257-259.

Percich, J. A., and Nickelson, L. J. 1982. Evaluation of several fungicides and adjuvant materials for control of brown spot of wild rice. Plant Dis. 66:1001-1003.

Percich, J. A., Bowden, R. L., Kardin, M. K., and Hotchkiss, E. S. 1983. Anthracnose of wild rice caused by a *Colletotrichum* sp. (Abstr.) Phytopathology 73:843.

Percich, J. A., Nyvall, R. F., Malvick, D. K., and Kohls, C. L. 1997. Interaction of temperature and moisture on infection of wild rice by *Bipolaris oryzae* in the growth chamber. Plant Dis. 81:1193-1195.

Watson, T. 1991. Smut of Manchurian wild rice caused by *Ustilago esculenta* in California. Plant Dis.75:1075.

Index